Retaining structures

Retaining structures

Proceedings of the conference *Retaining structures* organized by the Institution of Civil Engineers and held at Robinson College, Cambridge on 20 – 23 July 1992

Edited by C. R. I. Clayton

Thomas Telford, London

Conference organized by the Institution of Civil Engineers

Organizing Committee: C. R. I. Clayton (Chairman), M. D. Bolton, R. Fernie, C. J. F. P. Jones, R. J. Mair, G. W. E. Milligan, D. M. Potts, W. Powrie, B. Simpson, I. F. Symons

First published 1993

Classification
Availability: Unrestricted
Content: Collected papers
Status: Refereed
User: Geotechnical engineers and researchers

Distributors for Thomas Telford books are
USA: American Society of Civil Engineers, Publications Sales Department, 345 East 47th Street, New York, NY 10017-2398
Japan: Maruzen Co. Ltd, Book Department, 3-10 Nihonbashi 2-chome, Chuo-ku, Tokyo 103
Australia: DA Books and Journals, 648 Whitehorse Road, Mitcham 3132, Victoria

A CIP catalogue record for this publication is available from the British Library

ISBN 0 7277 1932 7

Published on behalf of the organisers by Thomas Telford Services Ltd, Thomas Telford House, 1 Heron Quay, London E14 4JD.

Printed in Great Britain by Galliards, Great Yarmouth

Contents

In situ embedded walls: observations

In situ embedded walls: design

Support systems to earth retaining structures

History of retaining wall design

J. KERISEL, Past President of the ISSMFE

Being always at the mercy of a new discovery, we shall never know for sure when the first retaining wall was constructed. As soon as man became sedentary, in fact, which was towards the end of the tenth millennium B.C., he set about building defences by raising obstacles that were difficult to climb. These defensive constructions were followed by monuments with a religious purpose inspired by the myth of the ascension towards the heavens.

It would appear that the first examples of the idea of providing a vertical support for a mass of materials are to be found among the megalithic monuments on the Atlantic coast of Europe: a typical example is the New-grange Cairn in Ireland, erected in the fourth millennium B.C. (Refs. 1 - 3). It was shaped like a drum with a diameter of 82 m (Fig. 1) and vertical sides of 4.20 m high formed of upright raised stones planted side by side in the earth up to a height of 1.20 m and surmounted by a masonry without mortar. This retaining structure enclosed a mass of rounded stones taken from the bed of a nearby river and resisted fro just over 500 years.

Builders were quickly made aware of the pressures at work on their walls and sought ways to reduce them. I see evidence of such efforts in two very different techniques to be found in monuments constructed thousands of kilometres apart as early as the fourth and third millennia B.C.: one is the insertion of horizontal mats and the other the technique of compartments.

Insertion of horizontal mats

I see the first use of this technique in the already mentioned Newgrange Cairn (Fig. 1). Between the layers of rounded stones the Irish builders laid mats of grass turf side by side. As we all know, pieces of turf like this weld together rapidly. Some experiments in traction (Fig. 2) based on mats of this kind and some rapid calculations have shown that they are capable of considerably reducing the pressure on the wall of the circular retaining structure which, without them, would have quickly collapsed under the lateral thrust of the rounded stones (Ref. 3).

Similar mats were also used in a few Mesopotamian ziggurats which time has spared. Ziggurats are heavy monuments shaped like superimposed coaxial truncated pyramids (Fig. 6) and are composed of successive courses of raw bricks made of silty clay and partially dried in the sun. The first edifices of this type were built at Sumer on the soft soil of the delta; they punched the

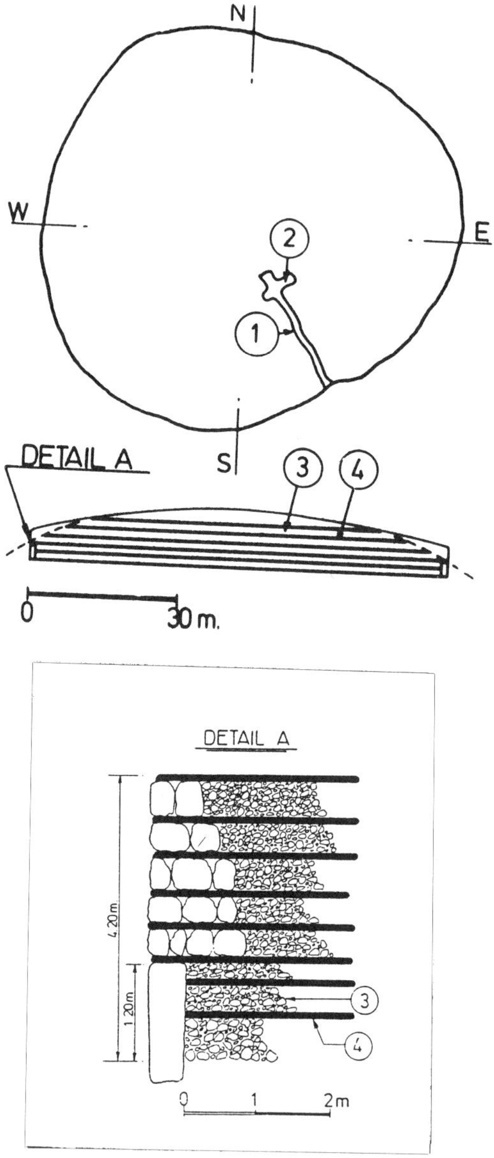

Fig. 1. Plan and cross-section of the Newgrange Cairn (3200 BC): 1 grave passage; 2 chamber (dotted line shows final shape after eventual overturning of retaining wall shown by detail A; 3 small rounded stones; 4 layers of turf

Fig. 2. Laboratory test to measure tensile strength of a square of turf (Simecsol, Paris)

soil and collapsed outwards. Someone then had the idea of inserting a mat of woven reeds laid in a bed of sand every five or six courses of brickwork. The mat neutralised the outward thrust and the sand drained away the water exuded by the raw bricks as the structure rose higher and higher. Cables made from woven plant fibres have also been found in the beds of sand.

System of compartments

The materials liable to exert outward pressure are placed in 'compartments', the walls of which form a grid. An example of this approach is the cairn of Gavrinis in the south of Brittany: the walls of the compartments are made of stones which, unlike the stones in the compartments, appear to have

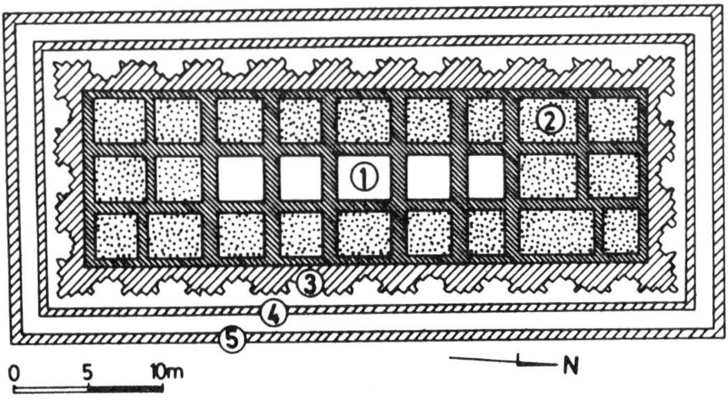

Fig. 3. Plan of tomb 3357 at Saqqarah (after Emery), dating from about 3000 BC: 1 subterranean burial chamber; 2 sand and small stones; 3 thick retaining walls; 4, 5 steps

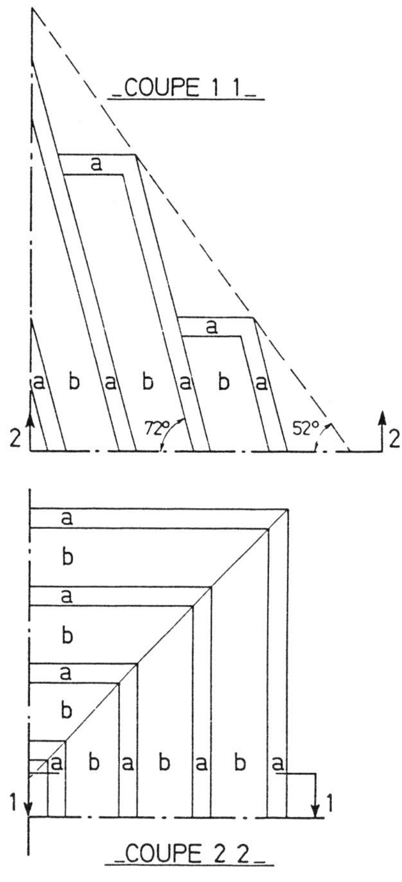

Fig. 4. First stone pyramids; principle of internal retaining walls invented by Imhotep at about 2700 BC: a walls dressed in stone; b chippings of stonecutters

been shaped with some care. We find the same technique in the mastabas, the tombs of important Egyptian dignitaries of the 1st and 2nd dynasties (Ref. 4): here the walls are of brick and the compartments filled with rubble and sand (Fig. 3). The height of these structures is still modest, only a few metres, but it was clearly this idea of compartments that later inspired Imhotep, the renowned builder of the pyramid of Saqqarah, the first major stone pyramid in history and about 60 m high. From the system of compartments Imhotep kept only the idea of walls that formed squares on the ground oriented NS and EW. These squares are centred on the centre of the pyramid, one inside the other, with their walls 5 Egyptian cubits (2.60 m) apart (Ref. 3 and Fig. 4).

Fig. 5. Step pyramid of Saqqarah; countermarches are simply tops of retaining walls

Fig. 6. The ziggurat of Ur, built about 2100 BC; ground level dimensions 62m x 42m; probable height 21m; the third level probably bore a temple

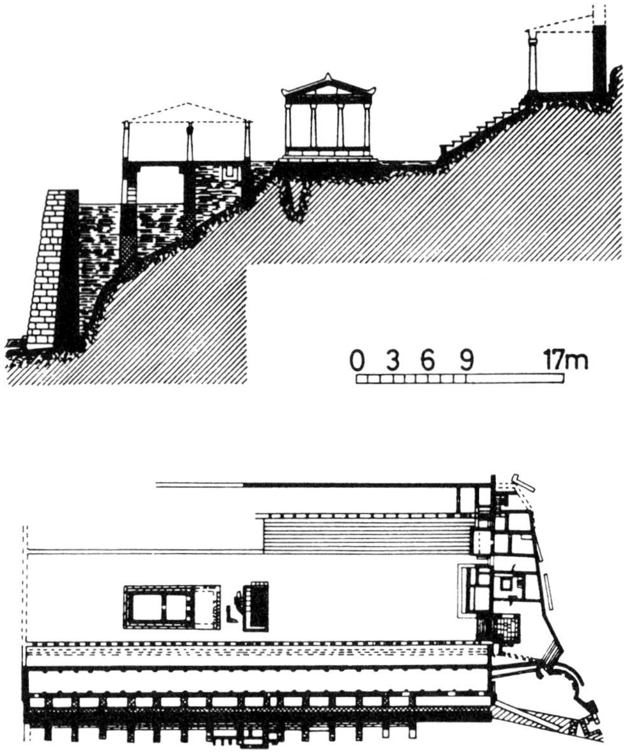

Fig. 7. Vertical and horizontal sections of retaining wall for the terrace of the Temple of Demester at Pergamon, 2nd century BC

We may note in passing that the intervals used for Meidum, a later pyramid, were 10 cubits (5.20 m). To prevent overturning the walls were inclined at 72° inwards; they were constructed of dressed stone and the spaces between them filled with the chippings of the stone cutters. The use of waste materials also economised on manpower (at a time when stone cutting tools were still very primitive). The pyramid (Fig. 5) has survived for nearly five thousand years and now stands as evidence of the intelligence of the first Egyptian builders.

Despite its ingeniousness, the technique of internal parallel walls imagined by Imhotep was used only in a few Egyptian pyramids. Quite quickly builders thought of constructing external buttresses as a means of preventing the retaining walls from overturning. Of this type is the wall of the terrace supporting the Temple of Demeter in Pergamon (2nd century B.C.), one of the finest retaining structures of the Ancient World (Fig. 7). The wall was 14 m high, 80 m long and supported by 15 external buttresses set just over 5 m apart.

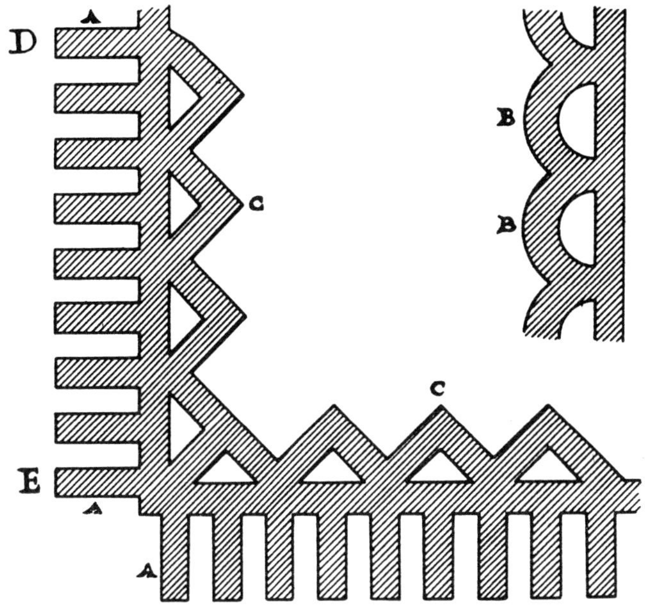

Fig. 8. The teaching of Vitruvius on the structure of walls; illustration by Perrault of the combs proposed by Vitruvius

As civilisations developed, more and more retaining walls were needed for fortifications. To withstand the onslaught of increasingly powerful weapons (catapults, battering-rams, etc) the walls were built higher and higher. The biggest mobile siege towers were about 60 metres high. The very vulnerable external counterforts were then replaced by internal ones or by much thicker walls. The treatises on fortifications of the School of Alexandria (Philo of Byzantium) lay a great deal of emphasis on the need for the thickness of walls to be in proportion to their height.

However, the first important book on retaining structures remains Vitruvius' De Re Architecutra (Refs 5 and 6). The use of internal buttresses is commented on at length: 'A series of supplementary walls should be built... to form the shape of the teeth of a saw or of a comb: by this means the earth is broken up into compartments and cannot push on the wall with such a great force' (Fig. 8). The ancients had also found ways of solving other problems relating to temporary retaining walls. We find evidence of this in De Re Architectura. Chapter 12 of Book V is entitled 'De portibus et struturis in aqua faciendis' - 'On building ports and other structures in the water'. For the construction of moles Vitruvius proposed alternative methods, one of which consisted of the construction of a double-walled cofferdam. Between the two walls, reed sacks filled with clay were let down to the bottom and the

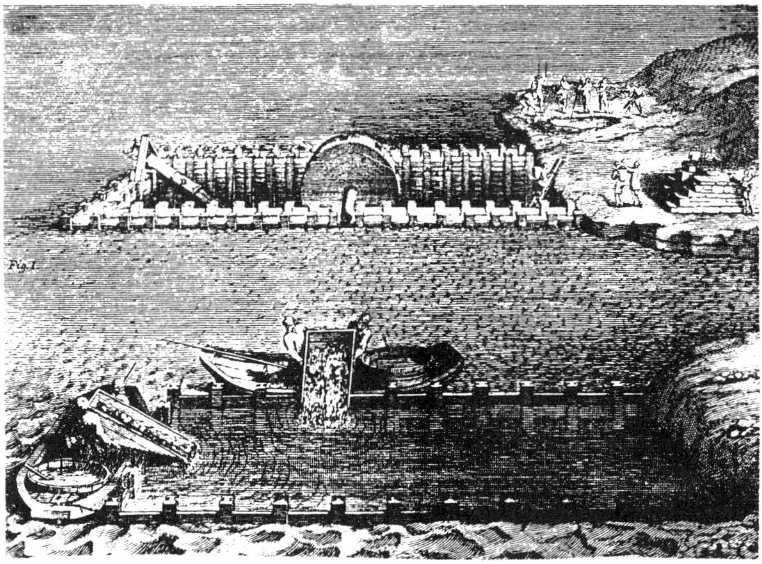

Fig. 9. Illustration for Vitruvius' treatise. Upper scene shows how the seabed near the shore was pumped dry inside a double-walled cofferdam; the space between the two timbered walls was filled with sacks of clay. In the lower scene the cavity is being filled with stone and pozzolana

inside of the cofferdam was then pumped dry with wheels and drums on the screw principle (Fig. 9).

In the middle ages and during the Renaissance there were a great many treatises on fortifications, one example being the study by Francisco di Giorgo in the 15th century. In 1684 Marshal de Vauban published (Ref. 7) a work entitled 'Table explaining the measures ... for building walls 10 to 80 feet high...based on the experience of the construction of 500,000 cubic fathoms (3,700,000 m^3) stonework in 150 forts, built on the orders of the King'. Vauban was in mid-career and, having so far encountered but a fairly limited range of soils, believed that standardisation was possible. He recommended the use of walls with internal counterforts with a centre to centre spacing of 5.40 m. The base of the external facade of the wall itself was slightly set back from the front of the foundations, an arrangement that provided a more effective resistance to stresses. The thickness of the walls at the base, as a function of the height H, was 0.20H + 1.48 (in metres) but the length of the counterforts added to the thickness of the wall came to almost half the height H.

The depth of the foundations is not given. Later, after having constructed fortifications in the north of France where the ditches in front of the walls cut into Sparnacian clay (from a geological level corresponding to London clay

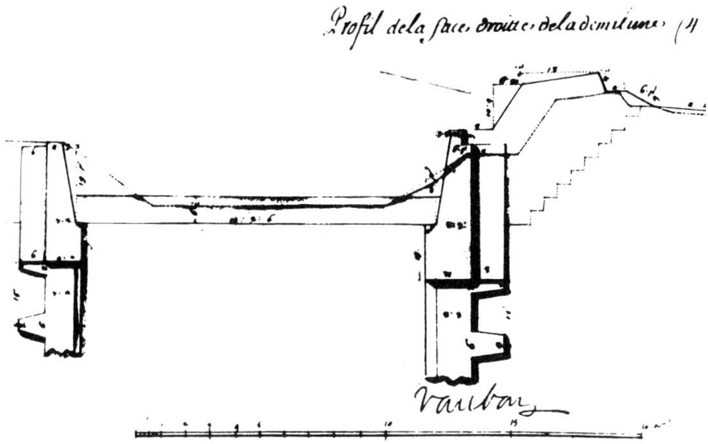

Fig. 10. Two retaining walls with internal buttresses founded in clay, designed by Vauban for the fortification of Ypres in 1699

but less heavily overconsolidated), Vauban was obliged to supplement his tables.

In an entry in his diary dated 25 June 1699 he wrote: 'On this fat soil slippery like soap..., the ground at the bottom of the ditch must be capable of containing the thrust of the covering when it is pushed by the soil of the ramparts'. Here we find the fundamental idea of using the passive resistance of soil by increasing the depth of the wall. In keeping with this idea, Vauban underpinned the walls of the fortification at Ypres which had been constructed before his time in 1678 (Fig. 11). Their foundations had settled and slipped forward quite a lot, with the result that the wall had rotated backwards under the action of its own weight and the friction force on the back face; the counterforts, moreover, had sheared off at their connection with the masonry. Where the wall had not moved too much, Vauban reduced the load on the back of the wall and then used temporary shoring to permit the addition of new masonry underneath.

All this is on the eve of the 18th century. That and the following century would see a great many studies on the question of retaining walls. Professor A. W. Skempton (Ref. 8) gave us a clear account of the subject at the Eleventh International Congress at San Francisco in 1985 in his History of Soil Properties 1717-1927. By observing the resistance of retaining walls and by subjecting scale models to experiments based on back analysis, a start was made on the development of a theory of earth pressure. The publications of Gauthier (1717), Belidor (1729), Sallonyer (1767), Coulomb (1773), Papocino (1781), Prony (1802), Mayniel (1808), Français (1820), Navier (1833), Poncelet (1840), Hope (1845), Rankine (1862), Darwin (1883), Boussinesq (1876, 1882 and 1883) are the main landmarks of the long history of research on retaining

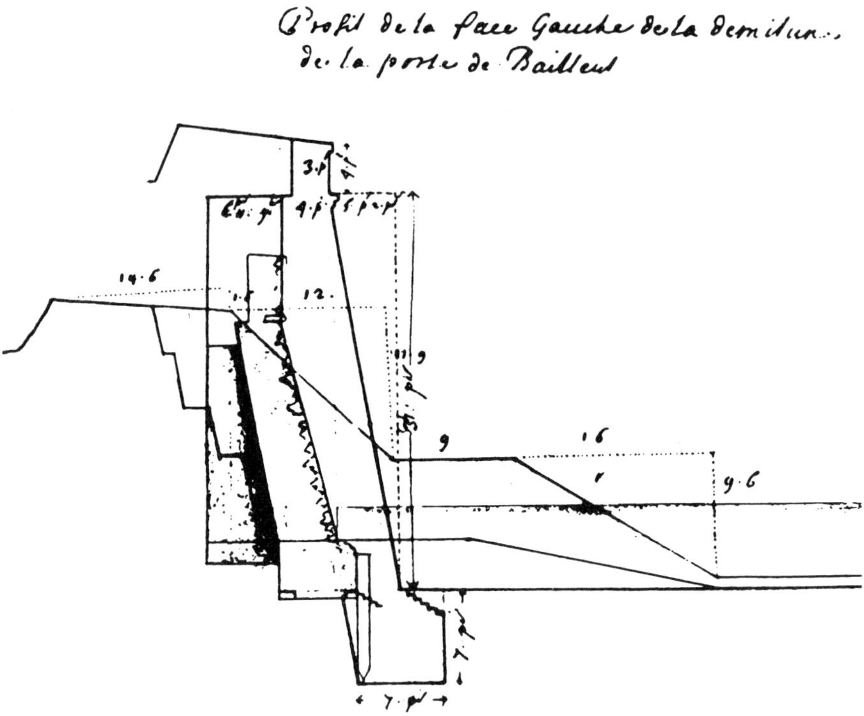

Fig. 11. *Under pinning by Vauban of the wall insufficiently embedded in clay at Ypres*

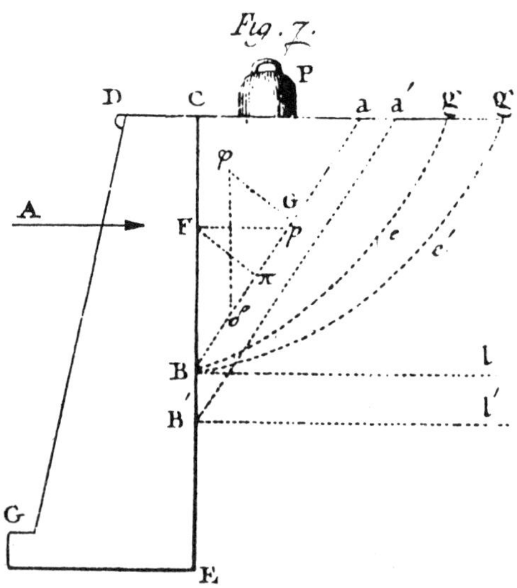

Fig. 12. The basic figure of Coulomb's memoir

walls (for references, see publications of Skempton and Heyman). I should like to make a few comments on just three of these authors - Coulomb, Rankine and Boussinesq.

The bicentenary of Coulomb's essay was celebrated in 1973 at the Seventh International Congress of Soil Mechanics in Moscow (Ref. 9). His life and work are described in detail in the book by Gillmore (Ref. 10), but it is in the book by Professor Heyman (Ref. 11) that the full significance of that essay, which goes far beyond the question of retaining walls, is brought out.

In your call for papers you reproduced one of the major figures from this essay (Fig. 12). 'The force A', writes Coulomb 'must be capable of supporting... all the surfaces CBeg bounded by any curve Beg, which, supported by its cohesion and its friction and acted upon by its own weight, would exert the largest thrust. It will be clear from this statement that if a force were applied at F infinitesimally smaller than that required to support the surface of greatest thrust, the mass of soil could cleave only along that curve, while all the other portions remained united by cohesion and by friction. Thus, to find the force A sufficient to support the whole mass, that surface must be sought, among all the sources CBeg, for which the thrust on the line CB is a maximum. Similarly, to find the greatest force which can act through F without disturbing equilibrium, another curve Be'g' is a minimum. The limits of the horizontal force that can be applied at F without moving the soil will be bounded by the limits A and A', where A will be a maximum and A' a minimum'. (We have used the translation by Professor Heyman). Here Coulomb opposes the active and passive cases and shows that the limits for the latter (far greater) are derived from a study of the minimum, and vice versa.

Coulomb assumes that the curve which gives A is a straight line, experience showing that the surface which breaks is very close to triangular. He does not make the same assumption for the passive case: the curves Beg and Be'g' which he draws by way of example do indeed bend slightly. In fact, the curve for the usual values of ϕ is even more rounded than the ones drawn.

Coulomb draws attention to the fact that the friction created upon the retaining wall has been neglected, and then writes: 'At the moment of rupture the soil is about to slide along CB, which diminishes A and at the same time increases the moment of the retaining wall'.

This problem of friction on the wall occupied researchers for the next two hundred years. Rankine (Ref. 12) applied his stress-field theory for cohesionless soils which leads to the familiar expression

$$Ka = tg^2(45 - \phi /2) \text{ and } Kp = tg^2(45 + \phi /2)$$

for a vertical wall and horizontal soil, confirming Coulomb's value for the active state.

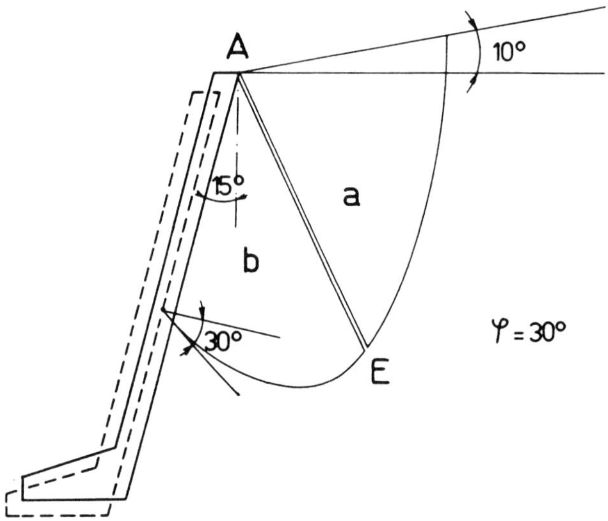

Fig. 13. Boussinesq introduces a discontinuity of the stress field through separating the Rankine equilibrium and the Boussinesq one; case of a wall that has settled

But the theory of Rankine (1857) is restrictive in that the angle of friction on the wall face is established a priori at failure being equal to slope of the free surface of the ground for a vertical wall.

The merit of Boussinesq (1876 and 1882) is to have clearly stated the equations governing the problem of active pressure (Refs 13 and 14) whatever the friction on the wall. The importance of his analysis lies in the introduction of a discontinuity within the stress field along a certain lane AE separating Rankine equilibrium and another one, his own (Fig 13). Moreover, in his note on 'Mr. Darwin's careful and extremely interesting experiments' (Darwin Ref. 15), Boussinesq (Ref. 16) accepts the concept that the angle ϕ within the mass of the sand is not necessarily the same all along the rupture line. As underlined by Professor Skempton, 'this analysis is one of the brilliant steps which, once they are made, seem simple and obvious'.

Starting from the (non-integrable) equations of Boussinesq we can now use computers to draw up numerical tables giving the value of the limiting states of passive and active pressures whatever the angle of friction on the wall, the slope of the ground or the inclination of the wall (Ref. 17).

Some 70 years later, Sokolowski (Ref. 18) made a little addition to the solution of Boussinesq: instead of having a plane of discontinuity between two equilibria they are separated by a fan OQR (Fig. 14). Nevertheless, as Professor Heyman points out, Sokolowski's solution only works if the surface is subjected to a uniform over burden.

In 1885, Boussinesq (Ref. 19) published his celebrated 'Application des potentiels' giving stresses and displacements in elastic foundations. Today the calculation of horizontal pressures on a retaining wall due to a concentrated surface load is still based on this work.

The 20th century put an end to the somewhat abstract research of the 18th and 19th centuries. As Professor Heyman emphasises, these cohesionless theories have been attacked repeatedly as poor models for the behaviour of soils. Professor Peck (Ref. 20) has even spoken of the 'numerical calculation subculture' to which too many theories lead.

The fact is that at the end of the 19th century, in the field of retaining walls with which we are concerned, mathematics was well in advance of physics. The 20th century saw an effort to catch up and there was big progress in our knowledge of the mechanical properties of soils. But the mathematical tool was not challenged for all that. Occasionally, however, certain factors were overlooked: for example, in the problem of horizontal pressures on a retaining wall due to a concentrated surface load, people took from Boussinesq only the stresses and did not examine whether the displacements of the wall at the corresponding point matched those given by that author, or, if they did so, they did not draw the consequences. As for cohesion, that was easy to take it into account as Caquot (1934) showed (Ref. 21) by introducing a uniform external stress $C/\mathrm{Tg}\phi$ on the free surface and behind the wall.

In fact, though progress was made, physics did not entirely catch up. In undisturbed sand said to be without cohesion, a low level of cohesion may develop quite rapidly as an effect of ageing. It plays a considerable role in the

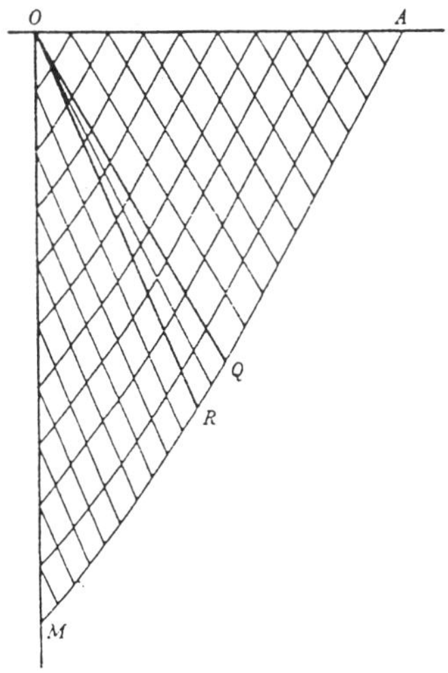

Fig. 14. Sokolowski's solution: a fan OQR separates the two regions; here $\phi = 30^{\circ}$, angle of friction on wall 15°, uniform burden 5C on the ground

calculation of active or passive pressures owing to the size of the multiplier of $C/Tg\phi$ though it is not at all easy to determine in the laboratory or in situ the value of this slight cohesion. Moreover many factors concerning the shear strength of clays are still unknown today.

With the emergence of concrete in the 20th century, gravity masonry walls gave place to concrete walls and reinforced cantilever walls, but there soon appeared a range of new retaining structures such as gabions, cribwalls, cantilevered or anchored steel walls and so on. Nor must we forget reinforced soils: the variety of these structures and the differences in flexibility led engineers to study the extent to which active or passive pressures are strain-dependent and the distribution of lateral pressures. This was the starting point for a considerable number of studies. Limiting myself to the main references prior to 1965, I wold begin by citing the work of Terzaghi (1934) and Rowe (1965) based on models incorporating large rigid screens. This research was extended to flexible models by Tschebotarioff (1949), Rowe (1952) and Terzaghi (1953), and to temporary retaining structures for deep excavations. This then provided an opportunity for Professor Peck (1943) and Bjerrum (1956) to make numerous earth pressure measurements in braced excavations and show their very large dependency upon the method of installation of struts or bracing frames.

Lastly Rowe (1965) (Ref. 23) described the mechanism of progressive rupture about which Boussinesq had had a certain intuition.

In spite of all this knowledge and further progress in the last twenty-five years, I note that your draft Code of Practice for earth retaining structures, prepared under the authority of the British Standard Institution, observes that 'there is still a considerable lack of reliable information upon the actual earth pressures encountered on walls in practice', and that the code is still 'limited to small and medium sized retaining walls which have a free standing height not in excess of approximately 15 m'. This gives some idea of the work before you at this Congress. However, in order not to end on a pessimistic note, I should like to stress that all this research now going on has at least made it possible to build in safety a large variety of structures which melt much better into the landscape through their architecture and perhaps through their decorative plant cover than the mass gravity walls they replace.

I shall bring to an end here this backward glance, which began so deep in the past at a time when our eyes are turned towards the future. It is never without interest to know what has already been accomplished and I have endeavoured to remind you of the most essential achievements.

References

1. O'KELLY M.M. *Newgrange, Archaeology, Art and Legend*, Thames and Hudson, London, 1982.

2. KERISEL J. A glimpse of geotechical engineering during the fourth and third millenia B.C. Proceedings of an international symposium organised by the Greek national group of IAEG. Balkema, Rotterdam, 1988, Vol 3, 1641-1653.
3. KERISEL J. La pyramide à travers les âges. *Mythes et religions.* Presses de l'Ecole des Ponts et Chaussées, Paris 1991.
4. EMERY W. B. *Archaic Egypt.* Penguin Books, London 1961, p 55.
5. VITRUVIUS. *De Re Architecutura.* 1st century B.C. Translated into French and illustrated by Perrault: the ten books on architecture. Coignard, Paris, 1674. English translation by Harper, Cambridge, Mass. and Heineman, London, 1955.
6. KERISEL J. The golden Jubilee book. *Proceedings of the Eleventh International Congress on Soil Mechanics and Foundation Engineering.* Part One: the history of geotechnical engineering up until 1700. Balkema, Rotterdam, 1985, pp 3-95.
7. VAUBAN. Tables publiées dans 'La manière de fortifier selon la méhode du Maréchal de Vauban'; par l'abbé du Roi, 1684.
8. SKEMPTON A.W. *Golden Jubilee Book* (see Ref. 6). A history of soil properties 1717-1927, pp 95-123.
9. KERISEL J. Bicentenary of the 1773 paper of Charles Augustin Coulomb. *Eighth International Conference on Soil Mechanics and Foundaiton Engineering*, Special Lecture, Moscow 1973.
10. GILLMOR C.S. *Charles Augusin Coulomb: Physics and Engineering in 18th century France.* PhD thesis, Princeton University, May, 1968.
11. HEYMAN J. *Coulomb's Memoir on Statics.* Cambridge University Press, 1972.
12. RANKINE W.J.M. On the stability of loose earth. *Phil. Roy. Soc.* 147, 9, 1857.
13. BOUSSINESQ J. Essai théorique sur l'élasticité des massifs pulvérulents et sur la poussée des terres sans cohésion. Paris 1876.
14. BOUSSINESQ J. Sur la détermination de l'épaisseur minimum que doit avoir un mur vertical ... pour contenir un massif terreux sans cohésion dont la surface supérieure est horizontale. *Annales des Ponts et Chaussées,* 6th series, 1882, pp 494-510.
15. DARWIN G.H. On the horizontal thrust of a mass of sand. *Minutes of Proceedings, Inst. Civ. Engrs,* 71, 1883, pp 350-378.
16. BOUSSINESQ J. Note on Mr. Darwins paper 'On the horizontal thrust of a mass of sand'. *Minutes of Proceedings, Inst. Civ. Engrs,* 72, 1883, p. 262.
17. KERISEL J. AND ABSI E. *Active and passive earth pressure tables.* 3rd Edition. Balkema, Rotterdam, Brookfield, 1990.
18. SOLOLOWSKI V.V. *Statics of Solid Media.* English translation: Butterworth Scientific Publications, London 1954.

19. BOUSSINESQ J. Applications des potentiels à l'étude de l'équilibre et du mouvement des solides élastiques, Paris, 1885.
20. PECK R. *Golden Jubilee Book* (see Ref. 6), The last sixty years, part 3, pp 123-136.
21. CAQUOT A. *Equilibre des massifs à frottement interne*, Gauthier-Villars, Paris, 1934, pp 10-13.
22. TERZAGHI K. Large retaining wall tests. *Engineering news record*, September 1932, February, March and April, 1934.
23. ROWE P. W. and PEAKER R. Passive earth pressure measurements, *Geotechnique*, Vol 15, No. 1, March 1965.
24. TSCHEBOTARIOFF G. P. *Large scale earth pressure tests with model flexible bulkheads*. Princeton University, 1949.
25. TERZAHI K. Anchored bulkheads. *Proc ASCE*, Vol 119, 1953, pp 1243-1324.
26. ROWE P.W. Ancored sheet pile walls. *Proc Inst. Civ. Engnrs*, 1952, Vol 1, p. 27.
27. PECK R. Earth pressure measurements in open cuts Chicago subways. *ASCE*, Vol 108, 1943, p. 1008.
28. BJERRUM L. AND EIDE O. Stability of strutted excavations in clay. *Geotechnique* 6, 1956, pp 32-47.

Introduction to Session 1
General principles

C.R.I. CLAYTON, University of Surrey, UK

Introduction

The design and construction of retaining walls requires a detailed knowledge not only of soil behaviour, in all its complexity, but also an appreciation of what types of wall are available, and what can go wrong (both during construction and when the wall is in service). This session considers these issues. The following broad issues should influence the engineer's thought as he starts to design his retaining structure:

- available wall types, and their geometries
- general design issues
- general analytical issues
- reliability of the construction technique and the finished structure
- the need for maintenance, and the possibility of providing it.

Design

Many engineers seem to believe that analysis is synonymous with design. But analysis is the calculation of the magnitude of specific variables, which in design is usually carried out to check that they do not exceed some pre-determined value. Design, however, is a much more complex process, normally including some element of analysis, but also involving the choice of materials and construction technique, the sizing of components and their spatial arrangement, etc., in order to fulfil a function (in this case, soil or rock retention).

As Figure 1 shows, design may take place without analysis (for example, when presumed bearing pressures are used for spread foundations) and analysis may take place without being part of a design process (for example, for research). Analysis may form a significant part of some designs, but it should not be mistaken for design itself.

The elements of design are shown in Figure 2. The circular nature of the design process shows why experience is such an important part of design; a knowledge of the performance of wall types on previous jobs, tempered by an appreciation that it is not always sensible to use the same solutions in every situation, helps improve the certainty of a design. The first stage of any design

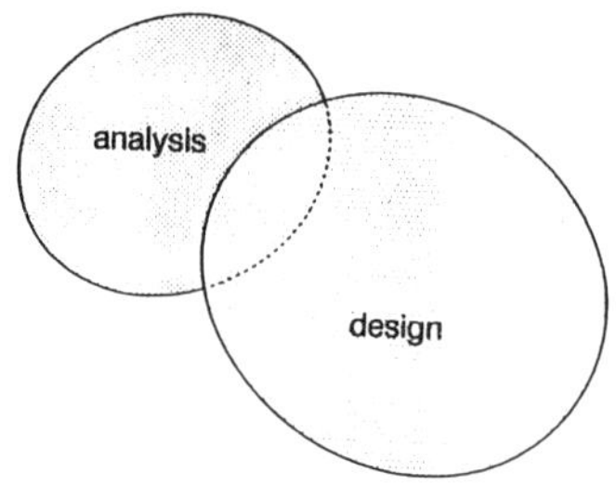

Fig. 1. Relationship between design and analysis

is the definition of need. In the case of retaining structures, this may be seem obvious - in fact the needs associated with retaining structures are often much wider than simply the retention of soil in order to provide a hole in the ground. In inner city locations, for example, the most demanding part of a retaining structure design is often created by the need to limit upward and inward deformation of the soil around an excavation, in order to prevent damage to structures and services around the proposed excavation, and tunnels beneath it.

Following definition of the need, it is important that research is carried out to stimulate thought on the various ways that the project requirements can be met. This involves not only site investigation to determine the ground conditions, but also an assessment of the different wall types that might be used, and of experience of construction of these wall types, and of the availability of plant and experienced contractors. Case records of previous walls are obviously very useful. Conceptualisation brings the fruits of research together, to demonstrate as many ways of achieving the needs of the project as possible. This is perhaps the most critical part of design, since it

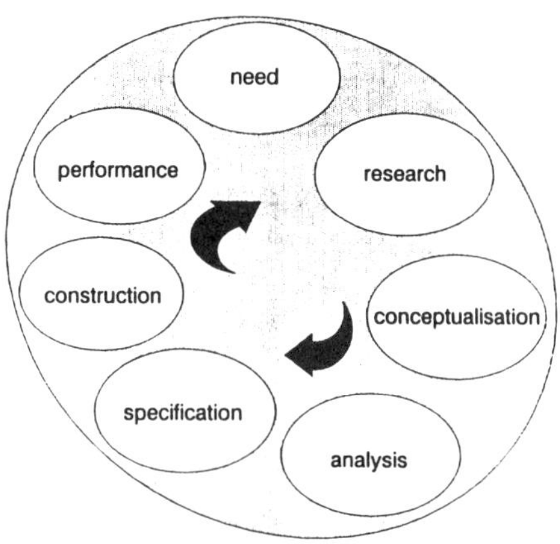

Fig. 2. Design process

will result in the selection of the wall types that are most favourable. It involves the assessment of as many options as possible, in terms of wall geometry and method of construction, in terms of:

- component cost
- buildability (performance during construction)
- construction cost (partly a function of buildability)
- reliability (post-construction performance)
- post-construction cost (maintenance).

Reliability is a difficult issue in any civil engineering design, and is even more important in geotechnical design. Unlike the manufacturing industry, the construction industry often has little opportunity objectively to assess the overall reliability of a design by component testing, since most components are formed in-situ. In the case of retaining structures, where soil-structure interaction is an important consideration, the properties of the soil and its geometry are only imperfectly known. Experience is very important, and comes in different forms - personal experience (the best kind), secondhand experience (for example, of colleagues in a firm), and literature.

Literature is often used to justify geotechnical design decisions, but we should remember the warning of van Weele (1989) that engineers only publish their successes. "Our young engineers, who start their careers by reading relevant literature,... are by definition misled. They get the impression that we, as experienced engineers, ... are able to solve the most complicated and exotic problems with great accuracy. This is however a wrong impression. Our failures are almost always considered unfit for publication."

The value of experience, then, is that it gives the engineer the missing component - a knowledge of what can go wrong. A "good" practising engineer is one with an extensive knowledge of what has failed to work elsewhere, and a sixth sense for what might go wrong on his own site. But even so, there is the problem of how to deal with the unexpected.

In the context of conceptual design, analysis is rarely of great importance. In civil engineering, analysis is normally used for the sizing of components, but this in itself is a relatively trivial part of geotechnical engineering. What is important is the correct identification of the limit states that are likely to occur, given the ground conditions and the retaining structure geometry. Once these have been identified, then avoidance (for example, by altering the geometry of the retaining structure, or by changing the type of construction altogether) is better than analysis.

Within the design process there are several strategies that can be adopted to avoid over conservatism while still minimizing risk. Experience of the behaviour of similar structures in similar ground conditions is obviously very helpful, and illustrates the positive aspect of publishing case records. Other strategies include:

- monitoring, and re-design during construction
- the observational method.

The need to design a structure to function wholly without maintenance would seem undesirable, but can maintenance normally be carried out on earth-retaining structures, and even if it could be done, is there any guarantee that the work would actually be carried out?

Analysis (as part of design)

Detailed analysis should only occur late on in the design process, once research has identified the many possible types and configurations of suitable retaining structures, and conceptual design has reduced these to a few likely options. Analysis is then used to investigate the relative merits and demerits of different design options, to optimize geometry, to check that the soil and structure are not overstressed, and in the more sophisticated forms of analysis, to predict the behaviour of the actual structure and its surroundings. It is. important to recognize the different levels of analysis used in retaining structure design, e.g.

- limit equilibrium
- beam on elastic foundations (Winkler spring)
- continuum (finite element/finite difference/boundary element)

since it is essential that the correct level of analysis is used for the particular problem requiring solution, and that the input parameters are appropriate. For example, Winkler spring models are suitable for determining the bending moments, shear forces and prop/anchor forces to be sustained by a diaphragm retaining wall. But if displacements around or below the excavation are to be predicted, then continuum analysis is required.

In the analysis of retaining structures, there are some basic issues which must be correctly addressed, and there are more sophisticated issues, which will improve (perhaps) the predictive accuracy of an analysis (Table 1).

Whilst it may seem somewhat extreme to state that the use of correct soil parameters is a "sophisticated" issue, it is clear that even if the correct soil parameters are used, failure to identify whether active, passive or K_0 conditions apply, will result in completely incorrect analysis. Similarly, misjudgment of ground water conditions may lead to huge underestimates of active total pressure, and also to a lack of appreciation of the likelihood of boiling and base heave in the base of excavations. In fact, many small retaining structures have been satisfactorily designed without any knowledge of the real soil parameters.

Table 1

"basic" issues	"sophisticated" issues
o identification of limit states	o selection of soil parameters
o water effects	o surcharge effects
o soil or structure placed first?	o installation effects
o mode and amount of wall movement	o dynamic (e.g. earthquake) effects

Papers

Table 2 lists the 5 papers that have been allocated to this session.

Grose and Toone consider the factors involved in the choice of wall type for a major retaining structure at Castle Mall, Norwich. These included "technical merit", construction safety risk, construction programme security, and cost.

In contrast, papers by Powrie and Kantartzi, and by Gunn,

Table 2. Papers allocated to Session 1

1. "The selection, design and performance of a multi-propped contiguous pile wall in Norwich"

   by W.J. Grose and B.H. Toone

2. "Installation effects of diaphragm walls in clay"

   by W. Powrie and C. Kantartzi

3. "Finite element modelling of installation effects"

   by M.J. Gunn, A. Satkunananthan and C.R.I. Clayton

4. "Surcharge and elasto-plastic computations of earth-retaining structures"

   by P.J. Schmitt and M.Gilbert

5. "Influence of seepage flow on the passive earth pressures"

   by A.H. Soubra and R. Kastner

Table 3. Papers of interest to Session 1: conceptual design

1.	Grose and Toone	Castle Mall, Norwich
2.	Puller and Puller	construction techniques - diaphragm walls
3.	Lee and Butler	relieving platforms
4.	St.John, Harris, Potts, Fernie	relieving platforms
5.	Tamaro and Gould	analysis and design - cast in-situ walls
6.	Porter, Lien, Abedi, Ramos	shaft sinking
7.	Bica and Clayton	component sizing - cantilever sheet walls
8.	Hagerty and Nofal	component sizing - anchored sheet walls

Table 4. Papers of interest to Session 1: analytical issues

1.	Gunn, Satkunananthan, Clayton	installation effects
2.	Powrie and Kantartzi	installation effects
3.	Soubra and Kastner	effect of seepage
4.	Schmitt and Gilbert	surcharge calculation methods
5.	Phillips, Ho and Pappin	long-term toe stability
6.	Butterfield	bearing capacity
7.	Iai and Kameoka	earthquake effects
8.	Whitman	deformations/earthquake effects
9.	Whittle and Hashash	deformations/propped diaphragms

Satkunananthan and Clayton, report the results of ongoing research into the effects of installation of in-situ (i.e. bored pile, or diaphragm) retaining walls. These researchers are attempting this difficult problem because almost all continuum analysis assumes that retaining walls of this sort are "wished in-place", i.e. installed without changing the stress conditions in the ground. Powrie and Kantartzi use the centrifuge to model this effect, while Gunn *et al.* used coupled-consolidation finite element analysis. In both cases, however, the problem was modelled as two-dimensional - it is questionable more than the general behaviour can be determined in this way, since the real problem is three-dimensional. Nonetheless, Gunn *et al.* show that installation effects are only likely to be an issue for propped walls, and where the water table is relatively low.

Schmitt and Gilbert highlight an important and difficult problem for designers using simple hand calculation methods of analyses, that of calculating the effects of surcharges applied over a limited area of the ground surface behind a retaining wall. They show how the use of elastic solutions, and Mindlin's "method of images" (Mindlin, 1936), lead to completely different solutions from those given by limit equilibrium methods (e.g. using Coulomb wedge analysis). While these problems exist when simple methods of calculation are used, they should not occur when using continuum analyses.

Soubra and Kastner present an important paper on the effect of upward seepage on passive resistance. This paper is again directed at users of relatively simple forms of analysis, this time for sheet-pile structures. They give earth-pressure coefficients as a function of upward hydraulic gradient, and show the dramatic effects of this. This paper will be essential reading for all who use limit equilibrium analysis for sheet pile wall design.

Tables 3 and 4 give lists of the papers in this conference (many of which have been allocated to other sessions) which contain useful information for conceptual design, and on the general analytical issues which should concern the designer.

Reference

VAN WEELE, A.F. (1989) "Prediction versus Performance" Keynote Lecture, Proc. 12th Int. Conf. Soil Mech. Found. Engng, Rio de Janeiro, vol. iv. Balkema, Rotterdam.

The selection, design and performance of a multi-propped contiguous pile retaining wall in Norwich

W. J. GROSE, Ove Arup and Partners, UK, and B. H. TOONE, ASA, Bovis Construction Ltd, UK

Introduction

The 'Castle Mall' shopping centre is currently being built in Norwich, Norfolk. It involves the excavation, up to 18 m deep, of an irregular shaped 2.5 ha site surrounded by roads and buildings, and subsequent construction of three levels of shopping malls alongside a five level underground car park. The majority of the 'roof' will be reinstated as parkland. The site is next to Norwich town centre, immediately south of Norwich Castle. Figure 1 shows the site layout relative to surrounding properties and roads. A diagrammatic isometric view southwards is shown on Fig. 2.

Whilst some of the nearby buildings are modern, others are considerably older and distorted due to foundation movement. In addition, the flint walled St. John the Baptist church is only a few metres away form the site boundary as are the Gate Houses and approach bridge to the Grade I listed Castle. The prediction and control of the effects of the new works on these buildings played a major part in the design process for the new wall.

The ground conditions at this site are described in detail elsewhere, see references 1, 2 and 3. In summary the stratigraphy is:

Made ground: 1 m to 10 m of archaeologically rich fill.
Norwich Crag: generally medium dense, fine to coarse sand with lenses of clay, gravel and occasional cobbles. A marine shoreline deposit.
Upper chalk: weak chalk (N = 10 to 20), weathering grade IV.

Chalk in this area is well known for solution features. These are irregularly shaped or inverted cone- shaped cavities in the chalk surface resulting from solution of the chalk by percolating groundwater. The voids thus formed tend to be filled with overlying soil.

The water table is at about +1 mOD, some 8 m below the lowest floor level and between 15 and 26 m below original ground level.

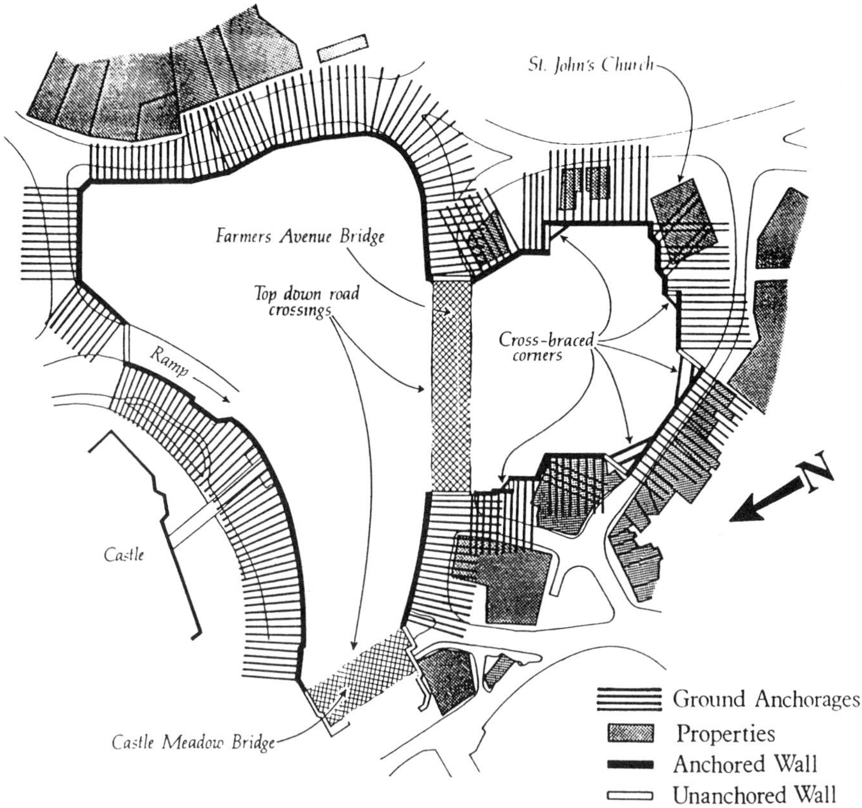

Fig. 1. Site plan

Wall selection

The layout of the project, with full-depth excavation over the whole plan area of the site, led to the early decision to use a concrete cast in situ flexible retaining wall system to provide both temporary and permanent support. Alternatives considered were diaphragm, secant and contiguous walling, as well as a king post and waling system. Factors affecting the choice were:

(i) the possibility of drilling fluid being lost into solution features
(ii) the low water table, watertightness therefore not essential
(iii) minimisation of ground movement
(iv) the ability to recognise and deal with solution features

These points led us to choose a contiguous piled wall, using 900 mm diameter piles at 1050 centres. Piles were bored using temporary casings, and solution features were overcome by lengthening the piles and permanently sleeving

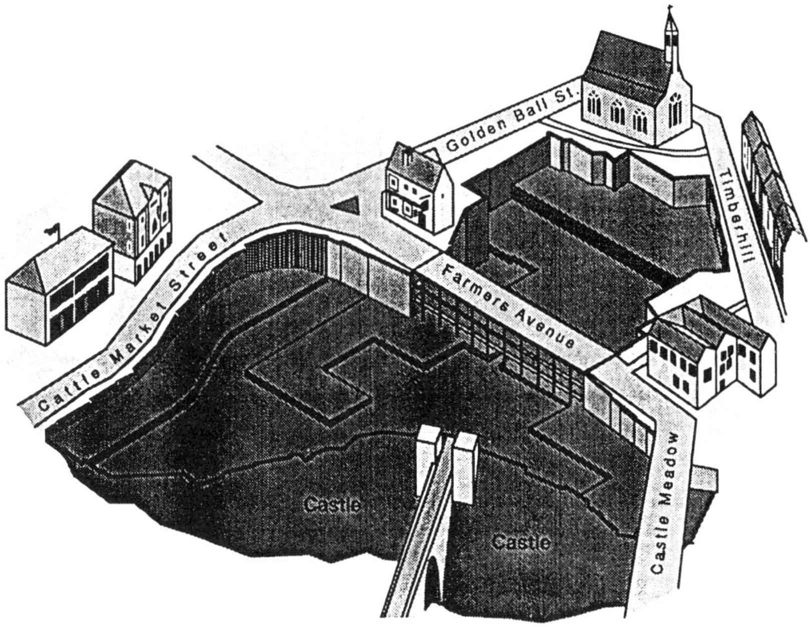

Fig. 2. Isometric sketch of excavation looking south

them to the underside of the feature. The retaining wall was some 800 m long, and was also designed to carry vertical load.

Support selection

Three alternative means of wall support were examined in terms of buildability, programme and cost:

Earth berm with top-down construction around perimeter

This scheme involved installing from original ground level the building's steel columns in open shafts, onto large diameter piles, for a width of about 20 m around the site perimeter. The open, centre section of the site is excavated leaving large berms, and the basement raft and superstructure constructed. The perimeter berm is than excavated in layers, and the building frame installed top-down until the remaining strip of basement raft is completed. This method minimises the temporary works but at the expense of programme flexibility. Elaborate sequencing is required to balance retained soil forces across the building, and column grids must be fixed early in the design.

Earth berm with raking props

A variation on the above, without the columns on large diameter piles but with raking props installed between the permanent raft and the wall as the 20m wide berm is removed. The principal disadvantage of this traditional construction method is the cost of the props.

Temporary ground anchorages

Ground anchorages are installed in layers as excavation proceeds. The large site allows for excavation to proceed in the centre in advance of anchoring at the perimeter. Other advantages are: an unobstructed site, both to following trades and the bulk excavation; column positions need only be fixed later in the construction programme; continuous basement raft not needed - pad footings can be used. A potential disadvantage is that wayleaves are not always given.

An appraisal of these three schemes was based on technical merit, construction safety risk, programme security and cost. Option 3 was significantly cheaper (half the cost of option 2 and one quarter the cost of option 1) and performed well in other comparisons, and was therefore the final choice. Figure 3 shows a typical cross section.

Design

Division of responsibility

The use of a permanent retaining wall to provide temporary support is not

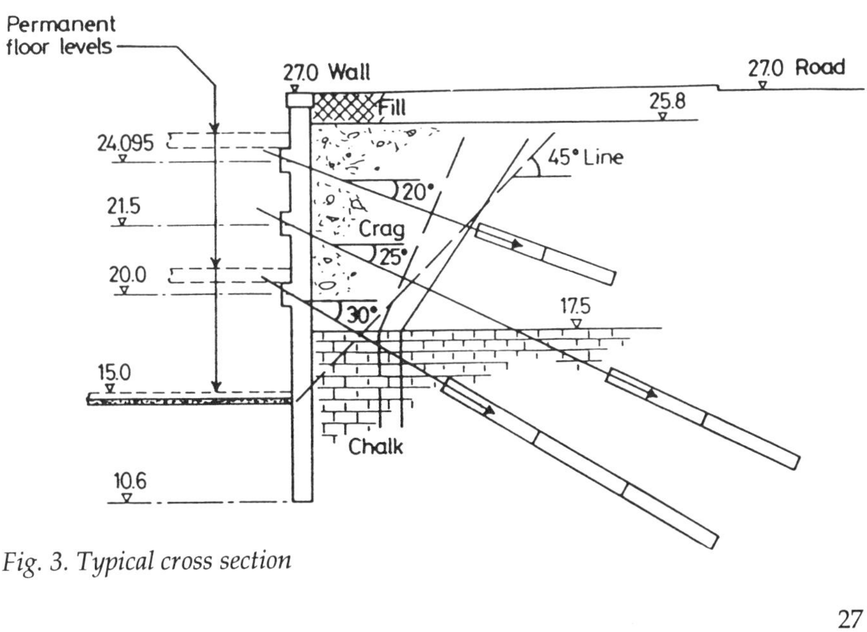

Fig. 3. Typical cross section

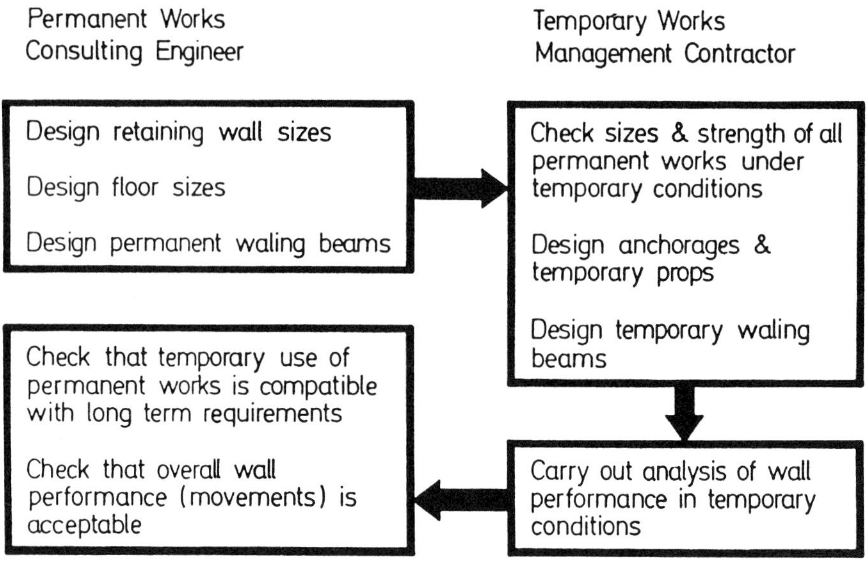

Fig. 4. Division of Design

unusual, but can lead to confusion over the duties and responsibilities of the Temporary Works Designer (in this case the Management Contractor) on the one hand, and the Consulting Engineer on the other. In order to achieve an efficient design process, while ensuring that all elements of the work were addressed, the division of responsibilities and activities relating to the wall design were agreed at the outset between the Engineer and Management Contractor. The flow chart on Figure 4 shows the principal design duties of each party, and their chronological order. This sharing of the design process aimed to combine responsibilities for design with control of the work: thus the temporary wall support (anchors, struts or top-down construction) was chosen, designed, programmed and specified by the Management Contractor, whilst the permanent works design, and its performance in the long term, were the province of the Engineer.

Wall design

Owing to the irregular shape of the site, and the difficulty of obtaining wayleaves all round the perimeter, it was not possible to use ground anchorages everywhere. Across internal corners, especially where close to re-entrant corners, diagonal bracing was used (see Figure 1). A total of 700 m of wall was retained using ground anchorages and, owing to the variation in ground level, stratigraphy and loading conditions around the site, sixty-three different cross sections were analyzed.

The forces taken into account in assessing the stability of the wall are

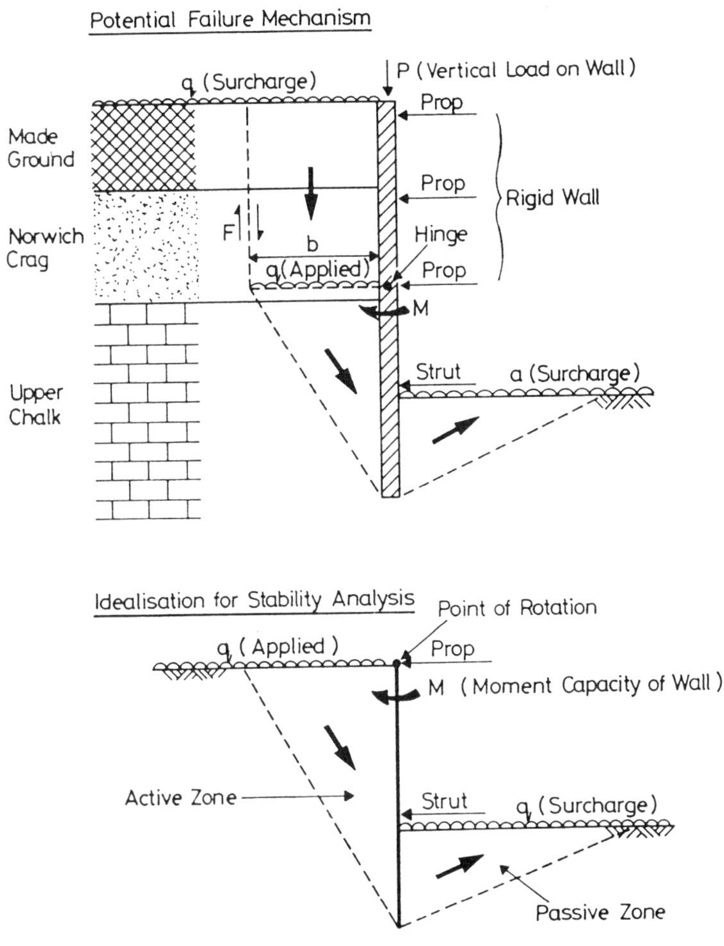

Fig. 5. Failure mechanism considered for wall stability

shown in Figure 5. In the general case, where there are several levels of lateral support, the system is structurally redundant and not amenable to a simple retaining wall analysis. Here, the mode of failure was assumed to be rotation about the bottom prop, the destabilising forces being (i) the vertical load on

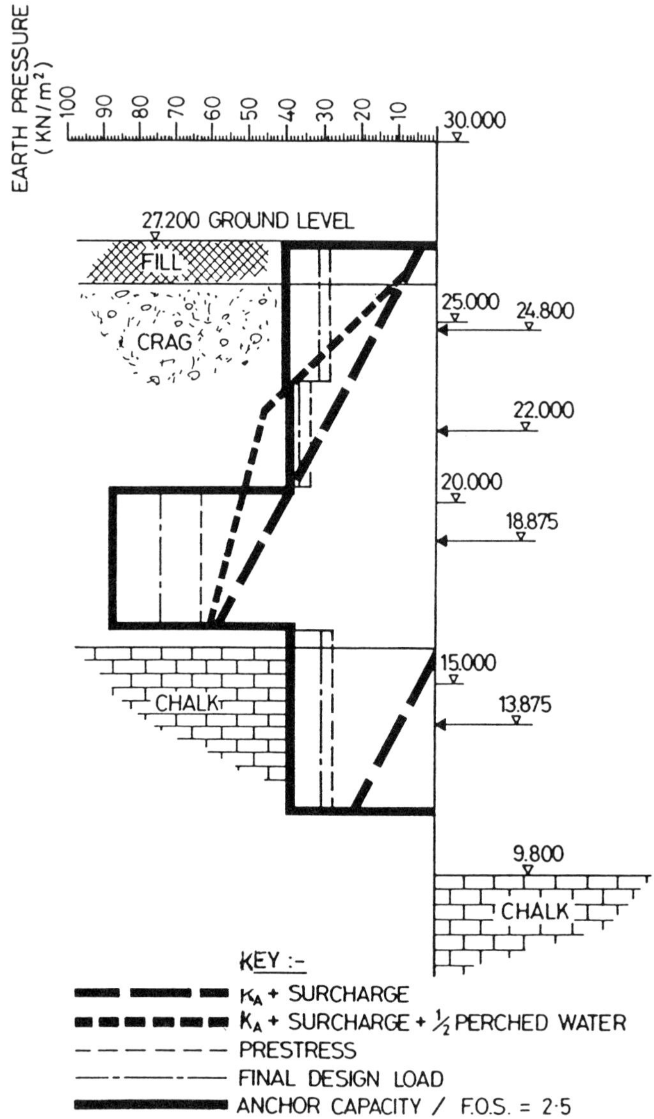

Fig. 6. Typical active pressure and anchor load diagram

the wall (ii) the weight of soil retained above prop level (modelled as a UDL at the lowest prop level), and (iii) the wedge of soil pushing the wall's toe on the active side. The resisting forces were (i) the passive soil resistance at the toe, (ii) the friction between the soil mass and the downward moving soil block and, (iii) the moment capacity of the wall. Generally, satisfactory results were obtained with a toe penetration of 4 m below excavation level or 4 m into chalk whichever was the greater. However, a critical condition was reached when high vertical components of anchor loads were combined with dead load from the permanent structure: this became a limiting condition when several rows of steeply inclined anchorages were needed.

For short term stability it was of course also necessary to ensure that the anchor forces were sufficient to resist the lateral forces on the wall. This was carried out by a simple graphical check on the active pressure diagram (see Figure 6).

Movement control

Of prime importance at this site was the need to limit wall, and hence ground, movements, as described before. Movements were estimated from an accumulation of effects: installation of the wall; excavation and anchoring; destressing of anchors. Whilst these were summed to give final movements, the largest single effect was excavation and anchoring and this aspect received the most attention.

Movement analyses were carried out using commercially available computer programs; the Management Contractor adopting WALLAP and the Engineer using the Oasys program FREW. Whilst these programs are superficially similar, the methods of analyses are not the same: WALLAP models the soil as a series of springs, the stiffnesses derived from a modulus of subgrade reaction, whereas FREW models the soil as an elastic continuum and derives an appropriate stiffness matrix for the soil on each side of the retaining wall.

In general, there were only slight differences in the input values for bulk density (γ_b) and effective angle of friction (ϕ'), although in FREW angles of wall friction (δ) of $2/3\phi'm$ and $1/2\phi'm$ were used on the active and passive sides respectively. For movement calculations FREW used Young's moduli of 5-15MN/m^2 for Made Ground, 20-50MN/m^2 for Crag and 200 to 400MN/m^2 for Chalk. WALLAP calls for a modulus of Subgrade Reaction and values between 2 and 3MN/m^2 were chosen for Made Ground and Crag, and 100MN/m^2 for Chalk.

For this project, the difference in the results produced by these programs were inconsequential, with the deflected shape, ground movements and final anchorage forces being broadly similar in most cases.

Analyses involved modelling the complete construction sequence of excavation, anchoring, excavation, 2nd row of anchors etc. to give deflection

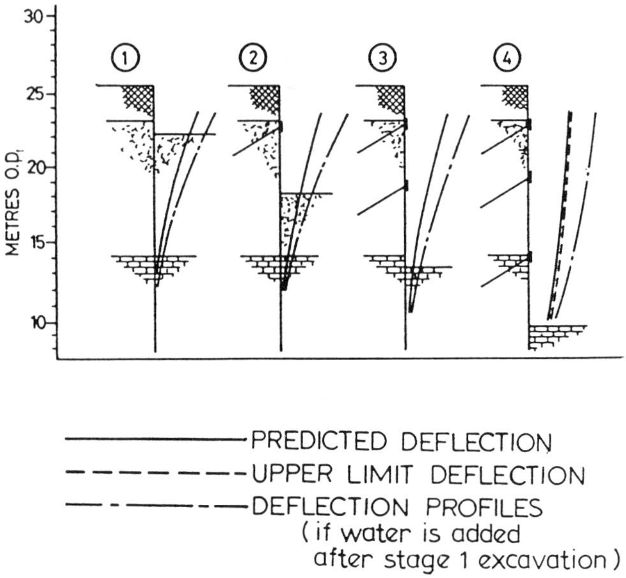

Fig. 7. Stage-by-stage deflection limits

profiles at all stages, see Figure 7. By applying full design loading (traffic loads, superimposed loads, notional water pressure) and using moderately conservative soil parameters, an upper limit of expected wall movement was obtained, and this was used on site as a limiting value for comparison with measured movements. These deflection profiles together with anchorage layouts and design load, waling design and relevant material parameters formed the principal parts of the temporary works design and were incorporated into design reports for issue and checking. Close attention to site monitoring, and comparison of observed performance with the design reports allowed a less conservative design philosophy to be used, thereby saving cost.

In order to be satisfied that the wall's performance was going to be acceptable, lateral movements were translated into vertical and horizontal components of ground surface movements away from the wall. From these, the effect on adjacent structures was assessed. In parallel with the 'moderately conservative' values, 'best estimate' movements were also obtained, using more optimistic soil parameters and not transient superimposed loads. These generally gave results between one third and one half of the 'moderately conservative' movements and later proved to be much closer to the observed values.

Water pressure

As stated elsewhere, the natural groundwater level occurred some 8 m below the deepest excavation level, and 4 m below the general toe level of the pile wall. It was not considered that the groundwater level would rise high enough to be problematic, but recognition was given to the likely sequence of events following a serious burst water main, as pipe diameters in the vicinity were quite large, ranging form 100 mm to 450 mm. The solution was to design the wall for no water table, then check this design with a perched water table in the fill, dissipating uniformly through the Crag to zero at the top of the chalk.

Anchorage failure

The robustness of the anchorage arrangement was tested against the supposition that an anchorage failed in service. Progressive failure obviously had to be prevented. Redistribution of loads to anchorages either side of, and above and below, the 'failed' anchorage was allowed as long as the adjacent anchorages could accept the load increase in the ultimate limit state. Waling beams were checked to ensure that they could span the extra distance, albeit with a lower factor of safety.

Berms

In order to speed up work on site, it was advantageous to excavate the central area of the site somewhat in advance of the anchor construction at the perimeter. The latter was therefore carried out from a platform at the top of a berm. It was thought that excavating in this way may have led to a reduction in the wall's stability and perhaps more movement. This matter was examined using the Oasys program STAWAL in conjunction with FREW. In the stability calculation, it was assumed that excavating to leave a berm would result in a reduction in the passive resistance of the soil adjacent to the wall. Reduced coefficients of passive pressure (Kp*) were derived at intervals down the buried part of the wall based on the most critical wedge or block failure passing through that point, see figure 8. The size of berms was established using a batter slope angle of 30%, approximately ϕ' for the soil, and making the crest width equal to the plan width of the batter slope. The calculations showed that stability was only marginally altered by excavation of berms in this way.

Performance

Considerable effort went into dealing with the concerns of neighbours. Before any site work took place, condition surveys were commissioned for fifty separate premises. Significant cracks were instrumented at this stage for subsequent monitoring. One of the first, and continuing, tasks of the Man-

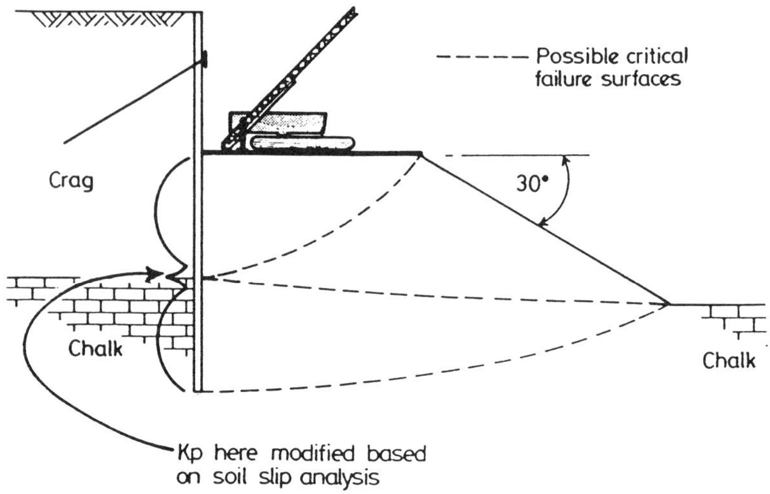

Fig. 8. Modelling of berms

agement Contractor was to survey the area around the site, using over seventy levelling stations in roads, paths and on buildings. During construction of the perimeter wall, some 30 inclinometer tubes were installed, about 25 m apart. Regular monitoring of all these instruments has provided a comprehensive step-by-step picture of the pattern of ground movements.

The results of monitoring which will continue for some time after excavation is complete, has shown ground and wall movements to be always less than 'moderately conservative' estimates and generally less than the 'best estimate' values. An example of the comparison is shown on Figure 9 for one of the deepest parts of the excavation.

Costs

The approximate cost of the wall was £3 m, or £250 per square metre finally exposed. The cost of the lateral support, including anchors, walings and braces is in the order of £3.6 m, or £300/m^2, about 25% of which is attributable to waling beams (including the temporarily used permanent walings).

Concluding remarks

This paper has described the design process for a retaining wall around an unusual-shaped excavation, in difficult ground conditions and with somewhat stringent performance requirements. The case history is an example of the successful combination of economic, safe design achieved within a relatively fast programme, as a result of clear division of responsibilities, close

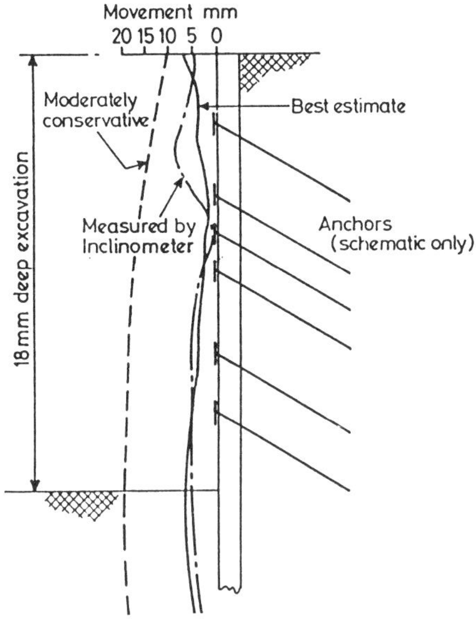

Fig. 9. Measured vs. predicted movement

Fig. 10. Completed wall, 18 m high, near the Farmers Avenue Bridge

cooperation between Engineer and Management Contractor, and a rigorous approach to design.

Acknowledgements

Projects of this magnitude depend heavily on team efforts, and the authors gratefully acknowledge the contribution of their colleagues in Ove Arup & Partners and Bovis Construction Ltd.

Client: Estates & General plc and Friends Provident
Consulting Engineers: Ove Arup & Partners
Management Contractor: Bovis Construction Ltd
Trade Contractors: Expanded Piling Ltd (contiguous pile wall)
Keller Colcrete Ltd (ground anchorages)
Geotechnical Instruments Ltd (inclinometer readings).

References

1. LORD J.A. (1989) Foundations in Chalk, Keynote address, International Chalk Symposium, Brighton.
2. TWINE D. AND GROSE W.J. (1989) Contribution to discussion, International Chalk Symposium, Brighton.
3. BARLEY A., EVE R. AND TWINE E. (1992) Design and construction of temporary ground anchorages at Castle Mall, Norwich.

Installation effects of diaphragm walls in clay

W. POWRIE and C. KANTARTZI, Queen Mary and Westfield College, UK

The factors influencing the changes in horizontal stress and pore water pressure and ground movements during diaphragm wall installation are discussed with reference to the results of a centrifuge model test.

Introduction

Designers of diaphragm-type retaining walls in clay soils are becoming increasingly aware of the importance of the ground movements and changes in lateral stress which may occur during construction. Although the horizontal effective stresses in an over-consolidated clay deposit will initially be high, they may be reduced significantly during wall installation. Finite element analyses (starting with the wall already in place) have shown that the influence of the assumed pre-excavation lateral stresses on the subsequent behaviour of a wall can be considerable (e.g. Potts and Fourie, 1984: Powrie and Li, 1991). Furthermore, the ground movements which occur during installation could be larger than at any other stage, especially if the wall is well-propped during and after excavation.

These issues are currently being investigated in a series of centrifuge model tests at the London Geotechnical Centrifuge Centre. In this Paper, the centrifuge modelling procedure is described; data are presented from a model test in which the excavation of a diaphragm wall trench under bentonite slurry was simulated; and the factors affecting ground movements and changes in stress during diaphragm wall installation are discussed.

Centrifuge modelling procedure

The centrifuge model, illustrated in Fig. 1, was formed from a block of speswhite kaolin clay, of plan dimensions 200mm x 550mm and depth 285mm, contained within an aluminium strongbox with a perspex viewing window in the front face. The sample was prepared by one-dimensional compression from a slurry of water content 100% to a vertical effective stress of 1250kPa. The sample was allowed to swell back to a vertical effective stress of 80kPa before being trimmed to the required depth of 285mm. The unit

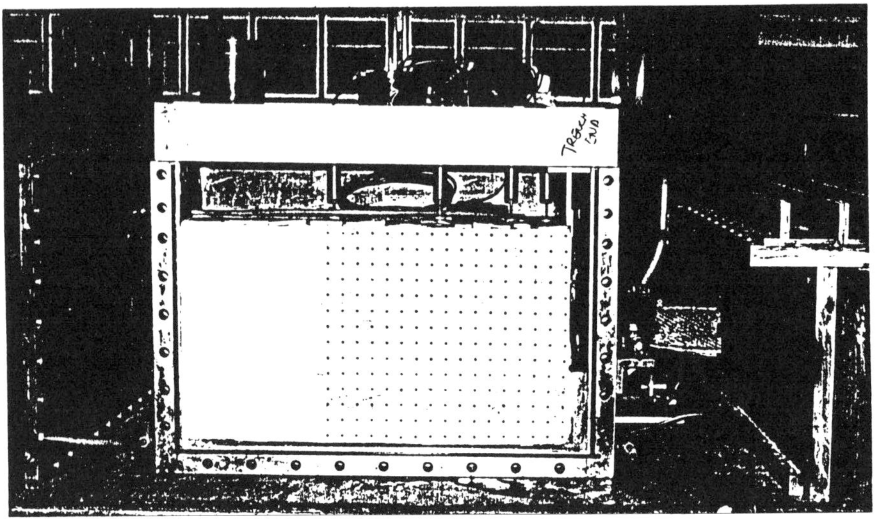

Fig. 1. View of centrifuge model in strong box

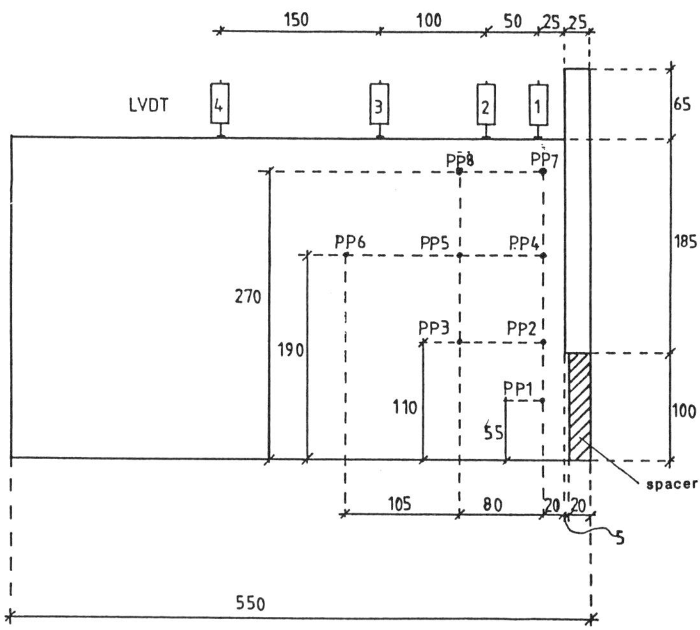

Fig. 2. Layout of centrifuge model and instruments (dimensions in mm)

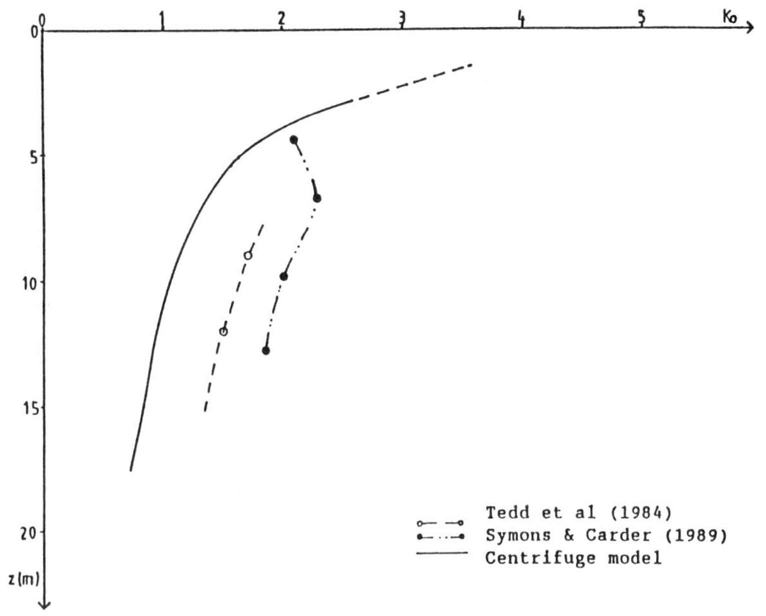

Fig. 3. Pre-excavation effective stress earth pressure coefficients

weight of the clay at this stage was 17.9kN/m^3. A 185mm-deep slot, simulating the excavation for the diaphragm wall, was cut at the right-hand edge of the clay sample, and the clay removed was replaced by a rubber bag filled to a level of 65mm above the soil surface with sodium chloride solution of unit weight 11.4kN/m^3. A rigid spacer below the slot was used to reduce the effective half-width of the trench to 5mm at model scale. At a scale of 1:100, the slot represented one half of a lm-wide diaphragm wall trench, 18.5m deep. The model was instrumented with Druck miniature pore water pressure transducers, and lvdts to monitor the movement of the soil surface, as indicated in Fig. 2.

The clay sample was reconsolidated in the centrifuge at l00g for a period of 2.5 hours until equilibrium conditions had been reached, with the pore water pressures steady and the rates of soil surface settlement negligible. During reconsolidation, water was supplied to both the soil surface and a sand drain at the base of the sample. The profile of earth pressure coefficient $K_o=\sigma'_h/\sigma'_v$ with depth in the soil adjacent to the trench at the end of reconsolidation is compared in Fig. 3 with field profiles for London Clay: it may be seen that earth pressure coefficients in the centrifuge model at the end of reconsolidation are significantly smaller, but their variation with depth is not dissimilar.

The changes in horizontal stress associated with the excavation of a diaphragm wall trench under bentonite slurry were simulated by draining the level of the salt solution in the rubber bag down to the soil surface, by means of a valve-controlled overflow pipe. During this process, the clay adjacent to the trench was subjected to a uniform reduction in horizontal total stress of 70kPa, and the resulting soil movements and changes in pore water pressure were measured. An attempt to reduce the horizontal stress still further by diluting the fluid remaining in the rubber bag with fresh water was not successful. The time taken to lower the level of the salt solution in the rubber bag was 5 seconds, which corresponds to 14 hours at full scale. Pore water pressures and soil movements were monitored for a further period of 146 seconds following excavation, corresponding to 17 days at field scale: this is likely to be beyond the upper limit to the time during which a trench would in reality remain open under bentonite before the concrete was placed. At all stages of the test, the fluid pressure in the rubber bag was generally sufficient to exclude water from the interface between the rubber and the clay sample. The face of the trench would not, therefore, have acted as a drainage boundary, except following excavation when the surface water could have penetrated the interface to a depth of approximately 20mm at model scale.

Changes in pore water pressure

Except where otherwise indicated, all results are presented at full scale, with linear dimensions and displacements multiplied by 100 and elapsed times multiplied by 100^2.

Fig. 4 shows the pore water pressures measured at various locations within the soil as functions of time before, during and after excavation. The pore

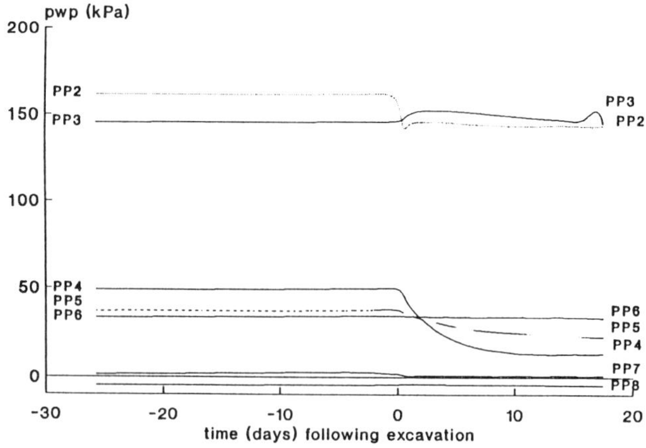

Fig. 4. Changes in pore water pressure during excavation

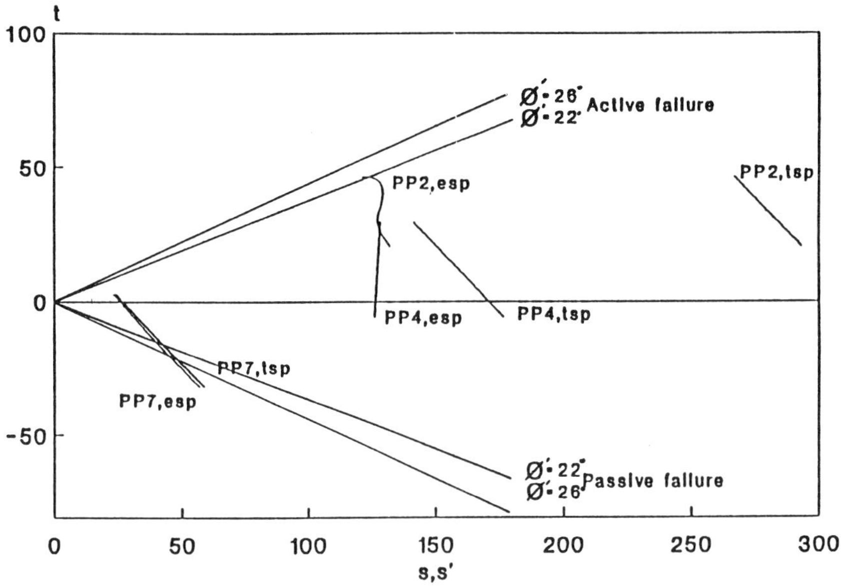

Fig. 5. Stress paths in t,s′ and t,s space

water pressures immediately prior to excavation corresponded approximately to hydrostatic conditions below a groundwater level just beneath the soil surface. The pore water pressures were sub-hydrostatic at mid-depth (PP4), and decreased slightly at each level with distance from the trench: however, the main point is that at the start of excavation, the pore water pressure regime was steady and reasonably well-defined.

During excavation, significant reductions in pore water pressure were observed at PP2, PP4, and PP5. The changes indicated by PP6, PP7 and PP8 were small, presumably because these transducers were either outside the zone of influence of the excavation (PP6), or very close to the drainage boundary (PP7 and PP8). The reason for the apparent increase in pore water pressure at PP3 is uncertain. The changes in pore water pressure recorded during the gradual reduction in centrifuge speed at the end of the test suggest that the delayed response of PP4 and PP5 was due to problems with the response time of these transducers, rather than a real effect: the overall magnitude of the changes in pore water pressure at these locations following excavation is nonetheless considered to have been reliably measured.

Assuming that the horizontal total stress in the soil at the positions of the pore water pressure transducers PP7 and PP4 is equal to the fluid pressure in the trench at the same level, the total and effective stress paths followed by soil elements at these locations may be plotted in t,s and t,s′ space (Fig. 5).

For the soil at PP2, near the bottom of the trench, there is more opportunity for stress redistribution, so that the reduction in horizontal total stress here will only be a proportion α ($\alpha < 1$) of that which occurs at the same level in the trench: the stress paths shown in Fig. 5 are plotted assuming α=0.75, a value calculated on the basis of an elastic stress distribution. ϕ' values correspond to peak (26°) and critical states (22°) reported by Bolton & Powrie (1987).

The soil response at PP7, 1.5m below the ground surface, was apparently drained, ie no significant excess pore water pressures were generated. This is just about plausible: taking c_v=2.5mm^2/s (Bolton and Powrie, 1987), excess pore water pressures would begin to dissipate at this depth after 7.5 seconds at model scale, assuming one-way drainage from a flooded soil surface. A higher value of c_v (due, for example, to an increased soil stiffness following a change in the direction of the stress path), and possible horizontal drainage from the top of the trench, would both tend to accelerate the rate of dissipation of excess pore water pressure. In a real situation, however, the soil surface is unlikely to be flooded, and unless the slurry in the trench acts as a source of water, the rate of excess pore pressure dissipation will probably be much slower.

The soil response at PP4 and PP2 is essentially "elastic", with changes in pore pressure occurring so that the average effective stress s' remains constant. In both cases, however, the soil is brought quite close to the active failure condition.

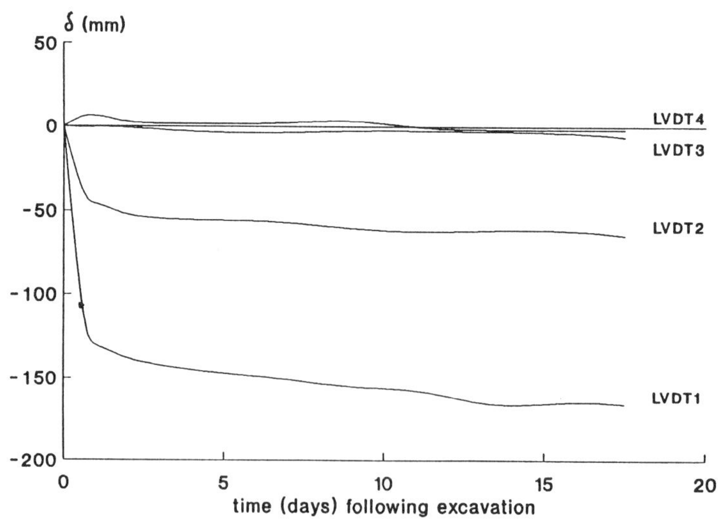

Fig. 6. Soil surface settlements during excavation

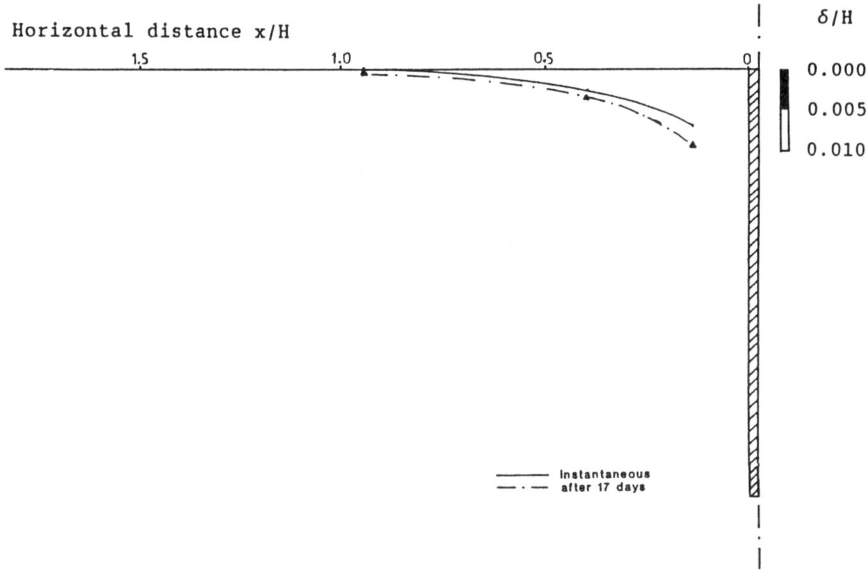

Fig. 7. Normalized soil settlement profiles following excavation

Ground movements

Figure 6 shows the development of soil surface settlements during and after excavation at distances from the trench of 2.5m, 7.5m, 17.5m and 32.5m. Figure 7 shows the soil settlement profile (normalized with respect to the depth of the trench H) immediately following excavation, and 17 days later. The maximum displacement is quite large (120mm, increasing after 17 days to 160mm at prototype scale), but the settlements beyond a distance of one trench depth are insignificant.

On a full scale site, with more favourable pore water pressure conditions and a stiffer soil; settlements would be expected to be rather smaller. There is no reason why the pattern of displacements, however, should be expected to be significantly different. The soil surface settlements would also be affected by the magnitude of the change in horizontal stress which occurs during wall installation: in a soil where the in situ horizontal stresses remained substantially unchanged, displacements might be expected to be very small.

Concluding discussion

Although it is not possible to draw firm conclusions from the results of only one test, the data presented in this paper indicate that significant soil settlements may occur within approximately one trench depth of the excavation during the installation of a diaphragm wall panel. Measured changes in pore water pressure suggest that the soil at mid-depth of the trench may experience a reduction in horizontal stress up to a similar distance sway, but that the distance of influence at the level of the bottom of the trench is much smaller.

The test reported in this paper must be viewed as a preliminary study only. The patterns of displacements and changes in pore water pressure are likely to depend on a number of factors, including:

1. The initial horizontal stresses, which may be higher or lower than those described in this paper, leading to more or less significant lateral unloading and deformation of the soil during excavation of the slurry trench.

2. The excavation geometry: the centrifuge test described in this paper modelled deformation in plane strain. The construction of a single diaphragm wall panel or a single circular pile (as part of a bored pile wall) would allow redistribution of the lateral stress reduction in the perpendicular horizontal direction, which would be expected to reduce the magnitude and perhaps the extent of ground movements.

3. The drainage path length, which would affect the rate of excess pore water pressure dissipation during wall installation, and hence the significance of ground movements due to volumetric strains.

It is intended to investigate these factors by means of further centrifuge model tests, and also to simulate the casting of the concrete wall in the centrifuge, which may result in an increase in the boundary stresses at the face of the trench, and corresponding changes in strain in the adjacent ground.

Acknowledgements

The Authors are grateful for the financial support of the Science and Engineering Research Council (grant no. GR/F/59278), and to Drs S. A. Jefferis and R. N. Taylor, and Messrs R Nelson, B Nicholson and H Skinner for technical advice and assistance.

References

BOLTON, M.D. and POWRIE, W. The collapse of diaphragm walls retaining clay. *Geotechnique*, 1987, **37**, no.3, pp 335 - 353.

POTTS, D.M. AND FOURIE, A.B. The behaviour of a propped retaining wall: results of a numerical investigation. *Geotechnique*, 1984 , **34**, no. 3, pp383 - 404.

POWRIE, W. and LI, E. S. F. Finite element analyses of an in situ wall

propped at formation level. Geotechnique, 1991, **41**, no.4.
SYMONS, I. F. and CARDER, D. R. Long term behaviour of embedded retaining walls in overconsolidated clay. *Instrumentation in geotechnical engineering*, pp 155 - 173, Thomas Telford Ltd, London, 1989.
TEDD, P. *et al* Behaviour of a propped embedded retaining wall in stiff clay at Bell Common Tunnel. *Geotechnique*, 1984, **34**, no.4, pp 513-532.

Finite element modelling of installation effects

M. J. GUNN, A. SATKUNANANTHAN and C. R. I. CLAYTON,
University of Surrey, UK

Introduction

The process of excavating a hole in the ground and filling it with wet concrete changes the ground stresses in the vicinity of the hole. These changes are not, generally, taken into account in the design of diaphragm, secant and contiguous bored pile walls. The structural design of these types of wall has normally been based on bending moments and prop or anchor forces derived from limit equilibrium calculations. These calculations can be criticised on the grounds that they do not take into account several factors which might be expected to have some effect on the wall bending moments. In particular the in-situ ground stresses, the stiffness of the ground, the stiffnesses of any props and the effects of the construction procedure do not appear explicitly in the limit equilibrium approach.

A soil structure interaction (SSI) analysis technique, using for example the finite element method, has potential to take into account all the above factors. Finite element techniques have generally been considered too time consuming and complex for routine design but, with continuing falls in the cost of hardware and software, their use is increasing. A well documented case of the use of the finite element method to supplement limit equilibrium calculations was the Bell Common cut and cover tunnel on the M25 (Hubbard et al, 1984). Here the analyses were elastic and the values at the bending moments were calculated to be up to ≈600kNm/m. These values were much lower than those from the limit equilibrium approach (3000kNm/m was obtained using CP2 and eventually a design moment of 1450kNm/m was adopted based on the Burland-Potts revised method (Burland et al, 1981)).

Potts and Fourie (1984) reported on some elasto-plastic finite element analyses of propped cantilever retaining walls of a similar geometry to the Bell Common wall, but where some simplifying assumptions had been made (for example the pore pressures were assumed to be zero everywhere). A later publication (Fourie and Potts, 1989) examined cantilever walls. An important finding from the elasto-plastic analyses of the propped walls was that the bending moments and prop forces were much higher for high K_0 soils (e.g. stiff, overconsolidated clays) than for low K_0 soils (e.g. lightly overconsolidated clays). In particular it was found that the bending moments for high K_o, soils were higher than those obtained from limit equilibrium analyses. These results were regarded as significant since the Bell Common tunnel was constructed partly in London Clay with K_o values in the range of 1.5 to 2.0.

The performance of the Bell Common tunnel was monitored (Tedd et al, 1984; Symons and Tedd, 1989) and it has been found that the higher bending moments suggested by finite element analysis have not been achieved in practice (a maximum bending moment of about 300kNm/m has been deduced from measured earth pressures and wall strains). One explanation of this is that the finite element analyses "wished" the wall in place. In other words they assumed that before excavation in front of the wall took place the wall was installed in the ground with no change in the in-situ stresses. As noted above one can expect significant installation effects which modify the ground stresses.

This Paper reports preliminary results of finite element analyses which include the effect of wall installation. The changes in pore water pressures during wall construction and the resulting time dependent flow of water through the soil were modelled using the coupled consolidation option of the CRISP program (Britto and Gunn, 1987).

Modelling assumptions

The analyses reported here were carried out for a wall with a similar geometry to the Bell Common retaining wall. The basic finite element mesh used in these studies is shown in Fig. 1. We have followed Potts and his co-workers in using an idealised geometry and set of soil properties so that comparisons can be made with limit analyses and previous work. Thus we have assumed that the wall is constructed in a material which yields according to the Mohr-Coulomb condition with a drained angle of friction of 25 degrees and a drained cohesion of zero.

The total height of the wall was 20m and its thickness lm. The excavation for the wall was modelled as a plane strain slot rather than the more realistic three-dimensional sequence of excavation used for real diaphragm or pile walls.The retained height of soil was taken as 8m and two cases were considered - in the first the wall was an unpropped cantilever and in the second there was a single rigid prop at the top of the wall. Three different positions of the ground water table have been assumed: at the ground surface, at the base of the retained soil (i.e. a depth of 8m) and at the base of the wall. The coefficient of earth pressure at rest was taken to be 2 throughout.

The bulk unit weights of the pore water, soil and concrete were taken as $10kN/m^3$, $20kN/m^3$ and $23kN/m^3$ respectively. The Young's modulus of the soil has been taken as being 250p' where p' is the mean effective stress at the same depth. This leads to three different distributions of modulus with depth

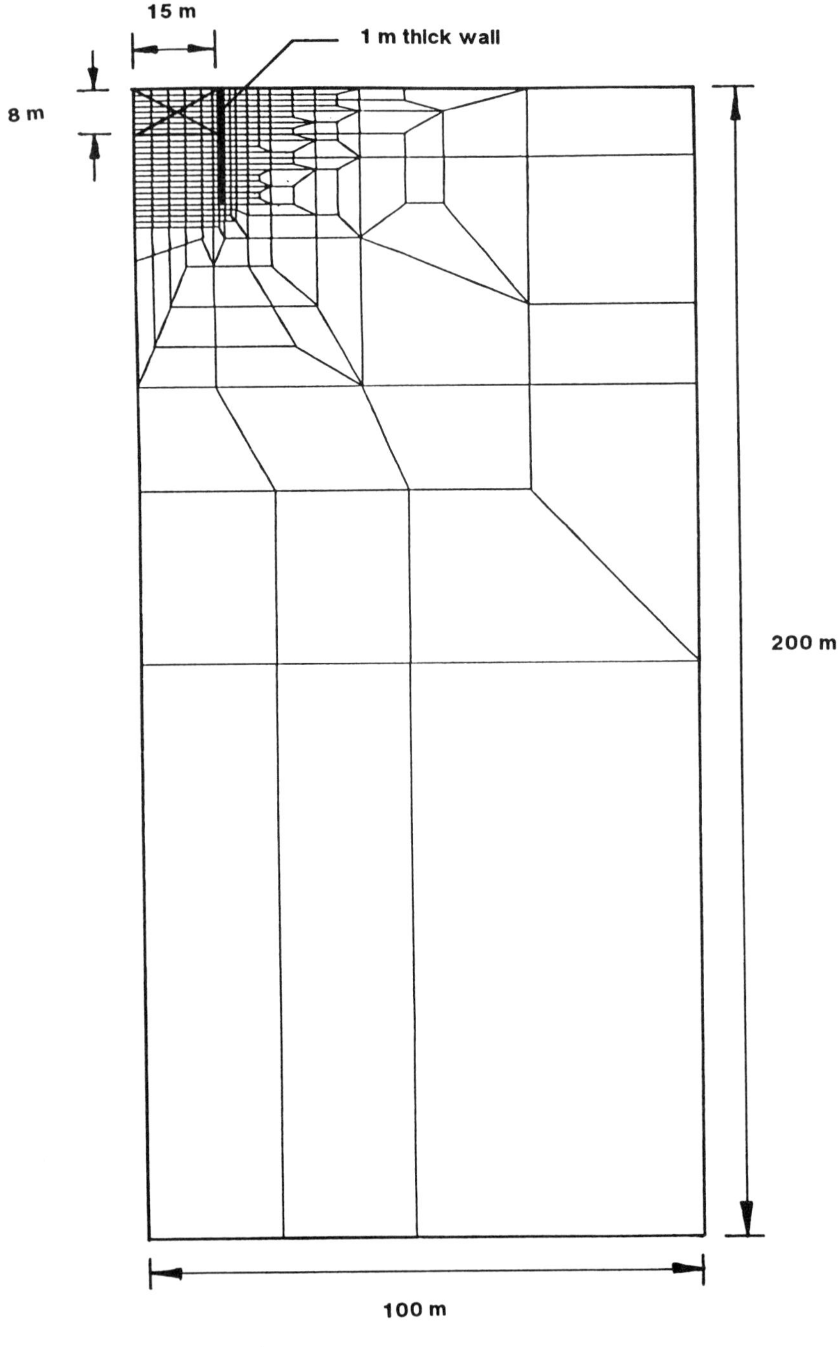

Fig. 1. Finite element mesh

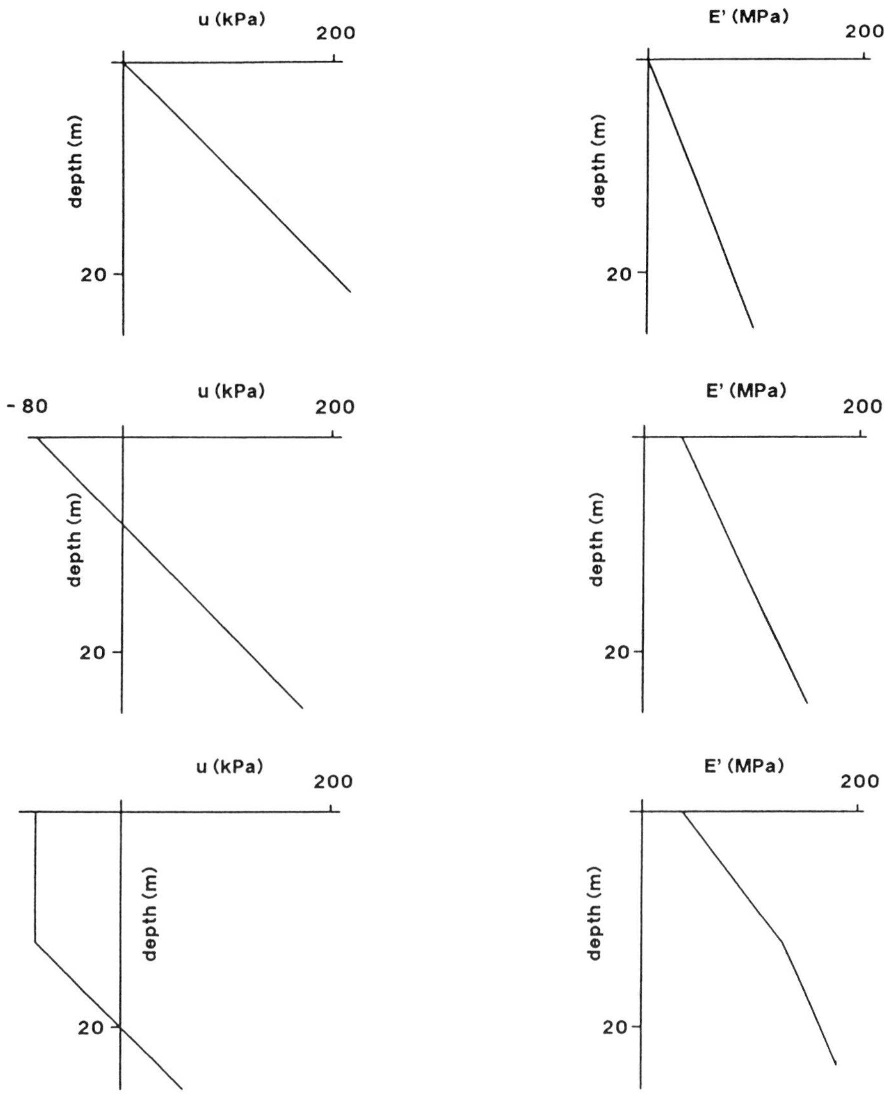

Fig. 2. Three assumed distributions of pore pressure and Young's modulus with depth

shown in Fig. 2 for the three different positions of the ground water table. The soil was assumed to have a drained Poisson's ratio of 0.2 and the wall had a Young's modulus of 28 GN/m^2 and a Poisson's ratio of 0.15.

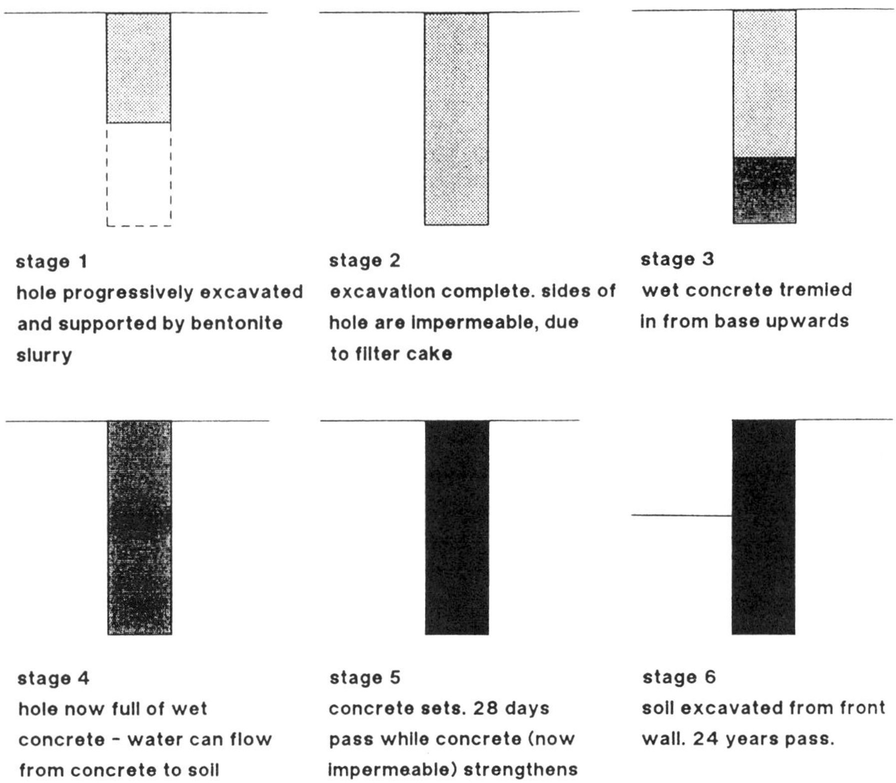

Fig. 3. Stages in analysis

Two types of analysis were carried out in order to ascertain the effect of wall installation. In the first type ("No Installation" or "NI") the wall is wished into place with no changes in the ground stresses and then the soil is excavated in front of the wall. In the second type of analysis ("Installation"or "I") the whole procedure of installing the wall in the ground is modelled according to the scheme shown in Fig. 3, and then excavation is carried out. Analyses were carried out for both cantilever and propped walls.

The bentonite support was modelled by applying the hydrostatic pressure due to the self weight of the bentonite (12 kN/m^3) to the sides of a hole which were assumed to be impermeable. The wet concrete also initially supports the hole with a fluid pressure, but here water flow was allowed to take place between the concrete and the soil for twelve hours, after which the concrete was assumed to have set. At this point the hydrostatic pressure was removed and finite elements representing the concrete wall were introduced. After 28 days (during which there was flow of water internal to the soil as excess pore

Table 1. Maximum bending moments

water table	no installation		installation	
	short term	long term	short term	long term
CANTILEVER WALLS				
at surface	560	2150	530	2220
at 8m depth	490	-350	-290	730
at 20m depth	640	430	350	760
PROPPED WALLS				
at surface	-1720	-2680	-1210	-2120
at 8m depth	-1810	-2630	-870	-1470
at 20m depth	-1760	-2230	-730	-910

Note: Positive moments correspond to concave curvature towards the excavation.
short term = after excavation in front of the wall
long term = after 24 years

pressures tended to equalise, but no flow between the concrete and soil) excavation of the soil in front of the wall was modelled. The final condition was either one of steady seepage around the wall (when the initial ground water level was at the surface) or otherwise hydrostatic, corresponding to the initial pore water pressure distribution.

Analysis

As may be deduced from the account given above, twelve analyses have been carried out in all: for three initial positions of the ground water table, two types of wall (cantilever or propped at the top) were analyzed according to the NI and I assumptions. We discuss below just one aspect of the results: the bending moment distributions in both the short term (after excavation in front of the wall) and long term (24 years after wall construction). Table 1 summarises the maximum moments in the short and long term for the twelve analyses.

Fig. 4 shows the bending moment distributions calculated for the cantilever walls. The NI and I analyses for the case with the water table at the ground surface are shown in Figs 4(a) and 4(b) respectively. The effect of wall installation is to increase the long term bending moments by a small amount (about 3%). The magnitudes of the moments (around 2200kNm/m) are relatively high, for example Fourie and Potts (1989) report maximum moments of approximately 1500kNm/m and 900kNm/m produced by excavating 8.85m in soil with K_0 values of 2 and 0.5 respectively. There are, of course, several differences in the assumptions made in our analyses and

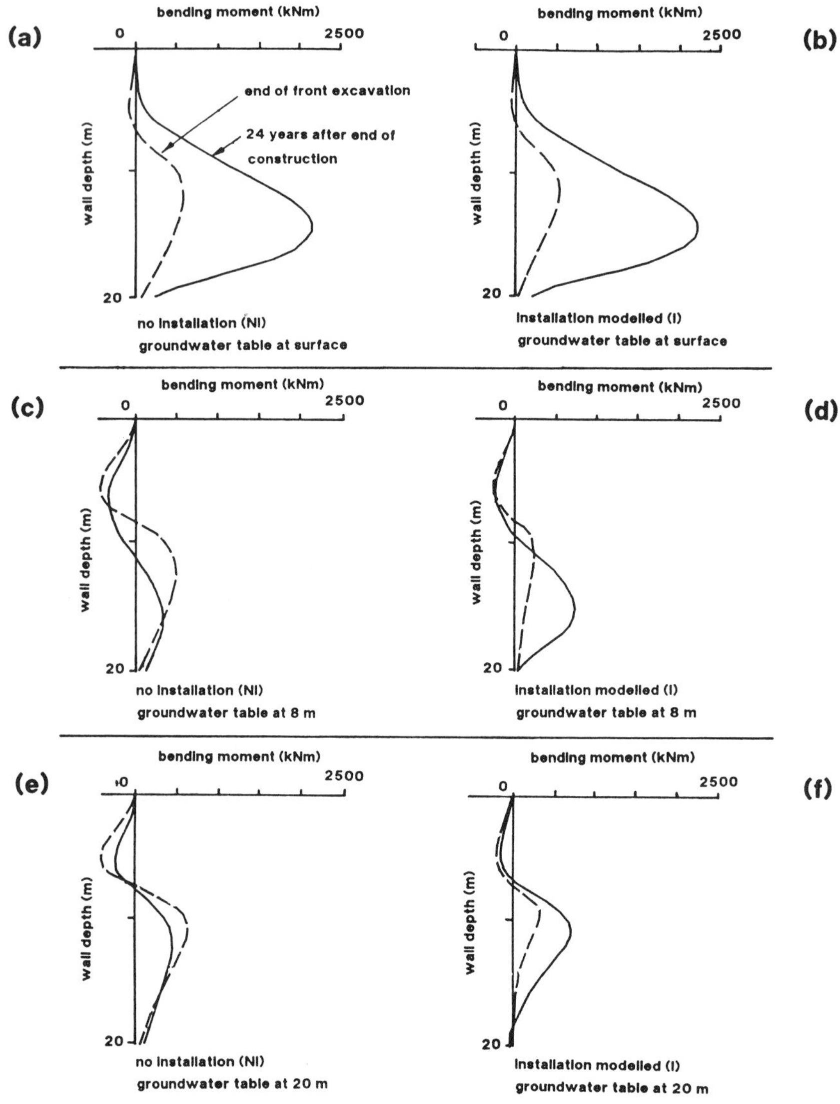

Fig. 4. Cantilever walls

those of Fourie and Potts. The most significant difference is that Fourie and Potts assumed that pore pressures in the soil were zero.

Investigation showed that the major contributing factor to our moments was, indeed, the water pressure behind the wall. This finding prompted the analyses with water table at 8m and 20m depth, the results of which are shown in Figs 4(c) to 4(f). As can be seen the lowering of the water pressure greatly diminishes the bending moments (see Table 1). A significant factor in all the long term results is the final pore pressure distribution acting on the wall. Corresponding to the bending moments in Figs 4(c) to 4(f) there is an area of negative total stress acting between the retained soil and the wall, i.e. the soil is "holding up" the wall. Of course this negative total stress,is due to the negative pore pressures that were assumed initially above dredge level and the continued existence of such a suction against a wall (as opposed to within the soil itself) would not be relied on in the long term.

Fig. 5 shows the bending moment distributions for walls propped at the top. The effect of propping was achieved in the analysis by fixing the corner node on the wall so the prop is being modelled as rigid. Considering first the NI analyses (i.e. Figs 5(a), 5(c) and 5(e)), different initial positions of the water table do not lead to significant differences in the maximum moments calculated for the short term condition (about 1800kNm/m). The long term bending moments for the water table at the surface and at 8m are similar (2680kNm/m and 2630kNm/m respectively). For the case of the water table at a depth of 20m, the maximum long term moment is less (the value is 2230kNm/m) but the difference with the other cases is not particularly significant. The bending moments for the NI analyses are broadly comparable with those obtained in the elasto-plastic analyses performed by Potts and Fourie (1984) - approximately 3000kNm/m for excavation to a depth of 9.26m when K_o=2.

Turning now to the corresponding I analyses (i.e. Figs 5(b), 5(d) and 5(f)), a marked decrease in both the short term and long term bending moments as a result of modelling wall installation is apparent (see Table 1). In the case where the water table is at the ground surface, the reduction in the long term moment is 21%, whereas the reductions are 44% and 59% for water tables at 8m and 20m depths respectively. The corresponding reductions in the short term moments are 30%, 52% and 59%.

Discussion and conclusions

For cantilever walls the results of our analyses predict that the effect of modelling wall installation is to reduce the short term bending moments and to increase slightly the long term bending moments. The walls are relatively free to move and the stresses acting on them tend to approach the active and passive limits. The most important influence on the final bending moments is the water pressures in the ground. The temporary reduction in ground stresses caused by wall installation has no lasting effect. This is essentially what is assumed when moments are calculated using the limit equilibrium approach - the initial conditions and the construction procedure seems to have little effect.

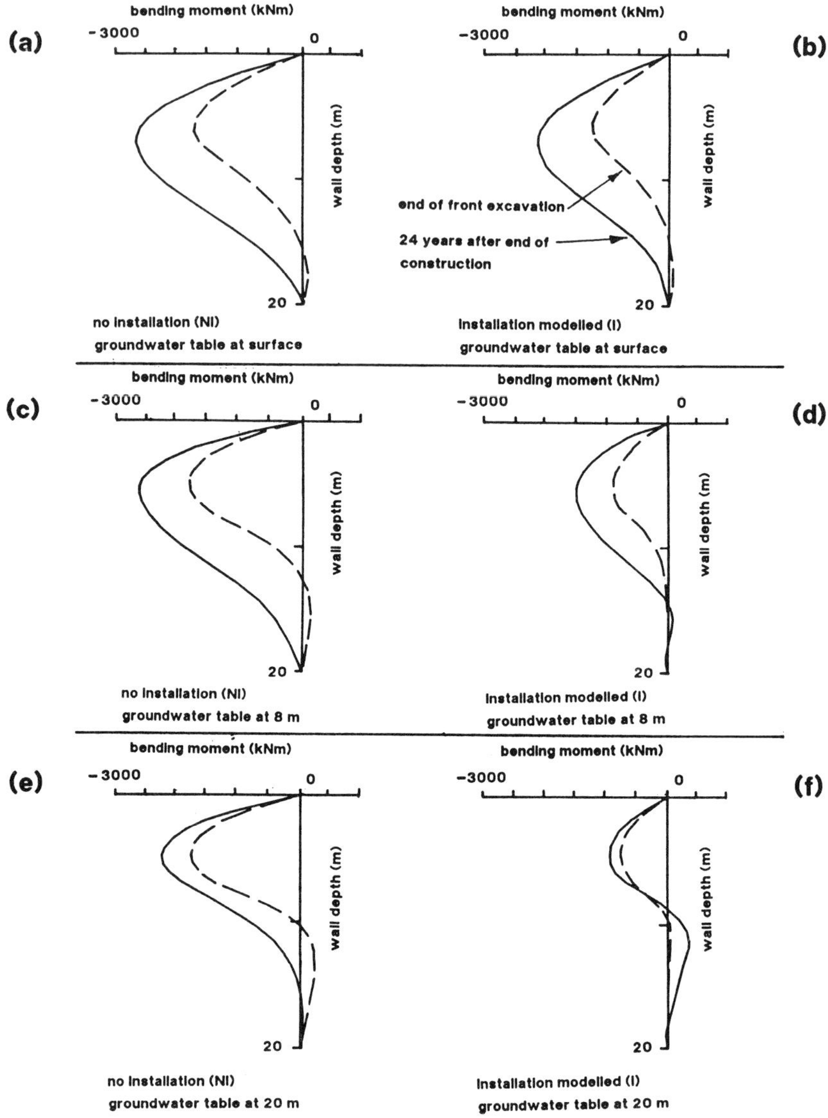

Fig. 5. Propped walls

In contrast, installation effects do seem to be significant when walls are propped. This can be understood as the propping action "locking in" the reduction in lateral stresses associated with wall installation - the soil around the wall does not have the same freedom to strain and approach the classical stress distributions. However, the magnitude of installation effects depends strongly on the initial position of the ground water table. Installation effects are relatively minor when the groundwater table is high, and become more significant as it falls below excavation level.

When considering the relevance of these results to design it is salutary to recall that in the case of the Bell Common wall that the maximum moments inferred from measurements are estimated to be about 300kNm/m. Not only are these much smaller than those used in design, they are actually of the opposite sign. As Hubbard et al (1984) point out, the designer is unable to predict all the events that might take place during the construction of a wall. This points to the importance of employing all types of predictive analysis to establish the possible bounds on the behaviour of the system being designed. A "best shot" type of analysis is probably unlikely to hit the target.

References

BURLAND, J. B. *et al* . The overall stability of free and propped embedded cantilever walls. *Ground Engineering*, 1981, **14** (5), pp 28-38.

BRITTO, A. M and GUNN, M. J. *Critical State Soil Mechanics via Finite Elements*, Ellis Horwood, Chichester, 1987.

FOURIE, A. B. and POTTS, D. M. Comparison of finite element and liniting equilibrium analyses for an embedded cantilever retaining wall. *Geotechnique*, 1989, **39**, 175-188.

HUBBARD, H.W. *et al.* Design of the retaining walls for the M25 cut and cover tunnel at Bell Common. *Geotechnique*, 1984, **34**, pp 495-512.

POTTS, D. M. and FOURIE, A. B. The behaviour of a propped retaining wall: results of a numerical experiment. *Geotechnique*, 1984, **34**, pp 383-404.

SYMONS, I. F. and TEDD, P. Behaviour of a propped embedded retaining wall at Bell Common Tunnel in the longer term. *Geotechnique*, 1989, **39**, pp 701-710.

TEDD, P. Behaviour of a propped embedded retaining wall in stiff clay at Bell Common Tunnel. ***Geotechnique***, 1984, **34**, pp 513-532.

Surcharge and elasto-plastic computations of earth retaining structures

P. J. SCHMITT and C. M. GILBERT, Soletanche Entreprise, France

Elastoplastic computation of earth retaining structures based on the coefficient of subgrade reaction is now common practice in the Geotechnical profession and has started to replace the Terzaghi and Peck earth pressure distribution rules for design. Simple and efficient algorithms which ease the designer's work are widespread and available. Nevertheless this method has some shortcomings as it mixes elastic and plastic theories when surcharge loads are present. Stress distribution generated by surcharge loads are either computed by the elastic theory (Boussinesq) or the plastic theory (Krey, Ohde) where the soil is in an elastoplastic state. Today no theory takes into account elasto-plastic behaviour of the soil to determine stress distribution due to surcharge load. This article reviews the assumptions made in these different theories of stress distribution for both elastic and plastic behaviour and gives comparative examples where the high variability of results, depending on the theory chosen (more than 100%), are stressed. Particularly, the so-called "theory of image" is analyzed and its validity is discussed. Understanding the assumptions and their consequences will help the designer choose the most appropriate method of design in each particular case.

Introduction

Surcharge loads and the influence on the stress distribution resulting from elastoplastic computations is very important. Particularly today where foundation Engineers have to build retaining structures adjacent to multi-story buildings. In this case, the pressure acting on the retaining structure comes mainly from the surcharge load and its determination is essential for a good and safe design. Underestimating these stresses may result in large displacements and cracking in the adjacent structure as well as in the retaining structure itself. On the other hand, excessive overestimation may result in useless expenses or uncompetitive projects. Faced with this problem, the Engineer has to make a decision as to the method to use, and it is with this goal in mind that the different methods of design are analyzed and compared. The problem that we analyze in this article is the computation of horizontal

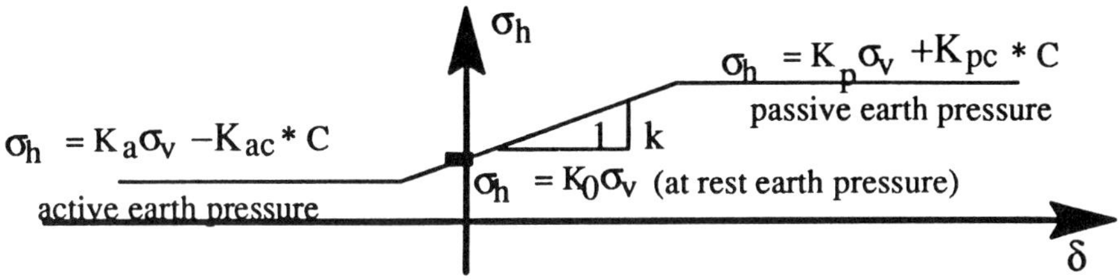

Fig. 1. Elasto-plastic earth pressure distribution

stresses acting on a vertical retaining structure due to vertical surcharge load applied on a horizontal ground. This analysis excludes inclined loads such as slopes or berms resting on a horizontal surface. Moreover, the means of analysis will be restricted to the two classical methods of computations (elastic and plastic) used in the design of retaining structures based on the modulus of subgrade reaction. Fig. 1 shows the classical stress-displacement case for computation based on this theory. The plastic states are the active and passive earth pressure and in between the stresses are computed based on the displacement, δ, of the soil .

The different loading cases investigated by the plastic and elastic methods of computations as well as their results are given in Fig. 2. The first method is based on an analysis similar to the one developed by Coulomb and generalized by Culmann. Ohde (1952) and Krey (1936) have applied this method to different cases of surcharge loading. The second method is based on the elasticity theory and the stresses are computed by means of the Boussinesq equations for the different loading cases.

These methods do not take into account the real behaviour of the soil-structure interface where some parts are in a plastic state (active or passive) while other parts are in an elastic state. These shortcomings have forced the designer to establish empirical formulations based on both theories but which are neither related by experiment nor by precise computations.

A large divergence between the results of these methods when applied to particular cases as shown in Figs 2a and 2c needs to be explained and analyzed. It implies a general analysis of the computational methods which are today common practice.

Methods based on plastic behaviour of soil

In this theory, the cases investigated by commonly recognized methods (Ohde, 1952; Krey, 1936) are shown in Fig. 2. Assumptions associated with these computations are:

(*a*) The soil-structure interface is smooth,

(*b*) The sliding line is straight and its angle is equal to $\pi/4 - \phi/2$ with respect to the vertical,
(*c*) The soil is imtropic, homogeneous, and verifies the Coulomb criteria on the sliding line.

The consequence is that the vertical applied load, Q, and the horizontal reaction, P, are connected by the elementary relationship:

$$P = Q * \tan(\pi/4 - \phi/2) \quad (1)$$

or in terms of average applied vertical pressure, q, and resulting average horizontal pressure, p, by :

$$p = q * \tan^2(\pi/4 - \phi/2) \quad (2)$$

Equation 2 is consistent with the result of Rankine's theory which applies in the case of an infinite band load.

Fig. 2 shows the most currently admitted stress distribution. For the line load (Fig. 2a), the stress distribution is triangular with a discontinuity above D*tan(ϕ) where the stress is zero. The stress distribution of a finite band load

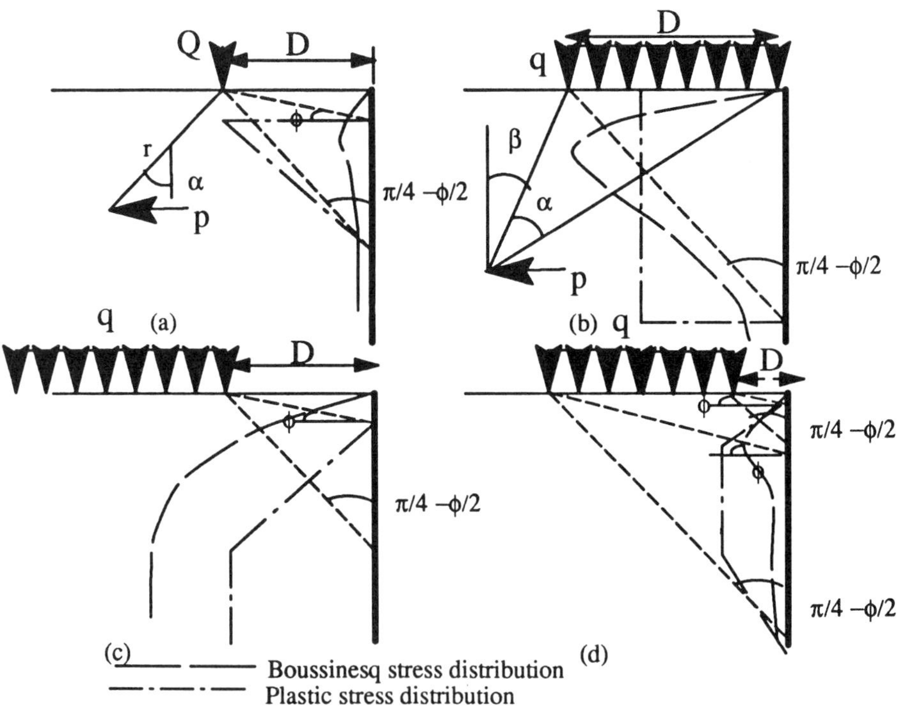

Fig. 2. Surcharge loading. (a) line load, (b) strip load, (c) semi-infinite load (d) strip load. Pressure distribution for elastic and active state (ϕ = 30°)

(Figure 2b) is uniform above the depth D*tan($\pi/4 + \phi/2$) and is equal to zero elsewhere. For a semi-infinite band load shown in Fig. 2c, the stress is found by Equation 2 below the depth D*tan($\pi/4 - \phi/2$) , it is linear between D*tan ϕ and D*tan($\pi/4 - \phi/2$) and is equal to zero above D*tan ϕ. Finally, the general stress distribution for the band load shown in Fig. 2d is explained in Houy (1986).

This method has some shortcomings as it neglects what happens below the sliding lines. In particular, in the active state, shear stresses appear along the sliding line and they are not taken into account in the computation. Also, the assumption of a smooth interface is not compatible with actual computations where the interface is assumed to be rough even in the active state. Values of the friction angle at the interface are often taken to be equal to $\phi/2$ or $2\phi/3$, which is not consistent with the assumption of a straight sliding line.

A comparison of stress distribution using plasticity theory and elasticity theory is shown in Figure 2 for the cases of a line load (case a) and a semi-infinite band load (case c). It appears that plasticity theory gives more concentrated stresses than elasticity theory for the line load and that the maximum horizontal stress is about five times larger for plastic than elastic distribution. Results are reversed for the band load where elastic theory gives larger stresses. This difference in stress distribution is important and can modify the design of the retaining structure depending on the theory chosen.

Design of a retaining structure often imposes limited displacement such that the active state is not reached. In this case the elastic stress distribution is to be considered rather than the plastic stress distribution.

At this stage it appears that the stress distribution obtained from plasticity theory should only be used when computation implies active and passive states. This kind of method will check the general equilibrium of the structure but does not find the internal stresses which can only be obtained by describing more closely the pressure distribution in relation with the state of stress in the soil (active, passive or elastic).

This lack of consistency for the interface conditions and the displacement conditions are specifically dealt with by writing the active pressure, p, as

$$p = Ka * \sigma_v \tag{3}$$

with

$$\sigma_v = p_1 / \tan^2(\pi/4 - \phi/2) \tag{4}$$

where p_1 is the active pressure found from the previous plasticity method represented in Fig. 2.

The same technique is adopted to generalize plasticity methods to elasto-plastic computations based on the modulus of subgrade reaction. From Figure 1, the elastic pressure is found by :

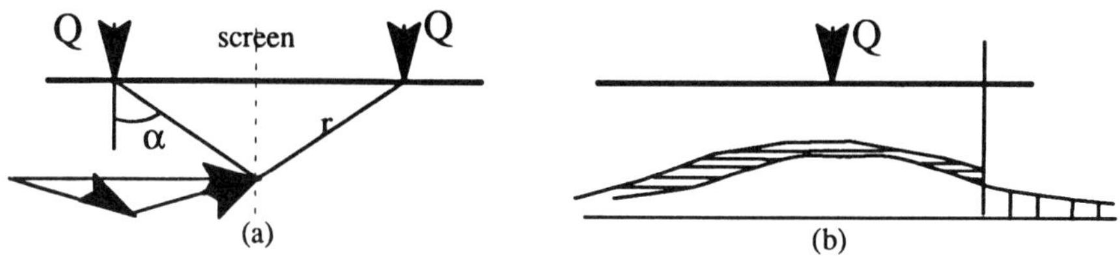

Fig. 3. (a) Image theory; (b) influence of smooth screen on stress distribution

$$p = K_0\,\sigma_v + k\,\delta \tag{5}$$

where σ_v is defined by Equation 4, K_0 is the at rest earth pressure coefficient, k the modulus of subgrade reaction and δ the displacement.

This method has been adopted by Engineers and is implemented in several codes but its validity has not been demonstrated and can be largely discussed as the elastic and plastic stress distribution cubes are not homotetic (Figure 2).

Methods based on elastic behaviour of soil

Today this method is classical and can be found in any Foundation Engineering manual. It is based on Boussinesq equations where the medium is assumed to be semi-infinite, weightless, isotropic, homogeneous and subjected to a vertical line load. Stress distribution resulting from surcharge loads defined in Fig. 2 are found by integrating Boussinesq equations. Computation of horizontal stresses, p, is done using Equations 6 and 7 given below for cases a and d of Fig. 2. Variables associated with these Equations are given in Figure 2. Cases b and c can be considered as limit of case d.

$$p = 2q \cos\alpha \sin^2\alpha/r/\pi \tag{6}$$

$$p = q/\pi\,(\alpha - \sin\alpha \cos(\alpha+2\beta)) \tag{7}$$

Despite the fact that the assumptions associated with this theory are not realistic in Soil Mechanics, it seems a little safer than the previous method, as the elastic pressure is always larger than the active pressure. Moreover, the elastic theory gives a homogeneous and continuous solution for the different loading geometry.

Nevertheless, this theory is not trouble free when applied to Soil Mechanics. Three different remarks can be made on its every day use

(*a*) Integration of Equation 6 over a semi infinite space for a uniform vertical load, q, results for the horizontal stress, p, in the value :

$$p = q/2 \tag{8}$$

which does not depend on elastic properties of the soil. Applying the image theory (Figure 3a) to simulate a uniform infinite load, the horizontal stress, p, is found by multiplying the value of Equation 8 by two such that:

$$p = q \tag{9}$$

while in Soil Mechanics this uniform loading over the entire space defines the earth pressure at rest whose value is given by :

$$p = K_0 * q \tag{10}$$

Equations 9 and 10 are different, while being used apparently for the same loading condition. Discrepancy between their results causes a problem when Boussinesq theory is mixed with K_0 theory in elastoplastic computations.

(*b*) Boussinesq development has been exercised for homogeneous media. Therefore the inclusion of an elastic structure inside the elastic media is not taken into account and the resulting stresses for this problem are obviously different from the stresses given by Equation 6.

(*c*) When comparing the value of total forces computed from elastic and plastic theory a paradox is found : results from the plasticity theory are larger than results from the elasticity theory (Figure 2a) which contradicts the minimum principle of plasticity theory discussed above. For example in the case of the line load of value P applied on a medium of internal frictional angle, ϕ, equal to 30 degrees, integration of Equation 6 gives a total horizontal force, Q, equal to :

$$P/Q=1/\pi=0.32 \tag{11}$$

while plasticity theory gives from Equation 1 :

$$P/Q = \tan(\pi/4 - 30/2) = 0.58 \tag{12}$$

Comparison of the results of Equations 11 and 12 shows that the horizontal active force is approximately two times larger than horizontal elastic force.

An attempt to explain the three kinds of discrepancies or non-conformity found in cases (a) through (c) will be made in the following sections of the paper.

Infinite uniform load

The at rest horizontal earth pressure used in Soil Mechanics is associated with a zero horizontal strain condition such that if the x and y directions are in the horizontal plane :

$$\varepsilon_x = \varepsilon_y = 0 \tag{13}$$

and the coefficient of at rest earth pressure, K_0, is related to the Poisson's coefficient, ν, by :

$$K_0 = \nu / (1-\nu) \tag{14}$$

When the Boussinesq equation is used for a uniform load over a semi-infinite domain, the horizontal earth pressure becomes p which implies that K_0 is equal to 1 or ν is equal to 0.5 (incompressible medium) and thus the horizontal stress does not depend on elasticity parameters any more.

The difference between Boussinesq and K_0 formulations comes from the fact that boundary conditions of these problems are different. While there is no strain at the boundary of the model for the K_0 formula, integrating Boussinesq formula over an infinite medium does not imply the same boundary condition as the uniform load becomes infinite over a domain whose order of infinity is larger (integration over an infinite domain of a finite load with boundaries going to infinity). Therefore the difference in boundary conditions for the two formulations is similar to the difference between an oedometer test where horizontal strain is equal to zero and a triaxial test where horizontal strain is free to develop.

Boussinesq formulation and K_0 formulation are only identical when the medium is incompressible (ν=0.5) as shown previously. This can be explained by the fact that theBoussinesq formulation gives isotropic stresses and thus isotropic strain ($\varepsilon_x=\varepsilon_y=\varepsilon_z$), and this result combined with the incompressibility condition ($\varepsilon_x+\varepsilon_y+\varepsilon_z=0$) implies zero horizontal strains, which is identical to the K_0 condition defined in Equation 13.

Besides the incompressible case, the two approaches are incompatible. Finite element method in this case does not give an answer to this problem as one must impose boundary conditions to the model, and thus implicitly defines the solution of the problem. Fixed boundary conditions gives the K_0 condition, while imposed stresses at the boundaries computed from Boussinesq equations give Boussinesq conditions.

From a practical point of view, the K_0 theory is relevant in the case of soil load since the load and the medium have the same order of infinity. Infinite surcharge loads are a simulation of large band load and in these cases Boussinesq theory is preferred.

Influence of the screen

Image theory takes into account the rigidity of the screen by multiplying by a factor equal to 2 the Boussinesq horizontal stresses as shown in Fig 3a. This is confirmed by experiments done by Matteoti (1970), Tcheng (1972), Kerisel *et al.* (1976) and Verdeyen *et al.* (1961) where it appears that this coefficient ranges generally between 1 and 1.5. Difference with the theory is explained by the fact that the screen is imperfectly rigid.

In elasticity, two problems have an identical solution if their boundary conditions are similar. Therefore the theory of image implies that along the axis of symmetry there is no horizontal displacement and no shear stress. These conditions are written as:

$$u_x = 0 \text{ and } \sigma_{xy} = 0 \tag{15}$$

Both conditions are met only when the interface soil-structure is smooth. Rough interfaces don't meet Equation 15 and therefore applying the theory of image is not correct.

Moreover, theory of image assumes that the soil is not loaded before building the screen. But the Engineer is faced with another problem: the surcharge load is present before construction and thus distribution of stress is changed because now Equation 15 is not true any more at the interface, as the displacement has already occurred in the soil due to the loading. In this case the stresses at the boundary are given by the Boussinesq equation.

Another unknown is the construction of the wall which could disturb the soil-structure interface. Driving a sheet pile or constructing a slurry wall perturbates the stress condition at the interface and does not allow the stress orientation to be guessed.

Comparison between active method and elastic method

Boundary conditions have an important effect on the results as shown in the previous section and can explain the paradox highlighted previously when we compared elastic and plastic horizontal resulting forces. With the elastic theory, the stresses are inclined along the interface while with the plastic method, the stresses are chosen perpendicular to the screen. In order for the elastic method to have stresses perpendicular to the interface, the theory of image has to be used. Therefore, Equation 11 becomes :

$$P/Q=2/\pi=0.64 \tag{16}$$

and can now be compared to Equation 12. The result is that the elasto theory, at least in the case of a smooth and perfectly rigid wall and for a load applied after construction, gives a larger horizontal resulting force than the active theory which is what is expected. But this remark shows the importance of interface conditions, as the stresses can be up to doubled as a function of the inclination on the wall. This is shown in Fig. 3b which compares vertical stress distribution before and after a smooth interface is inserted in an elastic medium. It appears that the increase of stress depends on the distance between the load and the smooth interface.

Thus it appears that the increase in elastic stresses comes from the interface condition and not from the rigidity of the retaining structure as is currently assumed. Interface conditions depend partially on the construction technique used. This point being unknown up to now, the use of an "ignorance coefficient " ranging from 1.5 to 2 is common practice in France and it has been exercised on many projects.

Elasto-plastic computations with elastic surcharge loading

Elasto-plastic computations with elastic surcharge loads are now common practice and we can classify the methods to take surcharge into account into two categories:

The first method consists of computing surcharge effects by the Boussinesq formulas in order to find the horizontal stresses, s(z), and applying a fudge factor, n, ranging from 1 to 2. The resulting horizontal stress is then written as:

$$q = n*s(z) \qquad (17)$$

Meanwhile, using this method, the effect of the surcharge is constant whatever the state of stresses in the soil. This term is then added to the active earth pressure before applying the cohesion term. Therefore, the effect of surcharge can disappear if the cohesion is large enough.

The second method, developed by Fages (1984) consists in adding an"equivalent" vertical surcharge computed from Boussinesq horizontal stress, s(z). It makes it possible to take into account the state of the soil (either elastic or plastic) for the influence of the surcharge. The method is implemented by computing the"equivalent" vertical stress, σ_v :

$$\sigma_v = s(z)/0.5 \qquad (18)$$

Then the at rest horizontal pressure, σ_h , for a uniform load is given by:

$$\sigma_h = K_0\,\sigma_v = K_0\ q/2\ 1/0.5 = K_0\ q \qquad (19)$$

which is compatible with the K_0 theory.

If the soil is in a plastic state, the horizontal stress due to a uniform sensi-infinite load is by definition equal to $K_a\sigma_v$, or $K_p\sigma_v$ and it is found using Equations 18 and 19. Nevertheless the same technique applied to any other type of surcharge load gives an earth pressure at rest equal to :

$$\sigma_h = K_0\ s(z) / 0.5 \qquad (20)$$

which is different from the horizontal stress, s(z), computed from Boussinesq theory for cases where K_0 is not equal to 0.5.

The paradox results in the fact that in this method the K_0 theory and Boussinesq theory are identical if K_0 is equal to 0.5 whereas it was demonstrated previously that only K_0 equal to 1 can give identical results. This arises from the fact that in this method we compare Equation 10 with the Boussinesq equation for a semi infinite load which gives p=q/2 (Equation 19) whereas the comparison should be made, to be consistent with the K_0 theory, with an infinite load which gives p=q. Thus, the use of Equation 19 is not correct but is tolerated because its result is compatible with the K_0 theory for an infinite uniform load and other types of load when K_0 is equal to 0.5.

In order to clear up a little bit the elastoplastic computations with surcharge loading, we propose the following procedure which contains three main steps :

(*a*) For uniform infinite loads, Boussinesq theory and Equation 18 will be preferably implemented in order to be consistent with the K_0 theory,
(*b*) On the contrary, for non-uniform loads the K_0 theory is not relevant and Equation 18 is replaced by :

$$\sigma_v = s(z) / K_0 \tag{21}$$

which gives the correct Boussinesq value s(z) when considering the at rest state.
(*c*) Stresses resulting from Boussinesq computation should be multiplied by a double coefficient $\alpha*\beta$ where α takes into account the boundary conditions at the interface soil structure and β takes into account the history of the surcharge load, ie surcharge applied before or after the retaining structure. Values of these coefficients will need to be determined based on experimental and theoretical works.

Conclusions

It appears clearly through this review of techniques that this subject needs experimental as well as theoretical studies. It is shown that stress distributions due to surcharge loads depends on interface conditions between the soil and the structure, state of the soil (elastic, plastic) and history of the surcharge (applied before or after construction of the retaining structure) . An existing method based on Boussinesq theory is explained and modified so that it is possible to take into account these three factors for any types of surcharges. Nevertheless experiments associated with finite element computations are needed to define a reliable method.

References

FAGES, R. Notice d'utilisation du Programme Rido 3. *Annexes Techniques*, 1984.

GRAUX, D. *Fondations et Excavations Profondes, Tome 1 : Geotechnique appliquee*. Fidons Eyrolles, Paris, 1967.

HOUY, A. *Dimensionnement des ouvrages en palplanches en acier*. Editions Gueblez, 1986.

KERISEL, J., TCHENG, Y and REIMBERT, M. Transmissions des surcharges a travers un massif. *Annales 1. t B. t p*, Janvier, 1976,

KREY, H. *Erddruck, Erdwiderstand und Tragfahigkeit des Baugrundes.* W. Ernst und Sohn, Berlin, 1936.

MATTEOTI, G. , Some results of quay wall model test on earth pressure. *Proc. Instn Civ. Engng,* 1970.

OHDE, *J.Zur Erddruck Lehre. Bautechnick,* 1952, Nov.

TCHENG, Y and ISEUX. J. Essai de butee en vraie grandeur et contraintes engendrees par une surcharge rectangulaire sur un mur vertical. *5eme congres Europeen Vol. II.* 1972.

VERDEYEN, J. and ROISIN, V. Relations entre les effets des surcharges exterieures concentrees et les caracteristiques des ecrans souples. 5eme congres international Vol. II, p 501, Vol. III, p316, 1961.

Influence of seepage flow on the passive earth pressures

A. H. SOUBRA, Ecole Nationale Supérieure des Arts et Industries de Strasbourg, France, and R. KASTNER, Institut National des Sciences Appliquées de Lyon, France

The variational limit equilibrium method is used for calculating the passive earth pressure coefficients in the presence of seepage flow. Numerical results are presented and discussed.

Introduction

When deep sheet piling structures are used in geotechnical engineering, we are frequently confronted by seepage flow around these structures as the excavation level is well below the groundwater table.

It is well known that the upward seepage forces will cause instability of the sheet piling structure. This is caused by the reduction of the passive earth force. Our aim in this paper is to propose an outline for the calculation of the passive earth pressures, taking into account the seepage forces.

Traditionally, the determination of these pressures is made using the classical method introduced by Terzaghi (ref.1). This Author has used an extension of Coulomb's sliding wedge theory in which seepage effects are taken into account by the addition of the pore water pressure due to the seepage flow on the plane sliding surface.

In this Paper, we present a more rational method which allows a rigorous solution of the earth pressure, taking into consideration the seepage forces. This method is based on a variational approach. It is a rigorous one in regard to the limit equilibrium method as it makes no assumptions with respect to the shape of the slip surface and the normal stress distribution along this surface.

The variational limit equilibrium method is equivalent to the upper-bound method in limit analysis for a rotational log-spiral mechanism (ref.2), hence the solution obtained is an upper-bound one for a rigid perfectly plastic material obeying Hill's maximal work principle.

Brief overview of the variational limit equilibrium method

The classical method introduced by Terzaghi (ref.1) is a limit equilibrium

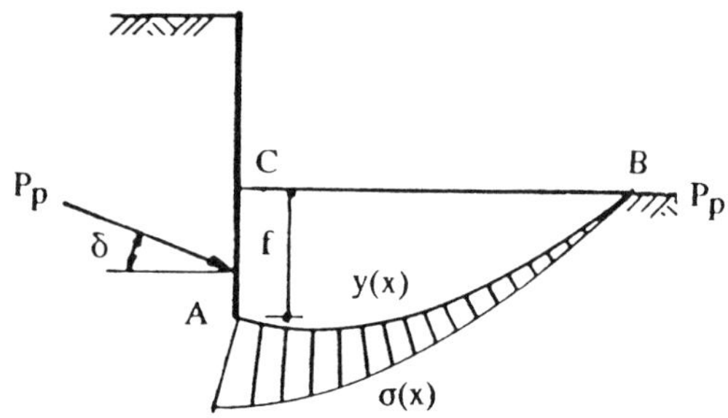

Fig.1. Slip surface and normal stress distribution for passive earth diagram pressure analysis

method giving unsafe solutions since it is based on Coulomb's approach which highly overestimates the passive earth pressure coefficients. This fact is due to the a priori hypothesis concerning the shape of the slip surface. In this paper, we look for the shape of the mechanism giving the minimum value of the passive earth force P_p, and for which the three limiting equilibrium equations are satisfied. This problem is formalized mathematically using a variational approach.

Mathematical formulation of the problem

It is well known that a rigorous limit equilibrium method is one for which the following conditions are satisfied.

(*a*) The shape of the slip surface y(x) and the normal stress distribution $\sigma_e(x)$ will give the minimum value of the passive earth force P_p.
(*b*) The three equations of the static equilibrium are satisfied for the soil mass ABC (Fig. 1).

Notice that a mass is in a state of limit equilibrium when the Mohr-Coulomb criterion is satisfied along the slip surface AB. Writing the three equations of the static equilibrium for the soil mass (Fig. 2), and combining these equations with the Mohr-Coulomb criterion, one obtains the three limiting equilibrium equations as follows

$$P_p . \cos\delta + U_1 = \int_{x_0}^{x_1} [\sigma_e (tg\phi + y') + u . y'] dx \quad (1a)$$

$$P_p . \sin\delta = \int_{x_0}^{x_1} [\sigma_e (1 - tg\phi y') + u - \gamma (f - y)] dx \quad (1b)$$

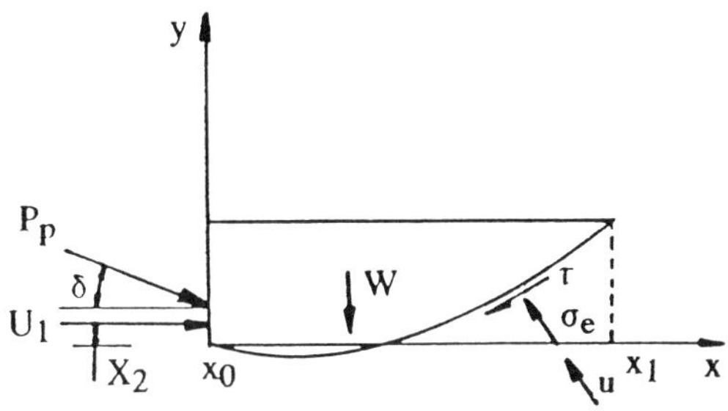

Fig. 2. Free body diagram

$$P_p.\cos\delta.f/3+U_1.X_2 = \int_{x_0}^{x_1}[\sigma_e(1-tg\phi y')x-\gamma.x(f-y)+u.x.$$
$$+\sigma_e(tg\phi+y')y+u.y.y']dx \qquad (1c)$$

where u represents the pore water pressure along y(x) and U_1 represents the resultant of the pore water pressure along the penetration depth f. A11 other parameters of these equations are defined in Fig. 2.

From these equations, it is easy to see that the passive earth force P_p is a functional of two functions y(x) and $\sigma_e(x)$. Thus, the rigorous passive limit equilibrium problem is a variational one of the isoperimetric type as follows

$$\text{Min } P_p = \int_{x_0}^{x_1} F(x, y, y', \sigma_e)dx$$

subject to

$$\int_{x_0}^{x_1} G_i(x, y, y', \sigma_e)dx = a_i \qquad (i=1,2)$$

where $F(x, y, y', \sigma_e)$ may be simply obtained through one of the equations (1). $G_i(x, y, y', \sigma_e)$ and a_i can be obtained from the two remaining equilibrium equations. It was shown (Petrov, ref.3) that the solution of such a problem is obtained using the Euler equations as follows

$$\frac{\partial H}{\partial \sigma_e} = \frac{d}{dx}\frac{\partial H}{\partial \sigma_e'} \qquad (2a)$$

$$\frac{\partial H}{\partial y} = \frac{d}{dx}\frac{\partial H}{\partial y'} \qquad (2b)$$

Where H is an intermediate functional given by

$$H = F + \lambda_i . G_i \qquad (i= 1,2)$$

Notice that H can be written as

$$H = \sigma_e . f(x, y, y') + g(x, y, y') \qquad (3)$$

Hence, equation (2a) is equivalent to f(x, y, y')=0. Solving this equation, one obtains the equation of the slip surface which is a log-spiral in the case of a constant ϕ (Fig. 4). Replacing this equation into equation (3), one can see that H becomes independent of the normal stress distribution. This result is a direct consequence of the shape of the slip surface. It was shown (ref.2) that any equation of the normal stress distribution having at least two degrees of freedom can satisfy the three equations of static equilibrium, and that only the equation of moments around the centre of the log-spiral is sufficient to calculate the passive earth force P_p. It is easy to see that the moment equation of the rotational log-spiral mechanism around the centre is identical to the work equation for the same mechanism in the upper-bound method in limit analysis. Thus, solving the passive earth pressure problem by writing the moment equation around the center of the log-spiral will give an upper-bound solution of the exact solution for an associated flow rule material. This method is detailed in the following section.

Calculation scheme of the passive earth pressures

As mentioned in the preceding section, the solution of the passive earth pressure problem by a variational limit equilibrium method is equivalent to saying that the equation of moment equilibrium must be satisfied for the soil mass bounded by the log-spiral AB and the ground surface BC. Notice that

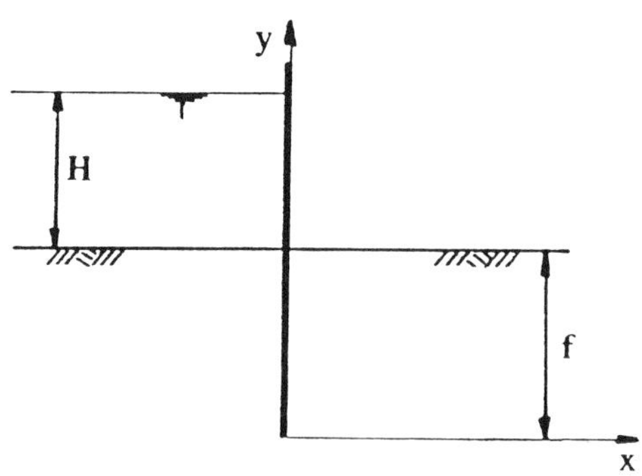

Fig.3. Illustrative example

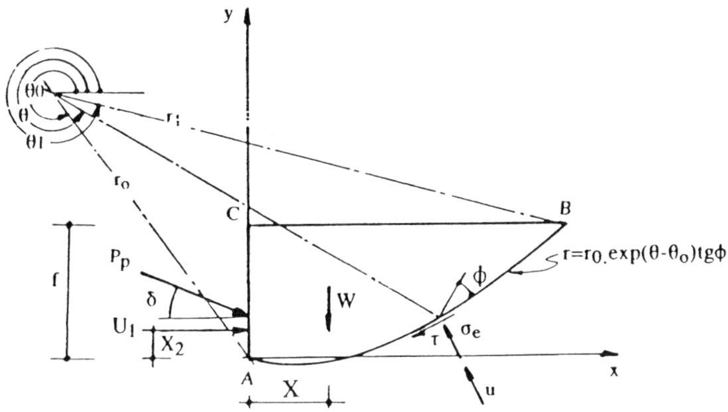

Fig.4. Log-spiral mechanism for passive earth pressure analysis

this conclusion is valid for any pore water pressure distribution in the soil mass. To investigate how the hydraulic head can affect the passive earth pressure coefficients, we present in the following section an illustrative example.

Illustrative example

As shown in Fig. 3 we consider a sheet pile whose penetration depth is equal to f and subject to a hydraulic head H. The soil is assumed to be homogeneous. To calculate the passive earth force, one must verify the equation of the moment equilibrium of the soil mass around the log-spiral centre. The external forces acting on the soil mass are shown on the free body diagram shown in Fig. 4. These forces consist of

(*a*) the weight of the soil mass between the log-spiral surface and the ground surface.
(*b*) the passive earth force which is inclined at δ to the normal of the sheet piling wall.
(*c*) the pore water pressures along the penetration depth and the log-spiral surface.
(*d*) the normal and tangential stress distributions along y(x).

Finally, the moment equation can be written as follows

$$W(r_0.\cos\theta_0+X)+P_p.\sin\delta(r_0.\cos\theta_0)=P_p.\cos\delta(-r_0.\sin\theta_0-\frac{f}{3})+M_1+M_2$$

where M_1 and M_2 represent the moments of the force U_1 and of the pore water pressure along y(x) respectively.

Details concerning the calculation of moments of the weight and the passive earth force are given elsewhere (ref.2). We will proceed now to the calculation of the terms M_1 and M_2.

Calculation of the moments of the pore water pressures

It is well known that the pore water pressure distribution in the soil mass can be obtained analytically in this case. The relation which governs the hydraulic head h at any point (x, y) in the coordinate system (ox, oy) shown in Fig. 3 is given by

$$\frac{(f-y)^2}{f^2.\cos^2\Phi} - \frac{x^2}{f^2.\sin^2\Phi} = 1 \qquad (4)$$

where

$$\Phi = \pi.\left(\frac{h_s}{H} - \frac{1}{2}\right)$$

and

$$h = h_s + f$$

M_1-value (along the penetration depth)

Based on equation (4), one can easily show that along the penetration depth (in the downstream of the sheet pile), the resultant of the pore water pressures is given as

$$U_1 = \frac{\gamma_w.f^2}{2} + \gamma_w.H.f\left(\frac{1}{2} - \frac{1}{\pi}\right)$$

This force acts at a distance Y away from the point O. Y is given as

$$Y = \left[\frac{H\left(\frac{3}{8} - \frac{1}{\pi}\right) + \frac{f}{6}}{H\left(\frac{1}{2} - \frac{1}{\pi}\right) + \frac{f}{2}}\right].f$$

Finally, the moment of the force U_1 is given as $U_1\,(-r_o\sin q_o - Y)$.

M_2-value (along the slip surface)

This moment can be obtained by first calculating the elementary moment acting on a surface element ds and then by integrating along the surface AB as follows

$$M_2 = \int dM_2 = \int_{\theta_0}^{\theta_1} (u.\sin\phi).\frac{r.d\theta}{\cos\phi}.r = tg\phi.\int_{\theta_0}^{\theta_1} u.r^2.d\theta$$

Substituting the equation of the log-spiral in this equation, one obtains

$$M_2 = tg\phi . r_0^2 . \exp(-2\theta_0 . tg\phi) . \int_{\theta_0}^{\theta_1} u . \exp(2\theta . tg\phi) . d\theta$$

where $u = (h-y)\gamma_w = (h_s+f-y)\gamma_w$.

Finally, the moment M_2 can be written as

$$M_2 = tg\phi . r_0^2 . \exp(-2\theta_0 . tg\phi) . \gamma_w . I$$

where $I = \int_{\theta_0}^{\theta_1} h_s . \exp(2\theta . tg\phi)\, d\theta + \int_{\theta_0}^{\theta_1} f . \exp(2\theta . tg\phi)\, d\theta - \int_{\theta_0}^{\theta_1} y . \exp(2\theta . tg\phi)\, d\theta = I_1 + I_2 + I_3$

While the integrals I_2 and I_3 are simples and can be obtained analytically, the integral I_3 however, cannot be obtained analytically and it needs numerical calculations (ref.2). Notice that this integral was calculated numerically using the gaussian quadrature method. The classical numerical methods (trapezoidal method) give insufficient accuracy in this case.

Minimization procedure

By writing the equation of the moment equilibrium around the centre of the log-spiral mechanism, one gets an expression of the passive earth force in terms of the θ_0 and θ_1 angles. The most critical P_p-value is obtained by minimizing with respect to θ_0 and θ_1 angles. The θ_0 and θ_1 angles at which the P_p-value is minimum determine the most critical sliding surface. A

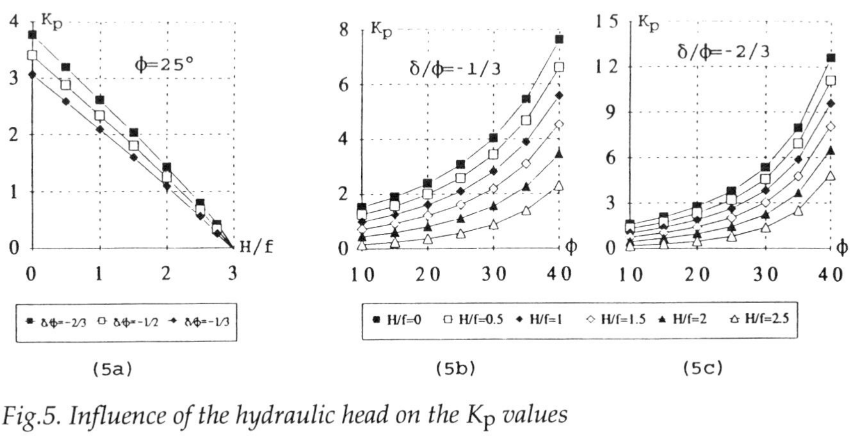

Fig.5. Influence of the hydraulic head on the K_p values

FORTRAN computer program has been developed for assessing the passive earth force and the corresponding critical slip surface.

Numerical results

Influence of the hydraulic head on the passive earth pressure coefficients

It is known that the seepage flow has the unfavourable effect of producing piping or heaving at the downstream of the sheet piling structure. Usually, when there is seepage flow, the sheet pile failure is considered to have taken place for high hydraulic head by heaving. The calculation of heaving risk is traditionally made by considering a rectangular mechanism adjacent to the sheet piling wall. This mechanism has a width equal to f/2. The vertical force equilibrium equation of this soil mass is then considered by neglecting the soil-structure force represented by the passive earth force. Kastner (ref.4), based on laboratory model tests, has shown that failure of the sheet piling structures is not only due to the heaving phenomena but it can be produced by reduction of the passive earth pressures. We present in Fig.5(a) one chart showing the influence of the hydraulic head on the passive earth pressure coefficient for an angle of internal friction ϕ equal to 25° and for three different values of δ/ϕ (δ/ϕ = -1/3; -1/2; -2/3).

It is easy to see that for zero K_p the corresponding H/f value is the same for different δ/ϕ values. This means that the angle of friction at the soil-structure interface has no effect on the H/f value causing failure by heaving. This fact can be explained as follows. When the passive earth force is vanishing, there is no interaction at the soil-structure interface and we have the traditional heaving phenomena. However, for small values of H/f for which the passive earth force does not vanish, this passive earth force depends on the δ value.

We present in Figs 5(b) and (c) some charts showing the variation of the k_p-value as a function of the nondimensional parameter H/f when $\gamma_{sat}\gamma_w$=2 and for δ/ϕ=-1/3, -2/3. It is easy to see that k_p decreases with the H/f increase. The reduction is about 62% when H/f increases from zero to 2.5 for ϕ=40°, δ/ϕ=-2/3.

Critical slip surface

The pore water pressures generated by seepage not only imposes extra loading to a soil mass but also shifts the sliding surface to less favourable positions. Consequently, in addition to the change in the passive earth pressure, the most critical sliding surface is also altered. The numerical results have shown that the slip surface approaches a planar surface due to the increase in the hydraulic head. Fig. 6 shows some typical changes in the critical sliding surface as the result of hydraulic head.

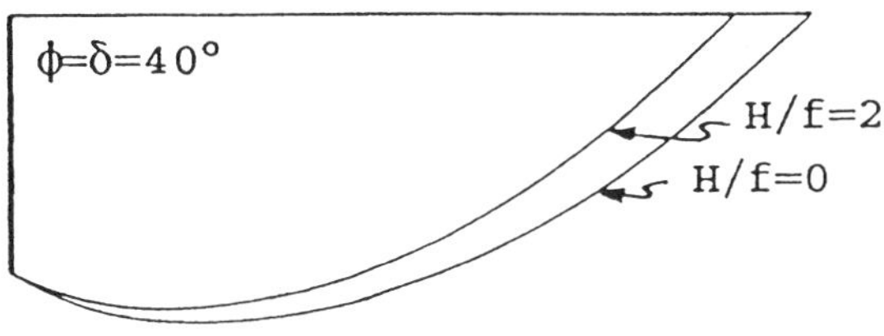

Fig.6. Critical slip surfaces

Analysis and discussion of results

Kastner (ref.4), based on laboratory model tests, has shown that for high hydraulic head, failure takes place for a planar slip surface. A previous theoretical work based on this hypothesis has been developed by Kastner et al (ref.5). The numerical results as obtained by the present variational approach are compared with the above mentioned results in the case (ϕ=25°, δ/ϕ=-2/3) as shown in Fig. 7. As is well known, the planar slip surface greatly overestimates the k_p-values especially for high ϕ and δ values. Hence, for small hydraulic head corresponding to high passive force, the triangular mechanism gives greater k_p-values than the log-spiral one. The difference in the k_p-value for the two mechanisms decreases with the H/f increase. For a

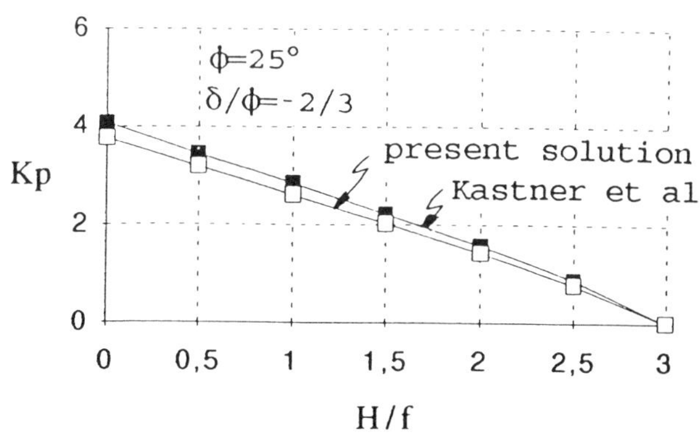

Fig.7. Comparison with the triangular mechanism (Kastner et al. 1987)

zero passive force corresponding to a failure by heaving, the two mechanisms give the same H/f value. This result confirms laboratory model tests of Marsland (ref.6) and Kastner (ref.4).

Conclusion

The variational limit equilibrium method is used to calculate the passive earth pressure coefficients in the presence of seepage flow. The numerical results as obtained by Kastner et al. (ref.5) concerning a planar slip surface are compared with the ones obtained with the above-mentioned approach. The present approach gives better solutions since we obtain less favourable mechanism giving less k_p-values.

References

1. TERZAGHI K. Theoretical soil mechanics, Wiley, New York, 510pp,1943.
2. SOUBRA A.H. Application de la méthode variationnelle au problème de détèrmination des pressions passives des terres. Influence des forces d'écoulement. Thèse de Doctorat, INSA Lyon, 200pp, 1989.
3. PETROV I. Variational methods in optimum control theory, Academic Press, New York and London, 216pp, 1968.
4. KASTNER R. Excavations profondes en site urbain: problémes liés à la mise hors d'eau. Dimensionnement des soutènements butonnés, Thèse de Docteur ès Sciences, INSA Lyon et Université Claude Bernard, Lyon I, 409pp, 1982.
5. KASTNER R., SOUBRA A.H. and DIDIER G. Etude de la butée en présence d'écoulement Proc. of the 9th Europeen Conf. on Soil Mechanics and Foundation Engineering, Dublin, 1987, 703-706.
6. MARSLAND A. Model experiment to study the influence of seepage on the stability of a sheeted excavation in sand. Géotechnique, 1953, vol. 3, 223-241.

Discussion

SESSION 1

C.W.W. NG, University of Bristol

During diaphragm wall construction under bentonite, the process of excavating a slot in the ground and back filling it with bentonite and subsequently with concrete will alter the lateral stress in the ground. For a given soil stratum and panel dimensions, the change of lateral stress in the soil adjacent to the diaphragm wall is governed mainly by the initial lateral stress in the ground and the lateral pressure exerted by concrete during construction.[1] Therefore, correct assessment of concrete pressure is paramount for modelling the wall installation process.

Full wet concrete pressure has often been assumed in numerical analysis along the entire depth of wall.[2,3] This assumption is contrary to theoretical considerations and the field evidence given below.

Theoretical consideration of lateral concrete pressure

According to extensive concrete research on the lateral pressure exerted by fresh concrete on vertical formwork,[4] it is widely accepted that the envelope of the lateral pressure with depth exerted by wet concrete is hydrostatic up to a certain limiting value (P_{max}) and thereafter remains constant with depth as shown in Fig. 1. Full fluid concrete pressure will not develop over the whole depth of the formwork since the concrete continues to set and harden as concreting progresses. The maximum pressure (P_{max}) on vertical formwork is governed by a number of factors such as the rate of placing, cross-section and depth of the form, temperature and concrete mix. The critical depth, h_c, below which the horizontal pressure is constant can be determined once the P_{max} value is known ($h_c = P_{max}/\gamma_c$) where γ_c is the unit weight of wet concrete. CIRIA report 108[4] provides general guidance for calculating the P_{max} value.

There are two major differences between concrete placed in vertical formwork and concrete placed under bentonite in a diaphragm wall trench. Firstly, the pressure increase due to concreting depends on the difference between the unit weights of the concrete (γ_c) and bentonite (γ_b) rather than the unit weight of concrete alone. Secondly, the envelope of lateral pressure with depth increases beneath the critical depth at a slope governed by the unit weight of the bentonite, rather than staying constant. Thus a theoretical

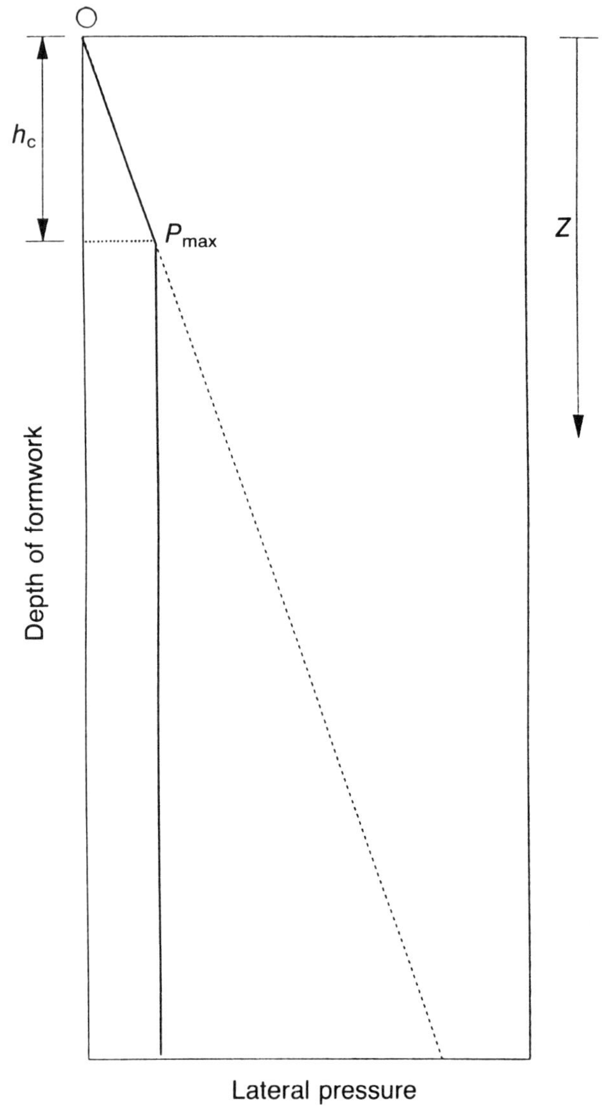

Fig. 1. Development of lateral stress in a vertical formwork during concreting

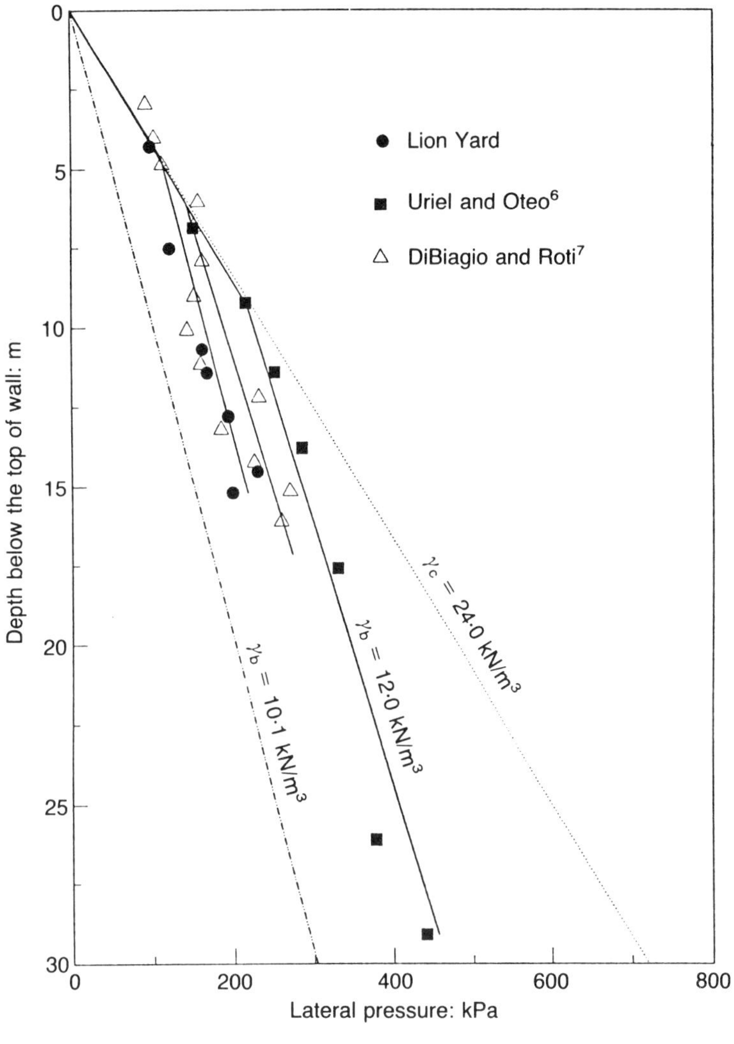

Fig. 2. Comparison of measured and predicted concrete pressure during concreting

bi-linear concrete pressure envelope can be derived as given below [5]

$$\sigma_h = \begin{cases} \gamma_c z & , \quad z \leq h_c \\ (\gamma_c - \gamma_b)h_c + \gamma_b z, & z > h_c \end{cases} \tag{1}$$

where σ_h and z are the total lateral pressure and depth below the top of a panel respectively.

Comparison of theoretical predictions and field observations

Theoretical predictions using the equation (1) are compared with three case records obtained during diaphragm wall construction: in stiff fissured Gault clay at Lion Yard;[5] in silty sand and gravel in Spain;[6] and in soft clay in Oslo.[7] Details of the interpretation of these field data using equation (1) are given by Ng.[5] The analysed results are shown in Fig. 2. It can be seen that the proposed bi-linear theoretical lines agree very well with the field observations. Clearly, full fluid concrete pressure did not develop over the lower two-thirds of the wall during concreting.

Conclusions

Based on theoretical considerations and the results of field observations obtained from the three different sites located in various deposits, it can be concluded that lateral pressures developed during concreting are substantially less than full fluid concrete pressures at the lower two-thirds of the wall. The h_c value appears to be approximately one-third of the depth of wall for all the three case records considered. Theoretical considerations lead to the conclusion that concrete placed under bentonite should exhibit a bi-linear pressure envelope. All three sets of site data correspond well with this theoretical pressure profile. It is therefore suggested that this envelope should form the basis of any future work where concrete pressures are required in analysis.

M. FUCHSBERGER, Technical University of Graz

There is no reference to retaining structures using prestressed diaphragm walls although the most extensive use in the UK is at Irlcom o' the Heights near Manchester. In this 1 km long vehicular underpass, 70% of the retaining walls are free cantilever prestressed diaphragm walls.

The principle is to use HT (prestressing) cables or bars, stressing them while the wall is embedded in the ground before excavation. This has the following advantages

(a) higher bending moment capacity

(b) only compressive stresses in the concrete
(c) greater unsupported height of the retaining wall leading to less support levels
(d) less total weight of steel reinforcement.

M. PULLER, Butler Puller Partnership

Regarding Mr Gunn's contribution (Paper 18) on the analysis of diaphragm wall panel excavation effects, I know of three field measurements of horizontal soil movement, attributable to panel excavation.

The first in London Clay was made at a diaphragm wall excavation for temporary works on the Victoria Line in Green Park. Only a few millimetres of horizontal soil movement were measured and this movement was coincidentally adjacent to the panel. In less overconsolidated clays and silts at Medinah, Saudi Arabia, measurements were made by inclinometer at varying distances from a panel approximately 40m deep. Again, the extent of horizontal soil movement over the depth of the panel and within a distance of 20m or so from the panel was very small and limited to a few millimetres. The third site in Central, Hong Kong showed appreciable horizontal soil movements due to panel excavations in a residual soil. It was considered that movement could be large in an expansive soil where groundwater was available and where the soil permeability allowed this water to migrate to the soil near the panel. The effect of concrete pressures during diaphragm wall panel installation was important: measurements reported some time ago in Travaux persuaded some that a minimum Ka of 0.5 should be adopted irrespective of soil strengths.

My conclusions regarding panel excavation are that soil movements are likely to be very small in clays and cohesionless soils, except expansive clays where groundwater can migrate towards the panel. Concrete pressures are important and can cause high pressures between soil and the panel face.

It is also my view that analysis by numerical methods of soil deformation adjacent to the panel must be made in three dimensions to give realistic results, and such analyses become more necessary as contractors seek larger panel lengths, i.e. up to 8m or more.

References

1. NG C.W.W., SIMPSON B., LINGS M.L. and NASH B.F.T., Modelling of 3-D effects of wall installation. 1992. In preparation.
2. HIGGINS K.G., POTTS D.M. and SYMONS I.F.Z. Comparison of predicted and measured performance of the retaining walls of the Bell Common tunnel. Transport and Road Research Laboratory, Crowthorne, 1989, TRRL Contractor report 124.

3. GUNN M.J., SATKUNANANTHAN A. and CLAYTON C.R.I. Finite element modelling of installation effects. Int. Conf. on Retaining Structures, Cambridge, 1992. The Institution of Civil Engineers.
4. CLEAR C.A. and HARRISON T.A. Concrete pressure on formwork. CIRIA, London, 1985, CIRIA report 108.
5. NG C.W.W. An evaluation of soil-structure interaction associated with deep excavation in a stiff fissured clay. University of Bristol, 1992, PhD thesis. In preparation.
6. URIEL S. and OTEO C.S. Stress and strain beside a circular trench wall. Proc 9th Int. Conf. Soil Mech. Fdn Engng, Tokyo 1977, 1, 781-788.
7. DIBIAGIO E and ROTI J.A. Earth pressure measurements on a braced slurry-trench wall in soft clay. Proc. 5th Eur. Conf. Soil Mech., Madrid, 1972, 1, 473-482.

Introduction to Session 2
Numerical and physical modelling

W. POWRIE, Queen Mary & Westfield College, London, UK

Introduction

Modelling is an important activity for almost all engineers. In addition to physical and numerical modelling, many of the traditional procedures in engineering design are models, in that they represent idealizations of real behaviour without which most problems would be too complex to analyze.

The essential feature of a model is that a decision is taken at an early stage as to the important aspects of the overall behaviour. These are separated out and analyzed, while the aspects which are judged not to be important are neglected. If this initial decision is wrong, the model may be inappropriate to the situation to which it is applied, and disaster may eventually ensue.

In theoretical (as opposed to physical) modelling, the behaviour of the soil must be described in an idealized form, by means of a constitutive model. The fact that soil strength is generally easier to define than the complete stress-strain relationship has led to the development of analytical models based on the theorems of plasticity and conditions at the collapse limit state, so that the calculation involves only the soil strength.

Limit-based models

Enoki, Yagi, Yatabe and Imamura describe a generalized limit equilibrium method, in which the soil construction being analyzed is divided into a number of rectangular and triangular blocks (Enoki *et al.*, Figure 1). The equilibrium equations for each block, in which the "factors of safety" on the base plane and the inter-block boundaries are included, are assembled and solved numerically. The geometry is optimized to find the most critical set of blocks. The end blocks must be triangular for the system to be statically determinate.

If the blocks constitute a kinematically admissible mechanism (i.e. if adjacent blocks have the same displacement rates perpendicular to their common boundary), then the generalized limit equilibrium method gives the same result as an upper bound plasticity solution, with the mechanism analyzed by statics rather than by a work balance. The solution can be refined by increasing the number of blocks into which the soil construction is divided. For the passive earth pressure problem, Sokolovski's solution is approached as the number of blocks is increased and the boundaries to the blocks become coincident with the planes of maximum stress ratio in the stress field (Enoki *et al.*, Figure 4).

Active thrusts for a wall with a battered back, and passive thrusts for a wall with a sloping backfill are derived. The method may also be used to estimate earth pressures on circular retaining walls under conditions of axial symmetry. The main significance of this Paper is perhaps that the calculations were carried out on a personal computer using a program consisting of 1000 lines of Fortran, including the optimization of the geometry to determine the most critical configuration of sliding blocks.

The Paper by Mortensen is concerned with the analysis of retaining walls by statically admissible stress fields, i.e. lower bounds. Stress discontinuities are used to create stress jumps at the wall, where the condition of the soil changes from active to passive. The technique is used in the analysis of propped walls: the appropriate stress field in any particular case will depend on the embedment depth, the soil/wall friction, and the load in the anchor, which may be pre-stressed (Mortensen, fig. 5). As with the previous Paper, the way in which the calculation is carried out is a major point of interest. The application of this style of calculation to design has not been widespread outside Denmark, but perhaps its incorporation into spreadsheet format will make it more accessible to a greater number of geotechnical engineers.

Modified plasticity models

Conventional plasticity models, being based on conditions at collapse, give no information on deformations. It is necessary to factor the calculation in some way to ensure that the structure as designed will neither collapse nor deform excessively in service. The application of this factor (conventionally if not quite appropriately termed the factor of safety) requires some thought. For an anchored sheet pile wall retaining sand, passive pressures are reduced by a factor F_p because the deformation of the wall which would be required to achieve fully passive conditions in front would be unacceptably large. Conversely, excavation in front of an embedded in situ wall in a highly over-consolidated clay could bring the soil below formation level to the passive condition without lateral movement of the wall: in an extreme case, it is perhaps the active stresses which should be increased by a factor F_a in order to estimate bending moments under working conditions.

An alternative approach involves the identification of the dominant class of behaviour of a serviceable structure, perhaps on the basis of experimental observations. Other aspects may then be ignored in the development of a partial or approximate solution. This is the method adopted by Maruoka, Aoki, Sato and Ikuta in their Paper on ground movements caused by the displacement of retaining walls. They present inclinometer data from an excavation in alluvial silt, supported by a retaining wall propped at two levels (Maruoka *et al.*, fig. 1). It is shown that the incremental displacement patterns during each stage of the excavation sequence are similar to those given by strain fields corresponding to idealized modes of incremental wall deformation (Maruoka *et al.*, fig. 10). Data are presented from a total of seven sites: despite some scatter, the measured soil surface profiles are generally close to the range given by the strain field calculation (Maruoka *et al.*, fig. 12). It should be noted that the strain fields correspond to undrained (i.e. constant volume) deformation, and that the approach is not truly predictive but enables the settlements of the retained soil surface to be estimated from a measured wall deflexion. The idealized strain fields suggest that significant soil settlements extend horizontally from the retaining wall for a distance of approximately twice the excavated depth.

The movements during the initial stage of the excavation are consistent with the idealized displacement field suggested by Bolton and Powrie (1988), on the basis of centrifuge model tests on unpropped walls. The lateral extent of significant soil movement in this case is equal to the overall wall height: however for walls which are eventually propped, this will in many cases be similar to twice the depth of the excavation.

Numerical modelling

With recent advances in computer technology, numerical modelling using finite element analysis is becoming an increasingly popular and powerful analytical tool. Where construction methods are complex, a numerical model may be
the only way in which the effects of construction sequence
on the behaviour of a retaining structure can be investigated. However, three dimensional effects, such as vertical loading from a bridge deck or its approach embankment applied behind a diaphragm wall over a short distance only (e.g. Powrie, Davies and Britto, fig. 1), are still rarely modelled fully.

The results of a numerical analysis may be influenced significantly by the constitutive relationship chosen to model the behaviour of the soil. A simple linear elastic/ perfectly plastic model is unlikely adequately to represent real soil behaviour, but with the judicious selection of soil parameters and in the context of a back-analysis rather than a prediction, it may be possible to obtain realistic results. The stress-strain behaviour of soils can be represented more closely by means of a kinematic soil model, or somewhat empirically using a curve which fits data from one particular stress path.

Construction sequence effects and soil constitutive relationships are addressed by Higgins, Potts and Symons, who analyze the M25 Bell Common retaining wall in two ways:

(*a*) using a linear elastic/perfectly plastic soil model, with parameters derived from the back-analysis of previous excavations in London clay
(*b*) using logarithmic relationships between G/p′ and axial strain; and K/p′ and volumetric strain, based on triaxial test data.

Higgins *et al.* conclude that it is in general possible to obtain reasonable results using either method, but that it may be necessary to model the construction sequence in some detail.

The use by Higgins *et al.* of an enhanced ϕ' value for the soil in front of the wall is perhaps open to question: the high stress ratios measured in monotonic tests representing conditions of vertical stress relief on the passive side of an in situ retaining wall do not appear to occur in tests in which the sample is subjected to small stress cycles, representing seasonal effects, during swelling (Stewart, 1989).

The value of an empirical non-linear soil model is diminished if the same stress-strain relationship is used for all soil elements. In the field, stress paths will differ widely, especially between elements on different sides of the wall. Furthermore, it is necessary to model the high stiffness of the soil when the direction of the stress path is reversed, for example by re-setting the strain datum to zero.

A second example of the use of finite element analysis is given by Whittle & Hashash, who present a parametric study of the effect of support spacing on the behaviour of braced in situ walls in lightly over-consolidated Boston Blue clay. The construction sequences investigated are shown in Whittle & Hashash, fig. 1. The results indicate that the settlement of the retained soil surface, and the movement of the wall above formation level, are dependent on the spacing of the bracing rather than on the overall length of the wall. Although wall length generally controls stability, insufficient bracing may result in a bending failure of the wall. The computed soil settlements extend a long way from the wall itself, to approximately ten times the excavation depth: this could be due to the constitutive model and/or the boundary conditions used in the analysis.

Finite element analysis has become much more accessible in recent years, and is no longer the preserve of specialists. Analyses which 2-3 years ago were left to run overnight on a mainframe computer can now be carried out in half an hour or so on a personal computer. The use of finite element programs, however, requires considerable care if misleading results are to be avoided - or at least identified, investigated and discounted.

The Paper by Woods and Clayton is useful in that it catalogues some of the problems experienced by users of a finite element program based on the concepts of critical state soil mechanics, CRISP. CRISP was originally developed as a research tool at Cambridge University. The code is widely available, and various modifications and special purpose elements have been added by research groups elsewhere.

Woods & Clayton discuss a number of points, including

(*a*) Modelling issues: the relative merits of empirical stress/strain relationships as opposed to kinematic soil models; the use of an associated flow rule, which leads to excessive dilation; the need for interface elements in problems where the wall tends to separate from the retained soil;
(*b*) Computational difficulties: the use of a tangent stiffness solution routine, which requires sufficient increments for numerical accuracy;
(*c*) Interpretation of results: the calculation of bending moments from the stresses within the wall, rather than the stresses in the adjacent soil.

Physical models

Physical modelling tends to be used for the investigation of basic mechanisms of collapse or deformation, parametric studies, and the validation or calibration of numerical codes. In principle, physical modelling techniques could be used directly in design, but this seems to be a less common application, at least in the UK. Budhu, Neelakantan and Richards have used a physical model to investigate critical collapse mechanisms of anchored retaining walls subjected to earthquake loading. Their tests were carried out on a shaking table (Budhu *et al.*, fig. 2). When the top of the wall was prevented from moving, the critical acceleration ratio was found to be 0.5, which is close to the value given by the Mononobe-Okabe equations assuming an angle of wall friction $\delta = 0.67\phi'$. However when the top of the model wall was restrained by an anchor designed for static loading conditions, the critical acceleration ratio was only 0.2.

From this, it is concluded that the seismic efficiency of anchored walls may be improved by increasing the capacity of the anchor to achieve a balanced design, i.e. one in which the acceleration ratio for anchor failure is the same as that for rotational failure about the anchor.

It has already been mentioned that in numerical modelling, it can be difficult to describe the stress-strain relationship of the soil in a way which is appropriate to the stress/strain paths being followed by all soil elements at all times. Small-scale physical modelling at normal gravity (lg) does not solve this problem, because stress levels due to self weight effects are too low. As stress-strain relationships depend on stress history, stress state and stress path, the stress-strain behaviour in the field will not be replicated faithfully in the model. Even if it is only conditions at failure which are being investigated, the low stress levels in lg tests in sands can lead to misleading results because the effects of dilation may be excessive. In clays, self weight stresses which drive failure (as in most retaining wall problems) may be far too small in relation to the undrained shear strength. These problems can be overcome

Fig. 1. Centrifuge model used to investigate installation effects of diaphragm walls in clay, showing hopper and fuel storage tanks

by testing a 1/n scale model in a geotechnical centrifuge in a radial acceleration field of n times normal gravity, so that self-weight stresses are the same at corresponding points in the model and in the field.

Powrie and Kantartzi describe the use of centrifuge modelling techniques to investigate the effects of diaphragm wall installation in over-consolidated clay. With the increasing use of in situ retaining walls and top-down methods of construction, geotechnical engineers are becoming more concerned with the effects of construction methods and sequences, both in their own right (for example ground movements during diaphragm wall construction) and because of their possible influence on the subsequent behaviour of the structure.

The centrifuge model represents the excavation for a 1 m thick, 18.5 m deep diaphragm wall at a scale of 1:100 (fig. 1). In the model shown, the trench occupies the full width of the clay sample, so that deformations will occur in plane strain and three dimensional effects, which may be important in the installation of an isolated panel, are eliminated. During reconsolidation of the clay sample in the centrifuge, a rubber bag which occupies the trench is filled with sodium chloride solution of unit weight 11.4 kN/m^3 to a height of 9 m or so (at field scale) above the soil surface. Pore water pressures at the end of the reconsolidation period are hydrostatic below the soil surface,

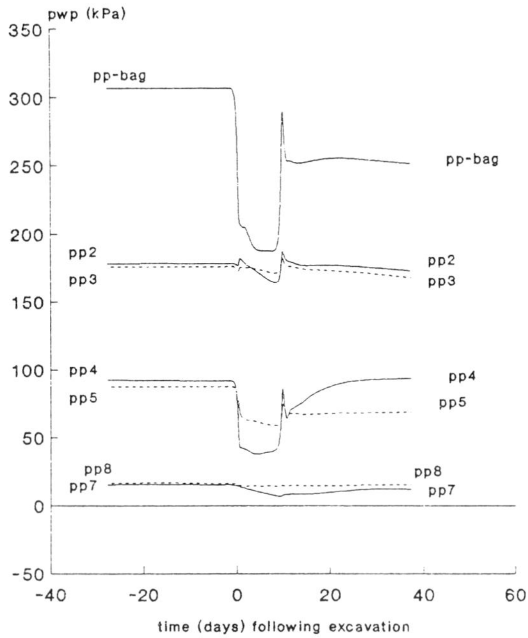

Fig. 2. Variation of pore water pressures during excavation and concreting

giving the initial profile of lateral earth pressure coefficient K_o ($=\sigma'_{h0}/\sigma'_{v0}$) with depth shown in Powrie and Kantartzi, fig. 3.

When reconsolidation is complete, the sodium chloride solution is drained down to the level of the soil surface through a valve-controlled waste pipe and the rubber bag is flushed through with fresh water, to simulate the reduction in horizontal total stress caused during excavation of a diaphragm wall trench under bentonite slurry. A pneumatic piston is then used to open a shutter at the base of a hopper mounted above the trench, allowing a mixture of iron powder, plaster of Paris, cement and fine sand to fall into the rubber bag. The mixture sets in approximately 10-15 minutes, modelling the concreting phase of diaphragm wall installation. Changes in pore water pressure and soil movements are monitored using pore pressure transducers and lvdt's respectively (Powrie and Kantartzi, fig. 2). Typical results are shown in fig. 2 (changes in pore water pressure during excavation and concreting) and fig. 3 (soil surface settlements). During excavation, some of the pore water pressures fall and the soil surface suffers considerable settlement. During concreting, the pore water pressures tend to return towards their initial values and the soil surface shows a very small recovery. These results are comparable with the field measurements presented by Symons and Carder. The influence of the groundwater level, the initial profile of

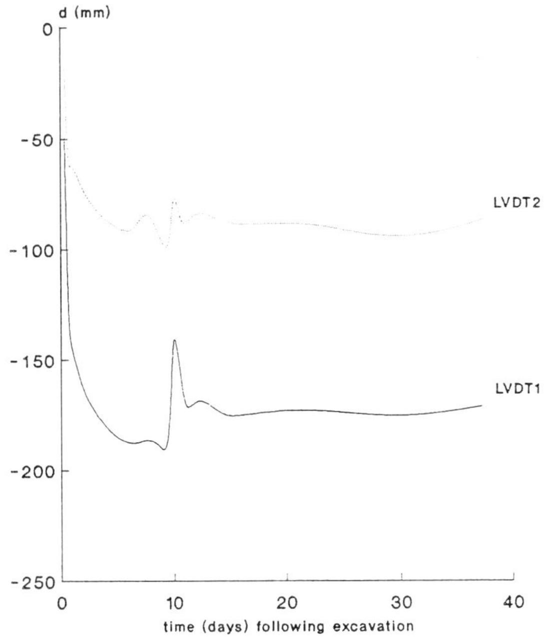

Fig. 3. Soil surface settlements during excavation and concreting

lateral earth pressure coefficient K_0, and three-dimensional effects will be investigated in future centrifuge model tests.

Conclusion

The Papers submitted to this Conference encompass a variety of different modelling techniques, and demonstrate their application to fundamental investigations of retaining wall behaviour, parametric studies, and design. The modelling method adopted in any given situation will depend on a number of factors including the problem which is being investigated, the purpose of the model, and constraints of time and cost. The various modelling techniques complement rather than compete against each other. Each has a contribution to make, and it falls to the geotechnical engineer to draw on the totality of these contributions both to reach an understanding of the problem in hand, and in the development of geotechnical engineering design.

References

All references are to Papers presented to this Conference, with the exception of the following:

BOLTON M. D. and POWRIE W. (1988): Behaviour of diaphragm walls in clay prior to collapse. *Geotechnique* 38 No 2, 167-189.
STEWART D. I. (1989) *Groundwater effects on in situ walls in stiff clay*. PhD dissertation, University of Cambridge.

The use of laboratory derived soil parameters for the prediction of retaining wall behaviour

K. G. HIGGINS, Geotechnical Consulting Group, D. M. POTTS, Imperial College of Science Technology and Medicine, and I. F. SYMONS, Transport Research Laboratory, UK

With the increased use of soil structure interaction analyses for design it is important to appreciate how such predictions compare with field behaviour. This paper uses parameters both from laboratory tests and from back analysis to predict the behaviour of the Bell Common Tunnel and compares the results with field measurements. It is concluded that laboratory derived parameters can be used to obtain a reasonable estimate of retaining wall behaviour.

Introduction

Soil structure interaction analyses are being increasingly used in the design of earth retaining structures to predict wall deflections and bending moments and to assess ground movements under working conditions. For embedded retaining walls in clay, consideration needs to be given to the likely behaviour both in the short term and in the long term when equilibrium is achieved under the new ground stress regime caused by the presence of the structure. This type of analysis requires information on the initial in-situ stress state in the ground, the type of structure and sequence of construction, and representative strength and stiffness properties of the soil. The selection of soil properties can be based on those previously obtained from comparisons of measured and predicted behaviour of similar constructions using similar analytical procedures. Where there has been little or no past experience of this type to act as a guide, the designer must rely solely on the results of site investigations when choosing parameters for use in analysis. The site investigations should incorporate a suitable programme of laboratory tests to determine the response of the soil to stress changes similar to those which are likely to occur in the vicinity of the proposed structure. It is therefore important to appreciate how analytical predictions made using parameters derived from this type of laboratory testing programme compare with actual measured performance.

 Retaining structures. Thomas Telford, London, 1993

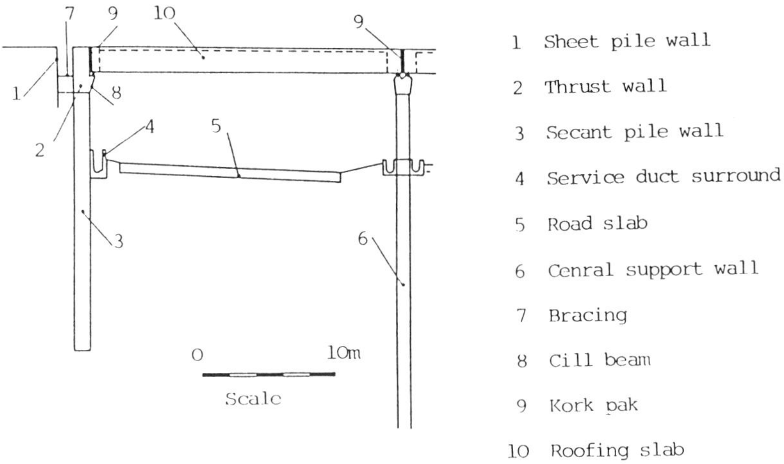

Fig. 1. Instrumented cross-section at Bell Common

This Paper aims to demonstrate the use of soil strength and stiffness parameters derived from laboratory tests in a soil structure interaction analysis to predict retaining wall behaviour. The Bell Common Tunnel, which carries the M25 London Orbital Motorway through the northern edge of Epping Forest in Essex, is used as a case study. The results of these analyses are compared with some published measurements of wall and ground behaviour.

The Bell common tunnel

The Bell Common Tunnel was constructed by the cut and cover method between 1982 and 1984. A section of tunnel was extensively instrumented to monitor the deformation and bending moments in the wall and the stresses and movements in the adjoining ground. Figure 1 shows a cross section through the Tunnel at this location. To limit the effects of thermal stresses in the roof slab, a compressible packing (Kork Pak) was installed between the walls and the roof slab. The road pavement provides no lateral support to the walls. Results of the observations made during construction are reported by Tedd et al (1984) and over the subsequent four year period by Symons and Tedd (1989).

Elastic finite element analyses were used for the design of the Tunnel (Hubbard et al (1984)) and these were extended to provide a Class 'A' (Lambe (1973)) prediction of behaviour at the instrumented section (Potts and Burland (1983)). The latter analysis modelled the soils as linear elastic perfectly

Table 1. Idealised construction sequence used by Higgins et al (1989)

Construction Activity	Stage (Tedd et al (1984))
Wall construction (sheet piles assumed 'wished in place' and secant pile installation simmulated in 2 dimensions)	I
Excavation (≈ 3.5 m) for and construction of cill beam and thrust wall	II
Excavation to approximately 5 m below original ground	IIIa
Construction of roof slab (represented by a horizontal spring of equivalent stiffness and equivalent vertical loads to cill beam and central wall)	IIIb
Excavation to approximately 8 m below original ground	IV
Application of vertical loads arising from protection to roof slab to cill beam and central wall and excavation to formation level	Vc
Backfill behind secant pile wall	VI
Construction of road slab and surface ducts and application of vertical loads due to 0.6 m of soil above roof to cill beam and central support wall	
100% dissipation of excess pore water pressures	

Table 2. Soil parameters used in Class A prediction

Soil Type	Cohesion c' (kN/m^2)	Angle of Shearing Resistance ϕ'	Angle of Dilation v	Drained Youngs Modulus (MN/m^2)	Poissons Ratio ν
Older Head	0	25°	0	9.143 y	0.2
Claygate Beds	0	35°	0	9.143 y	0.2
London Clay	0	25°	0	64+4.8*(y-7.0)	0.2
Backfill*	10	25°	0	9.143 y	0.2

y = depth below ground surface, *assumed parameters

plastic materials. In both cases the soil stiffness profile used was derived from back analyses of similar structures in the London Clay.

After construction of the Tunnel, a programme of laboratory testing using the most up to date techniques was undertaken on high quality samples of materials from the site. This work was reported by Hellings et al (1990) who used the results to derive strength and stiffness parameters. Further finite element studies of the tunnel were then carried out by Higgins et al (1989) to examine the effect on the predictions of more closely following the actual construction sequence, of allowing a softer response from the prop and of modelling the installation of the wall. These analyses used soil parameters as employed in the Class A predictions as well as parameters derived by Hellings et al (1990), without further interpretation. It was concluded that simulation of wall installation was necessary to reproduce the observed behaviour. The results from two of these analyses, which incorporated wall installation, are considered in more detail and compared with the measured behaviour in the subsequent sections.

Post construction finite element analysis of the Bell Common Tunnel

Table 1 shows the simulation of the construction sequence used for both analyses. In the first analysis, linear elastic perfectly plastic constitutive soil models were used. The soil parameters are given in Table 2, and were those used for the Class 'A' prediction. As previously noted the stiffness parameters were derived from back analysis of similar constructions in the London area.

In the second analysis a non-linear elastic perfectly plastic constitutive soil model was used. The high quality laboratory data were used to define the

Table 3. Non-linear elastic parameters after Hellings et al (1990)

Soil Type	A	B	C (%)	α	γ
London Clay	570	557	0.00007	1.47	0.453
Claygate Beds	633	647	0.0005	1.36	0.62
	R	S	T (%)	δ	λ
London Clay	520	495	0.0006	1.8	0.44
Claygate Beds	1300	1238	0.0006	1.8	0.44

Table 4. Strength data, after Hellings et al. (1990)

Soil Type	Cohesion c' (kN/m²)	ϕ'	ϕ' in Passive Stress Relief	Angle of Dilation v
Older Head*	0	32.7°	–	0
Claygate Beds	0	32.7°	–	0
London Clay	–	30° at p'= 83 kPa, 22° at p'= 258 kPa, 20° at p'= 525 kPa, 19° at p'= 703 kPa	35°	0
Backfill*	10	32.7°	–	0

* assumed parameters

soil properties (Hellings et al (1990)). To represent the non-linear elasticity, logarithmic functions of the form described by Jardine et al (1986) were used. The secant expressions which when differentiated give the tangent expressions used in the analyses are as follows:

$$\frac{G}{p'} = A+B\cos[\alpha(\log_{10}(\frac{\epsilon}{C}))^{\gamma}], \qquad \frac{K}{p'} = R+S\cos[\delta(\log_{10}\frac{|(\epsilon_v)|}{T})^{\gamma}]$$

where G is the secant shear modulus, K is the secant bulk modulus, p′ is the mean effective stress ($p' = 1/3(\sigma_1' + \sigma_2' + \sigma_3')$), ε is the axial strain observed in undrained triaxial tests and is related to the strain invariant E, used in the analysis by the expression $E = (3)^{1/2}\varepsilon$ where

$$E = 2(^1/_6((\varepsilon_1 - \varepsilon_2)^2+(\varepsilon_1 - \varepsilon_3)^2+(\varepsilon_2-\varepsilon_3)^2))^{1/2}$$

ε_1, ε_2, and ε_3 are principal strains, ε_v is the volumetric strain and A, B, C, R, S, T, α, γ, δ, and λ are all constants. The values of these constants are given in Table 3 while the derived strength parameters are summarised in Table 4. For the London Clay behind and beneath the secant pile wall a non linear failure envelope was adopted, but for the London Clay immediately in front of the wall a higher value of ϕ' was used, based on the results of special "passive stress relief" tests. Initial stresses in the soil were re-interpreted by Hellings et al (1990) and their modified distribution was used in the second analysis. For further information on soil properties see Hellings *et al* (1990).

In both analyses the temporary sheet pile and secant pile walls were assumed to behave linear elastically and have Young's moduli of 117×10^6kN/m² and 28.0×10^6kN/m² and Poissons Ratios of 0.2 and 0.15 respectively. The presence of the compressible packing between the roof slab

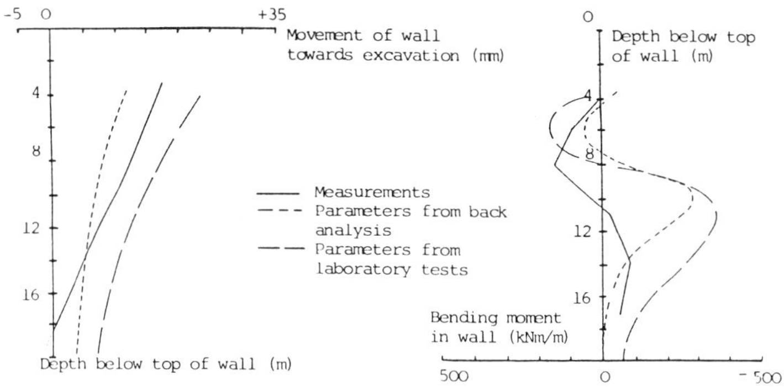

Fig. 2. Horizontal wall movement and bending moments in wall: stage VI

and wall made it difficult to assess the propping stiffness of the roof slab and in the original Class 'A' prediction a value of 29MN/m/m was used. However, observations made during tunnel construction (Tedd et al (1984)) suggested a much softer response. A value of 10MN/m/m has been adopted for both analyses. For further details see Higgins et al (1989).

Results

The wall deflections and bending moments at stage VI of the construction sequence are shown in Fig. 2. The analysis based on the 'back analyzed' soil stiffness underpredicts, while the analysis based on the laboratory data overpredicts the observed wall deflections. Both analyses show some movement of the base of the wall which is not indicated by the measurements. However, an inclinometer placed 3.25 m behind the secant pile wall did indicate movement at the level of the bottom of the wall. The analysis based on the laboratory derived soil parameters produced larger bending moments. In both analyses the maximum bending moment is greater than that measured and occurs at greater depth.

Similar plots of wall deflection and bending moments predicted after all excess pore water pressures had dissipated are presented in Fig. 3. Measurements at the Bell Common site are only available up to 4 years after construction and the values associated with the end of this period are given in Fig. 3. Measured wall deflections have already exceeded those predicted by the analysis based on parameters from 'back analyses' but are less than that predicted by the analysis based on laboratory derived soil parameters. The bending moment distributions predicted by the analyses are similar in shape with the analysis based on laboratory derived parameters producing larger values. The shape of these distributions does not as yet compare well

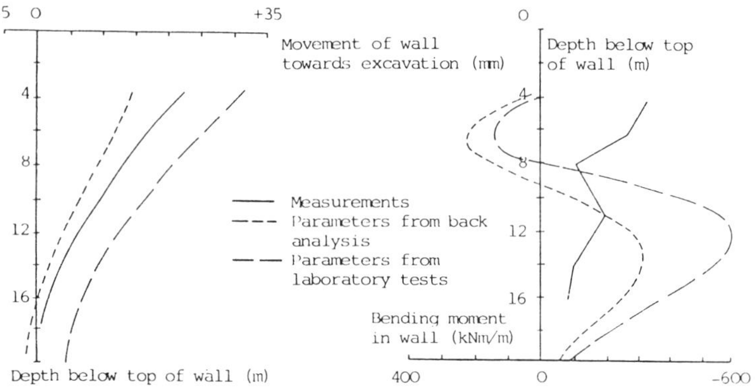

Fig. 3. Horizontal wall movements and bending movements predicted for the long term together with measurements at Stage V1 + 4 years

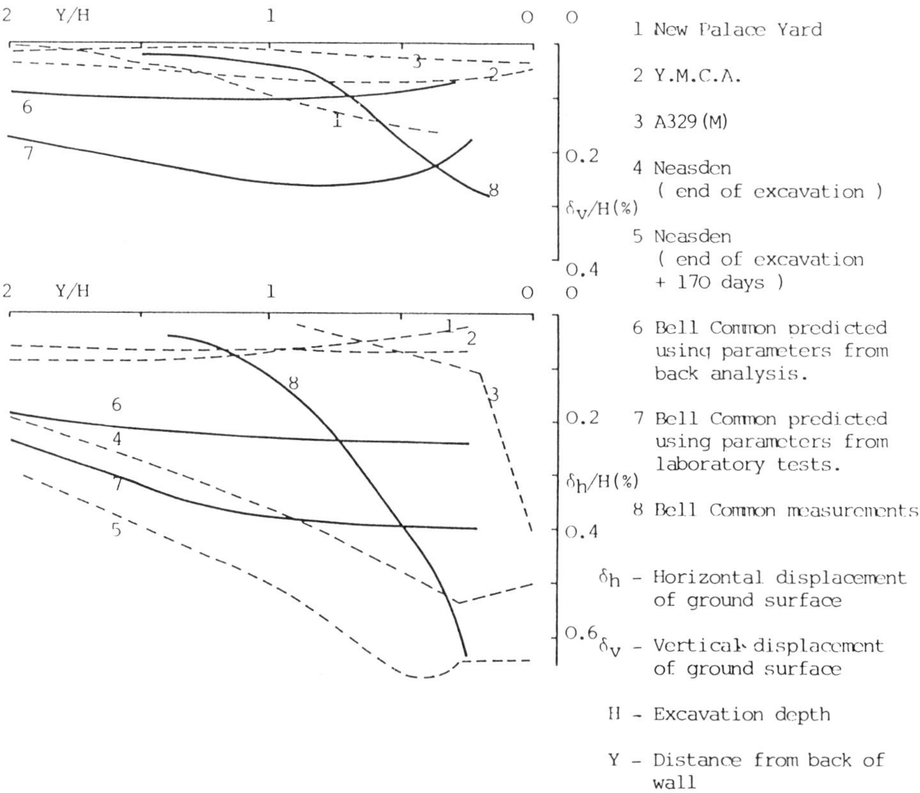

Fig. 4. Comparison of Bell Common data with other excavations in London Clay

with the measured distribution. However, it is difficult to reconcile the measured distribution, and in particular the large negative value apparent towards the top of the wall, with estimated loads and likely earth pressure distributions. It is to be expected that as the wall deflects further, bending moments will increase, although it is unclear how the currently measured distribution will change.

The effective stiffness of the propping system has a major influence on the wall behaviour. While the observed behaviour shown in figure 2 indicates that the roof may be providing lateral support during the later stages of construction, such behaviour is not so evident 4 years later. This is probably due to the compressibility of the Kork packing, which as mentioned was extremely difficult to estimate reliably.

Figure 4 shows ground surface displacements at stage VI of the construction sequence in which displacements have been normalised by dividing by the excavation depth allowing comparison with measurements from other schemes constructed in the London area. The New Palace Yard and YMCA walls were propped, the Neasden wall was supported by ground anchors and the A329(M) Reading wall was an unpropped embedded cantilever wall (St John (1975)).

Close to the excavation, the results of the analysis using 'back analyzed' stiffness values show reasonable agreement with the measurements at other sites but beyond a distance of one excavation depth from the wall, the movements are overestimated. This is thought to be a consequence of using a linear elastic representation for soil stiffness. When the soil stiffness was being derived by back analysis of other constructions only the wall behaviour was considered, it is therefore, perhaps not surprising that the analysis provides values of displacements near to the wall which are in reasonable agreement with other measurements. The Bell Common measurements show more movement than predicted by this analysis close to the wall, the maximum measured displacements occurring directly behind the sheet pile wall. The measured distribution of horizontal displacement is of a similar form to that of the cantilever wall on the A329(M). This could indicate a lack of support from the roof slab. Alternatively, the enhanced movements of the ground may be a result of the method used to form the walls at Bell Common together with the effect of the temporary excavation behind them. In either case, the fact that the observed behaviour differs from the behaviour of other propped walls shows that care must be exercised when extrapolating from one construction to another.

The analysis based on laboratory derived soil parameters shows larger movement than the analysis based on 'back analyzed' soil stiffness. It may be that the stiffness at low stress levels has been underestimated by the laboratory data giving a softer response of the superficial material. To improve the predictions the test data may need to be reinterpreted which was not an intention of this study. However, this analysis does show some reduction in movement with distance from the wall, which is more consistent with field measurements. Both analyses assumed fully undrained behaviour of the soils. Drainage occurring during construction may have influenced movements, an effect not accounted for in the analyses.

Conclusions

Soil parameters obtained from back analysis of existing structures in similar ground conditions can be used to provide a reasonable estimate of a retaining structures performance. To obtain the necessary parameters requires experience to have been gained through careful monitoring and back analysis of a sufficient number of relevant case histories. In the absence of such information, high quality laboratory testing can also be used to provide soil parameters for an approximate estimate of structural behaviour.

At Bell Common, the use of parameters obtained from high quality laboratory tests gave a somewhat better estimate of the maximum short term ground movement close to the wall than obtained when using parameters from back analysis. Measured movements at this site are likely to have been influenced by the temporary excavation immediately behind the wall. Further from the wall both predictions gave surface movements substantially greater than measured. Possible explanations are that the soil is stiffer than modelled in the analyses, and that drainage had occurred during construction that influenced the measurements.

Acknowledgements

The work described in this paper forms part of a research programme of the Ground Engineering Division of the Transport and Road Research Laboratory and the paper is published by permission of the Director. The views expressed are not necessarily those of the Department of Transport.

References

HELLINGS, J. E., BURIAND, J. B. and SYMONS, I. F. *Testing of soil samples from Bell Common Tunnel.* TRL Contractors Report 172, Transport Research Laboratory, Crowthorne, 1990.

HIGGINS, K. G., POTTS, D. M. and SYMONS, I. F. *Comparison of Predicted and Measured Performance of the Retaining Walls of the Bell Common Tunnel.* TRL Contractors Report 124, Transport Research Laboratory, Crowthorne, 1989.

HUBBARD, H. W. et al. Design of the retaining walls for the M25 cut and cover tunnel at Bell Common. *Geotechnique*, 1984, **34**, 4, pp 495-512.

JARDINE, R.J. *et al.* Studies of the influence of non-linear stress-strain characteristics in soil-structure interaction. *Geotechnique*, 1986, **36**, 3, pp 377-596.

LAMBE, T.W. Thirteenth Rankine Lecture: Predictions in soil engineering. *Geotechnique*, 1973, **23**, 2, pp 149-202.

POTTS, D. M., AND BURIAND, J. B. *A numerical investigation of the retaining walls of the Bell Common Tunnel.* TRL Supplementary Report 783, Transport Research Laboratory, Crowthorne, 1983.

ST JOHN, H. D. *Field and theoretical studies of the behaviour of ground movements around deep excavations in London Clay.* PhD thesis, University of Cambridge, 1975.

SYMONS, I. F. and TEDD, P. Behaviour of a proposed embedded retaining wall at Bell Common in the longer term. *Geotechnique*, 1989, **39**, 4.

TEDD P. *et al.* Behaviour of a propped embedded retaining wall in stiff clay at Bell Common Tunnel. *Geotechnique*, 1984, **34**, 4.

The application of the CRISP finite element program to practical retaining wall problems

R. I. WOODS and C. R. I. CLAYTON, University of Surrey, UK

Introduction

Finite elements may be used for practical retaining wall design in one of two fundamentally different ways

(*a*) The 'what if' approach, where the sensitivity of a particular calculated value is examined so that a design can be improved in a qualitative way, or critical design parameters can be identified.
(*b*) The 'absolute' approach, where values are required to be totally realistic and accurate, used for example when attempting to predict the displacements around a diaphragm wall.

There is an increasing trend towards using finite element analyses in the second way. There is a general optimism amongst programmers that finite element analyses always yield the 'truth', but many practising civil engineers have developed an extreme mistrust of finite element output. It is certainly true that it is very easy for an inexperienced engineer to produce results which appear ridiculous to any person with real experience of the problem being solved.

When using finite element packages for retaining wall analysis there are many factors which potentially can affect a desired result. What is required is

- a knowledge of the factors, and how they influence the results
- awareness of when problems are known to have occurred
- strategies for testing finite element output, to ensure as far as possible that the results are reasonable

Retaining wall analysis is often particularly difficult because

- geometry is often complex
- both short and long term solutions are required
- there are high stiffness contrasts
- for design purposes, accurate predictions of soil displacements, wall bending moments, and prop/anchor forces are required

Retaining structures. Thomas Telford, London, 1993

This Paper attempts to draw attention to the major issues involved and to highlight the potential pitfalls for the novice. Many of the examples cited in this paper have been encountered whilst using the CRISP package (Britto and Gunn, 1987) but all finite element analyses could, potentially, have similar problems.

Potential and observed problems

A number of problems and unexpected results have been reported when finite elements have been used in practical retaining wall analyses. The problems can be divided into the following categories

(*a*) Geometric modelling and discretization
(*b*) Constitutive modelling and parameter selection
(*c*) Modelling of installation, excavation and pore pressure equalisation
(*d*) Computational difficulties
(*e*) Obtaining required design output.

Geometric modelling and discretization

For all analyses (not only those involving retaining walls) the finite element method will always give an approximate solution, the accuracy of which will be influenced by the mesh used. At the outset, the analyst must decide how far away the remote boundaries should be located, how many elements (and what type) should be used, and how the element size should be varied across the mesh (grading). Each of these decisions will potentially affect the results, and the inexperienced user is usually offered little guidance.

Element-specific problems have been encountered in CRISP with bar elements used as ground anchors (Swain, 1989); it is possible to misinterpret the way in which the pre-stressing forces are applied. Interface elements to allow relative wall-soil slip and separation have been known to transmit residual tensile stresses after soil and wall have separated (Powrie and Li, 1990).

Constitutive modelling and parameter selection

Finite element packages such as CRISP generally offer the user a number of different constitutive models. These can range from simple elastic models to highly sophisticated elasto-plastic strain hardening/softening models, and the choice is closely linked with the selection of appropriate soil parameters. The issue facing the designer is, quite simply, how much of this complexity is required in order to ensure a realistic result. This section addresses some of the salient issues.

Non-homogeneity

For most soils, mean effective stress (and hence stiffness) increases with depth. Non-homogeneous stiffness models of the form $E = E_0 + m.z$ may cause problems (e.g. stress oscillations) when $E_0 = 0$. Difficulties with CRISP are known to have arisen in plate test analyses, where the "structure" is in full contact with this surface of theoretically zero stiffness (Hillier, 1992), and to a lesser extent with a retaining wall (Swain, 1989). Potts and Burland (1983), however, did not report any problems when analyzing the Bell Common Tunnel walls, with $E_0 = 0$.

For those elements at the soil surface, integration points just below the top boundary have a very low modulus E compared with integration points at the bottom boundary. Within a single element it is thus possible to have a large stiffness ratio (see below). Although there are relatively few soils which truly exhibit $E_0 = 0$, it is thought that some investigators have used a low E_0 to minimize the effect of the retained soil holding back the wall.

Anisotropy

Many natural soils exhibit anisotropy, as a result of stress history. For retaining walls embedded in such soils it would seem logical to use anisotropic elasticity as a constitutive model (e.g. Simpson et al., 1976; Creed, 1979). However, stress induced anisotropy depends on OCR, which is not itself constant but will vary (usually decreasing) with depth. So, to model stress induced anisotropy rigorously would require a fairly complex variation of E_h and/or E_v with depth. Care is also required in the selection of Poisson's ratio for anisotropic soils under conditions of undrained loading (Bishop and Hight, 1977).

Small-strain behaviour

The extent and magnitude of ground movements around retaining walls are often over-predicted using simple elastic models together with stiffness parameters measured in conventional laboratory tests. This is not only because conventional laboratory tests over-estimate strains by including bedding effects in their measured values, but also because the strain levels around walls are typically very small (see Jardine *et al.*, 1986). Two approaches are currently in use. In the first, back analysis of similar structures in similar ground conditions is used to obtain parameters (typically linear elastic cross anisotropic), and these are then used for forward prediction. In the second approach, small strain triaxial testing is carried out on samples obtained from the actual site to obtain stiffness parameters. Both approaches have potential problems associated with them. In the first approach, it is implicit that both geometry and ground conditions are sufficiently identical to allow satisfactory forward prediction, whilst the second approach assumes a simplified soil behaviour, as well as a knowledge of a suitable average operational strain level.

Linearity

The overall pattern of displacements predicted using a simple linear model (e.g. the settlement profile behind a wall) may be incorrect (Burland and Hancock, 1977). The principal reason for this has been shown to be the non-linear nature of most natural soils (Simpson *et al.*,1979), even at relatively small strains (Jardine *et al*, 1986, 1991). As with other codes (for example, ICFEP) a non-linear model based on a power law has been devised (Gunn *et al*, 1992) and used successfully within CRISP. It would be perfectly possible to implement an empirical stiffness-strain model of the form proposed by Jardine *et al* (1986). Many different testing strategies have been used to obtain the required parameters for such analyses, and they are typically relatively complex (see, for example, Jardine *et al.*, 1991). Simpler, bi-linear, elastic models have been developed for CRISP (e.g. Leach, 1985), and work is under way to develop kinematic models to mimic the changes of stiffness typically observed when soils are subjected to changes in stress path direction.

Yielding

All soils exhibit non-recoverable (i.e. plastic) behaviour above certain levels of stress or strain. For effective stress analysis, the Mohr-Coulomb yield criterion appears to be well established and is widely used in CRISP and other finite element programs. For total stress formulations, the equivalent Tresca yield criterion may be used (e.g. Jardine *et al*, 1986).

Closely allied to the selection of the yield criterion is the specification of in-situ stresses. As observed by Burland (1978), there is little point in carrying out a sophisticated non-linear elasto-plastic analysis if the initial stresses are incorrect. In this respect the at-rest earth pressure coefficient K_0 can be particularly difficult to measure or estimate over the full depth of interest.

Normality

When the stress state of an element of soil reaches (and remains on) the yield surface of an elasto-plastic material, plastic straining will occur. The relative components of plastic volumetric and plastic shear strain are determined by the plastic potential. If the flow rule is associated and normality applies, then the yield surface and plastic potential are coincident, and the plastic strain vector is normal to the yield envelope. This assumption is notorious for giving excessive dilation or, where drainage is prevented, negative excess pore pressures at yield/failure (for example, see Simpson *et al.*, 1979).

Associated flow is considerably easier to implement and use in a finite element package than non-associated flow, and is therefore commonly used. But such strong dilation (i.e angle of dilation = ϕ') is unrealistic, and can lead to over-enhanced strength and stiffness around the wall (Ponnampalam, 1990; Powrie and Li, 1990). There is little consensus on the most suitable angle of dilation ($\phi'/2$ is used by some) or what aspects of retaining walls are most affected. Potts and Burland (1983) studied it and concluded that wall movements were largely unaffected. It seems likely that earth pressure distributions would be more sensitive.

Modelling of installation, excavation and pore pressure equalisation

When analyzing the installation of a diaphragm-type wall with a finite element program there are three different strategies that one might adopt

(*a*) Begin the analysis with the wall already installed (the so-called "wished-in-place" wall),
(*b*) Begin with original undisturbed ground and then swap the relevant soil and concrete elements in the same time increment, or
(*c*) Begin with original undisturbed ground and then simulate the excavation of soil under bentonite slurry support, the placing of wet concrete via tremie pipe, and the subsequent hardening of the concrete.

Higgins *et al* (1989) compare strategies (*a*) and (*c*) in a re-analysis of the Bell Common tunnel, using undrained analysis in the short term and imposed pore pressure changes to model the long term. Gunn *et al* (1992) give further details of the numerical aspects of strategy (*c*), but using a coupled-consolidation approach. One significant anomaly which has been observed when wall installation is modelled is the spatial and temporal oscillation of effective stress and pore pressure in the lateral direction. This was first reported for CRISP by Kutmen (1986) and subsequently by Ponnampalam (1990). Similar oscillations were observed by Naylor (1974).

In contrast to most other finite element codes, users of CRISP are able to model the whole sequence of installation, excavation and pore pressure equalisation in one continuous analysis, using the coupled consolidation (Biot) approach. Whilst this is a convenient and elegant approach, the choice of time step is important if numerical problems are to be avoided (Woods, 1986). The method by which excavation is modelled can also be a problem - incorrect formulations lead to solutions which even for elastic cases are increment dependent (Ishihara, 1970; Gunn, 1982; Brown and Booker, 1985).

Computational difficulties

Disparate stiffnesses

There is a limit to the difference in stiffness that can be tolerated between (adjacent) elements in a mesh, before computational problems arise. This has implications for all soil-structure interaction problems where soil and concrete (or steel) are being modelled. The problem is well known and is caused by adding numbers of different orders of magnitude in a finite precision computer. The net result is equilibrium error in the analysis and possible ill-conditioning.

Equilibrium errors due to disparate stiffnesses have been reported by Woods and Contreras (1988), Vaziri (1988), and Hillier (1992) for rigid steel plates in contact with soil. Exactly the same problem can be expected to occur in the analysis of a thin steel sheet pile wall. These types of problem are associated both with the level of precision (i.e. single or double) invoked by the code, and also by the precision used by the computer's c.p.u. - they can therefore vary from machine to machine, even when using identical code.

Effective stress method

Most geotechnical finite element codes (including CRISP) use the effective stress method (Naylor, 1974) for both drained and undrained loading. The bulk modulus of the pore fluid K_W, is set either to zero (= drained) or to an arbitrarily large number (= undrained). It is generally believed that the actual magnitude of K_W used for undrained analysis is unimportant, provided it is much greater than the bulk modulus of the soil skeleton. However, Ponnampalam (1990) has shown that horizontal stress distributions resulting from wall installation can be very erratic when K_W, is large. The problem is analogous to that of selecting a value of Poisson's ratio close (but not exactly equal) to 0.5 for an undrained analysis in terms of total stresses.

If the value of K_W, is too low, then the behaviour modelled is similar to a partially saturated soil ($B < 1$), with some load being taken in effective stress even in the short term. Clearly K_W must be neither too high nor too low, relative to the drained modulus of the soil, but it is not clear that there will always be a satisfactory range between these limits, for all combinations of geometry and soil properties.

Horizontal stress distributions

Soil elements closest to the wall on the excavated side (just below final dig level) have been known to exhibit high horizontal effective stresses, after removal of elements to simulate excavation. It appears to be possible to obtain stress states outside the specified failure envelope near soil surfaces where the soil has undergone swelling. This phenomenon has been observed with CRISP by Clarke and Wroth (1984), Ponnampalam (1990), and Powrie and Li (1990). It has also been observed by Rodrigues (1975) and Creed (1979) for other finite element codes.

On the retained side, tensile stresses may develop behind the top of the wall as it attempts to move away from the soil. According to Powrie and Li (1990), these are particularly pronounced for elastic-perfectly plastic models and for high soil stiffness, but may be reduced by the use of interface elements (or setting by $E_0 = 0$, as noted above).

Stress oscillations have already been mentioned, and whilst these oscillations are clearly unrealistic, their cause remains uncertain.

Solution scheme

CRISP uses an incremental (tangent stiffness) solution scheme. For elastic perfectly plastic models, corrections are applied to elements that have yielded in order to bring the stress state back to the yield surface. For critical state based models no such corrections would be carried out as they would be considerably more complex. It is therefore relatively easy to produce an invalid analysis unless increment size is sufficiently small. For a single stage of loading on a Mohr-Coulomb material, about 50 increments are typically required. In contrast, other codes (for example, ICFEP) use different solution strategies, and can provide adequate solutions with only a few increments per loading step.

Obtaining required design output

Wall and ground deformations

There is no problem in obtaining displacements as they are the primary unknowns solved directly by the finite element method. As discussed above, however, there are a host of influences which will affect the reliability of both the magnitude and pattern of displacements.

Stress distributions

Stress distributions may be required for comparison with classical design methods, or perhaps with in-situ measurements. However, they should not be used as data for bending moment calculations, and may be unreliable as a basis for simple hand calculations of equilibrium. Unexpected patterns can occur on both the retained and excavated sides (see above), potentially confusing the inexperienced analyst. It must be remembered that the finite element method only ensures equilibrium of nodal forces, and that in general there will neither be local equilibrium within an element nor equilibrium of stresses across element boundaries. Creed (1979) describes a method of inferring approximate pressure distributions from nodal loads.

Wall bending moments

The calculation of wall bending moments may be based on

(*a*) transverse stress distributions in the wall elements
(*b*) horizontal soil pressures acting externally on the wall elements
(*c*) nodal forces acting between wall elements

(*d*) nodal forces from the soil acting externally on the wall elements

The first two methods are intuitively reasonable and also quite convenient as they make use of stresses calculated at integration points in the relevant elements. However, as noted earlier, stress distributions can be far from reliable; furthermore, method (*b*) multiplies (possibly) inaccurate stresses by an increasing lever arm as one moves down the wall. Comparisons of these two methods have shown significant discrepancies (Swain, 1989; Gunn and Ponnampalam, 1990; Powrie and Li, 1990).

Methods (*c*) and (*d*), on the other hand, make use of nodal forces which can be expected to be rather more reliable. As these forces are not normally output to the user, some program modifications may be necessary.

Prop and anchor forces

Temporary and permanent props are frequently modelled by solid 2D elements, and load distributions can be readily obtained from compressive stresses at integration points. Anchors may be modelled by bar elements, and once installed it is relatively easy for the program to calculate the excess force over the initial (pre-stress) value. The total force at any stage is thus readily computed.

Nodal forces may be used to make simple (hand) checks that the output satisfies force equilibrium; for example, to detect equilibrium errors due to inadequate precision.

Conclusions

Finite element analysis of retaining walls is potentially particularly problematic; perhaps more so than any other geotechnical application of finite elements. The most serious problems are choice of constitutive model and associated soil parameters, the modelling of installation and excavation, numerical irregularities, and the derivation of design output. Finite elements cannot be used "blindly" for retaining wall analysis. Validation against real problems is required, and in this respect there is a great need for more published case histories and guidelines for the inexperienced user.

References

BISHOP, A. W. and HIGHT, D. W. The value of Poisson's ratio in saturated soils and rocks stresses under undrained conditions. *Geotechnique*, 1977, **27**, No. 3, pp 369-384.

BRITTO A. M. and GUNN M. J. *Critical State Soil Mechanics via Finite Elements*. Ellis Horwood Ltd, Chichester, 1987.

BROWN, P. T. and BOOKER, J. R. Finite element analysis of excavations. *Computers and Geotechnics*, 1985, **1**, No. 3, pp 207-220.

BURLAND, J. B. and HANCOCK, R. J. R. Underground car park at the House of Commons, London: geotechnical aspects. *The Structural Engineer*, 1977, **55**, N.o 2, pp 87-100.

BURLAND, J. B. Application of the finite element method to prediction of ground movements. *Developments in Soil Mechanics* (ed. C.R. Scott), Applied Science Publishers, London, 1978, pp 69-101

CLARKE, B. G. AND WROTH C.P. Analyses of the Dunton Green retaining wall base on results of pressuremeter tests. *Geotechnique, 1984,* ***34****, No. 4, pp 549-562.*

CREED, M. J. *Analysis of the deformations associated with anchored diaphragm walls*. PhD thesis, University of Surrey, 1979.

GUNN, M. J. How not to analyze excavations. *Workshop on Critical State Soil Mechanics in Finite Elements, University of Cambridge*, 1982.

GUNN, M. J., HILLIER, R. P. and WOODS, R. I. The influence of nonlinearity on the interpretation of soil parameters from field plate load tests. *Proc. 4th Int. Symp. Numerical Models in Geomechanics*, (eds. Pande and Pietruszczak), 1992 .

GUNN, M. J. and PONNAMPALAM, A. Calculation of bending moments from CRISP output in soil-structure interaction problems. *Proc. 4th CRISP Users Workshop, University of Surrey*, 1990.

GUNN, M. J., SATKUNANANTHAN, A. and CLAYTON, C. R. I. Finite element modelling of installation effects. *Proc. Int. Conf. on Retaining Structures, Cambridge*, 1992.

HIGGINS, K. G, POTTS, D. M. and SYMONS, I. F. *Comparison of predicted and measured performance of the retaining walls of the Bell Common tunnel.* TRL Report CR 124, Transport Research Laboratory, Crowthorne, 1989.

HILLIER, R. P. *The interpretation of the field plate load test on clay soils via finite element analysis*. PhD thesis, University of Surrey, 1992.

ISHIHARA, K. Relations between process of cutting and uniqueness of solutions. Soils and Foundations, 1970, **10**, No. 3, pp 50-65.

JARDINE, R. *et al.* Studies of the influence of non-linear stress-strain characteristics in soil-structure interaction. *Geotechnique*,1986, **36**, No. 3, pp 377-396.

JARDINE, R. *et al.*Some practical applications of a non-linear ground model. *Proc 10th Eur. Conf. Soil Mech. and Foundation Engng*, 1991, **1**, pp 223-228.

KUTMEN, G. *Influence of the construction process on bored piles and diaphragm walls : a numerical study*. MPhil thesis, University of Surrey, 1985.

LEACH, G. *Abi-linear elastic model for CRISP*. Internal report, British Gas, 1985.

NAYLOR, D. J. Stresses in nearly incompressible materials by finite elements with application to the calculation of excess pore pressures. *Int. J. Num. Meth. Engng*, 1974, **1 8**, pp 443-460.

PONNAMPALAM, A. PhD transfer report, University of Surrey, 1990.

POTTS, D. M. and BURLAND, J. B. *A numerical investigation of the retaining walls of the Bell Common Tunnel.* TRL Report SR 783, Transport Research Laboratory, Crowthorne, 1983.

POWRIE, W. and LI. E. S. F. Design study of an in situ wall propped at formation level. *Workshop on Advanced Geotechnical Analysis, University of Cambridge*,1990.

RODRIGUES, J. S. N. *The development and application of a finite element program for the solution of geotechnical problems.* PhD thesis, University of Surrey, 1975

SIMPSON, B., O'RIORDAN, N. J. and CROFT, D. D. A computer model for the analysis of ground movements in London Clay. *Geotechnique,* 1979, **29**, N.o 2, pp 149-175.

SWAIN, A. Personal communication, 1989 .

VAZIRI, M. *Screw plate testing : experimental and theoretical observations.* PhD thesis, University of Surrey, 1988.

WOODS, R. I. Finite element analysis of coupled consolidation. *Proc. 2nd Int. Symp. Numerical Models in Geomechanics*, 1986, pp 709-717.

WOODS, R. I. and CONTRERAS, L. F. *Numerical modelling of field plate loading tests - Part II.* Research Report GE/88/8, City University, 1988.

Statically admissible earth pressure and bearing capacity calculations

K. MORTENSEN, Danish Geotechnical Institute, Denmark

A calculation method for plane earth pressure and bearing capacity problems is presented. The method is based on the theory of plasticity and statically admissible figures of failure, which may however more appropriately be labelled stable stress figures. The numerical calculations are carried out as finite-difference calculations applying Kötter's equation. Use is being made of curves of discontinuity with the normal stresses acting parallel to the curve not being identical on both sides of the curve. The method is applicable for investigation of failure problems for a great number of geotechnical structures. So far a spreadsheet program has been developed for the design of free and anchored sheet walls (Fig 7). The method is considered to be a more sound approach than the kinematically admissible methods dominating the standard calculation procedures used in geotechnical practice.

Stable stress fields and stress figures

In geotechnical calculations according to the theory of plasticity, distinction is generally made between kinematically admissible and statically admissible solutions.

Kinematically admissible solutions represent a possible, but not necessarily the most critical way of failure. The corresponding methods of calculation therefore generally indicate safety margins larger than the actual ones, i.e. they are on the "unsafe"side.

Statistical admissibility requires that the stress conditions in principle are known, that the conditions of equilibrium are fulfilled and that the failure condition is respected everywhere in the soil as well as in the structural material. Such a state of stress does not generally represent a possible state of failure, but rather a possible state of stability. Consequently, the labels statically admissible ways/figures of failure are deceptive. More appropriate terms would be stable/statically admissible stress fields and stress figures.

If the condition of normality (the angle of dilatancy = the angle of friction) is fulfilled in the soil as well as in the interface between the soil and the structure, statically admissible solutions will generally indicate safety margins smaller than the actual ones, i.e. they are on the "safe" side. Even when this condition is not fulfilled, statically admissible solutions will often be on the safe side, and in all cases they will be less on the unsafe side than the corresponding kinematically admissible solutions. Basically, statically admissible methods of calculation are therefore a more sound approach than the kinematically admissible methods often applied in ordinary geotechnical practice.

Main cases.

Earth pressure and bearing capacity calculations based on Coulomb's failure criterion may be distinguished according to 3 main categories

- Q-cases, dealing with a weightless, cohesionless soil carrying a surcharge q on the surface.
- G-cases, where the soil has weight (unit weight γ), but no cohesion and no surcharge.
- C-cases, where the weightless soil has a cohesion c, but no surcharge.

Superposition

In the Q- and G-cases it is generally on the safe side to superpose two statically admissible (compressive) stress fields, because the inclination of the frictional part of the stress on all sections will be smaller than or equal to the max. inclination in the two cases. On the other hand it is on the unsafe side to deduct one stress field from another, if the failure lines do not coincide, i.e. if the stress figures are not identical.

The C-case is generally coped with first by adding an isotropical principal stress $c\cot\phi$, then carrying out the calculations as a Q-case and finally subtracting $c\cot\phi$. Doing so, Coulomb's failure criterion will be satisfied, but the subtraction may result in unacceptable tensile stresses. Therefore the C-case requires special considerations, which will not be dealt with in the present paper. Consequently, only the Q- and G-cases are considered.

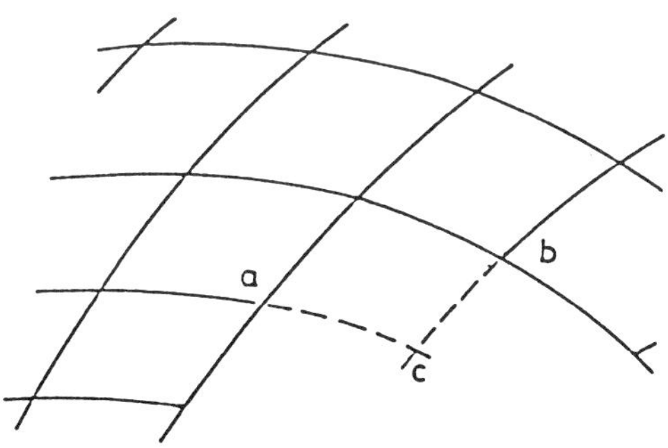

Fig. 1. Failure lines

Calculation method

The stress fields are generally described by failure lines along which Coulomb's failure criterion is satisfied. The points of intersection between the failure lines are characterized by the x-v co-ordinates, the inclination of a characteristic direction and a characteristic stress as for example the size of the minor principal stress. If the numerical values of these 4 parameters are known at points a and b in fig 1, their values may be determined for point c by means of the geometric conditions and Kötter's equation.

At the surface and at the structural interface (the wall or footing) there is an unequivocal connection between the size and direction of the characteristic stress in the stress field and the size and direction of the exterior stress (surcharge, earth pressure or foundation pressure). Consequently, when the boundary conditions are known, the whole stress field may be determined by finite-difference calculations based on Kötter's equation, provided that failure lines of the same category do not intersect. This method of calculation has been explained in detail in (ref.l).

Curves of discontinuity

The potential of this method of calculation may be greatly improved by introducing curves of discontinuity. On both sides of a curve of discontinuity identical total stresses act on the curve, whereas the normal stresses acting parallel to the curve are not identical on both sides of the curve.

Curves of discontinuity may be used to prevent undesired intersection of rupture lines and may also to a certain degree be used to separate layers with different angles of internal friction. The potential and shortcomings of curves

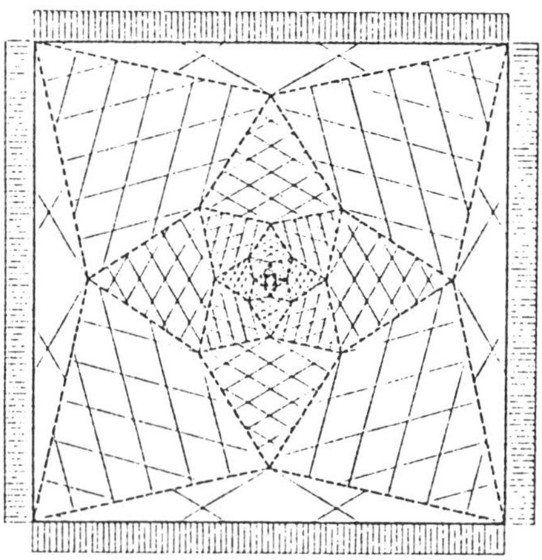

Fig. 2. Stable stress figure

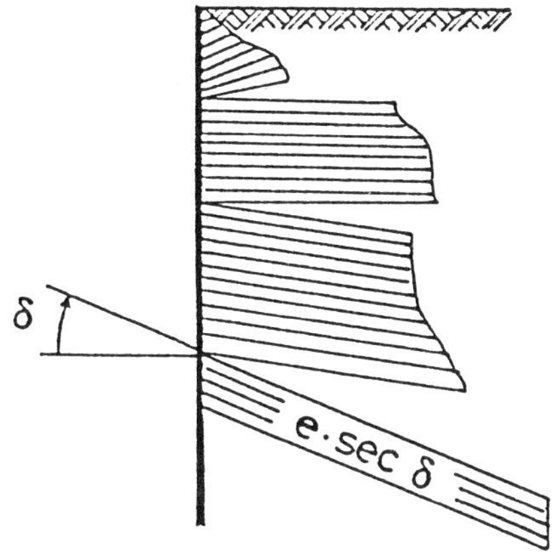

Fig. 3 Earth pressures

of discontinuity may be illustrated by the rather curious example shown in Fig 2. Present here is a square volume of soil with identical vertical and horizontal principal stresses. By the introduction of the lines of discontinuity shown dotted in the figure, it is possible as indicated to identify stress fields satisfying the failure condition everywhere, although the stress condition is the most stable one that can be imagined. This confirms that the stable stress figure in no way indicates a possible way of failure, but a state of stability which in this case is rather peculiar.

The potential of the method of calculation is demonstrated below by considering the earth pressure on a vertical wall. The surface of the soil is horizontal with a constant surface load q. The soil is cohesionless, frictional soil with constant unit weight γ and angle of internal friction ϕ.

Stress distribution

The theoretical determination of the earth pressure may be considered to be a process of integration, beginning at the top of the wall. When determining the earth pressure e*secδ acting on a small element of the wall as shown in Fig. 3, the stress distribution and principal stress direction are known above the element. The unit pressure on the element may correspondingly adopt any value between an upper boundary e_{max} and a lower boundary e_{min}. To reduce the vast number of statically admissible stress distributions, only extreme cases with e_{max} and/or e_{min} and δ_{max}, δ_{min} and/or $\delta=0$ are considered. δ is considered positive as shown in Fig. 3, i.e. when the resulting earth pressure acting on the soil is directed downwards.

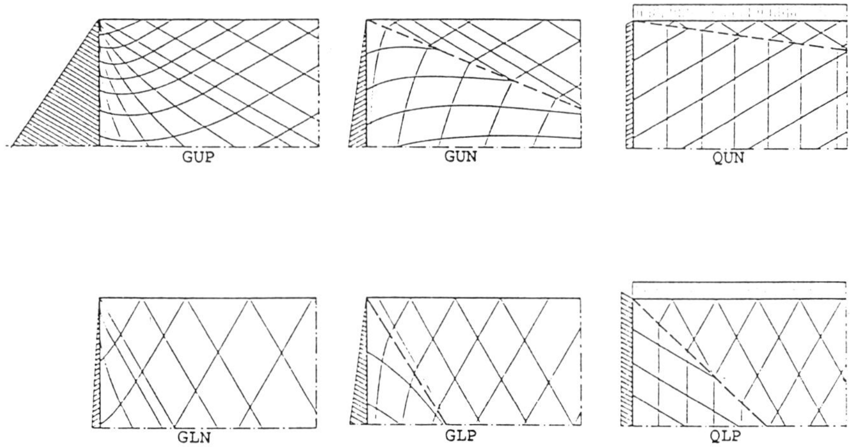

Fig. 4. Stable stress figures without pressure jump

Extreme earth pressures without pressure jump

The simplest earth pressures correspond to e_{max} or e_{min} with the same inclination along the whole height of the wall. The cases considered may be characterized by 3 letters

- Q or G referring to the main case.
- U or L referring to e_{max} (upper) and e_{min} (lower), respectively
- P, O or N referring to δ_{max} (positive), δ=0 and δ_{min} (negative), respectively.

Of the corresponding 12 stable stress figures, 6 are shown in fig 4 for ϕ = 30°. The remaining 6 are the well known figures with logarithmic spirals (Prandtl-zones) and/or straight lines (Rankine-zones). The upper three figures in the upper part of fig 4 correspond to e_{max} and the lower ones to e_{min}. The two G-cases to the left with the wall as envelope to the failure lines are generally known, whereas the remaining 4 with dotted discontinuity lines have possibly not been identified before.

Extreme earth pressures with pressure jump

Pressure jumps are defined as points at the wall, where the boundary values considered changes from e_{max} above to e_{min} below the point or vice versa. Stable stress figures with extreme pressures and one jump can be characterized by 5 letters

- G or Q as above
- U or L and P, 0 or N as above, referring to the conditions above the jump
- The same, but referring to conditions below the jump.

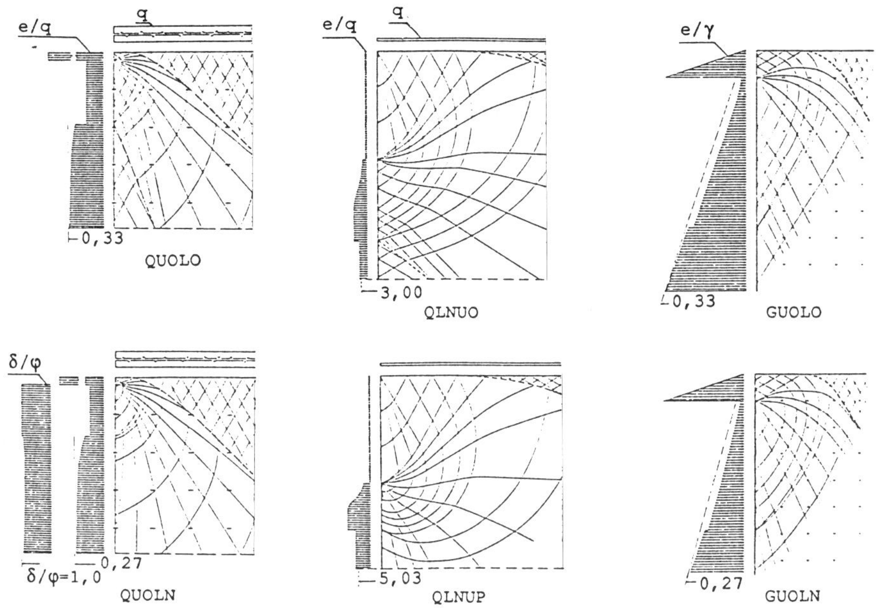

Fig. 5. Stable stress figures with one pressure jump

Correspondingly, 72 different stress figures exist. 6 typical examples are shown in Fig. 5. They have some characteristics in common

- Above the jump they are identical to the figures without jump (Fig. 4).
- Where the boundary failure line for this part of the figure reaches the surface, a dotted curve of discontinuity begins, delimiting an exterior Rankine-zone.
- Below the jump a singularity occurs at the wall, causing a sudden change in the pressure, which, however, is not a pressure jump of the type defined above.
- Below the singularity it is not statically possible to fulfil $\delta = \pm\phi$ when the same condition is also valid above the singularity. As an example, the variation of δ is shown in fig QUOLN. For δ=0 the corresponding phenomenon makes it necessary to introduce a curve of discontinuity as shown in fig QUOLO and QLNUO.
- Below the singularity the earth pressure approaches asymptotically the earth pressure corresponding to the exterior zone rupture, which actually surrounds the whole figure.

Example

As an example of the application of the method of calculation an anchored

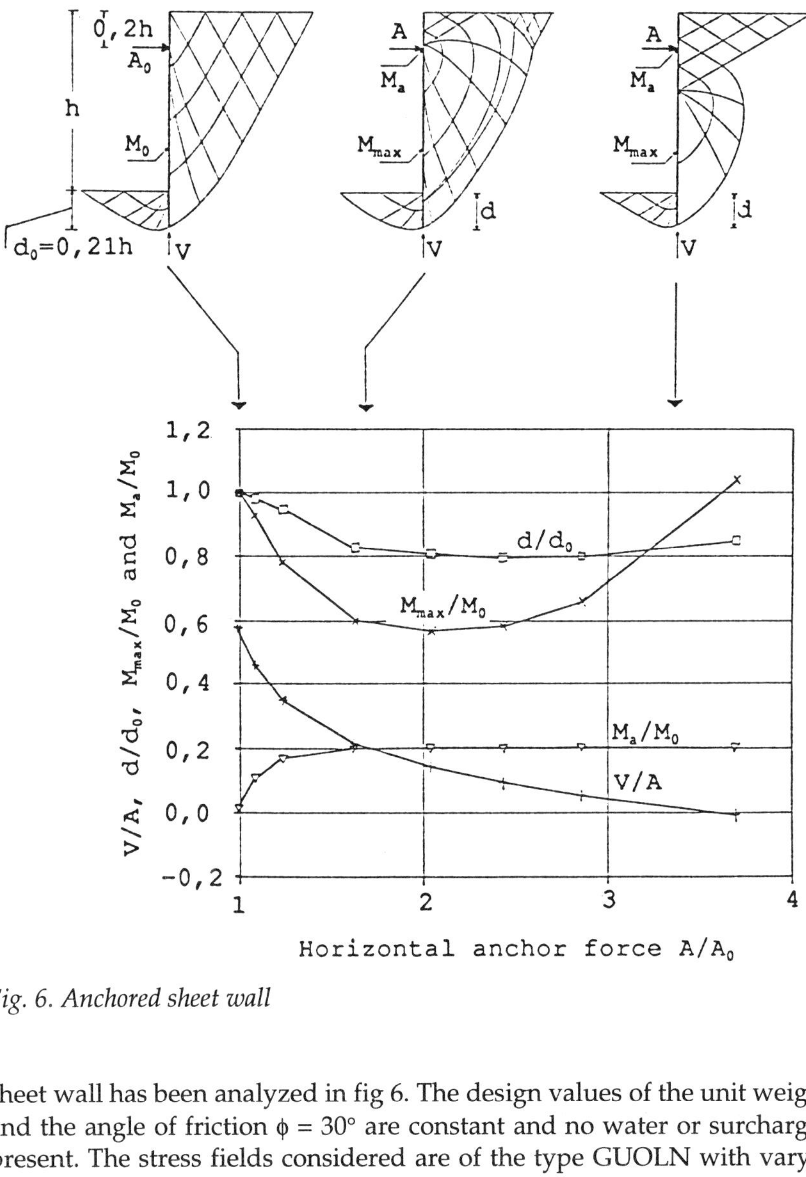

Fig. 6. Anchored sheet wall

sheet wall has been analyzed in fig 6. The design values of the unit weight γ and the angle of friction $\phi = 30°$ are constant and no water or surcharge is present. The stress fields considered are of the type GUOLN with varying positions of the pressure jump on the active side and of the type GUP on the passive side.

The diagram shows how the vertical anchor force V (or point resistance), the driving depth d and the max and min bending moments M_{max} and M_a vary with varying horizontal anchor force A. The left-hand end of the diagram corresponds to a free earth support solution, i.e. with depth 0 to the pressure jump, and the corresponding values are marked with index o. The bending moment M_{max} and the driving depth d have minimum values close to $A/A_0 = 2$ and 2.5, respectively. These two cases are very similar to the design methods proposed in (ref. 2).

For driving depths greater than the minimum value two solutions seem to exist, one on the left-hand side of the minimum corresponding to the wall acting as an ordinary anchored sheet wall, and one on the right-hand side with the wall acting as an anchor slab, representing the max horizontal anchor force which can be resisted with the driving depth and vertical anchor force considered.

For a certain interval of the horizontal anchor force equilibrium of the wall may be obtained also for negative driving depths, i.e. without any passive earth pressure. However, these solutions are statically impossible because the stable stress fields cannot be continued below the wall. For the solutions shown in the diagram this can be done however by means of curves of discontinuity.

The problems associated with the vertical equilibrium will not be dealt with in the present Paper.

Practical applications

The method of calculation described is applicable to a number of geotechnical structures such as

- free sheet walls or sheet walls anchored at one or more levels with or without fixation in the soil and with horizontal or vertical load, inclined surfaces, strip surface load, differential water pressure etc.,
- retaining walls with the stress fields for the earth pressures and the bearing capacity combined by means of curves of discontinuity to one entire, stable stress field,

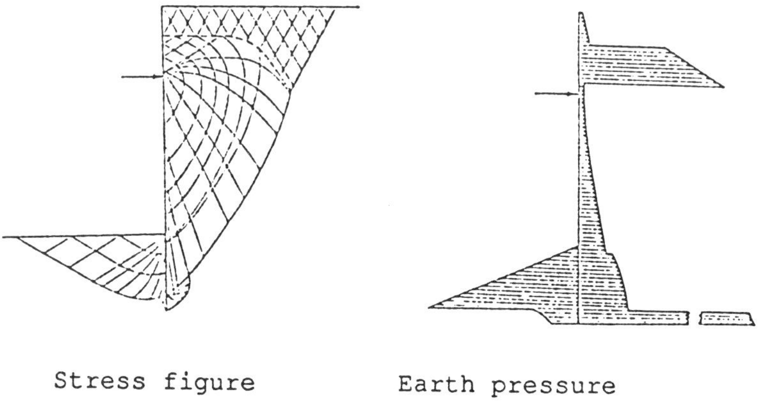

Fig. 7. Sheet wall design

- anchor walls, vertical or inclined, reaching the surface or buried,
- double sheet walls with differential water pressures,
- shallow footings, horizontal or inclined with vertical or inclined load acting centrally or eccentrically.

For all these cases the system of partial factors of safety may be applied. So far a spreadsheet program has been developed by the author for free and anchored sheet walls as shown in Fig. 7. On the active side three pressure jumps have been introduced and often a singularity also occurs, whereas only one pressure jump is considered on the passive side. The results of the calculations are the driving depth, the anchor force and the three bending moments, M_a at the anchor, M_{max} and M_{min} close to the tip of the wall. The three pressure jumps may be selected freely within certain limits in order to optimize the dimensions of the wall. The program allows for taking into account varying unit weight and strength parameters in layered soils, differential water pressures and the equilibrium in vertical direction.

Similar programs of calculation may be developed for other types of structures such as the ones mentioned above. The main problems in this connection are

- firstly to choose between the great number of statically admissible solutions, which exist for the various types of structures and
- secondly to establish acceptable approximations necessary to make the method applicable in ordinary geotechnical practice.

Acknowledgements

The writer wishes to thank Professor H.Lundgren, Dr. N.Krebs Ovesen, Managing Director of the Danish Geotechnical Institute, and Professor J.S.Steenfelt for valuable discussions and helpful suggestions in relation to this paper.

References

1. LUNDGREN, H. and MORTENSEN, K. Determination by the theory of plasticity of the bearing capacity of continuous footings on sand. *Proc. Third Int. Conf. Soil Mech.*, Vol. 1, Zurich, 1953.

2. HANSEN, J. B. *Earth pressure calculation.* Teknisk Forlag, Kobenhavn 1953.

Ground movements caused by displacements of earth retaining walls

M. MARUOKA, M. AOKI, E. SATO, and Y. IKUTA, Takenaka Corporation, Japan

This paper describes surrounding ground movements caused by earth retaining wall displacements in cohesive soil and its predictive method. The relationships between displacement patterns of the earth retaining wall and the patterns of zero-extension lines in the surrounding ground are discussed based on in-situ measurements and analysis using a rigid body spring model. The discussion leads to the conclusion that movements of the surrounding cohesive soil can be obtained on the assumption that the dilatancy angle ν equals 0°, and that the shape of the strain field is formed from a combination of straight lines and circular arcs.The ground surface settlements and the ground movements were calculated from earth retaining wall displacements for seven construction sites based on the specified strain field. These calculated values agree well with the measured values, thereby verifying this predictive method.

Introduction

The environmental conditions of building construction works in urban areas are usually rigorous and demanding. For instance, the decisions taken in the planning and execution of excavation works in urban areas are primarily based on an evaluation of the effects of the proposed excavation on nearby structures. The factors that are influential in determining the extent of the ground displacement resulting from excavations include: the displacement of earth retaining walls, ground heave, lowering of the groundwater, etc. The effect of earth retaining wall displacement is the main factor which needs to be investigated at the initial stage of planning the excavation work.

This paper describes examples of in-situ measurements and a method of predicting the ground movements caused by displacements of earth retaining walls resulting from excavation.

In-situ measurements of ground movement

An example of in-situ measurements of the movements of the surrounding ground caused by displacements of retaining wall is shown in Fig. 1. The

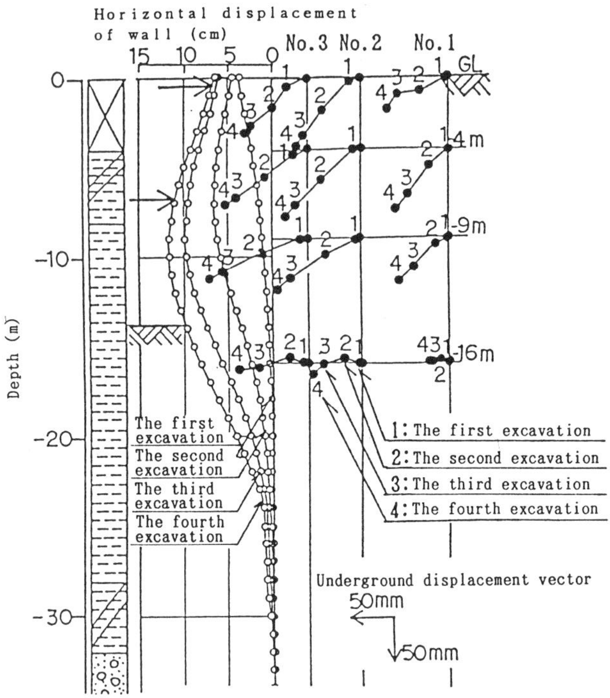

Fig. 1. Displacements of earth retaining wall and ground displacement vectors

outline of the soil profile and measurement section are shown in Fig. 2 . The approximate size of the excavation was 70m x 90m in area and 13.75m in depth. The excavation work was done by the top-down method in combination with steel struts , using a soil cement pile wall (800 ϕ) 24m in length reinforced with steel of H-beams 588 x 300, positioned at the centre of each pile to form the earth retaining wall. The soil profile of the site consisted of a layer of backfill to a depth 4m and an alluvium silt to a depth 32m overlying a gravel bed.

Measurements were made for the purpose of evaluating the movement of the ground caused by the retaining wall displacement in the sections where there are no neighbouring structures or buried objects. The displacement of the retaining wall itself and the horizontal and vertical movements of the ground were monitored.

Fig. 1 shows the measured horizontal displacements of the retaining wall and the displacement vectors in the ground. The displacement vectors between the ground surface and GL -9m showed a linear trend inclined 20-45° from the horizontal, while the displacement vectors at GL -16m indicated almost horizontal movements.

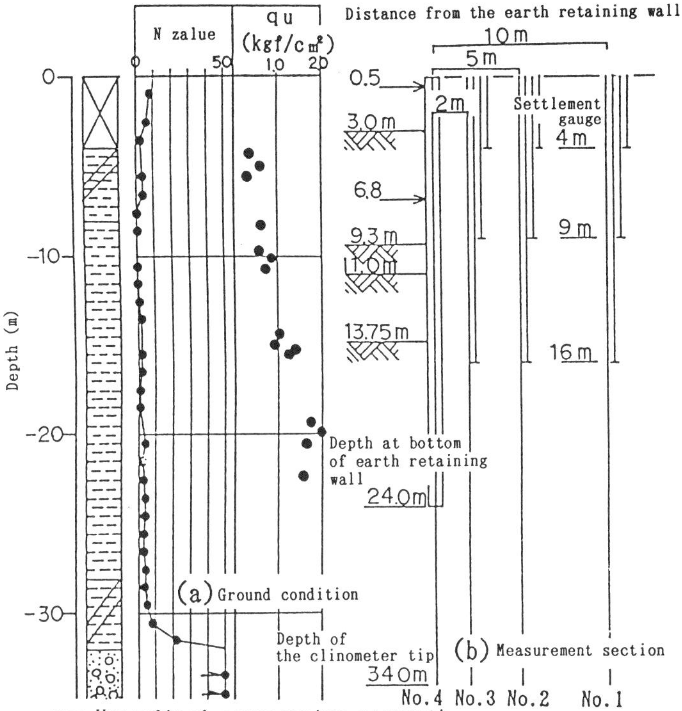

qu : Unconfined compressive strength

(1kg/cm² = 98.04KPa)

Fig. 2. Soil profile and measurements

From the measurements it is possible to define the strain field in relation to the displacement in the ground. The zero-extension lines are used later in a method of estimating the movement of the surrounding ground resulting from the retaining wall displacements .

Since no leakage was observed throughout the retaining wall, the observed ground movements were considered to be the result of the displacement of the earth retaining wall and the heave associated with the excavation. This assumption is implicit in Eqn. (I) .

$$\delta H = \delta H_1, \delta V = \delta V_1 + \delta V_2 \quad (1)$$

Where,

δH : Horizontal displacement in the ground
δH_1 :Horizontal displacement caused by the retaining wall displacement
δV : Vertical displacement in the ground
δV_1 :Vertical displacement caused by the retaining wall displacement

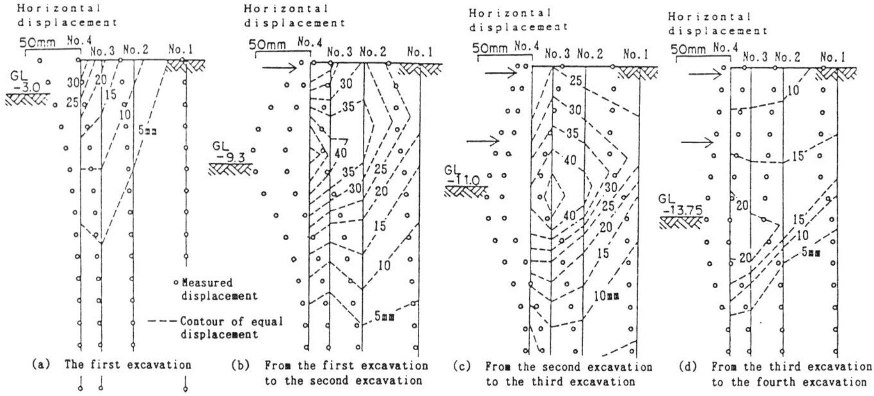

Fig. 3. Contours of measured equal incremental horizontal displacement

δV_2 :Vertical displacement resulting from the ground heave

The contours of equal horizontal ground displacements after the first excavation and the incremental horizontal displacements measured after each excavation are shown in Fig. 3. The contour lines after the first excavation are linear, after the second excavation they show a characteristic circular arc shape.

Since the ground being studied is cohesive alluvium soil the internal frictional angle ϕ and the dilatancy angle ν are both assumed to be 0°. In Fig. 4 the shape of the zero-extension lines is shown as a combination of a straight lines and a circular arcs. The dotted lines in the figure are contours of equal horizontal displacement.

Comparison with Fig. 3 shows that the actual behaviour for the first excavation is close to type A , having linear contours of equal horizontal displacement, after the second excavation it is close to type C or D , with circular contours of equal horizontal displacement near the retaining wall. Also, as the direction of the measured displacement vector (Fig. I.) on the ground surface is linear at nearly 45° to the horizontal, the actual behaviour

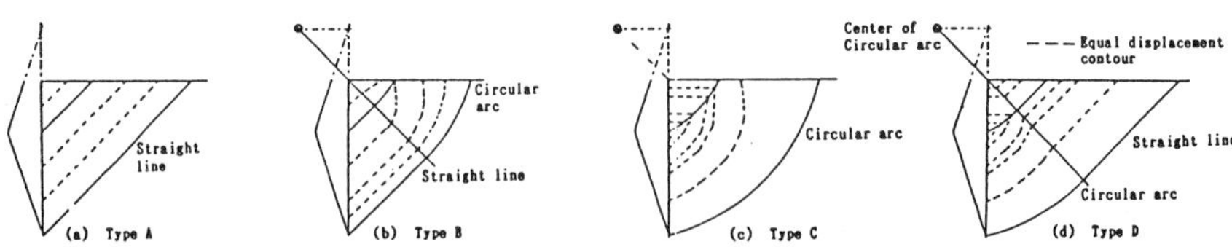

Fig. 4. Assumed zero-extension lines and distribution of equal horizontal displacement contours

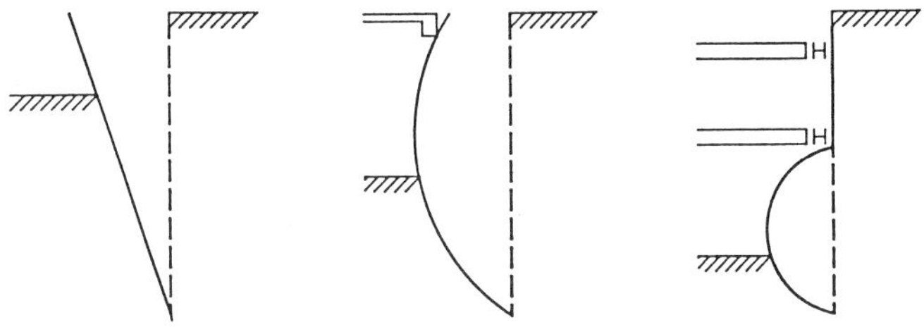

Fig. 5. Assumed incremental displacement patterns of earth retaining walls

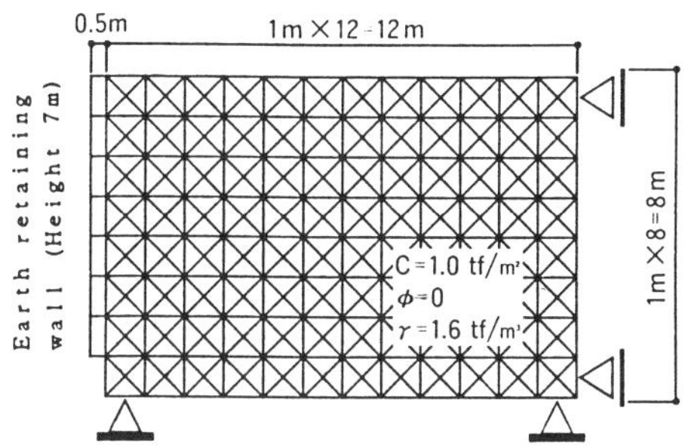

Fig. 6. Divided element and boundary condition (RBSM)

is considered to be close to the type A zero-extension line for the first excavation and to the type D after the second excavation.

Analysis using a rigid body spring model (RBSM)

A study was made using a Rigid Body Spring Model (RBSM). Three patterns of displacements of the retaining wall, as shown in Fig. 5 were used for the study. These patterns of displacement were assumed to be the incremental displacement for each stage of excavation .

Fig. 6 shows an elemental division and the boundary conditions of the models. The wall friction at the retaining end was assumed to be zero. The analysis was undertaken by increasing the soil weight per unit volume in order to determine the mechanism of failure and the pattern of the subsequent ground displacement.

Fig. 7 and 8 show the results of the analysis for two modes of displacement as pattern 1 and pattern 2 . Significant plastic displacement of the ground near the retaining wall was identified for both patterns 1 and 2 associated with the failure condition . The calculated soil displacement patterns corre-

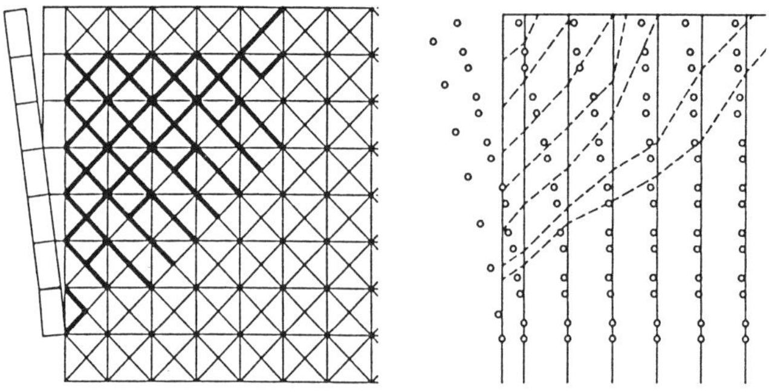

(a) Illustration of failure line (b) Equal horizontal displacement contours

Fig. 7. Result of the analysis in pattern 1

sponds well with the actual ground behaviour (Fig. 3) for both patterns .

Figure 9 shows the results of analysis of a third mode of displacement, pattern 3. Judging from the generating condition of the failure line, the soil near the displaced section of the wall is in a plastic condition. A horizontal failure line has been generated at the top of this section. The zero-extension line in the section where the earth retaining wall is displaced is considered to be close to type D, Fig. 4. Also, in the section where the wall is fixed, the zero-extension line can be considered a vertical line with zero horizontal displacement because most of the partially generated failure lines are on a vertical surface .

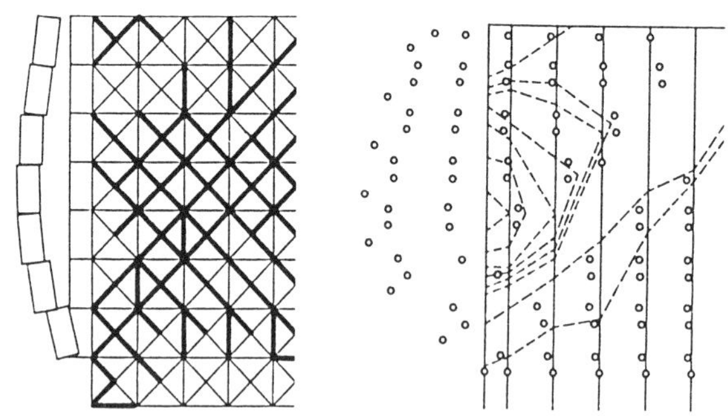

(a) Illustration of failure line (b) Equal horizontal displacement contours

Fig. 8. Result of the analysis in pattern 2

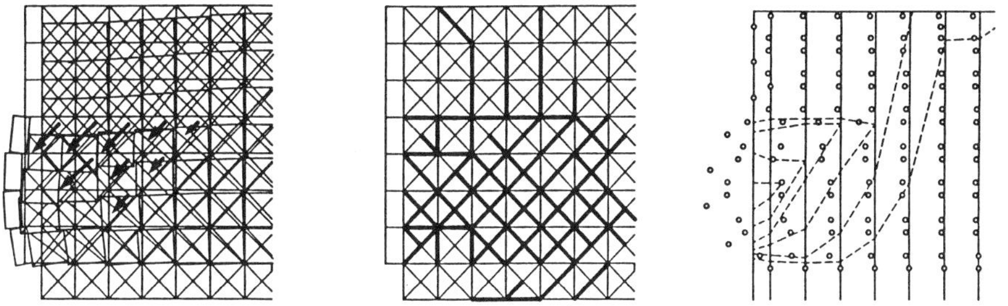

(a) Deformation (b) Illustration of failure line (c) Equal horizontal displacement contour

Fig. 9. Result of the analysis in pattern 3

Comparison of the calculated displacements with the measured values.

The strain field associated with the measured displacements in cohesive soil can be idealized in Fig. 10 (a - c) .

If the dilatancy angle $\nu = 0$, the following relations between the two velocities, V_1 and V_2 along the zero-extension lines can be identified

$$dV_1 + V_2 \cdot d\theta_1 = 0 \quad \text{along a } k_1 \text{ line} \qquad (2)$$

$$dV_2 - V_1 \cdot d\theta_2 = 0 \quad \text{along a } k_2 \text{ line} \qquad (3)$$

θ_1, θ_2 :Angle of inclination of zero-extension line, k_1 and k_2, to the vertical

It is assumed that the soil remains rigid outside the lowest k_1 line ABC. As the k_2 lines are straight, i.e. $d\theta_2 = 0$ along each k_1 line, Eqn. (3) reduces to $dV_2 = 0$. We can say that along ABC $V_2 = 0$ and so $V_2 = 0$, and $V_1 = $ const. everywhere.

In Fig. 11 the measured ground displacements are compared with the calculated values using the assumed strain field. The ground displacements were calculated only from the measured earth retaining wall displacements.

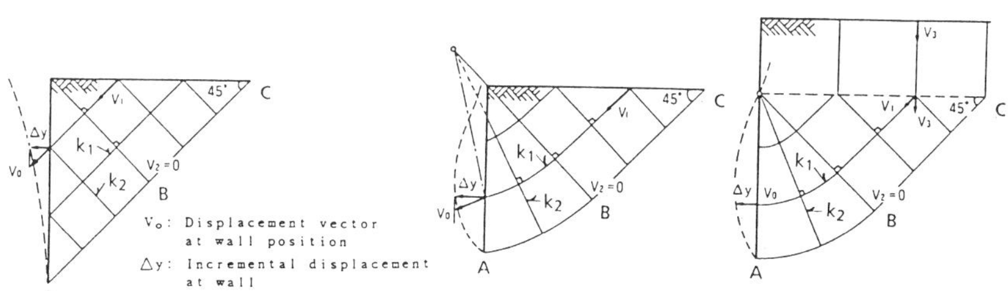

(a) Displacement pattern-1 (1st excavation) (b) Displacement pattern-2 (after 2nd excavation) (c) Displacement pattern-3 (after 2nd excavation)

Fig. 10. Assumed strain fields

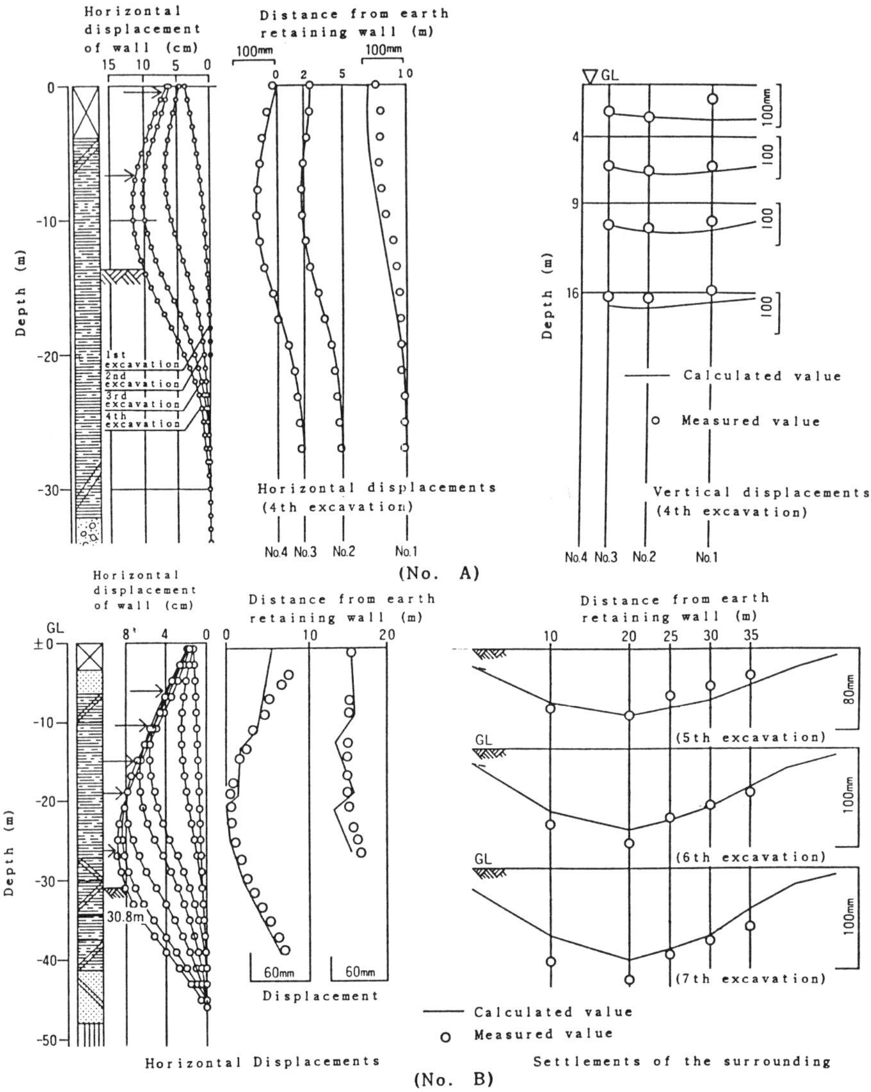

Fig. 11. Comparison between calculated values and field measurements of horizontal displacements and vertical displacements

The incremental displacement of the earth retaining wall, for the purpose of calculating ground displacements, was based upon

$$\delta(y) = \delta(y_1) + \sum_{i=2}^{m} \delta(y_i - y_{i-1}) \tag{5}$$

Where,

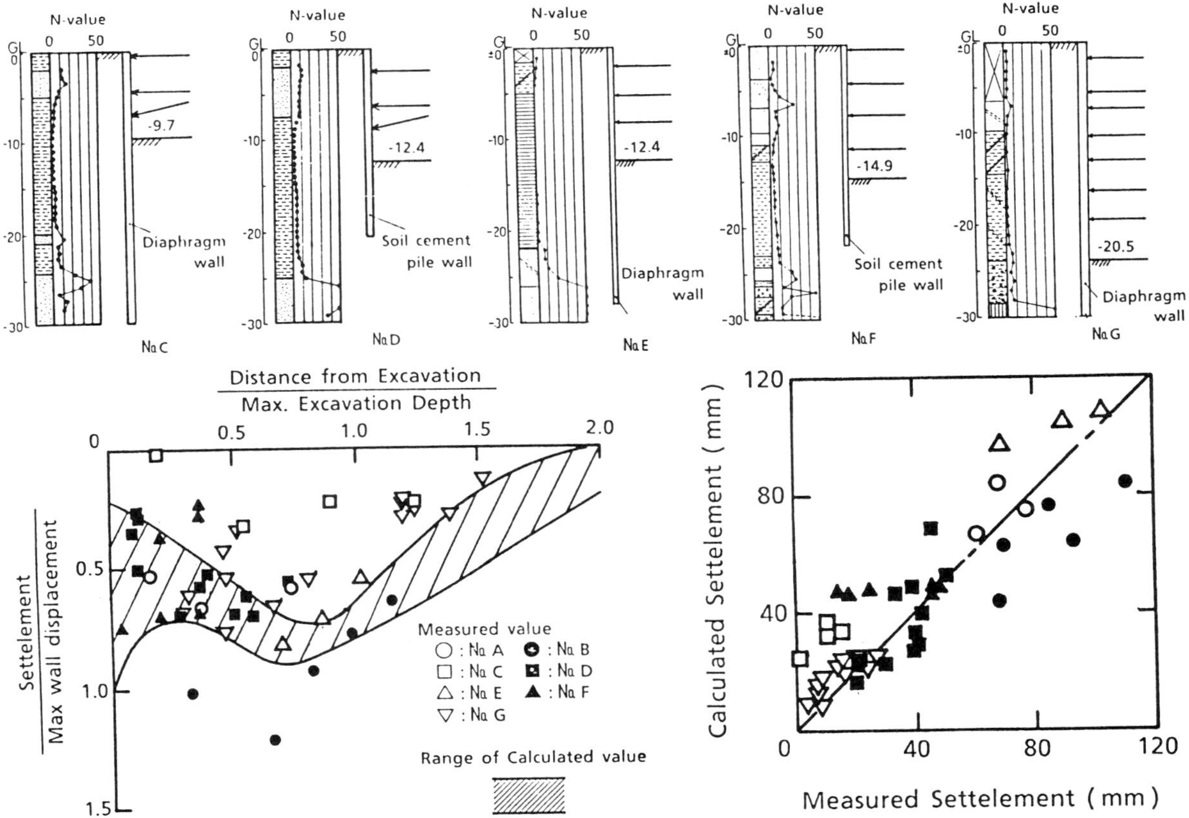

Fig. 12. Summary of measured and calculated settlements for excavations in soft cohesive soil and loose sand

m: order of excavation for which the displacement is to be calculated
y_i: retaining wall displacement after "i" th excavation
$\delta(y)$: ground displacement due to retaining wall displacement "y"

The calculated values agree closely with the measurements for both walls A and B, Fig. 11.

Figure 12 summarizes the measured and calculated settlements for excavations in soft cohesive soil and loose sand.

These measurements were obtained from sites where the lowering of the ground water level of the surrounding area was negligible. The calculated values agree comparatively well with the measured values, Fig.12

Conclusions

This paper presents in-situ measurements of ground movements caused by retaining wall displacements in cohesive soil and together with a method for predicting ground displacements. The conclusions are

(*a*)The strain field patterns in the cohesive soil near the earth retaining wall can be identified by a combination of straight lines and circular arcs .
(*b*)A simple method of predicting ground displacements which is based on zero-extension lines has been shown to correspond well with in-situ measurements.

References

1. ROSCOE, K. H. The influence of strains in soil mechanics. *Geotechnique,* 1970, **20** , No. 2 , pp 129 - 170.

2. KAWAI, T. *Analysis of dynamical problems of continuum shown in physical model* . SEIKEN Seminar Text, 1980 .

3. KAJIGAYA, M. *et al.* Behaviour and Analysis of Deep Excavation in Soft Ground . *The 23th Japan National Conf. on S. M. F. E.*, 1988, pp 1551 - 1562.

Analysis of the behaviour of propped diaphragm walls in a deep clay deposit

A. J. WHITTLE and Y. M. A. HASHASH, Massachusetts Institute of Technology, USA

Modern techniques of excavation, including the use of structural diaphragm walls and top-down construction, provide effective methods to minimize deformations in the surrounding soil. However, for deep excavations and walls embedded in deep layers of soft clays, it is difficult to control soil movements. In these situations, reliable predictions of deformations are an important consideration in the design of the earth support system. This paper describes results of numerical experiments, based on non-linear finite element analyses and incorporating an effective stress soil model which describes observed aspects of soft clay behaviour including small strain non-linearity and anisotropic stress-strain-strength. The analyses consider the undrained deformations around a braced diaphragm wall in a soft clay deposit. Principal parameters considered in the study are the wall embedment depth and support spacing. A simplified framework is developed to interpret wall deflections and ground movements with excavation depth.

Introduction

Excavations up to 20 to 25m deep, for an underground highway through downtown Boston (referred to as the Central Artery/ Tunnel project, CA/T), will be supported by structural diaphragm walls using techniques of top-down construction. In some locations, the soil conditions include very deep layers of soft Boston Blue Clay and it is not economical to extend the diaphragm walls down to the underlying till and rock layers. Reliable predictions of soil deformations are particularly important in the design of the lateral earth support at these sections especially due to the proximity of adjacent structures. Previous observations (ref.1) show that large lateral wall deflections and surface settlements, in the range of 2 to 3% of the excavation depth, can be expected for walls embedded in soft clays. However, most of the existing data were obtained from excavations less than 15m deep, supported by relatively flexible, sheet pile and soldier pile walls. This paper describes results of numerical experiments, using non-linear finite element analyses, which have been performed in order to investigate the effects of

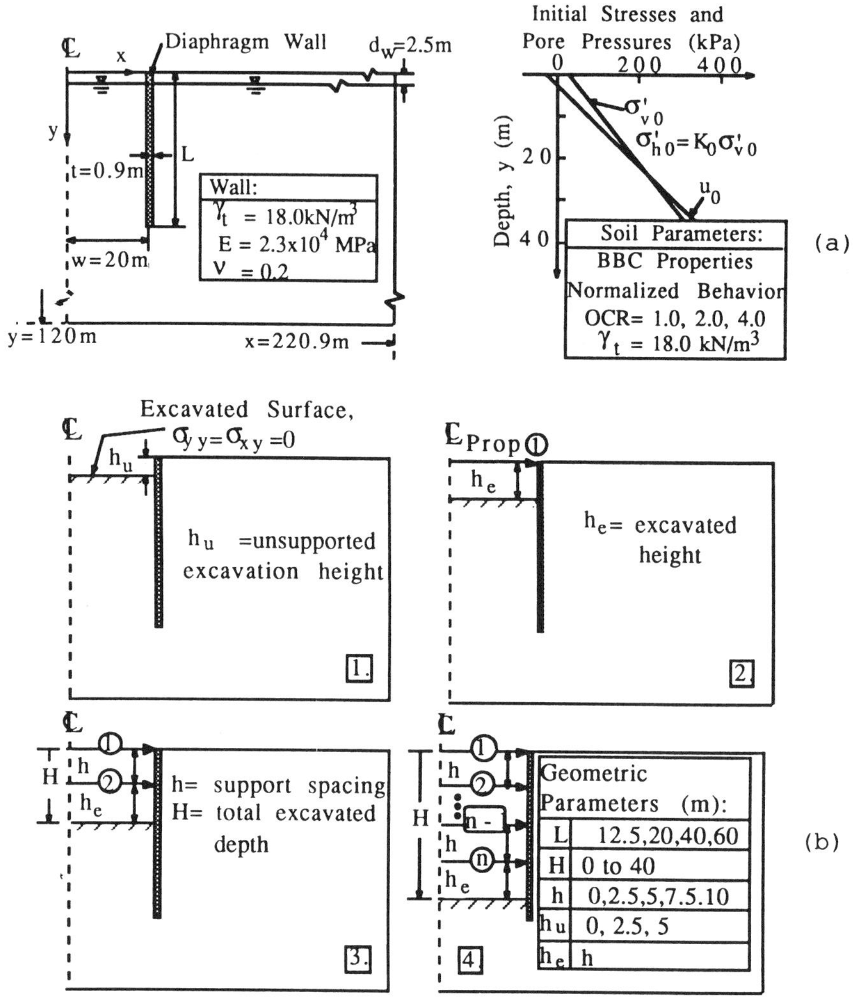

Figure 1. Initial Conditions and Model Excavation Sequence

wall length and support conditions on predictions of wall deflections and ground movements.

Scope of parametric study

Compilations of measured data and case studies emphasize the large number of factors, relating to construction activities and soil conditions, which can affect the development of ground movements for braced excavations. In contrast, the design of numerical experiments attempts to limit the number of parameters in order to facilitate the interpretation of the analyses, while still maintaining sufficiently realistic conditions that the results can have practical application. The following assumptions have been used in this study

(*a*) For deep layers of low permeability clay, the excavation process occurs sufficiently rapidly that there is no migration of pore water, hence the soil is subject to undrained shearing.
(*b*) The study focuses on a plane strain geometry with a fixed excavation half-width, w=20m, and standard diaphragm wall thickness, t=0.9m, (Fig. la) typical of conditions expected for the CA/T project. It is assumed that wall installation does not disturb the surrounding soil (i.e., the wall is wished-in-place) and that the full shear strength of the soil can be mobilized at the soil-concrete interface.
(*c*) The excavation is braced by incompressible (rigid) supports spaced at equal intervals, h, as shown in Figure lb. Installation of rigid bracing represents an ideal condition in which there is no further lateral deflection of the wall at the elevation of the support. The analyses do not consider support compressibility or other complex aspects of soil-structure interaction associated with different bracing systems, such as shrinkage and expansion of cast in-situ floor systems (ref.7) or prestressing of cross-lot bracing (ref.3).
(*d*) The analyses assume a simplified construction sequence (Fig. lb) comprising the following steps

- soil is initially excavated to a depth, h_u, without lateral support
- the wall is propped at the surface and excavation proceeds to a depth, h_e
- a second level of support is installed at a spacing, h, and soil is then removed over a further interval, h_e
- step 3 is repeated until failure occurs. By varying the parameters h, h_e and h_u it is possible to study a wide range of top-down construction sequences.

(*e*) The clay is saturated and exhibits normalized engineering properties. The numerical experiments then focus on simplified soil profiles in which the overconsolidation ratio (OCR = σ'_p/σ'_{vo}) is constant with depth, and hence the engineering properties are proportional to the initial overburden stress, σ'_{vo}. It is assumed that the pore pressures are initially hydrostatic with the groundwater table located at a depth, d_w = . 2.5m (typical of conditions in Boston) and full capillarity in the overlying clay.

The effective stress-strain-strength properties of the soil are characterized

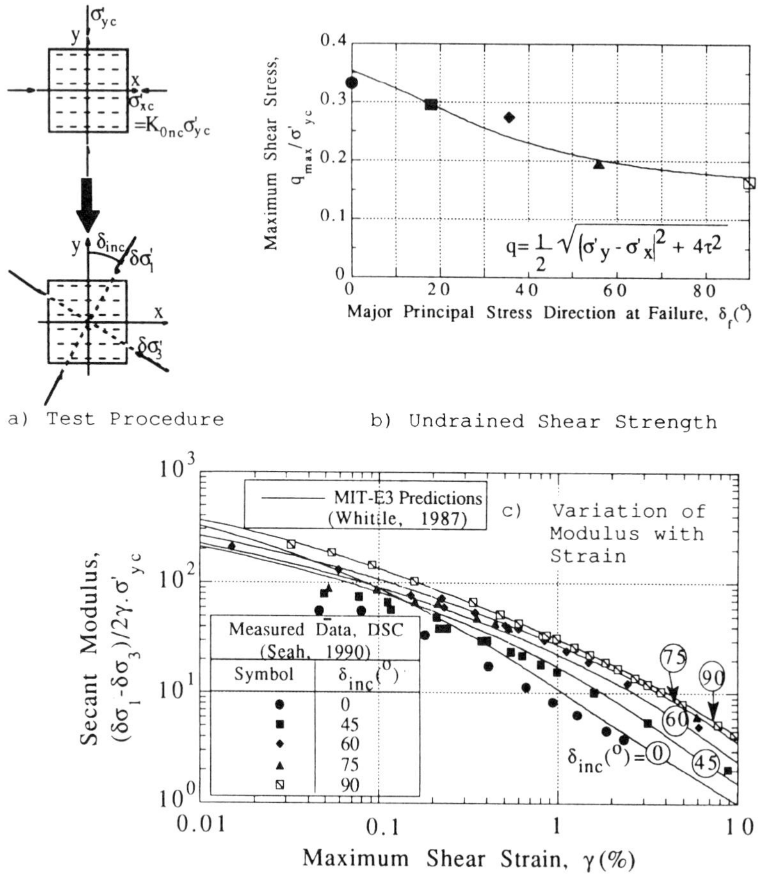

Fig. 2. Evaluation of the MIT-E3 model for undrained plane strain shearing with principal stress rotations: (a) Test procedure; (b) Undrained shear strength

by the MIT-E3 model (refs 5-6) using input parameters previously selected for Boston Blue Clay. The MIT-E3 model describes many aspects of the observed behaviour of K_0- normally and lightly overconsolidated ($1 \leq OCR \leq 8$) clays including

(*a*) small strain non-linearity,
(*b*) anisotropic stress-strain-strength; and
(*c*) hysteretic and inelastic behaviour due to cyclic loading.

The model has been extensively validated at the element level through comparisons with laboratory test data (refs 5-6). Fig. 2 compares model predictions with laboratory data from undrained plane strain shear tests on K_o-normally consolidated Boston Blue Clay measured in the Directional Shear Cell (DSC; ref.4). In these tests (Fig. 2a) the sample is sheared by applying incremental principal stresses oriented at, an angle δ_{inc} to the direction of the principal consolidation stress (σ'_{yc}). The data at $\delta_{inc} = 0°, 90°$ correspond to plane strain 'active ' and 'passive ' modes of shearing, while intermediate values of δ_{inc} produce continuous rotations of principal stresses (σ'_1 is oriented at δ to the y-axis). Fig. 2b shows that the model predicts accurately the measured changes in undrained shear strength ratio (i.e., q_{max}/σ'_{yc}) with the direction of the major principal stress. The predicted strength ratio ranges from $q_{max}/\sigma'_{yc}= 0.34$ at $\delta = 0°$ to 0.17 at 90° and illustrates the importance of anisotropic undrained strength properties for K_o -consolidated Boston Blue Clay. Measurements of the secant modulus (defined in Fig. 2c) show that the clay exhibits non-linear stress-strain behaviour for shear strains, $\gamma \geq 0.5\%$. Model predictions agree well with the data for tests with $\delta_{inc}= 60°, 75°$ and 90° but generally overpredict the stiffness at $\delta_{inc} = 0°, 45°$. These results illustrate that MIT-E3 is well suited to modelling the complex stress-strain history of the soil induced by excavation processes, including principal stress rotations.

Reference 2 describes the incorporation of MIT-E3 in the finite element code ABAQUS and gives details of the procedures used to perform numerically accurate, effective stress analyses of excavations in non-linear soils.

Typical results

Figures 3 and 4 show typical predictions of lateral wall defections and surface settlements for excavations in a deep layer of normally consolidated Boston Blue Clay. Fig. 3 compares results for two walls with lengths, L=20m and 40m, which are both braced at intervals h=2.5m over the full depth of the excavation (the initial unsupported depth h_u=2.5m in both cases). The effect of length on the deflected shapes of the walls is immediately apparent: For the short wall (L=20m), the maximum lateral deflection develops at the toe even for relatively shallow excavation depths (H≤10m). In contrast, for the 40m long wall, the maximum deflection is associated with a point of contraflexure which occurs 8 to l0m below the base of the excavation.

Figure 3 also includes the excavation geometry and deflected wall shapes at the stage immediately prior to failure (i.e., at H=15m and 22.5m for the 20m and 40m long walls, respectively). The actual failure depths are difficult to estimate precisely as the analysis models finite excavation increments of 2.5m. However, detailed interpretations of the data (ref.2) show that the mechanism of failure can be well defined from the development of shear strains in the soil and is controlled by anisotropic properties of the MIT-E3 model. The failure mechanism is affected both by the embedment of the wall

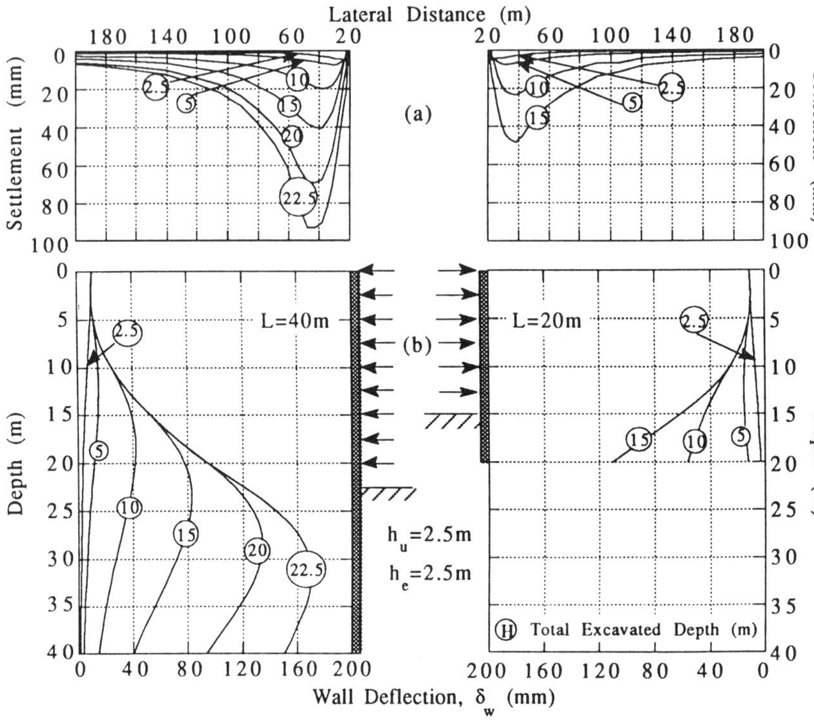

Fig. 3. Effect of wall length on predicted ground movements

and by the assumed excavation width (w= 20m; Fig. la). The results in Figure 3 show that as the wall length increases from L = 20m to 40m, the failure depth increases by 50%. This additional stability is not considered in conventional calculations of base heave used for braced excavations.

Although there are significant differences in the deformed shapes of the two walls, the maximum lateral wall deflections, δ_{max}, are very similar at early stages of excavation, and differ by less than 30%, even for conditions close to failure (i.e. at H=15m for the L=20m wall). The results in Figure 3a also show that wall length has practically no effect on the development of surface settlements. In both cases, the maximum settlement occurs at a lateral distance 15m to 20m behind the wall, with magnitudes approximately 40 to 60 percent of δ_{max}. The lateral distribution of settlement is strongly influenced by non-linear properties of the soil at small strain levels.

Figure 4 compares the behaviour of 'perfectly supported' (h=0m) and 'minimally supported' (h=10m), 60m long walls. For the perfectly supported case, wall deflections only occur below the current excavation level. The deflected shape is similar to that reported in Fig. 3b for a 40m long wall, with maximum lateral deflections occurring approximately 10m below the base of the excavation. Failure of the perfectly supported wall at H=35m is associated both with a deep seated mechanism in the soil (H= 32.5m to 35m; Fig. 4b) and with a vertical 'bearing' failure of the wall itself (observed from the wall settlements; Fig. 4a). Large deformations (in the range 100 to 150mm) develop in the soil prior to failure. In contrast, for the minimally supported wall, large lateral deflections develop over the excavated height, with maximum deformations located close to the base of the excavation. Failure in the soil is resisted by the bending of the wall. At the computed failure depth (H=20m; Fig. 4), the wall sustains a maximum bending moment, M_{max}= 5MNm/m, which is significantly larger than the ultimate moment (M_u) for even a heavily reinforced diaphragm wall of these dimensions (typically, M_u=1.5 to 2.0MNm/m for t=0.9m walls). Overall, the results demonstrate the importance of support conditions on the development of lateral wall deflections, and their corresponding implications for control of soil deformations induced by excavations in deep clay layers.

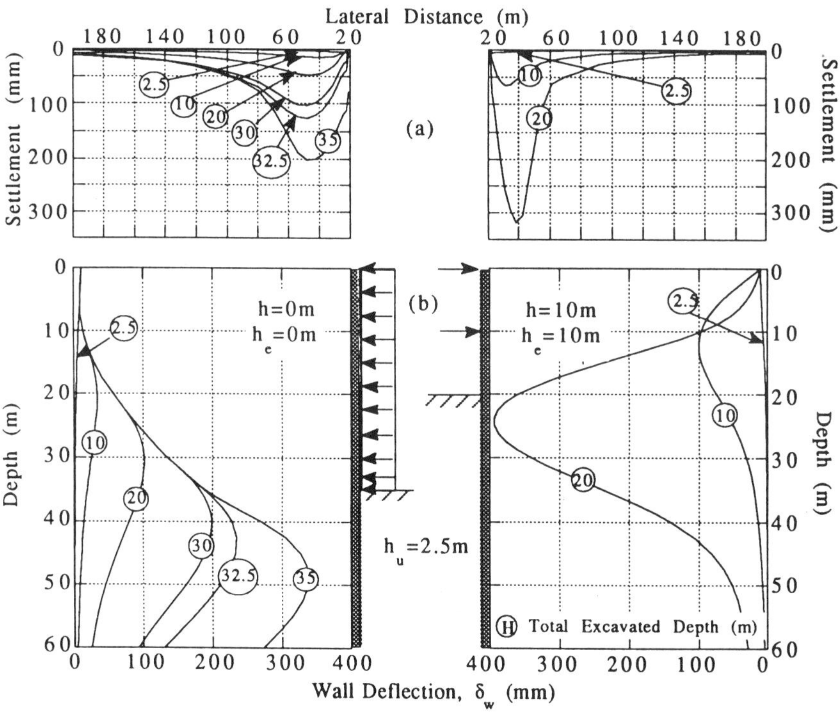

Fig. 4. Effect of support spacing on predicted ground movements

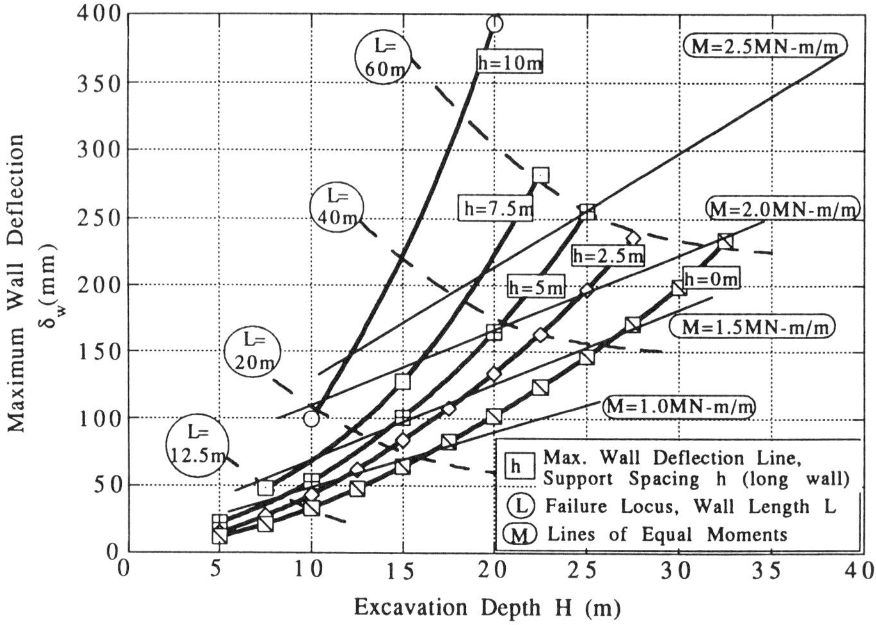

Fig. 5. Interpretation of maximum wall deflection from analyses

Framework for interpreting maximum wall deflections

Extensive parametric studies described in ref.2 confirm that wall length has a very minor effect on the pre-failure soil deformations and maximum lateral wall deflections (δ_{max}). Fig. 5 then summarizes the numerical predictions of δ_{max} for excavations in normally consolidated BBC as a function of the excavation depth, H, and support spacing, h. The figure includes loci of failure conditions obtained for walls of different lengths, L, ranging from 12.5 m to 60m, and implicitly incorporates observations of different failure mechanisms in the soil. Failure of the diaphragm wall can be evaluated from contours of the maximum moment compared to the capacity of the section (M_u).The figure shows the following

(*a*) For excavations which are perfectly supported, the deflection to depth ratio, (δ_{max}/H)≥0.5%. For minimally supported walls, this ratio increases to (δ_{max}/H) = 1.5 to 2.0%.

(b) Excessive bending moments in the wall are the most critical design condition for long walls (L≥40m) and lightly braced walls (h≥m).

The results in Fig. 5 should be used with caution for shallow excavations (H≤l0m), where wall deflections are significantly affected by details of the top-down construction sequence (e.g., initial unsupported height, h_{ur}; ref.2).

Conclusions

Predictions of wall deflections and soil deformations are an important consideration in the design of lateral earth support systems for excavations in deep deposits of soft clay. This paper has demonstrated the use of parametric studies, based on non-linear finite element analyses and incorporating effective stress modelling of soil behaviour, for interpreting the performance of braced diaphragm walls embedded in Boston Blue Clay. Results of the study indicate that support spacing is critical in controlling the development of soil deformations, while embedment length primarily affects the excavation stability. For these analyses, predictions of maximum wall deflections can be summarized as functions of the excavation depth and support spacing.

References

1. CLOUGH, G. W. and O'ROURKE, T. D. Construction induced movements of in situ walls, *Proc. ASCE Conference on Earth Retaining Structures, Ithaca, NY*, 1990, pp 439-470.
2. HASHASH, Y. M. A. *Analysis of deep excavations in Clay*. PhD Thesis, in progress, MIT, Cambridge, MA, 1992.
3. MANA, A. I. and CLOUGH, G. W. Prediction of movements of braced cuts in clay. *ASCE J. Geotech. Engng*, 1981, **107**, No. 4, pp 759-778.
4. SEAH, T. H. *Anisotropy of resedimented Boston Blue Clay*. PhD Thesis, MIT, Cambridge, MA., 1990.
5. WHITTLE, A. J. *A constitutive model for overconsolidated clays*, MIT Sea Grant Report, MITSG 90-15, 1990.
6. WHITTLE, A. J. Evaluation of a constitutive model for overconsolidated clays, to be published, *Geotechnique*, 1991.
7. WHITTLE, A. J. *et al.* Analysis of a deep excavation in Boston, submitted for publication, *ASCE J. Geotech. Engng*, 1991.

Balanced seismic design - improving the seismic efficiency of anchored retaining walls

M. BUDHU, G. NEELAKANTAN, University of Arizona, Tucson, and R. RICHARDS, State University of New York, USA

In this Paper, a mechanism generated by inertial forces which trigger large displacements of anchored retaining walls during moderate to strong earthquakes is examined. Through this mechanism, an investigation is carried out to determine whether anchor failures can precede liquefaction. A limit equilibrium analysis, using the Mononobe-Okabe (1926, 1929) seismic earth pressure equations, is performed to determine whether anchor failures lead to the general failure of anchored retaining walls during seismic events. Results of shaking table tests on aluminum walls with a dry cohesionless soil as the backfill confirm the analytical methodology. Based on the limit analysis, a balanced seismic design concept for the optimal design of tied-back retaining walls is presented. It was found that the balanced seismic design enhances the seismic resistance of tied-back retaining walls with only a small expense for strengthening the tie-backs. A typical design example to compare the balanced seismic design procedure with current static design practices, is presented.

Introduction

Large permanent displacements of anchored retaining walls such as quay walls, sheet pile walls and bulkheads are often reported in the literature (Werner and Hung *et al.*, 1982; Kitajima *et al.*, 1979) after moderate to strong seismic activity. These displacements or failures are often ascribed to possible liquefaction of the soil. In almost all reported cases of seismically induced large displacements of anchored retaining walls, anchor failure was observed. It is quite possible, however, that the anchor could have failed prior to liquefaction; liquefaction only intensified the displacements. It is postulated here an alternative mechanism of failure that is generated by inertia forces. This postulate, using a limit equilibrium analysis of anchored retaining walls supporting dry sand, is examined here.

Analysis

In this analysis, two possible modes of failures are considered: (a) rotation about the anchor (RA) and (b) rotation about the bottom (RB) of the wall. It is assumed that the retaining wall is structurally sound against bursting, yielding or bending but may become unstable due to increases in soil pressures. Pure sliding or horizontal translation is an unlikely mode of failure since there must be no net moment on the wall. The first mode of failure corresponds to a design where the anchorage is outside the active failure zone whereas the second mode is compatible with a design in which the anchor is swept up in the active zone or breaks and becomes ineffective.

The following assumptions are used

(*a*) straight failure surfaces occur on the active and passive sides and
(*b*) Mononobe-Okabe (1926, 1929) dynamic active and passive earth pressure coefficients, K_{AE} and K_{PE}, given in equations (1) and (2) are valid.

$$K_{AE} = \frac{\cos^2(\phi_A - \theta)}{\cos\theta\cos(\delta_A + \theta)\left[1 + \sqrt{\dfrac{\sin(\phi_A + \delta_A)\sin(\phi_A - \theta)}{\cos(\delta_A + \theta)}}\right]^2} \tag{1}$$

$$K_{PE} = \frac{\cos^2(\phi_P - \theta)}{\cos\theta\cos(\delta_P + \theta)\left[1 - \sqrt{\dfrac{\sin(\phi_P + \delta_P)\sin(\phi_P - \theta)}{\cos(\delta_P + \theta)}}\right]^2} \tag{2}$$

Where $\theta = \arctan(k_h/(1-k_v))$, y = the unit weight of the soil, H_A = height of the active wedge, ϕ_A, ϕ_p = angles of internal friction of the soil on the active and passive sides, δ_A,δ_p = angles of friction between the wall and the soil on the active and passive sides, k_h and k_v are the horizontal and vertical acceleration coefficients. For rotation about the anchor, the factor of safety (F) is (with reference to Fig. 1)

$$F = \frac{P_p\cos(\delta_p)\left(H_A - \dfrac{D}{3} - A_h\right) - P'_{PE}\cos(\delta_p)\left(H_A - A_h - \ell_p\right)}{P'_{AE}\cos(\delta_p)\left(\dfrac{H_A}{2} - A_h\right) + P_A\cos(\delta_A)\left(\dfrac{2}{3}H_A - A_h\right)} \tag{3}$$

where P'_{PE} and P'_{AE} are the passive and active dynamic components of the earth pressures and given as

$$P'_{PE} = \frac{1}{2}\gamma D^2(1 - K_v)K_{PE} - P_p \tag{4}$$

and

$$P'_{AE} = \frac{1}{2}\gamma H_A^2(1 - K_v)K_{AE} - P_A \tag{5}$$

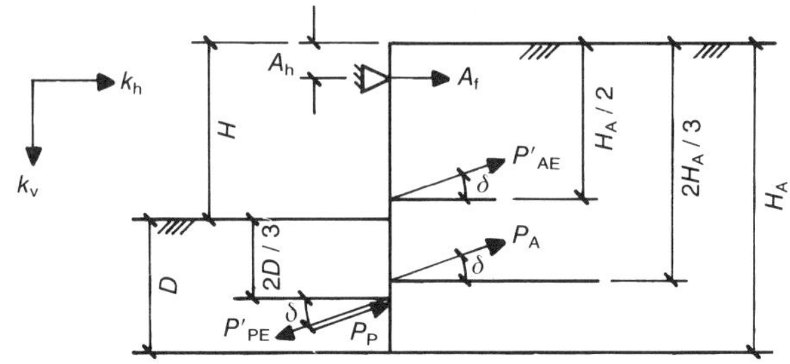

Fig. 1. Earth pressures on anchored wall.

where P_P and P_A are the static passive and active earth pressures. It was shown by Seed and Whitman (1970) that the point of application of the P'_{AE} is approximately one half the active height (H_A) of the wall. Neelakantan *et al.* (1990) have shown that the point of application of P'_{PE} is approximately two-thirds the passive depth. The retaining wall system will fail by rotation about the top when the value of F in equation (3) reduces to 1 and the acceleration ratio (tan θ) reaches a critical value tan θ^*_{RA}.

The primary concern in the seismic design of tied-back walls must be the rational design of the anchor system. The interplay between the various dynamic forces and the resistance offered by the anchor must be taken into account along with economic considerations and construction difficulties. Neelakantan et al. (1990) showed that a statically designed anchored retaining wall can be made to sustain a high level of acceleration by designing the anchorage to provide sufficient resistance to force failure, if it were to occur, to be by rotation about the anchor. The anchor must, therefore, provide the necessary equilibrating force throughout the seismic loading to avoid rotation about the bottom. To achieve a 'balanced design' an anchor system must be proportioned and positioned so that the failure acceleration ratio corresponds to that of RA failure mode.

Considering the equilibrium of forces on the wall in the horizontal direction, the required anchor force is

$$A_f = \frac{1}{2}\gamma\,(1 - K_v)\,((H + D)^2\,K_{AE} - D^2K_{PE}) \qquad \textbf{(6)}$$

In order to force RA failure, the maximum anchor force A_f must occur when the critical acceleration ratio reaches $\tan\theta^*_{RA}$.

For an anchorage system consisting of anchor plates, the passive resistance in front of the anchor plates will decrease as $\tan\theta$ increases. The ultimate load that can be sustained by an anchor plate of height 'h' is

$$P_{ult} = \frac{1}{2}\gamma h^2 (K_{PE} - K_{AE}) \tag{7}$$

Thus, for RA failure mode

$$P_{ult} \geq A_f \tag{8}$$

and consequently

$$h = \sqrt{\frac{2A_f}{\gamma(K_{PE} - K_{AE})}} \tag{9}$$

where K_{PE} and K_{AE} are computed for an acceleration ratio $\tan\theta^*_{RA}$.

It is recommended that the anchors be located such that the dynamic active wedge of the retaining wall does not interfere with the dynamic passive wedge of the anchor plate. Using this criterion, the distance of the anchor plate from the wall can be expressed as

$$d_a = (H + D)\tan(\alpha_a) + h\tan(\alpha_p) \tag{10}$$

where α_a and α_p, the dynamic active and passive wedge angles, are

$$\alpha_a = (\phi - \theta) + \tan^{-1}\left(\frac{\sqrt{(1+\tan^2\xi)[1+\tan(\delta+\theta)\cot\xi]} - \tan\xi}{1+\tan(\delta+\theta)(\tan\xi+\cot\xi)}\right) \tag{11}$$

and

$$\alpha_p = (\theta - \phi) + \tan^{-1}\left(\frac{\sqrt{(1+\tan^2\xi)[1+\tan(\delta+\theta)\cot\xi]} + \tan\xi}{1+\tan(\delta+\theta)(\tan\xi+\cot\xi)}\right) \tag{12}$$

where $\xi = \phi - \theta$.

Experimental set up and results

A series of experiments on aluminum walls (6.35 mm thick, 882 mm wide of various lengths) supporting a dry sand was conducted. The wall for a given test was mounted in the middle of a test box 0.9 m x 0.9 m x 5.2 m long. The test box was then placed on a shaking table where the desired motions were

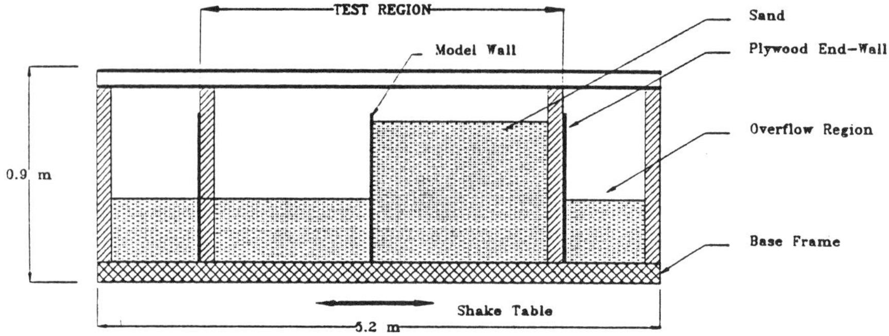

Fig. 2. The sand box

imparted. In one series of tests (test RA), the test wall was hinged at the top simulating a wall system in which the anchor is sufficiently strong to allow rotation to take place about its location. In a second series (test ANC), two 762 mm long steel rods of diameter 6.4 mm were connected to a steel plate (130 mm high x 880 mm long x 18 mm thick) and to the wall to simulate a statically designed plate anchor wall system. The size of the walls in both series was the same. A schematic drawing of the test box is shown in Figure 2.

In a typical set up, the sand was pluviated by a hopper and then densified by shaking the box on the shake table with the test wall restrained. Ramped horizontal accelerations were then imposed on the walls with the restraints removed. A frequency of 3Hz was used. Accelerometers were mounted on the test wall and the shake box, and linear variable transducers were used to measure the displacements of the test wall.

It was observed, during test RA, that at a horizontal acceleration level of 0.5 g (Fig. 3a), the wall began to rotate, and, since no vertical acceleration was imposed, the critical acceleration ratio for this test was 0.5. The wall continued to rotate for horizontal accelerations higher than 0.5g and ceased rotation when the horizontal acceleration dropped below 0.5g. Assuming a value of wall friction of 2/3 ϕ (a value shown by Neelakantan et al., 1990 to be reasonable), the critical acceleration ratio, computed using equation (3) for test RA, is 0.49, which is in excellent agreement with the laboratory measurements. Thus, the Mononobe-Okabe equations are quite satisfactory for the design of anchored retaining walls which can be simulated by the experiments described here.

For test ANC, the critical acceleration was 0.2g (Fig. 3b). The anchor plate, which was designed for static load, was capable of providing sufficient anchorage to withstand horizontal accelerations up to 0.2g. By comparing the results of test RA with test ANC, it appears that the wall in test ANC could withstand a higher level of acceleration (about 150% more than its failure

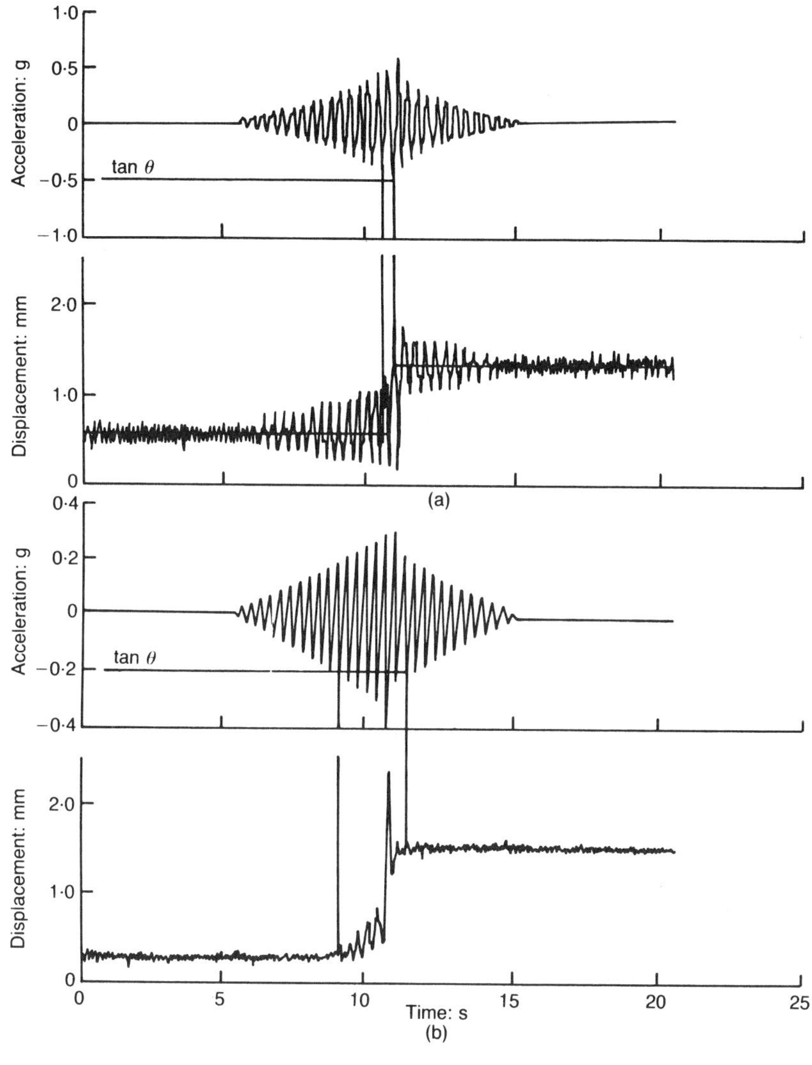

Fig. 3. Records of acceleration and deflection for: (a) Test RA at a peak input acceleration of 0.6g; and (b) Test ANC at a peak input acceleration of 0.4g

acceleration of 0.2g) for the same embedment depth if the anchors were adequately designed (equation 6) so that if failure were to occur, it would be by rotation about the anchor.

Table 1. Illustrative design example.

1	2	3	4	5	6
CASE	D/H	P_{ult} (static)	$\tan\theta^*_{RA}$	$\tan\theta^*_{ANC}$	P_{ult} (bal. Design)
(a)	0.48	178.5 kN/m	0.32	0.23	393 kN/m
(b)	0.35	149.9 kN/m	0.13	0.08	213 kN/m
(c)	0.58	178.5 kN/m	0.41	0.30	540 kN/m

Illustrative balanced seismic design example

Design an anchored retaining wall to retain 10 meters of dry cohesionless soil and evaluate its seismic stability. The soil's properties are as follows: (*a*) the angle of internal friction $\phi = 35°$, and (*b*) the dry unit weight $\gamma = 20$ kN/m^3.

The anchor ties are connected to the wall at 1m below ground level and the effects of wall friction are to be neglected.

The embedment depth ratios and the ultimate anchor forces, calculated following recommendations of the CP2:1951 and the Canadian Foundation Engineering Manual (1978) (case (a)), the USS Steel Sheet Piling Design Manual (1975) (case (b)) and the NAVFAC DM-7.2 (1982) (case (c)), assuming no scouring occurs), are shown in Table 1, columns 2 and 3. The acceleration ratios $\tan\theta^*_{RA}$ to initiate rotation about the anchor position is found by solving equation (3) for F = 1. The acceleration ratio required for anchor failure, $\tan\theta^*_{ANC}$, is found by equating equations (6) and (7) and solving for $\tan\theta$. The values of $\tan\theta^*_{RA}$ and $\tan\theta^*_{ANC}$ for each case are shown in columns 4 and 5 of Table 1.

In the cases examined, the acceleration ratio required for anchor failure is less than that required for failure by rotation about the anchor position. Consequently, the anchor will fail first as has been observed in practice. For any of these current design procedures, the seismic efficiency of this system can be increased by following the balanced design procedure proposed here. For example, for the given D/H ratios, the anchor may be designed from equation (7) by putting $\tan\theta = \tan\theta^*_{RA}$. The design anchor forces thus obtained are presented in column 6 of Table 1. These forces are much greater than those found by the conventional static methods. Thus, for the same D/H ratio the seismic stability of an anchored retaining wall can be enhanced by designing the anchor(s) to carry the balanced anchored forces as given (typically) in column 6 of Table 1.

Conclusions

The Mononobe and Okabe (1926, 1929) limit equilibrium approach appears to yield satisfactory practical results when applied to anchored retaining walls supporting dry cohesionless soils. Conventional static design of anchored retaining walls can sustain a moderate level of acceleration (generally about 0.2 g). These walls can, however, be designed to withstand a much higher level of acceleration, without increasing the embedment depth. The balanced design procedure suggested in this paper provides a method whereby the additional forces required to be carried by the anchor to improve the seismic stability of the retaining wall system can be calculated.

Acknowledgements

The research described here was supported by the U.S. National Science Foundation under Grant ECE86010345. The authors thank Dr. Cliff Astill for his support for this research.

References

MONONOBE, N. and MATSUO, H. On the Determination of Earth Pressures During Earthquakes. *Proc. World Engineering Conference*, 1929, **9**, pp 177-185.

NEELAKANTAN, G., BUDHU, M. and RICHARDS, R. Mechanics and Performance of a Tied-Back Wall Under Seismic Loads. Earthquake Engineering and Structural Dynamics, 1990, **19**, pp 315-331.

OKABE, S. General Theory of Earth Pressures. *J. Jap. Soc. of Civil Engineers*, 1926, **12**, No.1

SEED, H.B. and WHITMAN, R.V. Design of Earth Retaining Structures for Dynamic Loads. *Lateral Stresses in the Ground and Design of Earth Retaining Structures*, ASCE, 1970, pp 103-147.

WERNER, S.D. and HUNG, S. J. *Seismic Response of Fort and Harbour Facilities*. R-9122-5395, Agbabian Associates, El Segundo, CA, 1982.

ISTRUCTE. *BS Code of Practice No. 2: Earth Retaining Structures*. Institution of Structural Engineers, London, 1951.

CANADIAN GEOTECHNICAL SOCIETY. *Canadian Foundation Engineering Manual*. The Foundations Committee, Canadian Geotechnical Society, 1978.

UNITED STATES STEEL INTERNATIONAL *USS Steel Sheet Piling Design Manual*, 1975.

Generalized limit equilibrium method and its application to earth pressure problems

M. ENOKI, N. YAGI, R. YATABE, Ehime University, Japan, and
M. IMAMURA, Nishimatsu Construction Co. Ltd, Japan

Introduction

There is the Limit Equilibrium Method (LEM), Slip Line Method (SLM), Limit Analysis Method (LAM), which includes Upper Bound Method (UBM) and Lower Bound Method (LBM), to analyze the plasticity problems, such as bearing capacity, earth pressure and slope stability. Among these methods for obtaining the earth pressure, the Coulomb method (ref. 1) one of the oldest methods of LEM, is usually used, and for obtaining the safety factor of slopes the slice method, one method of LEM, is usually used. However, as LEM has vagueness in its theory, it is important to clarify the theoretical meaning of LEM and to develop a more versatile and theoretical LEM.

In this Paper, generalized LEM (GLEM) is proposed, generalizing LEM on the following points

- quadrangular or trianglular blocks as well as slices can be treated
- safety factors are defined both on the main slip surface and on the interblock planes
- non-circular slip surface can be treated
- every type of plasticity problem (slope stability, bearing capacity and earth pressure) can be expressed in a single formulation. Two formulations of GLEM will be introduced, one based on the limit equilibrium of all blocks and the other based on the energy equilibrium.

Then, it will be clarified that LEM can give the same solution as that obtained by SLM, LBM or UBM, according to the assumption made on interblock forces. Some examples of GLEM solution for the practical problems will be demonstrated, and GLEM will be extended to obtain the active earth pressure on a cylindrical wall. GLEM is useful for obtaining the solution of classical earth pressure problems with rough or smooth walls, as well as for obtaining the solution of complicated practical problems with accuracy.

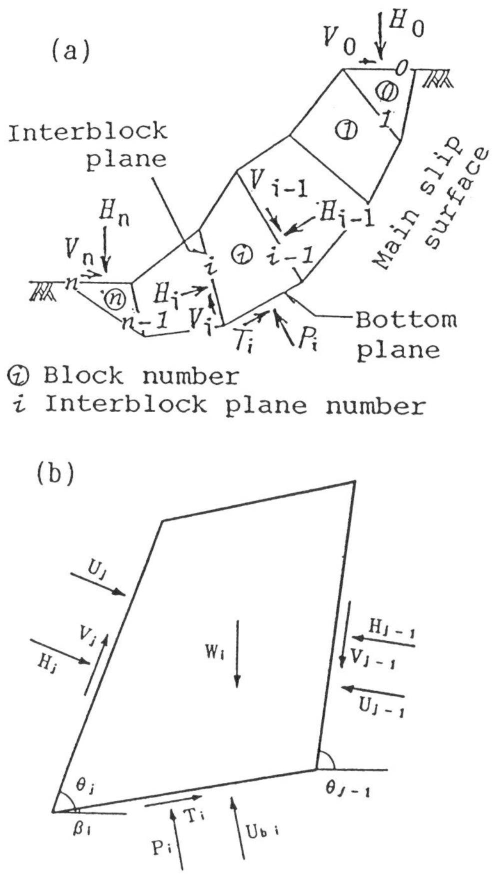

Fig. 1: (a) A block system; (b) limit equilibrium of a block

Formulation of GLEM

Limit Equilibrium Formulation (ref. 2)

For the limit equilibrium of *i*th block, as shown in Fig. 1(b), surrounded by ith base plane, jth and j-lth interblock lanes, the following five equations can be obtained

Equilibrium toward normal direction to base plane

$$(H_j+U_j)\cos(\theta_j-\beta_i)-V_j\sin(\theta_j-\beta_i)-(H_{j-1}+U_{j-1})\cos(\theta_{j-1}-\beta_i)+V_{j-1}\sin(\theta_{j-1}-\beta_i)+W_i\cos\beta_i=P_i+Ub_i \qquad (1)$$

Equilibrium toward tangential direction to base plane

$$-(H_j+U_j)\sin(\theta_j-\beta_i)-V_j\cos(\theta_j-\beta_i)+(H_{j-1}+U_{j-1})\sin(\theta_{j-1}-\beta_i)+V_{j-1}\cos(\theta_{j-1}-\beta_i)+W_i\sin\beta_i=T_i \qquad (2)$$

Definition of safety factors on base plane and interblock planes

$$T_i = k(P_i \tan\phi + c l_i)/Fs \quad (3)$$

$$V_j = m_j (H_j \tan\phi + c l_j)/Fs_i \quad (4)$$

$$V_{j-1} = m_{j-1}(H_{j-1} \tan\phi + c l_{j-1})/Fs_i \quad (5)$$

Here, all the variables are as given in the Fig.l(b). U_j, U_{j-1} and U_{bi} are the pore pressure forces on jth and j-1th interblock planes and on the ith base plane. k and m_j are the parameters for defining the directions of tangential forces on the bottom and interblock planes. They have the same characteristics as a parameter appears in SLM. The values of k and $M_1,..,M_{n-1}$ equal to +l for the slope and active earth pressure problems, and -1 for the bearing capacity and passive earth pressure problems.

By eliminating P_i, T_i, V_j, V_{j-1} from these equations, the following equation is obtained.

$$-H_{j-1} f_i + H_j g_i = h_i \quad (6)$$

Here, f_i, g_i and h_i can be determined only by the geometry of ith block and the safety factors on base and inter-block planes as given below.

$$f_i = \left\{-\cos\alpha_{j-1} + \frac{m_{i-1}}{Fs_i}\tan\phi \sin\alpha_{j-1}\right\}\frac{k}{Fs}\tan\phi - \left\{\sin\alpha_{j-1} + \frac{m_{i-1}}{Fs_i}\tan\phi \cos\alpha_{j-1}\right\} \quad (7)$$

$$g_i = \left\{-\cos\alpha_j + \frac{m_i}{Fs_i}\tan\phi \sin\alpha_j\right\}\frac{k}{Fs}\tan\phi - \left\{\sin\alpha_j + \frac{m_i}{Fs_i}\tan\phi \cos\alpha_j\right\} \quad (8)$$

$$h_i = U_j\left\{\sin\alpha_j + \cos\alpha_j \frac{k}{Fs}\tan\phi\right\} - U_{j-1}\left\{\sin\alpha_{j-1} + \cos\alpha_{j-1}\frac{k}{Fs}\tan\phi\right\} + \frac{m_i c l_i}{Fs_i}\left\{\cos\alpha_j - \sin\alpha_j \frac{k}{Fs}\tan\phi\right\} - \frac{m_{i-1} c l_{i-1}}{Fs_i}\left\{\cos\alpha_{j-1} - \sin\alpha_{j-1}\frac{k}{Fs}\tan\phi\right\} - W_i\left\{\sin\beta_i - \cos\beta_i \frac{k}{Fs}\tan\phi\right\} + \frac{k c l_i}{Fs} - Ub_i \frac{k}{Fs}\tan\phi \quad (9)$$

where, $\alpha_j = \theta_j - \beta_i$ and $\alpha_{j-1} = \theta_{j-1} - \beta_i$.

Suffix j=0 or j=n is included in Eqs 7 to 9. The roughness of footing or wall can be introduced into the analysis by the terms C_n and ϕ_n. As Eq. 6 can be written for every block shown in Fig. 1(a), the following simultaneous equations can be obtained.

$$\begin{aligned} -H_0 f_1 + H_1 g_1 &= h_1 \\ &\vdots \\ -H_{j-1} f_i + H_j g_i &= h_i \\ &\vdots \\ -H_{n-1} f_n + H_n g_n &= h_n \end{aligned} \quad (10)$$

After substitution and elimination of $H_i,...,H_{n-1}$ the following equation can be obtained.

$$H_n = \frac{1}{g_n}h_n + \frac{f_{n-1}}{g_n g_{n-1}}h_{n-1} + \frac{f_{n-1} f_{n-2}}{g_n g_{n-1} g_{n-2}}h_{n-2} + \dots + \frac{f_{n-1} f_{n-2} \dots f_1}{g_n g_{n-1} g_{n-2} \dots g_1}(h_1 + H_0 f_1) \quad (11)$$

If the value of F_{Si} is properly assumed, the unknowns in the above equation are H_0, H_n and F_S. In slope problems, H_0 and H_n and are given, and F_S is required. Then, Eq. 11 is an equation having an unknown F_S under the given H_0 and H_n. In bearing capacity or earth pressure problems, F_S is 1.0 and H_n is required under the given H_0. Then, Eq. 11 has an unknown H_n under given F_S and H_0. As the Eq. 11 is nonlinear including only one unknown in every problem, it is easy to obtain the solution numerically for the given F_{Si}.

Upper Bound Formulation (ref. 3)

This formulation of GLEM, using the displacement rate field and the principle of virtual work, is similar to that of UBM. Therefore the formulation of the displacement rate field for a block system in UBM is studied first. The difference between GLEM and UBM will be mentioned later. Chen's formulation (ref. 4) of displacement rate field for a simple block system of *c* material, is extended for a general block system shown in Fig. 1(a) and for c - ϕ materials as well. A displacement rate field for two adjacent blocks is shown in Fig. 2. It is determined under the following conditions.

Every block must slip parallel to its base slip plane. The normal displacement rate of the right block to the interblock plane must be equal to that of the left block. Thus the following equations can be obtained for two adjacent blocks.

$$w_i \cos\{\pi/2-(\theta_j-\beta_i)\} = w_{i-1}\cos\{\pi/2-(\theta_j-\beta_{i-1})\} \tag{12}$$

where w_i and w_{i-1} are the velocities of the right and the left blocks along the base planes, β_i and β_{i-1} are the inclinations of the base planes of the blocks, and θ_i is the inclination of the interblock plane.

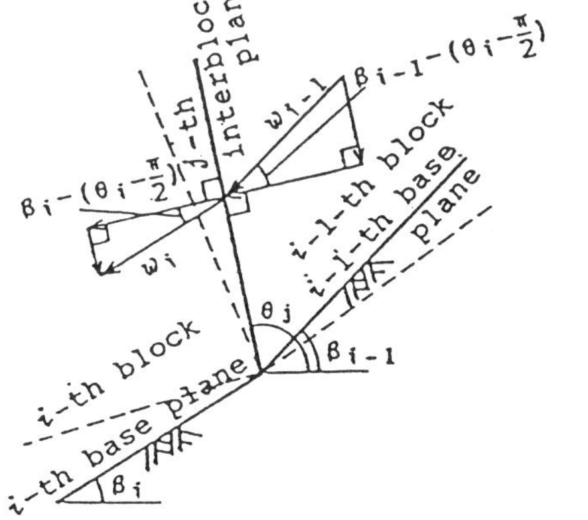

Fig. 2 Velocity field for adjacent two blocks

The relative displacement rate r_i along the interblock plane is the sum of the tangential components of w_i and w_{i-1} to the interblock plane, as follows

$$r_i = w_i \cos(\theta_i - \beta_i) - w_{i-1} \cos(\theta_i - \beta_{i-1}) \tag{13}$$

The horizontal and vertical velocities of the ith block are derived from w_i as $u_i = w_i \cos\beta_i$ and $v_i = w_i \sin\beta_i$. These equations contain only the inclination angles, β_{i-1}, β_i and θ_i, which are determined by the geometry of a given block system.

It is known that the internal energy dissipation for c-ϕ material can be obtained by assuming the displacement rate of a block forming the angle ϕ with the actual plane. Then β_i, β_{i-1} and θ_i in Eqs 12 and 13 must be replaced with the following apparent inclinations, $\beta_i = \beta - \phi$, $\beta_{i-1} = \beta_{i-1} - \phi$, and $\theta_i = \theta_i + \phi$, as shown by dashed lines in Fig. 2. Assuming a unit displacement rate for the first block, ($w_1 = +1$ for bearing capacity and passive earth pressure problems and $w_1 = -1$ for active earth pressure and slope stability problems) the whole displacement rate field can be constructed step by step.

Internal energy dissipation due to slipping, Es, is the sum of the dissipation on the interblock planes, Esi, and the dissipation on the base planes, Esb. It thus follows that

$$Es = Esi + Esb = \Sigma r_i c h_i \cos\phi + \Sigma w_i c d_i \cos\phi \tag{14}$$

where d_i and h_i are the areas of the base and interblock planes.

External work Ee, is the sum of the works by the load, P, on the first block, the weight Wi of the ith block and the vertical surcharge q, on the surface area li, and is expressed as follows

$$Ee = P \times 1 + \Sigma W_i v_i + \Sigma q_i l_i v_i \tag{15}$$

Equating Es to Ee, the force, P, on the first block can be obtained as follows

$$P \times 1 = \Sigma r_i c h_i \cos\phi + \Sigma w_i c d_i \cos\phi - \Sigma W_i v_i - \Sigma q_i l_i v_i \tag{16}$$

This formulation using the displacement rate field is a characteristic of UBM, but it can be applicable to GLEM with a little modification. In GLEM the safety factors Fs defined on base planes, must be used together with the safety factors Fs_i defined on interblock planes. Then, the parameters c and ϕ for base planes and for interblock planes implemented into the above formulation must be replaced with c/Fs and $\mathrm{Tan}^{-1}(\tan\phi/Fs)$ and c/Fs_i and $\mathrm{Tan}^{-1}(\tan\phi/Fs_i)$ respectively. For bearing capacity and earth pressure problems, Fs on base planes is set equal to 1.0. For slope problems Fs is obtained by trial and error so that Eq. 16 is satisfied under given P and q. If Fs and Fs_i are assumed to be 1.0 in the bearing capacity or earth pressure problems, or if Fs is assumed to be equal to Fsi in the slope problems, it is evident that the solutions are identical to those by UBM. This means that GLEM always provides an upper

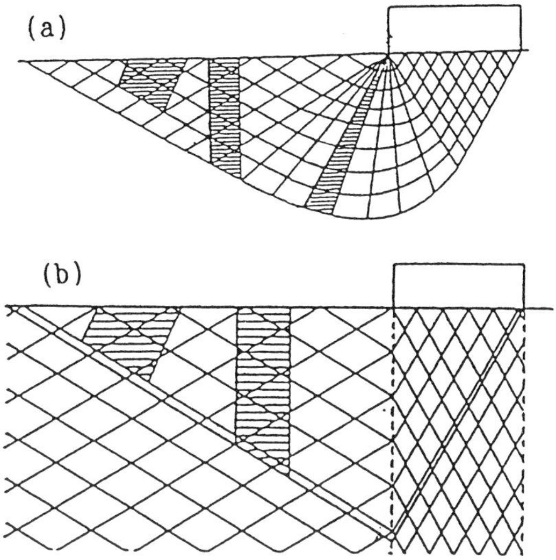

Fig. 3. Relation of LEM with SLM or LBM: (a) Blocks cut out from SLM solution; (b) Blocks cut out from LBM solution

bound of bearing capacity or passive earth pressure or safety factor of a slope, if all the safety factors are assumed to be 1.0 or equal to each other.

Optimization for geometry of block system

In the above sections, the safety factor or the bearing capacity or the earth pressure is formulated for the geometry of a given slip surface. In the actual GLEM analysis, the safety factor or the bearing capacity or the earth pressure must be minimized for a geometry of main and interblock slip surfaces. In the case of active earth pressure problems, since the displacement rate of the wall is negative, the resulting force on the wall is also negative. Minimizing a negative value is a kind of maximizing its absolute value. So, active earth pressure problem is also a minimizing of problem. The geometry of a block system is expressed by connecting the lines between adjacent nodes. Then, the optimization for geometry is able to be done in terms of the coordinates of all nodes. Though many methods for optimization are available, the authors have used classical Newton method to optimize function value.

Characteristic of GLEM solution

LEM obtains the force field which satisfies both the equilibrium condition of every block and the yield condition generally written by Coulomb's criterion on the bottom plane of every block. SLM obtains the stress field which satisfies the equilibrium condition written in terms of partial differential equations and the yield condition. SLM is characterized by a slip line field in the usual solution. LBM obtains an admissible stress field which satisfies the equilibrium condition and never exceeds the yield condition. If the yielding region is included in the obtained stress field, slip line field can be drawn in the region.

Suppose that a block is cut out from a failing zone of SLM or LBM n such a manner that the bottom plane agrees with slip line (Fig. 3 (a) and (b)). Then all the conditions for LEM are always satisfied for the block. These features mean that LEM is the necessary condition for SLM and LBM. In addition, GLEM, a special method of LEM, is evidently coincident with UBM as stated above. The matter which method of SLM, LBM or UBM, is equivalent to LEM, depends on the assumption on the interslice or interblock forces in LEM.

Application of GLEM to earth pressure problems

Passive earth pressure on rough wall

From slip line fields known for some simple problems of earth pressure triangular blocks having interblock planes passing through the top of the wall, are adopted in the following analyses. Fs and Fs_i are 1.0. Formulation for passive earth pressure is identical to one for bearing capacity and k=-1, $M_1=...=M_{n-1}=-1$.

It is recognized that for a rough wall, Coulomb's method slightly under-

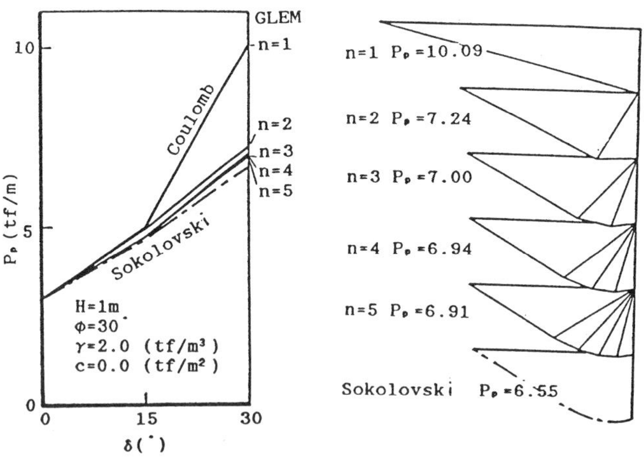

Fig. 4. Passive earth pressure problem on a rough vertical wall: (a) Passive earth pressure; (b) Slip surfaces and block systems

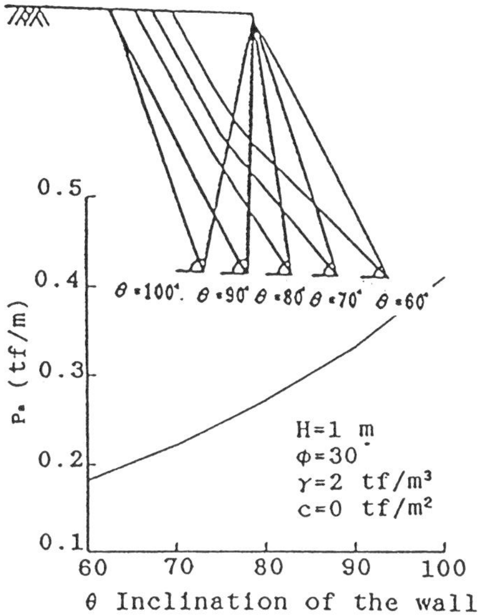

Fig. 5. Active earth pressures on an inclined wall

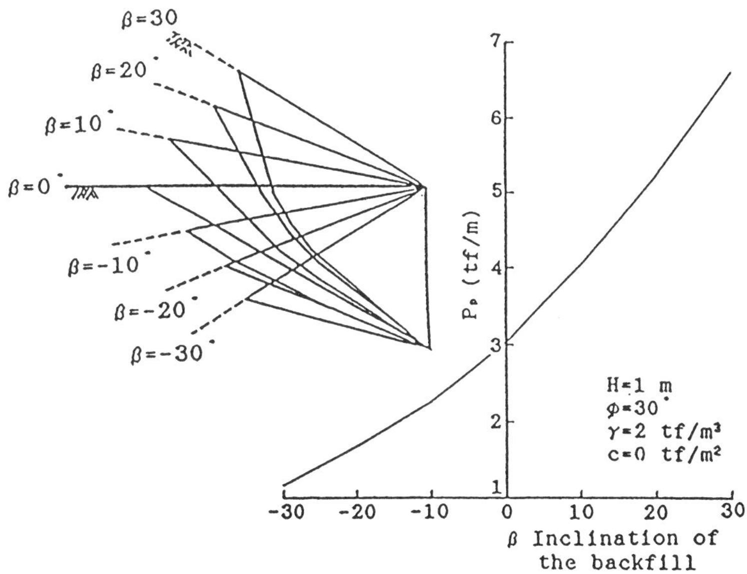

Fig. 6. Passive earth pressures on a wall having inclined surface of backfill.

estimates the active earth pressure and extremely overestimates the passive earth pressure. In Fig. 4(a), passive earth pressures on rough vertical wall (δ=0, 15° and 30°) calculated by the methods of Coulomb, Sokolovski (ref. 5)

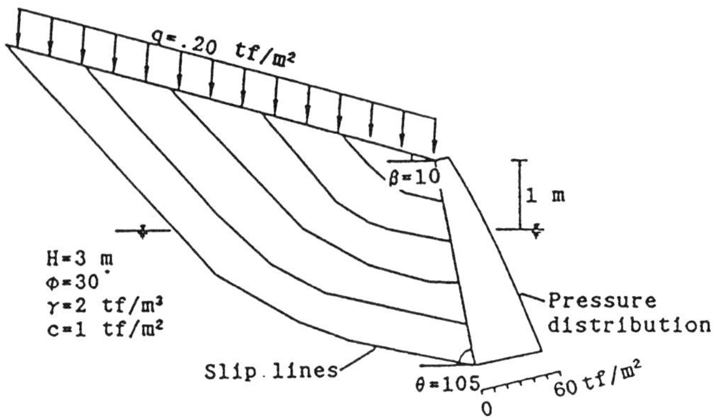

Fig. 7. A practical problem of passive earth pressure.

and GLEM, are compared.

When a single block is used, earth pressure obtained by GLEM completely agrees with Coulomb's, but the result of GLEM becomes close to Sokolovski's with an increasing number of blocks. In Fig. 4 (b) the block systems used in the analysis and the slip surfaces obtained, are shown. In this analyses, earth pressure is minimized only for radii of interblock planes. These results suggest that there is no fundamental difference between the nethods of Rankine and Coulomb to obtain the earth pressure, because the Coulomb method is the simplest case of GLEM.

Complicated earth pressure problems

As GLEM is relative to Coulomb's method, it is easy to treat practical problems with complicated conditions. Active earth pressures on inclined walls are obtained by GLEM as shown in Fig. 5 with corresponding slip surfaces, and pasive earth pressures on a wall having an inclined surface of backfill are also obtained, as shown in Fig. 6.

It is known that active earth pressure problems are common in practical retaining wall designs. Even though the Coulomb method gives a comparable result for active earth pressure problems, for passive earth pressure problems it gives a larger solution compared to SLM. Therefore a practical problem of passive earth pressure is solved as shown in Fig. 7. In the problem, the surface of backfill is inclined, the rough retaining wall is also inclined, and the distribution of pore pressure is given.

Active earth pressure on cylindrical wall

Active earth pressure on a cylindrical wall is analyzed using the combined

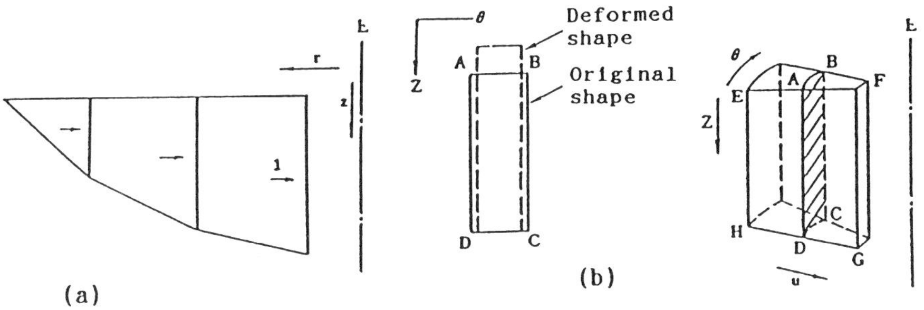

Fig. 8. Schematic diagram to obtain the earth pressure on cylindrical wall: (a) slices velocities; (b) Internal energy dissipation within a slice.

method of GLEM with the upper bound method. In this method, a slice system (the slice means a kind of ring in this case) is adopted in r-z plane as shown in Fig. 8 (a), and the internal energy dissipations along the slip planes and within the slices are obtained separately.

The displacement rate of every slice can be obtained in the same procedure as GLEM under the unit displacement rate of wall. The safety factors on the base planes and the interslice planes are assumed to be 1.0 and ∞ (infinite), respectively. When a slice has the horizontal displacement rate u, the circumferential strain rate ε_0 which has the magnitude of u/r is induced within the slice as shown in Fig. 8 (b), and this strain rate is a maximum principal strain rate. Then, the internal energy dissipation per unit volume within the slice is obtained as follows

$$D=\frac{2c\sin\phi}{1-\sin\phi}|\varepsilon_{max}| = \frac{2c\cos\phi}{1-\sin\phi}\left|\frac{u}{r}\right| \qquad (17)$$

By integrating this energy dissipation over the slice volume, the internal energy dissipation within the slices is obtained. This energy dissipation seems to correspond to the arch action which is effective around the hole. Though this method is not strict from the viewpoint of plasticity theory because it neglects some small internal energy dissipations, it is one of the methods to obtain approximately the active earth pressure on a cylindrical wall.

The result of the analysis obtained by this method is shown in Fig. 9. It is obvious that the active earth pressure decreases with the decrease of r/h where h is the height and r is the radius of cylindrical wall.

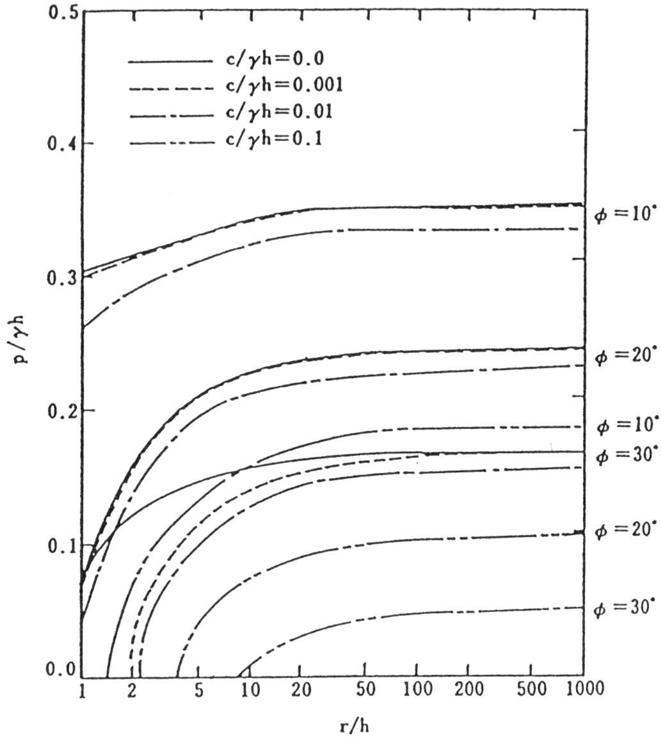

Fig. 9. Earth pressure on cylindrical wall

Conclusions

The following conclusions have been obtained.

(*a*) Limit Equilibrium Method (LEM) can give an approximate solution of Slip Line Method (SLM) or Lower Bound Method (LBM) or Upper Bound Method (UBM), according to the assumption made on the interblock forces.
(*b*) LEM is generalized with respect to the block geometry, interblock force assumptions etc., and Generalized LEM (GLEM) is proposed.
(*c*) Two formulations of GLEM based on the limit equilibrium of blocks and on the energy equilibrium, are possible.
(*d*) GLEM is powerful for obtaining earth pressures in practical problems.
(*e*) GLEM can be extended to analyze axi-symmetric problems such as earth pressure applied on a cylindrical wall.

Though GLEM requires knowledge and experiences for the engineer to construct an appropriate block system for a given problem, it has proven as powerful as the ordinary LEM. More significantly, it provides more accurate solutions than the ordinary LEM does. The program to analyze the problems demonstrated here, has about 1000 lines in FORTAN including optimization routine (250 lines) . All the results were obtained using a 16 bit personal computer system (NEC PC-9801 with 8087, 8MHz) within a few minutes calculations.

Acknowledgments

The Authors wish to thank H.J. Edirisinghe, M. Kubo, and M. Shioma, graduate students of Ehime University, for their assistance in these calculations.

References

1. COULOMB, C. A. Sur une application des regles de maximis et minimis a quelques problemes de statique relatifs a l'archtecture. *Acad. R. Sci. Mem. Math. Phys. par Divers Savants*, 1773, **7**,pp 343-382.
2. ENOKI, M. *et al.* Generalized limit equilibrium method and its relation to slip line method. *Soils and Foundations*, 1991, **31** ,No. 2,pp 1-13.
3. ENOKI, M., YAGI, N., YATABE, R. Relation of Limit Equilibrium method to Limit Analysis Method. *Soils and Foundations*, 1991, **31**, No 4, pp 37-47
4. CHEN, W. H. *Limit analysis and soil plasticity*, Elsevier, 1975, pp 49-55.
5. SOKOLOVSKI, V. V. *Statics of soil media*, Butterworth, 1960.

Discussion

SESSION 2

C.W.W. NG, University of Bristol

Recently, I have been involved in the back-analysis of the 10 m deep multi-propped underground carpark constructed top-down in Gault clay at Lion Yard Cambridge.[1] The site was unusual in that there was the opportunity to monitor both the wall deflections and the strut loads, as well as pore pressures, ground movements and total stresses on the face of the diaphragm wall which allows a comparison between the observations and retrospective predictions. A cross-section of the site is shown in Fig. 1.

The back-analysis was carried out using the non-linear finite element Brick model,[2] which has the capacity to mimic the soil behaviour at small strains. Careful thought was given to modelling the wall installation, and it is believed that the non-linear soil behaviour is very significant at this stage of the analysis.

The simulation of the multi-propped excavation was started by modelling the assumed geological history of the site, resulting in K_0 of around 1.5. The

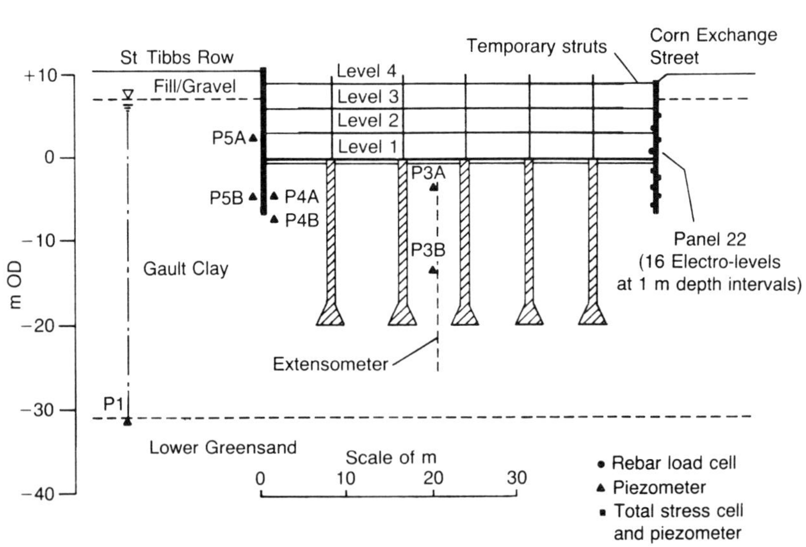

Fig. 1. Lion Yard, Cambridge: cross-section through site

Retaining structures. Thomas Telford, London, 1993

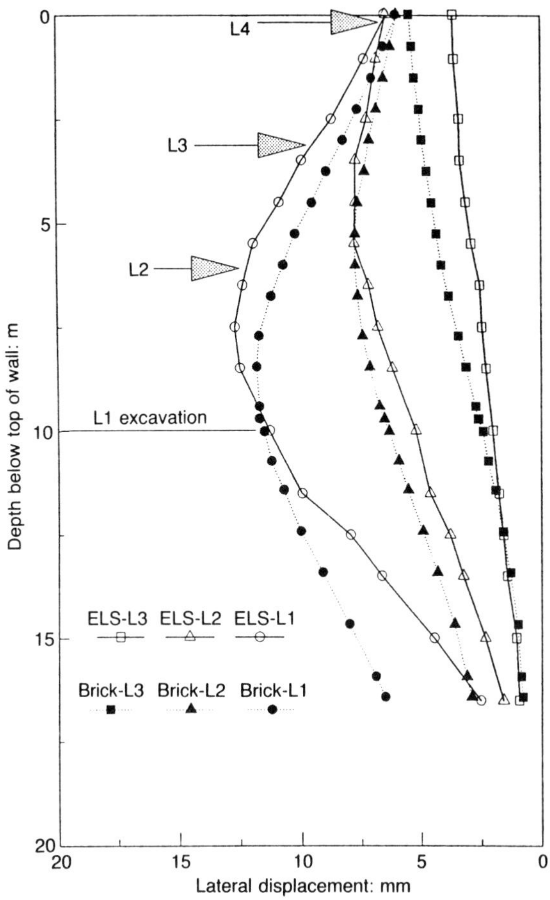

Fig. 2. Comparison of measured and computed displacements during main excavation

wall installation was modelled by imposing the measured earth pressures on the sides of the trench. Then the analysis of the three stages of the main excavation was proceeded as usual. As shown in Figs 2 and 4, the predicted wall displacements and strut loads agree well with the measured values.

To study the effects of wall installation, an additional analysis was carried out assuming a 'wished-in-place wall'. To isolate the effects of wall installation all other aspects of the analysis were kept the same.

Figure 3 shows the comparison between the measured and predicted wall displacements at the end of excavation for the two analyses. It can be seen that analysis with wall installation modelled, resulted in a somewhat closer match of predicted wall deflection to the measured deflection profile than the

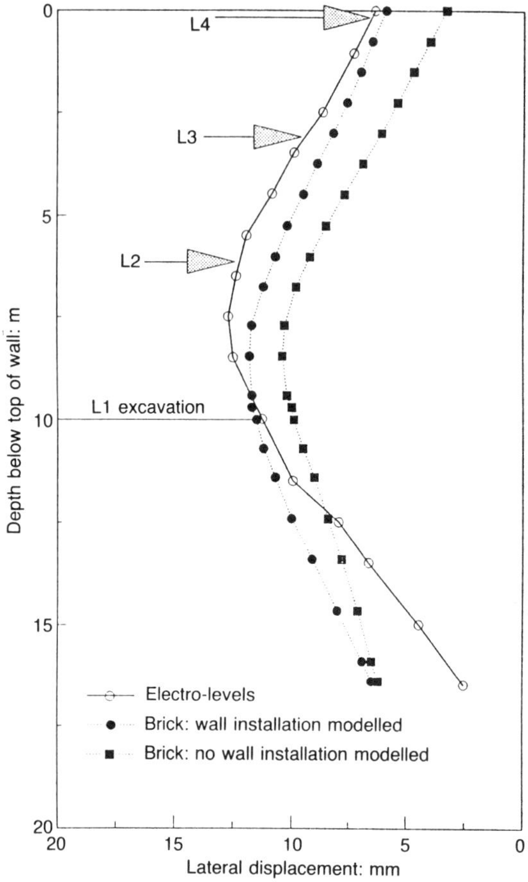

Fig. 3. Comparison of measured and predicted displacements at end of excavation

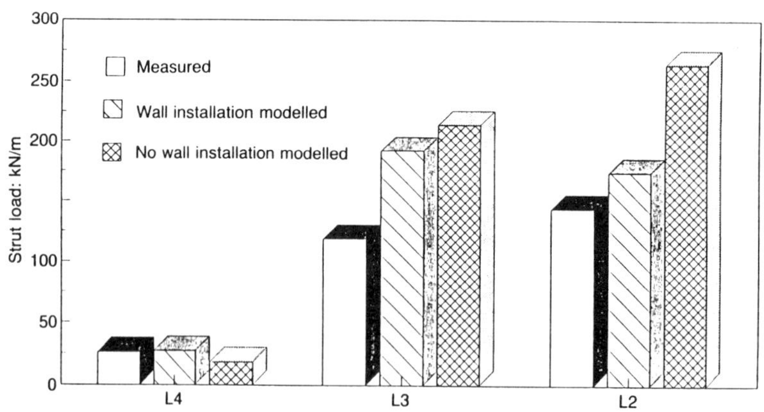

Fig. 4. Measured and predicted strut loads at end of excavation

simulation with the 'wished-in-place' wall. Of more significance is a comparison of the strut loads as shown in Fig. 4; analysis with the 'wished-in-place' wall overpredicted the strut loads by up to 80%.

In the analysis with wall installation, large stress reductions are imposed at that stage, leading to reduced prop loads at the main excavation. Ground movements are limited by the high initial soil stiffness during wall installation, but the resulting strains reduce the soil stiffness, which is then operative during the main excavation.

With the analysis of the 'wished-in-place' wall, the high initial lateral stresses are present in the soil leading to high prop forces and bending moments during the main excavation. Although the incremental displacements of the wall at the last stage of excavation computed in this analysis are larger than the analysis with wall installation modelled, the overall displacements at the end of excavation are somewhat smaller than those when wall installation is considered on account of the high initial operative soil stiffness. These results are opposed to the findings by using a linear elastic perfectly plastic soil model.[3]

In conclusion, the analysis using this non-linear soil model indicates that modelling the excavation using a 'wished-in-place' wall results in an underprediction of the wall deformations and overprediction of the strut loads.

M.L. LINGS, University of Bristol

At Bristol University, we have been attempting to simulate the 3-D effects of diaphragm wall installation (referred to by Dr Potts) by carrying out two separate 2-D analyses: one a horizontal plan section through the wall; the other a vertical cross-section normal to the wall.

The horizontal analysis assumes plane stress, and has an axis of symmetry along the centre-line of the wall. The initial K_0 conditions are first reduced to the bentonite pressure and then increased to the wet concrete pressure. As explained by C. Ng, this is not the full fluid concrete pressure, but rather a pressure based on the bi-linear pressure envelope. The calculated displacement profile is shown in Fig. 5.

The resulting stress distribution on section X-X after concreting one panel is also shown. The pressure over the panel itself is the wet concrete pressure, while the adjacent pressure is greater than the initial K_0 pressure, demonstrating horizontal 'arching'.

This process is then repeated for other panels, leading to a similar pattern of displacements. The key output from this analysis is the average displacement of the soil/wall boundary. As an aside, the analysis shows the importance of relatively short panels: if the length is doubled, it results in double the displacement. We also confirm Dr Potts's result that this analysis shows no overall stress reduction.

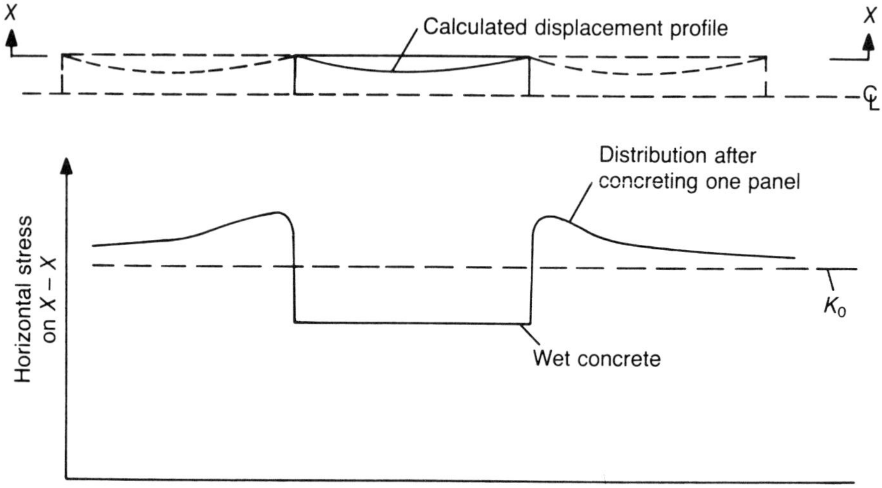

Fig. 5. Horizontal plane stress analysis

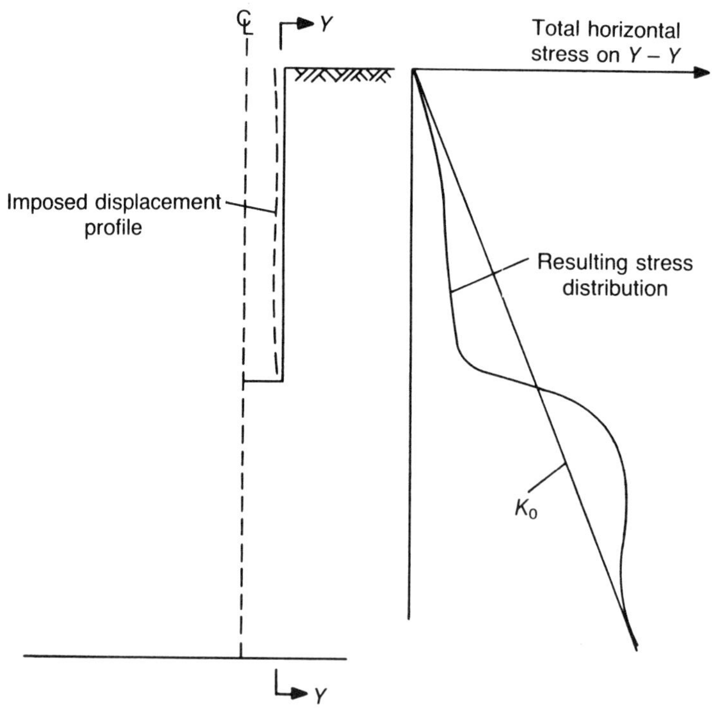

Fig. 6. Vertical plane strain analysis

The vertical analysis assumes plane strain, and has an axis of symmetry down the centre-line of the wall. The initial pressures represent K_0 conditions, but there are no bentonite and wet concrete stages, as this would merely simulate an infinitely long wall panel. Instead the calculated average displacement from the horizontal analysis is imposed on the wall. The resulting stresses on section Y-Y are indicated in Fig. 6, and these show a substantial reduction over the depth of the wall, and a substantial increase beneath the toe of the wall. Thus it appears that significant stress reduction can occur because of wall installation, but only through a vertical shearing mechanism.

M. PULLER, Butler Puller Partnership

Did the analysis of Professor Whittle and Mr Higgins examine the effect of wall stiffness on horizontal soil movements behind the wall and the maximum horizontal distance behind the wall within which such horizontal soil movements occurred.

We were taught many years ago that wall stiffness was not an important factor in influencing the extent of such soil movements. We were then advised that soil movements occurred over a distance from the back of the wall equal to approximately 2½ to 3 times the excavation depth.

It is my opinion that with cellular diaphragm walls or walls with plane forms of similar increased stiffness (compared with a straight wall of successive panels), that wall movements are reduced behind the wall. It is only when one's Client demands minimum soil movements behind the diaphragm wall and permits the extra cost of walls of increased stiffness (cellular walls, tee units, etc.) that the designer and analyst are called to give a very realistic answer, to justify such expense.

DR. M. BUDHU, University of Arizona

With regard first to Dr Powrie's concerns on the 1g shake table test, and secondly to address three issues: the applicability of the Mononobe-Okabe equation to the seismic design of anchored retaining wall; the effect of excavation depth - embedded depth ratio; the effect of wall friction.

Dr Powrie had two concerns, namely the effect of dilation on the test result in the 1g test and the low stress levels that are used. My response to the first concern is that dilation does not enter in the limit equilibrium calculation; indeed, it is stresses that are important. In response to the second concern, the low stress level is taken care of by using the appropriate angle of friction. Usually, a higher angle of friction (the peak angle of friction) than the critical state value is used.

There is concern among some engineers as to whether or not the Mononobe-Okabe analysis is satisfactory for seismic design of anchored retaining walls. I suggest that for certain wall configurations e.g. a wall supported by tie rods and anchor plates, the Mononobe-Okabe analysis is satisfactory (see Budhu et al., Paper 10). However, in the case of a wall configuration in which the anchors interfere with the potential active failure surface, one should not expect the Mononobe-Okabe analysis to give good predictions of the failure acceleration.

In the paper by Budhu et al., the statically designed anchored retaining wall can sustain a horizontal acceleration of 0.5g, instead of 0.2g for anchor failure, by increasing the load-carrying capacity of the anchor. This can be done without increasing the excavation/embedment depth ratio. Indeed, there is very little stability to be found by increasing the static excavation/embedment depth ratio.

Interface friction is particularly important for passive resistance. Neelakanton et al. [4] showed that soil-wall interface friction increases significantly the passive seismic coefficient but has little effect on the active seismic coefficient.

References

1. NG C.W.W. An evaluation of soil-structure interaction associated with deep excavation in a stiff fissured clay. University of Bristol, 1992, PhD thesis, in preparation.
2. SIMPSON B. Thirty-second Rankine Lecture: Retaining structures: displacement and design. Geotechnique, 1992, in print.
3. HIGGINS K.G., POTTS D.M. and SYMONS I.F. Comparison of predicted and measured performance of the retaining walls of the Bell Common tunnel. Transport and Road Research Laboratory, Crowthorne, 1989, TRRL Contractor report 124.
4. NEELAKANTAN G., BUDHU M and RICHARDS R. Mechanics and performance of a tied-back wall under seismic loading. Earthquake Engineering and Structural Dynamics, 1990, 19, 315-331.

The analysis of earth retaining structures

D. M. Potts, Imperial College, London, UK

Summary
This lecture considers the analysis of earth retaining structures. Firstly design requirements are presented, fundamental theoretical considerations are discussed and the various methods of analysis categorised. Secondly the basic problem of the mobilisation of earth pressure is examined. The objectives of the lecture are to categorise the types of retaining structure, provide a framework in which methods of analysis may be compared and indicate the potential of advanced numerical analysis.

Introduction

The purpose of an earth retaining structure is generally to withstand the lateral forces exerted by a vertical or near vertical surface in natural ground or fill. The structural system usually includes a wall, which may be supported by other structural members such as props, floor slabs or ground anchors. Alternatively, or additionally, the wall may be supported by ground at its base or into which it penetrates. In most situations the soil provides both the activating and resisting forces with the wall and its structural support, providing a transfer mechanism.

The design engineer must assess the forces imposed in the wall and other structural members and the likely movements of both the structure and retained material. Usually these have to be determined under both working and ultimate load conditions. Design of retaining walls has traditionally been carried out using simplified analyses or empirical approaches. Methods have been developed for free-standing gravity walls, embedded cantilever walls (fixed earth support) or embedded walls with a single prop or anchor (free earth support). These are described in BSI (1951) and Padfield and Mair (1984). Because of the statically indeterminate nature of multiple propped walls, these have often been dealt with using empirical approaches such as Peck (1969). The introduction of inexpensive but sophisticated computer hardware and software has resulted in considerable advances in the analysis and design of retaining walls. Much effort has been made attempting to model the behaviour of walls in service and to investigate the mechanisms of soil-structure interaction.

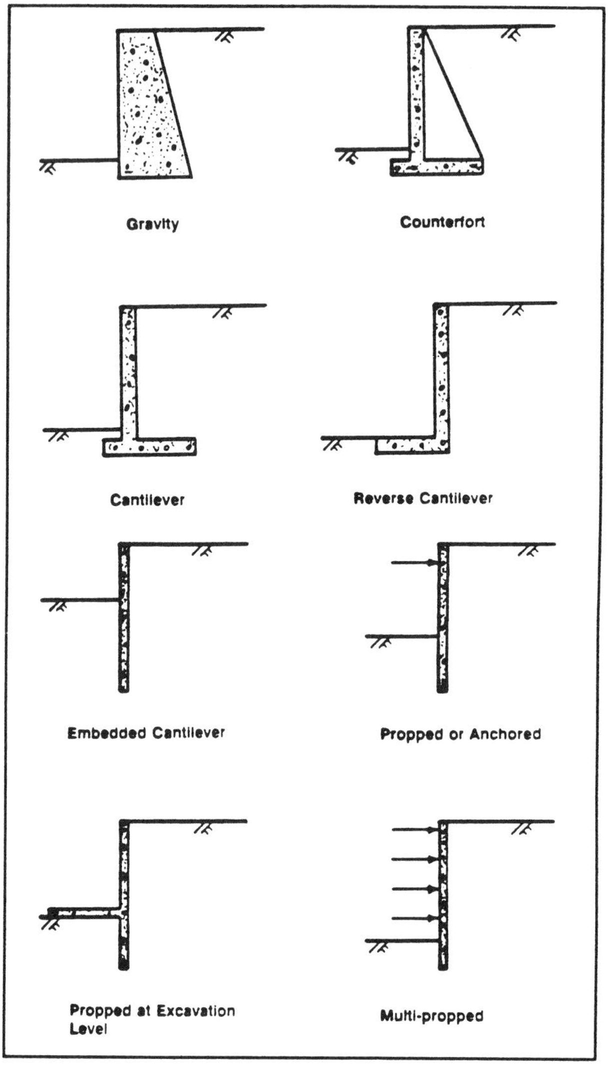

Fig. 1. Types of retaining structure

This lecture is essentially in two parts. In the first part types of retaining wall are categorised, design requirements discussed, fundamental theoretical considerations presented and a framework is provided in which the various design calculations can be compared. In the second part application of advanced numerical analyses to the basic earth pressure problem of mobilisation of active and passive pressure is considered.

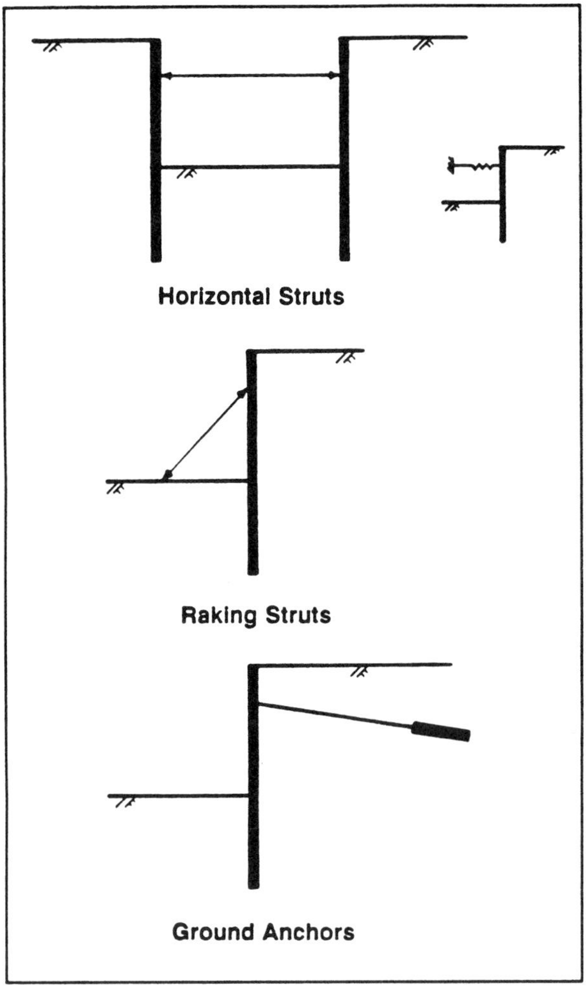

Fig. 2. Types of structural support

Types of retaining structure

The complexity and uncertainty involved in design increases with the degree of soil-structure interaction and thus depends on the type of structure to be employed. It is therefore appropriate to categorise the types of retaining structure on the basis of the soil-structure interaction problems that arise in design. In Figure 1 the main wall types are shown in order of increasing complexity of soil-structure interaction.

Gravity, counterfort and cantilever walls are stiff structures for which the soil-structure interaction is relatively simple. For overall stability, the earth forces on the back of a wall have to be balanced by normal and shear stresses at its base. Movement arising from deformation of the ground beneath the wall is often negligibly small, but exceptions occur if the wall and backfill are constructed on a deep layer of compressible soil. Rotation of the base may lead to large movements at the top of the wall.

When soil movements are important and/or construction space is limited, embedded cantilever walls may be used, with or without props or anchors. To maintain stability the walls rely on the resistance of the ground below excavation level and on the forces provided by any props or anchors employed. The flexibilities of embedded walls vary within a wide range, and this has a considerable effect on the distribution of earth pressures. A flexible wall may be constructed using driven sheet piles and much stiffer walls may be formed from bored piles or using the diaphragm-wall technique. The more flexible walls often have smaller bending moments in the structural elements but may lead to significantly larger deformations, particularly for embedded cantilevers with no props or anchors.

The complexity of the soil-structure interaction increases with the number of rows of props and/or anchors and hence with structural redundancy. The type of support system can also affect behaviour. When horizontal struts are used to support a wall, see Figure 2, analyses are often undertaken in which the struts are either assumed to be rigid with the wall restrained from any horizontal movement at the position of the struts or represented by springs suitably grounded (fixed in space) at some distance from the wall, see insert in Figure 2. Such assumptions imply symmetry and therefore that an identical retaining structure exists as a mirror image to that being analyzed. Such a condition rarely exists in practice as soil and loading conditions are likely to differ. However, even if such a situation did exist, the construction process is unlikely to be carried out in a symmetrical manner. For example the walls themselves are likely to be constructed sequentially imposing nonsymmetric conditions which could effect subsequent behaviour, Finno *et al.* (1991).

Raking struts, see Figure 2, also promote complex soil structure interaction. Load from the retained soil is transferred via the strut to be resisted by soil at formation level. The magnitude of the strut load is therefore dependent on the interaction between the wall and the soil both behind and at formation level. Depending on the position of the raking struts, additional soil pressures may be generated on the front of the wall.

Ground anchors, see Figure 2, involve a complex system of soil interaction as both the load in, and resistance of, the anchor are derived from the retained soil. Consequently the behaviour of an anchored wall is likely to be different to that of a strutted wall.

The type of connection between wall and support can also affect behaviour and increase the complexity of any design analysis. Three types of connection

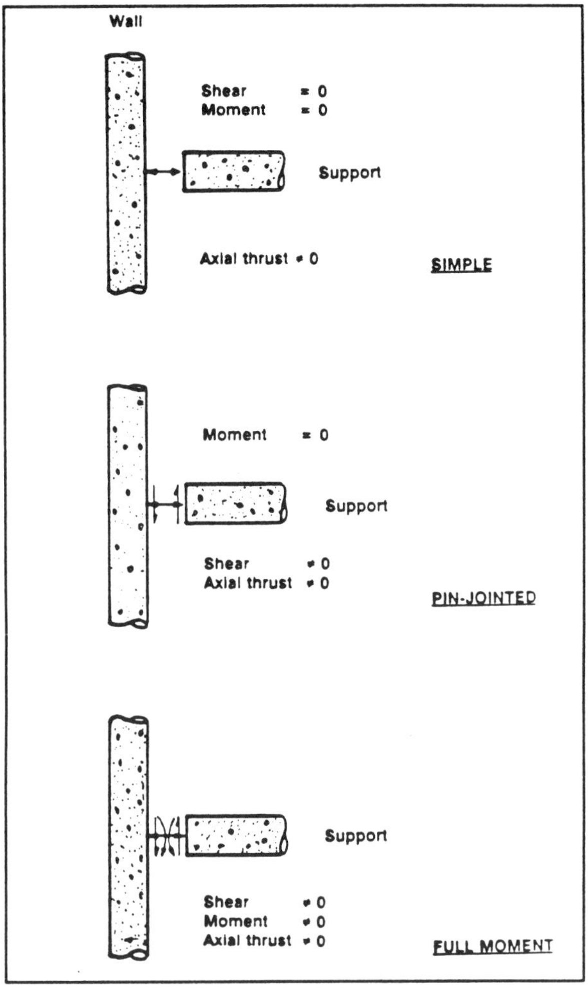

Fig. 3. Types of wall support connection

are shown in Figure 3. Most current design procedures often implicitly assume a 'simple' connection whereas in reality most connections are either 'pin-jointed', 'full moment' or somewhere in between.

It should be noted most of our design guides have been based on experience with the simpler retaining structures usually of a backfilled type. However recent and present trends are towards the more complex structures which involve excavation.

Design requirements

When designing any retaining structure the engineer must ensure that it is stable. Stability can take several forms. Firstly the wall and support system must be stable as whole. There must be no danger of rotational or translation failure. Secondly there must be no potential for excessive base heave due to either undrained shear failure (soft clays) or to piping as a result of high pore water pressure gradients. Thirdly overall stability must be established. For example, if the retaining structure supports sloping ground, the possibility of the construction promoting an overall slope failure should be investigated.

Structural forces (bending moments, shear forces and axial forces) must be obtained both under working and ultimate loading conditions. Estimates of movements, both of the structure, retained soil and soil below formation level may be required. This is particularly important if there are adjacent buildings and/or sensitive services.

Methods of analysis

A review of the literature indicates that there are many different methods of calculation currently available for analysing retaining structures. Unfortunately application of these methods to a particular retaining structure can lead to very different estimates of the main design parameters, see Day and Potts (1989). This is clearly unsatisfactory.

Fundamental considerations assert that for an exact theoretical solution the requirements of equilibrium, compatibility, material behaviour and boundary conditions, both load and displacement, must all be satisfied. It is therefore useful to review the broad categories of analysis currently in use against these theoretical requirements.

Current methods of analysis can be conveniently grouped into the following categories: simple, beam-spring approaches and full numerical analysis. Each of these categories will be considered in turn.

Simple methods of analysis

Limit equilibrium, stress field and limit analysis are in this category. Coulombs wedge analysis is an example of a limit equilibrium calculation, whereas a Rankine stress field and the work of Sokolvski (1960) are examples of stress field approaches. Upper and lower bound theorems of plasticity are used in limit analysis, Chen (1975). All methods essentially assume the soil is at failure but differ in the manner in which they arrive at a solution.

The ability of these methods to satisfy the basic solution requirements is investigated in Table 1. Inspection of this table clearly shows that none of the methods satisfy all basic requirements and therefore do not necessarily produce an exact theoretical solution. All methods are therefore approximate and it is perhaps not surprising that there are many different solutions to the

Table 1. Basic Solution Requirements Satisfied by the Approximate Methods of Analysis

METHOD OF ANALYSIS		SOLUTION REQUIREMENTS				
		Equilibrium	Compatibility	Constitutive Behaviour	Boundary Conditions	
					Force	Disp.
Equilibrium		Satisfied in the large	Not Satisfied	Rigid with a failure criterion	Satisfied	Not Satisfied
Stress Field		Satisfied	Not Satisfied	Rigid with a failure criterion	Satisfied	Not Satisfied
Limit Analysis	Upper Bound	Satisfied	Not Satisfied	Ideal plasticity with associated flow rule	Satisfied	Not Satisfied
	Lower Bound	Not Satisfied	Satisfied		Not Satisfied	Satisfied

same problem.

As these approaches assume the soil to be everywhere at failure they are not strictly appropriate for investigating behaviour under working load conditions. They do not distinguish between different methods of construction (i.e. excavation versus backfilling) nor account for in situ stress conditions. Empirical factors have to be used to account for wall flexibility, no distinction is made between anchors and props, and surcharges have to be treated in an approximate manner. Information is provided on wall stability but no information on soil or wall movements is given and separate calculations are required to investigate overall stability and base heave.

Notwithstanding the above limitations, simple methods form the mainstay of most design approaches. Where they have been calibrated against field observation their use may be appropriate. However it is for the more complex soil-structure interaction situations where calibration is more difficult that they are perhaps less reliable. Because of their simplicity and ease of use it is likely that they will always play an important role in the design of retaining structures. In particular they are appropriate at the early stages of the design process to obtain first estimates of both stability and structural forces.

Beam-spring approaches

In these approaches the retaining structure problem is reduced to that of analysing a single wall in isolation, see Figure 4. A major approximation is the assumed soil behaviour and two approaches are commonly used. The soil is either approximated by a set of unconnected vertical and horizontal springs

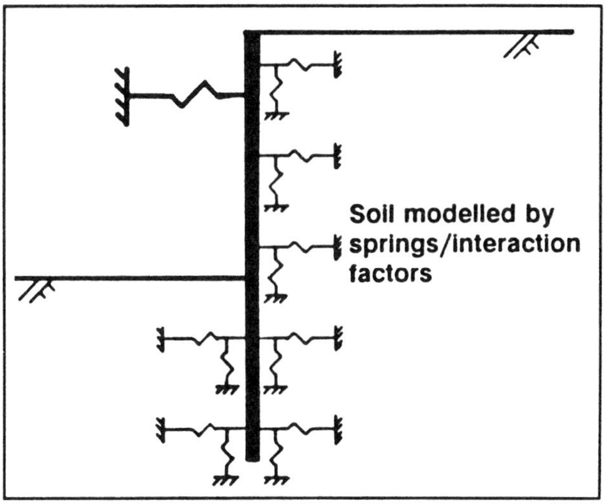

Fig. 4. Beam-spring approach

(Borin, 1989) or by a set of linear elastic interaction factors (Papin *et al*, 1985). Any structural support such as props or anchors are represented by simple springs. To enable limiting pressures to be obtained on each side of the wall 'cut offs' are usually applied to the spring forces and interaction factors representing soil behaviour. These cut off pressures are usually obtained from one of the simple analysis procedures discussed above. It is important to appreciate that these limiting pressures are not a direct result of the beam-spring calculation but are obtained from separate approximate solutions and then imposed in the beam-spring calculation process.

Having reduced the retaining structure to the behaviour of a single isolated wall and made gross assumptions as to soil behaviour, a complete theoretical solution to the problem is sought. Due to the complexities involved this is usually achieved using a computer. The wall is represented using either finite differences or finite elements and a solution that satisfies all the basic solution requirements obtained by iteration. Sometimes computer programs which perform such calculations are identified as finite difference or finite element programs. However it must be noted that it is only the wall that is represented in this manner and these programs should not be confused with those that involve full descritization of both the soil and structural members by finite differences or finite elements, see next section.

As solutions obtained in this way include limits to active and passive earth pressures, they can provide information on wall stability. This is often indicated by failure of the program to converge. However numerical instability may occur for other reasons and therefore a false impression of wall instability may be given.

Solutions from these calculations include structural forces in the wall, strut or anchor forces and movements of the wall. They do not provide information on overall stability, base heave or of movements in the adjacent soil. It is difficult to select appropriate spring stiffnesses and to simulate realistically the effects of soil berms. It is also difficult to reproduce stress redistribution arising as a result of wall flexibility. Most programs do not even consider this, whereas others use approximate procedures (Papin *et al*, 1985). Programs using interaction factors to represent the soil have problems in dealing with wall friction and often neglect shear stresses on the wall or make extra assumptions to deal with them. As a single wall is considered in isolation and structural supports are represented by simple springs fixed at one end (grounded), it is difficult to account for realistic interaction between structural components such as floor slabs and other walls. This is particularly so if 'pin-jointed' or 'full moment' connections are appropriate. As only the soil acting on the wall is considered in the analysis it is difficult to model realistically the behaviour of raking struts and ground anchors which rely on resistance from soil remote from the wall.

Full numerical analysis

This category includes methods which attempt to satisfy all theoretical requirements, realistic soil constitutive models and incorporates boundary conditions that realistically simulate field conditions. Because of the complexities involved and the nonlinearities in soil behaviour all methods are numerical in nature. Approaches based on finite difference and finite element methods are the most widely used. These methods essentially involve a computer simulation of the construction of the complete retaining structure.

Their ability to accurately reflect field conditions essentially depends on the ability of the constitutive model to represent real soil behaviour and correctness of the boundary conditions imposed. The user has only to define the appropriate geometry, construction procedure, soil parameters and boundary conditions. Structural members may be added and withdrawn during the numerical simulation to model field conditions. Retaining structures composed of several retaining walls interconnected by structural components can be considered and because the soil mass is modelled in the analysis, the complex interaction between raking struts or ground anchors and the soil can be accounted for. The effect of time on the development of pore water pressures can also be simulated by including coupled consolidation. No postulated failure mechanism or mode of behaviour of the retaining structure is required as these are predicted by the analysis. The analysis allows the complete history of the retaining structures behaviour to be predicted and a single analysis can provide information on all design requirements.

Potentially the methods can solve full three dimensional problems and suffer none of the limitations discussed previously for the other methods. At present the speed of computer hardware restricts analysis to two dimensional plane strain sections. However with the rapid development in computer hardware and its reduction in cost the possibilities of full three dimensional simulations are imminent.

It is often claimed that these approaches have limitations. Usually these relate to the fact that they need detailed soils information or a knowledge of the construction procedure. In the Author's opinion neither of these are limitations. If a numerical analysis is anticipated during the design stages of a project then it is not difficult to ensure that the appropriate soils information is obtained from the site investigation. It is only if a numerical analysis is an after thought, once the soils data has been obtained, that this may present difficulties. If the behaviour of the retaining structure is not sensitive to the construction procedure then any reasonable assumed procedure will be adequate for the analysis. However, if the analysis is sensitive to the construction procedure then clearly this is important and it will be necessary to simulate the field conditions as closely as possible. So far from being a limitation, numerical analysis can indicate to the design engineer where, and by how much, the retaining structure is likely to be influenced by the construction procedure. This will enable him to make adequate provision within the design.

Full numerical analyses are complex and should be performed by qualified and experienced staff. The operator must have engineering experience of retaining structures, understand soil mechanics and in particular the constitutive models that the software uses and be familiar with the software package to be employed for the analysis. Nonlinear numerical analysis is not straight forward and at present there are several algorithms available for solving the nonlinear system of governing equations. Some of these are more accurate than others and some are increment size dependent. There are approximations involved in these algorithms and errors associated with descritization. However these can be controlled by the experienced user so that accurate predictions can be obtained.

Full numerical analysis can be used to predict the behaviour of complex field situations. They can also be used to investigate the fundamentals of soil/structure interaction and calibrate some of the simple methods discussed above. Analyses have been used in this manner by Potts and Fourie (1984, 1985 & 1986), Fourie and Potts (1988 & 1989) and Potts and Day (1991) to investigate such effects as soil properties, wall flexibility, construction procedure, strut levels and initial soil conditions.

The advantage of beam-spring methods over full numerical analyses are that they are more economical in terms of computer resources. However with the rapid developments in computer technology this advantage will quickly disappear. Future design procedures are therefore likely to involve a combination of the simple methods and full numerical analysis. The simple methods being used to obtain approximate structural dimensions with the full numerical analysis being used to check stability, obtain structural forces and estimate movements.

Earth pressures

The mobilisation of earth pressures is fundamental to all retaining wall analyses. It is therefore instructive to review the application of advanced numerical analysis to this problem. The objectives being two fold, firstly to indicate the power of the full numerical approaches and secondly to investigate the accuracy of the simple methods of analysis which form the main stay of current design procedures.

Active and passive pressures

As a wall moves towards a collapsed state, zones will be formed in the adjoining ground within which the soil is in a state of plastic equilibrium. Within these zones the direction of shear is such that the soil thrust on the wall is a minimum within the active zones where movement of the wall is away from the soil, and a maximum within the passive zones where movement of the wall is towards the soil. The shapes and extent of these failure zones are greatly influenced by adhesion or friction developed between the soil and the wall, by the inclination of the wall and the slopes of the ground.

For uniform soils in simple situations it is often assumed in design that active and passive earth pressures increase linearly with depth. In this case the effective stresses acting normal to a wall at depth z can be calculated using equations of the following form:

$$p'_a = K_a(\gamma z + q - u) - c'K_{ac} \qquad \text{for } p'_a > 0 \qquad (1)$$

$$p'_p = K_p(\gamma z + q - u) + c'K_{pc} \qquad (2)$$

where: γ is the bulk unit weight of the soil
q is any uniform distributed surcharge pressure
u is the pore water pressure at depth z
K_a, K_{ac} are the active pressure coefficients
K_p, K_{pc} are the passive pressure coefficients

On this basis, the limits on total stress normal to the wall may be written as:

$$p_a = p'_a + u \qquad (3)$$

$$p_p = p'_p + u \qquad (4)$$

Values of the earth pressure coefficients for use in these equations are given in standard references and codes of practice (Caquot & Kerisel, 1948;

BSI, 1951: and NAVFAC, 1971). The values are dependent on soil properties and also on relative movements between soil and wall which govern wall adhesion and friction.

Current design practice relies on simple methods to estimate the active and passive earth pressure coefficients, e.g. limit equilibrium, stress field solutions, limit analysis. As noted above, all of these methods fail to satisfy at least one of the basic solution requirements and therefore the values quoted in the literature are not theoretically exact. In addition many of the solutions only furnish values of limiting force on the wall P_a(active side) or P_p(passive side). To obtain the coefficients K_a and K_p, a linear variation of earth pressure is assumed and these forces are divided by $\gamma H_2/2$.

Implicit in these conventional methods is the assumption that the angle of shearing resistance, ϕ', is fully mobilised and that the angle of mobilised wall friction is both constant and can be specified. No account is taken of the mode of wall displacement on the resultant earth pressure, and no indication of the distribution or magnitude of earth pressure prior to ultimate failure is given. The estimation of structural forces in the retaining structure under working load conditions is therefore extremely difficult. Although a great deal of experience has been gained using the approximate methods, the above short comings restrict their usefulness, particularly in solving unusual problems.

Mobilisation of earth pressure

An alternative approach is afforded by the finite element method of analysis. Potts and Fourie (1986) have used this method to study soil-structure interaction of a rigid wall embedded in an initially horizontal, uniform soil deposit. Three modes of wall deformation were studied, namely horizontal translation, rotation about the wall top and rotation about the toe. The mode of deformation of a real retaining wall is complicated, and depends, in part, on both flexibility and the type of wall. For embedded cantilever walls, the mode of displacement is essentially one of rotation about the toe. Alternatively, if a prop or anchor is installed near the top of the wall, the mode changes to one of rotation about the top. Horizontal translation is more applicable to a gravity type wall.

A 5 m deep by 1 m thick wall embedded in a uniform elasto-plastic soil was modelled, see Figure 5. The behaviour of both smooth and rough walls were investigated, as well as the effects of changing the coefficient of earth pressure at rest, K_o. As an example of the results obtained from this investigation, the development of the active and passive pressure coefficients with increasing wall displacement are shown in Figure 6. The results are for a rough wall and with an initial coefficient of earth pressure K_o=2. The equivalent coefficient K (active or passive) is defined as $K=2.P/(\gamma.H^2)$ where P is the resultant force exerted by the soil on the front (passive) or back (active) of the wall, H is the depth of the wall and the bulk unit weight of the soil.

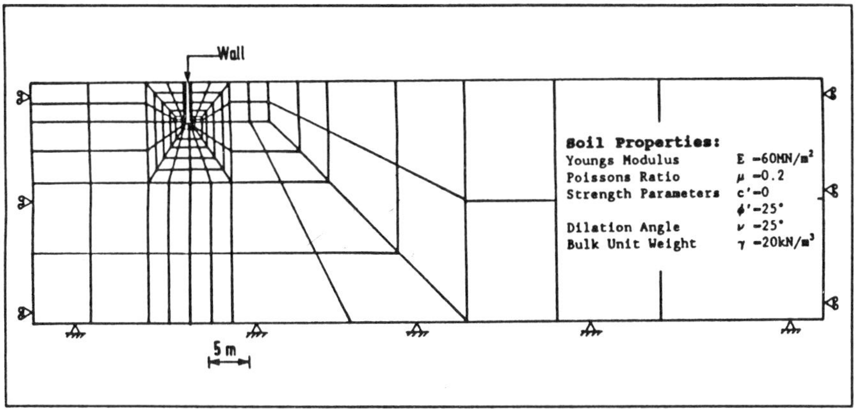

Fig. 5. Finite element mesh for studying earth pressure mobilisation

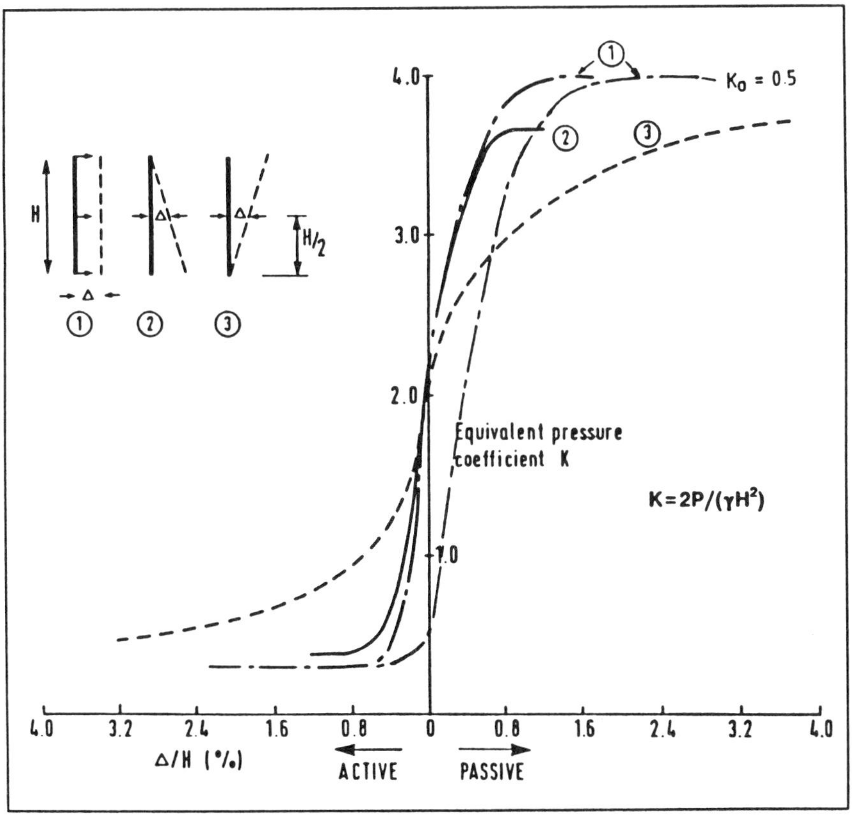

Fig. 6. Development of earth pressure coefficients with increasing wall displacements (rough wall)

Table 2. Earth pressure coefficients for φ′ = δ = 25°

AUTHOR(S)	METHOD OF SOLUTION	ACTIVE EARTH PRESSURE COEFFICIENTS	PASSIVE EARTH PRESSURE COEFFICIENTS
Coulomb	Limit Equilibrium	0.32	5.07
NAVFAC	Limit Equilibrium	0.36	4.20
Sokolovski	Advanced stress field	0.33[1]	3.86[1]
Lee and Herington	Advanced stress field	----	4.10[1]
Caquot and Kerisel	Limit Equilibrium	0.33	3.89
Chen	Limit Analysis (Upper bound)	0.33	4.19
Packshaw	Limit Equilibrium	0.32	3.9
Potts and Fourie	Finite element	0.32[2] 0.39[3] 0.45[4]	3.99[2] 3.68[3] 3.73[4]

[1]Interpolated value [2]Translation [3]Rotation about top
[4]Rotation about bottom

Clearly the magnitudes of these displacements are dependent on the elastic stiffness assumed for the soil. However, as the same stiffness was assumed in all three analyses, the results provide a relative measure of the amount of movement required to mobilise limiting conditions on either side of the wall. Also included on Figure 6 are results for a rough wall with K_o=0.5 subjected to horizontal translation.

Rotation about the base requires significantly more displacement to obtain failure conditions than do other modes of displacement. For high K_o values, active and passive conditions are mobilised at similar displacements. However for low K_o values active conditions occur before passive conditions. Clearly displacements necessary to mobilise active and passive conditions are dependent on the value of K_o as well as the mode of deformation. The limiting values (active and passive) of K are less sensitive to the mode of displacement and the K_o value. The values are compared with those based on conventional approximate analyses in Table 2.

The effect of the mode of wall displacement on the distribution of earth pressure is summarised in Figure 7. The development of the active and

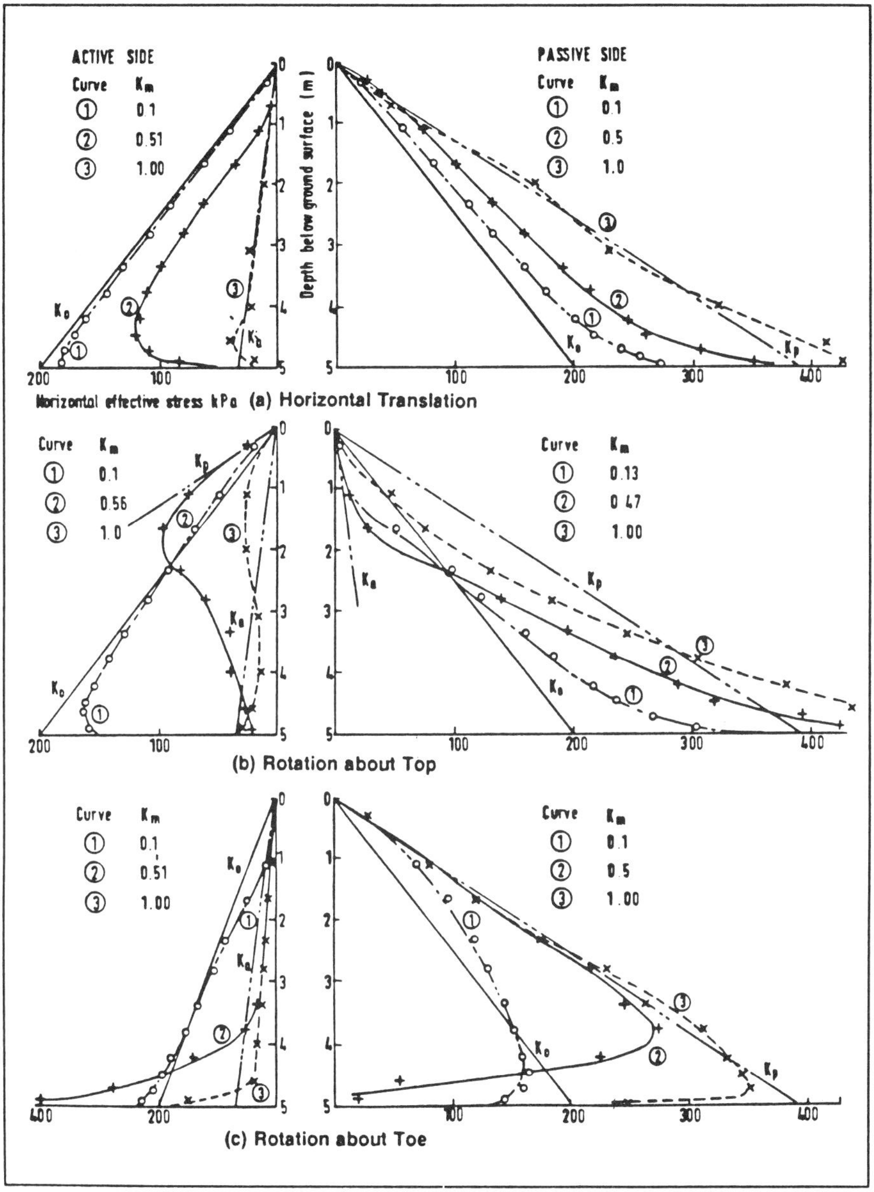

Fig. 7. Earth pressure distributions (rough wall)

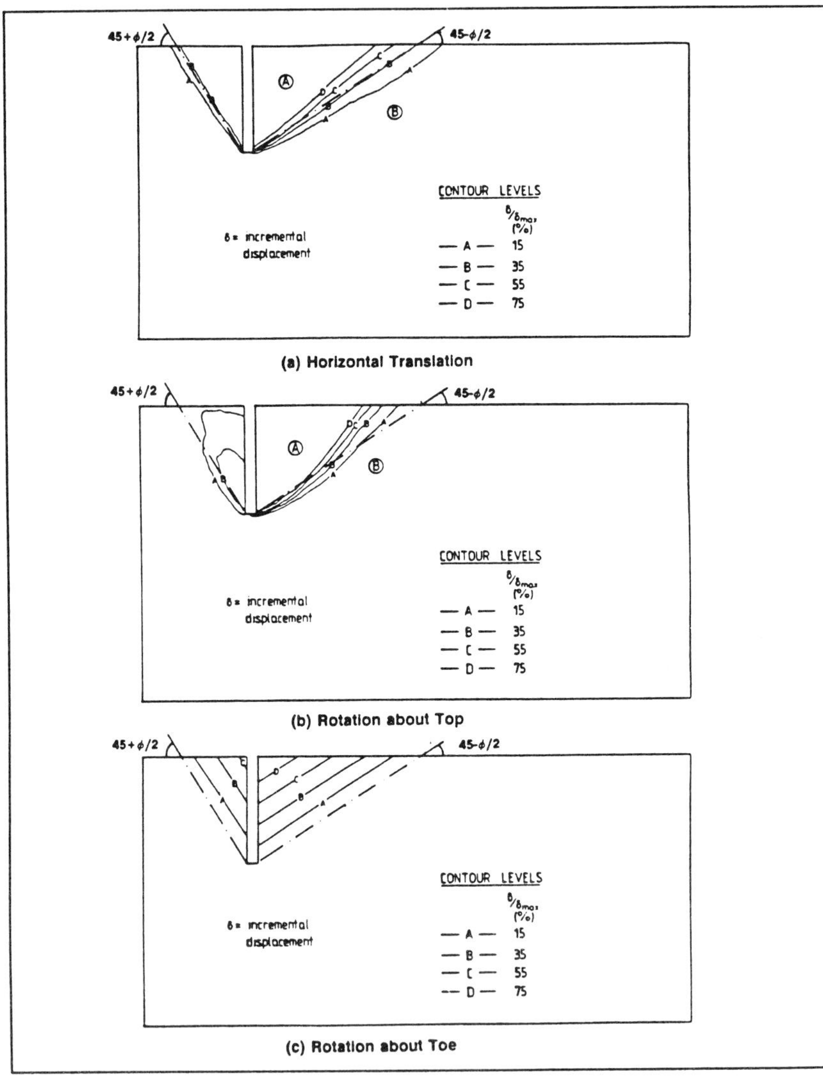

Fig. 8. Contours of normalised incremental displacement at failure (smooth wall)

passive pressures for all three modes of deformation are shown for various values of the pressure coefficient K_m where $K_m=(K-K_o)/(K_p-K_o)$ for passive pressures and $K_m=(K-K_o)/(K_a-K_o)$ for active pressures. K_m can therefore be seen to be equivalent to a load factor. The lines labelled K_a and K_p shown on these figures correspond to the Caquot and Kerisel (1948) values. For a wall rotating about its top or bottom, the distributions are far from the linear distributions commonly assumed in design and given by equations 1 to 4. For example, for a wall rotating about its top, the soil on the passive side at failure is in an active condition near the top and exceeds the classical passive value lower down. The pressure distributions also change during the mobilisation of a limiting condition. Such vastly different pressure distributions imply that wall bending moments will be highly dependent on the mode of wall deformation.

The above numerical predictions are in agreement with the experimental observations made by Bros(1972), who carried out laboratory model retaining wall tests in which the three modes of displacement were simulated.

From this study it is clear that the development of limiting earth pressures is complex. For modes of wall deformation similar to rotation about the toe and horizontal translation, approximate analyses which result in equations 1 to 4 give reasonable predictions for the limiting active and passive pressures. However for modes of deformation similar to rotation about the top and for all modes of deformation prior to reaching a limiting condition the stress distributions are not linear with depth.

Contours of normalised incremental displacement at failure are shown in Figure 8 for a smooth wall subject to the three different modes of wall deformation. Inspection of the figures for a wall rotating about its top and translating horizontally indicate that there are zones where the soil has a normalised displacement greater than 85% (zone A) or less than 15% (zone B). Soil in these regions have similar displacements and therefore are moving as blocks. Contours close together show high gradients of displacement and therefore indicate potential failure surfaces. For the wall that translates the failure surface, both the active and passive sides of the wall are planar and in agreement with the Rankine planes of maximum stress obliquity (inclined at $45 \pm\phi/2^{o}$ to the horizontal) that pass through the toe of the wall. For the wall rotating about its top the failure surfaces are curved.

A different pattern of displacement is indicated for the wall rotating about its toe. There are no zones of block movement adjacent to the wall and no indication of failure surfaces. Instead the contours are equally spaced, indicating that the soil in front of and behind the wall is subjected to either uniform compression or extension respectively. Clearly larger absolute wall movements would be required to mobilise plastic failure in these zones than in the narrow failure zones indicated for walls rotating about their top or translating. This difference in mode of failure explains the differences in displacements between the different modes of wall movement in Figure 6. It also helps explain the large differences in displacements associated with embedded cantilever walls without a prop, (rotation about a point below formation level) and those with a prop (rotation about prop position).

Conclusions

By considering the basic fundamental requirements of a theoretical solution, a means for categorising the various methods of analysis has been provided. The simple methods, such as limit equilibrium, stress fields and limit analysis do not satisfy these basic requirements and this explains why there are many different procedures currently available for analysing retaining structures. In addition these simple methods do not yield the same values of design parameters nor provide information on all the design requirements.

Beam-spring approaches provide an improvement over the simple methods allowing structurally redundant retaining structures to be analyzed and estimates of wall movements to be obtained. However, these methods involve severe approximations in modelling soil behaviour. In addition, complex retaining structures have to be reduced to analysing a single isolated wall and much of the important interaction between structure and soil is not modelled.

Full numerical analyses do not suffer from any of the limitations of the above approaches. They satisfy the basic theoretical requirements, can represent real soil behaviour and realistically simulate soil-structure interaction. Their ability to predict field behaviour depends on the accuracy of the soil constitutive models and the correctness of the boundary conditions. A single analysis provides information on all design requirements. Such analyses also can provide improved understanding of the effects of soil-structure interaction and can therefore be used to calibrate some of the simple methods of analysis.

In this lecture advanced numerical analysis in the form of finite element analysis have been applied to the basic problem of the mobilisation of earth pressures. The results have been compared with the assumptions and predictions from simple methods of analysis. This has enabled the limitations of the simple methods to be quantified and in particular has shown that they are unlikely to provide realistic predictions of stress distributions especially under working load conditions. The effects of different modes of wall movement has also been established. As the simple methods have been shown to have deficiencies, even when considering this basic earth pressure problem, their ability to analyze more complex retaining structures is questionable.

Future design procedures are likely to involve a combination of one or more of the simple methods combined with advanced numerical analysis. For simple retaining structures simple methods, suitably calibrated, may be sufficient. However for the more complex structures simple methods may be used to obtain approximate estimates of structural dimensions with the full numerical analyses being used to check stability, obtain structural forces and estimate movements.

References

1. BORIN, D. L. WALLAP - computer program for the stability analysis of retaining walls. GEOSOLVE, (1989).
2. BRITISH STANDARDS INSTITUTION - CP2 Earth retaining structures CP2: 1951 (currently under revision).
3. BROS, B. The influence of model retaining wall displacements on active and passive earth pressure in sand., Proc. 5th European Conf. on Soil Mech. and Found. Eng., Madrid, (1972), 1, pp 241-249.
4. CAQUOT, A. & KERISEL, J. Tables for the calculation of passive pressure, active pressure and bearing capacity of foundations., Paris, Gauthier-Villars, (1948).
5. CHEN, W. F. Limit analysis and soil plasticity, Developments in Geotechnical Engineering, (1975), Vol. 7, Elsevier.
6. DAY, R. A. and POTTS, D. M. A comparison of design methods for propped sheet pile walls., Steel Construction Institute, SCI publication 077, (1989).
7. FINNO, R.J., HARAHAP, I. S. and SABATINI, P. J. Analysis of braced excavations with coupled finite element formulations, Computers and Geotechnics 12, pp 91-114.
8. FOURIE, A. B. & POTTS, D. M. The estimation of design bending moments for retaining walls using finite elements., 6th Int. Conf. on Num. Meth. in Geomech., Innsbruck, A. A. Balkema, (1988), pp 1101-1107.
9. FOURIE, A. B. & POTTS, D. M. Comparison of finite element and limiting equilibrium analyses for an embedded cantilever retaining wall, accepted for publication in Geotechnique, (1989).
10. LEE, I. K. & HERINGTON, J. R. Soil mechanics new horizons, edited by I. K. Lee, Newnes-Butterworths, (1974)
11. NAVFAC DM-7 Design manual: Soil mechanics, foundations and earth structures, U. S. Department of the Navy, Washington, DC. (1982).
12. PACKSHAW, E. Earth pressure and earth resistance, Jnl. Inst. Civ. Engrs., (1946), Vol. 25, pp 233-256.
13. PADFIELD, C. J. & MAIR, R. J. The design of propped and cantilever walls constructed in stiff clay, CIRIA Project 317, London, Construction Industry Research and Information Association, (1984).
14. PAPIN, J. W., SIMPSON, B., FELTON, P. J. & RAISON, C. Numerical analysis of flexible retaining walls, Proc. Numerical Methods in Eng. Theory and Applications 85 Conference, Swansea, (1985), pp 789-802.

15. PECK, R. B. Deep excavations and tunnelling in soft ground, Proc. 7th int. conf. SM&FE, Mexico, (1969).
16. POTTS, D. M. & DAY, R. A. Use of sheet pile retaining walls for deep excavations in stiff clay., Proc. Instn Civ. Engrs, Part 1, (1990), 88, DEC., pp 899-927.
17. POTTS, D. M. & FOURIE, A. B. The behaviour of a propped retaining wall : results of a numerical experiment, Geotechnique 34, (1984), No. 3, pp 383-404.
18. POTTS, D. M. & FOURIE, A. B. The effect of wall stiffness on the behaviour of a propped retaining wall, Geotechnique 35, (1985), No. 3, pp 347-352.
19. POTTS, D. M. & FOURIE. A. B. A numerical study of the effects of wall deformation on earth pressures, Int. Jnl. for Num. and Anal. Meth. in Geomech. 10, (1986), no. 3, pp 383-404.
20. SOKOLOVSKI, V. V. Statics of soil media, Butterworth, (1960).

Introduction to Session 3 In situ embedded walls: observations

I. F. SYMONS, Transport Research Laboratory

Introduction

Of the 61 papers submitted to this Conference well over one third are concerned with in situ embedded walls of the diaphragm, bored or sheet pile types. This reflects their widespread use in basement works for buildings and their increasing role in the construction of road and rail schemes in urban areas.

In addition to the ten papers assigned to Session 3 there are three papers in other Sessions which include results of field measurements on embedded walls. These are the paper in Session 2 by Maruoka, Aoki, Sato and Ikuta describing ground movements caused by displacement of earth retaining walls, the paper in Session 4 by Tse and Nicholson on the design construction and monitoring of the basement diaphragm wall at Minster Court London and Newman and Covil's paper to Session 5 on a temporary secant mini-pile wall propped by a deep jet grout raft. In total the information presented in these thirteen papers represents a very substantial addition to the existing data base on the behaviour of in situ walls.

Table 1 illustrates how these papers can be categorised in terms of the primary purpose of the observations. Four of the papers report research studies where the primary objective is to obtain information which will assist in improving understanding of behaviour. Field measurements on full scale structures represent a key component in any coordinated programme of research in geotechnics since it provides the only means of investigating the effect on behaviour of construction activities. Site Specific Monitoring is reported in six of the papers where the main purpose of the measurements is either to provide confirmation of design predictions or assumptions, or to study innovative forms of construction or construction processes. The three remaining papers are concerned with field measurements where the results are used for the more specific purposes of construction control as exemplified by the Observation Method described by Peck (1969).

Table 2 lists the main types of measurement made and the number of papers in which each type of measurement is reported. Ten of the thirteen papers report measurement of structural deformation, using inclinometers

Table 1

PRIMARY PURPOSE OF MEASUREMENT			NUMBER OF PAPERS
RESEARCH	-	IMPROVED UNDERSTANDING OF BEHAVIOUR	4
MONITORING	-	CONFIRMATION OF PREDICTIONS OR DESIGN ASSUMPTIONS	6
	-	INNOVATIVE FORMS OF CONSTRUCTION OR CONSTRUCTION PROCESS	
CONSTRUCTION CONTROL	-	INTEGRATED DESIGN AND CONSTRUCT APPROACH (PECKS OBSERVATIONAL METHOD)	3

or electrolevels to assess the deflected shape and survey techniques such as tape extensometer measurements, precise levelling, theodolite offsets and electronic distance measurements, to determine the absolute movement of the walls. Similar techniques are used to determine the magnitude and distribution of movements in the adjoining ground and results are included in seven of the thirteen papers. Six papers refer to measurements of porewater pressure using standpipes or piezometers to determine changes caused by wall construction and the same number of papers mention measurements of structural loading, either in the wall or in the support system, using load cells, strain gauges or electrolevels. Measurements of earth pressure represent the most difficult and uncertain type of measurement to undertake and results from pressure cells installed on the wall or in the adjoining ground are given in only three of the papers.

In terms of the predominant soil type, six of the papers are concerned with

Table 2

TYPE OF MEASUREMENT	NUMBER OF PAPERS
STRUCTURAL DEFORMATION	10
GROUND MOVEMENT	7
POREWATER PRESSURES	6
STRUCTURAL LOADING (WALL AND/OR SUPPORTS)	6
EARTH PRESSURES	3

embedded walls in heavily overconsolidated sedimentary clays, four in soft clay, two in predominantly sandy or clayey silt and one in Coal Measures overlain by drift.

Review of papers

The paper by Wallace, Long and Ho Describes the behaviour of diaphragm walls for a three level basement of a 66 storey tower block in Singapore. The diaphragm walls, up to 40m deep, pass through a 30 metre thick layer of soft marine clay. An interesting feature is the monitoring of a trial panel to determine the ground response and test the feasibility of the proposed construction method. The paper gives comparisons between measured wall movements and design predictions based on FREW, an interactive spring program. The measured lateral movements of the wall were only about 50-60% of the design predictions. Much closer agreement was obtained with results from back analyses in which only very small changes were made in the assumed undrained strength and stiffness profiles of the marine clay. Fig 6 of the paper gives measured wall movements against system stiffness on the plot suggested by Clough and O'Rourke (1990). The authors include data from their own and from other sites in Singapore. According to this plot the wall movements would still have been below the design prediction for a fourfold decrease in system stiffness, corresponding to a reduction in thickness of the diaphragm wall from 1.2m to 0.8m. It would be interesting to learn if the authors consider that a thinner wall would have performed adequately.

Brooks and Spence's paper is concerned with the design and measured performance of a secant pile wall for an underground car park in West Croydon England. The walls were formed from 750mm diameter bored piles, had a maximum depth of 16m and were constructed through London Clay overlain by 5m of gravel and made ground. The excavation had a maximum depth of 11½ metres and was supported by temporary props at two levels. The primary purpose of the monitoring was to enable action to be taken if the measured bending moments were excessive compared with the design capacity of the piles. The paper also gives comparisons between the measured deformation profiles and those calculated using a subgrade reaction program WALLAP and a finite difference program FLAC with equivalent input parameters. Both of these programs overpredicted the measured movements particularly over the upper part of the wall and the authors conclude from further back-analyses that this was largely due to the adoption of conservative strength parameters for the overlying gravel and made ground. This resulted in an overestimate of the movements during the initial cantilever stage and a consequent underestimate of the bending moments in the propped condition. Cracking was observed in the walls at a position which agreed well with the predicted maximum bending moment, and from a parametric study the authors show that the actual factor of safety on the ultimate moment of resistance was at least 1.4.

Any design method requires adequate calibration against field observations and this paper illustrates the dangers inherent in transferring input parameters from one numerical method to another.

The paper by Ward is also concerned with the design and measured performance of a secant pile wall for a basement at Holborn Bars in London. The walls were formed from 880mm diameter bored piles and initially two levels of temporary propping were proposed. The authors report that the adoption of a single level of propping resulted in a 25 per cent reduction in propping costs. The main aim of the monitoring was to check on the actual wall movements since the structure had a low calculated factor of safety under fully drained conditions in the London clay. The paper gives comparisons between the deformation profiles of the wall obtained from inclinometer measurements at various stages of construction, and design predictions obtained using WALLAP and an early version of FREW. These predictions overestimate the observed wall movements by a factor of at least two and the author reports that a three fold increase in horizontal modulus of subgrade reaction is required to obtain good agreement with the measurements.

The paper also draws attention to the difference in magnitude of prop forces determined from limit equilibrium calculations and from subgrade reaction programs. The author suggests that subgrade reaction programs generally predict higher prop forces than are observed in practice. In this regard the actual drainage conditions in the clay will exert a major influence on the prop loads. In addition an underestimate of soil stiffness is likely to result in overestimation of the prop loads; so that adoption of less pessimistic stiffness profiles could well narrow the gap between both the predicted and measured prop loading and the predicted and measured wall movements.

Li, Nyirenda and Pickles describe the measured performance of diaphragm walls for the International Rail terminus at Waterloo Station in London. The purpose of their measurements was to enable the actual behaviour to be compared with design predictions obtained using the finite difference program FLAC. The paper includes comparisons between the calculated displacements of the wall and measurements from inclinometers obtained during the early stages of construction. The authors point out that adoption of worst credible loading conditions in the design is likely to lead to an overestimate of the actual wall movement, and mention a number of factors which can account for the disparities between the predicted and observed behaviour. The measurements given in the paper indicate that bending moments and wall deflections were significantly smaller than predicted. It would therefore be of interest to learn if the authors consider that, in the light of the measurements, the diaphragm walls on the eastern and northern sides of the site could have been replaced by more flexible sheet pile walls as used on the other two sides.

The paper by Curtis and Mirzabaigan compares the measured movements of a contiguous bored pile basement wall with predictions obtained from the program FREW. The site was located in Barnsley in North East England and the ground conditions consisted of up to 4½m of Drift and made ground overlying Middle Coal Measures. Ground anchors and temporary berms were used to support the walls during excavation for the 10m deep basement. A particular feature was the presence of a fault which traversed the site and caused a significant difference in ground water levels on either side. Predictions of behaviour of the wall were therefore made using a high and a low water level together with two ground stiffness profiles based on an average and a lower bound to standard penetration test results. In view of the highly variable ground conditions the overall agreement obtained between measured and predicted wall displacements was surprisingly good. The authors point out that closer agreement was obtained from analyses incorporating the higher ground stiffness profile based on the average of the SPT results. The paper provides a valuable case record for future construction in similar ground conditions.

Abedi, Porter, Lien and Ramos describe the construction of the foundations for a grit chamber at a wastewater treatment plant in Detroit USA. The ground conditions comprised 22m of soft to very soft low plasticity clay and the foundations were formed by driving three rows of sheet piles and excavating between them with temporary supports. Existing sludge thickening tanks and an ash lagoon had to remain operational and instrumentation was therefore installed to monitor the ground movements during construction at three sections. Extensive finite element analyses employing a non-linear soil model were also carried out to predict the magnitude and pattern of ground movements. At the section adjacent to the sludge thickening tank the measured horizontal ground movements were small and compared well with predicted values. By contrast at the two sections within the excavation zone the observed horizontal movements were substantially greater than the design predictions. At all three sections the vertical movements were in reasonable agreement with the predictions and the authors attribute the differences in horizontal ground movement to the method of construction and the effects of trafficking by construction plant. This paper is unique to the Session in that measured movements were substantially greater than design predictions.

The paper by Saito, Aoki and Ikuta describes an innovative retaining wall system used for the foundations of a 25 storey building with a three level basement, up to 17m deep, in Tokyo. The ground conditions comprise 29 metres of very soft alluvial clay overlying gravel. An arrangements of T shaped diaphragm walls connected by flat diaphragm panels served as temporary works for the basement construction and as the permanent foundation system.

Measurements were made of earth and water pressures, stresses in the reinforcement bars and wall deflections. The paper gives comparisons between observed and design predictions of behaviour. The measured earth pressures on the back of the walls showed very little change during basement construction, which the authors attribute to the high flexural stiffness of the T shaped wall panels. Good agreement was obtained between bending moments in the wall determined using measurements of stress in the reinforcing bars and calculated from measured earth and water pressures on the back of the T shaped panels. The authors were also able to show from their measurements, that the connection between the T shaped and plane panels of the outer perimeter wall is effectively rigid.

The paper provides an excellent example of the value of field measurements in evaluating the performance of an innovative construction system for basement works in very soft ground.

Kastner and Ferrand's paper describes the measured performance of the diaphragm walls for the western terminus of the Lyons underground system in France. The walls are 10½m deep and constructed in sandy silt. The excavation between the walls was carried out to a depth of 8m with two levels of temporary support. Measurements were made of the deformation of the retaining walls, the porewater pressures in the ground and the loading in the temporary struts. Of particular interest is the observed asymmetric behaviour of the two walls, attributed by the authors to the movement of construction plant and the construction of a shallow fill on one side. Wall displacements were calculated using a program which employed a spring model for the soil. Initial input parameters derived from the site investigation considerably overestimated the wall displacement during the early stage of excavation and revised parameters were derived which gave better agreement with the observed movements. However the measured loading in the upper struts, particularly during the later stages of construction, were substantially greater than calculated.

In view of the risk of failure by hydraulic uplift of the excavation floor, it would be useful if the authors could give the results of their measurements of porewater pressure.

Symons and Carder's paper is concerned with the changes in in situ stresses in stiff clay caused by the installation of embedded retaining walls. Results of measurements in London Clay are given for three sites where a contiguous bored pile wall, a T shaped diaphragm wall and a diaphragm wall composed of planar panels were installed.

At the sites where the T shaped diaphragm wall and the bored pile wall were installed, measurements of the in situ lateral stresses and porewater pressures indicated initial earth pressure coefficients of between about 1.5

and 2.0 in the London Clay. After wall installation the measurements from push in pressure cells suggest a reduction in the average earth pressure coefficient of about 10 per cent for the bored pile and 20 per cent for the T shaped diaphragm wall. The stress reductions are clearly dependent on the detail of the Symons 7 construction technique, and the field measurements given in the paper provide useful data for comparison with numerical and physical modelling of wall installation effects, as reported in two papers to Session 1.

The paper by Lings, Nash and Ng is concerned with an assessment of the measured earth pressures in the light of measured prop loads and wall rotations made on one panel of a multi propped diaphragm wall in stiff overconsolidated clay. The wall panel is 0.6m thick and forms part of the perimeter wall for an underground car park in Cambridge. The monitoring was carried out during construction and the comprehensive instrumentation included electro levels within the wall panel, earth pressure cells and piezometers on both faces, and load cells on the reinforcing bars and on the temporary props. The authors have constructed nett earth pressure diagrams, which together with the measured prop loads satisfy horizontal force and moment equilibrium and also, after integration, show agreement with the measured wall rotations. These nett earth pressure diagrams have then been compared with the measured earth pressures at two stages during construction. In particular these comparisons suggest that the measurements from pressure cells located at shallow depth on the excavated side, were significantly underestimating the in situ horizontal total stresses. High negative excess porewater pressures might be anticipated in this region as a direct result of the excavation, and it would be interesting to learn if the authors consider that these could have affected the readings from the pressure cells.

Concluding remarks

Embedded retaining walls involve complex soil structure interaction which serves to limit the accuracy of predictions of behaviour made at the design stage. This is illustrated in a number of the papers to this Session in which the predictions of wall deformations during construction are substantially greater than indicated by field measurements. This stems from the adoption in design of worst credible loading conditions in conjunction with conservative estimates of the strength and stiffness properties of the ground. Considerable scope therefore exists for economies if advantage can be taken of the improvements in understanding of soil behaviour and the enhanced numerical capability now becoming available.

The way forward may well lie in the wider adoption of Pecks Observational Method, in which the design and construction processes are more closely integrated, with site specific field measurements providing the essential feedback loop. The role of analysis would then shift from that of ensuring a conservatively safe design to cover all eventualities, to that of establishing the range of possible behaviour, and the required contingency action in the event that conditions proved less favourable than anticipated.

References

1. CLOUGH W.G. and O'ROURKE T.D. Construction induced movements of in situ walls. Proc. Conf. Design and Performance of Earth Retaining Structures. Geotechnical Special Publication No. 25, 1990. Am. Soc. Civ. Engnrs. New York 439-470.
2. PECK R.B. Advantages and limitations of the Observational Method in applied soil mechanics. Ninth Rankine Lecture, Geotechnique, Vol. 19, No. 2, 1969, 171-187.

Retaining wall behaviour for a deep basement in Singapore marine clay

J.C. WALLACE, C.E. HO, Arup Geotechnics, Singapore, and
M. M. LONG, Arup Geotechnics, UK

The new headquarters of the United Overseas Bank in Singapore, UOB Plaza, currently under construction, includes a 3 level basement. The site is underlain by up to 30m of soft marine clay over hard bouldery clay. The sides of the excavation are supported by diaphragm walls up to 1.2m thick and the basement was constructed using three levels of propping. This Paper describes the design and construction of the retaining walls. A comparison is made between wall and ground movements predicted using the computer program FREW and those actually measured using inclinometers and conventional surveying. The results of an instrumented trial panel excavation are described.

Introduction

The site is located between Chulia Street and Boat Quay in the Central Business District adjacent to the Singapore River. The project comprises a 66 storey tower block, podium and basement car park next to the existing 30 storey UOB building (see Fig. 1). The foundation design and construction has been described elsewhere (refs 1 & 2).

The basement excavation was generally 10m to 12m deep and up to a maximum of 16m beneath the tower core. The sides are supported by reinforced concrete diaphragm walls, of total plan length 320m, which are 0.8m to 1.2m thick and up to 40m deep.

Ground conditions

The soil conditions were investigated in three stages in 1981/B2, 1983 and 1987/88. The second stage included boreholes at approximately 10m intervals along the line of the perimeter of the diaphragm wall. In-situ vane and piezocone tests were carried out to establish the shear strength of the soft materials. Ground levels at the site varied from approximately +101.5m in the west to +102.5m in the east. The water level was located close to the ground surface. The investigations revealed the site to be underlain by a variable succession of strata as illustrated in Fig. 2.

Table 1 . Soil Design Parameters

Material	SPT 'N'	γ	K_o	c_u	E
	blows/0.3m	kN/m³	-	kN/m²	kN/m²
Made Ground (MG)	5	18	0.53	-	10,000 (13,500)
Peaty Clay (PC)	-	17	0.7 (0.6)	22 (22)	450c_u
Upper Marine Clay (UMC)	1	16	0.7 (0.6)	22 (25)	450c_u
Desiccated Clay (DC)	6	18	0.7 (1.2)	65 (70)	450c_u
Lower Marine Clay (LMC)	4	17	0.7 (0.6)	50 (55)	450c_u
U/I/L Sand (S)	5	18	0.53	-	12,500 (25,000)
Residual Soil (RS)	<100	22	1.0	375	450c_u
Bouldery Clay (BC)	>100	22	1.5	700	225,000

Marine clay

The behaviour of the retaining walls is largely governed by the properties of the marine clays. Data on these materials from the site investigations are shown on Fig. 3. Generally, the properties of Singapore marine clays are well

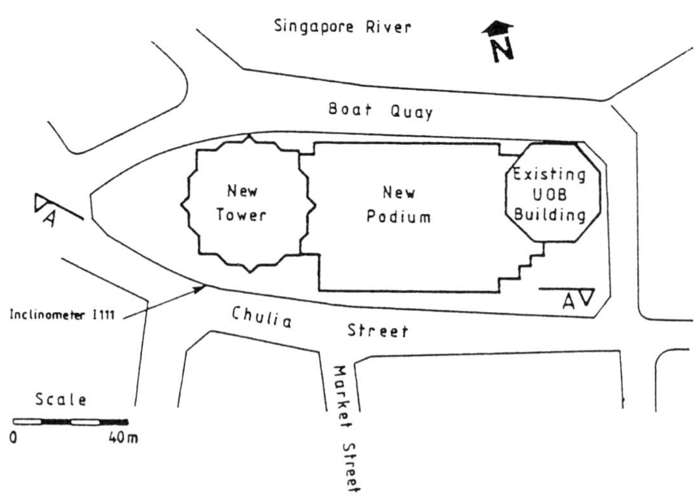

Fig. 1. Site location

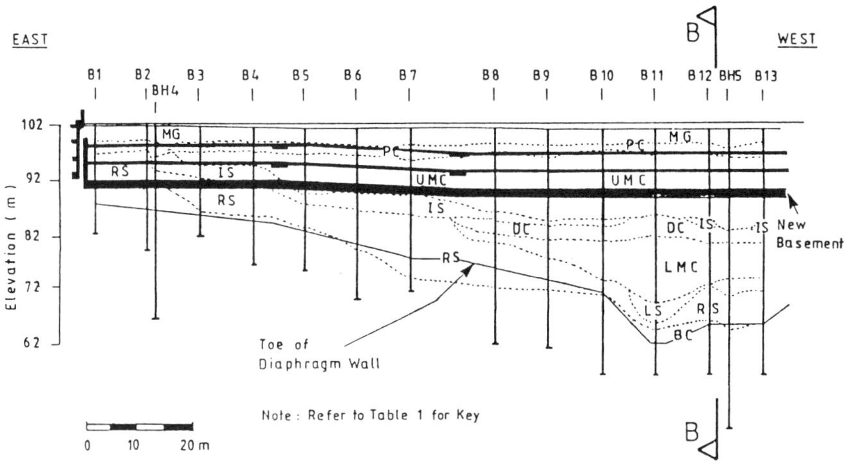

Fig 2. Geological cross section A-A (Section B-B is at location of Inclinometer I111)

documented elsewhere (refs 3 & 4). The soil parameters used in the analysis are given below. The values in brackets are those used subsequently in the back analysis.

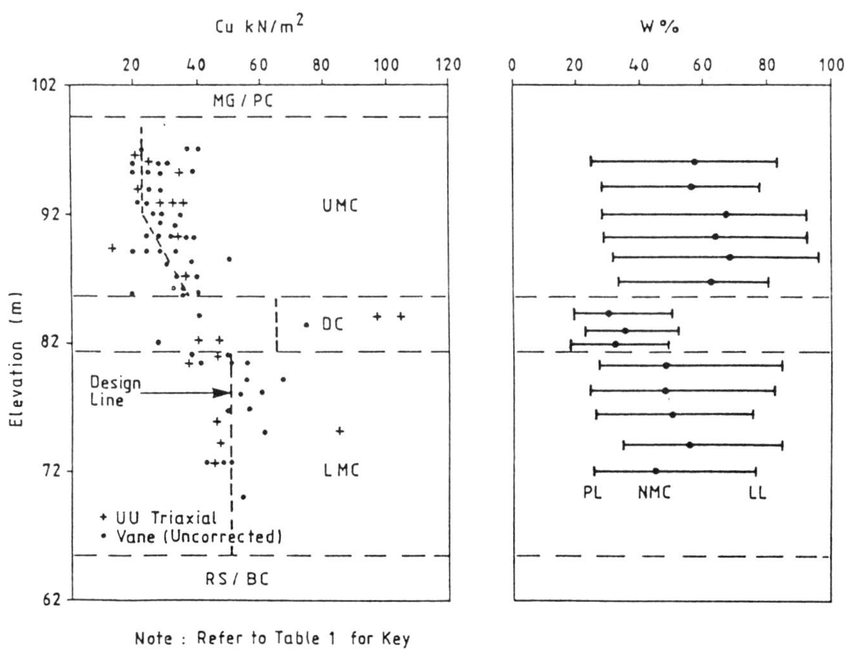

Fig. 3. Properties of Marine Clays from site investigation

Retaining wall scheme

Diaphragm walls were adopted. Sheet piles were rejected because of the large ground movements associated with their use. Contiguous bored pile walls were considered insufficiently watertight. A secant pile wall was not considered in detail because there were no contractors in Singapore with experience of this technique. There were several contractors available who had gained considerable experience of diaphragm walling during construction of the Singapore Mass Rapid Transit (MRT) system.

In order to enhance the factor of safety against bearing capacity failure of the base of the excavation and thereby minimise ground movements and to provide a deep groundwater cutoff, it was decided to found the diaphragm wall either in the Residual Soil or Bouldery Clay.

Method of analysis

The retaining wall was divided into nine different sections for analysis. This was because of the variability of the ground conditions and the geometry of the structure. Thus, it was necessary to adopt a fast and flexible method of analysis. The problem was too complicated for hand calculations. Finite element analysis was deemed too time consuming and cumbersome. Hence, it was decided to use the Oasys computer program FREW.

FREW is a quasi-finite element program. In FREW, the soil is represented as an elastic solid with the soil stiffness matrices being developed from prestored matrices calculated using the Oasys finite element program SAFE. FREW analyses each stage of the basement excavation and construction sequence. At each stage it calculates the lateral force at each node of the wall imposed by that stage and calculates wall displacement and soil stresses using the stiffness matrices. If the soil stresses are outside the active or passive limiting pressures, correction forces are applied and the problem is solved iteratively. Allowance is made for arching within the soil body. Allowance is also made for friction on the wall, by applying a vertical force at each node up to a limit specified by the user. This force is proportional to the relative vertical displacement between the soil and the wall. The program estimates relative vertical displacement using simplified method of calculation. Some details of the assumptions and analysis method are given elsewhere (ref 5).

The analysis was carried out in 5 stages as follows. Undrained soil parameters were used throughout. Hydrostatic water pressures from 1m below a ground level of +102m were assumed on the active side. On the passive side it was assumed that the water pressure varied from zero 0.5 m below excavation level to the hydrostatic value at the toe of the wall.

Stage 0: Initialisation, Soil at K_o pressures
Stage 1: Apply surcharge of $5kN/m^2$ outside, Dig to +100.5m

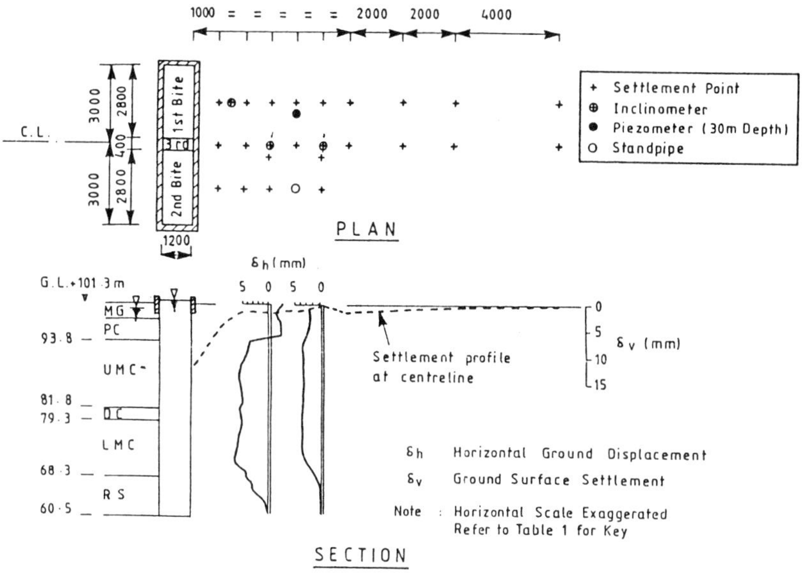

Fig. 4. Trial panel excavation

Stage 2 : Insert temporary prop T1 at +l0lm, Dig to + 95.8m
Stage 3 : Insert T2 at +96.2m, Dig to + 92.2m
Stage 4 : Insert T3 at +92.9m, Dig to + 89m.

The bending moments and shear forces in the wall predicted by FREW were multiplied by a factor of safety and then applied in the structural design. The forces in the props were also determined from the FREW analysis.

Construction

The diaphragm wall was installed with raised guidewalls using conventional techniques. For the basement, a bottom up construction method was used with three levels of temporary steel propping. The props were preloaded to 8% to 28% of their design load. There was very little seepage into the excavation. Water, mainly from rainfall, was expelled by pumping from sumps. After final formation level was reached, a sand blanket and a "no fines" concrete drainage layer were placed for permanent dewatering. The basement slabs were then installed progressing upwards towards ground level.

Piezometers installed in the Intermediate and Lower Sands surrounding the site registered an insignificant drop in water level indicating that the cut-off was effective. Seepage through joints in the wall was minimal.

Trial panel

Prior to construction of the contract diaphragm walls, a trial panel was excavated to determine the ground response. The panel was 6m long, 1.2m wide and 40m deep. The grab used was 2.8m x 1.2m and the trench was excavated in 3 bites, see Figure 4. After excavation the trench was left open for 17 hours with the bentonite slurry kept 1.5m above ground water level. The slurry level was then lowered by O.5m and the trench left to stand for another 5 hours during desanding prior to backfilling with mass concrete.

The layout of the instrumentation and the ground response to the excavation are shown on Fig. 4. The maximum lateral movement recorded was about 6.5mm, on the panel centreline 3m from the edge, reducing to 4mm after concreting. The total average settlement of the guidewall was about 11.5mm though this was probably largely due to impact from the grab. The vertical movement dropped to less than 1.5mm, 2m from the edge of the excavation. The trial panel excavation confirmed that the proposed construction method was feasible.

Monitoring

Monitoring during basement excavation was by standpipes and piezometers, inclinometers and ground settlement stations. The movements measured for a typical section of the diaphragm wall, at inclinometer I111, are discussed below. The inclinometer was installed to the toe of the diaphragm wall which was embedded about 7m into residual soil and hard bouldery clay. Thus it is reasonable to assume that movement at the bottom of the inclinometer was zero.

Discussion of results

Wall movements

The measured movements for a typical diaphragm wall section are shown on Fig. 5. The relatively low level of movement is believed to be primarily due to the high stiffness of the retaining wall system. This is demonstrated by plotting movement versus system stiffness as suggested by Clough & Rourke (ref 6), see Fig. 6. The plot contains information for various sites in Singapore and the trend broadly follows the parabolic relationship suggested by Clough & Rourke. The movements were also low because the construction sequence was clearly defined and closely supervised. This included careful control of the excavation, insertion of the props promptly and applying a preload to the props.

Back analysis

Figure 5 shows that, although FREW predicted the pattern of movement

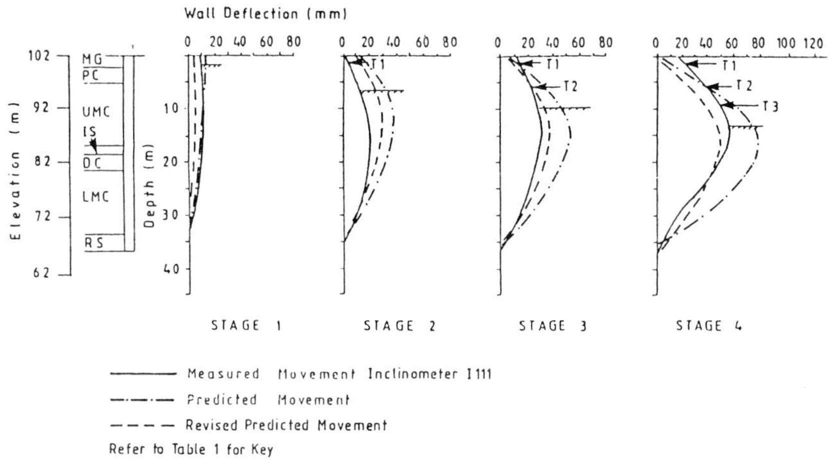

Fig. 5. Measured and predicted wall movements

accurately, it over predicted the magnitude of the movement. A parametric analysis revealed that predicted wall movement is most sensitive to choice of soil strength and stiffness values. This has also been found by others (refs 7 & 8).

The soil strength values assumed in the analysis were conservative values, see Fig. 3 and Table 1. If a more optimistic shear strength profile is assumed as detailed in Table 1, a better fit of the predicted and actual movements is obtained, see Fig. 5. There remain two unexplained differences between the

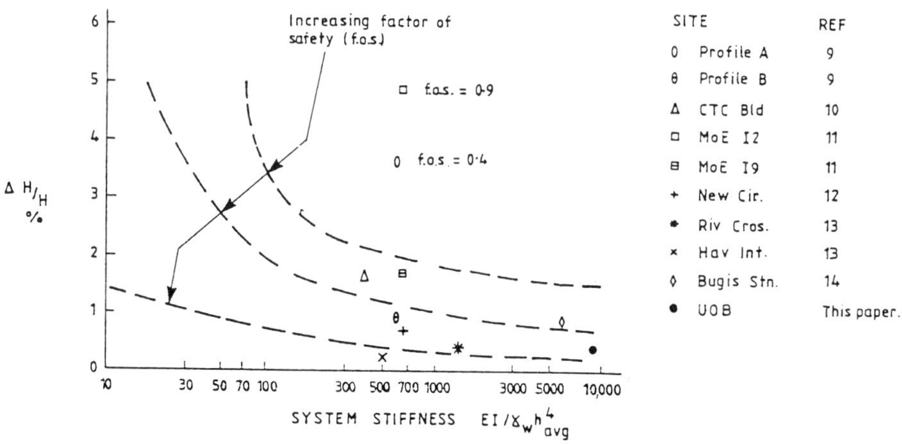

Fig. 6. Wall movement versus system stiffness, after Clough and Rourke (ref 6). ($\Delta H/H$ = max. measured movement over depth of excavation, EI = wall stiffness, h = average prop spacing)

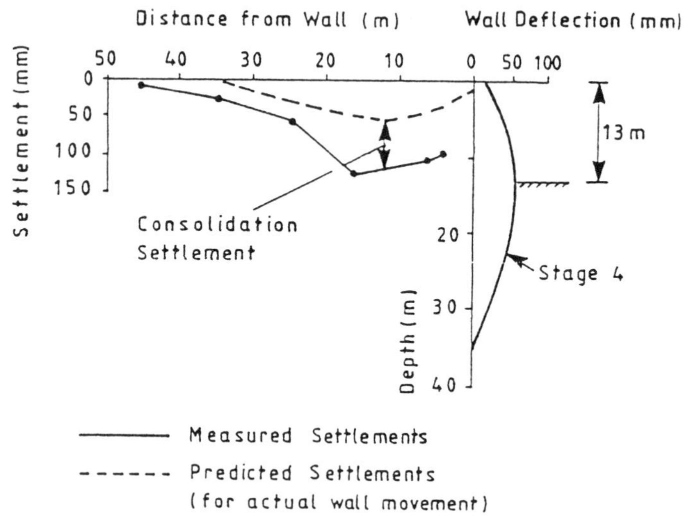

Fig. 7. Ground movements

actual and predicted movements. At Stage 2 the inclinometer data showed the wall to have been pushed away from the excavation by some 6mm to 8mm. This might have been due to the preloading of prop T1. However it was not predicted by the computer analysis. Between Stages 3 & 4 the inclinometer profile showed a 'body' movement of the wall by some 20mm. FREW shows no movement at the prop levels once the props are in place.

Ground movements

The settlements behind the retaining wall at the location of the inclinometer discussed above are shown on Fig. 7. The figure shows that the movements measured are greater than those which would have been predicted using a simple velocity field and assuming undrained behaviour. Clough & Rourke (ref 6) collated measured ground movements for a large number of deep excavations. They showed that, for soft to medium clays, the movement measured beyond two to two and a half times the maximum excavation depth was negligible. It is therefore considered that the considered that the measured movements reflecting settlement additional to that basement excavation. At an early stage of the project the road level adjacent to this part of the site was raised by about 0.5m.

Consolidation resulting from this surcharge is likely to have caused the additional settlement.

Conclusions

The problem presented shows that, with a good method of analysis and a good analytical model, reasonably accurate predictions can be made of retaining wall behaviour in deep deposits of soft clay. The wall movements and corresponding bending moments were over estimated because the designers judgement was to use soil parameters which were toward the conservative end of the estimated range. The predictions were shown to be improved considerably by choosing more optimistic parameters.

With the benefit of more documented case studies, designers confidence in their models and chosen parameters will increase. In this way the design process can be progressively refined.

Acknowledgements

The Authors are grateful for the assistance of their colleagues in Arup Geotechnics and in particular to Brian Simpson.

References

WALLACE, J. C., HO, C. E. and BROMS B. B. UOB Plaza Singapore Foundation Design and Construction. *Proc. Conf. Deep Foundation Practice, Singapore*, CI Premier, 1990, pp 257 -273.

WALLACE, J. C. *et al.* Large Diameter Caissons for Sixty-six Storey Building in Singapore. *Proc Int. Conf. Deep Foundations, Paris* , 1991.

TAN, S.B. and LEE, K.W. Engineering Geology of the Marine Member of the Kallang Formation in Singapore. *Proc. Int. Symp. on Soft Clays, Bangkok*, 1977, pp 75 - 88.

TAN, S. L. Geotechnical properties and laboratory testing of soft soils in Singapore. Int. Sem. on construction problems in soft soils in Singapore, 1983.

PAPPIN, J. W. *et al.* Numerical Analysis of Flexible Retaining Walls. Sym on Computer Applications in geotechnical Engineering. *The Midland Geotechnical Society*, April 1986.

CLOUGH, G. W. and ROURKE, T. D. Construction Induced Movements of In-Situ Walls. *Proc. ASCE Conf. on Design and Performance of Earth Retaining Structures, Cornell*, ASCE Sp Pub No. 25, pp 439-470, 1990.

WONG, K.S. A Method to Estimate Wall Deflection of Braced Excavations in Clay. *Proc. 5th Int. Geo. Seminar on Case Histories in Soft Clay, NTI, Singapore*, 1987.

MANA, A. I. and CLOUGH, G. W. Prediction of Movements for Braced Cuts in Clay. *Proc. J. Geo. Div. ASCE, 1981*, **107**, No. GT6, June, pp 759 - 777.

DAVIES, R. V. and WALSH, N. M. Excavations in Singapore Marine Clays. *Proc. Int. Sem. on Const. Probs. in Soft Soils, Singapore*, 1983.

LEE., S. L. *et al.* Developments in Soft Ground Engineering in Singapore. *Proc. 111th ICSMFE, San Francisco,* 1985, pp 1661-1666.

TAN, S. B., TAN, S. L., and CHIN, Y. K. A Braced Sheet Pile Excavation in Soft Singapore Marine Clay. *Proc. 11th ICSMFE, San Francisco,* 1985, pp 1671 - 1674.

NICHOLSON, D. P. The Design and Performance of the Retaining Walls at Newton Station. *Proc. Conf. Singapore MRTC,* 1987.

GONIN. H., DEPAUW, J. and N'GUYEN, M. Singapore's Expressway: Temporary Retaining Structures. *Proc. 10th ECSMFE, Florence,* 1991, pp 805 - 808.

HULME, T. W., POTTER, J and SHIRLAW, N. Singapore MRT System: Construction. *Proc. Instn Civ. Engrs, Pt. 1,* April, 1989.

Design and recorded performance of a secant retaining wall in Croydon

N. J. BROOKS and J. SPENCE, Stent Foundations, UK

Introduction

This Paper describes the design and recorded performance of a secant piled basement retaining wall in West Croydon. Two commercially available computer programs, WALLAP 3.12 and FLAC 2.25 were used at the design stage to calculate pile deflection and bending moment profiles.

Despite having successfully used WALLAP, a subgrade reaction program, for over five years the decision had been made to obtain a continuum analysis program in order that ground movements could be estimated. Before being able to use FLAC to calculate ground movements in a situation where such movements were critical, it was decided to take the opportunity to test the combination of program and parameters at a site where movement was not a problem.

The selection of soil strength parameters used in the design were based on experience. The spring stiffnesses used by WALLAP to represent the complex stress-strain behaviour of the continuum were those that had, for similar materials, given good predictions of wall movement in the past. Without the benefit of experience in the use of FLAC it was decided to use shear and bulk moduli directly related to the spring stiffnesses used with WALLAP, although it was recognised that this would underestimate soil stiffness. The calculated deflection profiles were significantly different in shape.

Field measurements were carried out to investigate the general accuracy of the deflection predictions, and to check that the induced bending moments were not excessive. The recorded deflections were similar to those calculated using WALLAP. It is concluded that the spring stiffnesses used in WALLAP were reasonable, but that their corresponding shear and bulk moduli were not suitable for use with FLAC.

A re-design was carried out and the amendments to the input data that produced a calculated profile similar to the recorded profiles are described.

The design and re-design assume a Ko value of 1.5 for the London Clay and Woolwich and Reading Beds. This value is considered to be a reasonable approximation, although it is appreciated that Ko will vary with depth and that Ko is a significant factor in the performance of retaining walls.

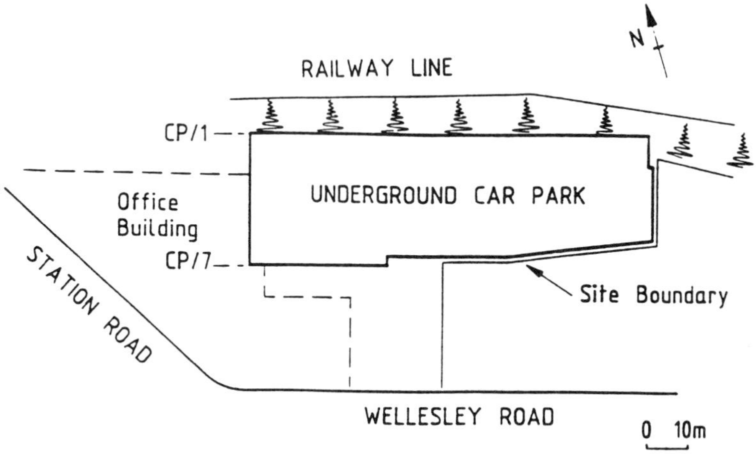

Fig.1. Site plan

Site conditions

A retaining wall was required to form the perimeter wall for a 4/5 floor underground car park, part of the Prospect West Office Development in West Croydon. The basement is surrounded on three sides by level ground with a 4 m deep railway cutting immediately to the north (Fig. 1).

The site investigation revealed Made Ground comprising grey brown clayey or silty sand of maximum thickness 2.8 m, overlying sandy gravel to clayey sand and gravel. Beneath the gravel, unweathered London Clay becoming very silty with depth was encountered overlying the Woolwich

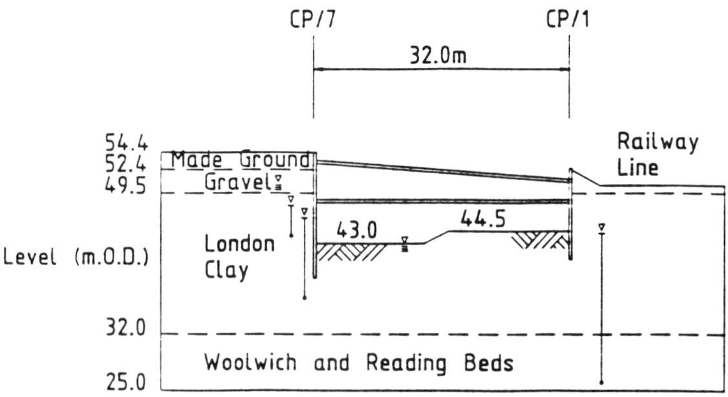

Fig. 2. Section across excavation

and Reading Beds at approximately 22 m depth. Fig. 2 is a cross section through the excavation with the soil strata and the estimated groundwater profile.

The recorded undrained shear strength profile of the London Clay increased from 100kN/m^2 at 5 m below ground level (bgl) at a rate of 6 kN/m^2/m to about 15. 5 m bgl, from where it increased more rapidly in strength with depth. The shear strengths were obtained from quick undrained triaxial tests on 38 mm diameter specimens.

Design

WALLAP is a subgrade reaction program, supplied by Geosolve, in which the wall and soil are modelled as a beam with springs, a check is made on active and passive pressures to ensure that the limiting values are not exceeded.

FLAC is an explicit finite difference code, supplied by Itasca, in which the soil is represented by zones or elements which form a grid that is adjusted to fit the shape of the problem being modelled. Each element follows a prescribed stress/strain law in response to the applied forces and boundary restraints. If stresses are high enough to cause the material to yield and flow, the grid deforms and moves the material it represents.

At the time of tender, WALLAP was used to determine the depth of wall and reinforcement requirements. On award of the contract, further analysis was carried out using both WALLAP and FLAC.

Figure 2 shows the difference in depth of excavation and ground levels across the site. WALLAP can only model a single wall and, therefore, it was necessary to calculate the force transferred to struts on the deeper wall and input these forces in the analysis of the shallower wall. FLAC is able to model the complete foundation cross-section.

The parameters assigned to each of the soils at the design stage are given in Tables 1a and 1b. The selection of the values used by WALLAP was based on experience gained from producing acceptable designs which had performed as expected. The spring stiffness used for the London Clay was approximately 350 times the measured undrained shear strength (Su) and 500 times the utilised Su. The same shear strength parameters were used with FLAC. The bulk and shear moduli given in Table 1b were obtained by taking the WALLAP spring stiffnesses as being equivalent to an elastic modulus. These values are significantly less than some results calculated from the back analysis of the performance of depth excavations (e.g. St John (Ref 1)). Wall friction was allowed for in each soil type. A value of 5e+5 kN/m^2/m was used for wall stiffness.

Table 1. Design parameters: (a) shear strength, (b) deformation

Soil Type	Undrained :SU (kN/m)	Drained: ϕ'	c' (kN/m2)
Made Ground		30	0
Gravel		35	0
London Clay Short Term	60 + 5.6/m depth		
Intermediate term retained side		20	10

Table 1a

Soil Type	Ko	WALLAP Subgrade Reaction	FLAC Moduli Bulk (kN/m2)	Shear (kN/m2)
Made Ground	0.45	2000 kN/m^3	1350 to 5360	800 to 3200
Gravel	0.43	5000 kN/m^3	20000 to 34000	2000 to 20000
London Clay Undrained	1.5	35000 kN/m^2 + 3267 $kN/m^2/m$	1e6 + 6.5e4/m	10000 + 900/m
Drained		25000 kN/m^2 + 2500 $kN/m^2/m$	14000 + 1350/m	10000 + 900/m

Table 1b

Figures 3(a) and 3(b) show the calculated deflection profiles for the retaining walls on Gridlines CP/1 and CP/7, based on undrained parameters for the London Clay up to completion of excavation. It will be noted that the deflected shape calculated by FLAC shows considerably greater movement of the base of the wall towards the excavation for both sections. This predicted movement reflects on the bending moment calculated by each analysis.

Figures 4(a) and 4(b) present the bending moment profiles for the same gridline positions. The lower maximum bending moments result in a significant saving in reinforcement in the secant wall; reinforcement suitable for the bending moment predicted by FLAC was employed.

Monitoring of the wall was undertaken firstly so that action could be taken should it become apparent that the actual bending moments, inferred from the deflection profiles, were excessive compared to the design capacity of the pile section, and secondly to determine which calculated deflection profile was more accurate, FLAC or WALLAP.

Method of construction

The secant wall was constructed using 750 mm outside diameter oscillator casing with pile spacing of 650 mm centre to centre. The characteristic

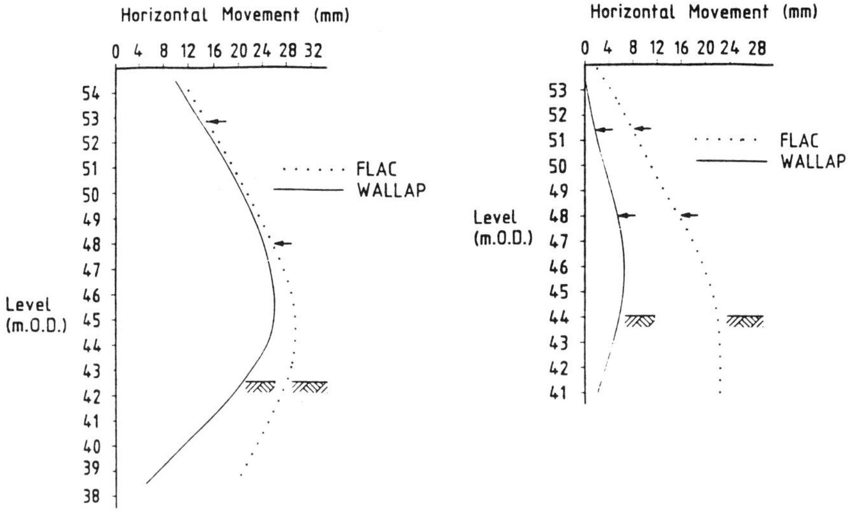

Fig. 3: (a) Calculated deflection profile, gridline CP/7; (b) Calculated deflection profile, gridline CP/1

strength of the concrete was 35 N/mm^2 at 28 days for the reinforced secondary piles and 25 N/mm^2 at 56 days for the unreinforced primary piles. Load bearing 1.2 m and 1.5 m diameter bored piles were constructed to support the substructure and to resist potential hydrostatic uplift acting on the underside of the basement slab.

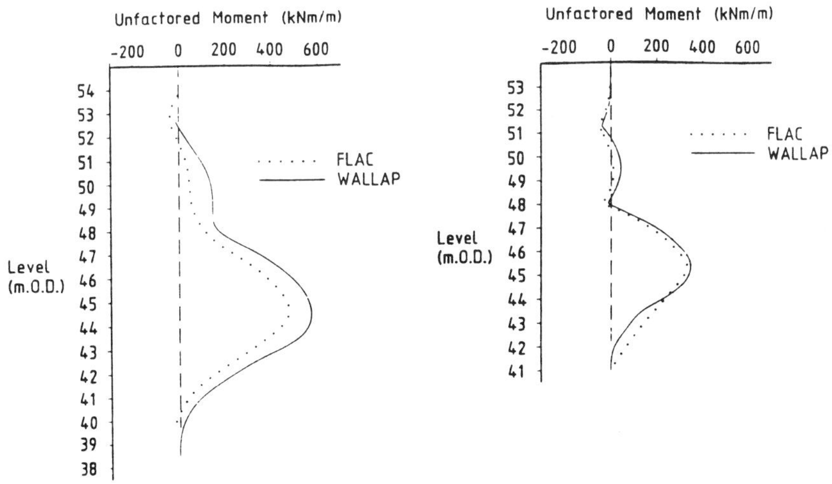

Fig. 4: (a) Calculated moment profile, gridline CP/7; (b) Calculated moment profile gridline CP/1

The excavation was carefully controlled and levels recorded. In general, the first and second levels of propping were installed after excavation to 50.5 m OD and 47 m OD respectively.

Instrumentation

Instrumentation was used to record wall movement and prop forces. Wall movement was measured in two ways. The top of the wall was monitored by precise survey techniques using a Geodometer 460 which gave a repeatable accuracy of 1 mm. Datums were established on the north west and south west corners of the wall. These, although on the wall itself, were considered to be the most stable position on the site, the site being surrounded by private land which was inaccessible and negated establishing datums outside the zone of influence. Cross checking of these stations showed them to be particularly stable.

Wall deflection was measured using an inclinometer. Geotechnical Instrument's biaxial torpedo was able to measure deflected shape to a reported accuracy of ±l mm. This correlated well with the precise survey when the excavation depths were shallow (2 to 3 m) and absolute wall movements only occurred at the top. Nine inclinometer tubes were installed, of which three were damaged prior to completion of the excavation.

Prop forces were to be estimated from strain measurements on the struts. Demec gauges were used for this purpose as they were considered to be reliable and cost effective. Anomalies have occurred in these readings giving rise to further investigation and the installation of SINCO vibrating wire strain gauges. Unfortunately the readings continued to be unreliable.

Recorded data

Precise surveying

The surveying found that the top of the wall moved into excavation by I to 3 mm during the free cantilever stage. After installation of the top prop and excavation to 47 m OD, top of wall movement into the excavation was approximately 5 mm on Gridline CP/7 and 2 mm on gridline CP/1. Following excavation to full depth, the position of the top of the wall on gridline CP/7 remained relatively constant, 5 mm into the excavation. On gridline CP/1 the top of wall tended to move out of the excavation; this may have reflected the effect of temperature variations on the steel frame. The outward movement reached a maximum of 5mm during August 1991 with the top of the wall returning to its original position in September.

Inclinometer results

Displacement increased at the point of maximum flexure by approxi-

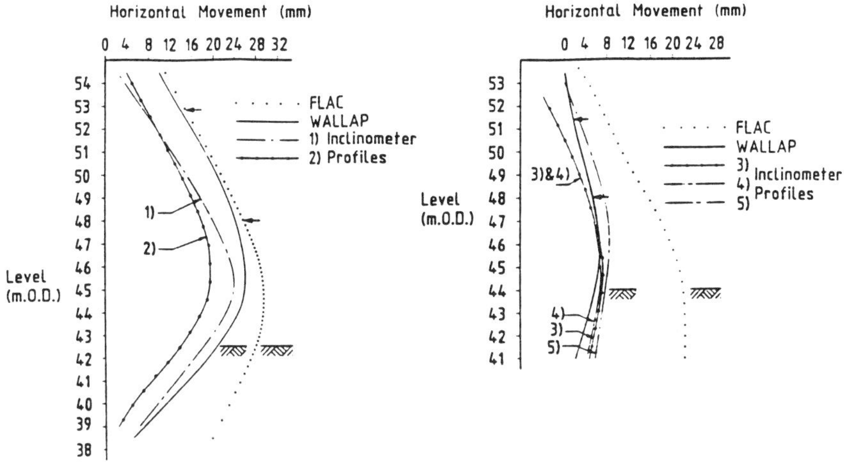

Fig. 5: (a) Inclinometer profile, gridline CP/7; (b) Inclinometer profile, gridline CP/1

mately 5 mm over a period of 4 months when the excavation was open to its full depth.

Figures 5(a) and 5(b) show the inclinometer profiles for the completed excavation immediately prior to the construction of the basement slab. The deflection at the top of the wall is that obtained by surveying.

It was expected that the inclinometers would show the toe of the two walls moving into the excavation more than was calculated by WALLAP; Figs 5(a) and (b) show that the toe of the wall on gridline CP/7 moved less than had been calculated using WALLAP, whilst on gridline CP/1 toe movement was 2 to 4mm greater than calculated by WALLAP. The profiles calculated by WALLAP were much closer to the recorded profiles than the profiles calculated by FLAC.

Re-analysis of retaining wall on gridline CP/7

With the large number of variables involved, there will be more than one combination that will produce the desired profile. It should also be recognised that some of the variables are not real soil parameters.

Only two changes were required to the input data for WALLAP in order to obtain a calculated deflection profile in close agreement with the two inclinometers. To reduce significantly wall head movement cohesion was allowed in the materials overlying the London Clay; $c' = 10 kN/m^2$ in the Made Ground and $c' = 5 kN/m^2$ in the gravel. The inclinometers and surveying had recorded movement of only 1 to 3 mm at the top of the wall during the free cantilever stage. To reduce movement at 48 m OD the compression of the prop was reduced by a third. The resulting profile is shown on Fig. 6.

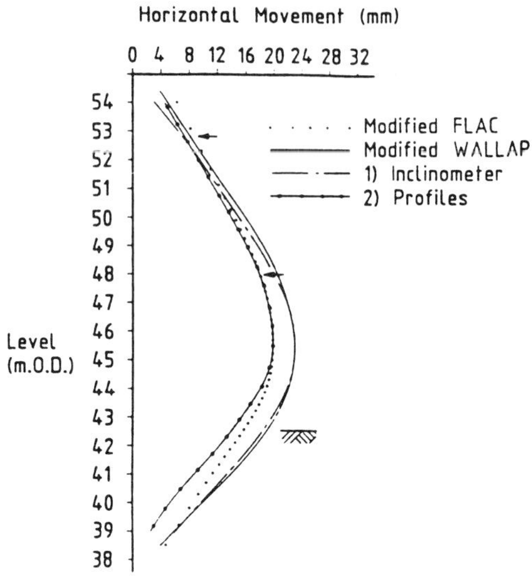

Fig. 6. Deflection profiles from revised analysis, gridline CP/7

The amended FLAC profile shown on Fig.6 was calculated after making the following changes to the input data

- The addition of cohesion for the materials overlying the London Clay
- Increasing the London Clay and Woolwich Beds shear modulus below 39.75 m OD to reflect not only an increase in shear strength, but also to take account of the small movements recorded at the toe of the wall and the greater stiffnesses associated with small strain, behaviour described by Jardine *et al.* (Ref 2).

Table 2. Shear moduli profiles for final stage of analysis, gridline CP/7Flac revised analysis

Shear Modulus (kN/m^2)	**Level** (mOD)
8400	49.5
17800	42.7 (dredge)
18700	42
20000	41
50000	39.75
70000	38.5 (toe)

The shear modulus depth relationship described for the final stage of excavation is given in Table 2. The bulk modulus was calculated using a Poisson's ratio of 0.4.

A 50% reduction in Su for the top 1 m of clay within the excavation prior to the installation of the second prop and after completion of the excavation.

Drained strength parameters for the London Clay on the retained side are $\phi' = 25°$, $c' = 0$.

The first two changes gave a deflected profile similar to that recorded immediately after excavation to full depth. The addition of the latter changes modified the calculated deflection profile so that it corresponded well with inclinometer profiles recorded approximately 4 months later.

Bending moments

One function of the inclinometers was to check that the wall reinforcement provided was adequate for the induced bending moments. Table 3 gives the moments calculated by running the analysis with different values of secant wall elastic modulus. The corresponding factor of safety on the ultimate bending moment capacity and the calculated toe movement are also presented.

The moments calculated by WALLAP and FLAC cannot be directly compared because, as Fig.6 indicates, the amended profile from FLAC is less curved than the inclinometer and WALLAP profiles. Table 3 confirms that the calculated moment is sensitive to the selected value of elastic modulus.

A single crack was observed in most of the piles on gridline CP/7 at approximately 45 m OD, i.e. at the point of maximum moment. The tensile capacity of the concrete had been exceeded and the reinforcement was providing the resistance to tension. A reduction in elastic modulus would

Table 3. Moments calculated for recorded profile CP/7

E Of Wall Section (kN/mm^2)	Calculated Max Moment (kNm/m)		Partial F.O.S On Ult Moment (852kNm/m)		Calculated Toe Movement (mm)	
	WALLAP	FLAC			WALLAP	FLAC
20	530	444	1.6	1.91	2	3.9
25	608	496	1.4	1.71	4	4.6
30	668		1.27		6	

have occurred at the time of cracking, a situation observed elsewhere by Nicholson (Ref 3).

It is concluded that the factor of safety on the actual bending moment is at least 1.4.

The initial design ignored apparent cohesion in the made ground and terrace gravels. It was thought that this would lead to a conservative design. However, ignoring cohesion in these materials leads to the calculation of greater movement during the free cantilever stage and a reduced maximum bending moment during subsequent propped conditions. Therefore, with respect to the determination of the maximum moment, in the case of propped walls where there is an initial free cantilever condition, the selection of lower bound shear strengths for the near surface materials does not lead to the calculation of an upper bound moment.

Conclusions

The monitored deflection profiles were in good agreement with the profiles calculated using WALLAP. It had been considered that WALLAP, by only allowing horizontal soil movement would under-estimate movement of the soil on the passive side, and hence be less reliable than FLAC in predicting wall movement and bending moments.

FLAC calculated a more accurate wall profile when the shear modulus was increased near the toe of the wall, and softening of the clay allowed. The increase in stiffness would reflect not only an increase in undrained shear strength but would also take account of the initial small strain behaviour.

An increase in modulus due to small strains was not allowed for with WALLAP. This omission may compensate for the assumption of one-dimensional soil movement.

The moduli used in the initial design were based on experience gained using WALLAP and appear to be appropriate for use with WALLAP. The results from this site indicate that FLAC requires higher moduli and the use of a non-linear stress-strain relationship. The changes to input data required for FLAC to produce the observed deflection profile are all reasonable which indicates that, with suitable data FLAC will give good quality predictions. With the recently updated version, FLAC 3.0, a non-linear stress-strain relationship can be prescribed using the built-in programming language.

The above comments are based on the results obtained from one site and will not be appropriate to all soil conditions, nor necessarily for free cantilever or propped cantilever walls. In particular the initial at rest earth pressures were a significant factor, a detailed consideration of which has not been included in this Paper.

The instrumentation was successful in monitoring wall movement and ensuring that the wall reinforcement was not overstressed. It has also provided information which should help in the design of deep basements in similar ground conditions.

The results have shown that, when using apparent rather than intrinsic soil parameters, the apparent parameters cannot be transferred from one method of analysis to another. Any new method needs to be tested against field observations or else with results from a computer model that utilises only intrinsic parameters, such as Imperial College's finite element program IPFEC.

Acknowledgments

The Authors wish to thank Speyhawk for their permission to publish this paper, and to thank the following companies for their co-operation, Waterman Partnership, Fairclough Building, Geosolve and Itasca. The authors are also grateful to the following individuals for their personal assistance, John Findlay of Stent Foundations and David Livings of Fairclough Building.

References

1. ST JOHN H.D. *Field and theoretical studies of the behaviour of ground around deep excavations in London Clay.* Phd thesis, Cambridge University, 1975.
2. JARDINE R.J., SYMES M.J. and BURLAND J.B. The measurement of soil stiffness in the triaxial apparatus. *Geotechnique* ,1984, **34**, No. 3, pp 323-340.
3. NICHOLSON D.P. Discussion session to use of sheet pile retaining walls for deep excavations in stiff clay. *Proc. Instn Civ. Engrs* (in print), 1991.

The design and performance during construction of the propped secant pile wall at Holborn Bars London

K. WARD, Sir Robert McAlpine & Sons Ltd, UK

This Paper reviews the design and performance during construction of the propped secant pile wall at Holborn Bars London. Engineering solutions to groundworks problems, particularly where these affect the permanent works, need to be acceptable to all the parties concerned. The design parameters for these temporary works were agreed between the Structural Engineer and the Management Contractor after both had made independent assessments of the requirements. A comparison of these is given, along with details of the propping system used and the excavation sequence adopted. Wall movements were monitored and are compared with those predicted. A back analysis has been carried out to check the validity of the initial assumptions.

Introduction

The development of a new head office complex for the Prudential Assurance Company Ltd required an 11 metre deep basement excavation to be formed through gravel into London Clay. The plan area measures 90 metres by 40 metres at its widest and is close to the architecturally significant Waterhouse buildings retained on part of the site.

In May 1988 Sir Robert McAlpine and Sons Ltd was appointed as Management Contractor 10 months before construction was due to begin. During this preconstruction period the permanent works design was finalized by the Structural Engineer Ove Arup and Partners, with McAlpine advising on buildability and designing the temporary works.

Ground conditions

From the site investigation the following strata were identified below the existing ground level of 20 metres A.O.D. (refer also to Fig. 3).

- Made Ground — up to 3.5 metres thick mainly granular fills
- River Terrace Deposits — 2 -5 metres thick medium dense sandy gravel
- London Clay — 16 - 20m thick stiff grey clay
- Woolwich and Reading Beds — 13-15m thick hard clays and very dense fine sand layers.

Groundwater level was at 14.1 metres A.O.D. indicating a perched water table in the gravel above the London Clay.

Basement design

The basement perimeter retaining wall is a secant wall formed from bored piles typically 880mm diameter at 755mm centres. The basement foundation is a 1.2 metre thick reinforced concrete raft slab. Below this a 200mm thick drainage layer of no fines concrete overlies the London Clay at formation level of 8.85 metres A.O.D. The top of the retaining wall is restrained by a concrete floor slab at ground level.

Assessment of the temporary propping requirements

The construction sequence proposed by the Structural Engineer at Tender stage indicated the conventional practice of excavation for a central portion of basement raft whilst berms were left around the perimeter to provide support to the secant wall. A series of raking props was then to be installed from the new slab, balancing the forces across the site. The berms would be reduced and a second level of props provided (except for the lower south wall). The permanent raft slab and ground floor slab would be cast and the props removed. Using this sequence, at the stage when excavation to formation would have been complete, the wall with two levels of props would not have relied on the passive resistance of the toe to prevent rotation about the top prop. It is usually more economical however to provide a single level of propping if possible. This seemed particularly relevant as the wall in its final state was to span vertically the full height of the basement plantroom. The moments for which the permanent wall would be designed would not need to be increased for the temporary state. A costing was carried out which demonstrated a 25% saving on propping costs if a single line of props could be used. Under these circumstances the stability and toe movement of the wall rely completely on the passive resistance of the ground in front of the wall. This ground is London Clay and difficulty arose with the determination of the operating shear strength of the clay whilst the wall was temporarily propped. The shear strength would diminish, from its undrained cohesive value immediately after excavation, to its residual frictional value in terms of effective stresses under fully drained long term conditions. Using the same depth of wall as for the two level prop case, gave a low factor of safety when considering rotation about a single prop under long term conditions. This softening of the clay is partially time dependent and designers are cautioned of the effect in the CIRIA Report R104 (CIRIA 1984). It was decided that these long term conditions would probably not be fully achieved during the time it would take to remove the berm and cast the raft slab against the wall. As a

Table 1.

Analysis		Prop Loads kN/m (horizontal)	
Active/Passive parameters		Raking prop at 16.3m AOD	Capping beams as prop at 18.6 AOD
McA	fully undrained	180	160
McA	modified undrained (25% reduction in strength)	247	191
McA	drained/undrained	331	222
OAP	undrained FREW	287	400
OAP	drained/undrained FREW	362	
OAP	drained/undrained STAWAL	318	
OAP	undrained STAWAL	160	190
McA	drained/undrained WALLAP *	344	216
Characteristic design load		270	285
Ultimate design load		380	400

* This analysis was carried out subsequent to the choice of prop forces.

contingency a small number of lower level props would be held on standby should monitored wall movements prove us wrong. Additionally the berm removal was restricted so as to minimize the amount of time the single propped wall would be in its most critical state. This did not prove too restrictive on programme as the excavation rate was also to some extent governed by the rate at which spoil could be transported off site.

Propping forces

With the principle of the scheme determined it was then necessary to establish the required propping forces and to detail the arrangement. A series of analyses was carried out by OAP and McAlpine allowing for various degrees of softening of the clay. As indicated in CIRIA Report R104 the retained clay is likely to experience tension cracks with water penetration and dilation due to release of horizontal stress. This softening effect was considered in the range of analyses by including drained conditions on the active side of the wall. Passive pressures were considered in the undrained state for the temporary condition. A summary of temporary prop forces from the different analyses is shown in Table 1. The lower figures are from stability analyses. The higher range are from different computer programs which model the wall and soil as a beam with spring forces acting within active and passive limits. (Pappin, 1986) and (Borin, 1988). Preload effects on the prop loads are included.

The propping forces chosen reflected the fact that it was considered that moderately conservative soil parameters had been used. At each end of the basement the widths reduce to 30 metres. To avoid an awkward arrangement of raking props, the proposed capping beams were widened and designed as two U frames on plan, with a single temporary prop across each open end.

Table 2. Soil Parameters used in Analysis

Made Ground OAP	Made Ground McA	River Terrace Deposits OAP	River Terrace Deposits McA	London Clay OAP	London Clay McA	
18	18	20	19	20	20	unit soil weight kN/m^3
35°	35°	35°	40°	25°	23°	ϕ'angle of internal friction
0	0	0	0	0	0	c'effective cohesion kPa
0	0	0	0	70+10z	80+10z	c_u undrained shear strength kPa
-	-	-	-	48+7z	40+5z	Euh undrained horizontal Young's Modulus MPa
22500	2500	37500	6000	400c_u	15000	E'h drained horizontal Young's Modulus kPa

(z is measured from the the top of the clay)

In this manner the capping beam became the top prop to the wall allowing full unobstructed access for excavation to formation level below. This construction sequence meant that the stiff capping beams would be cast early. This limited the inward movement of the wall and is reflected in the higher propping forces indicated in Table 2.

Predictions of wall movements

In order to ensure that excessive settlements of the surrounding area would not occur, the amount of inward yielding of the wall was predicted allowing for the sequence of basement construction. OAP used Arup Geotechnics with their in house program FREW. Wall movements predicted are shown in Figure 4. A substantial part of the predicted movement can be seen to occur during the initial berm formation. McAlpine did not carry out any prediction of wall movement during the berm formation stage. The berm sizes were judged to be sensible based on previous experience of deep excavations. Some inward movement was expected. Predictions for the performance of the propped wall were carried out using WALLAP as a check after the propping arrangement had been detailed.

Propping arrangement

The props used to the north wall were slightly larger than specified being 490mm OD circular hollow sections with a 30mm wall thickness. Typical spacing was at 5.5 metre centres. Walings were twin 610 x 229 x 140 Kg/m Universal Beams. These were continuous over the full length of wall and fixed perpendicular to the raking props. The triangular space between the secant wall and the walings was filled with in situ concrete over the full length of the wall. The south wall was similar but with twin 457 x 191 x 82Kg/m walings and 323mm OD circular hollow section props of 10mm wall thickness at 4 metre centres. Concrete thrust blocks were cast on the central portion of

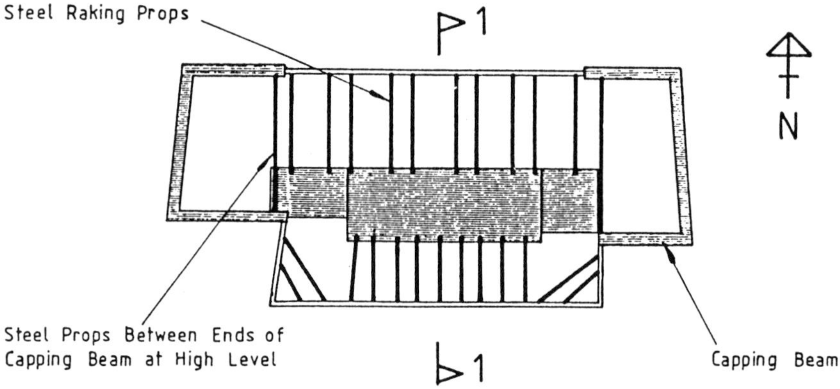

Fig. 1. Plan showing temporary propping arrangement

basement raft and these incorporated provision for jacking preload into the props (see Fig.1).

Preload and temperature effects

Despite the in situ concrete detail minimising take up under load, it was agreed that the props be preloaded. Approximately 40% of their presumed working load was jacked into the props to the north and south walls.The props were installed during the summer and the effect of temperature was noticed during their installation. Props installed at midday would lose most of their prestress by the evening. (Props to the north wall would contract by lmm for every 5°C fall in temperature).

Excavation sequence

The excavation immediately preceded the raft pour sequence. Berm removal was not permitted more than two bay lengths ahead of completed raft pours except below the capping beams at each end. Props to the south wall were released when the raft slab had been cast against it (see Fig. 2).

Monitoring of wall movements prop forces and settlement

Vertical inclinometer tubes were installed in 14 locations within the 260 linear metre perimeter secant wall. Reading were taken at weekly intervals during the critical excavation stages and monthly thereafter.

Global movement was recorded by relating inclinometer tube positions back to the original survey co-ordinates after each set of readings. Observed

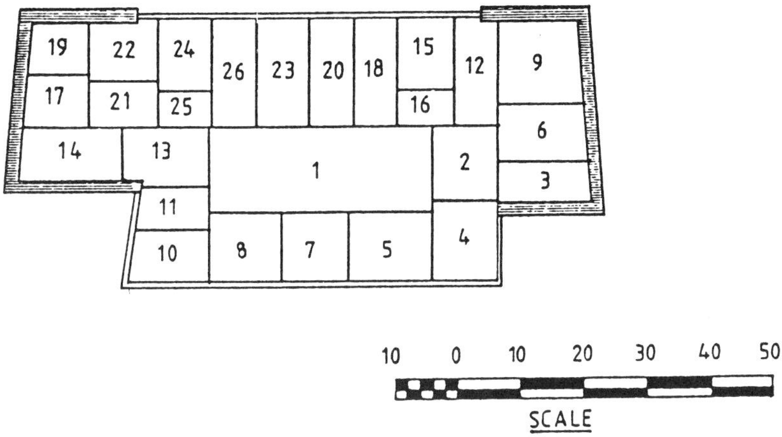

Figure 2. Plan showing basement slab construction sequence

movements are plotted in fig.4. Prop forces were not monitored generally. Three props to the north wall had their compressive forces measured immediately prior to release. At this stage the ground floor slab was complete. Two props recorded 660kN compression each (113kN/m horizontal). The third registered a negligible value.

From level surveys the maximum surface settlement measured outside the excavation was 2.9mm.

Comparisons of recorded wall movements with predictions

Predicted and measured movements are shown plotted in Fig. 4. It can be seen that the movements predicted by OAP for the unpropped wall at stage 1 did not occur.The top of the walls actually moved a nominal 2mm away from the excavation, probably as a result of rotational heave. Subsequent stage 2 analyses were similarly exaggerated having used the stage 1 predictions as their starting point.In order to look at second stage movements in their own right a revised prediction was produced by OAP.This was done by deducting the differences between the predicted and measured displacements for stage 1 from the predicted displacements for stage 2.

Maximum inward movements of 13mm were recorded for the walls with permanent structure providing restraint (stage 3). Smaller amounts were observed whilst temporary props were in place (stage 2). In all cases the deflected shapes were as predicted but with the actual magnitude of movement less than predicted.

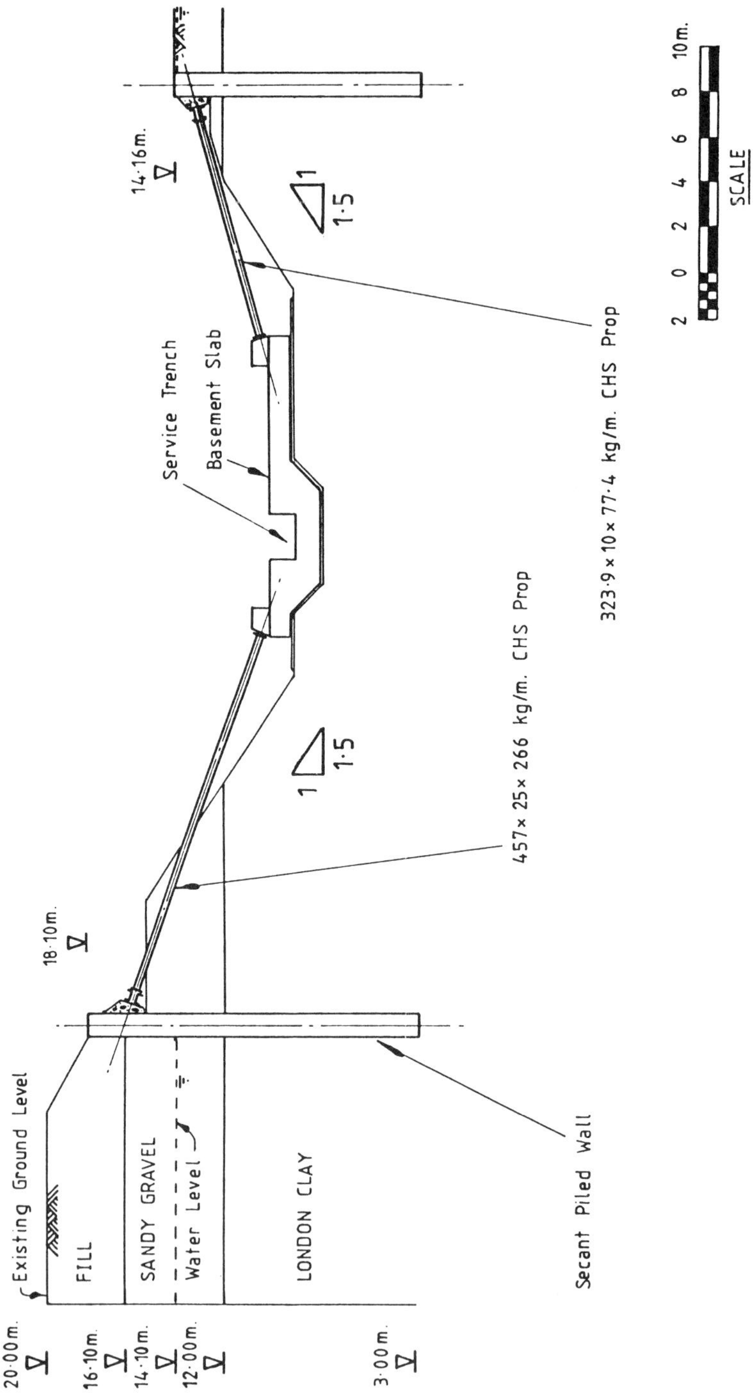

Fig. 3. Section 1.1 showing propping to secant piled walls

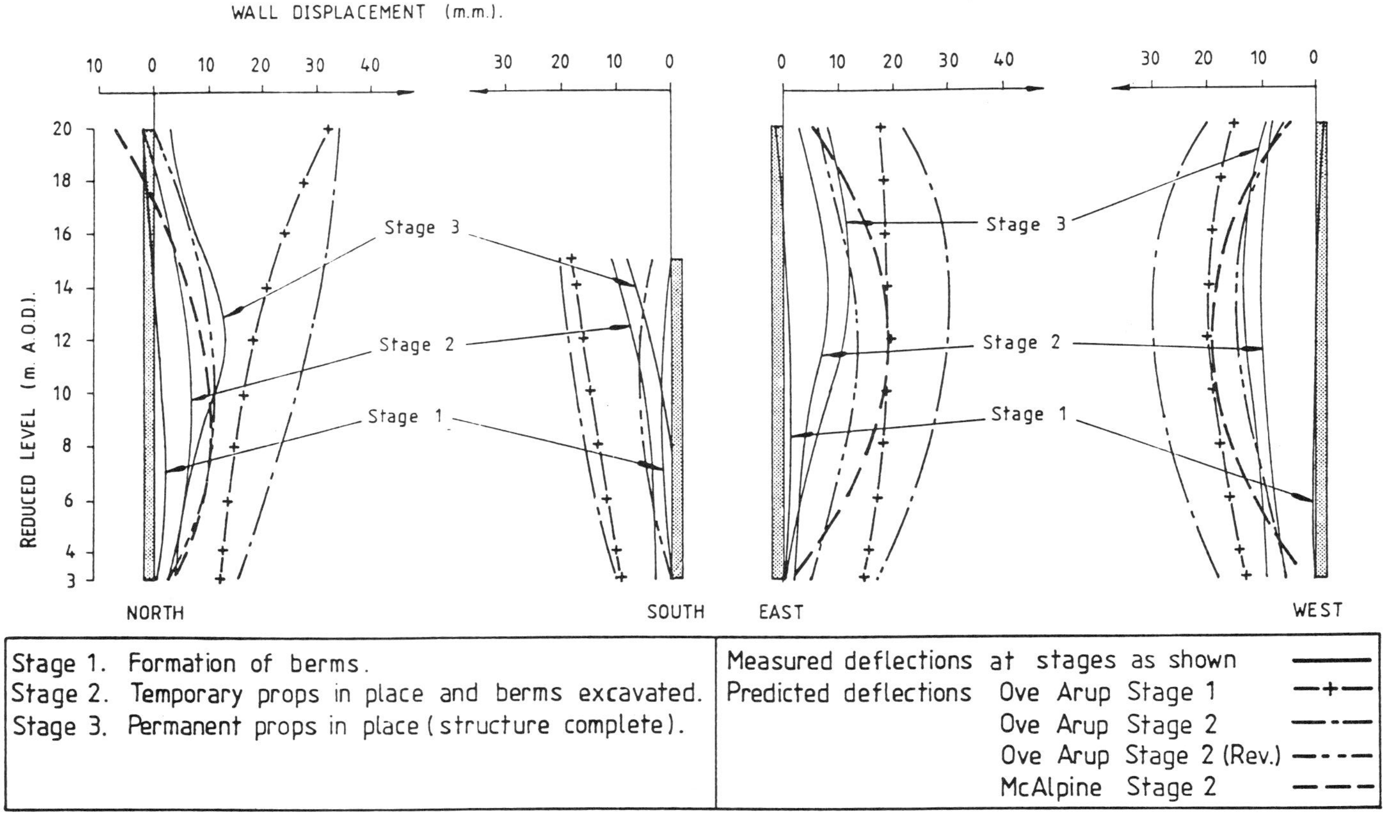

Fig. 4. Predicted and measured wall movements

Back analyses

OAP looked at the overestimation of movements predicted using FREW for stage 1. The program was compared with two finite element programs.The significant difference was that FREW assumed a non-friction boundary between the soil and wall. This was shown to have a major effect on predictions of cantilever wall movements in low K_0 (< 0.5) soils, such as the overlying terrace gravels at Holborn. FREW has since been revised to include the effects of these stresses.

The most likely cause of the overestimation of wall movements was thought by McAlpine to be the choice of horizontal modulus of subgrade reaction from the soils. The original values were increased in a series of back analyses but a factor of 3 had to be applied before observed movement could be reproduced.

Although prop loads were not monitored comprehensively it is possible to make some tentative assessments of their working value. Comparing relative waling plan positions for the various stages of construction indicates that the north wall struts lost 800kN each prior to the 660kN loads being measured. This equates to a maximum propping force of 250kN/m which is in good agreement with the design figure of 270kN/m.

Comment

The propped secant pile wall at Holborn Bars has performed well. The settlement/maximum depth of excavation ratio is 0.026%. This lies well below the upper bound of settlements from excavations in stiff to hard clay and within the limits associated with high horizontal stiffness propping. (Clough and O'Rourke, 1990).

Although the walls forming the perimeter support were analysed as four separate walls, the propping arrangement makes them interdependent. In particular the active forces are greater on the north wall than the south. It can clearly be seen that when the props to the south wall were released, after casting the raft slab against it, the south wall was pushed back some 3mm by the north wall props. This and the temperature effect could well explain the low recorded loads in the three north wall props as the back analysis presumes. In addition the ground pressure behind the north wall was less due to battering back at 45° from capping level.

With the general low magnitude of recorded wall movements the small effects of shrinkage of in-situ concrete slabs and temperature variation in props can be responsible for a large proportion of movement. Similarly in the east west direction the two ends of the southern wall rely on the raft slab cast against them and in the temporary case the pile wall cantilevering in plane to restrain the north-south capping beams. None of these effects was modelled in the wall analyses and it would have been unnecessarily involved to have made the attempt. These omissions do help to disguise the fact that wall movements were overestimated by a factor of about 2. This is acceptable with such low values of movement but does have an effect on the economical design of the temporary propping.

The design of this temporary propping has highlighted a common problem faced by designers. In the determination of the required amount of propping force for a single propped wall there is a choice of analysis methods which can generate quite wide variations in values.

Stability analyses, which may be hand calculations, can be carried out by the various methods summarised in CIRIA R104 and the Draft revision to CP2. These determine the embedment of the wall for the required factor of safety when considering rotation about the top prop. The prop forces are then determined from the difference between the active and passive forces on the wall assuming the embedment required for limiting equilibrium.

The prop force so determined is usually increased by 25% to allow for the effects of arching and stress redistribution on the back of the wall.

Many commercially available computer programs carry out a stability analysis as for the hand calculations but also have the option to model the wall and soil as a beam with spring forces acting between active and passive limits. Using this option, with the values of Modulus of Horizontal Subgrade reaction which appear appropriate from current thinking, invariably predicts higher prop forces and suggests larger wall movements than are observed in practice.

This is often explained by suggestions that the stiffness of the propping system does not allow the soil to move sufficiently to mobilize its shear strength and reduce pressures on the back of the wall to those of the idealized active state. The difficulty in attempting to design for a more flexible propping system is that the prop length is determined by site geometry and without an extremely flexible waling system (which may allow the soil to arch to the props regardless) higher movements of the wall can only occur with overstressed props.

It would seem that the prop forces chosen for design were reasonable for the raking props but somewhat high for the capping beams.

The general principle of a compromise between prop loads derived from stability analyses (plus 25%) and those predicted by beam/spring programs (considered as close to ultimate values) appears to have produced a reasonable solution.

Acknowledgement

The Author is grateful to the staff of Ove Arup and Partners and Arup Geotechnics for their assistance in compiling this Paper.

References

CONSTRUCTION INDUSTRY RESEARCH AND INFORMATION ASSOCIATION. *Design of retaining walls embedded in stiff clays.* CIRIA Report R104, 1984.

OVE ARUP & PARTNERS. *Predicted and measured movement associated with the secant pile retaining wall during construction of the deep basement.* Internal Report, 1991.

BRITISH STANDARDS INSTITUTION. *Draft code of practice for earth retaining structures (revision to civil engineering Code of Practice No. 2)* BS I, 1987.

PAPPIN J. W. *et al.* FREW - computer program for the analysis of flexible retaining walls. *OASYS*, 1986.

OVE ARUP & PARTNERS. STAWAL - Computer program for the stability analysis of retaining walls. *OASYS*, 1985.

BORIN D. L. WALLAP - computer program for the analysis of anchored and cantilevered retaining walls. *Geosolve*, 1989.

CLOUGH G. W. and O'ROURKE T. D. Construction induced movements of In-situ Walls. *Proc. ASCE Conf. on Design and Performance of Earth Retaining structures*, 1990.

Stress changes in stiff clay caused by the installation of embedded retaining walls

I. F. SYMONS and D. R. CARDER, Transport and Road Research Laboratory

Abstract

For embedded retaining walls the structural loading in service is likely to be influenced both by the initial in situ stress state in the ground and by the changes caused by the construction. Only limited information is currently available on the effect of wall installation, and field measurements are therefore being carried out at a number of sites where both bored pile and diaphragm walls are under construction. This paper gives results of field monitoring on three such schemes where embedded walls were installed in London Clay.

Introduction

Embedded retaining walls are now widely used in the construction of deeper sections of retained cutting and for cut-and-cover tunnels on urban road schemes. For walls of this type the structural loading in service is likely to be governed by the relative flexibility of the wall, the type of support, the initial stress state in the ground and the changes caused by construction (refs 1-3).

Many urban areas in the UK are founded on heavily overconsolidated clays where the in situ lateral stresses substantially exceed the vertical stresses. In such deposits the construction procedures used for the installation of the wall are expected to have a considerable influence on the stress regime in the ground close to the structure. Field measurements are therefore being carried out by the TRRL to determine the stress changes and ground movements caused by the installation of embedded retaining walls on a number of schemes. This paper gives some of the results from three sites where bored pile and diaphragm walls were installed in London Clay.

Measurement of earth and water pressures

Push-in spade shaped pressure cells, each fitted with an integral high air entry pneumatic piezometer, were installed at two of the sites both behind and in front of the proposed line of the retaining wall. Each cell was pushed a further 0.5m into the undisturbed soil beyond the bottom of a borehole using a lorry mounted drilling rig to provide the necessary reaction. During installation, the blade of each cell was carefully oriented to measure the total lateral stress acting towards the retaining wall. The boreholes were then backfilled with bentonite pellets to form an impermeable seal above each cell. In some boreholes additional pneumatic/vibrating wire piezometers were installed within 100mm long sand cells to give further data on porewater pressures.

The push-in action initially generated high stresses locally around each spade cell and for this reason a period of up to 3 months was allowed for readings to stabilise prior to any site construction activities. Even when readings have stabilised, the disturbance to the stress field caused by the presence of a cell is such that a correction to the readings is necessary to determine the true soil pressure. Tedd and Charles (ref.4) suggested a simple empirical relation that the over-reading is equal to one half of the undrained shear strength (C_u). Throughout this paper corrections have been made on this basis.

Bored pile retaining wall (Site A)

A field study was carried out to monitor stress and porewater pressure changes in the adjoining ground during the construction of a contiguous bored pile wall at this site. The piles, which were 1500mm in diameter and spaced at 1700mm centres, penetrated to a depth of 24m. The ground conditions comprised up to 3m of made ground and 3m of Claygate Beds overlying the London Clay.

Construction sequence

Steel guide tubes were driven 3 to 3.5m into the ground before auguring to the required depth using a crane mounted rig. The steel reinforcement cage was then lowered into the excavation and the concrete poured. A construction rate of 2 to 3 piles per rig day was achieved in the instrumented area.

Lateral stress changes during wall installation

Fig 1 shows the corrected total lateral stresses in the ground determined from measurements taken on five spade cells located on one side of, and at 1.5m away from, the wall during and shortly after installation of the piles. Large reductions in lateral stress occurred during the construction of the three piles nearest to each cell, and this was followed by a partial recovery to values lower than the initial stress which existed before pile installation.

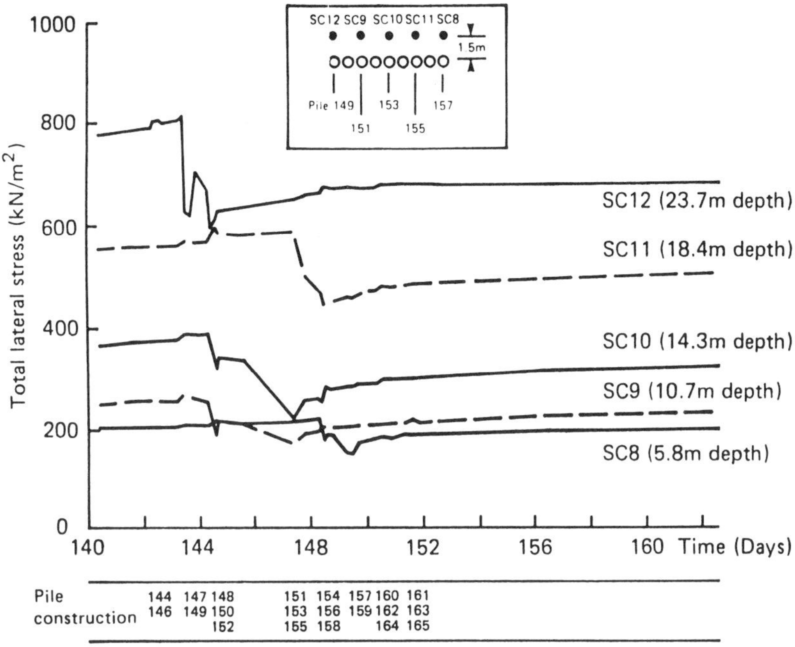

Figure 1. Stress changes during wall installation at Site A

The measured distribution of total lateral stresses before construction, together with the changes recorded on each of the ten spade cells installed on both sides of and at 1.5m from the wall, are shown in Fig 2. Fig 2a shows reasonable agreement between the initial total lateral stresses from spade cells, from Marchetti dilatometer tests, and as derived measurements of soil capillary pressure. The lateral stresses from the dilatometer tests have been calculated from the empirical relations given by Marchetti (ref.5) together with the measured distribution porewater pressure, and tend to represent lower bound to the results from all three methods. Measurements of soil capillary pressure (P_k) were obtained from triaxial tests on 100mm diameter specimens from thin walled tube samples, using the procedure described by Burland and Maswoswe (ref.6).

Fig 2b gives the overall reductions in total lateral stresses caused by installation of the pile wall, the average porewater pressure distribution from piezometer measurements before construction, and the corresponding lines of earth pressure coefficient (K). The initial K values determined from the individual spade cell readings before pile installation ranged between 1.3 and 2.3 with a mean of 2.1 over the depth of the wall. After pile installation the corresponding mean value reduced to 1.9. The reduction in total lateral stress

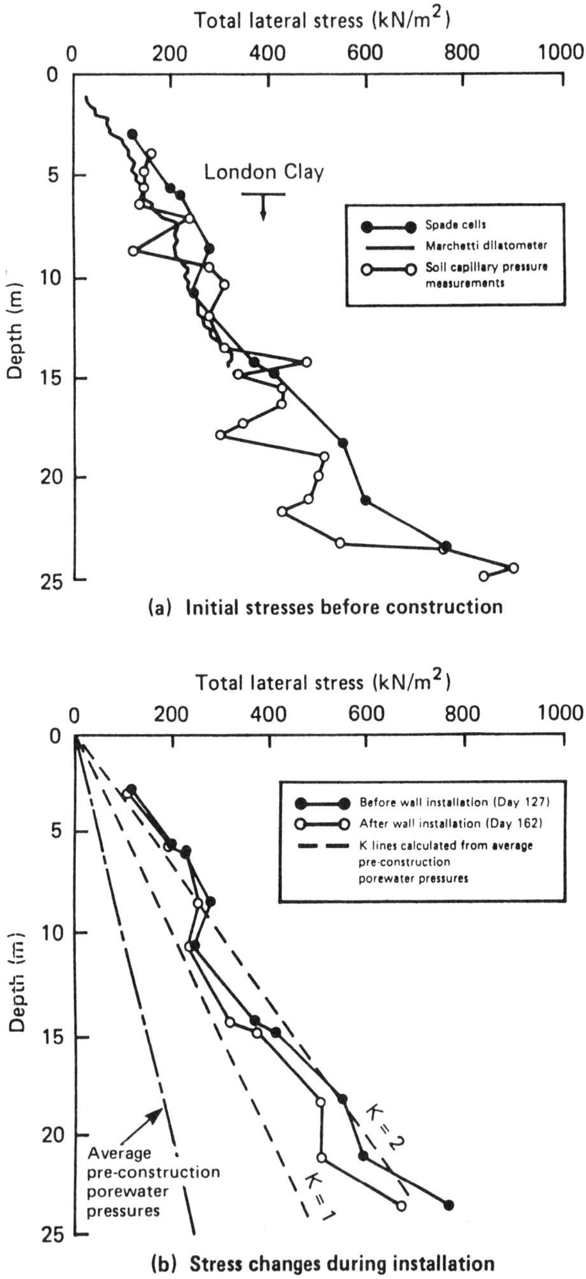

Figure 2. Total lateral stress distribution at Site A

caused by wall installation showed a near linear increase with depth below the top of the London Clay.

Porewater pressure changes during wall installation

At Site A the porewater pressure measurements before construction indicated a near hydrostatic distribution with depth from ground level (Fig 2b). The porewater pressures generally showed a slight rise as pile construction approached each piezometer location, followed by a rapid fall of up to 150 kN/m^2 as the three piles closest to the instrument position were constructed. The porewater pressures then rose steadily and within a few weeks were close to the values measured before piling took place.

Diaphragm retaining walls (Sites B and C)

Field monitoring has been carried out at two sites where diaphragm walls were constructed. T-shaped panels consisting of a 4.0m by 0.8m front section with a 2.7m by 0.8m counterfort were installed to a depth of 13.5m at Site B, while at Site C planar panels 4m in length by 0.6m in width and 15m in depth were constructed. At Site B the London Clay is overlain by about 2.4m of gravels and made ground. At Site C soft alluvial deposits were found to 6.9m depth below which a 2.2m layer of alluvial gravel overlies the London Clay. At Site B both total lateral stresses and porewater pressures were measured, while at Site C only porewater pressures were monitored.

Construction sequence

At Sites B and C the excavation for each wall panel was carried out using a mechanical grab operating between concrete guide walls. Throughout the period when the trench for each panel was open, a bentonite slurry was used to provide support. Following excavation, the reinforcing cage was lowered into the trench and concrete tremied to the bottom with the bentonite slurry being progressively displaced.

At Site B generally between one and three days were required for the excavation of each T-shaped panel with concreting taking place the following day. The simpler planar panels at Site C were generally excavated and concreted on subsequent days.

Lateral stress changes during wall installation

The corrected total lateral stresses measured on three spade cells at 1.5m from the front face of the wall during excavation and concreting of the T-shaped wall panels at Site B are shown in Fig 3. Spade cell SC8 was positioned directly in front of panel 83 and falls of 80 kN/m^2 in total lateral stress were

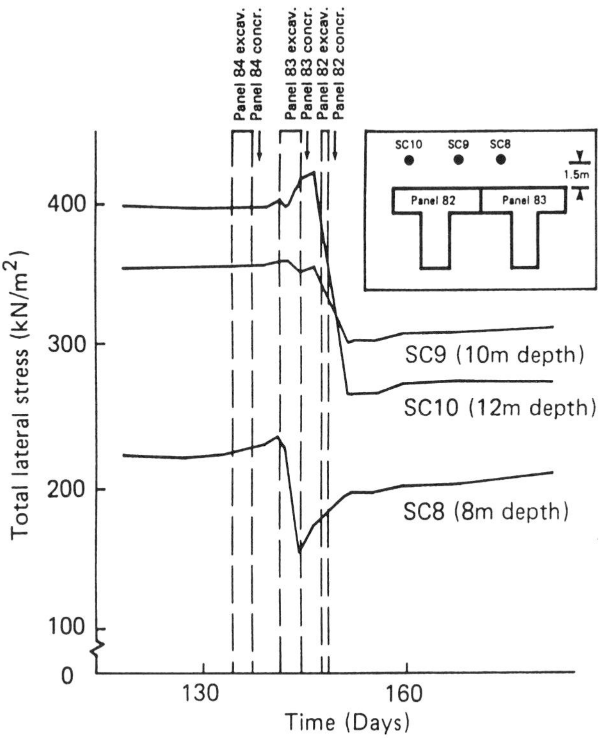

Figure 3. Stress changes during wall installation at Site B

observed during excavation. Concreting of this panel resulted in an increase in lateral stress with further rises measured after construction of the adjacent panel 82. Spade cells SC9 and SC10 showed only small changes in total lateral stress during construction of panel 83. The installation panel 82 adjacent to these cells resulted in reductions in total lateral stress of about 50 kN/m^2 and 160 kN/m^2 on cells SC9 and SC10 respectively. Four spade cells located at 1.5m behind the back counterforts of the T-panels showed a similar pattern of behaviour during wall installation, although the lateral stress changes were appreciably smaller.

The initial total lateral stresses in the ground from all ten spade cells installed at this site together with the results from self-boring pressuremeter tests and values derived from measurements of soil capillary pressure are summarised in Fig 4. Agreement between the total lateral stresses from the spade cell measurements and from pressuremeter tests was reasonably close. These results generally indicated an earth pressure coefficient (K) of between about 1.5 and 2 over the depth of the wall, based on the average distribution of porewater pressure from the piezometer measurements before construc-

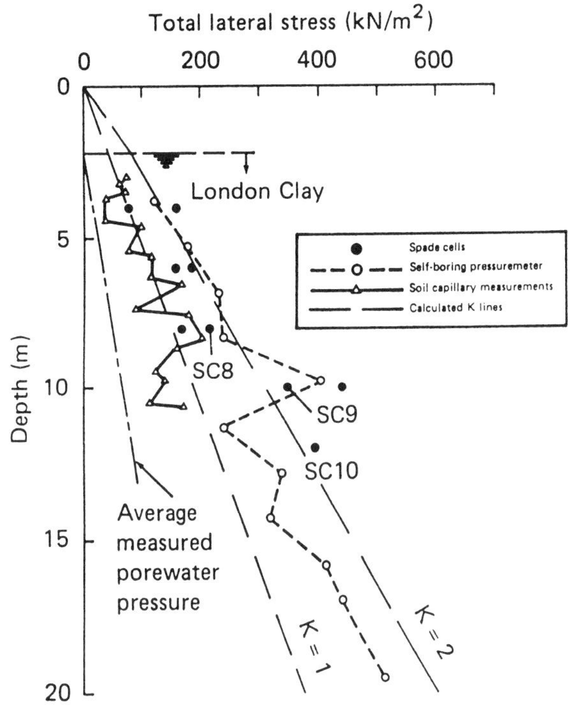

Figure 4. Initial stresses before construction at Site B

tion. The total lateral stresses derived from laboratory measurements of soil capillary pressure were significantly lower than those obtained from the in situ measurements.

Table 1 summarises the stress relief which occurred during wall installation for the cells at 1.5m in front of the panel and shows that the largest change was measured on the deepest cell (SC10). For these three spade cells the mean K of 2.0 prior to wall installation fell to a value of 1.6 after wall installation.

Table 1. Effect of wall installation at Site B

Spade cell	Depth (m)	K-value before wall installed	K-value after wall installed
SC8	8	1.6	1.5
SC9	10	2.3	2.0
SC10	12	2.2	1.3

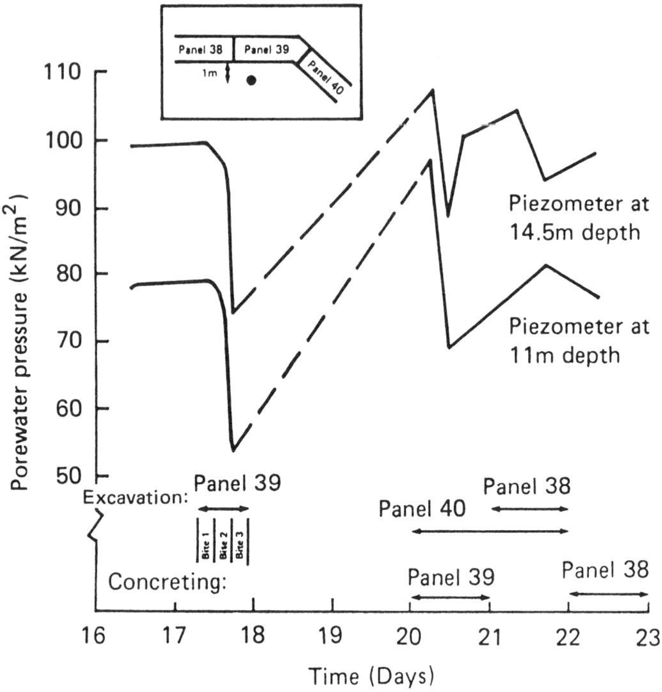

Fig. 5. Piezometer results at site C

Porewater pressure changes during wall installation

At Site B the porewater pressure measurements before construction showed a linear increase with depth below the surface of the London Clay which was slightly below that of a hydrostatic distribution and suggested that under-drainage may be taking place in this area (Fig 4). Reductions of up to about 20kN/m^2 were measured on the closest piezometers during excavation under bentonite for each panel, with rapid rises to above the pre-construction values then occurring during concreting. Over the next few months, the piezometer measurements gradually stabilised to give porewater pressures similar to those existing before wall installation.

More detailed information on the changes in porewater pressure during excavation and concreting of the planar diaphragm wall panels at Site C were obtained from piezometers installed at 11.5m and 14m depth in each of two boreholes located at lm from the wall. Prior to the start of construction the readings on these piezometers indicated stable porewater pressures of about 80kN/m^2 and 100kN/m^2 respectively. Fig 5 shows the changes in porewater pressures measured on the two piezometers in one of these boreholes measured over the period of wall installation. At this site each panel length was excavated under bentonite in three full depth bites of the grab, and the panel immediately adjacent to the instrumented boreholes was constructed as the starter panel. Excavation of this panel commenced at the end furthest from the instruments and only small reductions in porewater pressure were recorded during this first bite. Excavation of the second bite with the grab in the centre of this panel induced an overall porewater pressure fall of about $25kN/m^2$ on both piezometers. Readings taken after concreting of this panel then showed rises of up to about $45kN/m^2$ under the loading from the wet concrete. Similar changes but of reduced magnitude were then measured on both piezometers as the adjoining panels 40 and 38 were excavated and concreted. The readings on the piezometers which were located at about 1.5m and 4m respectively below the top of the London Clay reached equilibrium at values slightly above the pre-construction levels about six weeks after installation of the wall.

Summary

(i) Field measurements made during installation of both a contiguous bored pile and a counterfort diaphragm wall in London Clay showed substantial changes in total lateral stress in the ground close to the walls, caused by the construction operations. For the bored pile wall, reductions in lateral stress took place as the three piles closest to the instruments were installed, while for the diaphragm wall large reductions occurred during excavation of each panel with subsequent increases during concreting.

(ii) Before construction the spade cell measurements indicated initial stress states in the London Clay at both sites corresponding to earth pressure coefficients generally in the range 1.5 to 2.0. After wall installation the measurements at 1.5m from the walls showed decreases in average earth pressure coefficient of about 10 percent for the bored pile and about 20 percent for the diaphragm wall. These values can be compared with a reduction in earth pressure coefficient of about 30 percent in the London Clay at a similar location, inferred from measurements during installation of the secant pile wall at Bell Common Tunnel (ref.7).

(iii) Piezometers installed at lm from the face of a planar diaphragm wall at the third site showed reductions in porewater pressure as excavation took place under bentonite with subsequent increases to above the pre-construction values after concreting. At all three sites after wall installation was completed the porewater pressures gradually stabilised at values similar to those that existed before construction commenced.

Acknowledgements

The work described in this paper forms part of the research programme of the Ground Engineering Division of the Structures Group of TRRL and is published by permission of the Director.

References

1. Potts D M and Fourie A B. The effect of wall stiffness on the behaviour of a propped retaining wall. Geotechnique, 1985, vol.35, no.3, 347-352.
2. Fourie A B and Potts D M. Comparison of finite element and limiting equilibrium analyses for an embedded cantilever retaining wall. Geotechnique, 1989, vol.39, no.2, 175-188.
3. Higgins K G, Potts D M and Symons I F. Comparison of predicted and measured performance of the retaining walls of the Bell Common Tunnel. Transport and Road Research Laboratory CR124, Crowthorne, 1989.
4. Tedd P and Charles J A. Evaluation of push-in pressure cell results in stiff clay. Proc.International Symposium on Soil and Rock Investigation by In-situ Testing, Paris, 1983, vol.2, 16-21.
5. Marchetti S. In situ tests by flat dilatometer. J.Geotech. Eng. Div. Am. Soc. Civ. Engrs., 1980, vol.106, GT3, 299-321.
6. Burland J B and Maswowse J. Discussion on the paper by Tedd and Charles (1981). Geotechnique, 1982, vol.32, no.3, 285-286.
7. Tedd P, Chard B M, Charles J A and Symons I F. Behaviour of a propped embedded retaining wall in stiff clay at Bell Common Tunnel. Geotechnique, 1984, vol.34, no.4, 513-532.

Design and measured performance of diaphragm walls at Waterloo International Terminal

E. S. F. LI, Z.M. NYIRENDA and A.R. PICKLES, Alexander Gibb & Partners, UK

Introduction

A new airport style international railway terminal is at present being constructed along the western side of Waterloo Station in London to service the opening of the Channel Tunnel in 1993. Construction of the new terminal requires demolition of an area of the existing platforms and sidings which are constructed on brick arches approximately 8m above ground level. A plan of the site is given in Fig. 1 and a cross section through the terminal is given in Fig. 2. The site is bounded to the north, east and south by the brick arches of the existing station. An office block whose height varies between 7 and 15 stories is situated approximately 15m from the western site boundary.

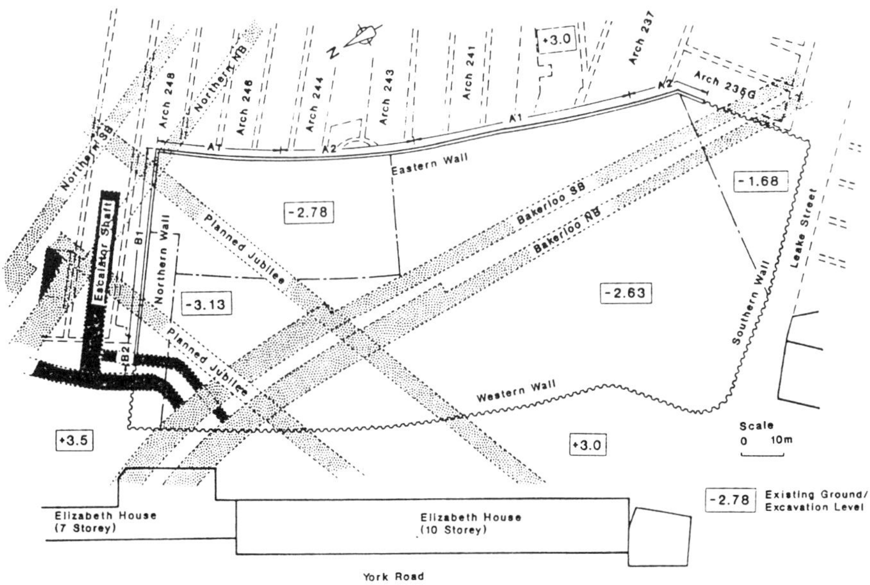

Fig. 1. General layout of Waterloo International Terminal

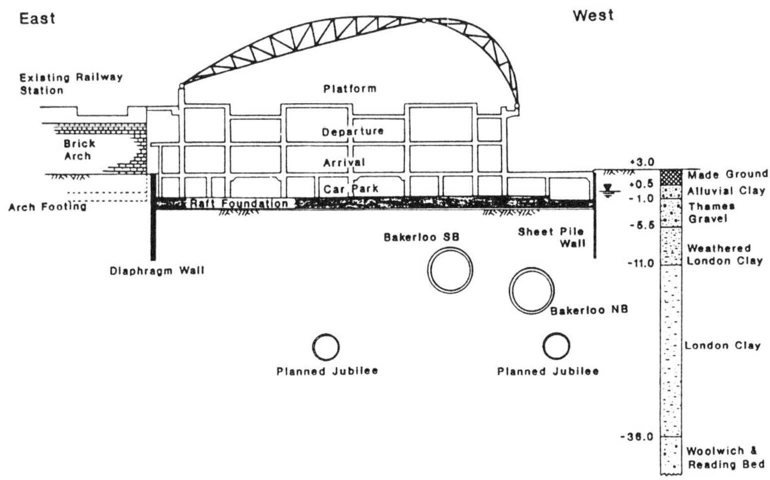

Fig. 2. Typical east-west cross section

Station and running tunnels of London Underground Limited's Bakerloo and Northern Lines pass directly beneath the site. A double bank of escalators from the existing rail station to the Underground is situated immediately adjacent to the northern boundary of the site. The alignment of the proposed tunnels for the Jubilee Line Extension also passes beneath the site. Excavation for the new basement along the north and east sides of the site is directly adjacent to the remaining arch foundations. The platforms along the eastern edge of the basement have remained fully operational throughout the demolition and excavation for the new terminal basement.

As a result of the close proximity of both above and below ground structures to the site, control of ground movements has formed one of the most important aspects of the design and construction of the new terminal. A diaphragm wall has been constructed along the north and east sides to minimise ground movements and to act as a water cut-off. Along the south and west sides of the site, where ground movements are less critical, a sheet pile wall has been installed. The terminal is founded on a 1.6 m thick raft which spreads the development loadings and smooths out differential ground movements. This Paper discusses aspects of the diaphragm wall design and compares the design wall movements with the actual movements measured to date on site. The differences between design movements and predicted movements, and design assumptions and actual field conditions are discussed.

Ground conditions

A comprehensive ground investigation has been carried out at the site. This comprised percussion boreholes with thin wall sampling, rotary coring of the London Clay and Woolwich and Reading Beds, trial pits to investigate existing foundations, and in situ and laboratory testing.

The typical soil sequence at the site is shown on Fig. 2. Approximately 2.5m of made ground overlies 1.5m of soft to firm alluvial clay which is underlain by approximately 4.5m of medium dense Thames Gravel. The arch footings are founded on the Thames Gravel. The Thames gravel is underlain by approximately 30m of London Clay which is weathered over the top 5m. Ground level is at approximately +3.0 mOD and the near surface water table is at approximately 0.0 mOD.

The in situ testing included piezocone probing, dilatometer profiling and pressuremeter testing. This testing was concentrated at five pre-selected locations in order to determine the effects of previous construction of the station and underground tunnels on the in situ stress state.

In the laboratory all triaxial testing was carried out using either mid-height or base mounted pore pressure probes on 100mm diameter thin wall or rotary core samples. In order to obtain further data on the in situ stress state all tests started with measurement of the initial soil suction. In addition filter paper suction tests were performed on the thin wall and rotary core samples.

Non-linear small strain parameters for the London Clay and the Thames Gravel were determined using sample-mounted strain measuring devices. The tests were carried out in both compression and extension after consolidation to the initial stress state following appropriate stress paths. The Thames Gravel samples were recompacted in the laboratory to their in situ density prior to testing.

A summary of the geotechnical parameters adopted for the design of the

Table 1. Soil parameters adopted for design

Soil	Bulk Unit Weight (kN/m^3)	Undrained Shear Strength Cu (kN/m^2)	Drained Cohesion c' (kN/m^2)	Friction Angle (deg)	Ko / In Situ Horizontal Stress (kN/m^2)
Made Ground	18	--	0	30	0.5
Alluvium	17	25	0	24	0.5
Thames Gravel	20	--	0	35	0.43
Weathered London Clay	20	50-100	5	23	125 at -5.5mOD 300 at -9 mOD
London Clay	20	125-260	15	25	360 at -11 mOD 700 at -36 mOD
Wooliwich & Reading Bed	22	500	200	27	1.0

wall are given in Table 1. WRB strength were based on peak values measured in high-quality cored samples.

Arch surcharge loadings

Generally the ground level at the site is at approximately +3.0 mOD. The final excavation levels vary across the site and are shown in Fig. 1. The remaining arches of Waterloo Station are supported on strip footings founded at the base of the soft alluvial clay. The existing arch surcharge loadings were estimated by considering the weight of the arches and the railway station above, plus the maximum design live loads which included full (Ru) train loading. For the analyses a value of approximately 900 kN/m was assumed at the footing level for the majority of the arch strip footings.

The arch foundations run parallel to the north wall and consequently for this wall the arch loadings on the retained side were modelled as discrete strip loads. Along the east side of the site the arch foundations run perpendicular to the walls. Taking into account the rigidity of the diaphragm wall panels, the capping beam and the temporary propping it was assumed that the arch loadings behind the wall could be modelled as being distributed uniformly at footing level. A 10 kN/m^2 construction surcharge at ground level was also allowed for in the analyses.

The arch foundations on the east side of the site were underpinned using minipiles prior to demolition and excavation. This was carried out in order to stabilise the arches during diaphragm wall panel construction and to decrease the magnitude of differential settlement of the arch footings during excavation for the basement. In addition, when the archistructures in front of the walls were demolished and the remaining station was redeveloped there was a re-distribution of load on the remaining footings. For the limit equilibrium and numerical analyses discussed later, it was assumed that the load increase within 5m of the wall would be carried by the underpinning. Beyond this zone the redevelopment loads increased over the next 5m to the fully re-distributed value.

Fig. 3 shows a typical cross section of the east wall (section A1 on Fig.l) at the end of excavation together with the assumed loadings.

Construction sequence

The construction sequence at the site is given below. For the numerical analyses discussed later the sequence can be conveniently separated into the 5 distinct stages as shown below.

Stage 1

(*a*) demolish arches within the site; the arches behind the north wall are

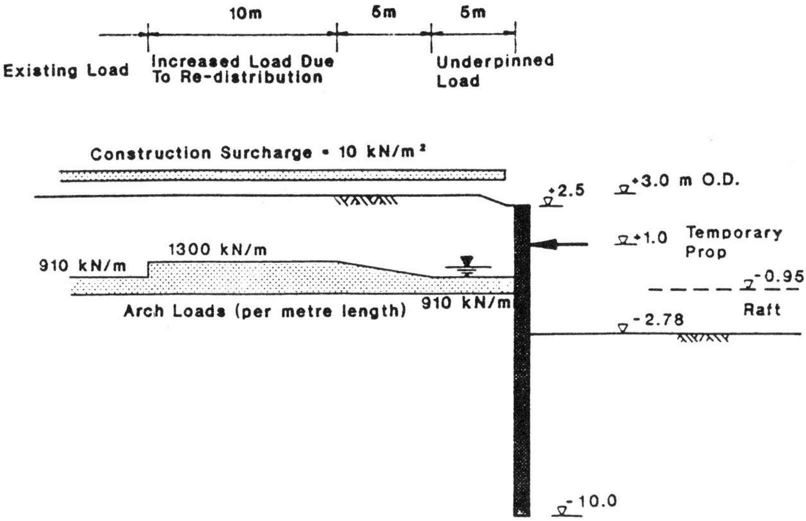

Fig. 3. Cross section of east wall A1

also demolished to springing level
(*b*) install diaphragm walls
(*c*) excavate to +0.5 mOD across complete area.

Stage 2 .

(*a*) dewater the site in front of the walls
(*b*) excavate central area to final formation level, leaving a 5m berm with slope of 2:1 (H=V) in front of the walls.

Stage 3.

(*a*) cast raft within berm
(*b*) install temporary props between the wall and the raft
(*c*) remove the berm.

Stage 4 .

(*a*) cast remaining raft
(*b*) construct station up to arrival floor level which acts as permanent prop
(*c*) remove the temporary prop
(*d*) apply final development loadings.

Stage 5.

(*a*) establish long term conditions after dissipation of pore water pressure.

Limit equilibrium design

The wall penetration depths were determined using limit equilibrium calculations as described in CIRIA 104 (Padfield & Mair, 1984). In view of the thorough investigation and high quality of sampling and testing, the recommended safety factors were slightly relaxed over those suggested in CIRIA 104.

When determining the required wall penetration an allowance was made for an extra depth of excavation for the removal of soft areas beneath the foundation raft. This allowance was needed at the southern end of the site where an extra metre depth of material had to be removed below raft formation level during construction. By taking this into account at the design stage the deeper excavation took place without the requirement for extra propping, which could have lead to construction delays.

The walls were temporarily propped at the +1.0 mOD level until the raft foundation, which acts as permanent prop, had been cast. Based on the limit equilibrium analysis results the toe level was generally at -10.0 mOD, however near Arches 243 and 235G on the east wall heavier surcharge loadings were imposed by the existing arches and a toe depth at -13.0 mOD is required, see Fig. 1.

Numerical modelling

The second stage of design is to assess ground movements, wall displacements and bending moments under working conditions. The major problem facing the design engineer is choosing the most appropriate set of design assumptions within the constraints imposed by the site conditions and the analytical techniques. The main difference between analyses carried out for design and those carried out to obtain class A predictions derives from the fact that the design engineer must always err on the conservative side. Design movements and bending moments should therefore represent upper bound values.

FLAC, a commercially available finite difference program, had been used to predict the design movements and bending moments. For the analyses presented in this paper an elasto-plastic sol model incorporating the Mohr-Coulomb yield criterion was used. A typical mesh is shown in Fig. 4 and the soil strength parameters are set out in Table 1. The initial stress state within the mesh had been established based on the results of the ground investigation and modified by modelling the construction of the existing station and underground tunnels on FLAC.

A 10 kN/m^2 construction surcharge at ground level and the arch surcharge loading behind the wall discussed above were included in the modelling, with the footing level assumed to be at -1 mOD. In reality it is doubtful if more than approximately 2 kN/m^2 construction surcharge and

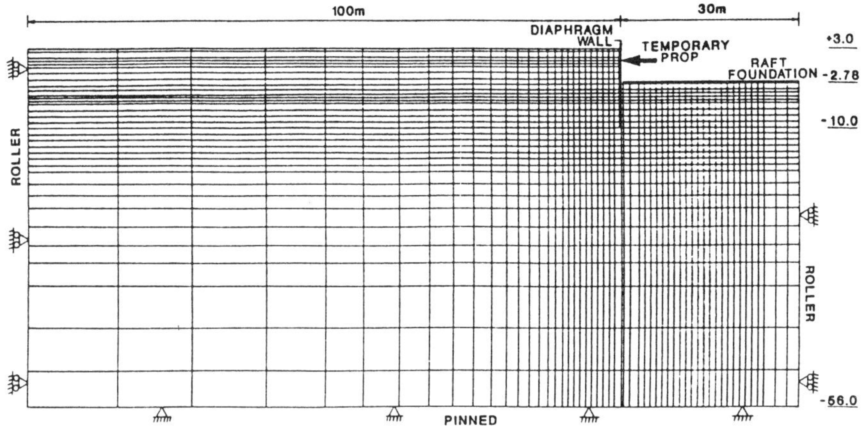

Fig. 4. Finite difference mesh

20% of the live load component on the arches were ever carried by the arch foundations at any time during construction. In addition the average footing level revealed during demolition at the site was approximately -2 mOD.

The construction of the basement took place in a relatively short period in comparison to the drainage characteristics of the London Clay and therefore undrained parameters were adopted for the London Clay strength and stiffness parameters. At the end of construction the London Clay was given effective stress strength parameters and a (variable) drained bulk modulus and the excess pore water pressures due to construction were dissipated. In reality some drainage will occur during construction of the basement and this was allowed for by modelling the top metre of the London Clay as being drained throughout the analysis.

It is well established that the stress-strain behaviour of soils is non-linear, particulary at small strain levels (Jardine *et al.* 1984). In addition, there is much published data which indicates that soils exhibit increased stiffness following a reversal in the loading path (Atkinson *et al.* 1990). The magnitude of the soil strains at Waterloo are both relatively small and variable and the construction sequence involves a number of loading-unloading cycles. Therefore both of the above phenomena are important for the accurate prediction of ground movements. This is particularly the case where the magnitudes of ground deformations at some distance from the wall need to be estimated, as at the Waterloo site.

One of the drawbacks of FLAC is that at the present time it does not incorporate the required non-linear stress strain models. Soil stiffness is modelled as being linear elastic until yield. It is therefore necessary to input variable shear moduli (G) to the soil elements within the mesh which reflect the differing degree of strain mobilisation. The effects on stiffness of stress

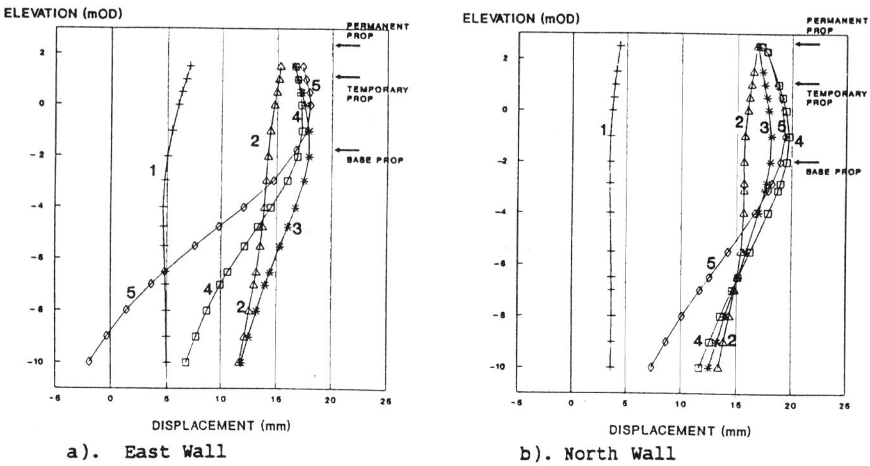

Fig. 5. Calculated wall displacements

path reversal must also be taken into account.

In order to establish appropriate stiffness values a cross section through the site was modelled by the Geotechnical Consulting Group using the Imperial College finite element program (ICFEP). The program incorporates the required non-linear stress strain models and the deformation moduli for different soil elements vary according to the mobilised strains. Based on the results of this analysis, pseudo non-linear stiffness values were derived for each element in the linear elastic FLAC analyses. The stiffness values were altered as necessary at each major change in the construction sequence.

The arch footings perpendicular to the east wall will act as stiff reinforcing strips within the soil mass. The footings were modelled in the analysis by increasing the stiffness and strength of the soil layers at the footing level. The mesh elements directly adjacent to the wall were given a reduced stiffness and no tensile strength so as to allow the wall to move away from the footing. With the exception of modifying the final footing load distribution as discussed above the effect of the underpinning piles was not modelled.

Discussion of calculated and measured performance

Figure 5 shows the wall displacements calculated by FLAC at various stages of construction for the east wall A1 south of Arch 242 and the north wall B1, see Fig. 1. The corresponding bending moment diagrams are shown in Fig. 6. The displacements and bending moments for the north wall are smaller than those of the east wall as the arches behind the north wall were partially demolished during construction of the terminal. The arch loading will be reimposed after completion of the basement.

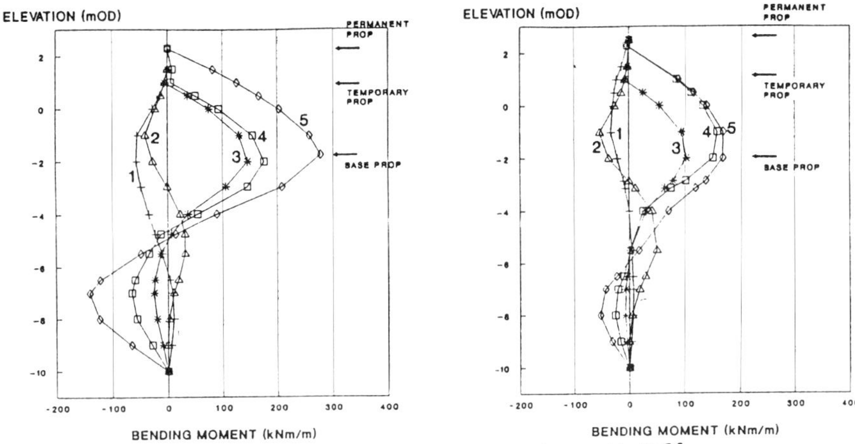

a). East Wall

b). North Wall

Fig. 6. Calculated wall bending moments

A comprehensive set of survey stations were installed on structures adjacent to and below the site to monitor movements of these structures during construction of the terminal. In addition 9 inclinometers were installed in the east west diaphragm panels, 4 inclinometers in the north wall and 2 combined inclinometer/extensometers between the north wall and the adjacent escalator shaft (Fig.7). Inclinometers installed within the diaphragm wall panels

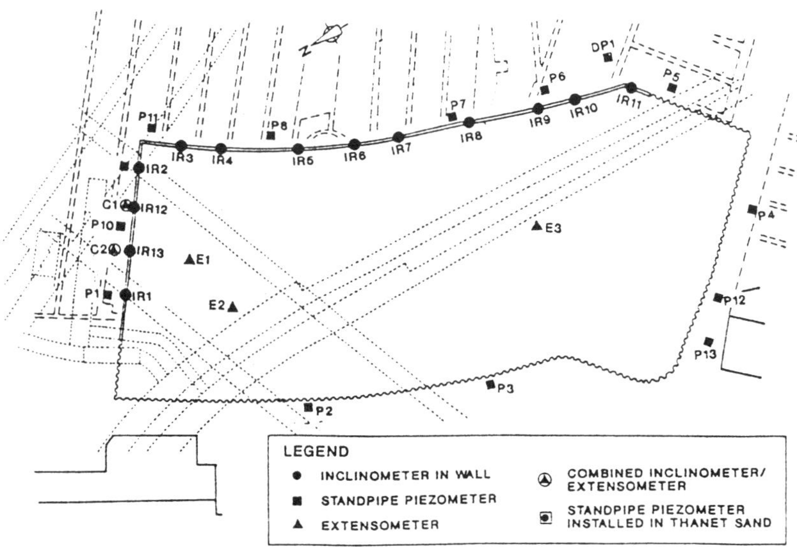

Fig. 7. Location of instrumentation

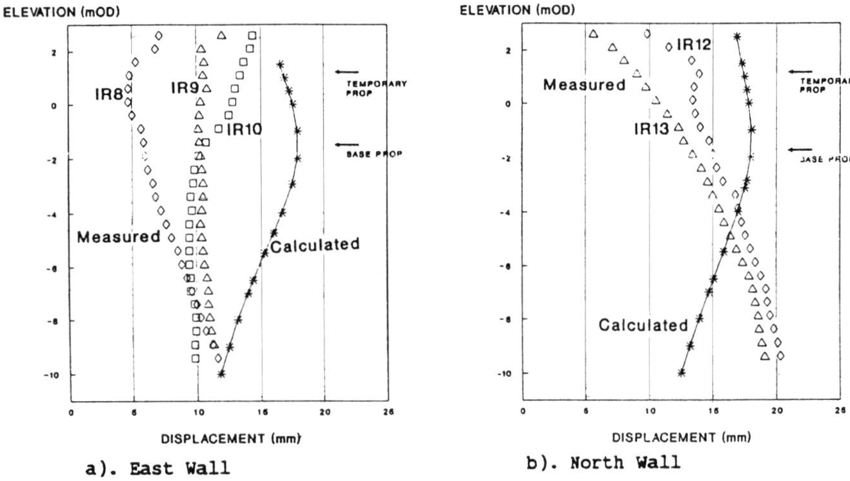

Fig. 8. Comparison of calculated and measured wall displacements

could only measure differential displacement of the wall, therefore the position of the top of the inclinometer tubes were surveyed in order to establish total movements. Standpipe piezometers were installed in the Thames Gravel around the periphery of the wall to detect any drawdown of water table outside the site boundary.

At the time of writing this paper the construction at the site was still in progress. The results for the inclinometers in the centre of the North wall and south of Arch 242 in the East wall at the final excavation level (Stage 3 of the construction sequence) are shown on Fig. 8 and are compared with the design deformations for this stage.

It can be seen that the actual movements are of the same magnitude but generally smaller than the design movements. This is to be expected as the design movements consider the worst credible loading condition behind the wall which, as discussed, did not occur in practice. However the profile of the actual wall movements differs markedly from the design profile. This is thought to be due to a combination of factors including: the underpinning, which was not modelled, transfers the footing load to ground below the toe of the wall thus increasing toe movements; and the arch footing acts as a more effective horizontal tie than allowed for in the model. Heavily reinforced concrete beams were constructed along the arch footings to tie the top of the underpinning piles together. These were not included in the original analyses.

Two other points which became apparent during monitoring are thought to be worthy of note. Firstly the load in the tubular steel temporary props were heavily influenced by temperature effects, the prop load doubling during hot weather as thermal expansion was restrained by the wall and base slab. Secondly there are notable time effects in the response of the made ground and Thames Gravel to changes in load. The use of fully drained parameters for these materials in the very short term, i.e. in terms of days, overestimated immediate wall movements.

When the basement has been completed, further results will be published to add to the body of data concerning real wall behaviour.

Acknowledgements

The Authors are grateful to the British Rail for their permission to publish the information.

References

ATKINSON J.H., RICHARDSON D. and STALLEBRASSE S.E. Effect of recent stress history on the stiffness of overconsolidated soil. *Geotechnique*, 1990, **4**, pp 531-540.

JARDINE R.J. *et al.* The importance of small strain behaviour in the analysis of soil-structure interaction. *Geotechnique*, 1984, **2**, pp 377-396.

PADFIELD C.J. and HAIR R.J. *Design of retaining walls embedded in stiff clay.* CIRIA Report 104, CIRIA, 1984.

Prediction versus actual movement of bored pile walls in middle coal measures

D. C. CURTIS and H. MIRZABAIGIAN, Ove Arup & Partners, Manchester, UK

For a new multi-storey shopping centre at an inner city site, a 10 m deep service basement was required. On two sides of the basement, temporary support of the surrounding ground including roads and buildings was necessary during basement construction. Contiguous bored pile walls, both acting as cantilevers and anchored back were selected for temporary support. These were constructed in the rocks of the Middle Coal Measures comprising interbedded mudstones, siltstones, sandstones and coal seams. The wall design was primarily based on parameters derived from Standard Penetration Tests carried out during the Site investigation. Lateral wall movements were predicted at design stage using a numerical analysis computer program. During pile construction, six inclinometer access tubes were installed in selected piles which were later used to monitor wall movements at different stages of basement construction. The predicted and measured movements of the wall are presented and comparisons made. It is demonstrated that ground parameters derived from Standard Penetration Test results used in numerical analyses methods can be used to predict retaining wall movements in the Middle Coal Measures lithology with reasonable accuracy.

Introduction

In 1987 a proposal was put forward to redevelop a partly derelict inner city site with a new multi-storey shopping centre and associated multistorey car park. The proposal included the provision of a basement, up to 10 m deep, to service the centre. The basement was flanked on two sides by roads and buildings which were to remain operational during construction of the shopping centre. Since the geology of the site was complex, there was an opportunity to investigate for, design and monitor retaining structures in interbedded and variably weathered Carboniferous rocks. This Paper covers the geotechnical design aspects of engineering retaining walls in a layered rock lithology together with comparing predicted wall movements with those measured.

 Retaining structures. Thomas Telford, London, 1993

The site

Location and history

The site is located in Barnsley in the north-east of England and is about 200 m long by 90 m wide at its maximum. The site is on a slope which results in a 12 m ground level difference across the longitudinal direction. A search of old records and maps indicated that the site had been built on since at least 1844. The search also suggested that cellars of previous buildings, up to 3 m deep, had been or were present.

Geology and ground conditions

On the geological map the site was shown to be on drift deposits of variable origin and thickness overlying rocks of the Middle Coal Measures. These Coal Measures included interbedded mudstones, siltstones, sandstones and coal seams. Two coal seams were known to sub-outcrop beneath the site. A third seam was also conjectured to sub-outcrop. As a result of the existence of coal seams, enquiries were carried out to ascertain whether mining of the seams had taken place beneath the site. It was established that the shallowest known worked seam was beneath the east of the site at about 140 m depth. A second deeper worked seam was at about 500 m depth.

Since the site was partly derelict, the site investigation had to be performed in two phases, before and after demolition of the remaining buildings, to ensure there was borehole coverage of the whole area. The investigations comprised rotary core drilling and probing together with machine dug trial pits and trenches. On completion of the second phase it was concluded that one major fault crossed beneath the site and that the shallow coal seams had not been mined. About 85% core recovery was achieved during drilling and the RQD's measured ranged from 0 to 100%. Standard Penetration Tests (SPT's) were carried out between core runs and piezometers installed in the boreholes to establish the groundwater regime. Laboratory tests were performed on core samples. The trial pit excavations allowed detailed examination to be made of the near surface rocks, faulting and made ground. The presence of a fault was confirmed trending north west to south east. Monitoring of the piezometers indicated that there were significantly different groundwater regimes on either side of the fault. Fig. 1 shows typical geological cross sections.

Brief summary descriptions of the strata encountered are as follows

- Made Ground (ground level to 1.3/4.0 m). Demolition rubble comprising brick, sandstone, metal and timber.
- Drift (1.3 to 2.2/4.6 m). Initially soft to firm silty clay becoming stiff with depth. The proportion of rock fragments also increased with depth.
- Middle Coal Measures (2.2/4.6 proved to 74.6 m).Interbedded mudstone and siltstone with subordinate sandstone horizons. Six discrete

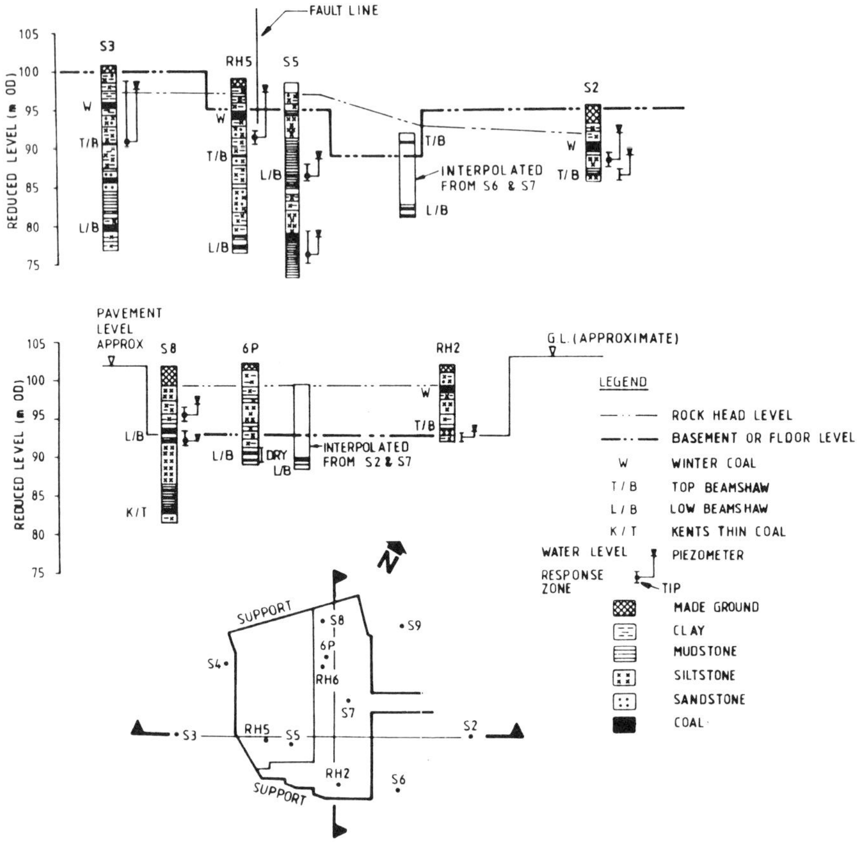

Fig. 1. Typical geological cross sections

coal seams with associates seatearths were revealed during the site investigation. Generally the top 3 m was completely weathered to silty clay with rock fragments. The effects of weathering decreased with depth until only discolouration was present on discontinuities. The results of the SPT's are plotted against depth in Fig. 2. The 'N' values ranged from 5 in the completely weathered zone to cover 300 at about 18 m below rockhead. Unconfined uniaxial compressive tests on mudstone cores gave strengths from 70 to 1800kPa. Point Load Index tests yielded Is(50) values of between 100 and 2500kPa for the mudstone, 100 to 4500kPa for the siltstone and 1500 to 4300kPa for the sandstone. The coal seams, varying in thickness from 0.2 to 1.3 m, comprised bright bituminous well fissured very weak to weak coal.

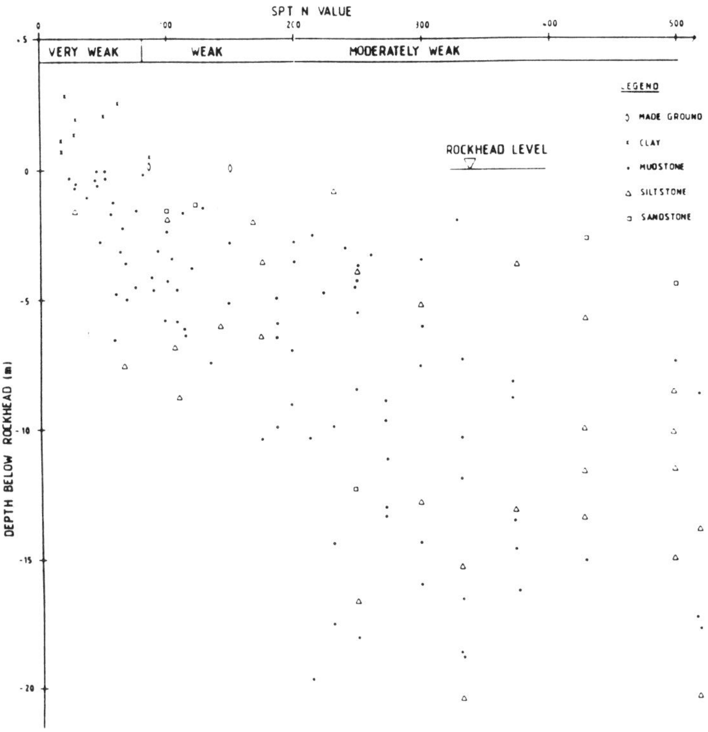

Fig. 2. Standard penetration test results

The project

The development comprises a multi-storey shopping centre and associated car park. The shopping centre, which is about 200 x 90 m, contains four floors. Due to the existing slope in ground level along the Site, some floors are below ground level. The service basement is totally below the existing ground level and is between 9 and llm below adjacent roads and properties. To minimise movement of the ground adjacent to the basement, upholding works were proposed to support the ground during basement excavation on two sides. The other two sides within the Site where space was available, were to be battered back. The basement itself would comprise a reinforced concrete box structure. A drainage layer along the sides and beneath the base of the box was to be installed to minimise uplift pressures from any rise in ground water level. In the permanent case, the upholding works were to be propped by floor slabs at existing ground and first floor basement levels. The introduction of these floors effectively introduced a lid to the basement box structure.

Basement retaining structures

General

Figure 1 shows the extent and depth of the basement together with the ground conditions. At the northern and southern ends of the basement, the ground had to be supported by retaining structures to limit ground movement at the adjacent public roads. After considering different types of retaining structure, contiguous bored pile walls were selected as being the most suitable for the ground conditions. Since bored piles were being used to support the Centre's structural columns, there was an added cost benefit related to piling rig mobilisation (Mirzabaigian and Curtis, 1991). The walls were to be formed with 750 mm diameter piles at 900 mm centres.

Temporary lateral support

With the basement excavation being about 10 m deep on average, it was not considered feasible for the piles to cantilever over this depth while restricting adjacent ground movements to magnitudes which would not damage roads and services. It was therefore necessary for there to be temporary support measures to the pile walls when the depth of excavation exceeded about 4 m. The type of temporary support would be governed by the main contractor's programme of working and the cost. Two alternatives were assessed; internal propping as the excavation proceeded or tying back the walls. The tie back option was selected primarily because it maximised the working space within the Site. Drilled and grouted rock anchors extending below the public roads formed the tie backs. Permission was obtained from the Highways Authority to carry out this operation.

Contiguous pile walls

Design

The first problem to be addressed was how the ground water would behave during the excavation. The site investigation had shown that the geological fault was acting as a dam with groundwater at 1.5 m below existing ground level on one side and no specific groundwater level on the other side.

At design stage, two cases were considered

(*a*) the retaining wall at the side of the fault with no specific groundwater level and in any case below the basement level.

(*b*) the retaining wall at the side of the geological fault with high groundwater level. The water level at the basement side was taken as at the excavation level at different stages of the operation with a three metre head behind the wall.

The length of the piles required for stability requirements were obtained using the OSAYS STAWAL computer program.

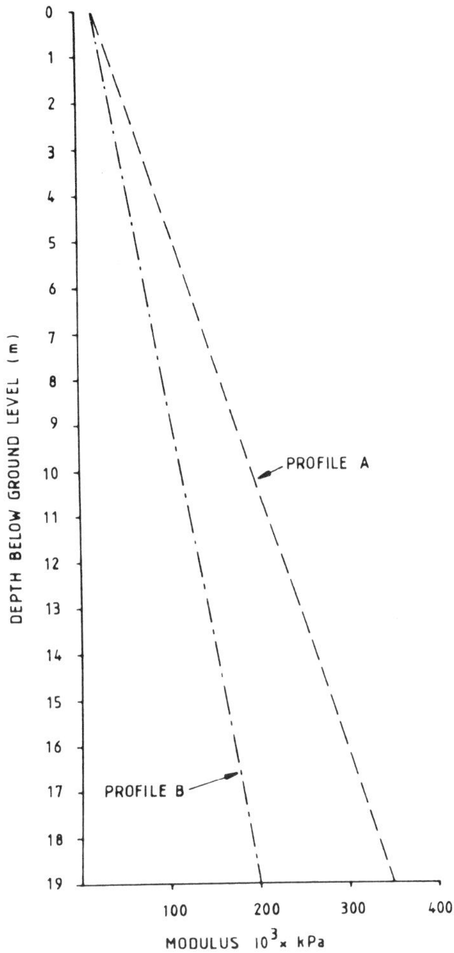

Fig. 3. Ground modulus profiles

The design analysis for the prediction of wall movement during and after excavation was carried out utilising the OASYS computer program FREW developed by Arup (Pappin *et al*, 1985). The input parameters for this program include modulus, earth pressure coefficients and density. From the laboratory tests on the rock core it was concluded that the core results gave the strength of the best intact material rather than the mass strength. The SPT's were considered to better represent the mass of the material and hence they were used in the derivation of modulus using the work of Stroud (1988). Since the predominant strata were mudstone and siltstone with SPT 'N' values generally less than 300 over the depth of the wall, it was considered that the relationships between SPT and modulus proposed by Stroud for overconsolidated clay could be safely adopted. As the basement construction programme was not of short duration, it was decided to use drained modulus values. The plot of SPT results against depth shows a considerable scatter which is not unexpected given the layered nature of the ground. Two SPT design lines were selected; one being regarded as an 'average' and the other as a 'lower bound'. Using a drained modulus E' of 130 x Cu where Cu was based on 5 x 'N' with these design lines, two modulus profiles, A and B, were produced as shown on Fig. 3. The earth pressure coefficients were derived from an assumed ϕ' range of 30 to 35° with C' of zero.

FREW models anchors as a prestress force with a stiffness. If the design anchor load was applied then once the berm had been removed, the wall near the point of application of the anchor would not move, resulting in high pressures behind the wall causing localised high moments and forces in the wall. The design anchor loads, which were variable due to different wall configurations and anchor levels, were determined from hand and computer analyses and lay in the range of 460 to 750kN. It was considered desirable to

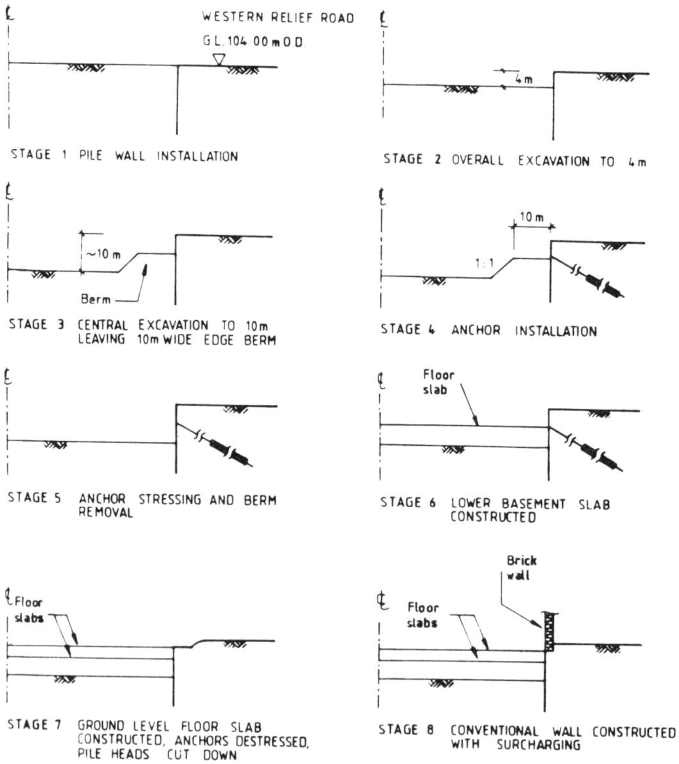

Fig. 4. Staged construction sequence for basement design

reduce the earth pressures behind the wall and consequently the forces and movements in the wall. This was achieved by only pre-stressing the anchors to 80% of their design load to allow some movement of an acceptable magnitude. To model this, the total anchor extension was calculated for the design load based on the stressed free length and then the extension compatible with the increase in anchor load from 80% to design load was calculated. This latter extension was used to estimate a realistic anchor stiffness.

Construction sequence

The staged construction sequence followed in the design is shown in Fig. 4.

The wall piles, 750 mm diameter and at 900 mm centres, were constructed with conventional rotary auger equipment with temporary casing to rockhead. Because of the presence of stronger siltstone and sandstone layers within the Coal Measures, it was anticipated that the auger could have penetration problems. The piling contractor was therefore instructed to provide core barrels for use wherever augering was not achieving progress. Full length reinforcing cages were installed in the piles with additional reinforcement being fixed local to the anchoring level.

The sequence of anchor construction was as follows.

1. Boreholes installed using TCH drilling rig equipped with a rotary-percussive head.
2. Cement grout injected (water cement ratio 0:45) using a tremie pipe to bottom of the hole.
3. When grout loss less than 500 mm per minute, along the borehole, the pre-assembled tendon installed.
4. Within 1 to 4 hours after the primary grout was placed, the secondary grout pipe was pressure grouted.
5. The anchors were stressed 7 days after grouting or after the grout had reach $30N/mm^2$.

Three anchors were subject to on-site suitability tests with all remaining anchors being tested for on-site acceptance test.

Monitoring of inclinometers

Six inclinometer access tubes were incorporated in selected piles. The monitoring, carried out by Arup, was most frequent following the completion of a construction stage. The monitoring equipment comprised the inclinometer torpedo, digital read out unit and data logger. For monitoring purposes, it was assumed that the toes of the inclinometer tubes were fixed. The position of the heads of the tubes were not surveyed due to the practicalities of this operation on the site.

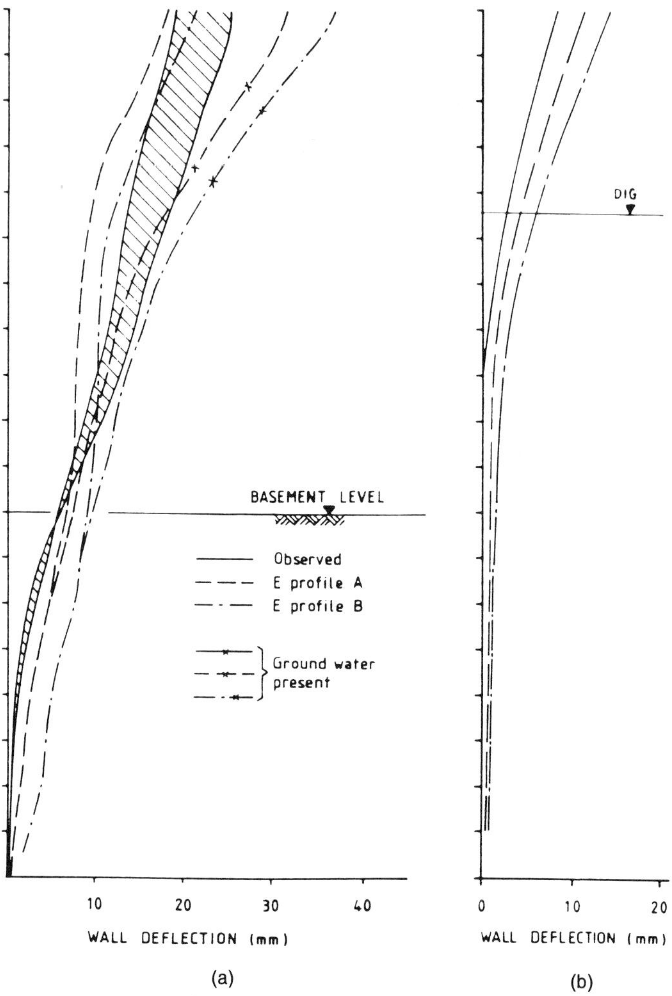

Fig. 5. Predicted and monitored deflections for: (a) Stage 5; (b) Stage 2

Comparison of predicted with actual pile wall movements

The predicted movements and those monitored for two inclinometer access tubes are shown in Fig. 5(a) for Stage 2 of construction with the wall as a cantilever and in Fig. 5(b) for Stage 5 which was the anchor stressing and removal of the berms. As previously mentioned, two groundwater regimes were analyzed to model the conditions on either side of the fault. The inclinometer results in these figures are for both sides of the fault. It is apparent that the monitored movements correspond more closely to those predicted using modulus profile A.

Conclusions

The following points have been concluded from the design, construction and monitoring of the contiguous bored pile walls.

1. Auger bored cast in situ concrete piles proved to be a suitable pile type for the layered strata present in the Middle Coal Measures rocks at the site.
2. Derivation of soil parameters from SPT values is appropriate in these ground conditions.
3. Numerical methods incorporated in computer programs such as FREW, used in this project, can provide good predictions of flexible wall behaviour.

Acknowledgements

The Authors would firstly like to thank Greycoat Shopping Centres Ltd for permission to publish this Paper. The piling and anchoring works were carried out by Bachy-Pigott Ltd and the site investigation by Geotechnical Engineering Ltd. The Authors would also like to thank our colleagues at Ove Arup and Partners (Manchester) for their support, in particular Andrew Morrison, who supervised the piling .

References

MIRZABAIGIAN H. and CURTIS D.C. The performance of bored piles, used as foundations and retaining walls, in the Middle Coal Measure. *Proc. 4th Int. Conf. on Piling and Deep Foundations, Stresa,* 1991.

PAPPIN J.W. *et al.* Numerical analysis of flexible retaining walls. *Proc. Midland Geotechnical Society on Computer Applications in Geotechnical Engineering. University of Birmingham,* 1986.

STROUD M.A. The Standard Penetration Test - Applications. General Report and State of the Art Review. *Conf. on Penetration Testing in the UK, Institution of Civil Engineers, Birmingham,* 1989.

Reliability of earth pressure measurements adjacent to a multi-propped diaphragm wall

M.L. LINGS, D.F.T. NASH and C.W.W. NG, University of Bristol, UK

Field monitoring data from one panel of a multi-propped diaphragm wall in a stiff fissured clay have been analyzed to investigate their internal consistency. Measured total earth pressures have been assessed in the light of measured prop loads and wall rotations. The raw earth pressure data are shown not to satisfy the fundamental requirements of equilibrium, and to fit poorly with measured wall rotations. However net earth pressure distributions have been constructed which do show consistency with all the other data. These imply that the measured earth pressures, while giving the right general picture, may be over or under recording by up to a factor of two.

Introduction

When carrying out field monitoring, it is desirable to record sufficient independent quantities to allow some degree of internal cross-checking on the measured values. This is important to avoid making deductions from one set of data whose validity has not been tested. Recently the authors have been monitoring a deep multi-propped excavation in Gault Clay situated in the centre of Cambridge. The 65m x 45m Lion Yard site has been developed as a three level underground car park with a hotel complex above ground. The structure is founded on large diameter bored piles, and the 10m deep excavation, retained by a perimeter concrete diaphragm wall, has been constructed top-down. A summary of the ground conditions, the instrumentation, the construction sequence, and the observed behaviour, is given by Lings et al (Ref.1). A cross section of the site is shown in Fig.1.

At Lion Yard, total earth pressures, prop loads, reinforcement stresses and wall rotations have all been measured, and it has therefore been possible to assess how well the measured earth pressures agree with the other data. Bolton and Powrie (Ref.2) drew attention to the importance of satisfying equilibrium when considering earth pressure measurements obtained from an in-situ device in the ground adjacent to an embedded wall. So far as we are aware, there have been few attempts to relate observed pressure distribu-

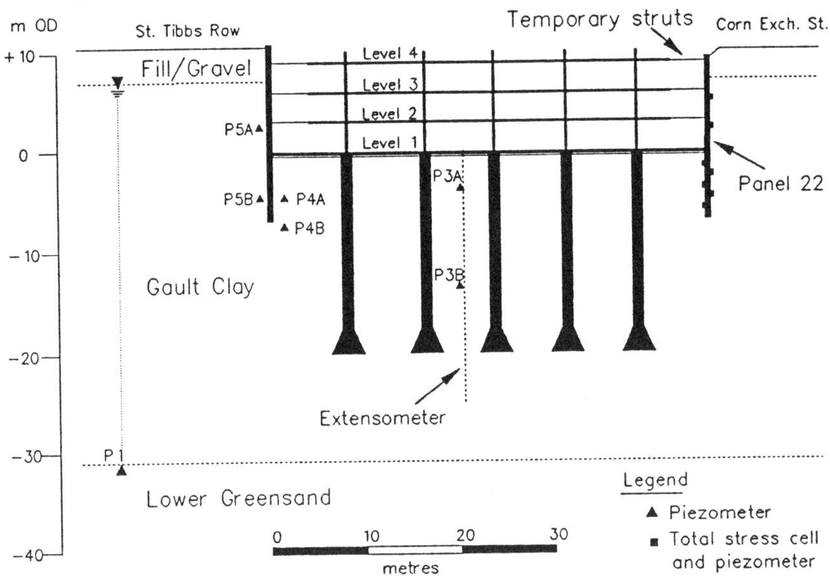

Fig. 1. Cross section through the site

tions to wall curvature.

Total earth pressures are particularly important, as they form the starting point for an assessment of the effective stresses in the soil adjacent to the wall. Information on the effective stress changes is needed if our understanding of soil-structure interaction is to be improved.

From beam bending theory it is well known that there are relationships between applied loading, shear force and bending moment. In each case, conversion is possible from one to the next by integration, provided the constants of integration can be determined from boundary values. At the bottom of a diaphragm wall, the shear force and bending moment diagrams must go to zero, thus ensuring that the applied pressures and loads satisfy force and moment equilibrium. In a similar way, curvature, rotation (or slope) and deflection can be derived from one another by successive integration. Conversion from bending moment to curvature requires a knowledge of the stiffness of the section. Thus a net pressure diagram can be converted in turn into shear force, bending moment and rotation diagrams. Initially it had been hoped that the wall bending moments would be known at the reinforcement load cell locations, and that this would provide an additional cross-check on any derived bending moment diagram. However a detailed analysis of the wall by Ng et al (Ref.3) has shown that there is considerable uncertainty in the interpretation of bending moments from field measurements on concrete walls.

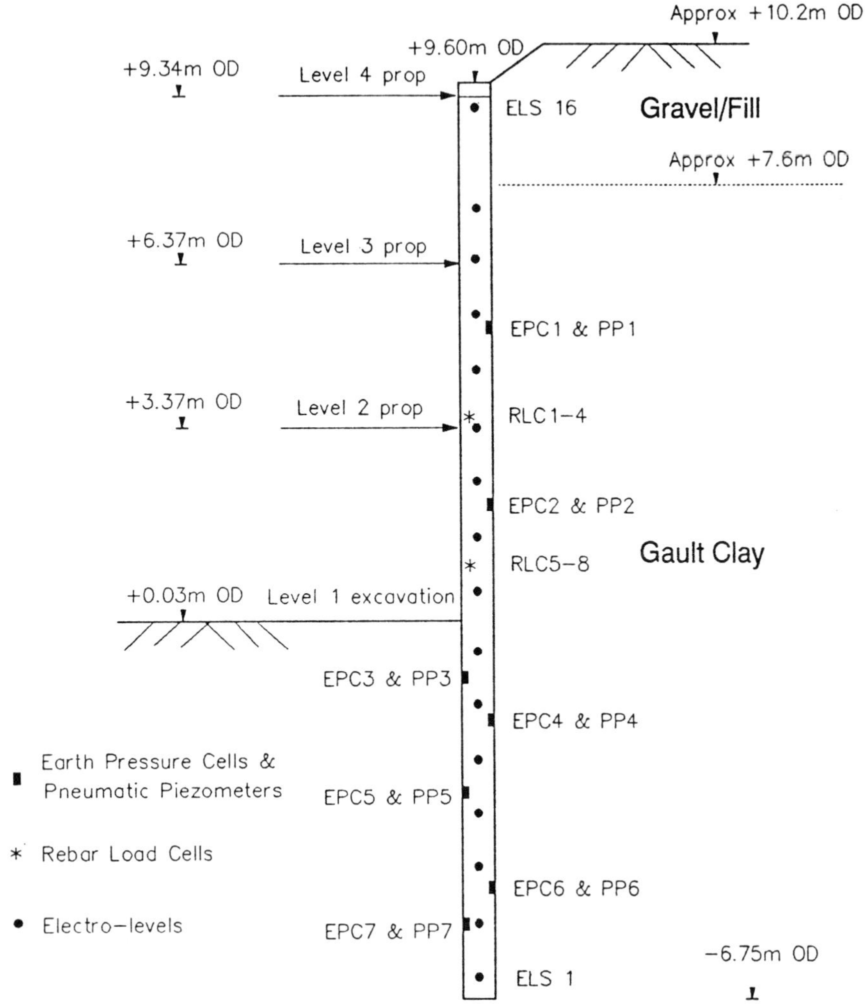

Fig. 2. Panel 22 showing the instrumentation

In order to investigate the relationships between various types of data, and assess the equilibrium of the wall under various earth pressure distributions, a numerical integration technique has been developed using a spreadsheet program. This allows an initial net earth pressure diagram to be constructed, from which shear force, bending moment and rotation diagrams can be derived. These different diagrams can all be viewed simultaneously, together with any values of raw data, and the net earth pressures adjusted progressively until all the boundary conditions are satisfied. In practice this means zero shear force and bending moment at the toe of the wall and a good fit with the measured rotation data.

Lion Yard instrumentation

One diaphragm wall panel (panel 22) near the centre of one long side of the site was chosen for instrumentation. Total earth pressure cells, together with pneumatic piezometers, were installed on both faces of the diaphragm wall. Temporary strut forces were measured by load cells at all levels of the excavation. Reinforcement load cells were incorporated in two sections of the wall, although since their data have not been used in this analysis they are not described here. A train of electro-levels was installed for recording wall rotations. The location of the monitoring equipment within the instrumented panel is shown in Fig.2.

Earth pressure cells

Seven 150mm x 250mm mercury-filled Glötzl earth pressure cells were installed on the wall/soil interface by attaching them to the reinforcement cage. When the cage had been lowered into the bentonite, they were jacked out onto the surface of the Gault Clay immediately before the panel was concreted. This technique has been used in the past in boulder clay (Ref.4) and in soft clay (Ref.5), but this is believed to be the first reported use in a diaphragm wall in stiff fissured clay.

Prop load cells

This was an unusual site in that loads could be measured in the temporary steel props placed across the relatively narrow openings at each slab level left for future concrete vehicle ramps. Each 8.5m long diaphragm wall panel adjacent to the ramp openings had five props at each car park level. Three out of five props at each level on panel 22 were fitted with vibrating wire load cells, mounted at the centroid of the I-section.

During the course of construction, significant load variations with time were observed apart from those due to excavation, and these can be attributed to changes in temperature. Subsequent tests on the load cells after removal showed their calibration factors to be very insensitive to changes in temperature, so that the variations recorded are considered to represent real changes in load. When the props were removed after completion of the concrete vehicle ramps, all cells recorded zero load. The measured loads are therefore considered to be reliable.

Electro-levels

A train of sixteen electro-levels was installed near the centre of panel 22 inside an inclinometer tube, which had been grouted into a duct cast into the diaphragm wall panel. Initially there were considerable problems with the stability of the electro-level data, some devices showing significant drift over a period of many weeks. This has meant that the rotation data are not very satisfactory leading to some difficulties in interpretation.

Assumptions of analysis

A general principle has been adopted here that best estimated values from the raw data should be used unless and until such values become impossible to reconcile with other data. When this has happened, earth pressure diagrams have been chosen as close as possible to the measured values, while still satisfying equilibrium. The distributions chosen are as far as possible smooth curves, with a minimum number of points of contraflexure.

The values of prop load used in the analysis are the average of the three measured values at each level. Other possible loads between the maximum and minimum recorded values were used to investigate the effect on the analysis, but this was found to result in only small changes in the earth pressures required to satisfy equilibrium.

The stiffness of the wall has been calculated assuming an uncracked section. This assumption may not be valid in some parts of the wall at the final excavation stage, but is considered to be a reasonable choice in the context of all the other uncertainties. A full discussion of the behaviour of the concrete wall is given by Ng *et al* (Ref.3). The contribution that wall friction makes to wall equilibrium can be significant, but its effects have not been included in this analysis. At Lion Yard there is uncertainty not only about the magnitude of the shear stresses acting on the wall, but also about their direction, since vertical loads on the panel from the superstructure gradually increased during construction. Trial calculations have indicated that the incorporation of wall friction into the analysis, acting in either direction, has a relatively minor influence on the derived earth pressures. In a similar way, base shear has also not been included in the assessment of equilibrium.

Calculation process

The data for level 1 excavation were looked at first, as this stage contained the maximum number of propping levels, thus providing the strongest check on equilibrium. Initially earth pressure distributions were chosen which fitted exactly with the raw data. From this starting point it was not possible to satisfy equilibrium. Reducing just one result substantially below its measured value (EPC1) meant that equilibrium could be satisfied, but the bending moments were much too high. Inclusion of full wall friction on both sides of the wall led to reduced bending moments, but they were still unacceptably high. At this point rotation diagrams were derived and com-

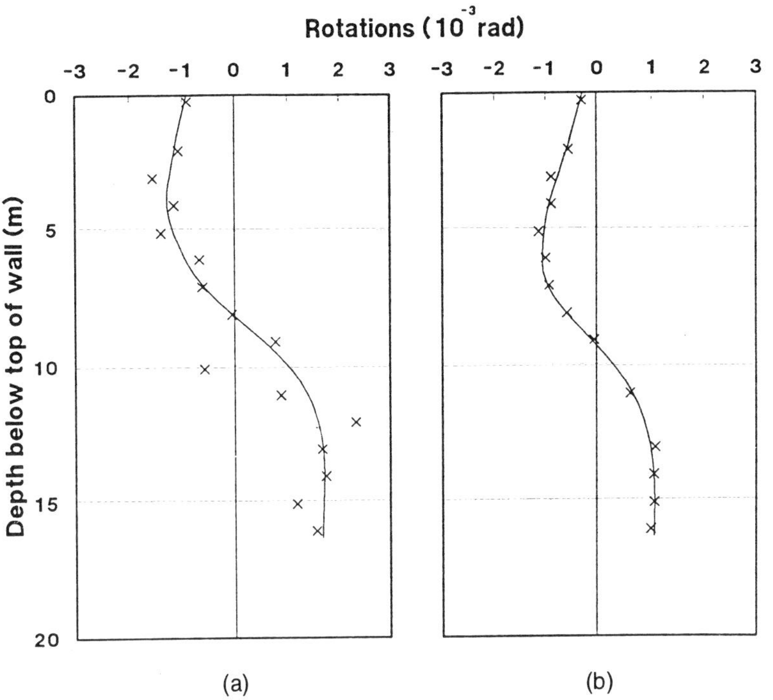

Fig. 3. Wall rotation data: (a) Level 1 excavation; (b) Change between level 2 and level 1 excavation stage

pared with the rotation data, and this showed poor agreement. A good fit could only be achieved if earth pressures were chosen which generally differed from their measured values.

Changes in earth pressure between the level 2 and level 1 excavation stages were looked at, as this enabled a comparison to be made with changes in rotation data over this period. Because of the relatively short time interval, and because the majority of the early electro-level drift had already taken place, a much smoother set of rotation data was available for curve fitting. Again the initial approach was to construct distributions of change in earth pressure which fitted the measured data, but although equilibrium could be satisfied, agreement with the rotation data was very poor. An acceptable fit could only be achieved if earth pressure changes were chosen which often differed considerably from the measured values.

Earth pressure diagrams for level 2 were derived by combining the level 1 earth pressures with the earth pressure changes between the level 2 and level 1 excavation stages. This approach was adopted because of the very

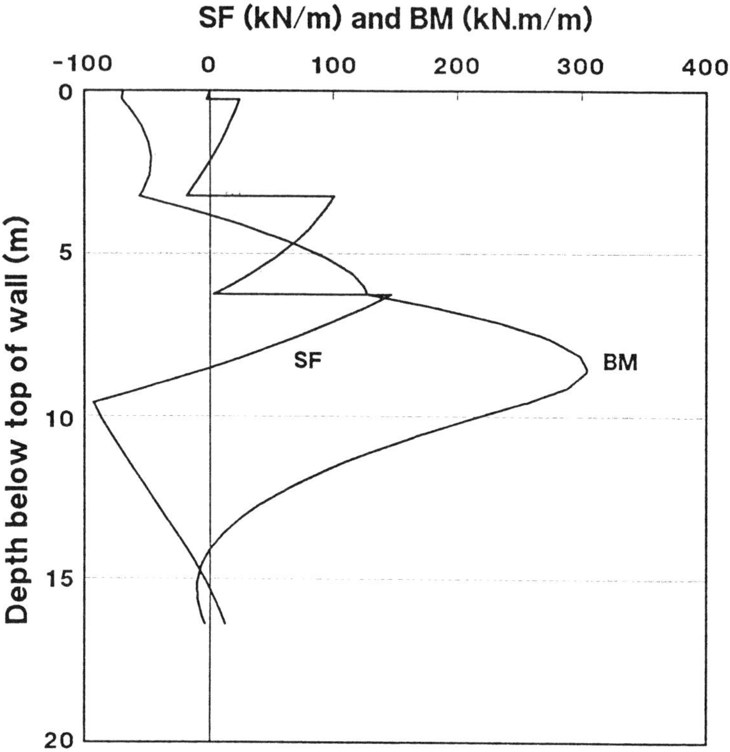

Fig. 4. Shear force and bending moment diagrams for level 1 excavation

scattered nature of the level 2 rotation data, which made curve fitting very unsatisfactory. Once derived, the level 2 earth pressure diagrams had to be reasonably smooth and contain no areas of negative pressure. This process required several iterations.

Results

The measured rotations for level 1, and changes in rotation between the level 2 and level 1 stages, are shown in Fig.3, together with the derived curves. Bending moment and shear force diagrams for level l are shown in Fig.4. In order to obtain satisfactory agreement with the measured rotation data for level l, it was found necessary to permit a significant continuity bending moment at the top of the diaphragm wall. This is considered to be reasonable, as the wall panel supports one of the major concrete columns for the superstructure, which has a full moment connection.

Net earth pressure diagrams for level l and level 2 are shown in Fig.5. A

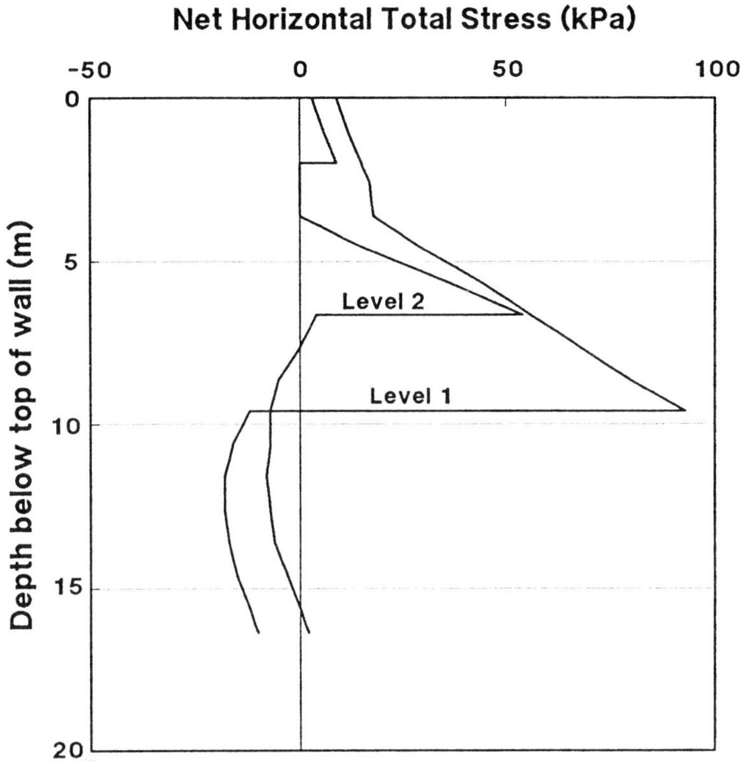

Fig. 5. Net earth pressure diagrams for level 1 and level 2

possible earth pressure diagram for the back and front of the wall for level 1 is shown in Fig.6, together with the measured values of earth pressure at the seven cell locations for this stage of construction.

Discussion

The technique of numerical integration is an inherently stable process and has been adopted here as a simple though somewhat tedious means of relating earth pressures to rotation data. The potential instability problems with the alternative technique of numerical differentiation are well known. More fundamentally, even if a good set of rotation data had been available, and a smooth curve had been fitted to it, the derived curvature/bending moment diagram would not have contained the necessary abrupt changes in slope at the prop locations. Without these, the shear force and net earth pressure diagrams could not be derived.

The chosen method of analysis takes as its starting point a net earth

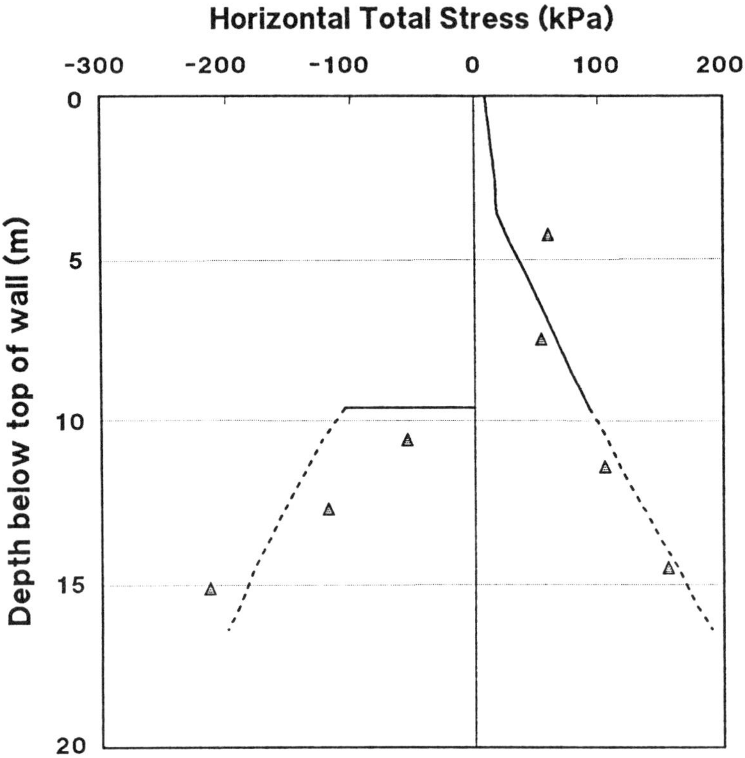

Fig. 6. Possible earth pressure diagram for level 1 showing measured values

pressure diagram. Below excavation level there is a whole range of pressures on the front and back of the wall which can yield the same net pressure diagram. With no means of determining the absolute values, the pressure distributions shown in Fig.6 have been chosen to give smooth curves that agree as closely as possible with the measured values. The earth pressures below excavation level have been shown dotted to emphasise their tentative nature. Nevertheless the chosen profile in front of the wall is not entirely arbitrary. The level 1 net earth pressure diagram (Fig.5) shows an abrupt step of approximately 100 kPa at excavation level. This was found necessary to satisfy equilibrium and give good agreement with the measured rotations. Thus the earth pressure in front of the wall at excavation level must be approximately 100 kPa to avoid an unrealistic step in the earth pressure diagram behind the wall. For level 2, the corresponding value at excavation level is approximately 50 kPa.

The earth pressure diagram (Fig.6) demonstrates that the performance of the earth pressure cells is somewhat variable. The highest cell behind the wall (EPC1) is clearly over-reading, while the highest cell in front of the wall (EPC3) is clearly under-reading, in each case by a factor of approximately two. On the other hand, when the installation technique is taken into account, namely that the cells were jacked out onto a roughly prepared surface in a stiff fissured clay, it is perhaps surprising to find as good a measure of agreement as is shown.

The net earth pressure diagram for level 1 (Fig.5) shows that the net horizontal total stresses below excavation level are small (approximately 20 kPa) but negative, that is the pressure in front of the wall is greater than that behind. This contrasts with the raw data, where the measured values in front of the wall are generally lower than those behind, apart from the lowest cell. In fact the raw data suggest that the pressures in front of the wall may approach zero at excavation level, whereas the equilibrium analysis clearly requires a substantial positive value.

A comparison of the two diagrams in Fig.5 reveals that, on carrying out the final stage of excavation, there is an increase in net horizontal total stress in front of the wall below excavation level. More significantly, there is an increase in horizontal total stress behind the wall above penultimate excavation level. This seems to imply that the wall is sufficiently flexible for some stress redistribution (arching) to occur, since during the final excavation stage the wall continued to move towards the excavation. It proved impossible to construct diagrams of change of earth pressure which both fitted the measured data and satisfied the internal checks. Some variation in nearly all the values was required, although the direction of stress changes appeared to be recorded correctly. This result is a little disappointing, as it had been hoped that the earth pressure cells might at least be recording changes in pressure correctly. This does not appear to be the case, which means that effective stress changes inferred from measured horizontal total stress and pore water pressure changes are not likely to be particularly accurate. A general problem with this type of analysis is that there are too many variables and uncertainties present to be able to put any realistic limits on the accuracy of the derived earth pressure diagrams. Undoubtedly the analysis would have been improved with better wall rotation data. This type of analysis would be worth repeating again in the future if an accurate set of prop load data and a reliable and comprehensive set of wall rotation data were available. Further work is required to assess the degree of precision required from the field data before it becomes possible to infer the net earth pressures with some confidence, rather than just carrying out an approximate internal check on the measured values as has been done here. However it is anticipated that the net earth pressure diagrams will be useful in assessing results from the ongoing back-analysis.

Conclusions

When considering earth pressure measurements in the ground adjacent to an embedded wall, it is important to check that equilibrium requirements are satisfied. An analysis of all the available field data recorded at a diaphragm wall panel installed in a stiff fissured clay has enabled an internal check to be carried out on the measured earth pressure values. Specifically, it has been possible to construct net earth pressure diagrams which, in conjunction with measured prop loads, satisfy the requirements of horizontal force and moment equilibrium, and after triple integration show agreement with measured wall rotations. These earth pressure diagrams have then been compared with the measured values of earth pressure.

Net pressure diagrams satisfying these requirements indicate values of horizontal total stress acting on the front of the wall at excavation level of around 50 and 100 kPa at penultimate and final excavation levels respectively. The net horizontal total stresses below excavation level are in both cases small, having values less than 20 kPa. The final stage of excavation shows an increase in horizontal total stress behind the wall above penultimate excavation level. This pattern of stress redistribution (arching) is consistent with the changes in earth pressure recorded by the instrumentation.

In general the earth pressure cells are not correctly indicating the earth pressures, since it has not been possible to construct earth pressure diagrams incorporating the measured values which satisfy all the necessary internal checks. However they do give broadly the right picture, which is perhaps surprising in view of the method of installation. No consistent pattern of over-read or under-read is discernible in the measurements, which differ from the derived diagrams by up to a factor of two.

It also appears that the earth pressure cells do not indicate changes in earth pressure with any great accuracy, although the direction of change is consistent with the derived earth pressures. This implies that changes in effective stress in the soil adjacent to the wall, inferred from measured total earth pressure and pore water pressure changes, are not likely to be particularly accurate. This study has demonstrated the importance of checking the internal consistency of raw field data, particularly where earth pressure measurements are concerned. An accurate knowledge of the earth pressures acting on a wall can help our understanding of soil behaviour: there is clearly a danger in taking field measurements at their face value.

Acknowledgements

The Authors gratefully acknowledge financial support from JT Design Build and the Science and Engineering Research Council, the help given by the Building Research Establishment, and the co-operation of JT Design Build during construction.

References

1. LINGS M.L. *et al.* The observed behaviour of a deep excavation in Gault Clay: a preliminary appraisal. *Proc. Tenth European Conference on Soil Mechanics and Foundation Engineering, Florence, Italy,* 1991, Vol. 2, pp 467-470.
2. BOLTON M.D. and POWRIE W. Discussion on Performance of propped and cantilevered rigid walls. Fifth Geotechnique Symposium in Print. *Geotechnique* ,1985, No. 4, pp 533-548.
3. NG C.W.W., LINGS M.L. and NASH D.F.T. *Back-analyzing the bending moment in a concrete diaphragm wall.* Submitted for publication to the Institution of Structural Engineers.
4. UFF J.F. In situ measurements of earth pressure for a quay wall at Seaforth, Liverpool. *Proc. Conf. on In Situ Investigations in Soils and Rocks, London,* 1969. British Geotechnical Society, 1970, pp 229-239.
5. DIBIAGIO E. and ROTI J.A. Earth pressure measurements on a braced slurry-trench wall in soft clay. *Proc. Fifth European Conference on Soil Mechanics and Foundation Engineering, Madrid, Spain,* 1972, Vol. 1, pp 473-483.

Large excavation with cantilever type continuous walls

E. SATO, M. AOKI and M. MARUOKA, Takenaka Corporation, Japan

Introduction

In order to construct the basements of buildings in city areas, generally we must build earth retaining walls, support the walls by temporary braces, and excavate the ground. However the supporting work of temporary braces is very consuming of time, money and manpower. In addition, the framed braces will prevent easy construction of basement at later construction stages.

In 1989 the construction of a 25-storied high-rise building in Tokyo, Japan was completed. In the 3-storied basement work of this building, a cantilever type of earth retaining structure, which consists of special shaped reinforced concrete continuous diaphragm walls, was used in order to retain the earth pressure with no supporting brace, and to excavate a rectangular area of 50 by 100m, down to the depth of 17.1m.

Intensive measurements were conducted during and after the construction. The displacement and stress of the wall, lateral pressure and water pressure on the walls were measured.

In this paper, the design for a new type of retaining structure and excavation works will be described. The measurement results during and after the excavation will also be shown in order to compare actual measured behaviour against the designed one.

Construction and ground condition

This office building has 25 storeys above ground, 3 storeys underground, and 137,200 m^2 total floor area as shown in Fig. 1. The construction site is located by the side of the Sumida River, one of the largest rivers in Japan.

The soil profile of the site, as indicated in Fig. 2, consists of an alluvial silty clay layer from the ground surface to GL-29m, a 1.5m thick diluvial clay layer, and a diluvial gravel layer underneath. The thick diluvial silty clay layer is made of soft clay with high natural water content exceeding the liquid limit, and with shear strength of 10-100 kN/m^2. The diluvial gravel layer is the bearing stratum of the foundations. The free groundwater table is at GL-1.8m, while the artesian head in the diluvial gravel layer is at GL-19.5m (Fig.2).

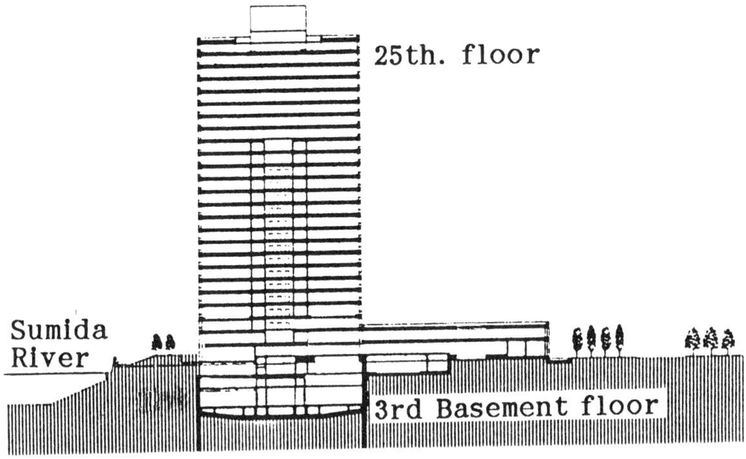

Fig. 1. Section of building

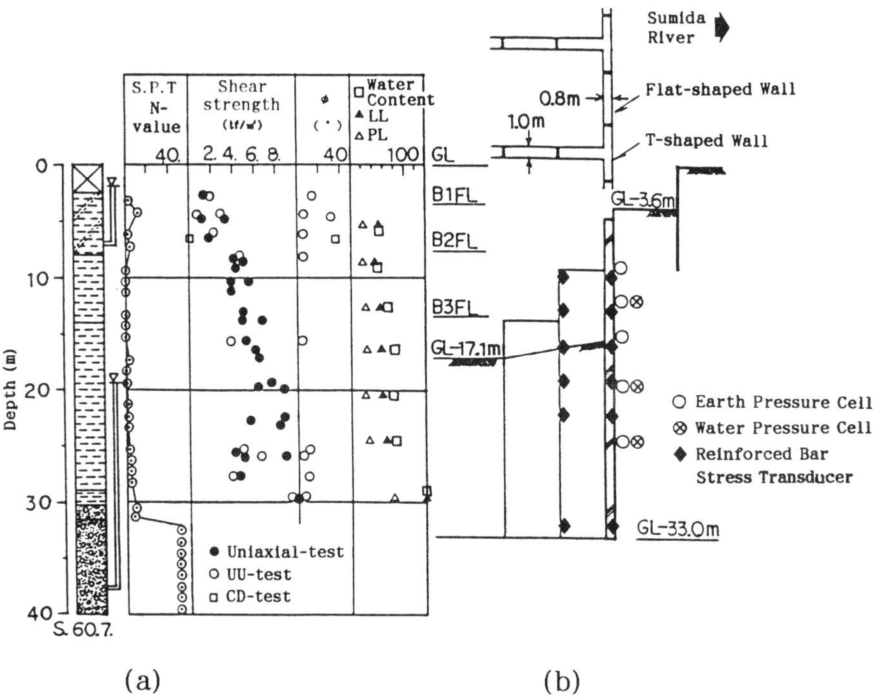

Fig. 2. Outline of construction and ground condition:(a) Ground condition;(b) Section of continuous diaphragm walls

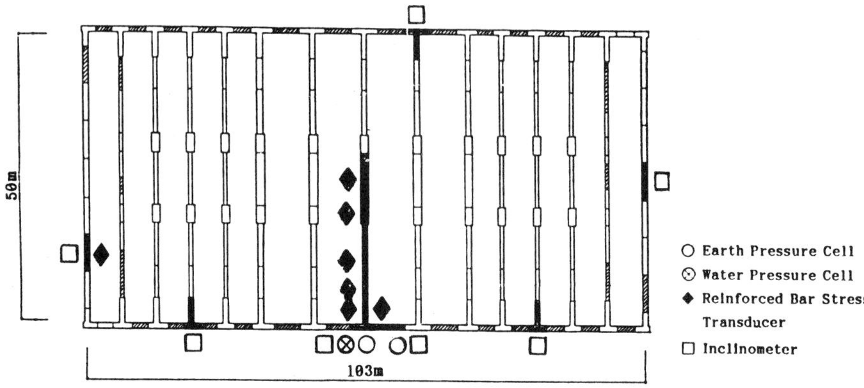

Fig. 3. Plane figure of continuous diaphragm walls

Designing of the basement structure

Outline of basement structure design

The following issues are raised in designing the basement structure

- The heavy vertical loads are concentrated on the core part of the superstructure. Therefore if circular piles are used, they should be very large, and not arranged more closely than the allowable distance.
- Due to the soft alluvial silty clay layer, which lies down to GL-29m, the

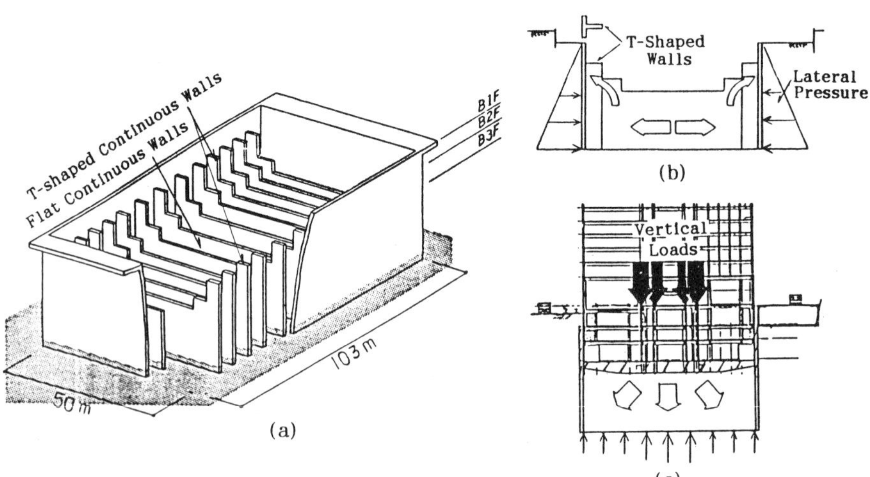

Fig. 4. Schematic continuous diaphragm walls

passive lateral pressure is considered to be small, and large stresses and deformation of the earth retaining wall are expected during the excavation work with ordinary earth retaining methods.

- As the excavation area is large (50x100m), bracing methods are unfavourable in terms of earth retaining structure safety, construction cost, and construction period.

Taking the above-mentioned problems into consideration, the continuous diaphragm walls (Ikuta, 1974) were selected, and 0.8, 1.0, & 1.5m thick walls were arranged as shown in Fig.3. T-shaped continuous walls are placed along the two long sides of the building and are connected to each other by flat continuous walls beneath the bottom of the excavation as shown schematically in Fig.4(a). Those walls are used as temporary retaining walls and also for permanent deep foundations.

T-shaped continuous diaphragm walls have distinctively high flexural rigidity in comparison with conventional flat continuous diaphragm walls. This structure forms a rigid cantilever type of earth retaining structure, therefore it enables excavation work to be conducted down to level of GL-17m without framing braces as shown in Fig.4(b). This method is suitable not only for the large scale excavation works but also the excavation works at the thick soft deposit, because it makes the earth retaining structure unaffected by the low passive lateral pressure.

On the other hand, this rigid foundation is able to bear the large loads from the superstructure, because the concentrated loads are dispersed by rigid foundations to the bearing stratum as shown in Fig.4(c).

Design of Earth Retaining Structure

The stress deformation analysis of the earth retaining structure is carried out by the elasto-plastic design method, which can deal with the elastic and plastic resistance of the excavated side ground (Endo, 1977). The design lateral pressure is set by the Rankine-Resal's equation, and it is set so as not to be lower than the assumed water pressure as shown in Fig. 5(a) . Flexural rigidity of the earth retaining wall is calculated by assuming the section as shown in Fig.5(b).

As the flat continuous diaphragm walls are placed beneath the excavated base, the coefficient of subgrade reaction per 1m of passive side is obtained by dividing the continuous by diaphragm wall stiffness by the intervals of the walls. Resistance of the soil was ignored because the soil was too soft.

Regarding the structure as a soldier beam-lagging structure shown in Fig.6, the horizontal stress is obtained for the horizontal steel bar design. These walls are connected by joints withU-shaped steel bars to transfer shear force.

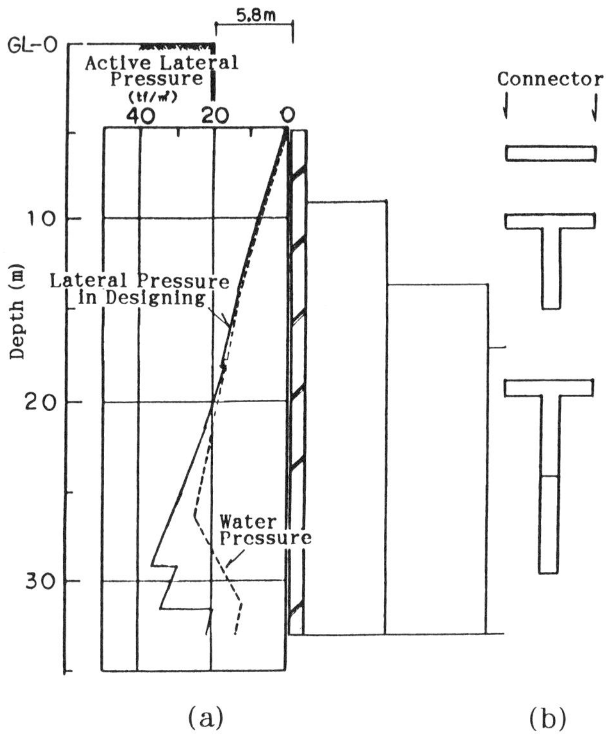

Fig. 5. Lateral pressure and flexural rigidity of wall in designing

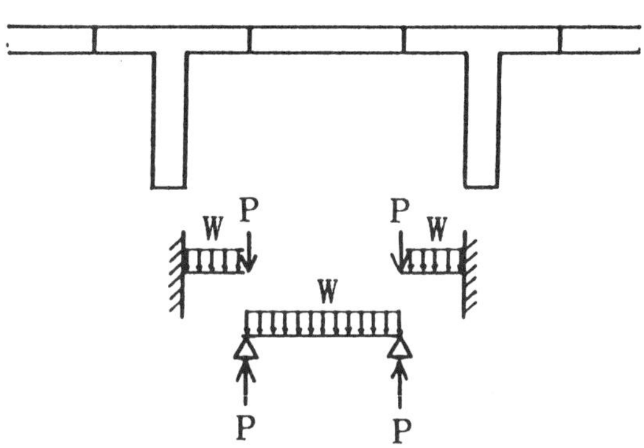

Fig. 6. Design of horizontal rainforcing bar

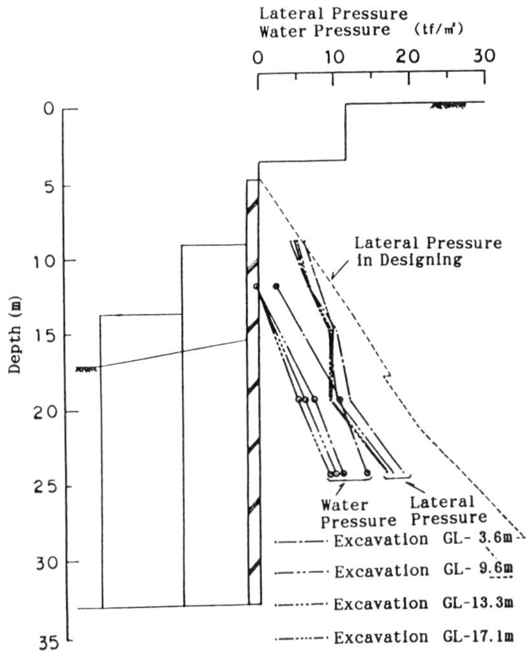

Fig. 7. Measured earth pressure and water pressure

Measurement results of the behaviour

During the excavation work, stress and deformation of the earth retaining structure consisting of continuous diaphragm walls was measured to utilize the data in safety monitoring. Some measurements have been continued after the basement work completion to observe the behaviour during the superstructure construction, and after the construction completion as well.

Measuring instruments such as earth pressure cells, pore water pressure cells, reinforcing bar stress transducers and roller-typed inclinometers, were installed on mainly the T-shaped continuous diaphragm walls, and on flat continuous diaphragm walls adjoining the T-shaped walls. In Fig. 2 and Fig. 3 the position of installed measuring instruments are shown.

In processing the measured data, the value measured before inserting reinforcing cages are set as the initial value for earth pressure and pore water pressure. And the values measured before the excavation work are set for reinforced bar stress and inclination.

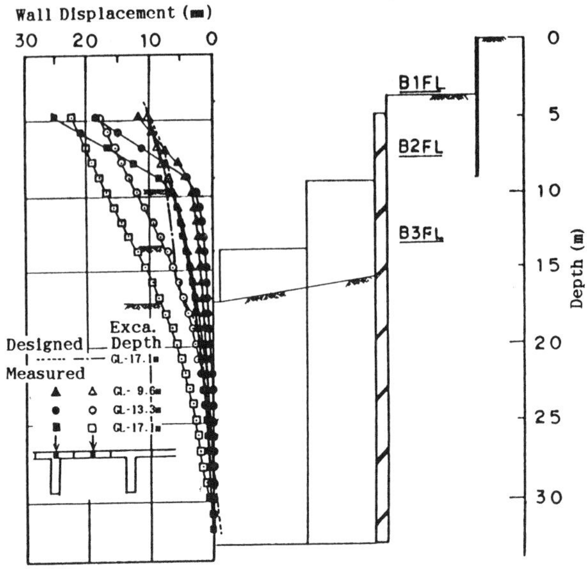

Fig. 8. Measured wall displacement

Lateral pressure and pore water pressure

Active lateral pressure and pore water pressure on the T-shaped continuous diaphragm wall are shown in Fig.7. There is no significant change of lateral pressure throughout the excavation. Equivalent coefficients of lateral pressure between the ground surface and GL-25m are almost constant at the value of K=0.4. And measured lateral pressure is smaller than the designed one at any depth.

Displacement of continuous diaphragm wall

Fig. 8 shows displacement of the T-shaped continuous diaphragm wall and the flat continuous diaphragm wall between T-shaped ones. The displacements of the T-shaped continuous diaphragm wall are small. The maximum displacement observed at T-shaped parts below GL-9. 1m is 8mm, corresponding well with the designed value. Meanwhile, the flat continuous diaphragm wall shows large displacement at the lower part as well as the upper part. It is distinguished from the behaviour of T-shaped wall.

Vertical stress of continuous diaphragm wall

The vertical reinforced bar stresses are shown in Fig. 9. Tensile stresses

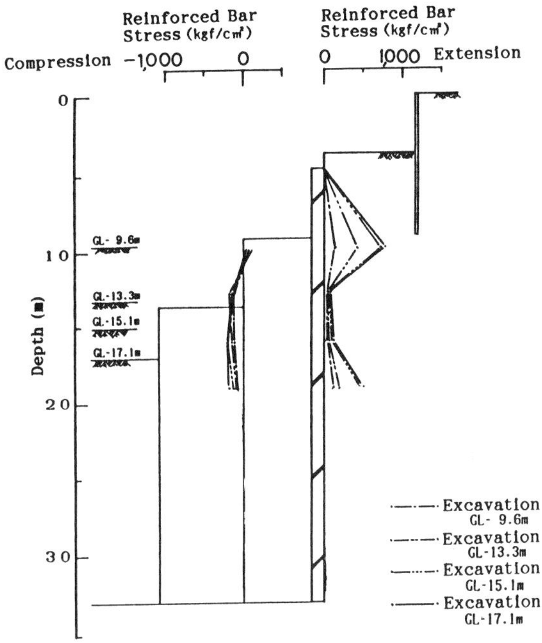

Fig. 9. Measured stress of vertical reinforcing bar

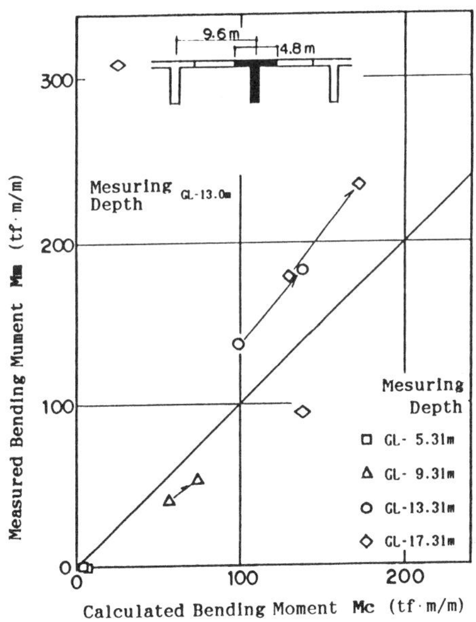

Fig. 10. Vertical bending moment of wall

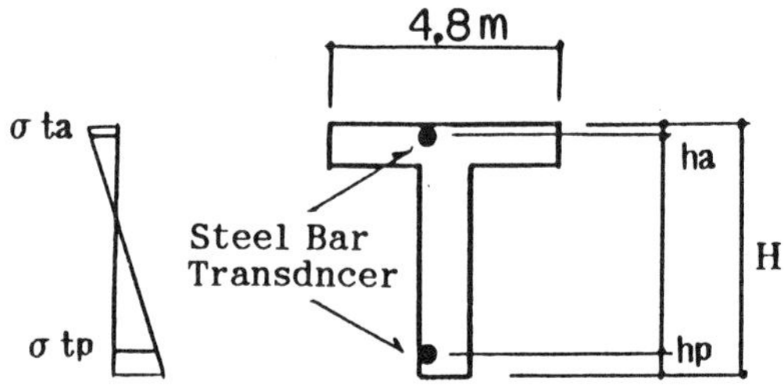

Fig. 11. Calculation of bending moment from reinforcing bar stress

occur in the active side reinforcing bars, reaching the maximum of 800 kgf/cm^2 at the depth where the section changes from the T-shaped to the flat walls. And compression stresses occur in the passive side reinforcing bars, those were lower than 200 kgf/cm^2.

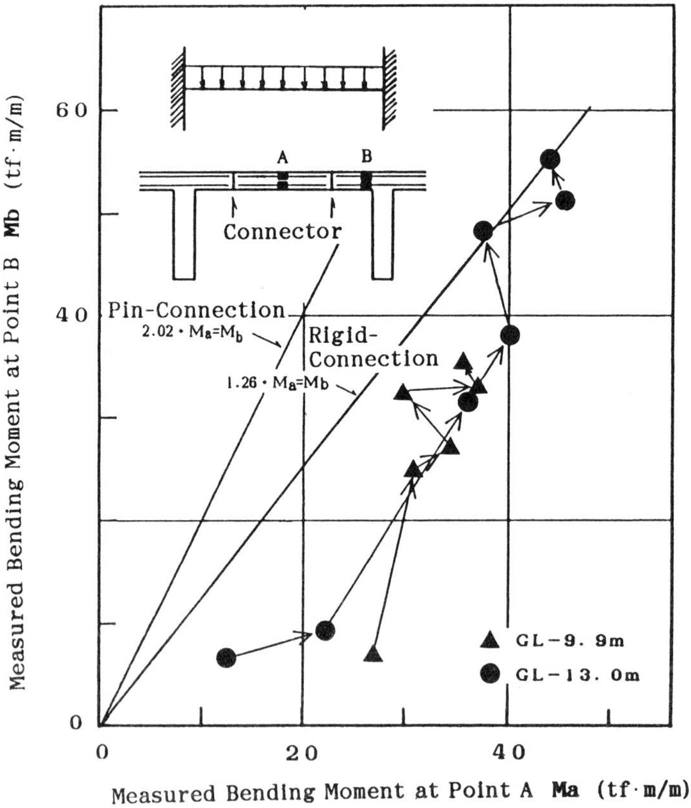

Fig. 12. Measured horizontal bending moment

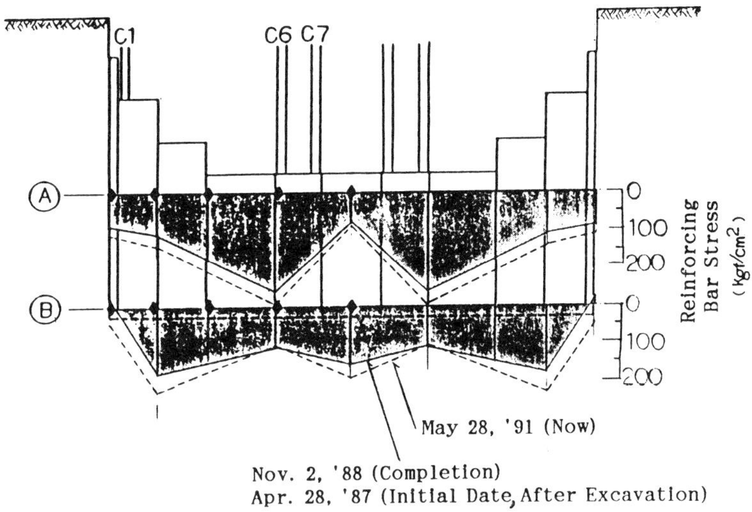

Fig. 13. Stress distribution of reinforcing bar in foundation

Fig.10 compares the bending moment Mm of the continuous diaphragm wall obtained by Eq. (1) with the measured steel bar stress, and the bending moment Mc of the wall calculated with measured lateral pressure.

$$Mm = (\sigma ta - \sigma tp)\, I / n\, (H - ha - hp) \tag{1}$$

where, σta, σtp: measured reinforced bar stress (Fig.ll), n : ratio of Young' s modulus, I : geometrical moment of inertia, H, ha, hp : refer to Fig.11.

In Fig. 10 the measured data in other constructions are also shown. The M m and M c correspond well. This correspondence shows that T-shaped continuous diaphragm walls performed as cantilevers.

Horizontal stress of the continuous diaphragm wall

The horizontal reinforced bars of the walls are designed with assuming the soldier beam-lagging type structure as shown in Fig.6. In order to measure the actual stress, reinforced bar stress transducers are installed on horizontal reinforced bar.

Fig. 12 compares the bending moment Ma obtained by measured horizontal reinforced bar stress at the point A of the flat wall, with the bending moment Mb at the point B of the T-shaped.

The ratio of Ma and Mb depends on the location of measuring points. In the case of the rigid connection between walls, the ratio is 1:1.26, whereas in the case of the pin connection, it is 1:2.02. The ratio of the measured bending moments agrees well with the former ratio. From this measured stress, the connectors are almost rigid.

Stress distribution of reinforcing bar in foundation

The stress distributions of the reinforcing bar in the foundation after excavation work are shown in Fig. 13. The stress were measured on two levels (A B) of foundation.

Although the stresses were measured on just one side, in Fig. 13 the distributions were written with the assumption that other side of foundation suffered the same stress.

The measured values after excavation completion are set as initial values. The stress under the column C6 on the level A is very large. But the stress on the level of B tend to be even. This means the rigid foundation transferred the concentrated loads downward evenly.

Conclusion

The cantilever type of earth retaining structure consisting of T-shaped continuous diaphragm walls was used for large excavation work in soft ground. The behaviours of the earth retaining structure were measured during the construction and those were compared with the assumptions in designing. From the results, it was verified that this construction method is effective for excavation works in soft thick ground, and for large area excavation works as well.

Reference

ENDO M. Introduction of Excavation work in Japan. *Ninth Int. Conf. on Soil Mechanics and Foundation Engineering*, 1977, Vol.4, pp 297-370.

IKUTA Y. Diaphragm walling in Japan. *Ground Engineering*, 1974, Vol.7, No.5, pp 39-44.

Performance of a flexible earth retaining structure in soft clays - comparisons between finite element method and field measurements

H. ABEDI, T. G. PORTER, B. H. LIEN and J. A. RAMOS, Metcalf & Eddy, Inc.,USA

Introduction

The City of Detroit Water and Sewerage Department has undertaken a major construction project currently ongoing in Southwestern Detroit. The project involves several excavations with different earth retention systems in very soft to soft clays. A major feature of the project is a new Grit Chambers Facility whose earth retention system is the subject of this report.

The Grit Chamber site, as shown in Fig. 1, is a 1.0 acre (4,000 m^2) area within the Detroit Wastewater Treatment Plant, which was formerly used as ash

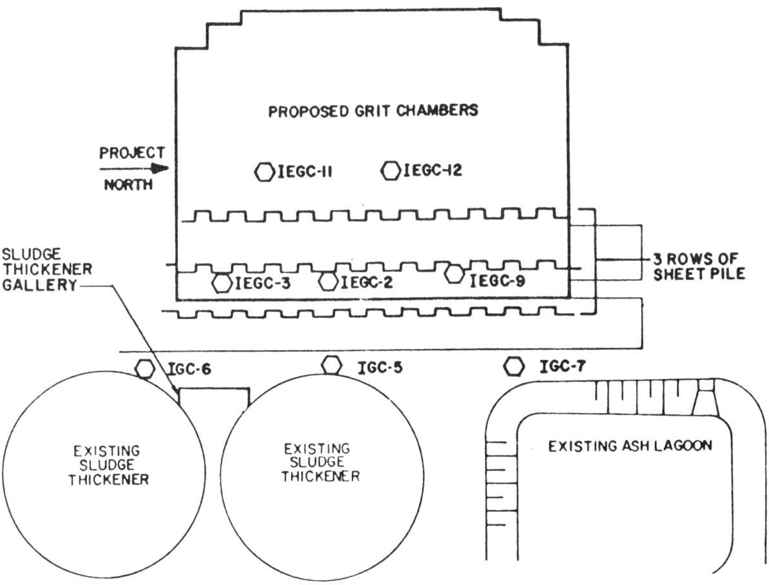

Fig.1. Project site and location of geotechnical instruments

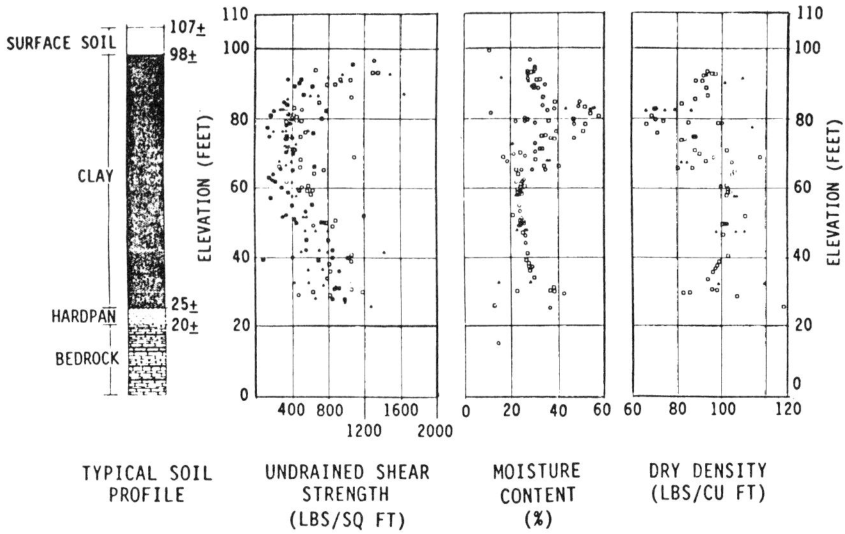

Fig. 2. Generalized subsurface profile and test results

lagoons to store ash from the plant incineration operations. The eastern portion of the site is the location of the sheetpile retaining structure for the construction of a portion of the Grit Chambers adjacent to the existing Sludge Thickener Tanks.

This Paper discusses the performance of the sheetpile retaining structure based on geotechnical instrumentation and compares the field measurements to the soil movements predicted by the finite element method. Effects of construction sequencing are also discussed.

Soil conditions

The soil conditions for the project were determined from three subsurface investigations. In addition, considerable soil data have been collected from previous investigations within the project site. In-situ vane shear tests and laboratory tests were performed within the clay layer. Fig. 2 shows the generalized subsurface profile and relevant test results.

The top of the clay stratum is approximately elevation 98 feet (29.9 metres) and extends down to elevation 25 feet (7.6 metres). The upper 10 feet (3 meters) of the clay layer is desiccated and consequently has a medium consistency with an undrained shear strength in the 750 to 800 psf (36 to 38 KPa) range. At approximately elevation 86 feet (26.2 metres) the consistency becomes soft to very soft, the undrained shear strength decreases to approximately 350 psf (17 KPa), the moisture content increases and the dry unit weight decreases.

The undrained shear strength gradually increases with depth reaching a value of about 700 psf (34 KPa) below elevation 48 feet (14.6 metres). The clay is normally consolidated or slightly over consolidated below about elevation 75 feet (22.9 metres). The clay is generally classified as low plasticity clay in the Unified Soil Classification System. However, occasional samples are classified as highly plastic clay based on Atterberg limits tests. Stratification or layering was observed in samples taken from the upper clay stratum. The layering typically consists of thin bands of silt or sand, and clay.

The clay is underlain by a glacial till, locally referred to as a hardpan, which consists of a very dense mixture of silt clay, sand and gravel. A limestone bedrock underlies the hardpan and is characterized as a massive, competent, gray limestone.

Excavation procedure

As shown in Fig. 1, three rows of PZ-35 sheetpiles were installed for the excavation adjacent to the existing sludge thickener tanks and the temporary ash lagoon. These tanks must remain operational after completion of the proposed Grit Chamber facility. Therefore, measures were taken to minimize any soil movements which may cause subsidence and/or cracks to the existing tank. An excavation sequence for the east end of the proposed Grit Chamber adjacent to the Thickener Tanks was designed and implemented during construction. Support of the Grit Chamber foundation consists of

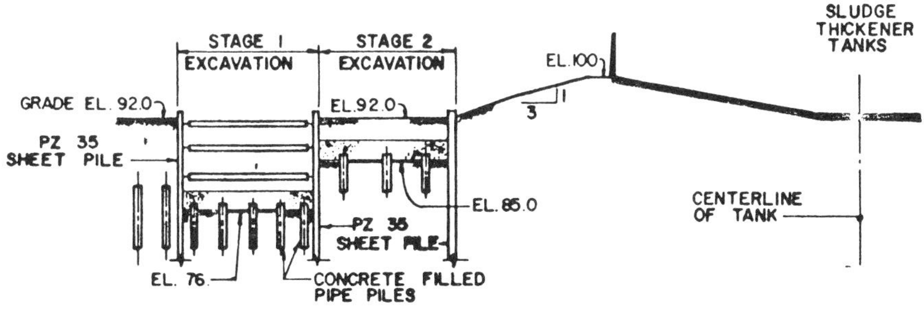

Fig. 3. Excavation sequence

concrete filled pipe piles installed prior to any excavation. These piles were driven to hardpan, filled with concrete and internally cut to the bottom of foundation level. The pile locations were predrilled to 10 feet (3 metres) above the hardpan to minimize disturbance due to soil displacements by pile driving operation.

The excavation sequence involved an eight-foot (2.5 metre) open excavation (pre-excavation) to elevation 92 feet (28 metres) followed by a two-stage braced excavation. Fig. 3 shows the general excavation sequence and excavation stages. The sheetpiles were driven prior to any excavation from elevation 100 feet (30.5 metres) to elevation 65 feet (19.8 metres) which is about 11.0 feet (3.3 metres) below the lowest level of excavation. Every fourth sheetpile was driven to the bottom of the clay layer. The major focus of the specified excavation sequence was to minimize soil movements which could affect the existing adjacent structures. Slope stability, the stability of the bracing system and deflection of the sheeting were all considered.

Instrumentation

The field instrumentation consisted of inclinometers, extensometers, and surface markers. Fig. 1 shows the Instrumentation Plan at the excavation site. Inclinometer/ Extensometer Nos. 3, 2 and 9 are located adjacent to the stage one braced excavation to measure horizontal and vertical movements adjacent to the sheetpiles.

Inclinometer Nos 5, 6 and 7 were located along the berm in front of the Sludge Thickener Tanks and the temporary ash lagoon to measure horizontal movements associated with potential slope stability failure. Inclinometer/Extensometer Nos 11 and 12 were located 30 feet (9 metres) to the west of stage one excavation to avoid conflict with the excavation and construction traffic.

Finite element method

The finite element method was utilized during the design stage to model excavation sequence and predict movements of the earth retention systems and the adjacent unexcavated soils and structures. Additional analyses were performed during the excavation phase to properly model the actual construction sequences. References cited (Ref. 1-8) are some of the available literature which utilized a finite element method to predict movements associated with excavation.

The analyses were performed with the program SOIL-STRUCT, developed by Clough and Tsui (Ref. 6), to study soil responses by simulating incremental construction and loading sequences under undrained, plane strain conditions. The program uses a nonlinear, stress dependent hyperbolic model for soil under primary loading. For soil under unloading and reloading, a stress dependent linear response is used. It also uses a bilinear stress-strain model for interface elements and a linear elastic model for structure elements. There also are bar elements, which take compression and tension only, to represent the installation of struts.

The basic soil parameters used in the analyses are listed in Table 1. Fig. 4 shows the employed finite element mesh. Construction sequences include excavation, strut installation, excavation equipment loading, and pile driving equipment loading used during pile installation. Note that the effect of previously installed pipe piles was not considered in the modelling. It was perceived that the pipe piles move with the clay and do not contribute significant resistance to lateral soil movements.

The PZ-35 sheet piles were treated as 2-foot (0.7 m) thick structural elements with equivalent flexural stiffness. Also, a bar element is attached between the tip of the sheet pile and the fixed bottom boundary to model a hinge at the bottom of the sheet pile. Such a bar element provides end restraint for the sheet pile. No interface elements are used in the analysis.

The finite element method served the following purposes

- To study magnitudes and patterns of the soil movement under different schemes of excavation sequences during the design phase.
- To simulate actual construction sequence and field loading conditions

Table 1. Basic parameters used in finite element analysis

PARAMETER	VALUE
Soil Unit Weight, Undrained Shear Strength, Su Profile)	See Figure 2 (Typical
Soil Initial Tangent Modulus	600 x Su
Coefficient of Lateral Earth Pressure at Rest	0.90
Equivalent Young's Modulus for Sheetpile Walls	113,000,000 psf (5,400 MPa) or 28,250,000 psf(1,350MPa) (Depends on Sheetpile penetration depth)
Young's Modulus of Concrete	432,000,000 psf(20,700 MPa)

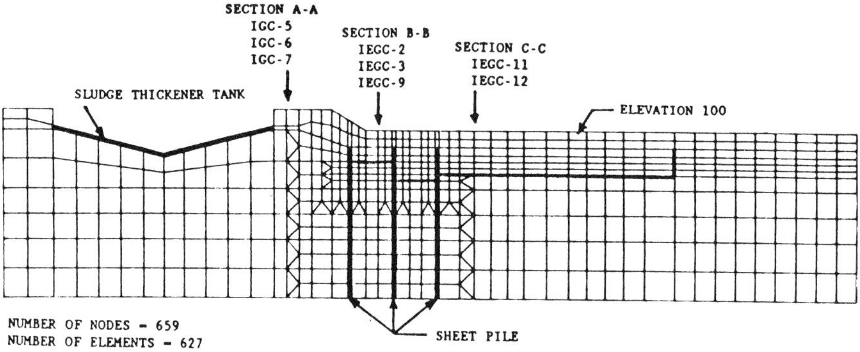

Fig. 4. Finite element mesh

during the construction phase. The results are compared with the observed field instrumentation data.

- To conduct a parametric study of different variables. The variables include soil shear strength, excavation sequence, surface loading conditions, and stiffness of braced system during the design and construction phase.

Discussion and conclusions

The predicted and observed horizontal and vertical movements of the actual construction sequence are shown in Figs 5 and 6, respectively. Fig. 5 shows the horizontal movements at three locations designated as Sections A-A, B-B, and C-C (refer to Fig. 4 for Sections). The sections coincide with the geotechnical instrument locations. The vertical movements are shown in Fig. 6 for Sections B-B and C-C. The predicted movements are those computed by the finite element method (referred to FEM in Figures). The observed movements were obtained from monitoring the geotechnical instruments during excavation. The movements are cumulative from pre-excavation through stages 1 and 2 excavations.

A comparison between the predicted and observed movements leads to the following observations

- In Section A-A, the predicted and observed horizontal movements are

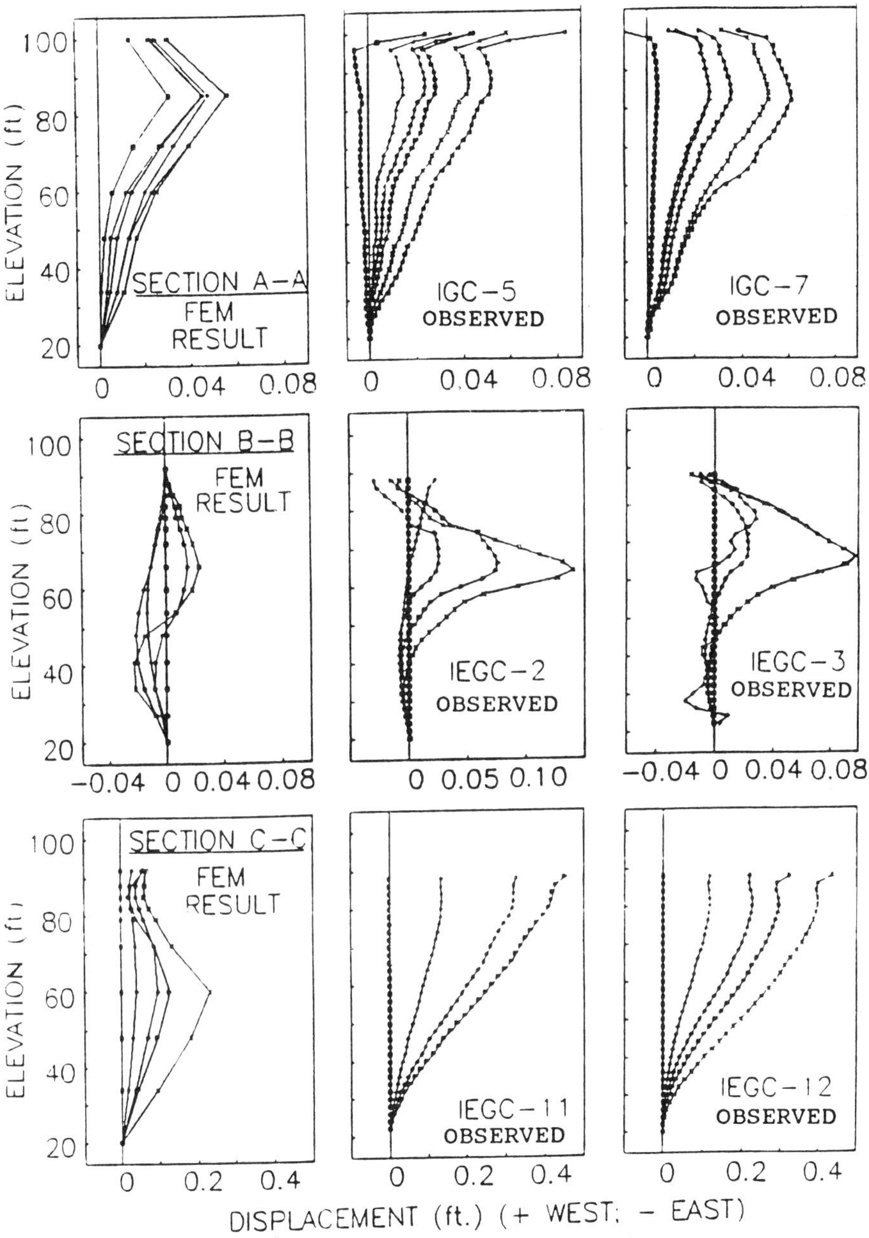

Fig. 5. Predicted and observed horizontal movements

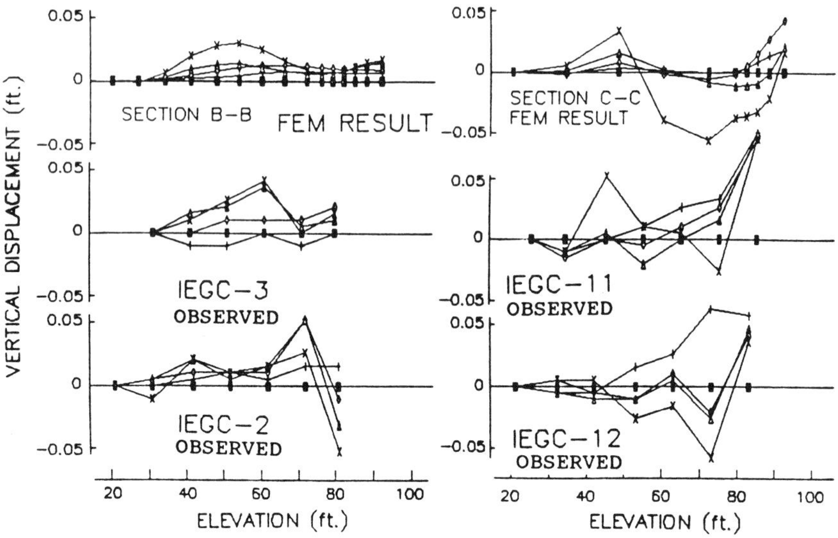

Fig. 6. Predicted and observed vertical movements

similar in both magnitude and pattern.

- In Section B-B, the observed maximum horizontal movements are three to five times the predicted movements, however the movement patterns are similar.
- In Section C-C, the observed horizontal movements near the ground surface are about four times the predicted movements. The movement patterns are similar for the lower clay layer, but are significantly different in the upper clay layer.
- In Sections B-B and C-C, the predicted and observed vertical movements are similar in both magnitude and pattern.

The discrepancy between the observed and predicted movements could be attributed to either one or a combination of the following.

- During pre-excavation and removal of top 8-foot (2.4-metres) of soil, the excavation was carried out in strips from east to west, and maintained almost vertical cut before reaching the bottom of excavation. Based on the instrumentation data, about 6 inches (150mm) of horizontal movement occurred at the top toward the east during this stage. This movement extended to the bottom of clay. The magnitude of this movement decreased linearly to the bottom of the clay layer. For this type of deep lateral movement any restraint provided by the struts, especially at the upper soil layer, would not be effective. A similar type of horizontal movement (cantilever excavation) has been discussed by

Clough et al. (Ref. 5).

- During all stages of construction, the excavation equipment was operating close to the edge of the cut and travelled back and forth frequently. Such activities could have disturbed the top layer of the soil, and induced additional horizontal movement.
- Soils parameters utilized are based on the laboratory and in-situ test results. The actual soil parameters could be different from those used in the finite element modelling.
- Exact simulation of the excavation behaviour requires three-dimensional modelling. The two-dimensional modelling by SOIL-STRUCT could contribute to discrepancy in the predicted movements.
- Sequence and method of construction are two important factors influencing the movements and must be accounted for in the analysis. For example in Section A-A, which is close to the Sludge Thickener Tanks, no excavation equipment was allowed and as previously stated the predicted and observed movements are essentially the same.

In conclusion, finite element predictions for the Grit Chambers excavation agreed reasonably well with the measured movement patterns and magnitudes, especially in Section A-A. Deviations in the sequence and method of construction is believed to account for most of the discrepancy between predicted and measured movements in Sections B-B and C-C. The accuracy of finite element predictions may be enhanced by a more detailed modelling of these factors.

Acknowledgements

The Authors would like to recognize the Detroit Water and Sewerage Department for their assistance during the design and construction of this project. The Department provided valuable input and recommendations during all phases of the project. The writers would like to thank the Department for permission to publish this article. The writers would also like to recognize T. Weber, M. Pochodylo, M. Cameron, and R. Sherman of Metcalf & Eddy, Inc. for their review of the paper, NTH Consultants, Ltd., Detroit for their assistance during the design and construction, Dr. G. W. Clough of Virginia Polytechnic Institute for his valuable input during the design of the earth retention system.

References

1. CLOUGH G. W. *et al.* Design and Observation of A Tied Back Wall. *Proc. ASCE Spec. Conf. on Performance of Earth and Earth-Supported Structures, Lafayette, Indiana,* 1972, Vol. 1, Part 2, pp. 1367-1389.
2. CLOUGH G. W. and MANA A. I. Lessons learned in Finite Element

Analyses of Temporary Excavations in Soft Clay. *Proc. Second Int. Conf. on Numerical Methods in Geomechanics, Blacksburg, Virginia*, 1976, Vol. 1, pp. 496-510.
3. CLOUGH G. W. *et al.* Prediction of Supported Excavation Movements Under Marginal Stability Conditions in Clay. *Proc. Third Int. Conf. on Numerical Methods in Geomechanics, Aachen, Germany*, 1979, Vol. 4, pp. 1485-1502.
4. CLOUGH G. W. and REED M. W. Measured behaviour of braced wall in very soft clay. *J. Geotech. Engng*, 1984, **110**, No. 1, pp. 1-19.
5. CLOUGH G. W., SMITH E. M. and SWEENEY B. P. Movement Control of Excavation Support Systems by Iterative Design. *Proc. ASCE Congress on Foundations Engineering: Current Principles and Practices, Evanston, Illinois*, 1989,Vol. 2, pp. 869-884.
6. CLOUGH G. W. and TSUI, Y. Behaviour of Tie-Back Walls in Clay. *ASCE National Meeting on Structural Engineering, Cincinnati, Ohio*, 1974.
7. MANA A. I. and CLOUGH G. W. Prediction of movements for braced cuts in Clay. *J. Geotech. Engng.*, 1981, **107**, No. GT6, pp. 759-777.
8. TAN S.B., TAN S.L. and CHIN Y.K. A Braced Sheetpile Excavation in Soft Singapore Marine Clay. *Proc. Int. Conf. on Soil Mechanics and Foundation Engineering*, 1985, Vol. 3, pp. 1671-1674.

Performance of a cast in situ retaining wall in a sandy silt

R. KASTNER, Institut National des Sciences Appliquees of Lyon (INSA) and J. FERRAND, Societe d'Economie Mixte du Metropolitain de l'Agglomeration Lyonnaise (SEMALY), France

Introduction

The western terminus of the D line of the Lyon subway was constructed by means of an open-air excavation protected by strutted cast walls, in a soil horizon of silt colluvial deposits with poor characteristics. A preliminary hydrological study showed that the groundwater level might rise when intersected by these cast walls. As a result, the retaining walls were designed with a reduced depth of embedment, and this in turn led to a heaving risk of the excavation floor. As the subway might later be further extended in the same soil horizon, the SEMALY entrusted the Geotechnical Laboratory of INSA with an instrument survey with the following main objectives

- to analyze of the behaviour of the strutted cast walls
- to check the dimensioning of the retaining walls in these colluvial deposits
- to evaluate of the heaving risk by hydraulic uplift of the excavation floor.

In this Paper, we present the behaviour of this retaining wall as well as a comparison between experimental results and calculations made according to the reaction modulus hypothesis.

Site description

Geotechnical section

Within the depth of the excavation, test borings showed that the soil consisted of a succession of sandy silt layers with fairly close identifying characteristics. Grain size analysis indicated that the sandy silt was fine (0.15 mm < D_{60} < 0.2 mm) and that the percentage of grain sizes inferior to 2 microns was 10 to 20%; some places on the surface contained 10 to 30% of coarser sands. The liquid limit of these silts varies between 25 and 30% with a plasticity index close to 5%. It should be noted that the natural water content of these silts is always close to their liquid limit, which explains why they are so sensitive to disturbance, especially during earthwork.

Mechanical behaviour is homogeneous within a depth of 9m. The characteristics, deduced mainly from triaxial tests, vane tests and pressuremeter tests, are as follows

Between 0m and 9m

Undrained shear strength: Cu = 70 to 90 kPa
Shear strength parameters with respect to effective stress: Φ'= 27° and c' = 0
Menard limit pressure: Pl = 0.2 to 0.4 MPa
Menard elastic modulus: Ep = 1 to 2 MPa

Beyond a depth of 9m, the soil has slightly better characteristics

Undrained shear strength: Cu = 100 to 1.50 kPa
Shear strength parameters with respect to effective stress: Φ' = 37 to 40° and c' - 0
Menard limit pressure: Pl = 0.7 to 0.9 MPa
Menard elastic modulus: Ep = 3.5 to 4 MPa

Excavation section

The excavation, 9 m wide and 8 m deep, is protected by two cast retaining walls of reinforced concrete 10.5 m high and 0.6 m thick. These are held by two series of passive struts made of metallic H-shaped sections. Before the

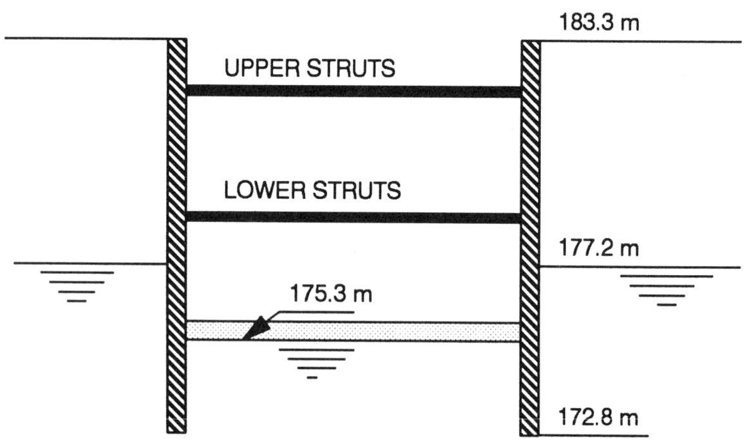

Fig. 1 . Section of the excavation

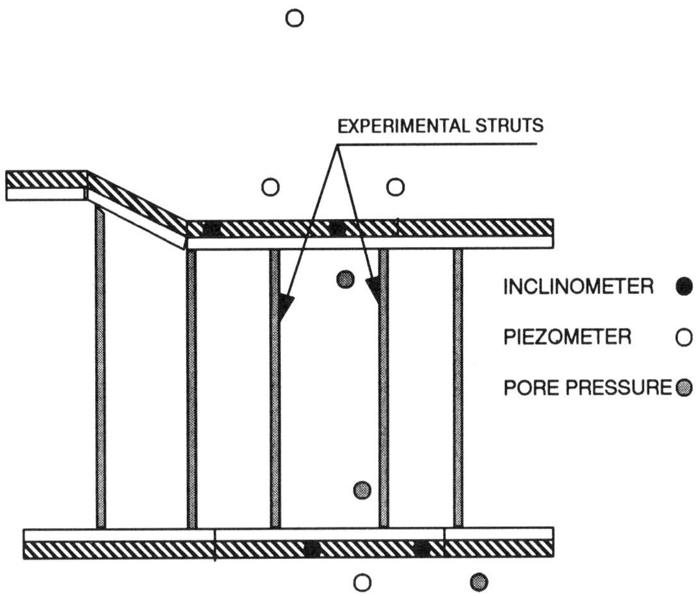

Fig. 2 . Plan view of the test zone

work, the water table varied between 4 and 5m below the natural ground level.

Experimental apparatus

The apparatus was used to measure the displacement and strain of the retaining walls, the stress in the struts, and the effect of the earthwork on the aquifer.

Displacement of the retaining walls

The deformation of the retaining wall was measured by inclinometry. Four inclinometer tubes were fixed to the reinforcing cages when they were set up in the excavation before concreting. Inclinometer readings were recorded at each stage of the work and were used to plot the displacement of the retaining walls. The curves were then readjusted: the movement at the head of the retaining walls was calculated from convergence measurements, using fixed points situated outside the zone influenced by the work.

Stress in the struts

All struts situated within the experimental section were equipped with 6 vibrating-wire strain gauges placed an the same H-shaped section, to determine the compression strain diagram. It might seem that a smaller number of gauges would have been sufficient, but compensations had to be made for the unavoidable failures taking place in situ; this apparatus gave precise measurements of stresses during the whole duration of work, while also leading to an estimate of the measurement uncertainties (ref.l).

Piezometric measurements

Changes in the groundwater level outside the excavation were followed on 4 open piezometers lined up at right angles to the excavation. To complement this, pore pressure measurements were carried out at 4 points at the level of the foot of the retaining wall to assess the foundation water pressure at the base of the excavation. Each measurement point was equipped with differential pressure cells linked to the open air by capillaries, this to compensate for variations in the atmospheric pressure.

The whole set of hydraulic pressure cells and vibrating-wire strain gauges, to which were added thermometric probes, was linked to an automatic measurement station. The analysis of the total set of measurements, read with a periodicity ranging from a few minutes to a few hours, was used in particular to assess the effect of temperature variations on the measurements, and thus correct them.

Stages of the work

After a first excavation to a depth of 1.8m, the first level of passive struts was placed 1.35m under the top the retaining wall. Excavation was then pursued in stages until a depth of 5.2m was reached; a second row of passive struts was placed at the -4.75m level.

When the excavation had reached the final depth of 8m, a geotextile was laid on the floor of the excavation, and a 30 to 40 cm gravel circulation layer was spread onto it. The excavation remained open in this state for 4 1/2 months before concreting of the lower floor, followed by construction of the tunnel.

Behaviour of the retaining wall

Displacement

Figure 3 shows the development of the retaining wall displacement as a function of the progress of the excavation and the placement of the struts.

The displacements remained small during the intermediary stages of the work; however, an important displacement of the foot of the retaining wall occurred during the last stage of the excavation: a likely hypothesis is that, at that moment, the very short depth of embedment of the retaining wall

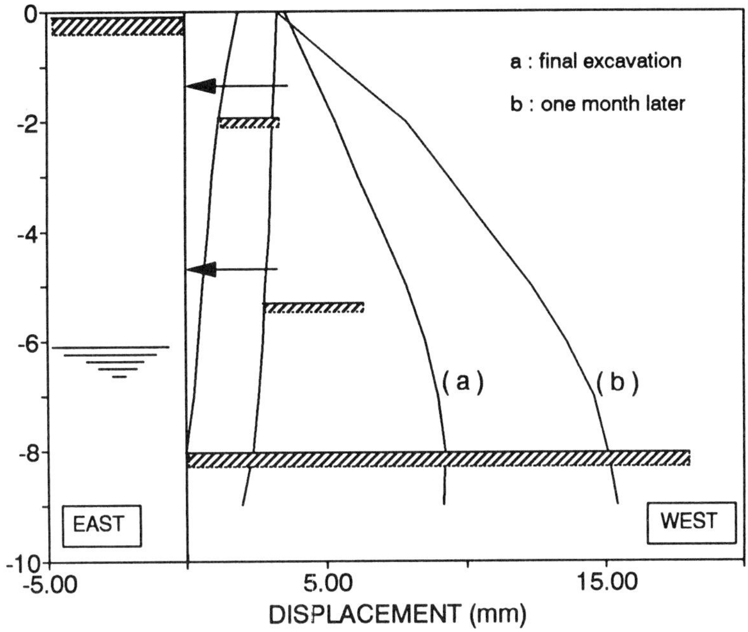

Fig. 3. Development of the displacement of the east retaining wall

produces a stress in the soil which is very close the passive pressure limit. The foot of the retaining wall is then very inefficiently held by the soil, especially as the bulk specific gravity of the soil is reduced by the vertical ascending flow between the two retaining walls, thus reducing the passive pressure available on the trench side of the embedment. A non-negligible creep of the soil appeared later, with an important evolution of the displacement during one month, varying at the base from 9mm to more than 15mm, with a subsequent stabilization.

When the east and west retaining walls are compared (Fig. 4), the asymmetry appears very clearly, with general displacement towards the west. This dissymmetry might be caused by the movement and parking of earthwork machinery, and by the construction of a 40cm thick embankment on the western side, very close to the excavation.

Stresses in the struts.

Figure 5 shows the evolution of stresses in the upper and lower struts as a function of the progress of the excavation. It appears that the stresses are not evenly spread between the two support levels, the stresses in the lower struts being inferior to those in the upper struts.

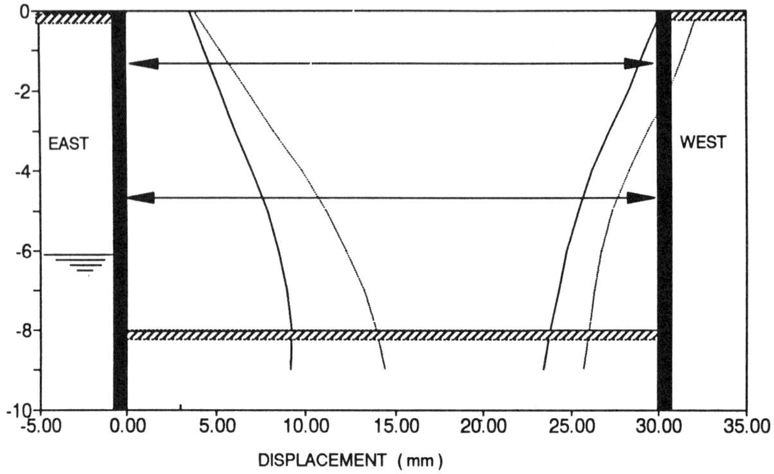

Fig. 4. Displacement of the east and west retaining walls during the final stage of excavation

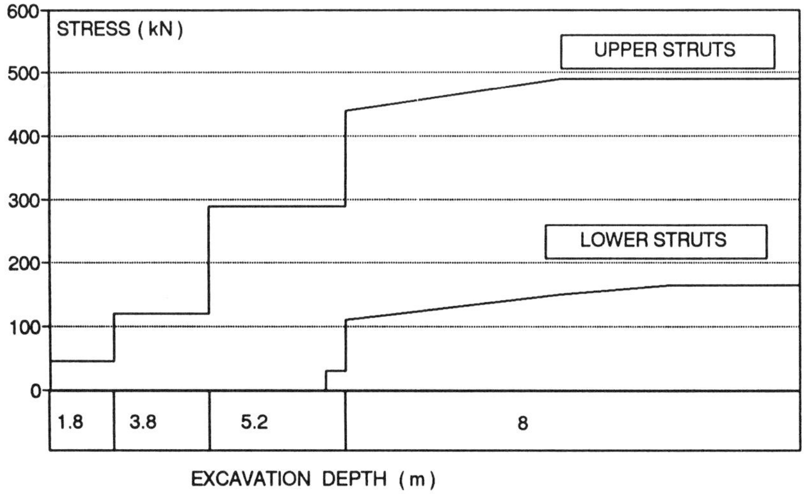

Fig. 5. Evolution of stresses in the struts

With the simultaneous measurement of stresses and displacements at the level of the struts, their real compressibility can be determined. Assuming that the solicitations are compressive, their theoretical stiffness ($\Delta F/\Delta l$) is equal to 8.10^5 kN/m. The experimental stiffness value for the upper struts is 8.10^4 kN/m and falls to 4.10^4: kN/m for the lower struts, i.e. 10 to 20 times smaller than the theoretical values. This difference has various causes

- the struts have to bear a bending as well as a compression stress
- the bearing between fender waling and the wall as well as the bearing between the strut and the fender waling introduce an additional compressibility.

When examining these results, one is led to think that the lower struts should have been prestressed to about 300 kN when being set up, which would have spread the stresses more evenly while also ensuring that the foot of the retaining wall was held more securely, thus making up for insufficient embedment depth. This prestressing might also compensate for the excessive compressibility of the strutting.

Comparison with calculations

The temporary retaining walls of the Lyon subway were dimensioned by using a method based on the reaction modulus hypothesis. The program used (RIDO) took separate account of the soil pressure on either side of the structure. The reaction law is linear, limited by the active and passive pressures, and with an irreversible behaviour once these limits have been reached (ref.2).

Calculations data

Since the drainage is fast, all calculations were made for a drained behaviour. Active and passive pressures were determined from the Caquot and Kerisel values, with the hypothesis that the soil-wall frictions are respectively equal to 2/3 and -2/3 of Φ. One of the sensitive points of these calculations is the choice of the reaction modulus: as a first step, we used the values recommended by the Laboratoire Central des Ponts et Chaussees (ref.3) calculated from the value of the pressuremetric modulus (i.e. kh varying from 1 000 to 4 000 kN/m^3 depending on the soil horizons in question). These values led immediately from the start of excavation, to displacement values considerably larger than those found in situ: for the first stage of earthwork, the movement calculated at the head of the retaining wall reached more than 12mm for a measured movement of only 2mm. It is possible that the LCPC recommended values for the determination of kh, calibrated on a few sites, are not well adapted to the soil in question. It is also possible that, with these stilts, which are very sensitive to disturbance, conventional pressuremetric tests lead to an important underestimate of the modulus, and that a self-boring pressuremeter might be better if a realistic estimate of the modulus of the undisturbed soil is wanted.

Following these first calculations, the values of the reaction modulus were modified so that the calculations would show displacement values compatible with those measured. The following characteristics were finally adopted

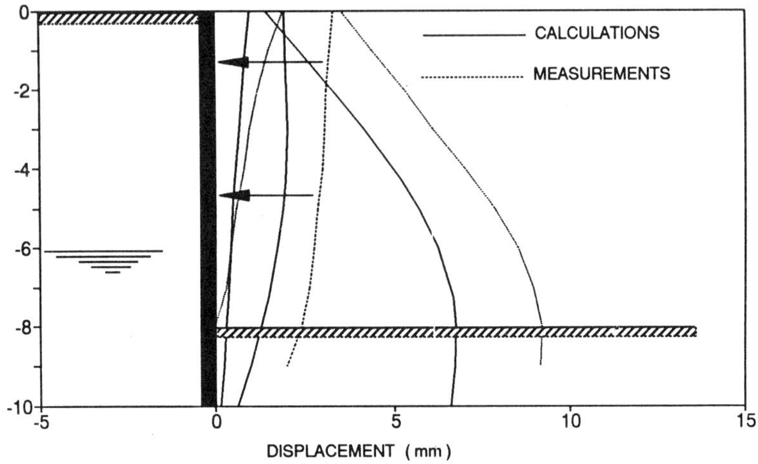

Fig. 6 . Comparison between calculated and measured displacements

Between 0 and 9 m

$\Phi = 27°$
$Ka\gamma.\cos\delta = 0.3$
$Kp\gamma.\cos\delta = 4.1$
$kh = 2.10^4$ kN/m^3

Between 9 and 10.5 m

$\Phi = 40°$
$Ka\gamma.\cos\delta = 0.15$
$Kp\gamma.\cos\delta = 10.7$
$kh = 4.10^4$ kN/m^3

Remark: below the aquifer, cohesion is nil; above, a capillary cohesion of 10 kPa was adopted.

In Fig. 6, the displacements calculated with those data are compared with the displacements measured on the retaining wall during the main stages of excavation. For the final stage, the value of the stress that can be summoned on the trench side of the embedment is reduced on an inclusive basis to take into account the effect of the vertical ascending flow. It should be noted that the displacement at the foot of the retaining wall is highly sensitive to this reduction, and as a result the movement of the retaining wall at this stage can only be very roughly estimated.

The results shown here correspond to a 40% reduction of the passive pressure at maximum excavation.

Table 1. Development of stresses in the struts

Depth of excavation		upper struts	lower struts
5.2 m	in situ	290 kN	
	calculation	172 kN	
8 m	in situ	440 kN	110 kN
	calculation	232 kN	112 kN
7.6 m	in situ	490 kN	165 kN
	calculation	226 kN	161 kN

It should also be noted that the calculated measured displacement curves are very similar in appearance with, however, a difference in translation reaching 3mm. This difference might be due to the general western movement of both retaining walls, of the same order in size and which cannot be simulated by the program used since the latter works under the hypothesis that the excavation is symmetrical.

Stresses in the struts

Table 1 shows the calculated and measured values of the compression stress in the struts. The values are fairly close for the lower struts; however, the calculations, to a large extent underestimate the stresses in the upper struts, while the difference increases as a function of the progress of the excavation. This phenomenon, already encountered on other experimental sites (ref.4), might be explained by the arching effect: the upper part of the retaining walls is fixed by the struts while the foot slips towards the excavation. This fixed point leads to a soil pressure concentration behind the retaining walls, pressure that has to be borne by the struts. Such pressure transfer cannot be described by the Winkler hypothesis, which does not take into account the relative displacements.

Conclusions

This experimental study allowed us to analyze the behaviour of the soil and the retaining walls, and to determine calculations hypotheses for future sites by showing in particular

- during the final stage, an important displacement of the foot of the retaining wall linked to the reduced embedment depth and to the reduction of the passive pressure that can be summoned when subjected to the vertical flow at the base of the excavation
- an effective stiffness of the strutting 10 to 20 times smaller than the theoretical stiffness, which might vindicate prestressing this strutting
- an underestimate of the reaction modulus by the traditional pressuremetric tests, which could well be replaced, in these sensitive silts, by tests with a selfboring pressuremeter
- an underestimate of stresses in the strutting when using the reaction modulus approach.

References

1. KASTNER R., PANTET H. and ONDEL C. Mesures en continu sur un soutenement du metro de lyon. *Conf. Int. Mesures et Essais en Genie Civil, Lyon* 1988, II, pp 222-230.
2. FAGES R. and BOUYAT C. Calcul des rideaux de parois moulees et de palplanches. *Travaux*, 1971, 439, pp 49-51.
3. BALAY J. *Recommandations pour le choix des paramstres de calcul des ecrans de soutenement par la methode aux modules de reaction.* Note d'information technique, 1984, LCPC, Paris.
4. KASTNER R. *Excavations profondes en site urbain. Problemes lies a la mise hors d'eau - Dimensionnement des soutenements butonnes*. These de Doctorat des Sciences, INSA Lyon et Universite Lyon 1 1982, p 409.

Discussion

SESSION 3

M. L. LINGS, University of Bristol

One of our main preliminary conclusions about the Lion Yard site was that installation of the diaphragm wall had caused a marked reduction of horizontal stress in the ground adjacent to the wall. Fig. 7 shows the readings of all seven earth pressure cells in panel 22 just before the start of the main excavation. Also shown are an initial stress line representing K_0 of 1.5, which is towards the lower end of the likely in-situ stresses, and active limit lines for no wall friction and full wall friction. The field measurements show the stresses before main excavation to be fairly close to active, and to be approximately half of their initial values. It is clearly important to know whether or not these field measurements are reliable.

There are two basic uncertainties in the earth pressure data.

(a) How accurate are the inital cell readings after installation? (For example, is there any over-read or under-read due to the installation

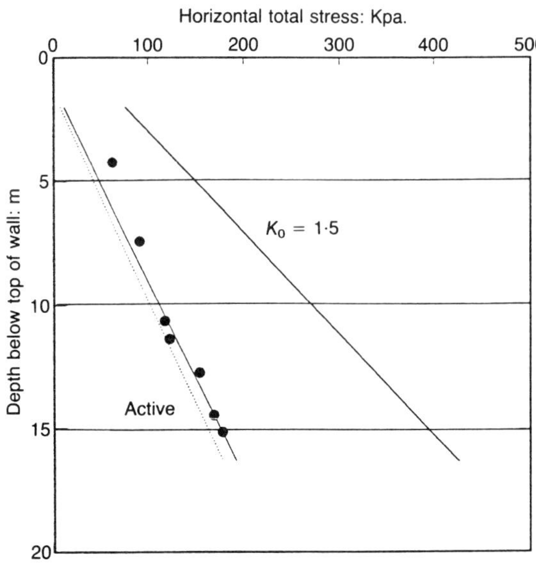

Fig. 7. Horizontal total stresses before main excavation

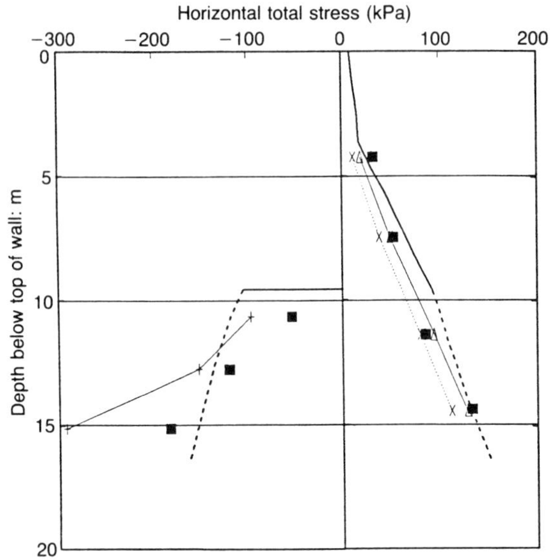

Fig. 8. Possible earth pressure diagram showing measured values (revised) and active/passive limits

method?)

(b) How accurate are the subsequent readings? (For example, what is the cell action factor when stress changes occur in the ground?)

My colleagues at Bristol and I have recently been studying the data acquired during the concreting of panel 22, and we feel we now understand more clearly what was going on during and immediately after cell installation. Without going into detail, what we have done is to calculate a revised set of pre-load values for the cells. For those not familiar with this type of cell, the pre-load value is the constant which needs to be subtracted from the raw data to give the actual pressure.

We have therefore revised Fig. 6 in the paper, the 'possible earth pressure diagram', to take account of these changes, and it is presented here as Fig. 8. Some of the measured values have changed, and the dotted lines have changed a little in sympathy, but the equilibrium analysis is unchanged, and therefore the solid lines are unchanged. We have also added active and passive limit lines, obtained using the measured pore pressures at that stage of construction.

As a result of this reappraisal, the worst cell is the highest one in front of the wall (cell number 3), which still appears to be under-reading by a factor of about two.

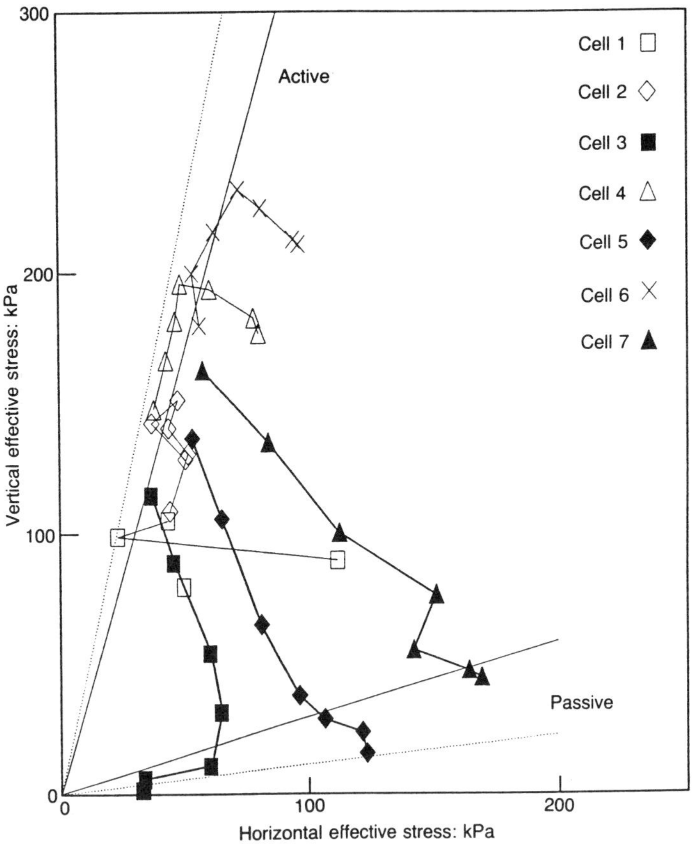

Fig. 9. Effective stress paths

Our conclusions about changes in earth pressure, and the reliability of the measured changes, have not been affected by the revised pre-load values. The worst cell in terms of changes in pressure is again cell number 3, which seems to have recorded a much larger pressure reduction than appears possible. This of course fits in well with it's under-reading by the time the lowest excavation level had been reached.

Returning to the installation process, by the time panel 22 had been fully concreted, the loads in the jacks were all very small; in other words, the cells were pushing into the soil with about the same pressure as the wet concrete. This is likely to mean that their initial performance was quite good. What seems to have happended subsequently is that they have all operated with different and rather variable action factors, so that the reliability of the measurements has decreased with subsequent load changes.

We have drawn effective stress paths for all seven cells, using the revised data, and these are shown in Fig. 9. Limit lines for active and passive states have been shown, the full lines representing no wall friction, the dotted lines representing full wall friction. The stress points plotted represent conditions just before main excavation, the end of each of the three stages of excavation, the removal of the temporary props, and then readings one and two years later. Soil pressures on the back of the wall remained close to active during excavation, and have subsequently increased. Soil Pressures on the front of the wall moved towards passive during excavation, but did not reach it. Subsequently, the soil at cell number 3 may have reached the passive limit. The consequence of cell number 3 reading approximately 50 kPa too low is perhaps not very serious given the slope of the limit line, and does not greatly affect this conclusion. The evidence of passive failure is as much the abrupt change in direction of the stress path.

Conclusions

An unusual opportunity to measure propping loads during main excavation has enabled a simple equilibrium analysis to be carried out on the values of earth pressure measured at the concrete/soil interface of a diaphragm wall in Gault Clay.

This analysis has shown that many cells were performing reasonably well at final excavation stage; the worst cell was under-reading by a factor of approximately two.

This important analysis has given us confidence that the lateral stresses just before main excavation were indeed very low. We believe the substantial reduction in stress from initial conditions was attributable chiefly to the construction of the diaphragm wall.

H. LONG, ESBI Atkins International Ltd

Dr Symons was concerned that, as our trial panel was only 6m long, it did not represent the behaviour of the ground during construction of the real wall. The purpose of the trial panel was to confirm that the proposed construction method was feasible. During construction of the real wall, ground settlements were monitored about 5m from the edge of the diaphragm wall. Settlements during wall installation were generally less than 12 mm, although in one area they were up to 16 mm. Their movements did not pose any problems.

Dr Symons queried why the predicted wall movements appeared very sensitive to the shear strength of the marine clay. This was because the computer program predicted that the soil had reached the active and passive limits over a considerable length of the wall, thus making predicted movements very sensitive to shear strength values.

'Frew' is not an interactive spring program. It is a 'quasi finite element program in which the soil is represented as an elastic solid with the soil stiffness matrices being developed from prestored matrices calculated using the Oasys finite element program SAFE.'

Dr. Symons wondered, on account of the low level of movements measured, whether, in hind-sight, the 1.2m thick diaphragm wall could have been replaced by an 0.8m thick wall. It should be noted that a contributing factor to the low movements was the careful control imposed on the construction sequence as detailed in the paper. It is not possible to confirm whether or not an 0.8m wall would have worked without detailed analysis. However, this further emphasises the point that more case histories are needed so that designers' confidence in their models and parameters can improve - only then can we progress.

Professor S. THORBURN, Director, Thorburn plc

I should like to pose three questions with regard to this session:

(a) Are our designs more economical?
(b) Do we have a better understanding of the real margins of safety?
(c) Do we have a better appreciation of the deformations patterns and the effects of construction?

DR M.M.H. AL-SHAIKH-ALI, DT Motorway Widening Unit, DTP MWU

In Fig. 7, Mr Lings showed the profile for K_0. He also modified markedly the values of the horizontal stesses. In Fig. 9, Mr Lings showed the profiles of k_a and k_p.

How did Mr Lings measure K_0 for Gault Clay? Did Mr Lings modify the values for K_0. Why was the K_0 profile not superimposed on Fig. 9? Why were Figs 7 and 9 not included in the Paper?

Professor R.V. WHITMAN, Massachusetts Institute of Technology

The garage at Post Office Square in Boston, Massachussets, has seven underground levels (extending to a depth of about 24 m) with a park on the roof at street level). The site-on irregularly shaped city block typically involves (from ground surface) 6-7m of fill, 13m of Boston clay, varying thicknesses of sand and glacial till, and, finally, weathered argillite at or somewhat below the bottom of the garage. The water table is about 3m below surface.

The 'top down' method was used to construct the garage. A continuous diaphragm was placed by the slurry-filled panel method; the thickness was nearly 1 m. The bottom of all panels extended at least 1.5 m into the argillite. Load-bearing elements (steel columns supported on concreted caissons) were likewise placed in slurry-filled slots, and the remaining slurry was hardened to the consistency of clay by an additive.

There were two major concerns and uncertainties, and the observational approach was adopted.

With the planned construciton schedule, settlements at adjacent buildings ranged up to 40 mm or 50 mm, which was judged to be the limit at which damage might begin to occur. A monitoring programme involving inclinometers and surveying points was established. Readings were taken weekly (or even twice-weekly) at some stages; results were displayed immediately and reviewed, frequently by a Geotechnical Monitoring Committee. The contractor understood in advance that adjustments to the construction schedule might be necessary.

Based on the best estimate for long-term seepage into the base of the garage, the design called for a thin slab on grade with an underdrain. However, with the worst case estimate, a pressure slab would be required. An alternate design - an anchored pressure slab that could be constructed with all overlying floors already in place - was prepared. Numerous piezometers were installed around the site, the seepage was monitored carefully as excavation progressed.

In the event, underseepage proved to be consistent with the best estimate. However, horizontal movements measured at several locations of the diaphragm wall during initial excavation were larger than expected and a cause for concern. The construction was required to slow the pace of excavation relative to the placements of the roof and first lower floor and to increase the width of beams. Subsequent movements were within anticipated limits, and displacements of adjacent buildings were actually smaller than expected.

The project produced a wealth of data, some of which has been published in two conferences held in the USA during June 1991: the ASCE Geotechnical Congress in Boulder, Colorado, and the ASTM conference on slurry walls in Atlantic City, New Jersey. Two items are of particular interest here.

At three locations, inclinometers were placed within a metre or two outside the yet to be trenched diaphragm wall, so as to record horizontal motions during trenching and concreting. Such movements were small and barely detectable - a few millimetres at most.

Subsequent to concreting of the roof, movements began at the diaphragm panels, then well away from excavation operations. As displacements at these panels were already of concern, these unexpected continuing movements were alarming. However, it was demonstrated that these movements were the result of rapid decreases of the ambient air temperatures (It was late

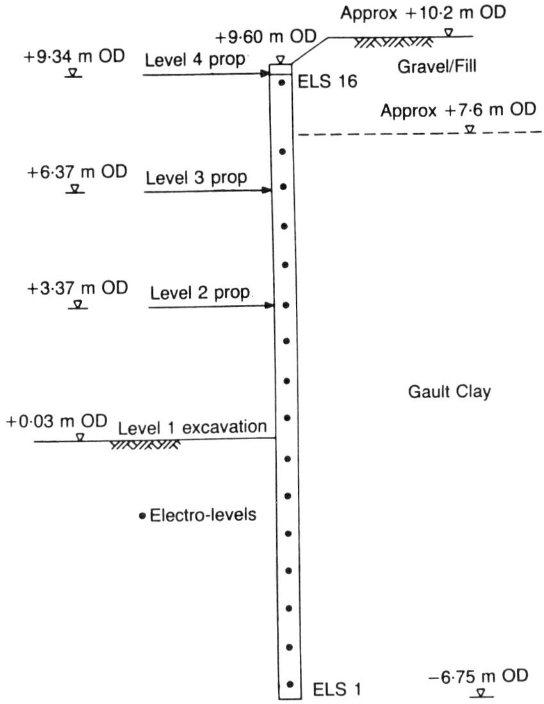

Fig. 1. Cross-section through wall

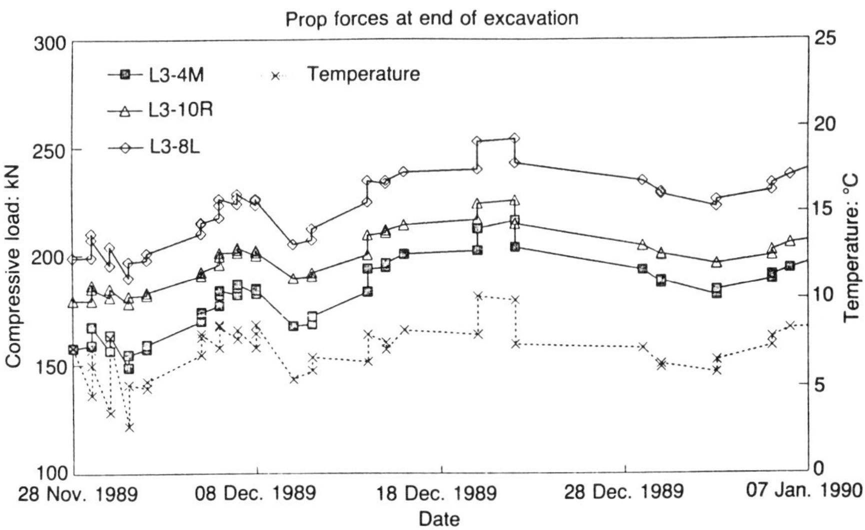

Fig. 2. Variation of strut load with temperature

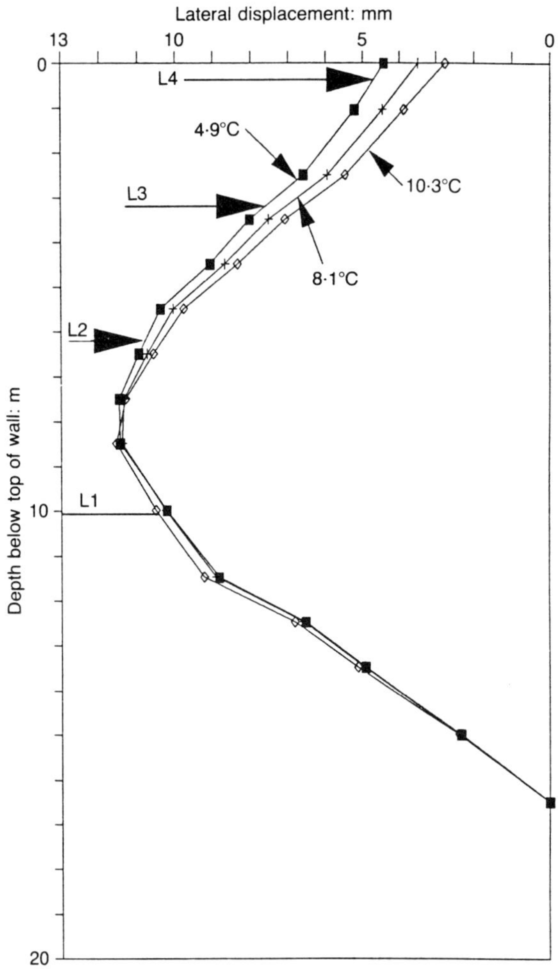

Fig. 3. Displacement of wall due to temperature effects

November in Boston!) During the late stages of construction, the upper portions of the diaphragm wall moved away from the excavation, apparently the result of warming during the summer months.

W.W.C. NG, University of Bristol

Load, displacement and time are three common physical quantities measured in a field monitoring program. Very often temperature effects are ignored since temperature may not be thought to be relevant or particularly important.

Recently, I have been involved in geotechnical and structural monitoring of a three-level underground car park constructed in stiff-fissured Gault clay at Lion Yard, Cambridge.[1] The 10 m deep underground car park was retained by a perimeter diaphragm wall. At the locations of two vehicle ramp openings, the wall was temporarily supported by steel props at three levels during the main excavation. The field monitoring program included measurements of the deflection of the wall and the strut loads using electro-levels and vibrating-wire load cells respectively. A cross-section through the diaphragm wall with its temporary steel props is shown in Fig. 1.

Temperature effects were ignored initially. However, it was found that the daily variation of strut loads could be as high as 30% of the average values. The meaning of these measurements became clear when temperature effects were taken into consideration as illustrated in Fig. 2. It can be seen that the strut load varied with temperature. As the temperature increased, the steel struts expanded and pushed the diaphragm wall panel away from the excavation as shown in Fig. 3. A rise of 5°C caused an outward movement of nearly 2 mm at the top of the wall (a 30% reduction of the maximum displacement).

At this time, it was not known for certain what proportion of the measured load variation was real, and what was attributable to zero shift of the load cells as the result of temperature fluctuations. An identical load cell was therefore placed adjacent to one of the struts for checking the zero shift. The results of this study is shown in Fig. 4. It was found that the variation in load of this control load cell was very small as compared with other load cells and

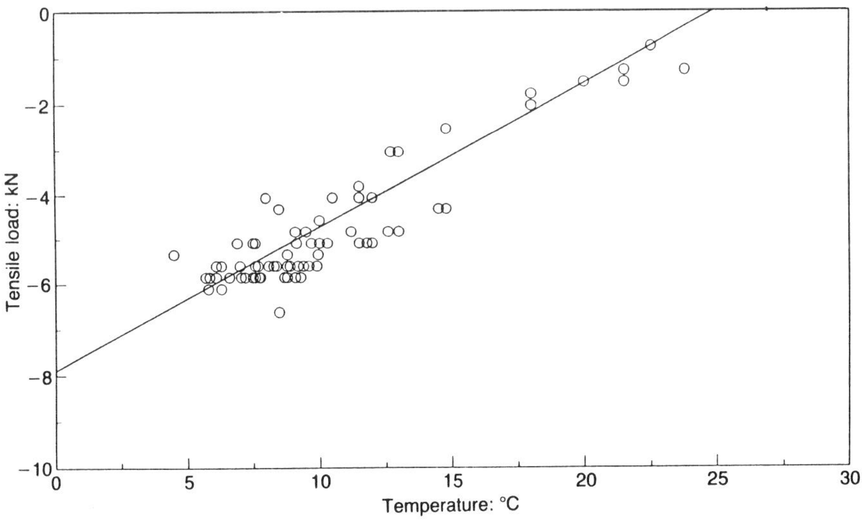

Fig. 4. Zero shift of load cell due to temperature effects

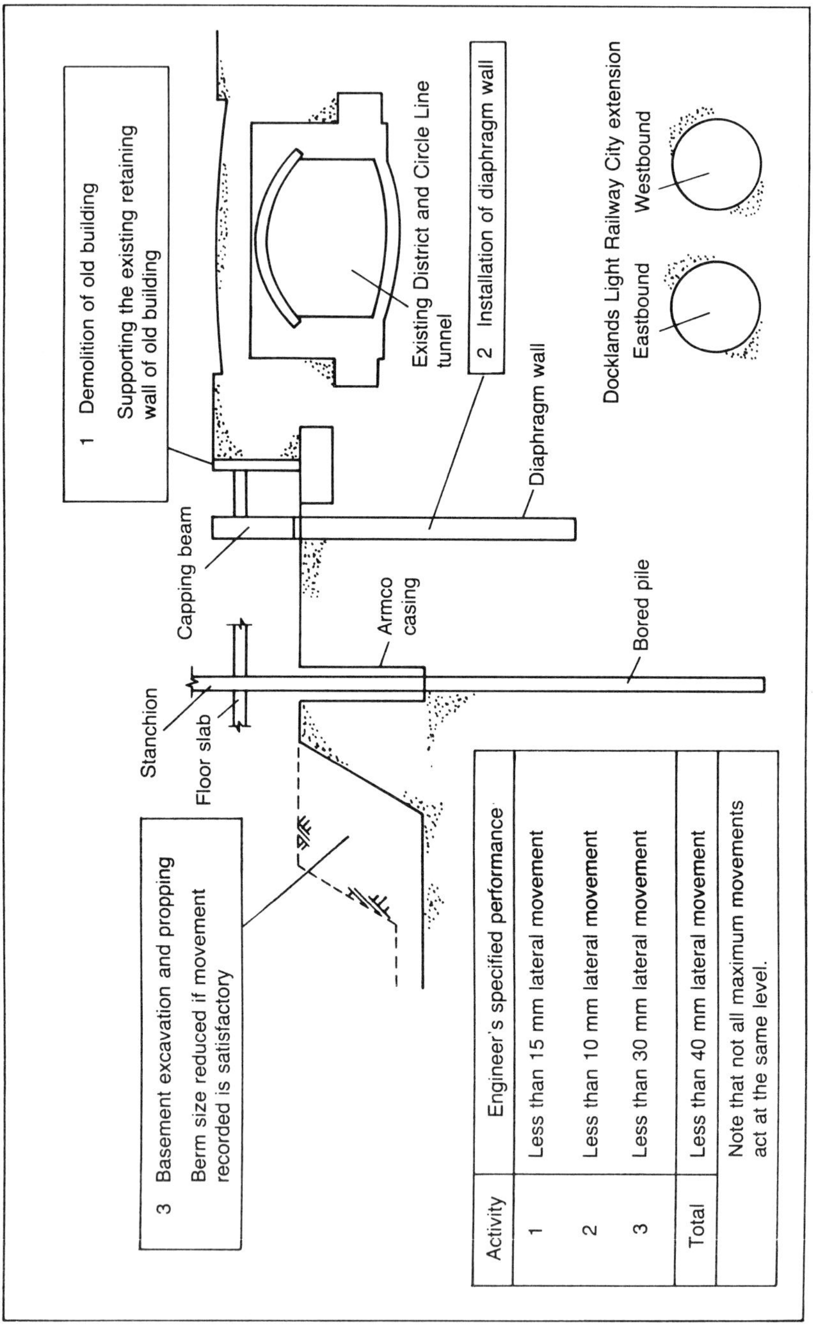

Activity	Engineer's specified performance
1	Less than 15 mm lateral movement
2	Less than 10 mm lateral movement
3	Less than 30 mm lateral movement
Total	Less than 40 mm lateral movement
	Note that not all maximum movements act at the same level.

Fig. 5. Minster Court, London

amounted to only 8 kN for a temperature change of 25°C. When the props were removed at the end of construction, all the load cells indicated zero loads. It was therefore concluded that the measured load variations in each prop were real.

Temperature effects on struts are significant

(a) when designing strutting systems, the effects of daily and seasonal temperature fluctuations should be taken into account, to avoid the risk of over-stressing the struts

(b) when interpreting strut load data (e.g. when carrying out a detailed equilibrium analysis of a wall as discussed by Lings et al. [2] in their Paper to this session).

CHE-MING TSE, Ove Arup and Partners

Instrumentation monitoring helps us to understand the interaction between the soil and the wall. It is sometimes viewed as an academic exercise. However, in some cases, instrumentation bears important contractual functions. I would like to illustrate this point using a case history I have written up for Session 4 of this conference.

The Minster Court project involved the construction of a diaphragm wall at approximately 5 m from the existing London Underground tunnel. Construction movements during demolition, wall installation and berm excavation were the main concern. The contractual set-up of the project was that the management contractor employed three different domestic sub-contractor to carry out these three activities. As Engineer to the project, Ove Arup and Partners specified the limits of movement for each subcontractorb to comply with (see Fig. 5).

During the final 'berm excavation stage', two berm sizes (a bigger and a smaller (see Fig. 5) had been designed by Ove Arup and Partners). The final choice of the berm size depended on the movements that occurred in the first two activities. Had large movements occurred in the first two activiites, the bigger berm would have been adopted. However, if the movements recorded at the end of the first two activities had been much smaller than those specified, the smaller berm would have been adopted.

Traditionally, temporary works, such as berm size design, are Contractors' responsibilities. In this particular project, the Engineer was actively involved in temporary works design and specification because the consequence of failure of the temporary works is disastrous.

In an informal discussion on 'The use of battered excavation in construction' at the Institution of Civil Engineers, my co-author Duncan Nicholson had advocated the classification of temporary works into three categories according to their risk, namely low, medium and high risks. For cases of high risk, Mr. Nicholson suggested that the Engineer should take the responsibility of design of temporary works. This is well illustrated by the case of Minster Court.

G. CHESHER, Posford Duvivier

Regarding comparisons between predicted and observed wall movements, I would like to generate comment on what, as a practitioner, I would call 'that blasted subgrade reaction value!'

Most of the case studies appear to have over-predicted wall movements by under-predicting soil stiffness by factors of up to 10. Routine designs are most commonly carried out using Winkler spring-beam wall models where subgrade reactions have to be estimated. Although I am very wary of predicting ground movements using these types of models, the wall deflections are directly related to bending moment.

Is it that subgrade reactions are not being assessed properly on account of

(a) inability to measure in the field or laboratory

(b) poor theoretical approximation in particular the modification of Terzaghi's subgrade reaction valve, which relates to reaction under footings which, for example, in WALLAP is divided by an arbitrary figure of 3 to model the unconfined excavated side of a retaining wall.

M. L. LINGS and D. F. T. NASH, University of Bristol

Dr. Symons has asked whether high negative pore water pressures may be put forward as an explanation for the under-reading of EPC3, the highest pressure cell on the excavated side at Lion Yard. On the day chosen to represent conditions at the final excavation level, negative pore water pressures were indeed present at EPC3, a value of 10 kPa below atmospheric pressure being recorded. However, at the same time, a pore pressure of - 11 kPa was recorded at EPC1, the highest pressure cell on the retained side. Yet this cell can be seen from Fig. 8 to be, if anything, slightly over-reading. We therefore do not think that negative pore pressures by themselves provide a satisfactory explanation for the under-recording of EPC3.

In reply to Dr Al-Shaikh-Ali's questions about K_0 values in the Gault clay, the self boring pressuremeter was used to assess in-situ horizontal stresses at the Lion Yard site (refer to Lings *et al.*, 1991 (ref. 1 in the Paper) for further details). An assessment of K_0 was also made from detailed considerationsof the geological history of the Cambridge Gault clay. These two approaches produce substantially different values for K_0 in the upper layers of the Gault, giving rise to uncertainties about the reliability of the SBPM measurements at Lion Yard. This only serves to highlight the difficulties that still remain in obtaining reliable information on in-situ stresses.

K_0 did not form part of the original Paper, which focused on the equilibrium analysis and the reliability of the earth pressure measurements at the end of main excavation. An initial stress line was only shown in Fig. 7 to illustrate the significant reductio in stress that occurred due to wall installation, and for that purpose an initial profile repesenting K_0 of 1.5 was chosen, which is towards the lower end of the likely in situ stresses. The Conference contribution concentrated on the low stresses recorded after diaphragm wall installation (Fig. 7) and our conclusion that they are reliable, rather than on the magnitude of the stress reduction that occurred.

Since K_0 at the site is rather uncertain, initial conditions were not shown on Fig. 9. As far as the exclusion of later figures from the Paper is concerned, Fig. 7 shows conditions before main excavation for which the equilibrium analysis has no direct relevance, and Fig. 9 was only produced after reappraisal of the data had given us greater confidence in the accuracy of the recorded earth pressure. The P_v values for four out of seven earth pressure cells were modified as a result of our reappraisal, leading to reductions in deduced stresses of between 20 and 30 kPa. Only in the case of EPC1, where the effect is to halve the stress, does this result in a marked modification to the original Paper.

References

1. NG, C.W.W. An evaluation of soil-structure ineraction associated with deep excavation in a stiff fissured clay. University of Bristol, 1992, PhD, in preparation.
2. LINGS, M.L., NASH, D.F.T. and NG, C.W.W. Reliability of earth pressure measurements adjacent to a multi-propped diaphragm wall. Int. Conf. on Retaining Structures, Cambridge, 1992

Introduction to Session 4
In situ embedded walls: design

R.J. MAIR, Geotechnical Consulting Group

Introduction

Of the 63 papers submitted to this Conference, 35 papers are of relevance to design of in situ embedded walls. While reading the papers, I have been impressed by their variety and quality. There is much material of interest to those involved in design of in situ embedded walls.

A convenient means of categorizing the papers is in terms of the principal soil types. 13 papers are concerned with stiff clays, 6 papers with soft clays, 9 papers with silts or sands, 2 papers with weak rocks, and 5 papers have general application.

Stiff clays

Of the 13 papers submitted concerning stiff clays, 6 are specifically related to London Clay. An interesting case history is presented by Tse and Nicholson concerning a 9 m deep basement constructed close to a London Underground tunnel. Ground movements were clearly of major importance for this project. Predictions were made and performance clauses were set specifying maximum movements for various stages of the construction. An interesting feature of this paper is the development of wall penetration criteria. The wall penetration was dominated by the design vertical loads, rather than by wall stability, and different adhesion alpha factors were specified, depending on how long the diaphragm wall panels were open prior to concreting.

Wall penetration in the context of long term toe stability is addressed by Phillips, Ho and Pappin in their paper on multi-propped walls. They consider the case where the lowest floor slab is wholly suspended. In 3 out of the 4 ultimate limit states considered by the authors, the penetration of the wall is of crucial importance to prevent the occurrence of excessive ground displacement in the long term. The authors consider a design situation where the soil and structure fail simultaneously and they postulate a failure mechanism to analyze this.

An informative case history is presented by Li *et al.* Their paper highlights the difference between design assumptions (which are necessarily conservative) and assumptions needed for the best Class A prediction of actual

Retaining structures. Thomas Telford, London, 1993

movements. The presence of heavy masonry arch footings, and underpinning of these structures, were difficult to model analytically. The authors also draw attention to the common problem of loads in steel temporary props being heavily influenced by temperature effects: the prop loads were found to double during hot weather, which is an important point to be borne in mind by designers.

In papers by Brooks and Spence and by Ward, case histories are presented of wall movements in London Clay. Brooks and Spence compare the observed wall deflection profiles with their predicted profiles. Better prediction was obtained with the beam-spring program WALLAP than was apparently obtained with the finite difference program FLAC, which models the soil as a continuum with elastic-plastic behaviour. Ward compares the observed wall deflection profile with predictions by the beam - spring program WALLAP and by the program FREW. In both cases significantly larger wall movements were predicted than were observed. It is evident that there is a real difficulty in deriving spring stiffnesses for a beam spring analysis and elastic moduli for a continuum analysis such that both sets of soil parameters are directly comparable. Ward also highlights the problem facing designers that the prediction of prop forces is invariably difficult. The different available methods of analysis can generate large variations in predicted values.

Comparisons between observed and predicted wall and ground movements are presented by Higgins *et al* using finite element analysis. The use of laboratory measured small strain stiffness parameters to predict the observed behaviour of the Bell Common tunnel was compared with the use of linear elastic stiffnesses derived from back analyses of measurements from various case histories. The authors conclude that soil parameters derived from high quality laboratory testing can lead to reasonable predictions of field behaviour, although a number of discrepancies are evident. Their paper highlights the importance of careful interpretation of small-strain stiffness data obtained from laboratory tests.

A good example of the effective use of finite element analysis for design is given by St John *et al.* Finite element analysis was used as a basis for a design study involving a relieving slab connected to a bored pile wall. It was found that a rear relieving slab reduces the required wall penetration, the wall deflection and the bending moment. The use of finite element analysis for unusual and complex soil structure interaction problems such as this is invaluable.

Another example of finite element analysis used for design is given by Powrie *et al*: predictions were made of the performance of a cantilever wall supported by a temporary berm. They highlight a common problem experienced by designers: the properties of the fill and gravels overlying the stiff clays can often dominate the predictions. Selection of appropriate parameters for these soils is particularly difficult. The authors also compare their finite element predictions with a simple approximate analysis.

The influence of berms on cantilever wall movements is explored in a parametric finite element study by Potts *et al.* The analyses indicate that the total weight of a berm is of more significance than its geometry. They also suggest that it is inaccurate to use an equivalent surcharge to model the effect of a berm.

A key question affecting most design methods concerning walls in stiff over-consolidated clays is the choice of initial horizontal stress prior to excavation. How important is K_o? How much are horizontal stresses modified simply by the process of wall installation? Should we take this into account? In a valuable paper, Symons and Carder describe field monitoring of the effects of wall installation for three schemes of embedded walls installed in London Clay. Changes of horizontal soil pressures were measured by spade cells. A decrease in average earth pressure coefficient K of about 10% was observed for a bored pile wall, and on two other schemes involving diaphragm walls K was found to reduce by about 20%. Estimation of K_o in stiff over-consolidated clays remains notoriously difficult. Even with high quality self-boring pressure meter tests, it is very unlikely that better than +20% accuracy is achieved in estimating K_o. Designers need to bear this in mind when considering wall installation effects.

An analytical study of installation effects is described by Gunn *et al.* They conclude from finite element analyses that installation effects only seem to be significant for propped walls, and not for cantilever walls. They also conclude that installation effects are relatively minor when the ground water table is high. Powrie and Kantarzi describe a centrifuge model test undertaken to investigate wall installation effects. As noted by the authors, a limitation of the test, and of the finite element analyses described by Gunn *et al*, is that the wall installation was modelled as a plane strain process - in other words as a long continuous slot. In reality the construction of a single diaphragm wall panel, or a single bored pile as part of a wall, would cause significant arching and redistribution of horizontal stress.

On the question of earth pressures acting on walls, the paper by Lings *et al* highlights the considerable difficulty in interpretation of wall bending moments from field measurements. The authors also showed that measured earth pressures using cells jacked out from a diaphragm wall trench into stiff clay may be over or under estimated by up to a factor of 2.

Soft clays

The 6 papers concerning soft clays illustrate a major difference from walls in stiff clays: the control of potentially very large wall movements and associated ground movements.

An interesting case history is presented by Wallace *et al* of a 12 m basement excavation in soft Singapore clay. The authors demonstrate the sensitivity of predictions to the assumed undrained shear strength profile: a 10-15% reduction in strength leads to a 50% reduction in predicted movement. This sensitivity to undrained strength is well illustrated by

Clough *et al*, 1989 (Ref.l). They showed that the wall movement is closely related to the factor of safety against base heave. When the factor of safety is low (as is often the case in excavations in soft clay), a small reduction in factor of safety can lead to a large difference in wall movement. This sensitivity of movements to the undrained shear strength of the soft clay highlights the importance of establishing the strength of the clay as accurately as possible. This requires high quality in situ testing, sampling and laboratory testing.

The use of finite element analysis for parametric studies is illustrated by Whittle and Hashash. They show that wall length has only a very minor effect on soil deformations and maximum wall deflections, which is consistent with field observations. Their results indicate that maximum wall deflections are primarily a function of excavation depth and support spacing. As noted by the authors, support stiffness also plays a major part in the development of wall movements, and this has not been included in this particular parametric study.

The difficulties in modelling important construction details are well illustrated in the paper by Abedi *et al.* They describe the performance of a sheet pile retaining structure and compare the field measurements to the soil movements predicted by finite element analysis. Deep excavations in soft clay were undertaken close to existing tanks and measures were taken to minimize ground movements which could cause damage to those tanks. Generally the measured ground movements were significantly greater than the predicted values, and this was largely attributed to the actual sequence and method of construction being different to those envisaged in the analysis. This is a common problem facing designers.

Comparisons of methods of analysing walls in soft clays (and in stiff clays) are given by Tamaro and Gould. They argue that simple beam - spring models for design have advantages over more sophisticated methods such as finite element analyses. They illustrate how wall bending moments, shear forces and prop loads are not particularly sensitive to the choice of sub-grade modulus values. However wall deflections (and therefore ground movements) are very sensitive to the assumptions of subgrade modulus.

An unusual and interesting case history is presented by Sato *et al.* A deep excavation in soft clay, extending to a depth of 17 m, was achieved without any bracing above excavation level by constructing continuous transverse diaphragm walls at an average spacing of about 7 m. The transverse diaphragm walls were also used to support the heavy vertical loads from the superstructure of the 25 storey building. Detailed measurements are presented of the wall movements, and stresses induced in the walls.

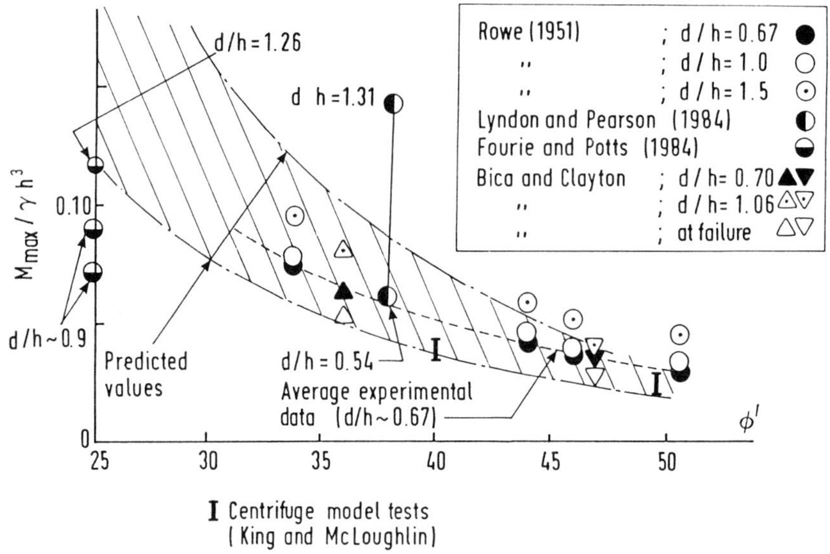

Figure 1. Experimental values of maximum bending moment in embedded cantilever walls in granular soil (after Bica and Clayton)

The same authors, Maruoka *et al*, in a separate paper present interesting methods of calculating ground movements caused by displacements of retaining walls in soft clays. They present several case histories showing very detailed sub-surface horizontal and vertical movements, and from these deduce a number of simplified ground movement patterns. As the excavation becomes deeper, the ground movement pattern becomes more circular near the base of the excavation, corresponding to the base heave mechanism identified by Clough *et al* (1989). The authors propose the use of assumed strain fields corresponding to these simplified patterns to predict ground movements for a given wall displacement.

Silts and sands

The major distinction between the behaviour of walls in silts and sands, as opposed to either stiff or soft clays, is that the soils are fully drained under static loading or unloading.

Bica and Clayton present design curves for bending moments in embedded cantilever walls in granular soil, as shown on Fig. 1. These are based on experimental results from model tests and also on various methods of prediction. The experimental values are shown as data points and the range of predictions is illustrated by the shaded area. This is of direct use to designers. King and McLoughlin present results of centrifuge model tests on embedded cantilever walls in sand. Tests were performed in a dense sand and a looser sand: the range of their measured bending moments close to failure of the walls is shown on Fig. 1. There is reasonable agreement with the other experimental values plotted by Bica and Clayton, particularly with those at or near failure.

Four papers to this Conference are devoted to tied-back walls in silts or sands. Hagerty and Nofal present a parametric study of free earth support method analyses of anchored walls in sands. They present normalized bending moments, anchor forces and wall deflections, which provide guidance for preliminary design.

Three of the papers on tied-back walls are concerned with seismic design. A dramatic illustration of the effect of an earthquake on a tied-back wall is illustrated by Iai and Kameoka. The loose backfilled sand liquefied during an earthquake, resulting in outward movement of the quay structure by up to 1.8 m, and ground surface settlements of the same order. A nearby quay structure of similar cross section suffered no damage, because the sand was significantly more dense. The authors carried out finite element analyses of the two structures and predict the difference in behaviour. Finite element analyses of dynamic response of tied-back walls are also described by Bangash *et al.*

A method of limit equilibrium analysis applied to seismic design of tied-back walls, based on the Mononobe-Okabe approach, is presented by Budhu *et al.* They show that the seismic stability of a tied-back wall can be enhanced by increasing the anchor capacity. Results of shaking table tests on model walls are also presented.

Newman and Colvil present a good case history on the use of jet grouting in a silt. A jet grouted raft was installed prior to excavation to act as an in situ prop to the retaining wall, as well as a means of cutting off ground water. Incorporation of the jet grout raft allowed significant reduction of the total length of wall, and ground movements were restricted to very small values. The use of jet grouting is widely used in continental Europe and in Japan; it is a versatile technique that should be considered more by U.K. designers.

Another case history of a propped embedded wall in a silt is described by Kastner and Ferrand. The authors present an interesting comparison between field measurements and calculations based on modulus of subgrade reaction. They draw attention to the overestimate of actual prop stiffness by 10-20 times. This is an important problem facing designers: practical effects such as waling flexibility and concrete shrinkage often lead to support stiffness being over-estimated. The paper also highlights the difficulty of estimating modulus of subgrade reaction, particularly from the initial modulus measured in a Menard pressuremeter test, in which there would inevitably be significant disturbance to the soil.

Lee and Butler describe comparisons of conventional design methods and finite element analysis of a quay wall structure with a relieving platform. They conclude that the finite element analysis shows the conventional design methods given in BS 6349 to be very conservative.

Weak rocks

Two papers were submitted on in situ embedded walls in weak rocks. Grose and Toone present an interesting case history of an in situ embedded wall in Upper Chalk. Monitoring of ground movements was a vital feature of the project and it was found that ground and wall movements were always less than the moderately conservative estimates. An interesting feature of the paper is the discussion of the division of responsibility between the permanent works consulting engineer and the temporary works contractor. The temporary wall support is selected and designed by the contractor, whilst the permanent works design remains the responsibility of the engineer. This is a very common feature of in-situ embedded walls, and it is important to be clear about the duties and responsibilities of the various parties.

Curtis and Mirzabaigian present a case history of a continuous bored pile wall constructed in the rocks of the Middle Coal Measures. The wall design was undertaken assuming linear elastic stiffnesses of these interbedded mudstones, siltstones and sandstones, derived from SPT values. Comparison with field measurements indicates that reasonably good predictions were obtained with this approach.

General application

Five papers submitted to this Conference have general application to the design of in-situ embedded walls.

Soubra and Kastner examine the influence of seepage flow on passive earth pressures. The authors undertake a rigorous analysis of the effect of seepage on passive pressure. It is important to note that the horizontal effective stress on the passive side of the wall is obtained by multiplying the authors' definition of K_p by the weight of the overlying soil using submerged density. This differs from conventional design approaches, e.g. CIRIA 104 (Ref. 2), where K_p is multiplied by the average effective vertical stress, taking seepage into account in calculation of the pore pressure.

Mortenson presents an interesting new calculation method for earth pressure problems, based on plasticity theory and statically admissible stress states. An example of an anchored sheet pile wall in sand is presented.

The treatment of line load and strip load surcharges often presents problems to designers: there is no well-established theory which takes into account the elasto-plastic behaviour of the soil and its influence on the stress distribution on a wall caused by the surcharge. In a valuable paper Schmitt and Gilbert review the various theories and design methods. They demonstrate that there are very large variations - up to 100% - depending on the theory chosen, and there is a need for further measurements and theoretical studies.

At any geotechnical conference concerned with design, particularly one on retaining structures, it would be inappropriate not to make mention of the contentious subject of partial factors. Pang describes the philosophy of adopting partial factors for limit state design of retaining walls in Hong Kong. He presents the proposed partial factors given in the draft revision of the Hong Kong design guide. The specified partial load factors are in fact very simple: all the factors on loads - with the exception of surcharge loads are 1.0. All the safety against an ultimate limit state of wall failure is embodied in the partial material factors. For an effective stress design, a single partial factor of 1.2 is specified to be applied to the soil shear strength parameters c' and tan . This is identical to the F_S approach described in CIRIA 104. A value of 1.2 for F_S is given in CIRIA 104 for 'worst credible' design, and 1.2 is also specified in the current draft of Eurocode EC7. Pang therefore demonstrates how in Hong Kong the potential minefield of partial factors is considerably simplified by the adoption of a single partial factor. This is analogous to the single factor of safety calculated for slope stability or embankment dam stability.

Puller and Puller describe developments in construction of diaphragm walls. The paper reminds us that new developments in construction techniques provide the possibility of innovation: design and construction cannot be divorced, and this is particularly true for in situ embedded walls.

Conclusions

It is apparent that the majority of the papers in this Conference relevant to design of in situ embedded walls are concerned with methods of analysis, although there are also several excellent case-history papers. Generally it can be concluded that 'beam-spring' or subgrade reaction methods of analysis are reasonably satisfactory for calculating bending moments and the deflected shape of the wall, but they are generally unreliable for calculation of absolute wall and ground movements. More sophisticated techniques, such as finite element analysis, are potentially of great value to designers - especially as both costs and time to undertake these are reducing rapidly.

References

1. CLOUGH G.W., SMITH E.M, and SWEENEY B.P. Movement control of excavation support systems by iterative design. Foundation Engineer-

ing: Current Principles and Practices, Vol. 2, ASCE, New York, NY, 1989, pp. 869-882.
2. CIRIA 104: Design of retaining walls embedded in stiff clays (authors C.J. Padfield and R.J Mair). Construction Industry Research and Information Association Report, 1984.

Design, construction and monitoring of the basement diaphragm wall at Minster Court, London

C. M. TSE and D. P. NICHOLSON, Ove Arup and Partners, UK

Introduction

Minster Court is a £400 million office development situated in the City of London (Fig. 1). It comprises three buildings and covers a total area of approximately 12 000 m^2. Building No. 1 is 14 storeys high; Building No. 2 has 12 storeys and Building No. 3 has 10. These three buildings are linked by a 9 m deep common basement. The foundation of the whole development is

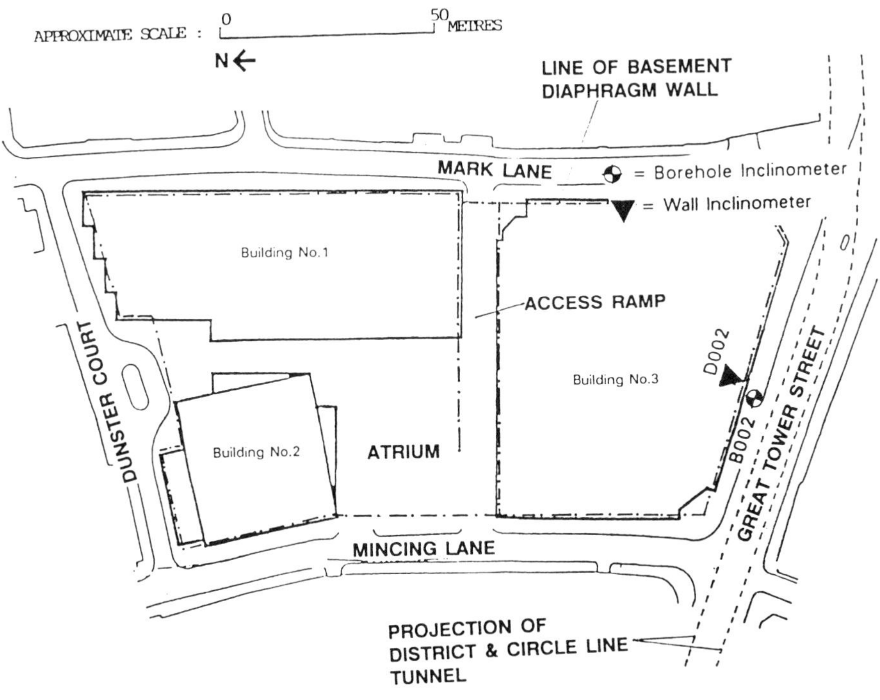

Fig. 1. Site plan

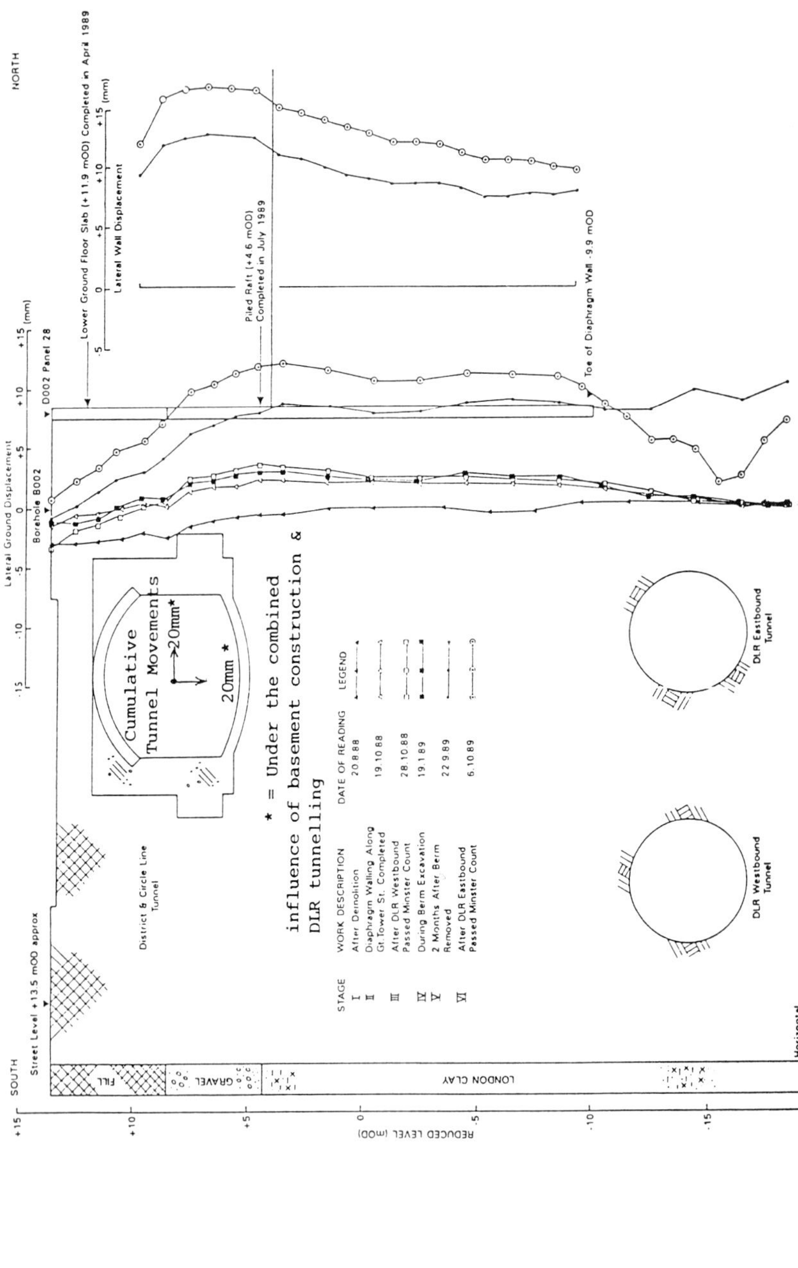

Fig. 2. Typical section and movement profiles along Great Tower Street

Table 1. Soil properties

Stratum	Top Level (mOD)	γ kN/m³	ϕ'	c' kN/m²	Ko	Euv kN/m²	Ev' kN/m²	Euh kN/m²	Eh' kN/m²
MADE GROUND	+13.1to+13.8	18	30°	0	0.7	-	10×10^3	-	10×10^3
TERRACE GRAVEL	+11.7to +6.7	20	35°	0	1.0	-	10×10^3	-	10×10^3
LONDON CLAY	+5.3 to +4	20	25°	0	1.0	500Cu	300Cu	1000Cu	600Cu
WOOLWICH & READING BEDS	-22.7to-30.7			[NOT USED IN DESIGN]					

a piled ground-bearing raft. The basement retaining wall is a 0.8 m thick diaphragm wall propped by the basement raft slab and one level of floor slab.

This paper concentrates on the diapragm wall behaviour along Great Tower Street (Fig. 1). This wall is approximately 5 m from the existing District and Circle Line tunnel (Fig. 2). Moreover it will be shown that the ground movements associated with the Docklands Light Railway tunnels construction below Great Tower Street (Fig. 2) have also influenced the diapragm wall movements at Minster Court.

The site

The site used to be occupied by four office buildings with one to two levels of basement. These buildings were constructed in the 1930's and 1950's and were demolished in 1988 before the construction of Minster Court.

The existing London Underground Limited District and Circle Line Tunnel is located to the south of the site (see Fig. 2). It comprises 5m high mass concrete walls with a 9m wide brick arch roof. The crown of the tunnel is approximately 1.5m below the Great Tower Street. The ground level of Great Tower Street is on average +13.5mOD.

Concurrent with Minster Court construction, the Docklands Light Railway bored two 5m diameter tunnels (Westbound and Eastbound Tunnels) for its City Extension. The crowns of the tunnels are at approximately 25m below the Great Tower Street (see Fig. 2). The wall of the Eastbound Tunnel is approximately 5m from the line of the diaphragm wall and is approximately 12m-15m directly below the District and Circle Line Tunnel.

Ground conditions

The stratigraphy revealed by the site investigation is summarised in Table 1. The soil properties are shown in Fig. 3. Also shown in Table 1 are the

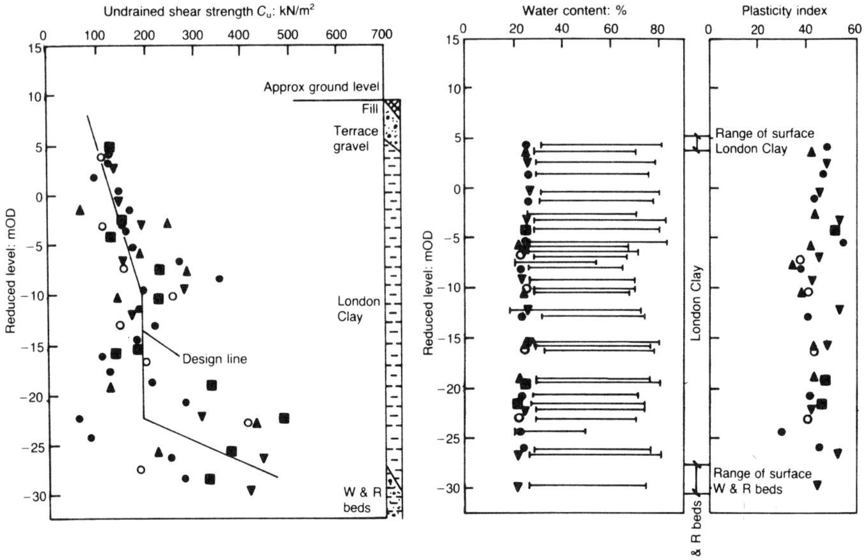

Fig. 3. Soil properties

parameters adopted for the diaphragm wall design. The ground investigations also revealed a perched water table at +4.75mOD in the Terrace Gravel.

Sequence of basement and tunnel construction

The basement was excavated leaving perimeter berms to stabilise the diaphragm wall. This enabled the raft and the ground floor slabs to be constructed using the "bottom-up" method over the centre of the basement. The Lower Ground Floor slab was used to prop the perimeter wall enabling the berm to be removed using the "top-down" excavation technique around the edge of the basement.

Summarised below is the basement construction programme for Minster Court and the Docklands Light Railway tunnels. This is also shown diagrammatically in Fig. 4.

Stage 1: Demolition of the old buildings. The basement retaining wall of the old building was supported by raking struts as demolition proceeded.
Stage 2: (a) Installation of diaphragm wall; (b) Constructing the piles in the edge row in front of the diaphragm wall.
Stage 3: Docklands Light Railway Westbound Tunnel passed Minster Court.
Stage 4: (a) Construction of capping beam on top of retaining wall; (b) The temporary raking support was removed. The load from the existing retaining wall was transferred to the diaphragm wall through a horizontal strut

connected to the back of the capping beam; (c) Excavation took place in the centre of the basement leaving behind a perimeter berm to provide passive resistance to the diaphragm wall.
Stage 5: (a) In the centre of the site, the raft was constructed. Steel columns were installed on top of the piles and the Lower Ground Floor slab was completed; (b) The Lower Ground Floor Slab was used as a working platform for constructing the super-structure while the berm was being excavated; (c) This is followed by the completion of the raft slab.
Stage 6: The Docklands Light Railway Eastbound Tunnel passed Minster Court.

Diaphragm wall construction

There are approximately 140 diaphragm wall panels at Minster Court, most of them are 3.75m-long rectangular panels. Special panels were also constructed, these included barrettes, V-shaped panels and cruciform panels. The panels were excavated in an alternating sequence, i.e. two "primary" panels were excavated first, followed by the excavation of a "secondary" panel in between the two primaries.

Due to the difference in ground levels between Building No.3 and the other two buildings, differential movements between buildings are anticipated. Movement joints have been constructed between buildings. A 25mm filled gap was constructed in between two connecting panels, a waterbar was provided for water tightness.

The diaphragm wall panels were excavated from the basement level of the old buildings at approximately +9.6mOD (Fig. 4, Stages 1 and 2). This working platform level is approximately 4m below the Great Tower Street level. The basement retaining wall of the old buildings, therefore, needed to remain and continued to retain the ground until the completion of the new diaphragm wall. It was propped by inclined struts as demolition proceeded. These inclined struts butted against concrete thrust blocks in front of the diaphragm wall (Fig. 4, Stage 1).

These struts were arranged so that initially they would not obstruct the construction of the "primary" panels. After the construction of the "primary" panels, the struts were repositioned to allow the construction of the "secondary" panels. The repositioning process continued until the completion of the wall. Preload jacking tests were carried out frequently to ensure that the load in the raking struts was maintained.

Diaphragm wall design

The diaphragm wall has three main functions: to carry the heavy vertical load from the super-structure, to retain soil and to control the ground

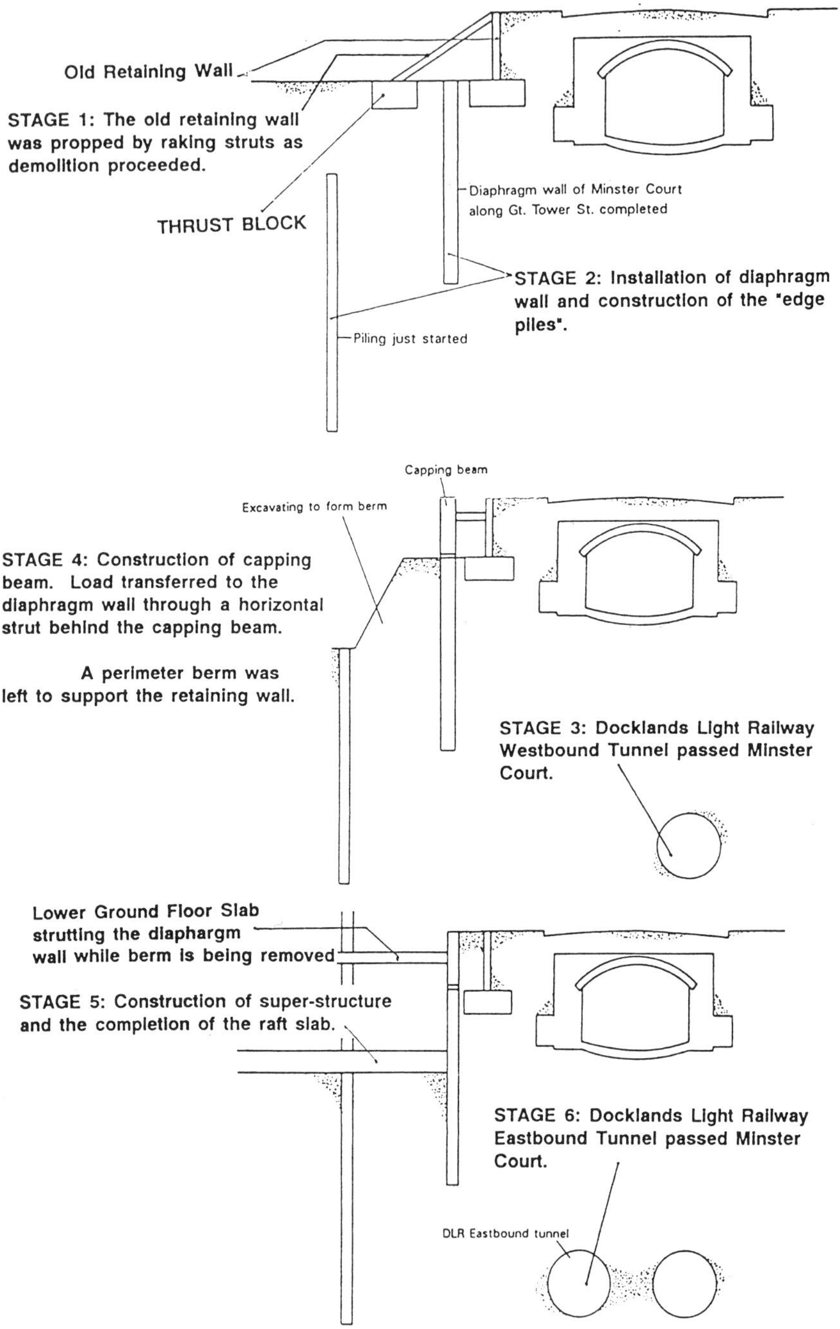

Fig. 4 . Construction sequence

movements throughout the construction. In the permanent condition, the perimeter diaphragm wall, internal piles and the basement raft combine together to act as a piled raft.

Vertical load requirements

In London Clay, the ultimate load capacity Q_{ult} of the diaphragm wall per metre run is given by

$$Q_{ult} = A_s \, \alpha \, \bar{c}_u + A_b \, N_c \, c_{ub}$$

where

$\bar{c}_u$=Average undrained shear strength over the embedment depth
A_s=Area of the wall in the clay to develop shear per m run
α = Shaft adhesion factor
A_b = Area of the base per m run
N_c = Bearing capacity factor, 7.5 for deep foundation
c_{ub}= Undrained shear strength at the toe of the wall

The embedment depths of the diaphragm wall panels are found to be governed by the vertical loading. The embedment depths of the panels along Great Tower Street ranged from 9m to 14m.

The panels have been designed with a shaft adhesion factor α of 0.35 and can be opened for up to 24 hours. It is considered that prolonged opening of the panels may lead to softening of the London Clay. This will reduce the load carrying capacity of the panels. It has, therefore, been specified by Ove Arup and Partners that the embedment depths of the panels need to be increased in case of prolonged opening. It is considered that α values of 0.3 and 0.25 should be used for panels open up to 48 and 72 hours, and the panels should be lengthened accordingly. The safe load capacity of an 800mm thick wall against embedment depth into London Clay is shown in Fig. 5, with a Factor of Safety of 3.0 and different α values.

This specified variation in panel embedment depth enabled the diaphragm wall contractor to programme his "excavation - concreting" cycle himself. It also covered the situations which panels could not be concreted within 24 hours, e.g. difficulty in ordering concrete in late evenings and weekends or unforseen site problems which delayed concreting.

Retaining wall requirements

The presence of the adjacent District and Circle Line Tunnel made it important to control ground movements throughout construction. These movements were associated with the following activities

(*a*) building demolition and propping of existing wall

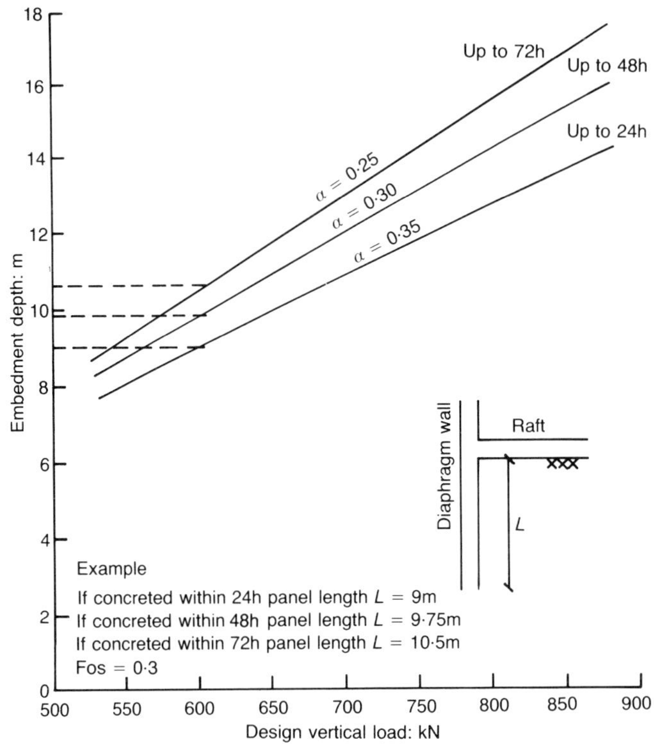

Fig. 5. Diapragm wall embedment depths

(*b*) diaphragm wall installation whilst propping existing wall, and
(*c*) basement excavation and propping.

It should be noted that three different sub-contractors were responsible for the above activities. The specifications contained performance clauses limiting the maximum components of lateral movements to 15mm, 10mm and 30mm for the respective activities. These specified limits were based on analyses using Arup Geotechnics computer program "FREW" and an assessment from case histories of similar basements in London Clay. These movements lid not act at the same level and the predicted maximum lateral ground movement was 40mm. This maximum lateral ground movement was agreed with London Underground Limited for limiting the damage to the District and Circle Line Tunnel.

Contingency plans were also developed which enabled the berm width (Fig. 4, Stage 4) to be varied to suit the measured ground movement. In this way an observational approach could be used to optimise conditions during construction.

Monitoring

It was anticipated that both Minster Court basement construction and the Docklands Light Railway (DLR) tunnelling would cause movements to the existing District and Circle Line Tunnel. As Consulting Engineer for Minster Court, Ove Arup and Partners were commissioned by the developer to monitor the movements caused to the building and to the District and Circle Line Tunnel by the DLR tunnels. A monitoring system was set up to record the ground movements along the southern boundary of Minster Court and inside the District Circle Line Tunnel. These included

(*a*) three borehole inclinometers and extensometers along Great Tower St
(*b*) wall inclinometers in three panels along Great Tower St
(*c*) street surface survey, and
(*d*) traverse, levelling and tape extensometer surveys inside the District and Circle Line Tunnel.

Typical lateral ground and wall movement profiles recorded by Inclinometers B002 and D002 (Fig.l) respectively along Great Tower Street are shown in Fig. 2 for various stages of construction. The maximum measured components of the lateral ground movement for the previously mentioned activities I to III were: -3mm for demolition (-ve means away from site), +4mm for diaphragm wall installation and +l0mm for basement excavation. These components were less than that specified and so they enabled a minimum berm width to be adopted during the final excavation stage. The cumulative lateral movement of the diaphragm wall at the end of basement excavation was 13mm.

It was noticed that the Docklands Light Railway Westbound Tunnel did not cause significant lateral movement to the Minster Court diaphragm wall. However, an additional 4mm lateral ground and wall movements occurred after the DLR Eastbound (nearest tunnel to Minster Court) passed Minster Court.

It was also noticed that the District and Circle Line Tunnel remained undistorted but moved laterally and settled 20mm as a result of the combined influence of the basement construction and tunnelling activities. DLR tunnelling accounted for 15mm of the vertical settlement.

Conclusions

This case history illustrates how specification allowed the diaphragm wall contractor to vary the wall embedment depths according to the length of time the panels had been opened. It also demonstrates how overall movements could be assessed by the consultant and assigned to contractors responsible for various activities. Monitoring provided a check on the movements and enabled appropriate construction sequence to be adopted.

Long term toe stability of multi-propped basement walls in stiff clays

A. PHILLIPS, Arup Geotechnics, UK, K. K. S. HO, Geotechnical Engineering Office, Hong Kong Government, and J. W. PAPPIN, Arup Geotechnics, UK

Introduction

Over the last few years, many deep basement excavations have been constructed, especially in London. These excavations are often in stiff clays and are commonly supported both in the temporary and permanent condition by propped diaphragm or secant pile wall construction. This paper concentrates in particular on failure mechanisms associated with the long term toe stability of such multipropped walls. The methodology described has been developed over the last few years on a number of projects.

Multi-propped walls work by the soil pressures acting behind the wall being balanced by forces in the props, and by resistance developed in the soil below excavation level in front of the wall. In their permanent situation the walls are generally propped by the basement floor slabs with a toe embedment beneath the lowest slab. This lowest slab in many cases is wholly or partially suspended to eliminate or reduce the swelling pressures that subsequently will arise. This is often done for "top-down" construction where piles are used as foundations for the superstructure.

For multi-propped walls embedded in stiff clays a relatively small toe embedment may be sufficient for stability during construction if undrained behaviour of the clay can be relied upon. However, as time progresses, the negative pore water pressures set up in the clay dissipate and the ground swells and loses strength. Even though the lowest slab is cast by this stage, stability of the wall soil system is likely to decrease. If instability of the wall-soil system were to be reached, excessive displacement below the slab would occur. This could lead to upward forces on the slab possibly resulting in its failure.

At failure, the soil pressures acting on the wall will not only depend on soil properties, but also on the influence of the wall on the soil. Relative vertical movements between the soil and the wall will control mobilisation of wall friction. The bending flexibility of the wall and the mode of deforma-

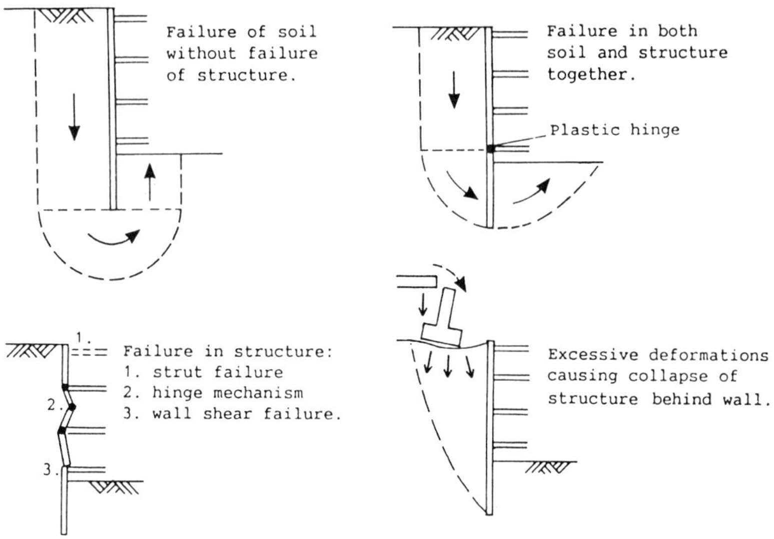

Fig. 1. Ultimate limit states

tion can result in pressure redistribution (arching) within the soil mass. These effects need to be considered in order to fully understand likely failure mechanisms.

Limit state design of multi-propped walls

Limit state design of geotechnical structures, including propped embedded walls, is advocated in the draft Eurocode on Foundations, EC7. Limit states are states beyond which the soil-structure system no longer satisfies the design requirements. They are generally classified into two types; namely serviceability limit states and ultimate limit states.

To comply with serviceability limit state requirements, deformations of all elements in the soil-structure system under normal "working" conditions must be satisfactory, and must not be such as to cause deterioration of the materials. To satisfy ultimate limit state requirements, it must be shown that there is an adequate margin of safety against collapse of any significant element of the soil-structure system, for the worst combinations of loading and material properties that can occur. In the case of an embedded wall, four types of ultimate limit states can be identified. These are illustrated in Fig. 1.

For a satisfactory design, the occurrence of each limit state must have an acceptably low probability.

Table 1. Factors of Safety for strength factor method

Recommended Range for Moderately Conservative Parameters (c',ϕ')	Recommended Minimum Values for Worst Credible Parameters (c'= 0, ϕ')
1.2 to 1.5	1.2

Methods of assessing toe embedment and structural forces

Cantilevered and singly-propped walls

There are well established methods for the design of cantilevered and singly-propped walls. They are generally limit equilibrium methods based on the assumptions of fully mobilised active and passive pressure distributions. Procedures for assessing toe embedment and structural forces are discussed by Padfield and Mair (ref. 1).

Toe embedment. Padfield and Mair recommend Factors of Safety for determining a stable wall geometry in stiff clays. For permanent works design, using the "strength factor method" they recommend the Factors of Safety for effective stress analysis shown in Table 1.

Application of such Factors of Safety can be considered to give "allowable mobilised" strengths which are suitable for design when considering ultimate limit states. Such "allowable mobilised" strengths are applied to the soil when assessing the embedment depth required for stability. The use of such values should limit soil movements. The structural forces (e.g. bending moments and shear forces in the wall, and prop forces in the slabs) corresponding to this ultimate limit state calculation are normally very conservative and consequently are ignored and not designed for.

Structural forces. One of the approaches currently used to determine ultimate structural forces, as discussed by Padfield and Mair (ref. 1), involves the computation of structural forces corresponding to the limiting condition with unfactored soil parameters. (The additional embedment depth associated with the ultimate limit state "toe stability" calculation is not accounted for in this "working condition" calculation.) The calculated "working" bending moments and shears are then factored up, for example in accordance with BS 8110 (ref. 2) by 1.4, to determine ultimate structural forces.

An insight into the actual behaviour of cantilevered and singly-propped walls can be obtained by parametric studies with numerical modelling using, for instance, finite elements. Potts and Fourie (refs 3 and 4) reported on a finite element study of a singly-propped 1m thick reinforced concrete wall in a stiff overconsolidated clay. The important conclusions are that for this type of rigid wall the initial in-situ soil stresses and passive soil yielding during excavation can have a significant influence on the structural loads. Much higher loads compared to those calculated by conventional limit equilibrium methods were computed. Similar conclusions were reached by Bolton *et al* (ref. 5) based on a theoretical approach and model tests.

Multi-propped walls

In the case of multi-propped walls, the assessment of the design structural forces is generally based on similar approaches adopted for singly-propped walls, with no account being taken of soil arching. Toe stability is conventionally assessed by considering the equilibrium of the free-ended span below the lowest strut. Rotation is assumed around the lowest strut and the plastic moment capacity of the wall can be included in the calculation.

However, the design toe embedment depth may be such that the soil strength associated with the lower part of the wall cannot be fully mobilised at collapse. This is because failure may be governed by the strength of the structure. Preliminary studies suggest that this is normally the case for embedded walls in overconsolidated clays where the design is based on the methods and the Factors of Safety recommended in Padfield and Mair (ref. 1) and BS 8110 (ref. 2). The apparent incompatibility of the Factors of Safety applied to the soil and those used for the design of the structure is also demonstrated by Simpson (ref. 6). It is therefore useful and logical to consider a design situation where the soil and the structure fail simultaneously. This idea is developed later in the Paper.

Analysis using the computer program FREW

To model adequately the behaviour of a multi-propped wall system, the sequence of excavation and floor construction must be considered; the addition of stiff propping elements (e.g. floor slabs) gives rise to complex soil-structure interaction effects. As a result of the relative stiffnesses of the floors, wall and soil, together with the strength limits on the soil and concrete, redistribution of earth pressures (arching) can take place.

The OASYS program FREW models the behaviour of a flexible retaining wall allowing for interaction between the wall and the soil. Details of the program are given by Pappin *et al* (ref 7). In the program the wall is modelled as a series of elastic beam elements between nodes, with the soil as blocks of elastic material on either side of the wall.

The program analyses each stage of the construction sequence in turn. At each stage the force imbalance at each node and stresses in the soil are calculated. If the stresses are outside specified active or passive limiting criteria, correction forces are applied and the problem is solved iteratively until the stresses are acceptable. It is often assumed that limiting earth pressures increase linearly with depth, giving, for example

$$p_a = K_a\ (\gamma z - u) \tag{1}$$

where p_a is the active effective pressure limit at depth z below ground, K_a is the active pressure coefficient, γ is the unit weight, u is the pore water pressure at depth z.

If the effective wall pressure at depth z is defined as p then ensuring that p is greater or equal to p_a at all depths will be sufficient for stability, but may be unnecessarily severe. If P is the effective force on the wall between depth z and the ground surface, an alternative condition would be

$$\int_0^z p\,dz = P \geq P_a = \tfrac{1}{2}K_a\ (\gamma z - u)z \tag{2}$$

This equation would allow the type of stress distribution indicated as p in Fig. 2a. Working from the top downwards, it implies that the area beneath the p_a line must never exceed the area above the p_a line.

However, although equation 2 is necessary, it is not sufficient, since, even for a cohesionless soil, it would allow a zero pressure on a finite depth of wall. This is clearly not admissible. There must be another limiting lower value to the pressures. In FREW this limit on pressure redistribution is given by the following criteria

$$\int_{zj}^{zi} p\,dz \geq \int_{zj}^{zi} K_a\ [\int_{zj}^{z} \gamma dz + K_{aj}p_j + u_j - u]\ dz \tag{3}$$

This equation is based on the principle that at any depth z_j the minimum vertical effective stress is K_{aj} times the horizontal pressure p_j immediately above depth z_j. The minimum horizontal pressure below z_j is K_a times the vertical effective stress. At depth z_j this is therefore equivalent to $K_{aj}2.p_j$.

Equation 3 is enforced between all pairs of nodes corresponding to depths

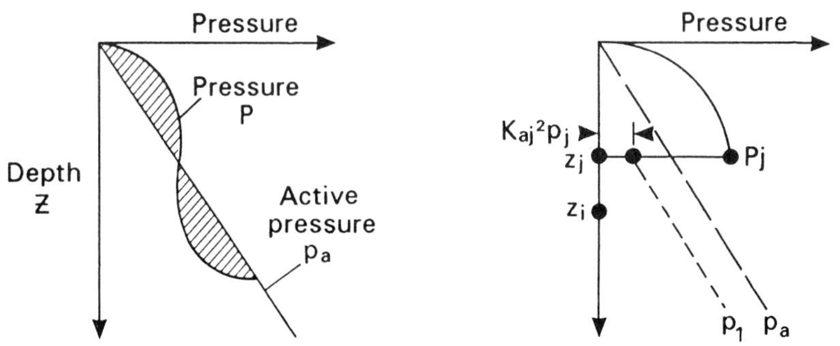

2a 2b

$$p_1 = K_a\ [\int_{zj}^{z} \gamma dz + K_{aj}p_j + u_j - u]$$

Fig. 2. Redistribution of wall pressures

z_i and z_j ($z_i > z_j$), as presented in Figure 2b. It is considered that this provides a good approximation to limits which are necessary and sufficient (ref 7).

Design of embedded wall using FREW

The computer program FREW can be used to examine serviceability limit states using most probable soil parameters, and to examine ultimate limit states using "allowable mobilised" parameters.

The design procedure uses the following steps

(*a*) Compute the embedment depth required for vertical load capacity or for cut-off purposes if appropriate
(*b*) Estimate a conservative embedment depth necessary for stability and use FREW with "allowable mobilised" soil strength and stiffness parameters suitable for the ultimate limit state to model the collapse condition. Ultimate structural forces are calculated directly. The overall stability of the soil-structure system is demonstrated by the ability of the program to reach convergence
(*c*) Where the program converges readily and the embedment depth estimated for step (*b*) is greater than that for step (*a*), the toe embedment assumed in step (b) can be progressively reduced (down to the limit of that corresponding to step (*a*)) until the system is just stable
(*d*) Check for serviceability using FREW with "most probable" parameters. The structural forces corresponding to the working condition are factored up (e.g. by 1.4) to obtain the ultimate values
(*e*) The ultimate structural forces are taken to be the envelope of those calculated from step (*c*) and step (*d*).

The procedure described above may appear to be quite involved, but it has the advantage that it may be generally applied. In certain design situations, for instance where there are large spans between floors, the conventional serviceability check (i.e. employing step (*d*) only) has proved to provide insufficient safety, and the full procedure has been required.

A typical result from a FREW analysis for a multipropped wall in stiff clay is presented in Fig. 3. The figure shows the ultimate limit state pressure distribution calculated for the wall in its final condition; the stiff clay being fully drained, and the pore water pressure distribution based on a linear drop of head around the wall. Behind the wall the active pressure near the toe of the wall has dropped well below the active pressure limit. In front of the wall the available passive pressures have been fully mobilised.

Checking of FREW solution

Simple failure mechanisms can be postulated and the embedment depth

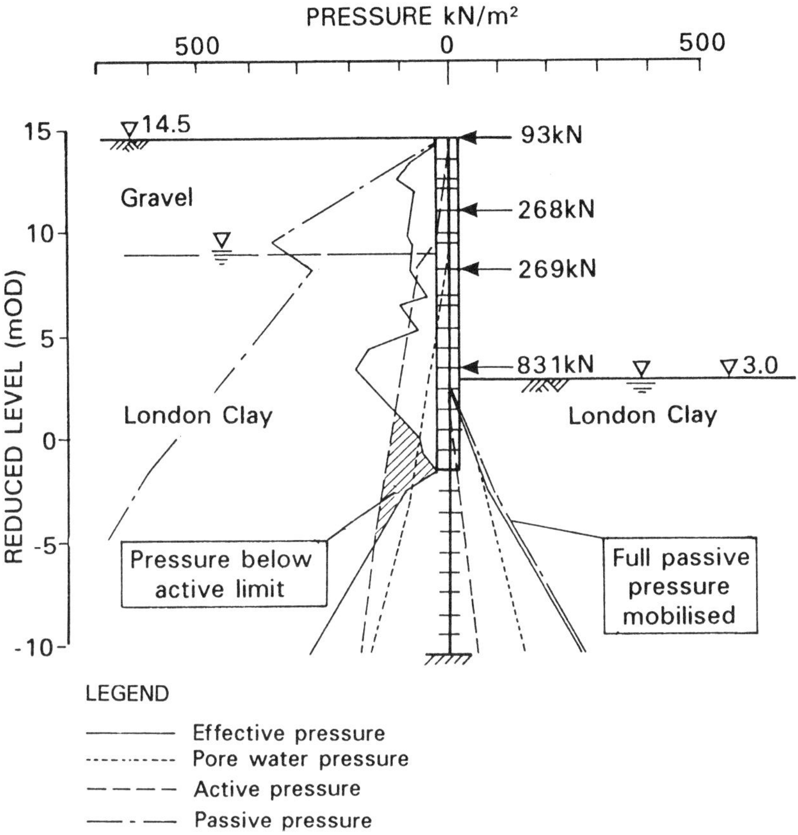

Fig. 3. Typical FREW results

assessed by hand calculations, to compare with results of stability analyses by FREW. A possible soil failure surface is shown in Fig. 4.

The soil being retained by the wall and any associated applied loads act to cause failure. It is assumed that the propping system is stiff enough to restrict horizontal wall movement into the excavation above the lowest prop, resulting in a vertical failure plane within the retained soil. The soil below the excavation acts to restrain the wall by way of conventional passive pressure, which may include for wall friction.

Finite element studies have been carried out where the strength of the soil in the long term is reduced incrementally and the development of ground movements observed. These analyses have confirmed that the postulated soil failure surface shown in Fig. 4 is credible. Two failure mechanisms are checked, one by the wall soil failure surface shown in Fig. 4 is credible. Two failure mechanisms are checked, one by the wall failing in shear below the

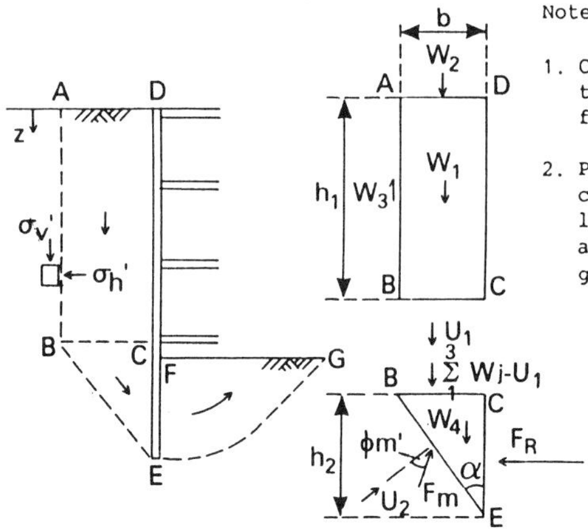

Notes:

1. Calculation to be done in terms of "allowable mobilised" friction angle ϕ_m'
2. Pore water pressure u can be calculated assuming uniform loss of head around wall; assume u constant along any given horizontal plane.

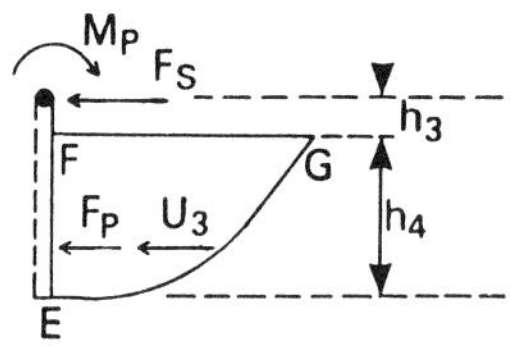

Stresses along AB:

assuming a vertical failure plane

$$\sigma_h' = \frac{1 - \sin^2 \phi_m'}{1 + \sin^2 \phi_m'} \sigma_v'$$

where $\sigma_v' = \gamma z - u$ and γ = bulk unit weight
z = depth below ground
u = pore water pressure

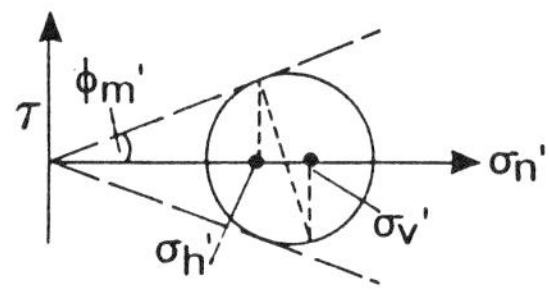

Magnitude of forces:

F_R = required force on BCE to maintain stability, calculated using "wedge" analysis; assumed to act at $\frac{1}{2}h_2$ below lowest strut.

F_P = available "effective" passive resistance of EFG ($\bar{\sigma}_{vp}'.K_p.h_4$); assumed to act at $\frac{2}{3}h_4$ below ground level.

F_s = shear force capacity of wall immediately below lowest strut.

M_p = plastic moment capacity of wall at lowest strut

Note: F_R and F_P are horizontal components of force

$W_1 = \gamma h_1 b$

W_2 = load applied on block ABCD from surcharges and structure.

$W_3 = \int_0^{h_1} \sigma_h' \tan \phi_m' dz$.

$W_4 = \frac{1}{2}\, b\, h_2\, \gamma$

$U_1 = \frac{u_B + u_C}{2} \cdot b$

$U_2 = \frac{u_B + u_E}{2} \cdot \frac{h_2}{\cos\alpha}$

$U_3 = \frac{u_F + u_E}{2} \cdot h_4$;

assumed to act $\frac{2}{3}\, h_4$ below ground level.

Stability requirements:

$$F_R \not> F_p + U_3 + F_s$$

$$F_R \cdot \tfrac{1}{2}h_2 \not> M_p + (F_p + U_3)\cdot(h_3 + \tfrac{2}{3}h_4)$$

Fig. 4. Postulated failure mechanisms used to check toe stability

lowest prop, and the other by the wall developing a plastic hinge and rotating about the lowest prop.

It is assumed that the required force on the wedge BCE, (Fig. 4), acts at the mid-height of the length of wall below the lowest strut. This may seem initially to be unconservative since generally earth and pore water pressures increase with depth, which results in the resultant force acting below the mid-height of the wedge. However, it is considered that arching will lead to effective earth pressures increasing near the strut level, and decreasing near the toe of the wall, as shown in Fig. 3. Therefore, the assumption of the force acting at mid-height of the wedge is considered realistic and generally conservative.

The wall bending moments and shear forces at the lowest strut level calculated using FREW and using the calculation method described in Fig. 4 have generally been found to be in reasonable agreement. For the case shown on Fig. 3, the FREW analysis calculated a bending moment of about 400kNm/m and a shear force of about 400kN/m, compared to values of 500kNm/m and 300kN/m calculated by hand.

Discussion

The stability of a multi-propped retaining wall is a complex soil structure interaction problem. It cannot be realistically studied using simple statically determinate mechanisms of the type assumed for cantilevered and singly-propped retaining walls.

A method is presented for using the computer program FREW to allow for the known soil arching conditions that will exist in the case of long term stability failure of the toe of this type of wall. The results appear reasonable and can be used directly within the limit state framework. A hand calculation method for investigating the toe stability mechanism is also given. This method is likely to be conservative because it is not kinematically admissible and the internal energy required to mobilise this postulated failure must therefore be underestimated. Nevertheless, it is much closer to the actual mechanism of this type of failure than ones which assume full active pressures to act behind the wall below the lowest slab level. When using the hand calculation method one must consider whether the assumptions about points of action of forces, as presented in Fig. 4, are appropriate.

The problem of toe stability must be addressed to prevent the occurrence of excessive displacement in the long term. If it is found that, even with use of the design methods described, excessive toe penetration is required, then other methods of ensuring toe stability could be used. The use of a ground bearing edge strip around the lowest slab, as used on the British Library (ref. 8), would greatly enhance toe stability. However this may lead to high uplift earth pressures acting on the slab. Fernie *et al* (ref. 9) have used a "pin-pile" system to enhance the passive resistance of the soil in front of the wall. All of these solutions are relatively expensive and require significantly more difficult calculations to demonstrate their effectiveness. In many cases it is beneficial to more accurately examine the toe penetration required in the long term. This paper gives methods which go some way in achieving this aim without having to resort to time consuming finite element analyses.

References

1. PADFIELD C.J. and MAIR R.J. *Design of retaining walls embedded in stiff clay*. CIRIA Report No. 104, Construction Industry Research and Information Association, London, 1984.
2. BRITISH STANDARDS INSTITUTION. *BS 8110. Structural use of concrete*. BSI, London, 1985
3. POTTS D.M. and FOURIE A.H. The behaviour of a propped retaining wall: results of a numerical experiment. *Geotechnique*, 1984, **34**, pp 383-404.
4. POTTS D.M. and FOURIE A.B. The effect of wall stiffness on the behaviour of a propped retaining wall. *Geotechnique*, 1985, **35**, pp 347-352.
5. BOLTON M.D., POWRIE W. and SYMONS I.F. The design of stiff in-situ walls retaining overconsolidated clay: Part II, long term behaviour. *Ground Engineering*, 1990, **23**, No. 2, pp 22-28.
6. SIMPSON H. Discussion on "Limit equilibrium design methods for free embedded cantilever walls in granular materials", by Bica A.V.D. and Clayton C.R.I. *Proc. Instn Civ. Engrs*, 1991, **90**, pp 209-229.
7. PAPPIN J.W. *et al.* Numerical analysis of flexible retaining walls. *Proc. Conf. Numerical Methods in Engineering Theory and Applications, Swansea*, 1985, pp 789-802.
8. STEVENS A. and RYALLS P.J. Design and construction of the structural works phase lA, British Library, St. Pancras, *The Structural Engineer*, 1990, **68**, No. 9, pp 159-166.
9. FERNIE R., St. JOHN H.D. and POTTS, D.M. Design and performance of a 24m deep basement in London Clay resisting the effects of long term rise in groundwater, *Proc. l0th Eur. Conf. on Soil Mech. and Fnd. Eng., Florence*, 1991, pp 699-702.

Analysis and design of cast in situ walls (diaphragm walls)

G. J. TAMARO and J. P. GOULD, USA

The Paper concerns four methods utilized to analyze cast-in-situ (CIS) walls which serve as both temporary and permanent retaining structures. Practice is tending toward computer-assisted analysis and programs are available for the continuous beam method, finite difference method (FDM), or finite element method (FEM). Both FDM and FEM require detailed knowledge of soil and wall properties and construction procedures. Properties of most soils are crudely defined but observed performance permits empirical analyses of soils' stress-strain characteristics. The authors discuss the effect of these uncertainties and strengths and weaknesses of the methods. The focus is to test alternative designs by comparing moments predicted by conventional methods against those predicted by the finite differences method.

Introduction

Increasing use of cast-in-situ (CIS) walls (slurry trench walls, tangent pile walls, secant-walls) requires methods to evaluate soil-structure behaviour. Usually, in the past the temporary earth retention system was isolated from the permanent basement wall, which was constructed independent of the temporary supports. The two could be designed independently disregarding the effects of one on the other; but use of CIS walls requires a revision in design philosophy in as much as CIS walls perform initially for temporary earth support and subsequently as the permanent wall. Stresses and movements in the temporary structure will remain after the temporary system has served its function and can add to those developing in the permanent condition.

Behaviour of CIS walls is complicated since wall properties are difficult to estimate (although better known than properties of the adjacent soils). Wall thicknesses may vary from point to point depending upon ground conditions and installation method. Behaviour of the concrete/steel wall reinforcement can be estimated, but will depend upon concrete stress level and whether the wall is cracked or uncracked. Soil properties on both sides of the wall can change due to dewatering, wall movement, consolidation, and construction procedures. Lastly, stresses in the wall will change with time due to shrinkage, creep, temperature and elastic shortening, all depending upon the type of support from the interior structure.

The Geotechnical Engineer/Structural Engineer interface

The geotechnical engineer must estimate ground pressures, deformations and movements of the retained soil; while the structural engineer must determine stresses in the structure and proportion it to support applied loads. Knowledge of ground properties is limited to tests of a few samples at a few locations. These properties can change with time, location, stress, water content, as well as the sampling process. Local knowledge of the ground and site is essential to predicting soil behaviour. The structural engineer, on the other hand, has more explicit knowledge and control of his structural materials since he has specified their manufacture. He deals with steel, concrete or their combination in elements which are usually linear or planar, performing in the elastic range.

The geotechnical engineer forms his expectation before start of work and uses the "observational method," hopefully having reliable instruments and good measurements to verify that the work accords with his assumptions. If variations from those assumptions occur, changes must be made in the work. The structural engineer, on the other hand, starts the work with more confidence (perhaps ill founded) in the precision of his design. The luxury of independent analyses by the geotechnical and the structural engineer is no longer possible. CIS walls require that the participants integrate their understanding of the wall's behaviour during construction and in the permanent condition. This paper outlines methods of achieving that understanding.

Various methods of analysis

Currently, four different procedures are available to analyze a CIS wall. Each was developed with a different goal and exhibits specific advantages and disadvantages. These procedures are summarized below and are elucidated subsequently. The first is the time-tested Terzaghi-Peck "enclosing" pressure diagram. The second is a "staged excavation method" where the structure is analyzed as a continuous beam with rigid or yielding supports and an equivalent reaction is assumed below the excavation subgrade. The third procedure is also a "staged excavation method," where the structure is analyzed using the finite difference method employing soil and support springs (Winkler Model) which may be elastic or elasto-plastic. This is the classic analysis of soil-structure interaction. The fourth procedure is a "staged excavation method" which uses finite elements to estimate loads, soil and structure behaviour. It is the most comprehensive since it considers performance of both soil and pore water in the surrounding cross section to a distant boundary.

Terzaghi-Peck procedure

This was developed using measured strut loads to compute an enclosing pressure diagram for design of temporary support with the water table below a stable subgrade. "Apparent pressure diagrams" are suggested for sands "soft-medium" clay, and "stiff-fissured" clay. They are altered by the geotechnical engineer to reflect variations in groundwater, soil strength and type, and "local knowledge." The structural engineer uses these pressures to design the temporary structure, assuming that the wall spans as a simple beam from support to support. Some structural engineers have "refined" it by assuming that the sheeting is continuous. An equivalent rigid support is taken at the excavation subgrade.

It is conservative for estimating brace loads; easy to use and effective for preliminary sizing of wall and struts; it is widely accepted for flexible sheeting; it only requires gross knowledge of soil properties. It was not intended for design of a stiff permanent wall and does not readily accommodate seepage and/or unusual variations in soil properties. It assumes a rigid support at subgrade and base stability must be assured for intermediate and final conditions. It does not accommodate the effect of strut preload and cannot be used to estimate wall and ground movements.

To design permanent walls some contemporary practice employs the "apparent pressure diagram", increased typically 10 – 30 % to account for greater wall and support stiffness. The wall constructor then must study each stage of excavation and provide supplemental reinforcing and/or supports for a safe excavation.

Staged excavation - continuous beam method (CBM)

The method was developed to account for the realities of a staged excavation. It assumes "active" soil pressure and water pressures applied outside the wall and "passive" soil pressures inside as the subgrade reaction. The wall is assumed to act as a continuous beam and an equivalent support is taken below subgrade, typically at a point of zero net pressure (See Fig. 1). The method requires active (or at rest) and passive pressure diagrams for each stage of excavation to accommodate the stylized geometry in order to perform a continuous beam analysis at each stage. It permits consideration of wall movements just before installation of supports, the "installation" of a support after the wall has moved during that stage of excavation and the deformation of the support for all subsequent stages of excavation.

The method is easy to use and generally accepted by structural engineers. It provides reactions, shear and moment envelopes for design, and can approximate the wall deformation from elastic performance of the structural system. Precise knowledge of the properties of wall or supports is not necessary and modifications easily accommodate changes in soil properties or geometry. The method yields conservative reactions and shears; conservative negative bending moments but unconservative positive moments compared to other methods (Kerr and Tamaro, 1990). Computations do not

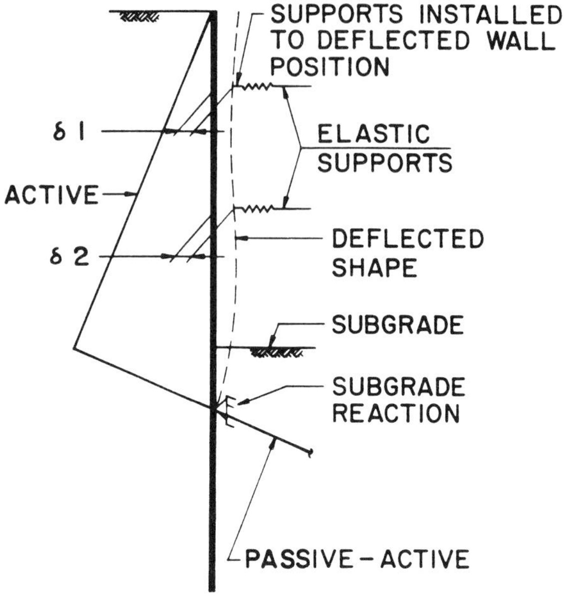

Fig. 1. Staged excavation - continuous beam method

proliferate and are easily reviewed. However, soil deformation is not considered except indirectly in judging an equivalent interior support point. This method is site-specific and computations must be redone for each soil/structure or support change. Separate analyses may be necessary to be sure of base stability or to evaluate seepage effects.

Staged excavation - finite difference method (FDM)

This analyzes the wall as beam-columns under applied forces and constraints. It assumes that soil adjacent to the wall has definable elasto-plastic properties (Halliburton, 1968). For example, prior to start of excavation, at-rest pressures are assumed on both sides of the wall. Then, stage by stage, soil is removed from inside the wall with concurrent installation of supports. As excavation is carried downward, pressures outside the wall reduce from at-rest to a limiting active condition, while pressures inside the wall below subgrade increase from at-rest to passive as a limit. Elastic behaviour of the soil during the transition from at rest to either active or passive pressure is represented by soil springs with stiffness either uniform with depth (cohesive) or increasing with depth (granular), (Terzaghi, 1955), with a limiting yield value (see Fig. 2).

Practical design has been simplified by the assumption that active pressure is achieved outside the wall, and soil springs are reacting only inside the wall

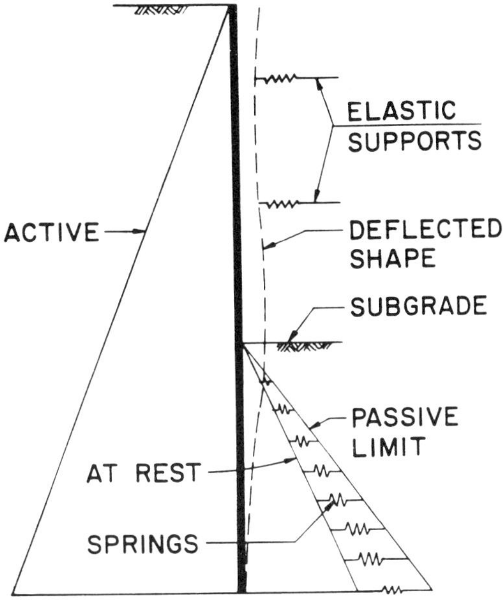

Fig. 2. Staged excavation - finite difference method

below subgrade. The wall is permitted to deform at each excavation stage with the struts represented by elastic springs. Supports can be preloaded and wall or brace deflections can be preset. The procedure requires full definition of the construction sequence, the development of resisting pressures for each stage of excavation, the definition of geometry and structural properties for each stage, an estimate of soil springs for each stage, and, lastly, an analysis by use of computer.

Refinements can be obtained by considering the deformed position of the wall prior to installation of a support, by considering changes in lateral pressures due to drawdown of water levels on either side of the wall and by performing sensitivity analyses by varying soil and support springs to determine a range of wall movements, reactions, and deflections.

It is comfortable for structural engineers; it allows for soil deformation; it can estimate lateral wall movement (and with some judgement, settlements outside the wall). Moments, shears and reactions are not highly sensitive to passive soil springs. Computers can be used to develop reactions, shear, moment and deflection envelopes for design. There is a limited output which can be easily reviewed.

However, the method is highly site-specific and the analysis must be rerun for changes in geometry or soil properties; wall deflections are sensitive to assumptions on passive soil springs and yield limits. The method overestimates positive moment when passive spring values are underestimated. The method cannot be used to estimate the effect of wall installation procedures. The methods assumes that wall and soil properties remain unchanged with time.

The finite element method (FEM)

FEM can be used to evaluate the behaviour of soil/wall systems during staged excavation. It assumes that the soil/structure acts as a series of connected elements and soil properties can change with strain and time. The procedure requires a detailed soil and groundwater description, selection of a particular soil model and a detailed construction procedure. Sophisticated definition of elasto-plastic soil behaviour and precise structural properties are required. A mesh is set up to define the geometry and analyzed by computer. The method can handle deflection of supports, soil changes with time and varying properties at the soil/structure interface. Drained or undrained soil conditions are accommodated..

The FEM outputs vertical and horizontal soil movements on both sides of the structure; it can be used to estimate soil behaviour at each stage of excavation and to perform a sensitivity analysis of assumed soil and structural properties. But the procedure yields masses of data for the engineer to review and comprehend; it cannot be modified easily for changes in geometry or soil properties; it yields lower wall shear, moments and support reactions than the foregoing methods and requires a sophisticated estimate of soil and structural properties prior to analysis. There is limited experience in its use for design. Results can vary significantly with assumed soil and structure properties.

Central artery examples

Two of the foregoing analytical methods, the continuous beam procedure (CBP) and the finite difference method (FDM), have been evaluated as tools for design and analysis of diaphragm walls for the Central Artery Project in Boston, Massachusetts. The methods have been compared for consistency of results.

The Boston subsurface profile comprises approximately 5 meters of variable but chiefly granular fill, 6 meters of desiccated marine clay overlying 6 meters of moderately overconsolidated marine clay, 4 meters of cohesive glaciomarine soil and 3 meters of very compact glacial till. Design undrained shear strengths are 100 kPa in the desiccated clay and glaciomarine soils and 50 kPa in the overconsolidated marine clay. The wall bottoms in argillite ranging from completely to moderately weathered. Soil spring values have been obtained from another site, Post Office Square Project in Boston (Ressi, Tamaro 1991) and have been back-computed from observed wall deflections using FDM, as modified by Mueser Rutledge Consulting Engineers from the Halliburton program. This permits continuous staged excavation analysis where all preceding wall movements are retained for use in subsequent computations.

For this study it was accepted that sufficient wall movement will occur so that exterior pressures will be midway between at-rest and active and that passive pressure will fully develop above the level of zero net pressure. Variations in exterior soil pressures do not have an important effect inasmuch as hydrostatic pressures dominate and the difference between active and at-rest are only 8 percent of the total driving load. As a further simplification the calculated pressures have been smoothed to straight-line relationships between changes in strata. Surcharge loads are estimated using conventional procedures and added to the basic diagram.

Passive pressures below subgrade inside the wall are calculated using conventional procedures. Wall friction is usually ignored, as is the effect of interior constraint imposed by closely spaced walls. The reduction in subgrade strength due to upward gradients attributable to uncontrolled dewatering, the indiscriminate use of berms and over excavation and the installation of bored piling or load bearing elements adjacent to the walls is not considered. Ignoring wall friction and the interior confinement is conservative, while ignoring disturbance of the subgrade can have a major consequence on the actual wall behaviour. These passive limits (calculated ignoring both wall friction and confinement) are not further modified. Safety factors are not applied to the passive pressures. Underestimations of subgrade soil strength will yield an overestimate of wall forces and deformations. Safety factors are applied by the structural designer when he selects wall properties, dimension, and reinforcement as well as when he sizes the bracing or other support system.

Modifications have been made in both analyses to reflect "weak" clay strengths as well as "strong clay strengths" wall stiffness, stiffness of the argillite and the passive subgrade modulus to evaluate the effect of these variables on strut loads, wall moments and deflections. There is general agreement between the two analyses with most of the difference resulting from changes in soil springs and assumed behaviour of the argillite (See Figs 3 & 4). As a result of these studies, both methods are being recommended for use as practical design tools. Performance data are not yet available for comparison to analytical results.

Federal Center Station example

In 1973 a diaphragm wall was constructed at the Federal Center Station of the Washington D.C. Metro subway. The wall was analyzed using modified Terzaghi-Peck for each stage of the excavation. Loadings and staging of

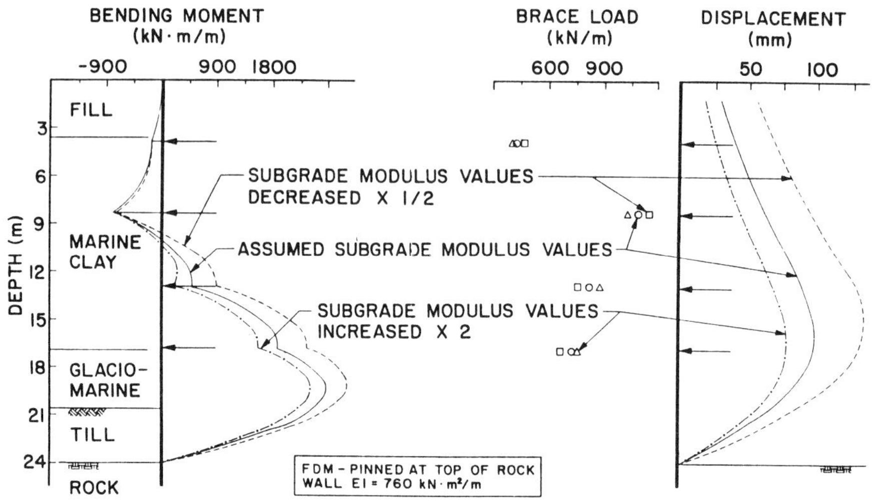

Figure 3. Central artery project excavation stage 5 soil spring sensitivity studies

excavation were defined by the owner and the wall was designed by the contractor following those criteria. Measurements were taken by the owner to verify wall behaviour during construction.

The subsurface profile consists of 3 meters loose to medium compact silty fill, 14 meters of Pleistocene terrace deposits, 7 meters very stiff plastic

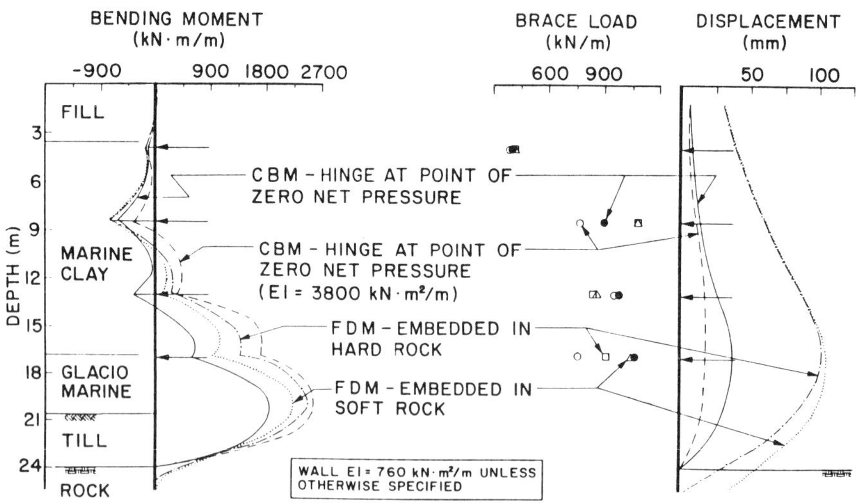

Figure 4. Central artery project - comparison of analytical methods

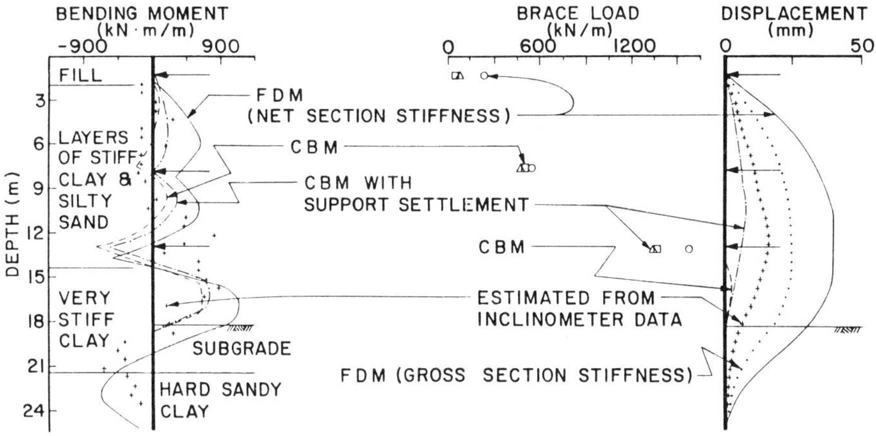

Figure 5. Federal center S.W. comparison of analytical methods

Cretaceous clay underlain by hard Cretaceous sandy clay. The terrace deposits range from overconsolidated clay lenses with strengths of 70 to 120 kPa to gravelly sand with friction angle equal to 38°. The upper Cretaceous clay has strength of 200 to 250 kPa and the lower clay strength varies from 200 to 300 kPa.

Inclinometer data were obtained during construction and recently evaluated (Tamaro, 1991). The measured deflection diagrams were not sufficiently sensitive to permit more than a crude estimate of wall bending moments. However, comparison of these data with recent analyses by the Terzaghi-Peck, Continuous Beam and Finite Difference Methods indicate that a reasonable correlation can be obtained by both the continuous beam and finite difference methods (See Fig. 5).

Modelling of the structure was made more difficult by incomplete records of the interior excavation. Some overcutting of the subgrade was permitted at the center of the excavation and some wall movements and weakening of the subgrade occurred as a result of this construction expedient. Observed wall movement below subgrade confirms this condition. Springs values ($Kh \approx 0.01 N/mm^3$) used in the calculations were initially obtained from Terzaghi, 1955 and then reduced 50% to acknowledge this weakening of the passive reaction. A 10% increase in maximum deflection and negligible moment increase resulted from this modification.

Conclusions

Design of cast-in-situ walls requires careful modelling to account for loading and deformations which occur during staged excavation, especially when the wall becomes part of the permanent structure. Stresses remain in the wall as a result of the construction. Construction procedures which vary from those considered during design could have a greater significance to wall behaviour than changes in the design assumptions. Four techniques are currently available to designers. The Terzaghi-Peck method appears to be superseded by two other more straight-forward procedures (continuous beam method or finite difference method). The relative accuracy of these two methods needs to be further verified by measurements. The finite element method may eventually be useful as a design tool when correlations with observed behaviour of the wall system are available. The writers believe that either the continuous beam method or the finite difference method can be used to reasonably portray the structural behaviour of the wall. The FDM can also be used to estimate wall deformations during construction; but empirical relationships and "local knowledge" are still needed to calibrate the model.

References

KERR W.G. and TAMARO G.J. *Diaphragm walls - update on design and performance design and performance of earth retaining structures.* ASCE, 1990, pp 576 - 594.

HALIBURTON T.A. Numerical analysis of flexible retaining structures. Proc. Am. Soc. Civ. Engrs, 1968, SM6, pp 1233 - 1251

TERZAGHI K. Evaluation of coefficients of subgrade reaction. *Geotechnique,* 1955, **5**, pp 297 - 326.

RESSI A. and TAMARO G.J. Post Office Box Square Garage, Boston, U.S.A. *Piling and Deep Foundations.* Deep Foundation Institute, 1991, pp 117 - 122.

TAMARO M.J. *Parametric study of methods of analysis of structural slurry walls.* Masters Thesis Lehigh University, 1991.

A limit state code for retaining wall design in Hong Kong

P.L.R. PANG, Civil Engineering Department, Hong Kong Government

Introduction

Due to the scarcity of flat land in Hong Kong, a large number of retaining walls have had to be constructed to provide space for development on hillsides. In 1977, the Geotechnical Control Office of the Hong Kong Government initiated an exercise to catalogue all significant slopes and retaining walls in the territory. More than 3 700 retaining walls (including retaining walls over 3 m high and cut and fill slopes incorporating retaining walls) had been identified at the end of this exercise in I978. Since then, over 600 permanent restraining walls have been built, up to mid-1991, averaging about 50 walls per annum.

In 1982, the Geotechnical Control Office published its first Geoguide, viz. *Geoguide 1 . Guide to Retaining Wall Design* (ref l), to give guidance on design of retaining walls in Hong Kong. Experience in the use of this Geoguide has shown that there is a need to revise the document. As a result, in late 1987 , preparation for revision of the Geoguide began.

In preparing the revised Geoguide reference has been made to various national codes and guidance documents(refs. 2 - 7). Also, account has been taken of the results of research on earth retaining structures, as well as the nature of the materials and the local experience available in Hong Kong.

This Paper outlines some of the principles adopted for the first draft of the revised Geoguide (ref . 8) and the background to its development. Some results of the studies carried out to check the recommended minimum factors of safety are also presented.

Types of retaining walls

Many types of retaining walls have been built in Hong Kong. Those covered by the draft revised Geoguide include the following

(*a*) Gravity retaining walls, which include mass concrete retaining walls, crib walls and gabion walls
(b) Reinforced concrete (RC) retaining walls, which include RC L-shaped and inverted T-shaped cantilever, counterfort and buttressed retaining walls

(c) Cantilevered retaining walls.

Temporary structures which are used to support the sides of deep excavations (e.g. cantilevered, strutted and anchored walls formed with sheet piles, soldier piles, grouted piles or diaphragm wall panels) are not covered. A review of the design methods for such structures is given in a separate GCO Publication No. 1/90 . *Review of Design Methods for Excavations* (ref 9). The design of reinforced fill structures is covered by Geospec 2 . *Model Specification for Reinforced Fill Structures* (ref 10), the background for development of which has been given by Pang (ref 11).

Design philosophy and limit states

A limit state approach incorporating partial safety factors has been adopted in the draft revised Geoguide. This is because this approach is considered to be more rational than the conventional global factor of safety approach, in that it provides a logical framework for applying appropriate partial factors to the relevant design parameters to cover uncertainties directly and individually (see next section). Thus the difficulties of defining a single lumped factor to cover all uncertainties are removed. Also, a more consistent level of safety margin than the global factor approach can be obtained for a wide range of wall and backfill geometries. It is worth noting that the partial factor approach is becoming widely accepted internationally. For example, both the Danish Code (ref 4) and the draft Eurocode 7 (ref 5) have adopted such an approach.

The draft revised Geoguide recommends that designers should consider both ultimate and serviceability limit states. Calculation models are given for various limit states for the types of retaining walls covered. Verification of safety and serviceability by other means, e.g. prescriptive measures, load tests and the observational method, are also discussed.

Factors of safety

The draft revised Geoguide recommends that 'factored values' of loading and material parameters should be used in design calculations. Factored values are defined as follows

$$F_f = F.\gamma_F \quad (1)$$
$$X_f = X/\gamma_m \quad (2)$$

where F_f, X_f = factored values of loading F and material parameter X respectively, γ_F, γ_m, = partial load factor and partial material factor respectively . The recommended minimum partial factors of safety for ultimate limit state design of retaining walls are given in Tables 1 and 2. To calculate 'ultimate' values of earth pressures, partial material factors are applied to the soil (or

Table 1. Partial load factors for use in design against ultimate limit states

Loading	Partial Load Factor, γ_F
Dead load due to weight of the retaining wall	1.0
Dead load due to weight of soil, rock and water	1.0
Surcharge on the retained ground	1.5[(1)(2)]
Earth pressure	1.0
Water pressure	1.0

Notes : (1) Alternatively, the 'ultimate value' of surcharge may be calculated by applying appropriate partial load factors given in the relevant structural code for any dead, live and wind loads which are assumed to act on the structure supported by the retaining wall.
(2) γ_F should be set to zero for those surcharges which produce a favourable effect.

Table 2 . Partial material factors for use in design against ultimate limit states

Material Parameters	Partial Material Factor, γ_m
Soil : unit weight γ	1.0
shear strength parameters c', tan ϕ', tan δ_b	1.2[(1)]
s_u	2.0
permeability k	1.0
Rock : unit weight γ	1.0
shear strength parameters c', tan ϕ'	1.2[(1)]
permeability k	1.0
Water : unit weight γ_w	1.0
Granular Filter and Drainage Materials : permeability k	10.0
Structural Materials : unit weight γ	1.0
strength parameters f_m	As per relevant structural code

Note : (1) The γ_m value should be applied directly to shear strength where a strength model other than a c' - ϕ' model is used in design.

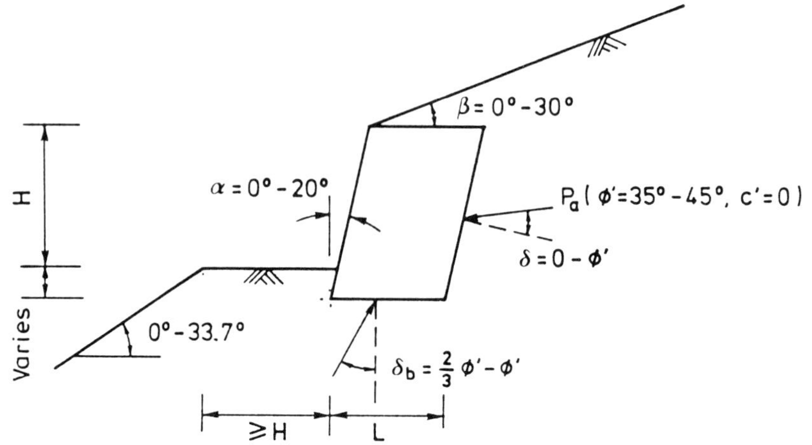

Fig. 1. Parameters covered in comparison exercises for gravity and RC retaining walls

rock) shear strength parameters, viz. tan ϕ' and c' (or s_u). These ultimate values are then used directly in the relevant analyses without further factoring, i.e. a partial load factor of 1 is applied to the derived earth pressures.

For serviceability limit state checks, the draft revised Geoguide recommends that all values of γ_F and γ_m should be set to unity. An exception is that the value of γ_F is set to zero for those variable surcharge loads which produce a favourable effect.

Calculations were carried out to check the partial safety factors in Tables 1 and 2 (see next section). It is recognized that the reliability of a geotechnical design depends not only on the factors of safety, but also on the methods of analysis, the calculation models, the way in which factors of safety are defined, the assessment of appropriate geotechnical parameters and the quality achieved in construction. Therefore, the draft revised Geoguide gives detailed guidance on these matters and recommends that the partial safety factors in Tables 1 and 2 should not be used out of context.

Comparison exercises

Gravity and RC retaining walls

For these walls, calculations were carried out to compare designs obtained using the partial factor γ_m on tan ϕ' with those obtained using the global factors of safety. The walls were modelled as monolithic blocks of uniform cross section, retaining fill for which a c'=0 soil strength model was used. Fig. 1 shows the range of parameters covered in the comparison exercises.

The results of the calculations to check against sliding failure indicated the following

(*a*) The relationships γ_m on tan ϕ' and F_s, the global safety factor against sliding, is approximately linear and is relatively insensitive to the ϕ' value of the fill for ϕ' between 35° and 45°. It is also relatively insensitive to the assumed angles of wall and base friction, and to the unit weight of the wall material.
(*b*) For a given F_s value, γ_m reduces with increase of both the back batter α of the retaining wall and the angle β of the retained sloping fill. The relationship is more sensitive to α than β.
(*c*) For α = 0-20° and β = 0-30°, F_s = 1.5 corresponds to g_m= 1.10-1.22.

Calculations were also performed, for the same parameter variations given in Fig. 1, to calibrate γ_m on tan ϕ' with F_o the global safety factor against overturning, and F_b, the global safety factor against bearing failure. These revealed that F_o = 2.0 corresponds to γ_m = 1.20-1.72 and F_b=3.0 corresponds to γ_m = 1.12-1.20.

As expected, the calculations carried out indicated that there is no unique one-to-one correspondence between γ_m and the various global safety factors. In order to achieve designs which satisfy the standard of the previous Geoguide 1 (ref l) using global safety factors, different γ_m values are required for different limit states. However, overturning is usually not the controlling mechanism, i.e. F_o = 2 does not govern the design. Subsequently, γ_m was set at 1.2 for all limit states. Further checks were then carried out, as outlined below.

Selected designs obtained using γ_m = 1.2 were compared with designs achieved by using global safety factors. This comparison showed that for a retaining wall of a given height, the difference in the wall widths obtained from the two different approaches is relatively small (generally less than 10%). Hence, use of γ_m = 1.2 would yield designs which are consistent with current experience for most cases. However, the wall width obtained using γ_m = 1.2 can become significantly larger than that achieved by using global safety factors when β approaches the factored ϕ' value. This indicates that there may be a relatively small margin of safety against uncertainties in soil shear strength for such cases when the global safety factor approach is used. It is considered that for cases where β is large, a cantilevered retaining wall is a more appropriate structural form than a gravity retaining wall.

Cantilevered retaining walls

Bica & Clayton (ref 12) have reviewed various limit equilibrium methods used in the routine design of cantilevered (sheet) walls embedded in granular materials. Fig. 2 shows the experimental data on limit equilibrium depths of

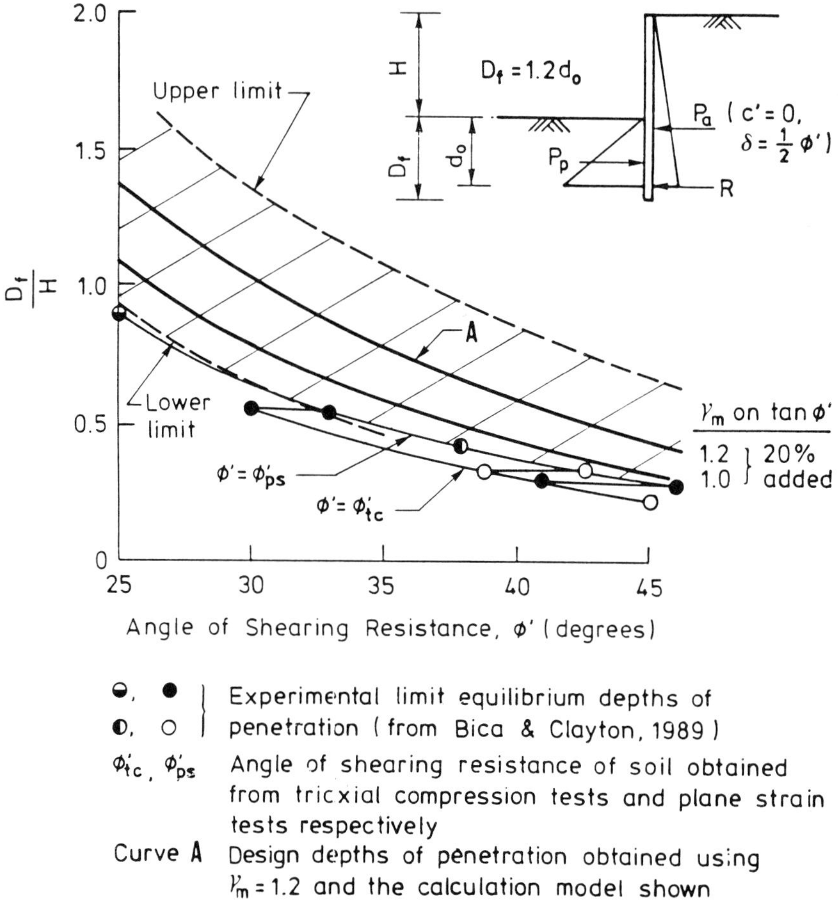

Fig. 2. Comparison of design depths of penetration and experimental values for cantilevered retaining wall

penetration extracted from Bica & Clayton, together with the upper and lower limits of design penetration depths obtained from the various design methods. Curve A, which gives the design depths of penetration D_f obtained using $\gamma_m = 1.2$ on tan ϕ', has also been plotted. To obtain the D_f values, d_o values were first calculated from a consideration of moment equilibrium of the walls, assuming that the earth pressures near the toe of the walls act as a concentrated force R. The calculated d_o values were then increased by 20% (i.e. $D_f = 1.2d_o$, see Figure 2) to compensate for the errors involved in the use of this simplified design calculation model. It can be seen from Fig. 2 that the design depths of penetration are approximately 1.8-1.9 times the experimental limit equilibrium depths which have been plotted against angles of shearing resistance of soil determined using triaxial compression tests.

The factor $\gamma_{m'} = 1.2$ on tan ϕ' is consistent with the safety factors recommended by Padfield & Mair (ref 13), who have presented results of a comprehensive study on cantilevered walls, and those given in the draft Eurocode 7.

Geotechnical parameters

Guidance is given in the draft revised Geoguide on the evaluation of geotechnical parameters relevant to retaining wall design. It is recognized that many geotechnical parameters are not true constants and it may be necessary to adopt different design parameters in different design situations. Also, different values are sometimes required for calculating loading and resistance.

The draft revised Geoguide discusses the use of the conventional c'- ϕ' soil strength model. It points out the problems which can arise with the use of the Least Squares Method for obtaining c' and ϕ' values, where the soil is heterogeneous and there are too few test results. A practical procedure for determining the values of c' and ϕ' of a soil from laboratory shear test results is suggested. Designers are advised to take account of the amount of data available in the relevant stress range, extrapolation required in the low stress region, curvature of the shear strength envelope and the likely variability of the soil in the evaluation of c' and γ'.

A $c'=0$ soil strength model is recommended for fill material which has undergone significant remoulding in the process of excavation, deposition, spreading and compaction. Such a method gives a conservative estimate of the shear strength of the backfill material and is analytically simple to apply in design. The draft revised Geoguide recommends the use of the minimum dry density specified in the contract for specifying laboratory shear tests on the fill.

Earth pressures

The relationship between wall movement and earth pressure is discussed in the draft revised Geoguide. Recommendations are given on the magnitude and distribution of earth pressures to be adopted for ultimate and serviceability limit state design. Detailed guidance is given on the evaluation of at-rest, active and passive earth pressures and also of compaction effects.

Designers' attention is drawn to research results on earth pressures. These include

(*a*) the progressive failure in dense sand close to passive failure
(*b*) the mobilisation of wall friction and its influence on earth pressures

(*c*) the effect of compressibility of wall material
(*d*) arching action which results in non-triangular earth pressure distributions, and
(*e*) the effect of initial stresses on retaining wall movements.

The application of the Winkler model for soil-structure interaction analysis and its limitations are also discussed.

Water pressures

The draft revised Geoguide recommends that design water pressures should be based on the worst credible water conditions that would arise in extreme events selected for design. Examples of extreme events given include severe rainfall, flooding and bursting of water mains.

In Hong Kong, severe rainstorms can occur in the wet season. The draft revised Geoguide differentiates between storm and seasonal water responses. Where the water level exhibits a storm response, the design water level is to be based on a storm rise corresponding to a 1 000-year return period rainstorm added to a typical groundwater level occurring before the storm. For a seasonally responding water level, a seasonal rise corresponding to a 1 000-year return period wet season is to be added to a typical groundwater level occurring at the beginning of the wet season.

Some groundwater levels exhibit both storm and seasonal response. In such cases, either a 1 000-year return period storm rise is to be added to a typical wet season water level, or a typical storm rise is to be added to a 1 000 year return period seasonal rise, whichever is the most appropriate.

The intensity of a 1 000-year return period rainfall has a probability of exceedance of about 11% in 120 years. The draft revised Geoguide suggests that this low probability level may be used as a guide for selecting other extreme events for design.

Guidance is given on the evaluation of water pressures for design. Both the groundwater seepage condition and the infiltration condition are discussed. The importance of providing adequate drainage behind retaining structures is emphasized in the draft revised Geoguide.

Conclusions

The draft revised Guide to retaining wall design (Ref 8) is a limit state code incorporating partial safety factors. Calculations have been carried out to check the recommended minimum partial safety factors. Detailed guidance is given on various aspects of retaining wall design, including selection of design geotechnical parameters, evaluation of earth and water pressures and the effects of surcharge. Design and construction aspects of various types of retaining walls are also covered by the draft revised Geoguide, as well as control of construction and maintenance. However, space limitations prevent discussion of these aspects in this paper. Nevertheless, it is hoped that the information presented herein will allow readers to gain an appreciation of the philosophy followed in the preparation of the draft revised Geoguide.

Acknowledgements

This Paper is published with the permission of the Director of Civil Engineering, Hong Kong Government but does not represent Government policy. W. K. Pun and P.C.Y. Yung performed the calculations described and contributed to the revision.

References

1. GEOTECHNICAL CONTROL OFFICE, HONG KONG. *Guide to Retaining Wall Design (Geoguide 1)*. Geotechnical Control Office, Hong Kong, 1982.
2. BRITISH STANDARDS INSTITUTION. *Draft Code of Practice for Earth Retaining Structures. (Revision of Civil Engineering Code of Practice No. 2). Technical Committee CSB/2 Earth Retaining Structures*. British Standards Institution, London, 1988.
3. CANADIAN GEOTECHNICAL SOCIETY. *Canadian Foundation Engineering Manual. (Second edition)*. Canadian Geotechnical Society, Ottawa, 1985.
4. DANISH GEOTECHNICAL INSTITUTE. *Code of Practice for Foundation Engineering*. DGI Bulletin No. 36. Danish Geotechnical Institute, Copenhagen, 1985.
5. COMMISSION OF THE EUROPEAN COMMUNITIES. *EUROCODE 7 . Geotechnics. (First Draft)*. Prepared for the Commission of the European Communities, 1989.
6. US DEPARTMENT OF THE NAVY. *Design Manual . 7.1 - Soil Mechanics, 7.2 - Foundations and Earth Structures*. Department of the Navy, Naval Facilities Engineering Command, Washington, D.C., 1982.
7. NEW ZEALAND MINISTRY OF WORKS. *Retaining Wall Design Notes*. Civil Division Publication CDP 702/C, New Zealand Ministry of Works, 1973.
8. GEOTECHNICAL ENGINEERING OFFICE, HONG KONG. *Guide to Retaining Wall Design (Geoguide 1) (Second edition). First Draft*. Geotechnical Engineering Office, Hong Kong, 1991.
9. GEOTECHNICAL ENGINEERING OFFICE, HONG KONG. *Review of Design methods for excavations (GCO Publication No. h/90)*. Geotechnical Control Office, Hong Kong, 1990.
10. GEOTECHNICAL CONTROL OFFICE, HONG KONG. *Model Specification for Reinforced Fill Structures (Geospec 2)*. Geotechnical Control Office,

Hong Kong, 1989.
11. PANG P.L.R. Design of reinforced fill structures in Hong Kong. *Proc. Int. Conf. Reinforced Soil , Glasgow*, 1990, pp 25-30.
12. BICA A.V.D. and CLAYTON C.R.I. Limit equilibrium design methods for free embedded cantilever walls in granular materials. *Proc. Instn Civ. Engrs, Part 1*, 1989, **86**, pp 879-898.
13. PADFIELD C.J. and MAIR R.J. *Design of Retaining Walls Embedded in Stiff Clay*. CIRIA Report No. 104, Construction Industry Research & Information Association, London, 1984.

Design study for a cantilever bored pile retaining wall with relieving slabs

H.D. ST JOHN, D.I. HARRIS, Geotechnical Consulting Group, D.M. POTTS, Imperial College of Science, Technology and Medicine, and R. FERNIE, Cementation Piling and Foundations Ltd, UK

Introduction

The construction of an underpass at an existing motorway interchange requires cutting below present ground level as well as placing 4m of fill to form a vertical retaining wall some 8m from top of wall to finished road level. A bridge passes over the cutting which is to be supported on the retaining walls and central piers. The underlying soil is Gault clay.

The Engineer's design comprised reinforced concrete walls cantilevering from a 2m thick concrete base slab spanning the entire width of the dual carriageway. This mode of construction would necessitate bulk excavation followed by placement of the slab and wall and subsequent backfilling. The bridge could only be placed once the road was virtually complete.

In order to reduce the construction period it was proposed that a bored pile retaining wall be placed prior to any excavation. The walls could then be extended upwards in cast-in-situ concrete, thus allowing bridge construction and backfilling to commence before bulk excavation. The head displacements of a pure cantilever are likely to exceed those which can be tolerated at the bridge abutments. Various options were considered to reduce this movement

- The addition of a single row of anchors.
- The provision of a narrow, 'relieving' slab in front of the wall placed after excavation, with a full moment connection to the wall.
- The provision of a 'relieving' slab behind the wall placed before backfilling, using the mass of the backfill to pull the wall back.

The use of permanent anchors was not examined in detail because of the possible loss of effectiveness in the long term.

Analyses were carried out, within a period of two weeks, to estimate design bending moments, ground movements and the required wall pene-

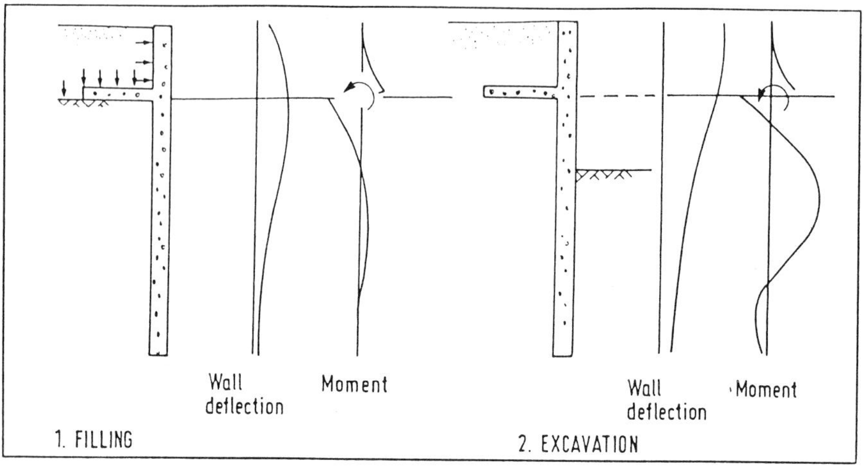

Fig. 1. Effect of rear relieving slab on bending moments and displacements

tration for long term stability. These studies were carried out using the Imperial College Finite Element Program, ICFEP. This paper describes the assumptions made in these analyses, the results obtained, and the conclusions drawn there from.

Principle of operation of relieving slabs

The general principle of a relieving slab is to reduce movement and/or bending moment in the wall by providing additional support, thus minimising the thickness of wall required to achieve a specified maximum displacement. Depending on the size of the slab it may only have a very marginal effect on overall stability and therefore on wall length. However, a small reduction in wall length may, in some circumstances, make wall construction feasible, e.g. if there is a water bearing sand near the base of a wall which could otherwise be constructed dry through clay. The principles of operation of front and rear relieving slabs are quite different and are discussed separately below.

Rear relieving slab

Such a system is only suitable where fill is to be placed behind a wall and the weight of the fill can be employed to bear on a horizontal slab. The beneficial effect of the slab is the input of a restoring moment to the wall prior to excavation, which reduces the peak bending moment and pulls the head of the wall back (Fig. 1). A secondary effect is the reduction of horizontal pressure behind the wall beneath the slab. Short and long term, stability is

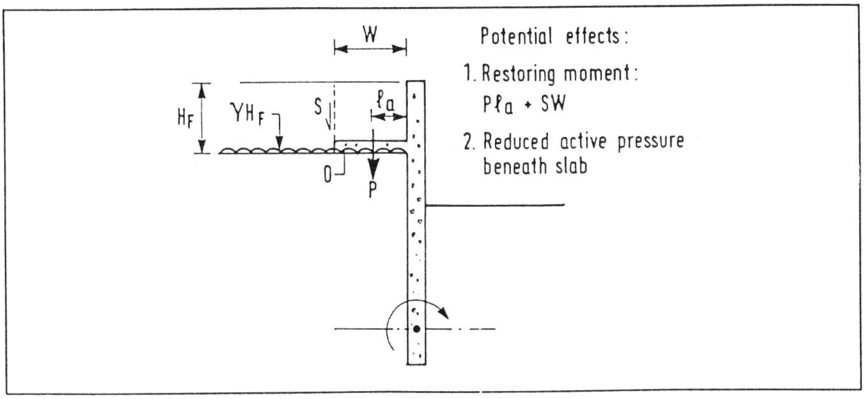

Fig. 2. Beneficial effects of rear relieving slab for stability

enhanced by both the reduction of local wall pressure and the addition of a restoring moment from the mass of the soil above the slab, when taking moments about the base of the wall (Fig. 2).

Front relieving slab

Since the front relieving slab can only be installed after complete excavation, its main benefit is in reducing the effect of changes that take place in the clay soil due to drainage. This will not be the case if the wall is supported by a temporary prop or anchors, the release of which will immediately bring the slab into action.

If the relieving slab is purely to reduce the effect of the swelling of the soil then its action is explained in Fig. 3. In the long term the wall rotates forward as the pressures on these wall increase and the ground at the base of the excavation swells and loses strength. This mechanism will cause a relative vertical movement between the slab and the soil beneath it causing an

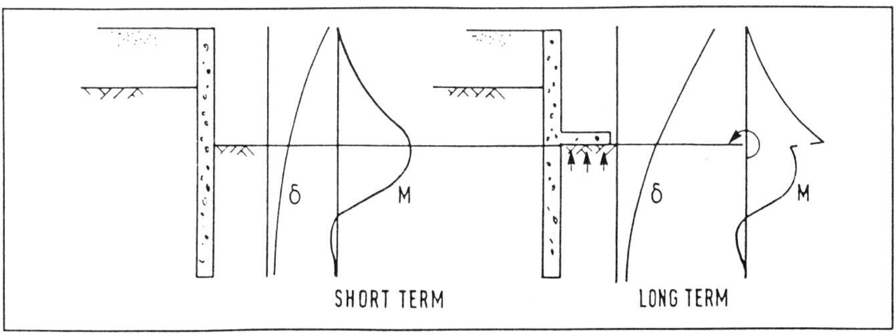

Fig. 3. Effect of front releiving slab on bending moments and displacements

Table 1

	c'(kPa)	φ'(°)	γ_{sat} (kNm^{-3})	K_o	E' (MPa)	ν'
Fill	0	32	19	0.5	40	0.1
Road Base	0	35	19	-	40	0.1
Cryoturbated Gault Clay	10	14	19	1.5	69	0.15
Gault Clay (I)	13	24.5	20	0-3 m 3.0 3-7 m 2.5	105 + 9.75z	0.15
Gault Clay (II)	20	24.5	20	2.0		
Folkestone Beds	0	40	20	1.0	500	0.1

increase in contact pressure and the development of a restoring moment. This, together with the local effect of increased horizontal pressure on the wall beneath the slab will reduce the maximum wall bending moment. Clearly the benefit derived depends on the size and stiffness of the relieving slab and the stiffness of the ground in front of the wall.

The integrity of the moment connection between the wall and the slab is of permanent importance, and the determination of its magnitude under both serviceability and ultimate limit states is of paramount importance. Because of the complexity of the behaviour it is essential that both limit states are examined using a numerical or physical model which accurately replicates the soil behaviour.

Description of the analysis

Soil parameters

The results of field and laboratory tests from four boreholes in the vicinity were available to the tenderers. These showed the ground conditions to comprise a small thickness of intermittent Made Ground over Gault Clay over the Folkestone Beds. The Gault Clay was around 23m thick, and was seen to be firm to stiff near the surface, becoming very stiff at depth. There is evidence that the top 3m of the Gault has been affected by cryoturbation. It was dry except near the surface. The Folkestone Beds were described as a dense silty sand with some gravel. Observations indicated that water was present in sub-artesian conditions in the Folkestone Beds, the head of water at the base of the Gault Clay being 5m.

The finite element analyses using ICFEP were all carried out all in terms of effective stress and, in this case, soils were modelled as elastic-plastic with a Mohr-Coulomb yield criterion. The maximum undrained strength is thus a function of c', φ', and the initial state of stress as well as the flow rule adopted when the state of stress is at yield. In order to attain a consistent and credible set of characteristics, the soil parameters adopted are shown in Table 1.

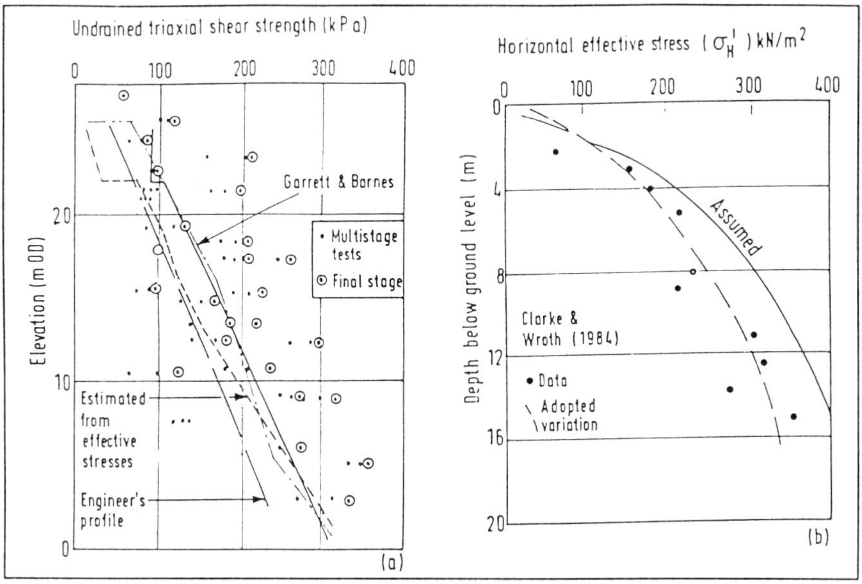

Fig. 4. Gault Clay: Undrained strength and in-situ horizontal stress

The undrained shear strengths estimated from these parameters are compared with the measured values and those used by Garrett and Barnes (1984) for the Gault Clay at Dunton Green (Fig 4a). The initial horizontal stresses are compared to those measured by Clarke and Wroth (1984) again at Dunton Green (Fig 4b).

Other assumptions

- It was assumed that currently and in the short term a perched water table exists 2.5m below the ground surface at +22.5m OD. Water pressures increased hydrostatically to a depth of 10m whereafter they become constant, dropping to 5m head from +5m to + 3m OD.
- In the long term, it was assumed that the perched water table would rise to the base of the fill at +26.0m OD. Below this and inside the cut, it was assumed that water pressures would be hydrostatic to a maximum pressure of 110 kPa, remaining constant below this depth.
- It was assumed that the fill placed would be granular, and would have a density of 19 kNm^{-3} and an angle of friction of 32°.It would be placed in 0.5m layers and compacted to give an initial K_o, of 0.5 (ν' =0.33).

Finite element analyses

A single mesh was used to analyze all cases, the basic geometrical features

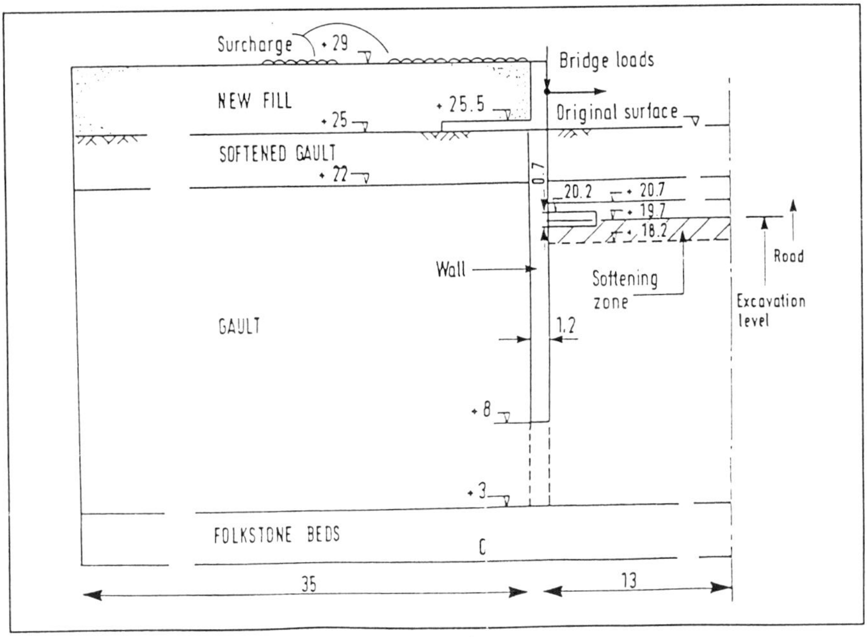

Fig. 5. Basic geometry for finite element analyses

of which are summarised in Fig 5. The mesh allows for

- a 1.2m diameter bored pile wall
- a 500mm thick rear relieving slab 5m wide
- a 1.0m thick cast in situ concrete wall above the relieving slab
- a 3.0m wide front relieving slab, 500mm thick
- a 1.0m thickness of road construction.

It was assumed that the excavation would extend to +19.7m OD and that 1.5m of clay below that would drain to a fully softened state during the construction period.

The sequence of construction modelled was as follows

- Construct bored pile wall 'wished-in-place'
- Construct relieving platform (if appropriate) & extend wall
- Construct fill in 0.5m thick layers
- Apply bridge loads
- Excavate to +19.7m OD
- Allow ground in front of wall to drain over depth of 1.5m
- Construct front relieving platform (if appropriate) and construct road to +20.7m OD
- Allow dissipation of pore water pressures to long term values.

Table 2. Long-term results

TABLE 2 Long Term Results	Run Number									
	1	2	3	4	5	6.1	6.2	7.1	7.2	8
Wall ρ_{max} mm	44	42	58	29	130	25	35	72	112	68
Wall BM_{max} kNm/m	543	492	826	599	10	372	491	13	9	855
Wall BM_{min} kNm/m	-434	-294	-746	-210	-1372	-244	-267	-1114	-1270	-744
Wall SF_{max} kN/m	266	251	319	246	313	296	317	341	392	334
Wall SF_{min} kN/m	-13	-21	-39	-40	-149	-511	-522	-656	-850	-33
Relieving slab BM_{max} kNm/m	566	519	815	392	-	429	551	-	-	840
Relieving slab SF_{max} kN/m	247	232	338	194	-	217	252	-	-	343
Front slab BM_{max} kNm/m	-	-	-	-	-	-286	-295	-348	-454	-
Front slab SF_{max} kN/m	-	-	-	-	-	220	236	262	338	-

Index to Analyses

Run 1	1.5m ϕ piles	Relieving slab		
Run 2	1.2m ϕ piles	Relieving slab		
Run 3	1.2m ϕ piles	Relieving slab		Factored shear strengths
Run 4	1.2m ϕ piles	Relieving slab		No bridge loads
Run 5	1.2m ϕ piles			
Run 6.1	1.2m ϕ piles	Relieving slab	Front slab	
Run 6.2	1.2m ϕ piles	Relieving slab	Front slab	Factored shear strengths
Run 7.1	1.2m ϕ piles		Front slab	
Run 7.2	1.2m ϕ piles		Front slab	Factored shear strengths
Run 8	1.2m ϕ piles	Relieving slab		Factored shear strengths & stiffness

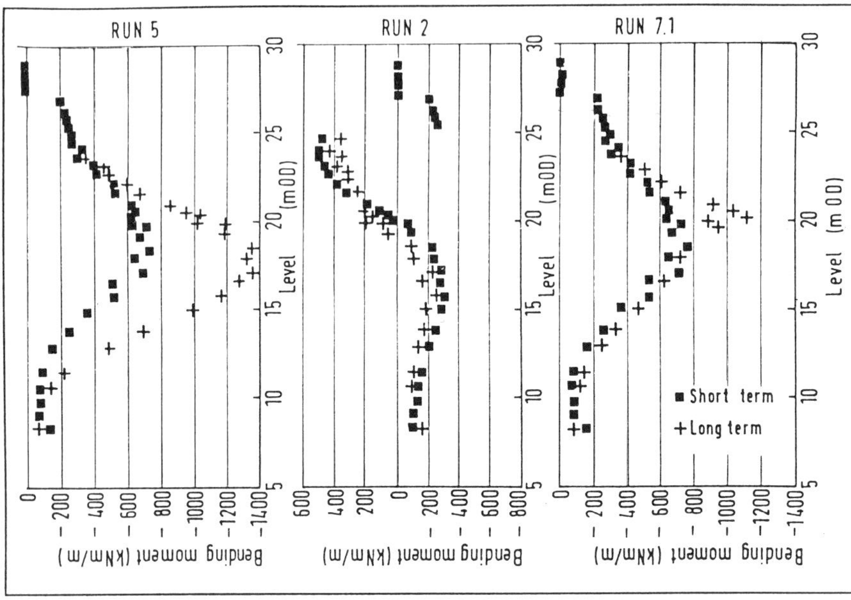

Fig. 6. Bending momemt distributions

All runs were carried out with a wall extending to 17m below original ground level. This is a minimum giving a factor of safety (F) of 1.4 including an allowance for the stabilising effect of the relieving slab. Table 2 gives a summary of the cases examined together with a summary of the critical values of movement, moment and shear. Runs 1, 2, 4, 5, 6.1 and 7.1 were carried out using unfactored shear strengths. For runs 3, 6.2, and 7.2 shear strengths were factored by 1.3 (on c′ and tan ϕ') in order to approximate an ultimate limit state for the wall. In run 8, both shear strength and ground stiffnesses were factored by the same amount. It was necessary to modify the initial effective stresses when factoring shear strengths in order to avoid initial stress states beyond yield.

Figure 6 shows the variation of short and long term bending moments in the wall for three different wall support conditions. Figure 7 shows the horizontal wall displacements for the three cases described above at critical stages.

Conclusions

The analysis showed that with no additional support, wall displacements were excessive. The addition of the rear relieving platform served to significantly reduce the maximum bending moment in the wall and reduce the wall head deflection by pulling the head of the wall back. This enabled the size of the piles to be reduced. It appeared also to have a significant effect on the stability of the wall.

The application of a vertical bridge load eccentric to the centreline of the wall and a horizontal load produces an additional moment in the wall. This increases the maximum (negative) moment for a cantilever or a wall with a front relieving slab; but reduces the maximum (positive) moment induced by the rear relieving slab.

The front relieving slab was not as effective as that at the rear, because it can only be installed at the end of an excavation, and because of its limited size.

It is interesting to compare the magnitudes of the bending moments and displacements for the different cases, particularly the serviceability and factored runs. It can be surmised that the difference between the moments depends on the amount of movement that has occurred in the unfactored runs.

This series of analyses allowed a suitable wall configuration to be chosen and detailed in the knowledge that wall deflections would not realistically exceed a specified maximum value.

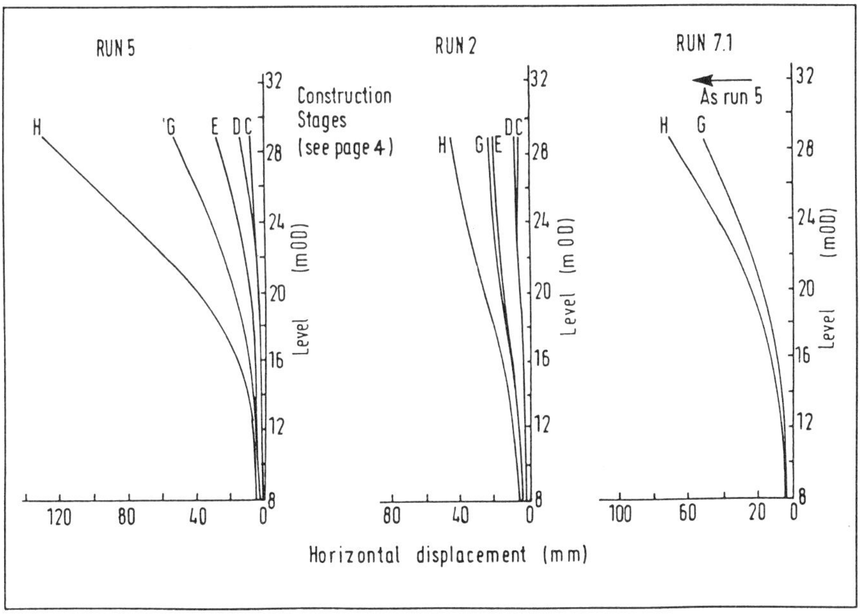

Fig. 7. Horizontal wall displacements

References

1. GARRETT C. and BARNES S. J. The design and performance of the Dunton Green retaining wall. *Geotechnique*, 1984, **4**, pp 533-548.
2. CLARKE B. G. and WROTH C.P. Analysis of Dunton Green retaining wall based on the results of pressuremeter tests. *Geotechnique*, 1984, **4**, pp 549-561.

Developments in structural slurry walls

M.J. PULLER, Butler Puller Partnership, UK, and D.J. PULLER, Ove Arup and Partners ,UK

Synopsis

The development of structural slurry wall techniques since the introduction of Icos walls in Europe in the late 1950s has been centred around both civil engineering and mechanical engineering innovations. Civil engineers faced with demand for environmentally acceptable means of deep excavation support have sought more effective use of slurry walls by introducing structurally efficient plan shapes and post tensioning and precasting methods. Mechanical engineering designers have developed sophisticated means of excavation using reverse circulation principles for walls in soil and rock. This paper describes these developments and their application.

Historical review

The earliest structural slurry trenches were constructed as a trial by Veder in 1948 but the first commercial applications were made in Italy by I.C.O.S. in 1950 during the construction of the Fedala Power Dam and at the hydraulic works at Venafro on the River Volturno. The first continuous reinforced concrete walls, excavated panel by panel and concreted by tremie were built only in the 1950s; in the U.K. the earliest structural wall (at Hyde Park Corner, London) was built by Icos in 1961-62. In Europe the method gained early acceptance; in Italy (for metro construction) and in Belgium and France for metro and highway cut and cover construction, but had to wait until 1959 for it's introduction to Japan. By 1991 the originators of the process had ceased to trade as specialist contractors (although Icos U.S.A. continued in the United States) and with no patent restriction, diaphragm walling was accepted and used for a wide variety of underground works in both the developed and Third World Countries.

The design and construction of a wide range of soil retaining structures followed the earliest walls for basement and cut and cover construction. In Europe, large diameter circular shafts, elliptical shafts and counterforted walls using T shaped panels were built (refs 1 and 2). In some cases innovative designs were used: at Bloomsbury, London, Icos used a large circular shaft for underground car parking with double helical ramps for the entrance and exit of vehicles, the ramps also provided structural propping for the wall. Elsewhere, where buttressed walls were built, vertical and inclined anchoring was used. The T shaped walls were concreted in a single pour but no advantage was taken of soil-structure interaction and the stabilizing effect of the soil weight between the tees. Barrettes were introduced in France to withstand heavily concentrated loads (and overturning moments) below high rise buildings, and in Paris the earliest examples of top-downwards construction used diaphragm walls as peripheral basement walls (ref.3).

Development in wall construction

The improvements made by both civil and mechanical engineering designers since the earliest walls built by Icos include:

* the development of structurally efficient plan shapes of diaphragm walls based on improved joints between panels
* use of post tensioned diaphragm walls, either precast or cast in-situ
* use of precast concrete diaphragm walls
* use of reverse circulation excavation equipment known as Hydrophaise or Trenchcutter rigs
* development of excavation equipment for work on congested urban sites
* use of polymeric slurries for excavation support.

Each of these items is reviewed in this paper.

Variations in plan shape

The use of wall plan shapes other than straight walls can provide greater flexural stiffness, and where combined with soil self weight, greater stability. In order to derive such shapes economically, however, joints between panels of wall are required to transmit tension or shear, or both.

The use of patented tension joints by Icos shown in Fig.1 in the late 1960s and early 1970s allowed the use of "wine glass" shaped cells for heavy quay wall construction. The cells (at Redcar) were 30m by 15m in plan and were formed with front and back lines of arches, supporting crane arms, and interconnecting cross walls. The diaphragm wall, up to 45m deep, provided 20m of water at berthing. The inverted arch forming the quay face was designed to be restrained by the tensile forces within the cross walls and were resisted by the rear arch. The weight of the soil enclosed within the cells was not used to restrain the wall in overturning because the panel joints could only transmit tension.

Recent developments in panel joint construction and methods of soil/structure analysis enable the construction of T shaped buttress walls of flexural strength derived from the whole T shaped section together with the stabilising effect of the weight of soil between the legs of the tees. Fig.2 shows

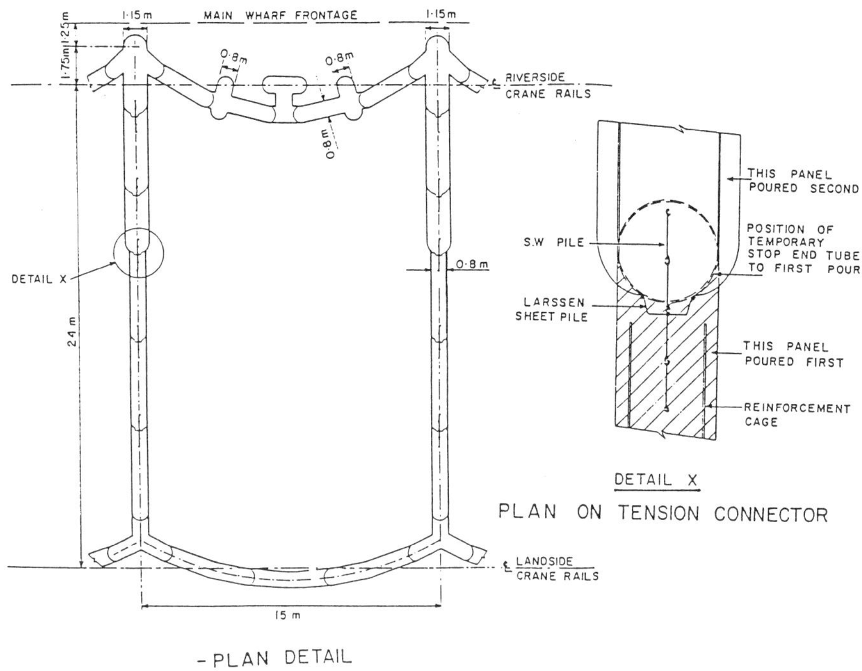

Fig. 1. Panel layout at Redcar Ore Terminal and detail of tension joint used at Redcar and Seaforth Dock by Icos

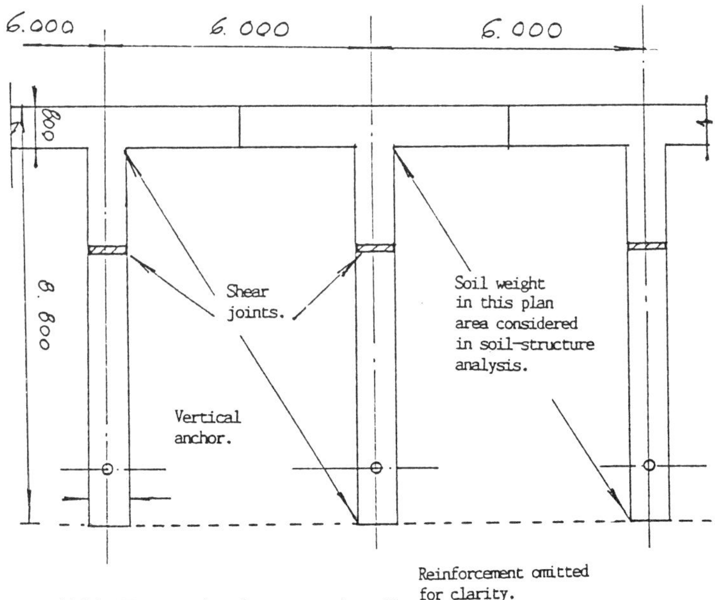

Fig. 2. Monolithic Tee section buttressed wall

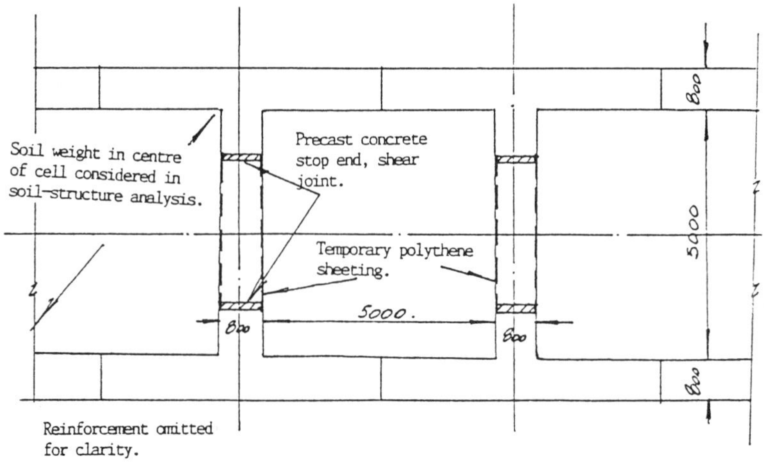

Fig. 3. Cellular wall at Medina

the location of a panel joint in a large buttressed wall designed in this way.

The development of joints between diaphragm panels has recently been significantly advanced during the construction of deep cellular walls as peripheral walls to a very large basement structure in Medinah, Saudi Arabia. The incentive to design fully monolithic cellular walls taking into account the weight of the enclosed soil was brought about by the proximity of religious buildings and the need to curtail soil movement due to bulk excavation. The cellular units, shown in Fig.3, were built with precast concrete permanent stop ends incorporated in the centre web panel of each H unit of the cellular wall. Concrete was poured into the central web excavation enclosed within a bag of plastic sheeting to avoid leakage of concrete around the stop end and thereby reduce the effectiveness of the shear and tension bond to the outer panels. The use of plastic sheeting as temporary measure proved most effective to retain all the concrete within the web panel. The depth of bulk excavation on this site was 17.5m throughout and where bedrock occurred at less than 35m from ground level overturning and sliding stability were improved by the use of vertical rock anchors at the rear wall of the cells. Over much of the site, however, the bedrock exceeded 35m depth and then no vertical anchorage was used, the wall standing as a cantilevered semi-gravity structure. Although the use of projecting reinforcement from one panel to it's neighbour has been practised in Japan for some years, and similar methods have been developed by Bachy called the 'Teba' joint, their application has been much restricted; the precast joints at Medinah, however, were used more than one thousand times on the site with construction difficulties occurring in less than five panels.

It is evident therefore that where design criteria stipulate minimal soil movement around very deep excavations, diaphragm walls can now be built to plan shapes which effectively resist movement by a combination of soil/structure mass and very considerable flexural stiffness.

Post tensioned diaphragm walls

Where design moments are excessive for the straight single wall, recourse can be made to improving the flexural strength of the section by prestressing. Two advantages exist in post tensioned diaphragm walls: the section remains relatively slender and induced moments due to soil pressure are thereby minimised. Tensile stresses due to stressing can be largely avoided within the wall section as the soil stiffness is used to restrict wall movement when the initial prestress is applied. The advantages in post tensioning are therefore not limited to reduction in reinforcement quantities. Both wall thickness and wall movement are reduced compared to the reinforced concrete wall section.

The use of post tensioning in the United Kingdom has at present been restricted to two contracts, the German Embassy basement in London and Irlam of the Heights underpass near Manchester. In Switzerland and Italy however, the method has been applied successfully on a number of basements (ref.4). Because of their improved bending resistance and the reduced moments induced, post tensioned walls can be used with greater unpropped heights of wall in the temporary condition before the permanent propping from the basement floors is applied.

Precast concrete diaphragm walls

Although precast concrete diaphragm walls have been used in France successfully by both Soletanche and Bachy using patented proprietary methods, the technique has not been applied in the United Kingdom. In the Far East, in Hong Kong and Thailand, thick heavily reinforced in-situ diaphragm walls appear to be preferred, there are relatively few examples in Japan and there are no known precast walls in the U.S.A.

The experience in France, particularly in metro cut and cover construction in Paris, Lille and Lyon only extended beyond French boundaries in the hands of the patent holders, Soletanche and Bachy. The former method, known as the 'Panosol' method used bentonite slurry excavation followed by a slurry change and replacement with a cementatious slurry for the permanent condition. The Bachy technique known as the 'Prefasil' method used a cementatious slurry for excavation without subsequent change. The development of these precast wall methods since their introduction in 1970 has shown several advantages compared with in-situ diaphragm wall construction:

* Reduction of site nuisance by more rapid execution. The sequence of panel excavation is simplified by successive panel excavation whereas in-situ walls frequently use primary, secondary and intermediate panel excavation

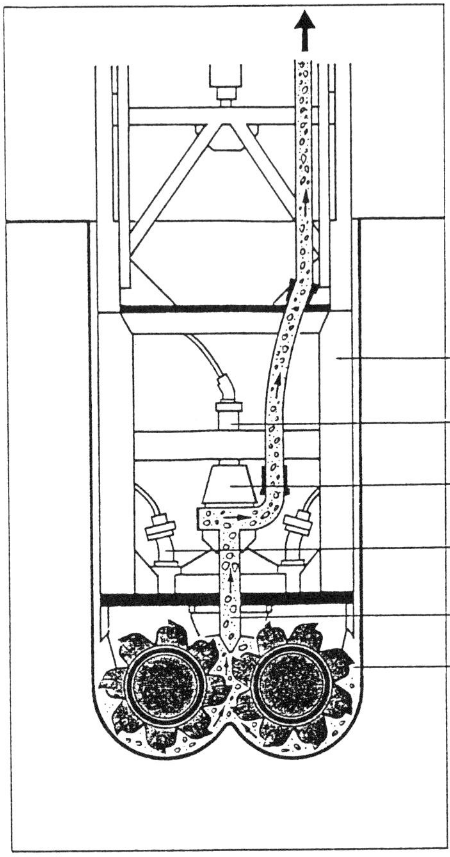

Fig. 4. Sectional view of cutter head to Trenchcutter reverse circulation equipment

sequences to allow hardening of concrete. Remedial works in breaking down walls to level or to profile are largely unnecessary.

* Reduction of nuisance by avoidance of site concreting operations and stop end extraction.

* In the permanent phase constructional thicknesses are reduced by the improved concrete qualities brought by precasting, wall thickness as slender as 400mm are permissible. Good wall finishes and better waterproofness by incorporation of water bars into the precast panels are possible.

Reverse circulation excavation equipment

The evolution during recent years of the trench cutter or hydrofraise as an excavation tool probably represents the single most important technical development to affect the method of construction of slurry walls. (Fig.4).

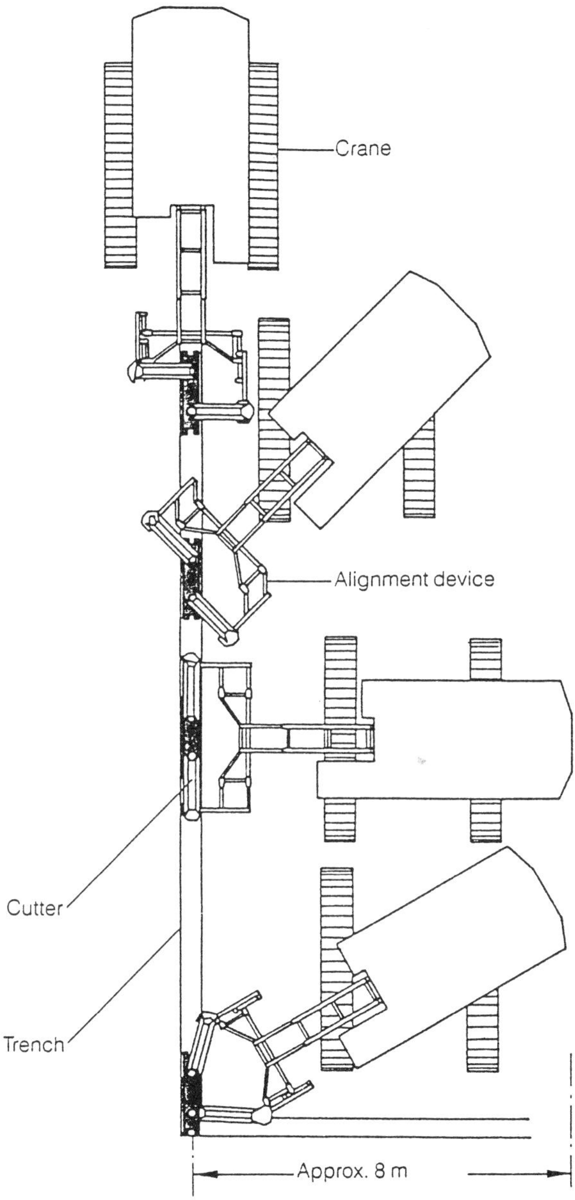

Fig. 5. Plan view of Trenchcutter machine positioned on site with limited space

There are three main advantages of the trench cutter method over traditional Kelly or cable - operated grabs.

* greatly reduced noise and vibration during the penetration of weak rocks (i.e. unconfined compressive strength $\sigma_c < 50$ MN/m^2)

*increased production rates (in excess of 400m^2/12 hour shift)

* improved trench verticality.

Furthermore it is the authors' belief that panel instability, overbreak and ground movement associated with excavation is significantly reduced using the trench cutter method because of the minimal disturbance caused to the trench walls.

Interestingly the original idea was a Japanese one realised by Tone Boring and Okumura in the early 1960s. A steel frame mounted with wheels powered by electric motors was constructed to enable continuous excavation of the soil by using the slurry both as a means of trench support as well as a medium for transportation of the spoil. The slurry pump used to remove the slurry-spoil suspension was located at the ground surface. In the mid 1970s the French contractor Soletanche developed the idea further by replacing the electric motors by more powerful hydraulic units. The slurry pump was also located immediately above the cutter wheels, housed within the body of the trench cutter.

Further refinement by Bauer of Germany led to a number of innovations which enhanced the equipment's power and range of application.

The inclusion of a feed cylinder within the body of the cutter enables the operator to vary the crowd force applied to the cutter wheels very precisely, up to a maximum equal to the dead weight of the cutter. The importance of this feature is realised particularly in variable ground conditions where the ratio of crowd to torque has to be varied according to the strata encountered to achieve maximum production.

The cutting end of modern trench cutters generally consists of two pairs of two wheels which rotate about horizontal axes. Each wheel consists of rows of cutting teeth. The Soletanche wheel contains teeth mounted on a spiral arrangement. Bauer found in stiff clays and weak rocks that a saddle of material was often left between the adjacent cutting wheels during excavation. The cutter tended to sit on this material, unable to excavate any further. A hinged or tip-aside tooth was developed so that a pair of teeth per wheel had the facility to tip inwards during each rotation hence ensuring excavation of any material located between the wheels see Fig 6.

If care is taken and ground conditions permit then verticality to within 1% can be achieved on panel excavation using a grab. However tolerances of less than 0.5% can be attained using modern trench cutters, with a digital read-out giving the operator the deviation from the vertical in both x and y planes as excavation proceeds. Such outstanding tolerances may be further improved by the introduction of hydraulic pads which are contained within the body

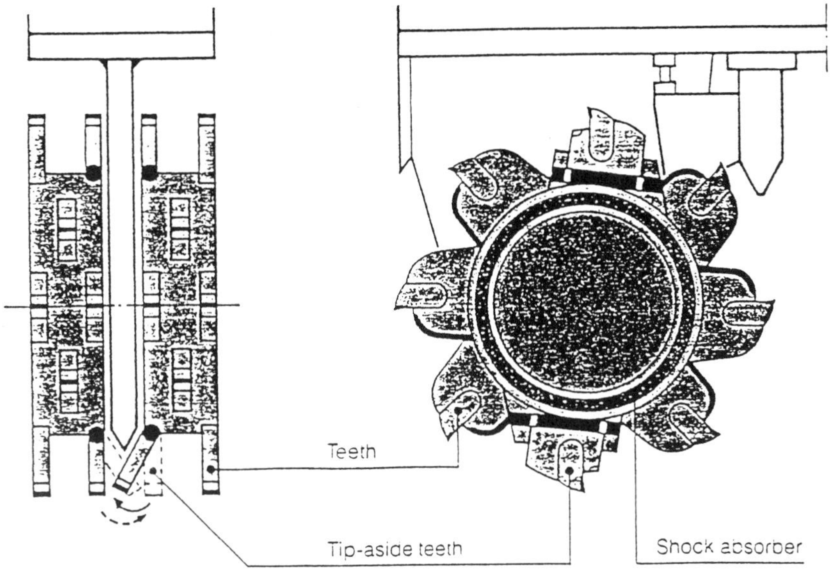

Fig. 6. Arrangement of tip-aside teeth on main cutter wheel of Trenchcutter

of the cutter frame. If the cutter starts to deviate from he vertical then the hydraulic pads are brought to bear against the walls of the trench until the alignment is corrected. Excavation to exacting tolerances is particularly relevant in the case of very deep basements and cut-off walls where the risk of windows between adjacent panels is greatest.

Wall thicknesses between 600 and 2800mm and depths exceeding 130m are now within the range of equipment currently available.

Traditionally trench cutters have been best suited to sandy soils. Indeed the early Japanese machines were designed particularly for such ground conditions. Modern cutters, having the benefit of more powerful hydraulic motors as well as an improved understanding of the optimum number and configuration of cutting teeth, are able to achieve high production rates in a wide range of ground conditions. Much of the wheel smearing associated with the excavation of clayey soils has been eliminated by the introduction of clearing plates which clean out the soil between adjacent rows of teeth. Excavation in weak rocks has become common-place, with trench cutters becoming an essential item of equipment especially on many inner-city sites, where the vibration and its effect on adjacent buildings associated with chiselling is considered unacceptable. Much recent effort has gone into extending the technique to strong rocks (i.e. $\sigma_c < 100MN/m^2$). Bauer has developed a trench cutter capable of excavating rock with an unconfined compressive strength exceeding $150MN/m^2$. Cutting teeth have been replaced by roller-bits which rotate individually as the cutter wheel itself rotates. The Hard Rock Cutter has been used successfully, in the excavation of a cut-off wall at a dam site in Japan in Andesite rock with an unconfined compressive strength of up to $200MN/m^2$. An average production rate of lcm/min was achieved.

Not surprisingly there are a number of disadvantages associated with the trench cutter method. As a general rule there needs to be something special about a project, for a trench cutter to be the cheapest suitable solution. For example where the program is particularly short and the site is a congested one or the wall must be embedded into sound rock and it is located close to a sensitive existing building or perhaps the water-tightness of a deep cut-off wall necessitates particularly onerous verticality tolerances. The cost of a trench cutter and the required desanding unit is considerable and therefore the potential cost of standing time is not insignificant. The useful life of slurry used during excavation is much shorter than in the case of a grabbed wall since the slurry is used as a transportation medium and it becomes loaded with fines more quickly, especially in the case of silts and fine sands. Particularly in the case of hard abrasive rocks teeth can become worn out in a matter of hours and their replacement is a time consuming operation. Whilst modern trench cutters are able to cope with most ground conditions, cobbles and boulders still cause problems since they frequently become jammed between the cutter wheels.

Construction: congested sites

During the last five years diaphragm walling equipment has been miniaturized to enable construction on increasingly congested sites. For example, modern two-tier desanding plants are being used in preference to row upon row of slurry ponds used in the past. Bachy's fully automated mud plant is able to provide $195m^3$ of slurry storage capacity in three tanks on the lower tier.

On the upper tier the desanding unit can desand up to $240m^3$/hour and the mixer batches up to $40m^3$/hour of fresh mud. Also located on the upper tier is a bentonite silo with a 30 tonne storage capacity. Surprisingly all of this equipment occupies a plan area of only 12 x 19 metres.

The first generation of trench cutters mounted as they were on large cranes were best suited to large open sites. However in 1987 Bauer developed the city cutter which is mounted on a much smaller base crane, is relatively compact and has extendable tracks which allow transport without special width permits. The crane has also been modified so that the machine can work at almost any angle to the wall axis. Where space is particularly limited it can even straddle the trench. (See Fig.5). In 1990 the Spanish contractor Terratest completed diaphragm walls for the extension of the Bilbao Metro using a city cutter. The site consisted mainly of two 800m long strips less than

8 metres wide. Over 23000m^2 of wall to a maximum depth of 20m was completed without affecting the continuous operation of the existing train services located only a few metres from the wall axis.

The latest development to affect equipment size is the Mini Cutter which at the time of writing is undergoing final testing. Bauer have constructed machine capable of excavating slurry walls to a maximum depth of 55m in conditions of very restricted space such as tunnels, existing basements or shafts. The dimensions of the unit are 5.2 x 4.1 x 5.0m (length x width x height). Mud hoses are stored on a drum which rotates about a vertical axis whilst the hydraulic hoses are stored on a separate vertical drum.

Polymeric slurries

Slurry walls in the U.K. are excavated almost exclusively under bentonite. However on the Continent the use of polymeric slurries is more widespread. Polymeric slurry behaves quite differently compared to bentonite acting as a pseudoplastic fluid without forming a filter cake. Bentonite on the other hand relies on thixotropic gel viscosity (ref.5) to provide trench support.

Although polymer is approximately ten times more expensive than bentonite, disposal costs of used bentonite are very high and are likely to continue to rise as suitable tips become more scarce. Polymer however is almost infinitely re-usable and therefore only a small quantity of polymer is required for an entire project. Furthermore disposal of the spent polymer can be carried out by the addition of an agglomerator followed by transport to a regular tip or by discharge directly into a sewer.

Summary and conclusions

The developments described in this paper have been made by mechanical and civil engineers employed by specialist contracting firms seeking to minimise the cost of soil retaining walls in either the temporary or permanent conditions or both. The use of slurry walls in building basements, metro construction, marine and industrial structures continues to expand internationally and in doing so new demands are made for technical and commercial improvement.

The structural improvements including the use of precast and post tensioned concrete, structurally efficient plan shapes and changes to joint design have been applied in some European countries but have failed to gain acceptance in others. As demand for slurry walling grows it is likely that more underspread use of the improvements will also occur.

The hydrophaise or trench cutter represents the most important technical development in terms of equipment enabling increased production rates, reduced noise and vibration levels, excavation within rock and improved trench stability and trench verticality. Application of the cutter has been extended to hard rocks as well as to sites with space or headroom restrictions. The relatively high cost of mobilization and high consumption of slurry limit the application of this equipment to larger sites at present.

For the future, developments in wall design and construction and equipment design will continue and as the cost and environmental consequences of bentonite slurry disposal increase, engineers must consider the alternative of polymeric slurries more seriously.

References

1. ICOS (1968) Underground Works. Icos Milan 1968.

2. XANTHAKOS P.P. (1979) Slurry Walls. McGraw-Hill Book Co., New York.

3. FENOUX G.Y. (1971) Le realisation fouilles en Site Urbain. Travaux Aug-Sep. 1971.

4. GYSI H.J., LINDER A., LEONI E. (1975) Prestressed diaphragm walls. Proc. Eur. Conf. S.M.F.E. Vienna 1975.

5. BERESFORD J.J. et al (1989) Merits of Polymeric Fluids as Support Slurries. Proc. D.F.I. Conference London 1989.

Discussion

SESSION 4

P. MURPHY, Wiltshier Construction (London) Ltd

As a designer of temporary support works for a building contractor, the primary results that I am looking for are that we build a suitable support wall system, which is

(a) stable in itself and without risk of jeopardising the construction site
(b) adequate safety to uphold adjoining buildings and structures without risk of collapse or of unacceptable movements.

I would therefore be happy with any simple, quick method that would give me reasonable and workable solutions for the above.

Dr. Main noted that any assessment of settlements of adjoining buildings would require a reasonably sophisticated finite-element analysis.

I note that a sophisticated anaysis seems to involve sophisticated variables and assumptions, giving an answer which would appear to come with little more confidence than a simple empirical approach. For our purposes, an empirical approach is adequate at least in the early stage.

R.P. THOMPSON, Allott and Lomax

Four new design standards are currently in draft form and will shortly be available. These are

Eurocode EC7 - Geotechnical design
International Standard, produced by ISO/TC 182/SC3
Hong Kong Geoguide 1
British Standard BS 8002

Each standard in the design of retaining structures gives certain advice regarding the derivation of earth pressure.

Eurocode EC7 in respect of passive resistance includes guidance that the ground level of the passive soil should be lowered by an amount - for cantilever walls - equal to 10 per cent of their height. The document also advises that in calculating the design earth pressures, partial factors should be used for the material properties and that design earth pressures should not be derived by multiplying characteristic earth pressures by partial factors. The International Standard makes no specific requirements in respect of passive or active pressures.

The Hong Kong Geoguide recommends the inclusion of a surcharge load of 10KN/m^2 when determining the active pressures. Finally, BS 8002 calls for both a minimum 10KN/m^2 surcharge loading in calculating the active pressures and for a minimum 'unplanned' excavation of not less than 0.5 m and not less than 10 per cent of the retained height for cantilever walls.

Regarding the derivation of the design value for material properties, all four documents obtain the design value by applying a partial factor to the characteristic value or a similar such termed value.

Eurocode EC7 gives the definition of a characteristic value as a value selected such that the probability of a more unfavourable value governing the occurrence of a limit state is not greater than 5%. The International Standard refers to the values as being based on a careful assessment of the range of values envisaged. The Hong Kong Geoguide states that selected values are commonly based on 'best estimate' parameters. BS 8002 refers to conservative estimates of the mass properties of the soil.

Table 1 is an outline table which compares the numerical values of the partial factor γm (materials) for the ultimate limit state condition.

Table 1

Design Code	ϕ' or $\tan\phi'$	c^1	c_u
Eurocode EC7	1.2-1.25	1.5-1.8	1.5-1.8
ISO/TC182/SC3	1.2	1.5	1.5
HK Geoguide 1	1.2	1.2	2.0
BS 8002	$1.2\phi'_{peak}$ $1.0\phi'_{crit}$	1.2	1.5

It can be seen that there is a marked similarity between the documents, particularly in the numerical value of the partial factor for the effective angle of shearing resistance. It is to be hoped that a common approach can be adopted for the derivation of characteristic values and the determination of active and passive earth pressures.

Codes, standards and design codes

M. D. Bolton, Cambridge University, U.K.

Background to codes in the UK: CP2

The Civil Engineering Code of Practice CP2 Earth Retaining Structures was published in 1951 by a Joint Committee of the Institutions of Civil, Municipal, Water and Structural Engineers. Code writing then passed to BSI where revision of CP2 has been under way for about twenty years. Some delay was caused by the apparent conflict between traditional "factor of safety" or "permissible stress" methods of evaluating designs and the new "limit state" approach advocated by structural committees, and widely accepted in Europe.

The old CP2 served simultaneously as a design standard, a guide to calculation procedures, and a record of what was considered good practice in detailing and construction. This paper will concentrate on essential design standards. It is first necessary to recall the mixture of safety provisions which were used in CP2. With regard to the possibility of a deep slip circle carrying away the whole facility, there was to be a factor of safety of at least 1.25, defined as restoring moment divided by overturning moment. For embedded walls, there was to be an arbitrary reduction factor of 2 applied only to passive resistances. This is tantamount to reducing both the self-weight and soil strength components of passive resistance, while maintaining full values on the retained side. This was shown by Burland *et al* (1981) to be illogical and, for the long-term analysis of walls in clay, impossible to implement. For gravity walls on spread bases a safety factor of 2 was to be applied to sliding resistance, but otherwise the approach was based on permissible states. The middle third rule was to be applied to the eccentricity of the base thrust, while the maximum bearing pressure at the toe was to be shown to be smaller than the allowable bearing capacity, which was not defined.

Limit stage design: The analysis of critical events

Bolton (1989) advocated the replacement of arbitrary rules and definitions of "factor of safety" by the introduction of two concepts compatible with limit state design methodology as it was emerging through the Eurocode process. Design situations are those combinations of harsh environmental circumstances, severe loads, and pessimistic soil-structure characteristics, which the facility is intended to survive. Limit modes are the independent ways in which the new facility (or a neighbouring facility or service) might come to grief in any one of those situations, distinguishing between ultimate limit states (uls) which endanger people, and serviceability limit states (sls) of excessive deformation affecting appearance, efficiency, cost or durability.

At the core of limit state design evaluation is the explicit checking of the various critical events - each of the possible limit modes in each of the design situations. Analysis of structures in normal working states is to be avoided. These precepts are now adopted in both the current draft of Eurocode 7 Geotechnics for the EC Standards organisation CEN, and in the current draft of BS 8002 (revising CP2) for BSI.

Setting standards

The objective description of design situations is a feature of the draft BS 8002 not coherently covered in CP2. Each design situation should be sufficiently severe, and the collection sufficiently varied, to encompass all foreseeable conditions during construction, normal operation, and foreseeable classes of accidental loading. Examples of recommended conditions are listed below.

(a) Environment: Groundwater levels should be the most onerous that are considered reasonable, taking into account storms, seasonal changes, trends, or accidental leakages. Allowance will be made for the adverse effects of upward hydraulic gradients in the passive zones beneath excavations.
(b)Geometry: Provision should be made for over-dig in excavations, no less than the larger of O.sm, or 10% of the vertical span concerned. Provision should also be made for a service trench in front of walls, which could eliminate the passive zone in front of gravity walls, for example.
(c) Unit weight: A cautious estimate of soil unit weight shall be made.
(d) Loads: Surcharge will be not less than lOkPa overall. The transient or dynamic effect of live loads must also be taken into account: it may be noted that the average ground pressure beneath the multi-axle bogey of a special transporter unit can be in the range 50 to lOOkPa over an area circa 3m by 4m.
(e) Soil conditions: Water-filled tension cracks will be invoked at a retained face in fine-grained soils, where the possibility exists of a significant inflow due to existing perched water, burst pipes, or storms.

Evaluation of both stability and deformation

Although the principle of limit state evaluation of critical events is now very widely accepted, the exact means of selecting critical parameters has been much more contentious. Structural code writers have generally quoted "characteristic values" for loads and strengths, and then specified "partial

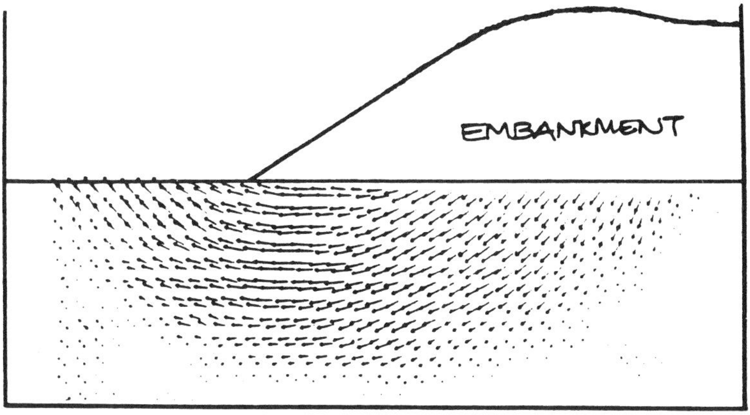

(a) Displacements increase smoothly towards clay surface, reaching circa 0.25 m prototype just prior to slip.

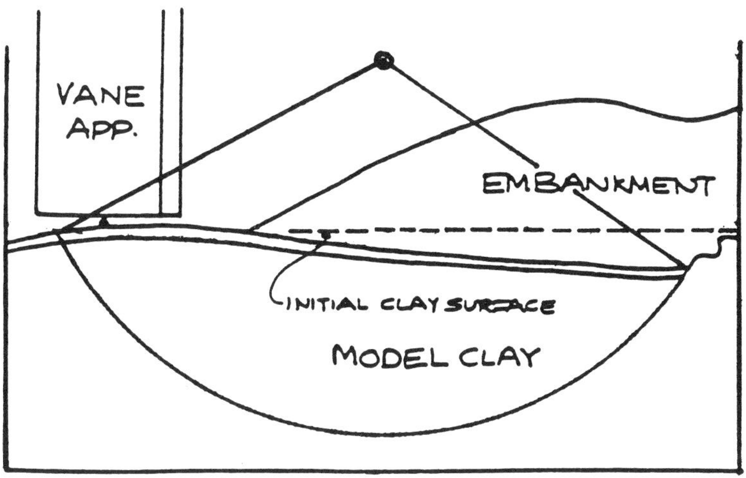

(b) Approximately circular slip surface, displacing circa 1.25 m beneath 6 m high embankment.

Fig. 1. Centrifuge tests on embankments over soft clay

factors" on all these parameters so that design values could be derived automatically by division or multiplication. Bolton (1981) asked that partial factors should be applied only to items where statistics can be applied to allow for variation within some experimentally determined frequency distribution. Geotechinical engineers recognise the overwhelming importance of inferences deduced from an experienced interpretation of site investigation data available only to the engineer on the job, not to the codewriters. Critical values of soil parameters should therefore be selected by the designer, under advice from the code. A specimen code for bridge abutments on spread bases was produced for TRL on this basis: Bolton (1991a).

Soil parameters should, in any event, be relevant to the limit state under consideration. Different parameters should be required for deformation (sls) checks and collapse (uls) checks. In the past, textbooks usually implied that equivalent elastic stiffness moduli should be used for deformation calculations while plastic strength parameters were to be used for collapse checks. Most practising engineers have, on the other hand, "used a safety factor" against collapse to guard against excessive shear deformation during construction, and have followed this up with a separate calculation of settlements due to consolidation. It is proposed below that a modification of this latter approach is actually more consistent with research evidence of deformations, and less subject to errors of judgement in selecting parameter values, than the "elastic" approach.

CP2 placed all its emphasis on stability (uls) calculations. Figure 1 shows two different episodes of embankment construction on soft clay, observed in centrifuge model tests (Sharma, 1992). The collapse event in Fig 1(b) is rather similar to the quay wall failure at Gothenburg Harbour, discussed in CP2 as a global slip circle failure. This class of event was to be prevented by the use of a factor on undrained soil strength of 1.25. But the existence of a small "factor of safety" in Fig 1(a) clarifies that as collapse is approached shear strains can accumulate to produce significant surface displacements (exaggerated in the figure). If the depth of soft soil affected by the plastic mechanism is D, and the mean shear strain mobilized within the mechanism is γ_{mob}, the horizontal surface displacement δ will be $\gamma_{mob}D$.

The following lessons must be appreciated: deformation mechanisms prior to collapse mobilize shear on the same family of lines as potential ultimate collapse mechanisms;whereas collapse involves a monolithic rotation, shear deformation can be characterized by a transition from far field displacements which may be negligible to near field displacements which increase towards the ground surface; the line of demarcation between the far-field and near field zones of pre-failure deformation coincides approximately with the ultimate slip surface; the magnitude of pre-failure deformation is proportional to the degree of shear strain permitted, which is easily related to the degree to which the strength of the soil is mobilized in an appropriate stress-strain test.

Strength reduction factors

A mobilization factor ML (defined as a reduction factor on peak strength) of about 1.5 is required to limit the undrained mobilization of shear strain to below 1% in a range of typical clays; M = 3 would be required to limit shear strains to about 0.25%. If the depth of the near-field plastic zone is 5 to 10 m, M = 1.5 would imply a displacement $\delta \approx 50$ to 100 mm. It is unlikely that a larger displacement would prove tolerable for an embankment carrying a structure such as a quay wall A larger value M = 3 might reduce lateral displacements to $\delta \approx 12$ to 25 mm.

The confusion engendered by the terms "lump factor of safety" and "partial factor on soil strength" can now, perhaps, be dispelled. The CP2 "lump safety factor" of 1.25 might best be regarded as a factor of safety on soil strength. It is likely that a value of 1.25 will be required to insureagainst collapse, considering the variability of the deposit, and dependence of undrained strength on the mode of deformation. The use of F = 1.25 on a slip circle might, in a limit state code, be described as a partial factor $\gamma_m = 1.25$ to be applied to some carefully defined characteristic element strength in order to derive a design value capable of assuring sufficiently against collapse. However, soil is a compliant material, and a much higher strength-reduction factor M between 1. 5 and 3 is required to assure serviceability on undrained clays. Serviceability is evidently the critical design consideration.

Terminology can be important. The old British practice of calling the typical strength-reduction factor of 1.5 on earthworks and 3 on more sensitive foundations a "factor of safety" is confusing, because safety is not the issue which has demanded factors as large as these. A lawyer who could show that the true strength-reduction factor was 1.3 not 1.5, or 2 not 3, could presently ask that safety was compromised whereas the real threat would be to serviceability. The new European practice of achieving the same reduction factor through the successive application of a chain of "partial factor" will be equally confusing if it is not made clear that the threatened limit state is sls, not uls. It is the certainty of deformation, not the uncertainty of strength, which generally controls the value of soil strength mobilized in geotechnical design. The reduction factor M should therefore be called a "mobilization factor" as recommended in the current draft of BS 8002.

Soil strength

CP2 (1951) offers the reader the opportunity to reflect on the slow rate of diffusion of fundamental ideas from the universities into engineering practice, especially in its treatment of the strength of soil. It invoked cohesionless soils (sands etc) which satisfied a simple constant -ϕ failure criterion, and cohesive soils (clays) which ideally possessed a constant cohesion. Most soils, and certainly silts, were thought of as intermediate between the two extremes, possessing both c and ϕ at failure on a Mohr-Coulomb diagram of total stress. The principle of effective stress had then been understood for twenty years, and the textbooks of Terzaghi and Taylor were in the bookshops. It should therefore have been possible in 1951 to explain the transition of fine grained soils from apparently cohesive ($\tau = c_u$) at constant water content during construction, to apparently frictional ($\tau = \sigma'\tan\phi'$) in the long term, as excess pore pressures dissipated.

However, the principle of effective stress was missing from the treatment of clay in CP2, probably because its authors were uncomfortable about transcending current practice. Textbooks teaching that soil strength could be written in total stress terms $\tau = c + \sigma\tan\phi$ were still being produced twenty years later, possibly because some university teachers were uncomfortable about transcending current codes. Stiff fissured clays were thought troublesome because it was necessary to estimate their softened strength: some values were quoted without much conviction. Fortunately, CP2 also introduced a long-stop in the form of a rule that active earth pressure in stiff clays should never be taken to be less than that which would be exerted by an equivalent fluid of density 30lb/ft^3. This has much the same effect as specifying an active earth pressure coefficient of 0.25 with zero pore pressures, a situation which would obtain with under-drained clayey silt attaining $\phi' \approx 32^o$ in a "fully softened" condition according to effective stress analysis. BS 8002 will emphasise the use of effective stress analysis -with an appropriate value of $\phi'$and carefully estimated pore pressures - irrespective of soil type, whilst permitting undrained strength calculations for fine-grained soils of low mass permeability prior to any pore water equilibration.

Strength parameters

It is necessary to distinguish between peak strength and ultimate strength, as shown in the effective strength envelopes drawn in Fig 2a. Peak strength is important because it terminates the rising branch of the stress-strain curve, upon which the normal working states of the soil should reside. It can therefore act as an asymptote or delimiter to states of acceptable deformation. Ultimate strength is important because it will be engendered during the large deformations associated with collapse. Here "ultimate" must be taken to imply the strength measurable after indefinitely large shear strains. Most soils deform uniformly up to a brittle peak and then strain soften in shear bands, about 5 to 10 particle diameters thick, as they dilate to a "critical state" in which further shear causes no further change of density. Soils with a predominant clay fraction exhibit a further phenomenon allied to mechanical polishing: the strength can drop even further to "residual" values on slip surfaces which continue shearing.

The successful back-analysis of large deformations at collapse in the field, or in centrifuge model tests, invariably demands the use of ultimate soil strengths. Because residual strengths require large relative displacements across shear bands, the problem effectively falls in two parts: recognizing pre-existing slip surfaces where they exist, and specifying critical state par-

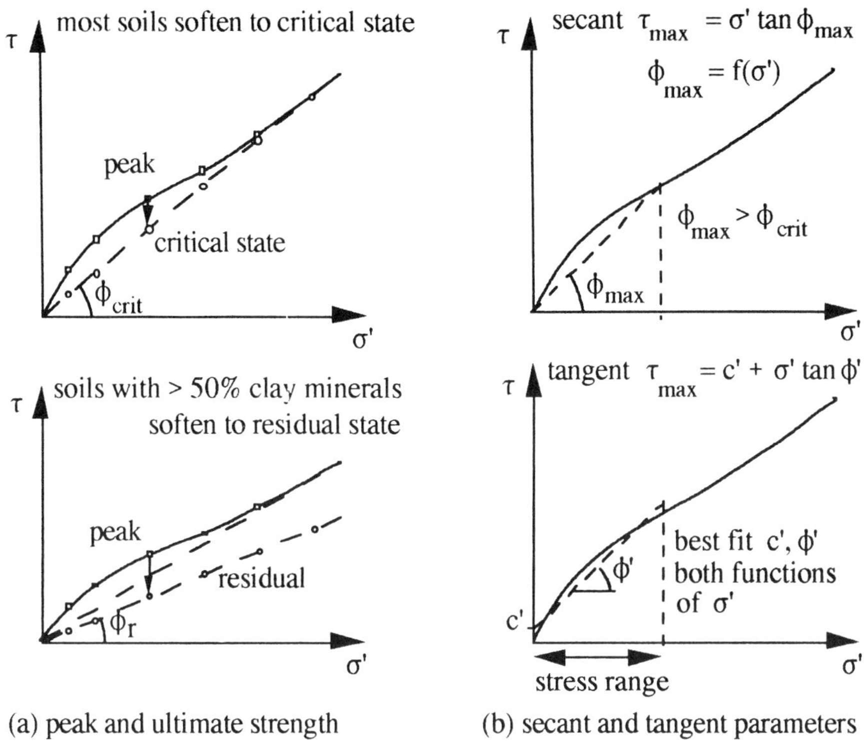

Fig. 2. Soil strength in terms of effective stress

ameters for all other uls checks. If there is then an attempt to restrict soil states to the pre-peak condition, it is generally held that residual slip surfaces will not initiate. A sufficiently close estimate of ϕ_{crit} can generally be found from the ultimate states of triaxial or shear box tests carried out on saturated samples reconstituted to be loose (if granular) or normally consolidated (if clayey). Reasonable estimates may also be obtained from the effective angle of shearing ultimately developed on rupture surfaces observed in shear tests on intact material.

While ϕ_{crit} controls the effective stress analysis of collapse, the role of the peak strength, described in Fig 2b either with a secant angle of shearing ϕ_{max} or a (c', ϕ') tangent envelope, is to act as a limit to normal working states. Although practitioners have become used to the tangent description of peak strength, the strong curvature of the envelope near the origin makes the secant definition much more convenient since the variation of only one parameter, ϕ_{max}, has to be accounted for. Furthermore, secant ϕ_{max} is a fundamental parameter related to the relative density and crushability of the aggregate, and is predictable from soil classification and relative density alone: Bolton (1986).

Where total stress analysis is admissible, with the short term undrained strength of saturated and relatively impermeable plastic soils, both the peak and ultimate strengths $(c_u)_{max}$, $(c_u)_{ult}$ should be determined. Here, the ultimate strength can best be defined as the peak strength of the clay remoulded at its in-situ water content. There are severe difficulties in selecting an appropriate ultimate strength for some soils which are highly sensitive, brittle, or cemented, and with cyclic loading events such as earthquakes which cause pore pressures in low plasticity soils to rise much higher than in monotonic undrained shear tests and which therefore cause a "degradation" in undrained strength. In any such case, expert geotechnical advice must be sought. Otherwise $(c_u)_{ult}$ should be used in collapse calculations, and $(c_u)_{max}$ can serve in serviceability checks following the strength-reduction approach.

Following the independent assessment of the peak and ultimate strengths of various soil elements, there must follow a process of decision-making. First, where a variety of values has been inferred for a particular region which is to be treated analytically as homogeneous, representative values of both peak and ultimate strength must be derived. Representative values should be cautious estimates of values which might govern performance in the field in a limit state. Usually, there will be little or no uncertainty in ϕ_{crit} governing collapse. Uncertainty in ϕ_{max} will be greater due to variations in density and stress level: the worst credible value will pertain to a representative element with least possible dilatancy. A conservative bound will be provided by simultaneously inferring the lowest relative density of sands, or the lowest preconsolidation of clays, together with the highest stress level expected in the region of interest.

More difficulty may be experienced in selecting representative undrained strengths. Excess pore pressures at peak and ultimate are a great deal more variable than the underlying effective stress parameters. However, for soils which are not especially sensitive or brittle it has been demonstrated earlier that deformation control is likely to be more critical than assurance against collapse. A conservative lower bound to the population of individual peak strength values could be used as the definition of representative value. It may, however, be possible for the experienced geotechnical engineer to discount certain low values which might be due to sample disturbance or swelling for example.

Mobilizable strength

The current draft of BS 8002 advises on the selection of a single design value of strength mob which will be mobilizable under all foreseeable conditions, taking both safety and serviceability into account. This requires two inequalities: $\tau_{mob} \leq \tau_{ult}$; τ_{max}/M, where M is the specified mobilization factor

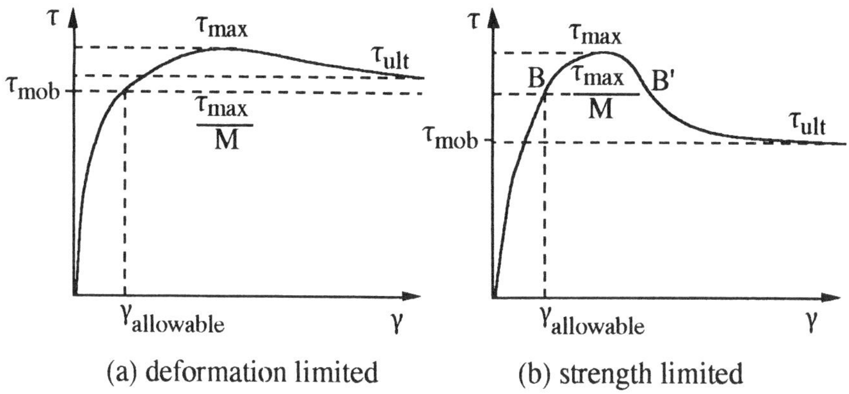

Fig. 3. Mobilizable strength deduced from shear stress-strain curves

to assure acceptable deformations . In total stress analysis M is set to 1.5 for isolated retaining walls, while in effective stress analysis M is set at 1.2. The intention of these values for mobilization factor M is that shear strains would be less than 1%, so that $\delta/H \leq 0.5\%$ in service, if the soil is not classified as soft or loose. Otherwise, stress-strain data would be required from stress-path triaxial or pressuremeter tests so that higher values of M could be chosen to achieve the required limit on strain.

The two possible cases of (a) deformation-limited and (b) strength-limited design values are shown in Fig 3. Each is intended to be conservative, and a more accurate value might be derived by advanced techniques of investigation, testing and analysis if the project warranted it. In (a) the conservatism comes from ignoring the possibility of favourable preloading, which might reduce the strains necessary to mobilize any given strength, compared with those offered in the virgin loading of triaxial tests. This is identical to the use of the proof stress (the stress at which the plastic strain on first loading exceeds a given value, e.g. 0.2%) in aluminium alloy structures. In (b) the conservatism comes from denying the possibility that states such as B are permissible. Any pre-existing softened zone, or unforeseen accident, could lead to progressive failure with the strength of the soil dropping below that required at B to maintain static equilibrium. Such an incident would lead to catastrophic dynamic collapse. Engineers generally go to some trouble to avoid the consequences of brittle failure, ignoring the upper yield point of low-carbon steels in favour of the long plateau of the lower yield strength, and then ignoring subsequent strain hardening on the grounds that brittle fracture would be too close for comfort if design relied on any plastic strength other than the lowest.

The outcome is a "permissible stress" approach to limit state design satisfying the partial factor format but never departing from an objective description of soil behaviour.

Shear strength of an interface

The available shear strength at the interface between the structure and the soil can best be determined by direct shear tests in which a sample of the steel or concrete material fills the bottom half of a shear box, and the appropriate soil is placed above and forced to slide over it. Where the surface texture of the structure is rougher than the d_{50} size of the soil, the interface will mimic the parent soil except that slicing at critical states is very easily induced: for these "rough" surfaces ϕ_{crit} for the parent soil may be used as the representative value in default of experimental measurements. Surfaces which are "smooth" relative to d_{50} may give lower values of ϕ dependent on the materials, which would then require measurement.

In total stress analysis, the shear strength of an interface may be affected both by friction and pore pressure deviations. Regarding friction alone, the representative undrained strength for sliding of a "rough" surface might be estimated as $(c_u)_{ult}$ for the parent clay, on the same grounds as before. However, where porous permeable materials are placed as drains between the structure and a clay soil, the undrained interface strength will never be relevant. The total normal stress σ in the clay will then be transformed into an effective normal stress at the interface, so that the shear strength becomes the lower of tan $_{crit}$ (for shear in the clay close to the drain) and c_u (for shear further from the drain). For all except the softest clays the former condition would be critical, and the equivalent a-value former/latter) could be significantly less than unity.

It would clearly be impermissible to attempt to mobilize an interface strength either greater than the representative strength of "smooth" surfaces, or greater than τ_{mob} in the neighbouring soil. A design limit $(\tau_{mob})_{wall} = 0.75(\tau_{mob})_{soil}$ can be set for wall faces to allow for the possibility that the wall is not aligned with the planes of shear in the soil. Vertical equilibrium must be demonstrated in every case.

UK designers should note that taking $(\tau_{mob})_{soil} = \tau_{max}/1.2$ for granular soils, with $(\tau_{mob})_{wall} = 0.75(\tau_{mob})_{soil}$, gives a total reduction factor $(\tau_{mob})_{wall}/(\tau_{max})_{soil} = 0.63$. The alternative inequalities set out above to limit (mob soil) will result in a somewhat smaller (variable) ratio. These new wall friction factors are similar in magnitude to those in current use, but are now expressed in objective terms.

Method of earth pressure analysis

The outstanding advantage of the concept of mobilizable strength is that the usual plastic methods of analysis are available, except that no further "safety factor" is to be used. The reduced strength τ_{mob} can be expressed in terms of ϕ_{mob} or c_{mob} as appropriate. Mobilizable active and passive pressures are then calculated as a function of mobilizable soil strength and wall friction. In effective stress analyses, earth pressure coefficients K_{mob} are the usual functions of ϕ_{mob} and wall friction angle δ_{mob}. In total stress analyses, the limiting earth pressures (taking safety and serviceability into account) are functions of c_{mob} and α.

Convenience of analytical techniques

The theory of plasticity has been much extended in its application to soil mechanics problems since 1951. Whereas CP2 specified earth pressure coefficients based on Coulomb's assumption of sliding wedges, the draft EC7 contains explicit formulae based on the method of characteristics, which satisfies equilibrium and obeys the strength limitation at every point in the soil, not just on an assumed failure surface. These new earth pressure theories offer safe bounds to the plastic solution if the values of soil strength and interface friction are themselves on the safe side. The degree of under-estimation which they offer is much smaller than the over-estimation caused by assuming planar slip surfaces, for example. They can also account for geometrical parameters such as sloping fill or battered walls.

Bolton (1991b) used similar techniques to derive thrust coefficients for the effect of load patches on the surface of retained granular fills. Effectively, Prandtl's bearing capacity solution was extended downwards and outwards to produce earth pressures on a nearby retaining wall. The earth pressures were then integrated to provide the magnitude, inclination and location of thrust on the wall.

Water-filled cracks

Confusion often occurs regarding the possibility of cracking in active zones of lateral spreading, and the thrust which can be exerted on walls retaining clay if such cracks initiate and fill with water. Although CP2 warned of the effect, the detailed advice it gave is typical of the worst sort of code of practice, comprising a misunderstanding of the mechanism, an over-elaboration of a completely mistaken formula, and an overlay of arbitrary conditions.

First, it is necessary to recognize that the conditions for opening a crack are that tension is applied to a material which has finite shear strength but small tensile strength and small fracture toughness. These are fulfilled best on the interface between a wall and clay which it is retaining, where the wall is moving away from the soil. The interface has a much smaller fracture toughness than the parent soil, and offers the perfect crack-raiser, as described by Bolton and Powrie (1987). Second, it is necessary to appreciate that if a crack fills with water, the pressure can extend the crack beyond the "dry" depth $2c_u/\gamma$ down to $z_w = 2c_u/(\gamma - \gamma_w)$. The hydrostatic pressure $\gamma_w z$ replaces any active soil pressure, and tends to push the wall outwards. The water follows the wall out, keeping the crack full if there is sufficient inflow. Below depth z_w, the "active" soil follows the wall and keeps the crack shut, whereas above that depth the soil is supported by the water pressure in the crack. The practical consequence of this is that in many overconsolidated clays the maximum possible depth of a flooded crack will exceed the depth of the wall. A crack will then open dawn to the point of zero lateral displacement, which is close to the foot of a rigid cantilever. Bolton *et al* (1989) documented centrifuge tests and charted the strength mobilized in a variety of cases. It is simply necessary to make a judgement of whether free water could be present ro flood the crack at the wall/clay interface.

Geo-structural mechanisms

Plastic soil deformations around a stiff cantilever

Fig. 4 shows an idealization for uniform soil shear strains around a free embedded cantilever wall, consistent with the usual assumption of uniform soil strengths. Wall rotation θ is shown to relate to mobilized soil shear strain $\gamma_{mob} = 2\theta$. The use of a mobilization factor $M = 1.2$ on $\tan\theta_{max}$ or 1.5 on $(c_u)_{max}$ is intended to restrict shear strains to within 1%, so that $\theta \leq 0.5\%$. If some different limiting displacement were desired, or if good quality stress-strain data were available, the factor M could be altered ta suit. Whatever value of ϕ_{mob} or c_{mob} is selected, the plastic equilibrium of the wall in terms of horizontal forces and moments provides the two conditions necessary to determine the point of rotation and the required length of the (assumed rigid) wall. This approach was used by Bolton *et al* (1989) to calculate and chart the minimum permissible lengths of a stiff wall consistent with the specified displacement criterion, under various soil conditions.

Walls rotating about props at their crest can be treated in the same fashion. A compatibility condition $\gamma_{mob} = 2\theta$ is maintained only when a frictionless horizontal surface is introduced at the foot of the wall: otherwise, an additional strain concentration propagates from the foot. In general, the region of soil immediately surrounding the wall mobilizes $\gamma_{mob} \geq 2\theta$ so that the assumption of equality causes the engineer safely to underestimate the degree to which soil shear strain (and therefore strength) will be mobilized.

Structural serviceability for a stiff cantilever

Design bending moments, shear forces and prop forces can be derived by equilibrium calculations from the linear design earth pressure distributions of the type shown in Fig 4, with ϕ_{mob} or with c_{mob} and pore water pressures. If the soil around the wall starts with zero mobilized strength (initial earth pressure coefficient $K_i = 1$), and if the same degree of shear strain is mobilized in both active and passive zones due to wall movement, the same strengths

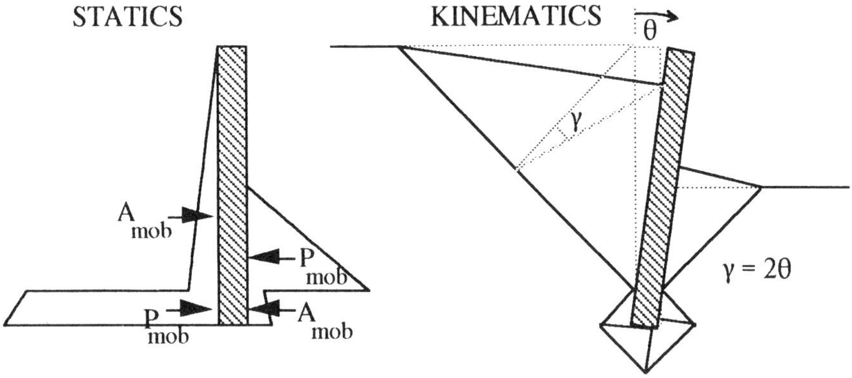

Fig. 4. Geo-structural mechanism for embedded retaining wall

should be mobilized. This is equivalent to assuming a uniform mobilization factor $M_a = M_p = M_{target}$, 1.2 or 1.5 as appropriate.

Higher bending moments will be induced if it then transpires that $M_p < M_{target}$ or $M_a > M_{target}$ so that $K_{p,mob} \rightarrow K_{p,\ max}$ and $K_{a,mob} \rightarrow 1$. The two conditions of higher earth pressure either side of the wall inevitably go together since equilibrium demands they balance. This can happen where there is a high value of K_i, such as in heavily overconsolidated soil, or where clay swells on the passive side due to excavation. It can also occur where compaction increases soil pressures in retained fill, or where large prop forces are jacked into place.

In reality, the mobilized strength around a wall will vary with position and time in a complex way. The objective of the designer is not to predict such variations but to allow for their extreme values. The consequence of failing to predict enhanced lateral pressure may be the premature yield or cracking of the wall. If the wall material is ductile, local yielding will relieve pressures by creating wall movements sufficient to achieve target mobilizations of soil strength.

It follows that steel will present few problems. The wall could never collapse if it has been designed properly to a target value of M, nor can it rotate significantly more than the target value of θ (eg 0.5%). It may, however, have yielded unexpectedly at some locations. Similar problems afflict steel frames due to lack of fit in construction: steel designers generally ignore such self-stressing effects. In the design of steel retaining walls, mobilizable earth pressure coefficients can be used in any arbitrary pattern, so long as the conditions of overall equilibrium are met.

Reinforced concrete retaining walls may be more troublesome if steel yielding causes concrete to crack which in turn hastens corrosion of the steel. It would be sensible, for example, to assume for the purposes of structural design that diaphragm or secant-pile walls in stiff overconsolidated clays ultimately develop $M_p = 1$, or $K_{p,mob} \rightarrow K_{p,max}$ beneath the excavation. The resulting bending moments could hardly be exceeded unless some gross error had been made, such as in the groundwater pressure distribution. No further "load factor" need be applied, therefore, prior to section design.

Wall flexibility effect

The relative bending deflections need to be calculated from the assumed earth pressure distributions. If they exceed 0.5% (or the selected target strain limit) the previous use of a constant M factor with depth needs to be re-examined.

Three stages of increasing flexibility might be recognised.

i) Stiff wall - any tendency to bend does not greatly influence the strains in the surrounding soil. Simple triangular distributions of earth pressures are unaltered.
ii) Flexible cantilever - greater soil strains are induced near the crest, reducing local earth pressures and thereby bending moments. Target mobilization factor M can not be reduced and the serviceability limit can not be relaxed, so account can not be taken of this effect.
iii) Flexible sheet with redundant supports - arching of earth pressures on to the supported sections, and encastre at the base with "fixed-earth" support. This class of soil structure interaction may benefit from numerical analysis, leading to economies in the provision of steel in sheet pile wall design following Rowe's method, for example. One aspect of this is considered below.

Flexible cantilever with single level of support

If a high-level prop is provided, the simple "statically determinate" calculation corresponds to "free-earth" support with a rotation about the prop. This sets the minimum permissible penetration and a conservative value for the required bending resistance. A greater than minimum length of penetration may be provided but it might not all be effective due to excessive flexibility. The effective extension will be that part which can develop significant net earth pressures, and which can therefore contribute extra resistance. Hetenyi (1946) developed solutions for the bending of a beam on springs in terms of the relative stiffness of the beam and the springs. The diffusion along the beam of the effect of an imposed load or displacement can be generalized in terms of a characteristic length, which in plane strain for a retaining wall embedded in an elastic medium might be written $\lambda = (EI_{wall}/E_{soil})^{1/3}$. An extra penetration of up to λ beyond the "statically determinate" length may be regarded as simply extending the "free earth solution", and therefore having little effect on the required bending resistance. An extension of 3λ, however, may be taken to provide full moment fixity and fixed earth support. Deeper penetrations will not be effective. A significant part of Rowe's moment reduction factor for walls with small EI values is captured by eliminating the wall below the effective depth, thereby mobilizing higher soil strengths, smaller active pressures, and larger passive pressures, over a smaller span.

Conclusions

New "limit state" codes are being written which set objective performance criteria for the safety and serviceability of retaining walls and other geotechnical structures. Critical events are to be forestalled by direct evaluation of the performance of the structure in certain well-defined design situations.

Eurocode EC7 is a design standard, specifying what must be checked without prejudicing the method. By avoiding calculation details as much as possible, it aims to cover the whole scope of geotechnical design in a single slim document. This effort deserves to succeed, and the BSI committee have aimed to make the new BS 8002 compatible with EC7.

The concept of mobilizable soil strength, c_{mob} or ϕ_{mob}, simultaneously satisfying safety and serviceability, is proposed. Careful attention to the selection of a strength parameter can lead to simple, conventional calculations of plastic equilibrium. This should be equally acceptable to "lump factor" or "partial factor" advocates, and to those who prefer to be guided to the direct selection of a design value taking both stiffness and strength into account.

If a more detailed analysis is sought, the method can be broken down into its basic components of plastic equilibrium ($K_{mob} = F(\phi_{mob})$) and plastic kinematics ($\gamma_{mob} \approx 2\theta$) , and it can then be enhanced through the use of high quality stress strain tests ($\phi_{mob} = f(\gamma_{mob})$). The term "geo-structural analysis" has been coined to cover this simplified treatment of soil as a jigsaw pattern of structural elements. The rate of advance in soil mechanics theory and geotechnical practice has been rapid. New textbooks, advanced courses, and computer programs will continue to emerge, as will new technologies in earth retention. The old BSI system of code writing has given way to the new process of European standardization. If European geotechnical engineers wish to have committees winnowing new ideas and practices, their professional bodies should establish them.

Acknowledgements

I wish to recognise the close collaboration of Dr Brian Simpson and Mr Ian Symons in our joint proposals to the BSI committee reviewing CP2 under the chairmanship of Mr Tom Ackroyd, although I take responsibility for this particular statement of views.

References

1. BOLTON M.D. Limit state design geotechnical engineering, Ground Engineering, 1981, 14, No.6, 39-46.
2. BOLTON M.D. AND POWRIE W.
The collapse of diaphragm walls retaining clay, Geotechnique, (1987), 37, No.3, 335-353.
3. BOLTON M.D.
The strength and dilatancy of sands, Geotechnique, (1986), 36, fmo.l, 65-78.
4. BOLTON M.D. The development of codes of practice for design, 12th ICSMFE, Rio de Janeiro, (1989), 3, 2073-2076.
5. BOLTON M.D., POWRIE W. and SYMONS I.F. (1989, 1990) The design of in-situ walls retaining overconsolidated clay. Part I, Ground Engineering, 22, No 8, 44-48; 22, No. 9, 34-40; Part II, Ground Engineering, 23, No 2, 22-28.
6. BOLTON M.D. (1991a), (1991b) Geotechnical stress analysis for bridge abutment design, TRL Contractor Reports 271, 270.
7. BULAND J.B., POTTS D.M. and WALSH N.M. (1981) The overall stability of free and propped cantilever retaining walls, Ground Engineering, 14, No 5, 28-38.
8. HETENYI M. (1946) Beams on elastic foundation, Ann Arbor Press
9. SHARMA J.S. (1992) Research report RESC4, TRL Contract N04900, A.N.S. 7 A., 9 Little St. Mary's Lane, Cambridge. (Confidential)

Introduction to Session 5
Support systems to earth retaining structures

R. FERNIE, Cementation Piling and Foundations Ltd, UK

Support to ground can have either a reactive or a pro-active effect on soil structural behaviour (and hence loading and deformation).

Reactive support can include:

. ground in front of wall toe
. berms (passive earth buttress)
. yielding props
. deadman anchorages.

Pro-active support can include:-

. props subject to expansive forces (jacks, temperature)
. stressed anchorages
. possibly props spanning sites with significant ground level differences opposite each other

Following this classification, of the nine papers published under the 'support systems' heading, three can be categorised as reactive and six proactive.

Reactive

. A temporary secant mini pile wall propped by a deep jet grout raft. Newman and Covil

. A cantilever retaining wall supported by a berm during temporary works activities. Powrie, Davies, and Britto

. The use of soil berms for temporary support of retaining walls. Potts, Addenbrooke and Day

Proactive

For reporting facility the proactive papers have been further (fairly arbitrarily) categorised into:

Case studies with a clear BS8081 theme:-

. Ground anchorages to the retaining walls at Pen Y Clip. Turner, Seeley

and Cooper
. Design and construction of temporary ground anchors at Castle Mall Development, Norwich.Barley, Eve and Twine

Use of stereo nets for anchorage force prediction:
. An anchored and tied retaining block structure at York Road car park Guildford. Barsby
. Stereographic projection methods used in the design of anchorage orientations for retention of rock mass slope. Maurenbrecher
Transfer of local technologies and expertise:-
. Rock anchorages in South Brazil - bond strength evaluation.Macedo, Bressani and Milititsky
. Regroutable ground anchorages. Sherwood and Harris

Proactive BS8081

Turner et al

The A55 is sandwiched between the North Wales mountains and the sea, and has some of the most demanding of geotechnical problems to overcome, not least a series of anchored structures approaching Pen Y Clip tunnel (see figure 1).

1600 permanent anchorages were placed in Glacial Till, Mircodiorite, Mudstone and Scree and their design/construction and performance controlled very much to BS 8081.

The Paper illustrates how a professional approach permitted successful tremie grouted and pressure grouted anchorage placement in a variety of ground conditions despite geographical, topographical, access and programme constraints.

With a proportion of anchorages fitted with "permanent" restressable heads for monitoring and some 'sacrificial anchorages' which can be exhumed at a later date (a TRRL contract) it is likely that this site will be revisited and will appear in the literature as a BS8081 monitor and case study for the next decade or so.

Barley et al

This is a companion paper to the already presented Castle Mall project in Norwich dealing with 1100 plus temporary anchorages to BS 8081. To compensate for the weaknesses of the anchoring media; a Soft CHALK with many solution features or a SAND (NORWICH CRAG) affected by solution feature collapse - the anchoring contractor (well experienced in these conditions) used the cost effective SBMA anchorages and pressure grouting (with some pre-grouted treatment) to support up to 18m high of piled retaining wall structure.

About 10% of the placed anchorages required special pre-grouting. As the job progressed (anchorages are temporary), it was possible to exhume some anchorages. This demonstrated that pressure grouting enhanced anchor diameters (in the weak material) and that 8081 double corrosion protection was satisfactorily installed.

The large costly anchoring project was supported by a well planned S.I. Also as many as 18 proving anchorages in 3 locations were carried out with a significant amount requiring pre-grouting (good cost/ benefit for S.I.) thus allowing the team to build confidence in the working procedure. The scheme illustrates the use of BS 8081 designer contractor co-operation to the jobs best advantage.

Barsby: stereonets

Presents a case study of an extension of a vertical cutting in Grade II/III Chalk with a requirement to underpin existing footings without disrupting the use of the adjacent and overlying car park.

Failure could disrupt the local car park area and possibly affect the global stability of the chalk cutting (amounting to an overall height of 20 m). Stereo-plots are used here to assess anchorage need in terms of daylight envelope and the necessity to increase frictional reserve.

The study describes design parameters, construciton procedures and apparent performance acceptability of the works.

The chalk rock slope is generally dry (facilitating a stereonet analysis) but would anyway perhaps have been drained by some air-raid shelter tunnels. The tunnels could also be used for anchorage fixity within the ground and a significant percentage of the anchorages were indeed tied through to the tunnels - an excellent example of engineering to a solution waiting for a problem.

Proactive: stereonets

Mr. Maurenbrecher's Paper should perhaps be read prior to Mr. Barsby's since it takes us through basic steronet literature (geotechnical). It reminds us that FEM is not the only elegant modelling available to Engineers.

A slant to the use of stereonets which is seldom discussed in the literature but shown here is: 'imposing a friction circle due to the inclination and azimuth variation of separate anchorages to cover any "stray" admissible failures in the daylighting envelope. This gives the user the ability to optimise general anchorage positioning and Mr. Maurenbrecher illustrates the technique on an actual bridge abutment support (founded on a rock slope)'.

As well as introducing this concept he reviews the toppling/spalling problem in rock slope design and discusses methods of choosing declination angles of anchorages.

For most practising engineers stereo-nets are useful for 3D introductions to rock slope problems, but they have insufficient regular "hands on" experience to use them confidently. Further, the gathering of pertinent information has to be left to experienced Engineering Geologists who can probably visually/intuitively carry out the analyses without masses of stereo-net data. This is I believe how most contracts are run in UK and overseas, we salute Mr. Maurenbrecher's theory but rely on him and his Engineering Geology colleagues practising it.

Proactive: technology transfer
Macedo et al

This is a bond strength evaluation on a dywidag type rock anchor in volcanic 'bedrock' in South Brazil comparing some in-situ pullout tests with a laboratory study.

There is a reminder that the use of Beniaskwi's RMR gives us an internationally applicable method of indexing rock. With RQD's of 95 and 90 and overall RMR's 60, this is a very competent volcanic rock in which one would expect good rock/anchorage bond.

Excellent core recovery allows a lab procedure involving UCS tests on grout core specimens cut at different angles, to produce the "shear bond". A test spec (rate, numbers, capping) is included, but the advantage of this method compared with a simpler shear box test on a cut sample is not apparent since neither gives an answer other than an INDEX.

Field trials show reasonable correlation with lab tests and indicate ultimate bond between 4 and 10 times the limit bond permitted by the local control authorities. The authors would welcome a BS 8081 approach (say using F of S = 3) to increase the permissible bond stress.

Proactive: technology transfer
Sherwood and Harris

If the Brazilians would like to use BS 8081 practice to increase their anchorage capacities in very good ROCK, ironically the French experience (and claim) is that if their IRS system is used BS 8081 is too conservative in CLAY.

This Paper is to be presented and needs little introduction, it relates about 20 years of French experience of this anchorage type.

In their presentation the Authors have been asked to address:

. whether the weaker clays of the Paris basin (cu - 90 kN/m^2) do not better respond to IRS than highly over-consolidated stiff clays (cu = 200/300 kN/m^2).

. whether other systems, including underreaming and SBMA anchorages do not achieve similar load carrying capacities (with the benefit of familiarity).

Reactive:

Innovation to UK

Newman and Covil

This Paper is to be the second presentation and involves a case study on the provision of a jet grouted raft and a mini-pile (300 mm) perimeter wall for temporary support for a structure in Jersey.

Jet grouting (first used by the British in Pakistan in the early 60's) has been further developed by the Japanese to provide a variety of ground solutions to problems in a range of materials throughout the world. This is one of the first uses in UK environs for a raft, although mix-in-place techniques have been used in similar ground conditions (at least three times) to provide the same end result. The Authors have been asked if mix-in-place techniques were considered and what governed the choice (access, cost innovation, familiarity?).

Powrie et al

case study

This is a case study and another of the A55 (North Wales) schemes.

In order to provide contractors access without too much local movement prior to placing temporary/permanent propping, a berm was used to buttress diaphragm wall support to a major rail and road causeway diversion.

The berm effect on inhibiting movement was modelled by:

. FEM (predicting about 4-6 mm movement at the head of the wall)
. the Bolton mob method (predicting 25+ mm movement at the head of the wall).

The FEM (CRISP) gave a better fit prediction of movement when allowance was made for relaxation due to diaphragm wall installation (see other papers in conference) and when the stiffness of the granular material was varied.

FEM was employed because of the limitations of CP2 in guidance on deformation. An empirical method would have suggested deformation less than lO mm and one wonders if pedantic authority forced the work? The Authors however we can thank for some first class information on ground conditions, lateral stresses, soil modelling information and FEM data in Irish Sea Boulder Clay/Glacial Lake deposits. This is a useful case study for the general data base.

Potts et al

FEM Study

This Paper will be presented in the discussion session. It is an important topic relating to an often utilised temporary construction technique (berm support) to limit Bending Moment and Deflection.

Despite the Authors inference to the contrary there are methods of predicting:

. BM reduction (via geometrical constructions to produce a 'lesser excavation depth' and a surcharge - see Fleming *et al* South African Code).
. Movement predictions have been addressed via FEM by Clough and Denby (1977) for soft to firm clays.

Stiff over-consolidated clays however have not previously been studied?

There are features of the paper which will perplex the practising Engineer, e.g.

. the relative values of the cantilever and prop BM's (a continuation of the FEM/limit equilibrdum debate?)
. the observation that berm geometry has no apparent effect.

The Authors have been asked to address these points in their presentation.

An anchored and tied retaining block structure at York Road Car Park, Guildford

EurIng R. D. BARSBY

Background

As part of the continued expansion of car parking facilities in Guildford, the Borough Council proposed to increase the number of available spaces by expansion of York Road Car Park. Built in the early 1970s in the floor of a disused chalk pit, the existing footings of the car park structure, adjacent to the quarry high wall, are founded on pad footings at first floor level on a chalk bench. The proposed extension involved the removal of this bench of chalk to a line coincident with the existing column foundations and the construction of a retaining structure to support the cut chalk face. This retaining structure had to be designed to maximise the number of extra car parking

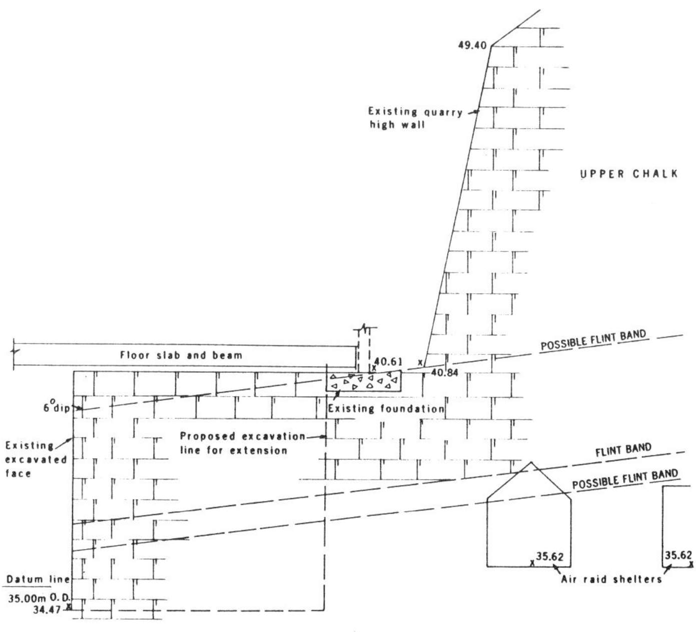

Fig. 1. Section through chalk bench

spaces and constructed with the minimum disruption to the car park operations.

The fulfilment of these requirements, whilst maintaining the integrity the existing structure, posed a number of engineering problems and the solution adopted had also to take account of the very restricted working space beneath the lower floors of the car park. Furthermore, the design had to . consider not only the long term loading from the footings and the active earth pressures from the chalk cliff above but also the short term stability of the cut slope beneath the loaded footings.

Geological investigations

A geological investigation was made involving an extensive discontinuity mapping survey along the face of the chalk bench. The regional mass discontinuity pattern of the Upper Chalk of the North Downs was confirmed at this site, being closely spaced, discontinuous, tight and planar. Superimposed upon this pattern is a less regular, more widely spaced joint system associated with regional faulting.

A visual examination of the weathering grade of the chalk based on the system proposed by Wakeling (ref.l) indicated the mass material to be primarily Grade II to III, and described as white to greyish white, blocky to rubbly, slightly weathered medium hard chalk, see Plate 1. Within the depth of the bench, three continuous bands of flints were recorded and throughout the mass, discrete flints, up to 300mm in cross-section, were occasionally noted.

The mass discontinuity data recorded in the survey is reproduced in Fig. 2 as a series of points on an equal area spherical projection, enhanced by statistical grouping to reveal zones of discontinuity concentrations.

Plate 1. Grade II to III Upper Chalk

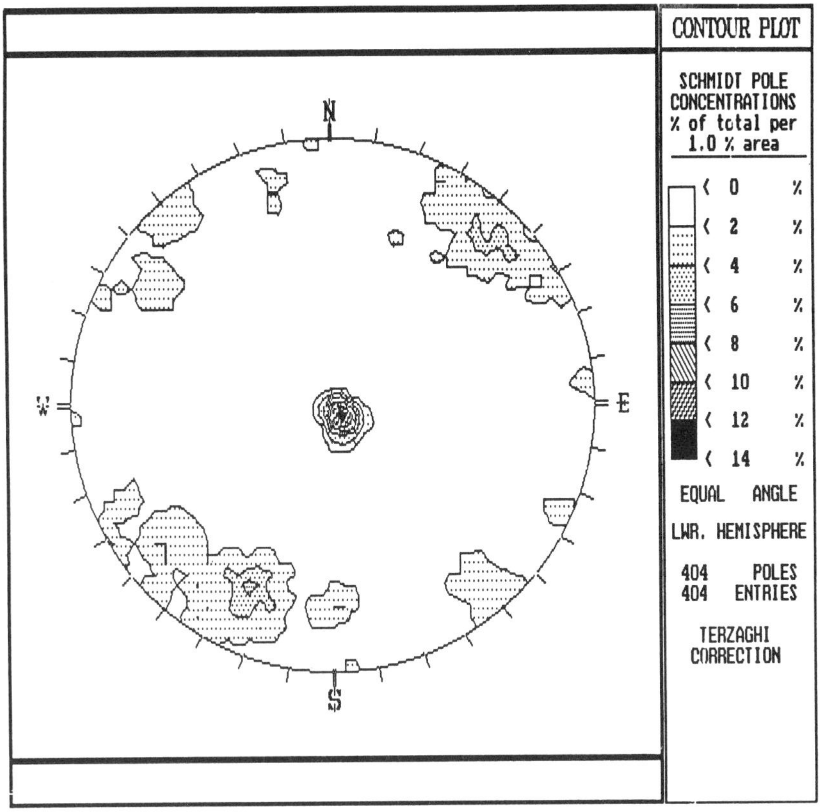

Fig. 2. Mass discontinuity survey data

Stability assessment

The mass strength and deformability of the chalk behind the retaining wall depends upon the shear strength along the discontinuities within the chalk mass rather than the strength of the intact chalk. Hutchinson (ref.2) showed by back-analysis that chalk exhibits a strain softening characteristic and that the mobilised shear strength along a failure surface was somewhat lower than the peak strength. Clayton (ref.3) found that some of the mechanical and physical characteristics of chalks are dependent on their porosity and reports that, for low porosity chalk, the peak shear strength may be several thousands of kPa. Because of the uncertainties regarding the exact failure mechanism and the shear strength mobilised along the failure surface, the stability of the vertical cut slope was assessed by considering the sensitivity of the factor of safety to shear strength for various potential slope failure mechanisms.

Planar and wedge analyses were represented by equal area spherical

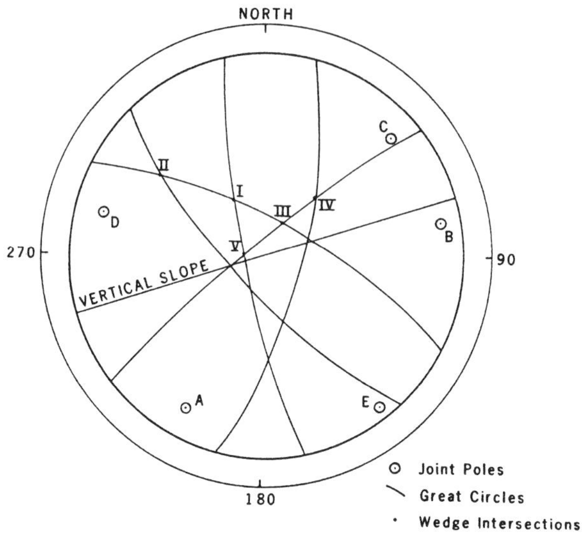

Fig. 3. Stereoplot wedge analysis

projection as depicted in Fig. 3. A number of potential wedge failures were recognised and analysis showed that, with the incorporation of a surcharge from the existing footings, the factor of safety against instability was approximately unity for conservative shear strength parameters (cohesion of $25kN/m^2$ and an angle of friction of 35 °) and assuming zero water pressure.

Circular or quasi-circular failure was represented by a failure surface developed through the rock mass along the discontinuity pattern. Although these failure representations are approximations, the calculated factors of safety (based on conservative strength parameters) were found to be close to 1.0.

It was concluded, therefore, that the integrity of the chalk bench beneath the existing foundations could be impaired if the cut was allowed to remain fully unsupported for the time required to construct a traditional cantilevered retaining structure.

Support alternatives

Alternative methods of support to the footings that were considered included traditional underpinning of the foundations, grouting and retaining walls.

Traditional underpinning methods rely on relieving the load from the foundation while the pins are constructed in phases beneath the footings. Such methods are satisfactory for lightly loaded strip foundations but for isolated pads complete relief of the load is essential. It was not considered practical to provide temporary props throughout the full height of the car park and raking props would have severely restricted the limited working space available.

As failure of the chalk would have taken place through pre-existing discontinuities, an improvement in the strength along these joints by grouting would allow a vertical cut to be made. However, the discontinuity survey showed that the jointing was primarily tight and the ability of a grouting programme to successfully enhance the mass strength characteristics of the chalk was considered low.

A cantilevered retaining wall was also considered but for the reasons expressed above, propping to relieve the structural loading from the footings during the construction of the wall would be a severe restriction in the confined working space of the lower floor of the car park.

To minimise disturbance to the pad footings, a retaining wall which comprised a series of levels of twin anchored and tied blocks offered advantages over the alternatives by both maximising the available parking spaces and minimising the risk of displacement of the existing foundations. A typical elevation of the block arrangement is shown in Fig. 4.

Block-anchor and tie design

Behind the quarry wall a network of tunnels were excavated during the Second World War for use as air raid shelters and storage facilities but have been long abandoned. A survey of their location showed that the upper two

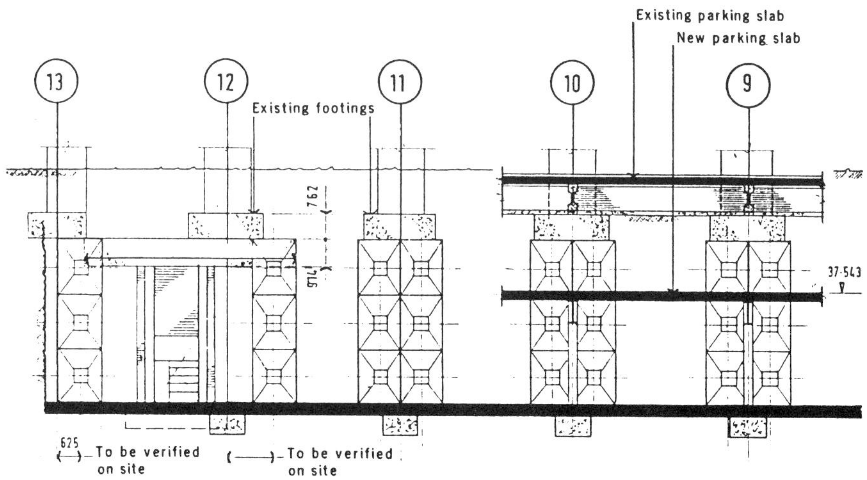

Fig. 4. Typical elevation of block arrangement

levels of blocks could be tied into the tunnel walls, whilst the lower row of blocks would require the use of a rock anchorage system.

The design of each block took into consideration a 300kN/m^2 load from the footing and the earth pressures associated with the 30 metre high near vertical quarry wall above. In predicting the stresses and deflections of the blocks two conditions were evaluated with respect to the block arrangement. The first assumed that the blocks formed a single retaining wall for the full height of the bench to be supported; the second assumed single blocks one third of the height of the bench. The results of the former condition were used in the assessment of the earth pressures acting on the wall, the latter to assess wall movements.

The parameters used to evaluate the earth pressures were based on drained conditions and zero water pressures. Whilst water pressures were incorporated into the design to assess their overall affect, any increase in the earth pressure resulting from a build up of water behind the blocks was evaluated as only transient and a temporary lower factor of safety was considered acceptable for the anchor and block design.

The earth pressure coefficients used to calculate active and passive pressures were assessed from classical soil mechanics formulae to be

k_a=0.25, k_{ac}=0.61, k_p = 7.0 and k_{pc}= 6.48

In the assessment of the earth pressure coefficients, consideration was given to the fact that the chalk face had been excavated and left exposed for many years prior to the construction of the retaining wall. Relaxation of the chalk within the active wedge over this period would have reduced the in situ stresses towards the active condition.

The force equivalent to the area of the earth pressure diagram over which an anchor or tie acts was taken as the prestress force and a value of 300kN/m^2 was applied to each tie and anchor. In addition, blocks, anchors and ties were designed to be capable of withholding the equivalent foundation load at each block level on the assumption that failure could occur along a pre-existing unfavourable joint or series of joints.

Estimates of the deflected shape of the wall were made by computer analysis using the program WALLAP licensed from Geosolve. This program utilises the modulus of subgrade reaction to represent stiffness and a value of 100 000 kN/m^2, assessed from published data, was considered appropriate. The computer analyses predicted movements of up to 5 mm for individual blocks compared with over 20 mm for a cantilevered retaining wall spanning the full height of the chalk bench.

A typical cross-section showing the arrangement of ties, anchors and blocks is presented in Fig. 5.

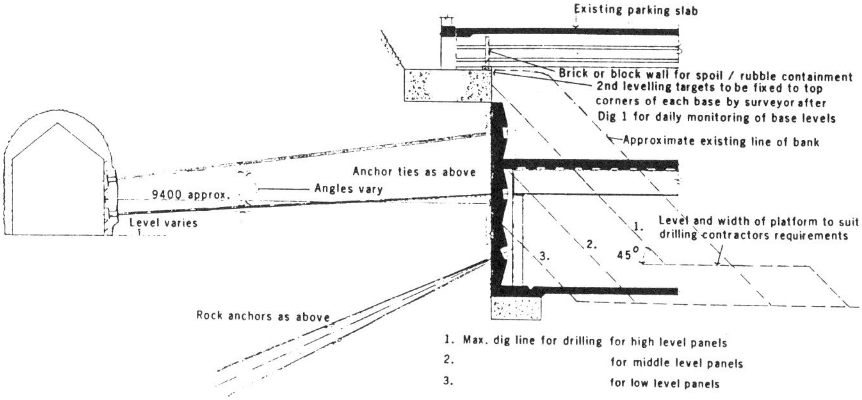

Fig. 5. Section of typical tie, anchor and block arrangement

Construction and performance

To ensure that the existing foundations were at all times adequately supported, and with the knowledge that the chalk could fail in a relatively sudden and brittle mode, rather than by plastic deformation, excavation in front of the footings was only allowed to be left unsupported for 48 hours before a block and anchor/tie was installed and stressed. Additionally no two adjacent blocks were allowed to be excavated and remain unsupported at any one time. There was, therefore, a detailed planning and logistical exercise necessary to comply with these restrictions.

To ensure excavation headings were left unsupported for the minimum period, a programme of work was planned so that half footing width headings could be excavated, the tie or anchor hole drilled and the pre-cast concrete block offered up and grouted in place all in one day. The grouting of the tendon was programmed for the second day with prestressing taking place at the start of the third day. The drilling contractor was programmed to complete two drillings per day and managed, with good programming and close liaison with the main contractor and the Engineer, three drillings on occasions.

The alignment criteria imposed on the drilling of the anchor and tie holes were particularly tight to ensure that adjacent blocks were not misaligned, and that sufficient tolerance was provided to allow the installation of the back-plate within the tunnels.

Special precautions were taken by the drilling contractor to satisfy the requirements of the National Rivers Authority.

Drilling fluids were discharged into the tunnels rather than into the local

Plate 2. Drilling operations

surface water drainage system where the high chalk fines content from the drilling may have caused siltation of the drainage network.

All excavations for headings were described by an engineering geologist in order to confirm that the discontinuity pattern encountered during the construction of the anchored and tied block retaining wall was consistent with the assumptions made during the design. No adverse conditions were identified and the excavation of the headings, the drilling, prestressing and grouting all followed the programme without disruption or delay.

The sensitivity of the car park structure to displacement of the foundations necessitated careful monitoring for movements of the foundations. Two survey stations were mounted on each footing and surveyed for position and level on a daily basis.

This monitoring showed undetectable movement of all footings. In the event that movements of the footings had been noted, precautionary measures included load bearing struts being stored on site in sufficient quantity to provide load relief throughout the complete height of the structure for at least two footings.

The rock anchor and tie design was made by Colcrete Ltd. and consisted of a double strand arrangement with a fixed length of 10 metres and a free length of 5 metres. To confirm the anchorage design, one proving anchor was installed and tested to 3 times its working load. All production anchors were tested to 1.5 times their working load before locking at 1.1 times their working load with load loss monitored after 1 day. In all respects the anchors passed the testing requirements of BS 8081: 1989 Ground Anchorages and the proving anchor showed minimal residual load loss after the 10 day creep test.

Conclusions

The proposed extension to York Road Car Park was considered to represent a relatively high risk for foundation movements. The solution to the stabilisation measures of the chalk face adjacent to the existing footings had also to address the potential for instability during construction. It was necessary, therefore, to adopt a construction sequence and method that minimised the risk and to implement monitoring procedures and contingency plans to safeguard the integrity of the structure.

A sequential excavation procedure provided a high degree of certainty of minimising foundation displacement and, by anchoring the support structure to caverns behind the quarry wall, stability of the chalk excavation could be assured.

The use of a rock anchorage system to restrain a vertical chalk slope supporting footings has proven highly successful. The use of ties anchored to the walls of tunnels excavated in the chalk helped keep the cost within economic limits whilst providing a low risk solution in the difficult and restricted working conditions in the base of the car park.

References

1. WAKELING, T. R. M. A comparison of the results of standard site investigation methods against the results of a detailed geotechnical investigation in Middle Chalk at Munford, Norfolk. *Proc. Conf. on In situ Investigations in Soils and Rocks, BGS, London,* 1970, pp 17-22.
2. HUTCHINSON. J. N. Field and laboratory studies of a fall in Upper Chalk cliffs at Joss Bay, Isle of Thanet. *Proc. Roscoe Mem. Symp. on Stress-strain behaviour of soils. Foulis, Henley-on-Thames,* 1972, pp 692-706.
3. CLAYTON. C.R.I. The Mechanical Properties of Chalk. Keynote address. *Proc. Inter. Chalk Symp., Brighton,* 1989.

Acknowledgements

The Author would like to thank Guildford Borough Council and Cooper and Withycoombe, Consulting Engineers for their kind permission to use extracts from the Contract Drawings. The main Contractor was Lampard and Partners and the ground anchorage system was installed by Colcrete Ltd.

A cantilever retaining wall supported by a berm during temporary works activities

W. POWRIE, Queen Mary and Westfield College, UK,
J. N. DAVIES, Mott MacDonald (formerly with Tarmac Construction), UK, and A. M. BRITTO, Cambridge University Engineering Department, UK

The effectiveness of an earth berm in providing support to an embedded cantilever retaining wall during temporary works activities is investigated with reference to a particular case study. Finite element analyses were carried out to assess possible movements of the wall during and for some months after excavation. The effect of the assumed lateral stresses acting on the wall prior to excavation was investigated. The results of the finite element analyses are compared with those of a hand calculation based on a modified limiting stress field and a conservative assumption concerning the effectiveness of the berm, and with the wall movements actually measured.

Introduction

The new A55 North Wales Coast Road passes beneath a causeway on the east side of the Conwy estuary in an underpass whose sides are retained by diaphragm walls. The causeway carries two twin-track railway lines in addition to the old A55 road. A plan view of the site is shown in Fig. 1 (a). Where the function of the diaphragm walls is earth retention alone, the panels are of conventional monolithic design and are installed to a depth of about 19m below ground level. The diaphragm walls supporting the causeway overbridge are formed of T-shaped panels, for additional structural stiffness.

In the permanent condition, the underpass structure is propped below carriageway level with reinforced concrete struts bearing onto cill beams cast against, and fixed into, the diaphragm walls. The specified construction sequence required the use of temporary props at two levels during excavation before placement of the permanent struts. In order to provide access between two parts of the site the contractor, Costain Tarmac Joint Venture, wished to excavate a corridor between the diaphragm walls along the line of the new A55 road. To enable the construction plant to pass beneath the soffits of the overbridge supported by the diaphragm walls, it was necessary to excavate

Retaining structures. Thomas Telford, London, 1993

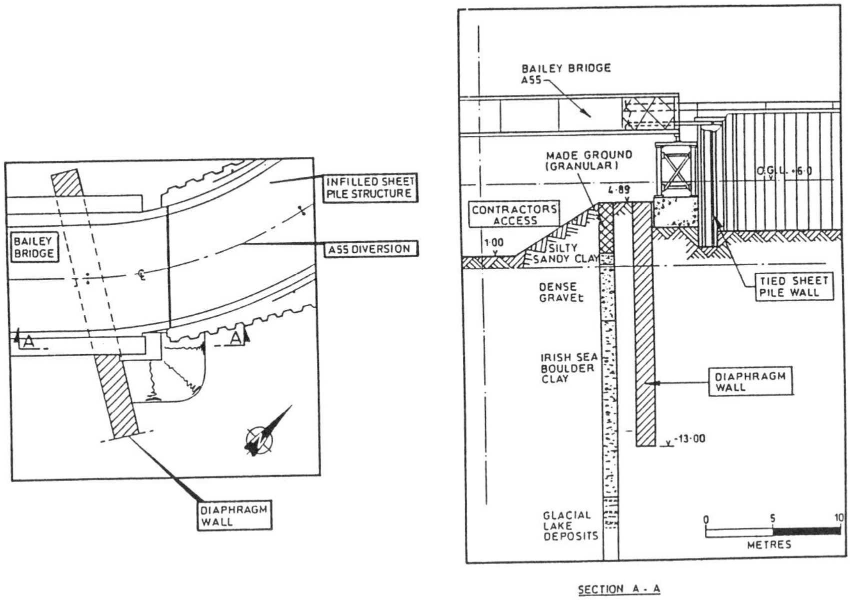

a) plan b) cross-section

Fig. 1. Plan view of site and design cross-section

locally below the level originally specified for the first row of temporary props. To overcome the need for temporary support, the contractor proposed to minimize wall deflexions and structural stress resultants by leaving wedges or berms of soil in place against the walls. The purpose of the analysis described in this Paper was to investigate the suitability of the proposed new excavation procedure.

Geometry

The analysis was carried out at the most critical cross section, adjacent to the north east pier base of the diversion structure provided for the original A55 road. At this location, imposed loadings from the pad foundation of the Bailey bridge supporting trestles and the retained fill of the approach embankment were closest to the rear of the western diaphragm wall. For the purpose of analyses, the foundation and fill loadings were modelled as surcharges acting at ground level, +4.9m AOD. The design section is shown in Figure lb, and the idealized model in Fig. 2.

Ground conditions

The ground conditions associated with the Conwy estuary form a com-

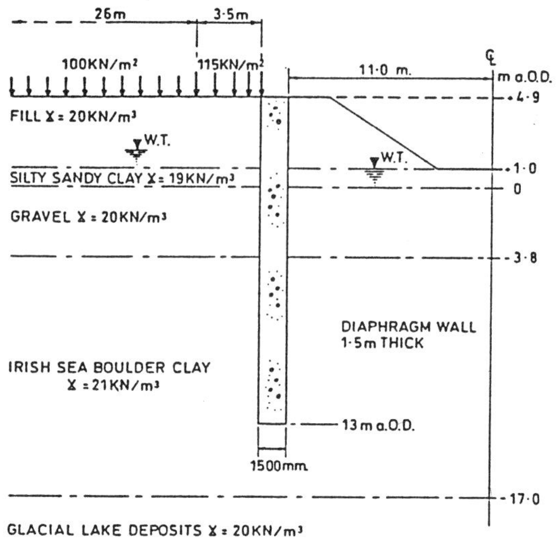

Fig. 2. Idealized geometry and soil profile

plex, interstratified sequence of sediments resulting from the direct influence of successive glaciations. These are overlain by Recent alluvial deposits which also exhibit considerable variation in soil type. Around the periphery of the estuary and beneath the causeway area, soft marine clays and silts have been displaced by Made Ground, which is generally granular in nature.

A typical succession of the soil strata in the area of the causeway underpass is fresh to slightly weathered rockhead overlain by stiff/dense, grey, gravelly North Wales Lodgement Till which is in turn overlain by Glacial Lake Deposits. This last stratum, locally in excess of 20m thick, is a laminated sequence of silts and clays with partings and thicker horizons of fine to medium sand. It generally exhibits a well defined varved fabric. The Glacial Lake Deposits are overlain by stiff overconsolidated brown Irish Sea Boulder Clay. These glacial soils are covered by a deposit of poorly sorted granular material, the origin of which is uncertain. Above the gravels, variable but generally granular Made Ground was encountered with an impersistent layer of silty, sandy clay at the base. This lower cohesive material is considered to represent a remnant of the estuarine deposits removed when the original causeway was constructed by Thomas Telford in 1826. The ground conditions are summarized in Fig. 2.

In situ lateral stresses

The geological stress history of the Irish Sea Boulder Clay and the Glacial Lake Deposits was believed to consist (following deposition) of the removal

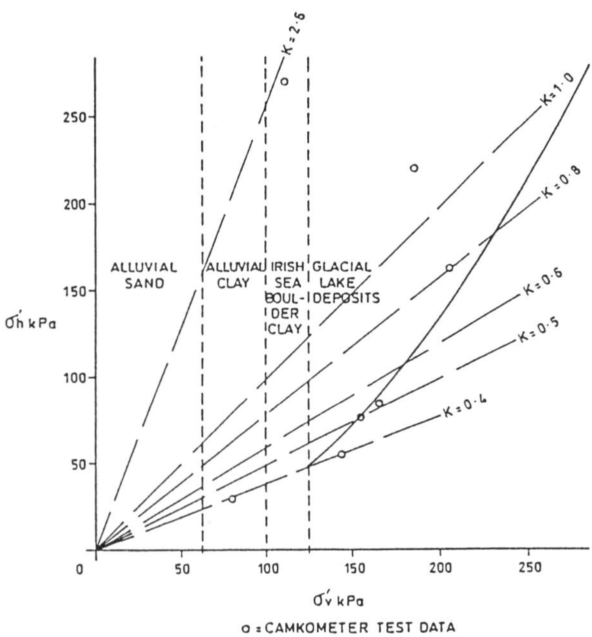

Fig. 3. In situ lateral effective stresses measured using a Camkometer

of an effective overburden of 500kPa. The overconsolidation ratio should therefore decrease with depth, and empirical expressions for the in situ lateral earth pressure coefficient K_O such as $K_O = (1-\sin\phi').OCR^{\sin\phi'}$ (equation 1: Mayne and Kulhawy, 1982) would suggest a corresponding decrease in K_O with depth. However, the in situ lateral effective stresses measured using a Camkometer indicate a more complex situation (Fig. 3). Although in the Irish Sea Boulder Clay the earth pressure coefficient is much as would be expected, there appears in the Glacial Lake Deposits to be an increase in K_O with depth from a value of approximately 0.4 at the top at the stratum. This may be due to the layered structure of the Glacial Lake Deposits, which results in significant cross-anisotropy. Variations to the general trend might arise from inhomogeneities within the stratum at certain horizons.

For the purpose of the analysis, the in situ earth pressure coefficients in the Irish Sea Boulder Clay were calculated on the basis of equation 1 with the removal of an overburden of 500kPa. In the Glacial Lake Deposits, a linear increase in K_O from 0.4 at the top of the stratum to 1.0 at the base was assumed. In the backfill K_O was taken as 0.5, which is approximately equal to $(1-\sin\phi'_c)$ for the granular material. The sensitivity of the analysis to the pre-excavation lateral stresses in the upper strata was, however, investigated.

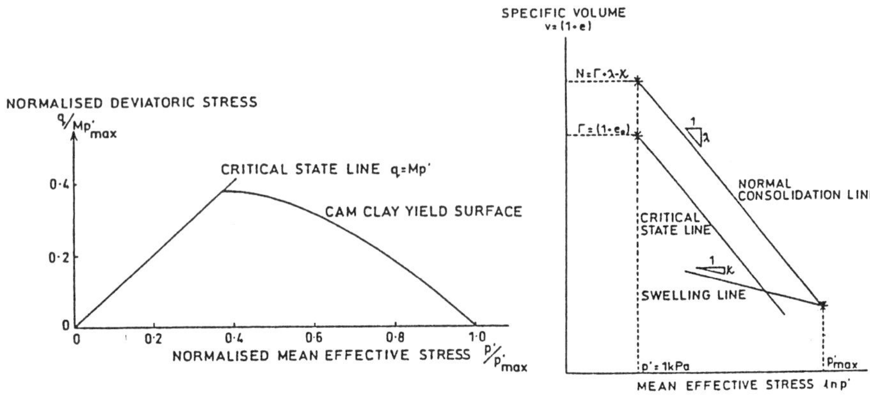

a) Yield surface in q:p' space

b) State paths in v:lnp' space

Fig. 4. Soil model used to represent Irish Sea Boulder Clay and Glacial Lake Deposits

Soil properties

Stress-strain data were not available for the granular fill, nor for the gravel. These strata were therefore modelled simply as elastic-perfectly plastic with a Mohr-Coulomb failure envelope having a critical state angle of shearing $\phi'_c = 32°$. Values of Poisson's ratio $\nu'=0.2$ and normalized shear modulus $G/p'=100$ and 200 were taken as typical of medium-dense granular materials over a shear strain increment of approximately 0.2% (eg Bellotti *et al*, 1989). Representative Young's moduli based on the initial average effective stress p' at the centre of each stratum were 6.25MPa and 15.5MPa for $G/p'=100$, and 12.5MPa and 31MPa for $G/p'=200$. The unit weight γ of both materials was taken as $20kN/m^3$.

The Irish Sea Boulder Clay and the Glacial Lake Deposits were modelled using Cam Clay, with the shape of the yield surface altered to prevent the computation of unrealistically high stress ratios on the dry side of the critical state (Fig. 4 (a)).

The parameters used in the analysis are given in Table 1, and were determined on the basis of laboratory and in situ test data as follows: λ, κ

Table 1. Values of parameters used for the Irish Sea Boulder Clay (ISBC) and Glacial Lake Deposits (GLD).

	λ	κ	e_o	ϕ'_c	G MPa	γ kN/m^3	k_h m/s	k_v m/s
ISBC	.028	.0065	.60	31°	30	21	$7x10^{-11}$	$7x10^{-11}$
GLD	.040	.010	.857	30°	35	20	$1x10^{-7}$	$5x10^{-10}$

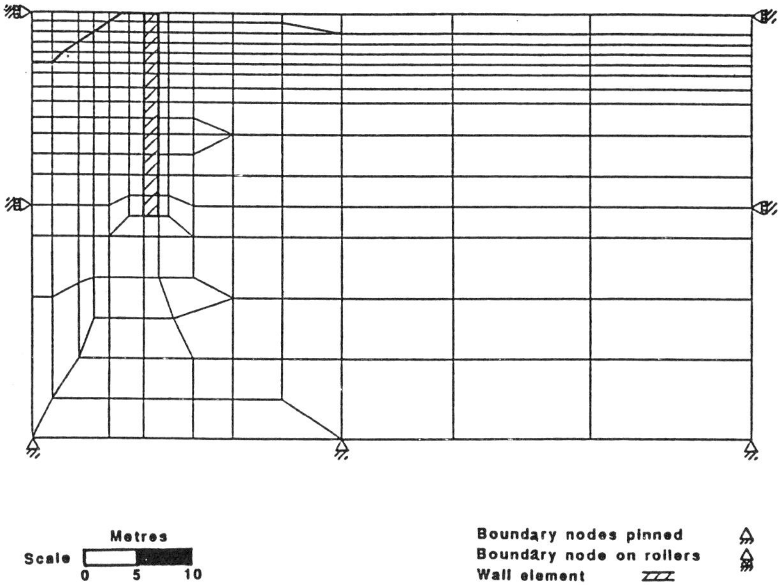

Fig. 5. Finite mesh element

and e_o (defined in Fig. 4 (b)) in one-dimensional compression and swelling from oedometer tests; ϕ'_c in triaxial compression; shear moduli G using a Camkometer; and the vertical permeabilities k_v from isotropic and one dimensional consolidation tests in the triaxial cell and oedometer. The horizontal permeability k_h of the Glacial Lake Deposits was estimated from the results of in situ pumping tests, but for the Irish Sea Boulder Clay it was assumed that $k_h = k_v$.

The back-filled clay was assumed to have the same properties as the Irish Sea Boulder Clay, except that the shear modulus was reduced to 5MPa and the unit weight to 19kN/m^3.

Finite element analysis

The idealized geometry and soil profile were shown in Fig. 2. The Bailey bridge foundation was modelled as a surcharge of 115kPa acting at the retained soil surface, and the 5m high approach embankment as a surcharge of 100kPa. The finite element mesh is shown in Fig. 5. The analysis was carried out using the program CRISP (Britto and Gunn, 1987).

The initial pore water pressures were assumed to be in hydrostatic equilibrium with a water table at +2mAOD. After excavation, groundwater levels were assumed to be at +2mAOD behind the wall and at +1mAOD (ie at formation level) in front. It was further assumed that in the steady state

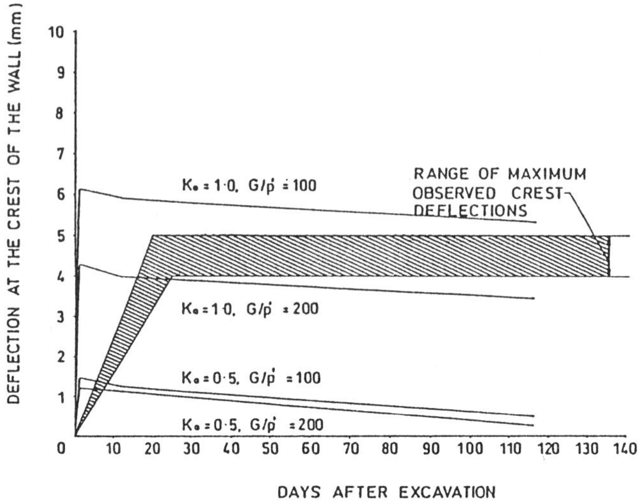

Fig. 6: Observed and computed crest deflexions as functions of time

flownet which would eventually develop, all excess head loss would occur in the Irish Sea Boulder Clay, with pore water pressures in the overlying strata in hydrostatic equilibrium below the appropriate groundwater level. This is reasonable in view of the high permeability of the granular materials, and the relatively small thickness of the layer of backfilled clay.

The 1.5m thick reinforced concrete diaphragm wall was modelled as an elastic material with Young's modulus 15.25×10^6 kPa and Poisson's ratio 0.2. In reality the wall was thicker over the top 2.89m, and the Young's modulus was increased to 175.84×10^6kPa in this region in order to achieve the correct flexural rigidity.

The starting point for the analysis was with the diaphragm wall in place and a ground level of +4.9mAOD. The temporary works activities were modelled using the following procedure

- Application of surcharges behind the wall to represent the Bailey bridge foundation and the approach embankment.
- Three months' dissipation of excess pore water pressures in the Irish Sea Boulder Clay and the Glacial Lake Deposits.
- The instantaneous removal of elements to simulate excavation in front of the berm to a level of +1mAOD.
- Ten months' dissipation of excess pore water pressures.

Throughout the analysis, excess pore water pressures in the fill and the dense gravel were assumed to dissipate immediately. In the Irish Sea Boulder Clay and the Glacial Lake Deposits, excess pore water pressure dissipation was coupled to the rate of consolidation of the soil.

Further analyses were carried out, in which the pre-excavation earth pressure coefficients above the toe of the wall were set to unity in an attempt to model the possible effect of diaphragm wall installation. The computed deflexions of the crest of the wall for all four analyses are compared in Fig. 6 with the maximum wall movements actually measured (using accurate surveying techniques) during the temporary works period.

Approximate analysis

In the approximate analysis (following Bolton *et al*, 1990), the mobilized soil strength ϕ'_{mob} required to maintain the wall in equilibrium was calculated on the basis of the idealized long-term pore water pressure distribution used in the finite element analysis and the assumed distributions of horizontal effective stress shown in Fig. 7. Assumptions had also to be made concerning the lateral pressure exerted by the berm in front of the wall above formation level, and the effectiveness of the berm as a surcharge: these have an important influence on the calculated mobilized soil strengths. In the current analysis, it was assumed conservatively that fully-active conditions would develop in the berm, with the active earth pressure coefficient K_a based on the critical state soil strength and zero wall friction, $K_a=(1-\sin\phi'_c)/(1+\sin\phi'_c)$. The effective surcharge was estimated as 35kPa, which is equal to the total weight of the berm divided by the half-width of the excavation. The stress-strain relationship was based on a consolidated-undrained triaxial compression test on a sample of Irish Sea Boulder Clay, starting from an effective cell pressure of 150kPa and an earth pressure coefficient of unity (Fig. 8).

The result of the approximate analysis is summarized in Table 2. The earth pressure coefficients were taken from Caquot and Kerisel (1948), assuming full wall friction ($\delta=\phi'_{mob}$ on both sides of the wall.

Discussion

It may be seen from Fig. 6 that the results of the finite element analysis are sensitive to the stiffness of the granular materials, and more especially to the pre-excavation lateral earth pressures. However the analyses with the pre-excavation lateral stresses set to unity in the soil above the toe of the wall seem to bracket the measured displacements. The discrepancy between the

Table 2. Result of approximate analysis

K_1	K_2	ϕ'_{mob}	shear strain γ	crest deflexion
0.502	2.17	15.5^o	0.4%	29mm

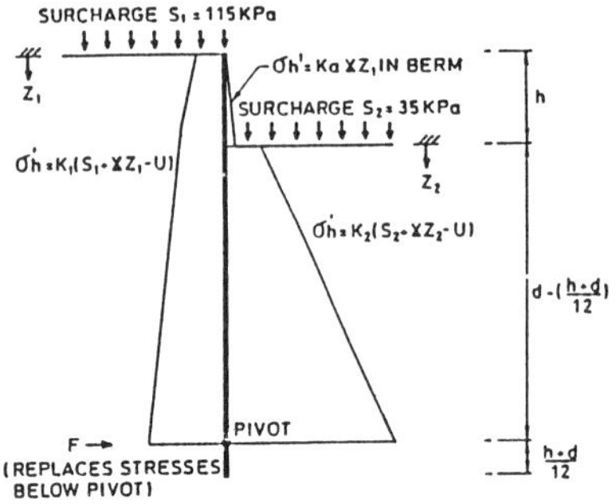

Fig. 7. Lateral effective stress distribution, approximate analysis

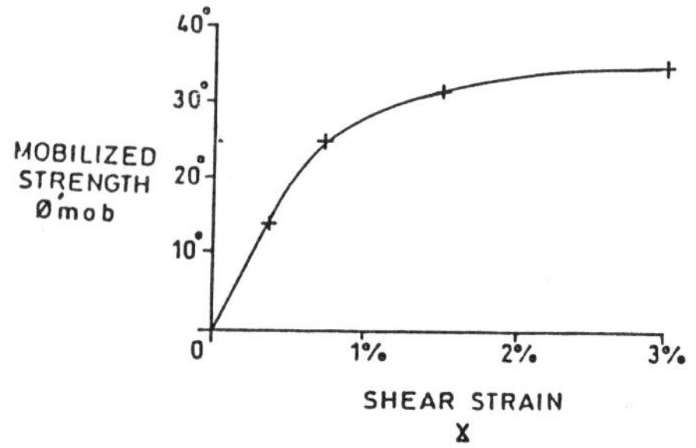

Fig. 8. Relationship between shear strain and mobilized soil strength

initial rates of movement probably arises because no attempt was made to model the actual timescale of excavation in the finite element analysis. All of the finite element analyses indicate a gradual movement of the wall back into the retained soil during the period following excavation. This is probably due to the continuing dissipation of excess pore water pressures in the Irish Sea Boulder Clay and the Glacial Lake Deposits following the application of the initial surcharge loads and excavation in front of the wall, together with the assumption of reversibly elastic soil behaviour.

The displacement of the wall was significantly over-predicted by the approximate analysis, due primarily to the use of a single stress-strain relationship for the Irish Sea Boulder Clay and the fill materials, which did not follow the correct stress or strain path for the soil on either side of the wall. Since the stress-strain relationship started from a mobilized strength of zero (ie K_O=1), the approximate analysis may reasonably be compared with the second pair of finite element analyses. The difference between the calculated displacements is due partly to the discrepancy between the shear moduli used in the two analyses. In the approximate analysis, an increase in shear modulus with depth is implicit in the use of a unique relationship between the mobilized soil strength and the shear strain. At a shear strain of 0.4%, the secant shear modulus according to Fig. 8, based on the pre-excavation average effective stress at a level of -8.4mAOD (ie at the centre of the part of the Irish Sea Boulder Clay above the toe of the wall), is approximately 11.5MPa: this may be compared with the shear modulus of 30MPa used in the finite element analysis. Furthermore, the effect of the berm was deliberately modelled conservatively: this would also tend to increase the calculated wall displacement.

The analyses described in this Paper were carried out after initial calculations based on CP2 (1951) had been completed. The initial calculations had indicated bending moments well within the structural capacity of the wall, but offered no guidance on the likely magnitude of wall movement. A more detailed treatment of the soil/structure interaction is often beyond the scope of much temporary works design, but was carried out in this case to support a revised construction technique which led to a considerable saving in cost.

Conclusions

The deflexion of the wall during temporary works activities was satisfactorily limited by the use of soil berms. The measured wall movements fell generally within the range indicated by finite element analyses in which it had been attempted to model the effects of wall installation. The results of the finite element analyses were dependent to some extent on the effective stiffness of the granular strata. They would undoubtedly also be sensitive to the other input parameters, whose values it was in this case possible to estimate with more confidence. The approximate hand calculation over-predicted the deflexion of the wall: this was due partly to the use of inappropriate stress-strain data, and partly to the deliberately conservative representation of the effect of the berm. The influence of the soil stiffness and of the pre-excavation lateral stresses on the results of either analysis cannot be over-stated.

The implications of the construction sequence in terms of programme and hence cost can be significant. Although unusual in traditional temporary works design, analyses of the type described in this Paper, together with appropriate performance monitoring, can be extremely cost effective as the basis of a "value engineering" approach to the optimization of construction methodology. It is vital, however, that the parameters used in analysis are determined with considerable care.

Acknowledgement

The Authors are grateful to Costain Tarmac Joint Venture for permission to publish this Paper.

References

BELLOTTI R. *et al.* Interpretation of moduli from self boring pressuremeter tests in sands. *Geotechnique,* 1989, **3**, No.2,pp 269-292.

BOLTON M.D., POWRIE W. and SYMONS I.F. The design of stiff in situ walls retaining overconsolidated clay: part II. long term behaviour. *Ground Engineering*, 1990, **23**, No.2,pp 22-28.

BRITTO A.M. and GUNN M.J. *Critical state soil mechanics via finite elements.* Ellis Horwood, Chichester, 1987.

CAQUOT A. and KERISEL J. *Tables for the calculation of passive pressure etc.* Gauthier-Villars, Paris, 1948.

INSTITUTION OF STRUCTURAL ENGINEERS. *Civil Engineering Code of*

Practice No. 2 : Earth Retaining Structures. Institution of Structural Engineers, London, 1951.

MAYNE P.W. and KULHAWY F.H. K2 -OCR relationships in soil. J. *Geot. Eng. Div. ASCE*, 1982, **108**, GT6 ,pp 851-872.

Design and construction of temporary ground anchorages at Castle Mall Development, Norwich

A.D. BARLEY, Keller Colcrete Ltd, R. EVE, Bovis Construction Ltd, and D. TWINE, Ove Arup and Partners, UK

Introduction

The Castle Mall Development in the centre of Norwich involves the construction of two storeys of shopping and a basement up to 18m below ground level, alongside a five storey underground car park. The majority of the ground surface of the site will be reinstated as parkland. A plan of the 2.5 hectare site is shown in Fig. 1.

Ove Arup and Partners' permanent works design consisted of a 900mm diameter contiguous bored pile perimeter retaining wall propped by the floor slabs. Management contractor, Bovis Construction Ltd, were responsible for the temporary support of the perimeter wall during construction and adopted a scheme requiring 1120 ground anchorages. This Paper describes the design and construction of these ground anchorages. A companion paper in this conference by Grose and Toone (Ref. 1) explains why ground anchoring was chosen and describes the overall design of the wall.

The site

The site topography varies widely from 17 to 28 mOD resulting in retained heights of 7 to 18m. A five phase site investigation comprised three inspection shafts up to 28m deep and 65 shell and auger boreholes up to 50m deep at the locations shown in Fig. 1. The general findings of these investigations have been presented previously (refs 2 to 4) and a summary is given below.

Description	Level (mOD)	Depth (m)	Thickness (m)
Made ground	17 – 29	0	0.1 – 11
Silty sands and gravels with lenses and pockets of silty clay (Norwich Crag)	14 to 26	0 to 5 (av. 3m)	0 to 10

Very weak to moderately weak partly weathered to unweathered white chalk with flint bands(upper chalk).	12 to 18	3 to 15 (av. 8m)	proved 45m

The groundwater table was about 1mOD, some 8 m below basement level. As part of the site investigation 24 boreholes were put down outside the site to establish the stratigraphy and confirm the crag and chalk were similar to

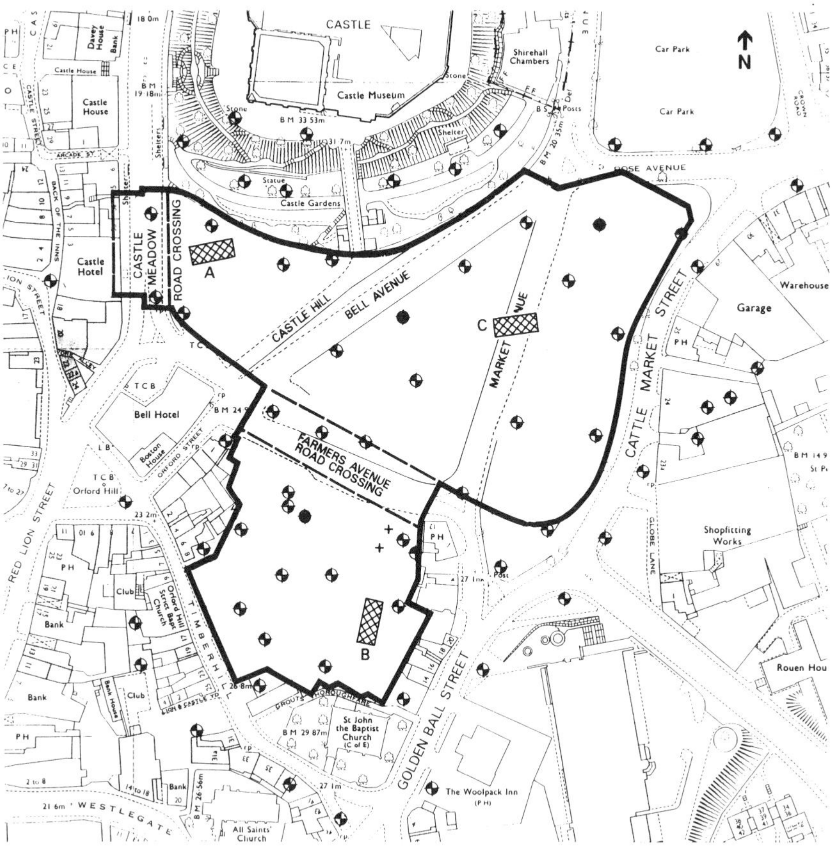

Reproduced from the 1974 Ordnance Survey 1:1250 Map with permission of the Controller of HMSO, Crown Copyright Reserved

LEGEND

- Perimeter Bored Pile Wall
- Borehole
- Trial Shafts
- + Instrumented Trial Pile
- Proving Test Anchorages

Fig. 1. Site plan

the materials investigated in more detail on the site. This investigation cost £30,000 and was considered cost-effective to minimise the risk of serious delays to the £3.5m anchoring contract caused by unexpected ground conditions in the fixed anchor zone.

The presence of solution features within chalk is a well-documented phenomenon and they were anticipated on this site. Not surprisingly, drilling and sampling anomalies attributed to solution features were recorded to depths of 30, (-3 mOD) in eight of the 65 boreholes. These sub-vertical, steep sided features tended to be loosely infilled, with possible voids, and the anchorage design had to overcome the problems their occurrence presented. Furthermore their location could not be predicted so they would only be discovered during anchorage construction.

Ground anchorage design

The responsibilities for the design of the anchorages were divided between the Designer and the Contractor as recommended in Table 1 BS 8081: Ground Anchorages (ref 5.) In this case, Bovis were the Designer and Keller Concrete Ltd were the contractor with responsibility for the design of the fixed anchor length. This reflects that specialist anchorage contractors

(*a*) are likely to have most experience in design of the fixed anchor length for his anchorage system
(*b*) offer different forms of anchorage construction, making it difficult for Designers to produce designs which can be used in competitive tendering.

Therefore, it was decided to tender the contract for the 1120 anchorages with the requirement that the contractor design the items given in Table 1 BS 8081 except anchorage spacing, orientation and loads and the anchorage monitoring system. The winning tenderer's anchorage design would be verified by on-site proving tests in advance of the contract anchorages. If these anchorages failed the contract would be terminated and the exercise repeated with another tenderer. Nonetheless, this approach required a post-tender assessment of the likely success of each tenderer's design, since failure of the proving tests would have incurred undesirable delays to the construction programme.

Double corrosion protection was specified and factors of safety of 1.6 and 2.5 on grout/tendon and ground/grout interfaces, respectively, were adopted in accordance with BS 8081.

Anchorage system and fixed length design

Analysis of the overall stability of the anchored wall (ref. 1) determined that a limited number of fixed anchor lengths would be located in the

Norwich Crag but that the majority would be founded within the underlying chalk stratum. BS 8081 provides clear guidelines for estimating the capacity of anchorages founded in granular materials, such as Norwich Crag. However, such guidelines did not exist with regard to anchorages founded in weak to very weak chalk. Littlejohn and Bruce (ref. 6) only recommended working and ultimate bond stresses at the grout to ground interface for chalks of Grades I to III. Littlejohn (ref. 7) had, within this range, recommended a relationship of ultimate bond stress equivalent to 10N in kN/m2 where N is the Standard Penetration Test value in blows per 0.3 m penetration.

Barley (ref. 8) had researched some 2000 anchorages founded in chalks ranging from strong to weak covering Grades I to V. He recommended that when using grout injection techniques ultimate bond stresses may range from 320 to 1100kN/m^2 and generally giving a relationship to STP N values of 20N to 30kN/m^2 (see Fig. 2). The N values measured in the fixed anchor zone ranged from 5 to 15 which indicated likely ultimate bond stresses of between 100 and 450kN/m^2. However, the most relevant data available tender stage was Keller Colcrete's report on four test anchorages (ref. 9) installed approximately 1.5km east of the site. They were founded in similar weak to very weak chalk (N ranging 14 - 20) and exhibited ultimate bond stresses of 270 and 290kN/m^2 . The other two anchorages recorded bond stresses of 340kN/m^2 without failure. In terms of SPT values, these approximate to 16N, 17N and 20N, respectively (considering average values of the range of N).

In order to attain the required maximum working load of 500kN utilising the maximum fixed length of 10m recommended in BS 8081, then an ultimate bond stress of 349kN/m^2 would be required in a 114 mm diameter borehole. Such a value was in excess of that attained in previous test anchorages in slightly better chalk and did not allow for any reduction due to the effect of progressive failure mechanism in such a long fixed length. Furthermore, there was the additional problem of the random location of solution features in the chalk.

The specification called for any anchorages which encountered such features within the fixed length to be extended beyond them. This requirement, together with the bond stress requirement, lead the contractor to propose the use of the novel "Single Bore Multiple Anchorage System (SBMA)", which is fully compliant with BS 8081. This system comprises a single borehole of 100 to 200mm diameter installed using normal anchor drilling techniques; generally cased or partially cased holes with a water flush system. Conventional permanent anchorage systems involve the installation of a double corrosion protected anchor tendon using a single encapsulation to transfer the load from the anchor tendon to the grout and then to the ground (Fig. 11 (a), BS 8081). The novel aspect of the SBMA system is that it involves the installation of a multiple of unit anchorages down the one borehole (Fig. 11(d), BS 8081). The unit anchorage encapsulations are located at staggered depths in the borehole and each transfers its load component in a controlled manner to a discrete length of the borehole. This allows a more efficient load transfer than a normal anchorage. It also allows the utilisation of an unlimited length of bore to which the load may be efficiently transferred. This length is limited in BS 8081 to only 10m in total with a normal system since there is little increase in anchor capacity above this length.

The SBMA system had already been well proven in a number of site conditions and demonstrated that the load transfer mechanism, and the efficient utilisation of long fixed lengths, achieved capacity enhancements ranging from 50% to 150% over and above normal anchorages in the same conditions. Furthermore, the system can provide a partial anchor load capacity should any of the unit anchors fail during stressing when the location of the low capacity soil would be identified. The load capacity could be downrated to that of the sound unit anchors. This had been satisfactorily demonstrated in other contracts and could clearly provide a benefit in reducing the failure of anchorages due to solution features. In the event of this occurrence with a normal anchorage, then total replacement would be necessary.

The capacity per metre fixed length for the anchorages in Crag was estimated as 150kN/m (419kN/m^2 bond stress) from recommended formulae and parameters in BS 8081. This allowed utilisation of unit anchor working loads of up to 175kN with unit fixed lengths of 3 m. Anchorages were made up from single, double or triple unit anchorages with maximum working loads of 175, 350 and 525kN respectively with total fixed lengths of 3, 6, or 9 m. The capacity per metre of fixed length in chalk of 110kN/m (310kN/m^2 bond stress) was taken directly from the trial anchorage results at the nearby Norwich site. This design concept of capacity per metre length was acceptable since drilling method, drill casing size, grout method etc proposed at Castle Mall were identical to those used at the earlier site and the SBMA load transfer mechanism can, providing unit anchor fixed lengths are not excessive (greater than say 5 m), provide linearity between load capacity and fixed length. In the chalk, unit working loads and number of unit anchors were identical to those in the Crag but longer unit fixed lengths of 4 m were utilised.

Proving tests on anchorages

Proving tests were specified to demonstrate, in advance of the contract anchorages, the quality and adequacy of the design in relation to the ground conditions, materials and workmanship. Eighteen tests were carried out at three locations chosen to reflect possible variations in ground conditions (see Fig. 1). Details of the anchorages are given in Table 1. The contractor intended using 15.2 mm diameter Dyform strand for the contract anchorages. How-

Table 1. Details of proving test anchorages

LOCATION	ANCHOR REFERENCE	GROUND CONDITIONS	FIXED LENGTH (m)	MAXIMUM TEST LOAD (Tw = working load)	TYPE OF ANCHORAGE
A	3M7	Crag	2	-	Single unit SBMA, with single corrosion protection
	1M7	Crag	10	2.4Tw	Four unit SBMA, with single corrosion protection
	2M7	Crag	10	2.4Tw	Normal anchorage (4.5m encapsulation) with single corrosion protection
	6M7	Chalk	2	-	Single unit SBMA with single corrosion protection
	7M7	Chalk	12	1.8Tw	Three unit SBMA with double corrosion protection (Production anchorage)
	4M7	Chalk	12	2.4Tw	Four unit SBMA, with single corrosion protection
	5M7	Chalk	-	-	Tendon dropped 3-4m during installation and was lost. New hole drilled and pregrouted (17 tonnes). Hole abandoned after redrilling
B	3T1	Crag	2	-	Single unit SBMA with single corrosion protection
	7T1	Crag	9	1.8Tw	Three unit SBMA with double corrosion protection (Production anchorage)
	1T1	Crag	10	2.4Tw	Four unit SBMA with single corrosion protection
	2T1	Crag	10	2.4Tw	Normal anchorage with single corrosion protection
	6T1	Chalk	2	-	Single unit SBMA with single corrosion protection
	4T1	Chalk	12	2.4Tw	Four unit SBMA with single corrosion protection
	5T1	Chalk	12	2.4Tw	Normal anchorage (4.5m encapsulation) with single corrosion protection
C	1M4	Chalk	2	-	Single unit SBMA with single corrosion protection
	4M4	Chalk	12	1.8Tw	Three unit SBMA with double corrosion protection (Production anchorage)
	2M4	Chalk	12	2.4Tw	Four unit SBMA with single corrosion protection
	3M4	Chalk	12	2.4Tw	Normal anchorage (4.5m encapsulation) with single corrosion protection

ever, this strand gave insufficient load capacity unit for proof testing and 18 mm strand was used for all test anchorages. The unit anchorage encapsulation was 2.5 m long except for the 2 m fixed length cases. Even with the 18mm strand the maximum test load for a three unit anchorage was limited by strand capacity to 1.8 times the working load (Tw). Therefore, to proof load the anchorages to a higher factor of safety (2.4Tw) four unit anchorages were also adopted.

To confirm the feasibility and workmanship of production anchorages, as well as design capacity, some three unit anchorages were also installed. These anchorages were identical to the proposed contract anchorages except they used 18mm diameter strand, instead of 15.2 mm. Normal anchorages were included in the test programme for both comparative purposes and as a contingency should the SBMA system not have proved satisfactory.

Review of test anchorages

Seventeen of the eighteen test anchorages were successfully installed using the proposed construction techniques which in some cases involved pre-grouting and re-drilling techniques to overcome the fissure/solution feature problems in the chalk. One tendon penetrated a cavity below the drilled length and was beyond reach to allow testing.

Of the 11 boreholes in chalk, 6 required pre-grouting and sand/cement grout take varied from 2 to 29 tonnes. Grout "take" during anchor pressure grouting at 3 to 6 bar generally ranged from 3 to 7 times the borehole volume. Such takes were not considered excessive since it was necessary to enhance the capacity of the chalk local to the fixed anchor. Of the boreholes in the Crag, none involved pre-grouting and grout take during pressure grouting at up to 10 bar ranged from 3 to 8 times the borehole volumes. These takes were on the high side, but it was considered likely that takes would reduce during production anchorage works as the ground tightened up during anchor grouting.

The short test anchorages in Crag attained approximately 300kN/m, equivalent to 840kN/m^2 bond stress, and the full scale anchorages which could not be failed confirmed that the design proposals were more than satisfactory. The short test anchorages in chalk attained approximately 300kN/m length of bore, equivalent to 840kN/m^2 bond stress, while the full scale anchorages carried 100kN/m length with no indication of failure.

The three and four unit SBMA anchorages were installed and tested satisfactorily in both chalk and Crag. They proved factors of safety in excess of 2.4 on a 525kN working load. None recorded load characteristics to indicate failure was imminent even at the maximum test capacity of the tendons, although unit anchorages installed in pre-grouted zones exhibited greater permanent deflections. Creep characteristics and elastic behaviour of all SBMA anchorages complied with requirements of BS 8081. The four normal type anchorages also performed satisfactorily with regard to elastic behaviour and creep characteristics, proving similar capacities to SBMA's and without failure.

Two anchor bores in Crag with lengths of 15 and 24 m were investigated for borehole alignment. Maximum deviations were 0.13° in the vertical plane and 0.30° in the horizontal plane. Four anchor bores were investigated in the chalk to lengths of 24 and 30 m. Maximum deviation was 1.64° in the vertical plane and 0.78° in the horizontal plane.

In all, the tests fully confirmed both the adequacy of the anchorage design proposals of the SBMA systems in both chalk and Crag, and, together with laboratory test data, the quality of the materials and workmanship. In addition, the problems of solution features were highlighted (e.g anchorage 5M7), particularly regarding loss of drill flush and rapid advance of the drill rods, and the need for pre-grouting was confirmed. All the anchorages installed in pregrouted zones performed satisfactorily albeit with larger permanent deflections. This indicated that where it was impossible to extend anchorages beyond a solution feature it was likely that acceptable anchorages could be installed in pregrouted feature material.

Contract anchorages

The installation of contract anchorages commenced July 1990 and continued with the use of four Hands England drilling rigs until May 1991 followed by a gradual reduction until completion in July 1991. In total, 1120 anchorages were installed which were made up of 2832 "unit" anchorages (one, two or three "unit" SBMA anchorages per main anchor). Of the total number of holes, 101 required a single pre-grouting and re-drilling operation whilst 23 required up to three pre-grouts. Pre-grouting utilised at 3.1 sand cement mix and quantities varies from 0.15 to 12 tonnes of cement per hole in the chalk to 0.15 to 10 tonnes of cement per hole in the Crag.

The large grout takes in both Crag and chalk were generally associated with solution features and were anticipated from the results of the proving tests. The presence of very loose Crag, highly fissured chalk and voids were identified during drilling by monitoring the advance of the casing or drill rods and by partial or total loss of water flush. It was imperative to identify such critical areas during drilling since a large ingress of water flush into a solution feature area could trigger collapse of the infill material and overlying ground. This could have resulted in extensive subsidence of adjacent roads, services and buildings and all reasonable steps had to be taken to prevent such an occurrence. With this aim in mind a detailed flowchart was developed to control the drilling and grouting of all anchorages. This construction procedure included the use of pregrouting techniques, as necessary, as preliminary ground stabilisation to permit anchor construction beyond the feature, or in exceptional circumstances within it.

The grouting technique during anchor construction consisted of end of casing injection grouting operations using either a neat cement or 1:1 sand-cement mix. In all anchors founded in a grout-tight stratum, the initial penetration of neat cement grout into the coarse granular material or into the chalk fissures or alternatively the compaction of strata adjacent to the anchor bore, during grout pressure application is beneficial to the anchor capacity. During construction of anchors founded in a stratum which freely allowed neat grout penetration at nominal pressure, the grout mix was changed to a sanded grout mix in order to reduce flow by blocking off fissure entry after nominal penetration. This procedure optimised grout penetration or compaction required to enhance anchor capacity with high grout takes associated with unnecessary bulk grouting of the surrounding strata. Total grout placed associated with pre-grouting works amounted to 1057 tonnes, whilst sand-cement mix and neat cement associated with anchor installation amounted to 780 and 1783 tonnes, respectively.

Approximately 4% of the total number of anchor bores were checked for deviation in alignment using either a fotobor or maxibor photo-checking system. The majority of bores recorded deviations not exceeding 1° (1:57) in the vertical plane and 1.5° (1:38) in the horizontal. These deviations did not exceed the tolerance requirement specified in the contract. Five holes that did exceed the tolerances exhibited maximum deviations of 2° (1:29) and 3° (1:19) in the two planes. After taking the maximum precautions during set up and then continuing the use of quasi-Duplex drilling techniques (advancement of drill rod and drill casing) with a "stiff" drill string, it is not possible to guarantee any more accuracy than that observed. Deviations generally resulted from non-homogeneity of drilling strata particularly the presence of flints and flint beds.

All double protected anchor tendons carried a dual layer of plastic protection throughout their length and fully complied with the requirements of BS 8081. All were prefabricated in the contractor's tendon manufacturing department at Wetherby, each carrying an identification tag in accordance with the in-house quality assurance system.

The stressing system utilised for SBMA anchors involves load increments applied by monostrand loading techniques due to the variation in elastic lengths between unit anchors. During the load test application the total anchor load was monitored by an annular load cell installed temporarily below the anchor head plate. Although some modification to the recommended load increments was required the system provided all data to satisfy the acceptance criteria requirement of BS 8081. A small number of unit anchors which did not comply with load extension requirements in the initial cycles, probably due to friction forces within the anchor free length, were recycled in accordance with BS 8081 requirements to confirm repeatability in behaviour.

The anchorages, where possible, were installed at levels such that their reinforced concrete walling constituted permanent floor slab support. In areas where solution features were encountered during bored piling operations additional levels of anchor support were necessary. This amounted to up to 7 rows of anchors supporting a retained height of up to 18m.

Exhumation of proving test anchorages

During site excavation works at Location A (Figure 1) it was possible to fully observe, monitor and record the exhumation of the fixed anchor lengths of three proving test anchorages in Crag. The brief for the investigation work involved

(*a*) the monitoring of the extent of grout penetration into the surrounding soil

(*b*) identifying any changes in condition of the soil surrounding the fixed anchor length

(*c*) the monitoring of the size and shape of the full grouted body of the anchor and comparison with the diameter of the original borehole
(*d*) assessing the integrity of the plastic sheathing and ducting which provided the corrosion protection layer.

To allow inspection of the fixed anchor length with minimum disturbance, a trench was excavated by a backacter machine down each side of the anchor. This then allowed local and controlled excavation around the anchor with pick-axe and spade and finally with a wire brush. On completion of shape identification and general observation, the centre grouted body and tendons of the fixed anchor were pulled from the soil by a tractor. This allowed further inspection and shape measurement of the grouted body and section cutting of the body to determine precise shape at critical sections.

The following conclusions were drawn from the exhumations and analysis

(*a*)There was no indication that the corrosion protection plastics in either normal or SBMA anchors were ineffective even after trial anchorage testing.
(*b*) The effect of drill flushing and pressure grouting was generally to replace all removed soil in and around the bore with bodies of pure grout. These grout bodies had only occasional uniform cylindrical shape smallest diameters 140 to 180mm) and were frequently grossly irregular in shape, varying in diameters from 135mm to as much as 750mm (equivalent to 1.2 to 6.6 times diameter for the drill casing)
(*c*) The effect of pressure grouting was to fill the drill casing, replace the removed soil and recompact the disturbed soil around the bore, but with little or no penetration of the grout into the granular material, of which the major constituent was fine sand. The presence of pebbles ranging from 10 to 70mm diameter was observed in the grout body above the original bore over a short fixed length, which may have resulted from wash out of fines and replacement by grout.

Although design formulae suggest the distribution of uniform bond along a cylindrical grout to ground interface, the gross irregularity of the shapes observed would suggest that pull out resistance of anchorages founded in such granular materials is grossly enhanced by the presence of an inordinate number of irregular areas loaded in end bearing mode.

Concluding remarks

The provision of temporary support to a 700lin.m long contiguous bored pile wall up to 18m high in ground containing numerous large solution features was overcome by using the 'single bore multiple anchorage system

(SBMA)'. A total of 1120 ground anchorages were installed of which 124 required up to three pre-grouting operations.

These ground anchorages were installed between July 1990 and July 1991. A monitoring programme for both the anchorages and the retaining wall has indicated that the anchorages are behaving satisfactorily to date.

Acknowledgements

Client: Estates and General plc and Friends Provident
Consulting Engineer: Ove Arup and Partners
Management Contractor: Bovis Construction Ltd
Ground Anchorage Contractor: Keller Colcrete Ltd

References

1. GROSE W.J. and TOONE B.H. The selection, design and performance of a multi-propped contiguous pile retaining wall in Norwich. *Proc.Int. Conf. on Retaining Structures*, 1992.
2. TWINE D. and WRIGHT R.H. Farmers Avenue Road Crossing, Castle Mall Development, Norwich. *Proc. 4th Int. Conf. on Piling and Deep Foundations. Balkema*, 1991
3. TWINE D. and GROSE W. J. Discussion: Session 1. *Proc. Int. Chalk Symposium, Brighton*, 1989.
4. TWINE D. and GROSE W. J. Discussion: Session III. *Proc. Int. Chalk Symposium, Brighton*, 1989.
5. *Code of Practice for Ground Anchorages BS 8081* - BSI Publication, 1989.
6. LITTLEJOHN G. S. and BRUCE D. A. *Rock Anchors - State-of-the-Art*. Foundation Publications Ltd, Brentwood, England, 1977
7. LITTLEJOHN G. S. Soil Anchors. *Proc. Conf. on Ground Engineering, ICE, London*, 1970, pp 33 - 44.
8. BARLEY A. D. Ten Thousand Anchorages in Rock. *Ground Engineering, September/ October/ November*, 1988.
9. BARLEY A. D. *Test Anchorages in Chalk at Colmans, Norwich*. Confidential Report, 1984.

The use of soil berms for temporary support of retaining walls

D. M. POTTS, T. I. ADDENBROOKE and R. A. DAY, Imperial College, UK

In this Paper non-linear finite element analyses are used to investigate the effectiveness of soil berms when used as temporary support to embedded retaining walls. The results of the analyses show that provided the berm exceeds a minimum height the volume, and therefore weight, of the berm is more important than its geometry. To provide a frame of reference two 'control' analyses, one with support at the top of the wall prior to excavation and the other with no support above excavation level, have been performed. By comparing the results from the berm analyses with those from the 'control' cases the effectiveness of the berms is quantified. The results provide some simple guidelines for use in practice. Analyses in which the berm is simulated by an equivalent surcharge load show significant differences from analyses in which the berm is modelled. It is concluded that such surcharges do not accurately represent the effects of a berm.

Introduction

A common construction technique for building deep basements in urban areas is the so called 'top down' method. In this approach the required foundations such as piles or caissons are installed from the original ground surface. Diaphragm or secant pile walls are then constructed to form the perimeter walls of the basement. A floor slab is cast near to the original ground surface (with minimal excavation) and then soil is excavated from beneath. When the excavation is sufficiently deep the next floor slab is placed and the process repeated until the required number of basement levels are completed.

This method has the advantage that support to the basement walls is provided at an early stage of construction thus minimising disturbance to any adjacent buildings or services. In addition it is possible to construct the superstructure of any building above ground level simultaneously with basement excavation therefore reducing the construction period. The disadvantage of the approach is that excavation has to be performed under severe restrictions and is an expensive operation.

In order to reduce costs some Contractors prefer to excavate to the first basement level below ground without support from a floor slab. To prevent

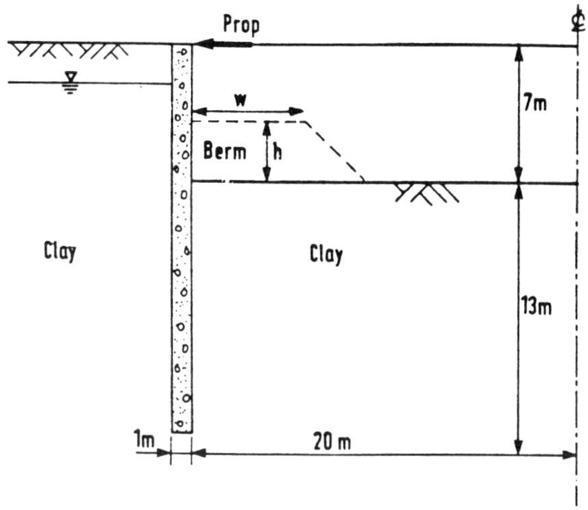

PROPERTIES

CONCRETE:		
Youngs modulus	:	$28*10^6$ kN/m^2
Poisson's ratio	:	0.15
CLAY:		
Bulk unit weight	:	19 kN/m^3
Youngs modulus, E'	:	6000z kN/m^2
Poisson's ratio	:	0.2
Angle of shearing resistance	:	25°
Cohesion	:	10 kN/m^2
Dilation angle	:	12.5°

where z is depth below original ground surface

Fig. 1. Geometry and material properties

excessive wall and soil movements soil berms may be left immediately adjacent to the basement walls. Unfortunately no simple design guidelines for such berms are available.

In this paper nonlinear finite element analyses are used to investigate the behaviour of soil berms. A basement construction situation typical of that found in the London area has been studied. Only the first excavation stage (i.e. first basement level) has been considered. Two 'control' analyses, one with support at the top of the wall prior to excavation (i.e. top down situation) and the other with no support to the wall (i.e. cantilever situation) have been performed to provide a frame of reference. In both these cases no berm was included, however, additional analysis with different berm geometries have been undertaken. In these analyses excavation is performed leaving a berm in place. A top support is then installed and the berm removed.

Analyses in which the berm is simulated by an equivalent surcharge load have also been carried out.

Problem geometry and material properties

The problem geometry and material properties assumed are given in Fig. 1. The 1m thick concrete wall was assumed to be linear elastic and the clay was assumed to be linear elastic - perfectly plastic with a Mohr-Coulomb yield surface. Throughout the analyses the clay was assumed to behave in an undrained manner. This was simulated by assigning a large bulk modulus (=l00K', where K' is the effective bulk modulus of the soil skeleton) to the pore water. The wall was assumed to have been installed without affecting

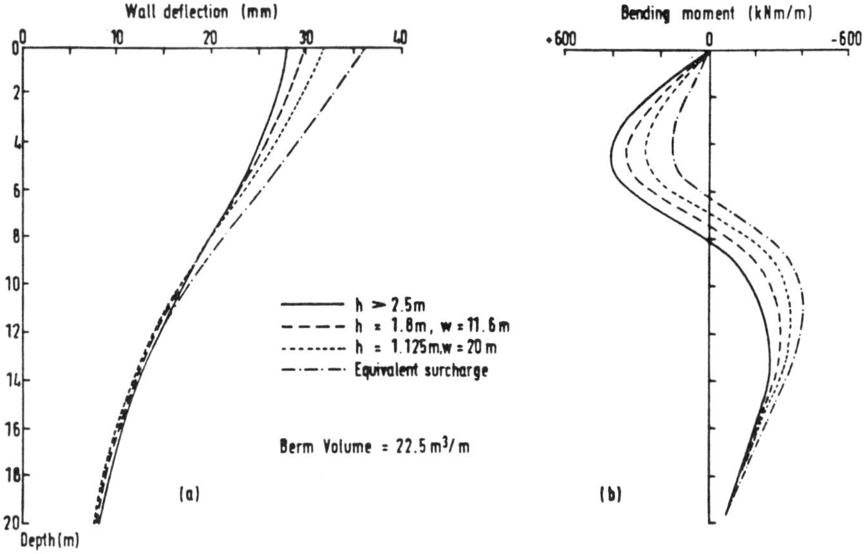

Fig. 2. Effect of berm geometry on wall geometry on wall behaviour: (a) Deflections; (b) Bending moments

the initial soil stresses (i.e. 'wished' in place). These initial stresses were calculated from the bulk unit weight given in Fig. 1 and by assuming hydrostatic pore water conditions with a water table 2m below original ground level and a coefficient of lateral earth pressure at rest, K_o = 2. The prop stiffness was taken as $50MN/m^2$.

All berms had 45° face slopes and are defined in terms of berm height, h and crest width, w, (Fig. 1). A berm was simulated in an analysis by carrying out the excavation while leaving the berm in place. A horizontal spring was then placed at the top of the wall to model the prop before the berm was excavated.

Plane strain conditions were assumed and eight noded isoparametric finite elements were employed with reduced integration. Several finite element meshes were used to enable the various berm geometries to be modelled. Typically these consisted of approximately 240 elements. All analyses were carried out with the Imperial College Finite Element Program, ICFEP (Potts and Fourie (1984), Potts and Day (1990)).

Different berm geometries

The results of several analyses carried out with berms of different geometries (i.e. different h and w values) but with the same sectional area (and therefore volume) are presented in Fig. 2. Also shown are results from an analysis which treated the berm as an equivalent surcharge loading. In this analysis both the magnitude and extent of the surcharge loading applied to the newly excavated surface increased with the depth of excavation. The results presented are for a berm with a volume of 22.5 m^3/m which represents 16% of the total material to be excavated. Berms with heights (h) up to 5m have been considered. The wall deflections and bending moments given in Fig. 2 correspond to full excavation (i.e. top prop placed and berm excavated).

Results from analyses with berm heights greater than 2.5m are so similar that they are indistinguishable when plotted. These results are therefore represented by single curves in Fig. 2. As the berm height reduces below 2.5m wall deflections increase and the bending moment distribution changes. The analysis in which the berm is treated as an equivalent surcharge produces significantly different results. The predicted movement of the retained ground surface behind the wall follows a similar trend to that of the wall deflection shown in Fig. 2. Similar results were obtained for berms with different cross sectional areas.

It may be concluded that for berms with h greater than 2.5m the behaviour is not sensitive to berm geometry. If the height h reduces below 2.5m the berm becomes less effective. An equivalent surcharge does not accurately simulate the effect of a berm.

Different berm volumes

As noted above for berm geometries with h greater than 2.5m, behaviour is not sensitive to the actual berm geometry. The behaviour is, however, dependent on the cross sectional area (volume) of the berm. This can be seen in Fig. 3 which shows the wall deflections and bending moment distribution for analyses with berm volumes of 6.0 m^3/m (4.3% of total excavation) and 42.5 m^3/m (30% of total excavation), h is greater than 2.5m in both cases. Also shown are the results from the 'control' analyses (propped and cantilevered walls with no berms). To avoid over complicating Fig. 3, results from analyses with intermediate berm volumes are not shown. Results from such analyses lie between those given. The 'control' analyses provide envelopes of behaviour within which the analyses with berms lie.

The deflected shape of the wall is dependent on the size of the berm left adjacent to the wall. Not surprisingly, the smaller the berm the larger the movement. The normalised maximum wall deflection, or berm efficiency, is plotted against berm volume in Fig. 4 to allow comparisons of the different cases to be made. Results for several different berm volumes are shown. Berm

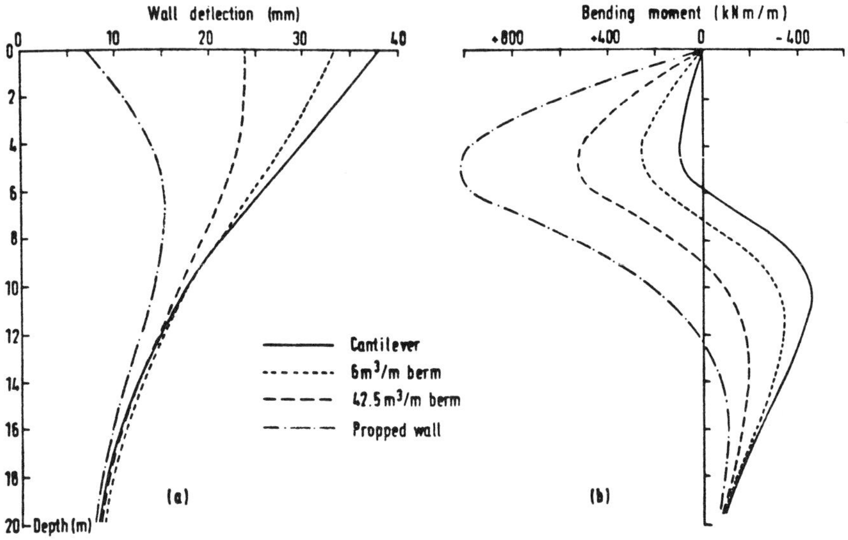

Fig.3. Effect of berm volume on wall behaviour: (a) deflections, (b) bending moments

efficiency is defined such that for a cantilever wall it equals 0% and for a propped wall it equals 100%. These represent the two 'control' cases. For the analyses incorporating berms the following definition of efficiency applies

$$\mathrm{Eff} = \frac{(\delta_{cant} - \delta)}{(\delta_{cant} - \delta_{prop})} \cdot 100$$

where δ refers to a maximum deflection and the subscripts 'cant' and 'prop' refer to the cantilever and propped 'control' analyses respectively. It is clear that the larger the berm, the greater the efficiency, but it should be noted that the largest berm analyzed (volume = 42.5 m^3/m), which represents 30% of the total material to be excavated, only has an efficiency of 61.5%.

In Fig. 5 the normalised maximum bending moment, defined as the magnitude of the maximum positive or negative bending moment divided by the maximum bending moment in the propped wall ('control') analysis, is plotted against berm volume. An optimum berm volume is apparent.

Profiles of vertical and horizontal movement of the retained ground surface behind the wall after completion of excavation are given in Fig. 6. The

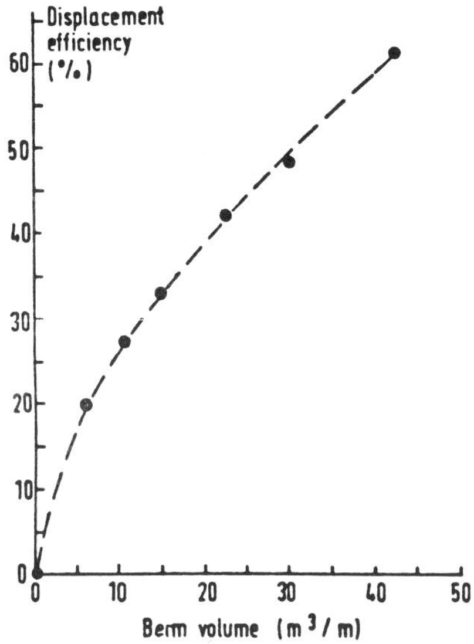

Fig. 4: Variation of displacement efficiency with berm volume

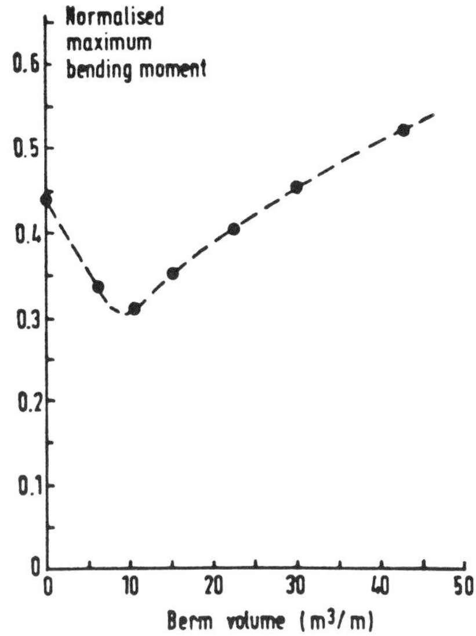

Fig. 5: Variation of normalised bending moment with berm volume

smaller the berm the more the results tend to that of the cantilever wall. For the largest berm displacements lie approximately midway between those from the two 'control' analyses. As this berm represents 30% of the total material to be excavated it is likely to be larger than anything that would be used in practice. It may therefore be concluded that the use of a berm for temporary support is likely to be less than 50% efficient in reducing soil movements in the retained soil. At distances greater than 15m from the wall the use of soil berms has little advantage over the use of a cantilever wall.

Conclusions

The analyses presented in this Paper have assumed simplified soil behaviour and have idealised the construction procedure. In particular the soil has been modelled as a linear elastic - perfectly plastic material and no account of the effects of small strain stiffness (Jardine et al (1986)) have been included. Such shortcomings must be appreciated when applying these results to practice. However, by making such simplifications it has been possible to investigate the effects of different berm geometries in a systematic manner. The following conclusions and guidelines are apparent from the

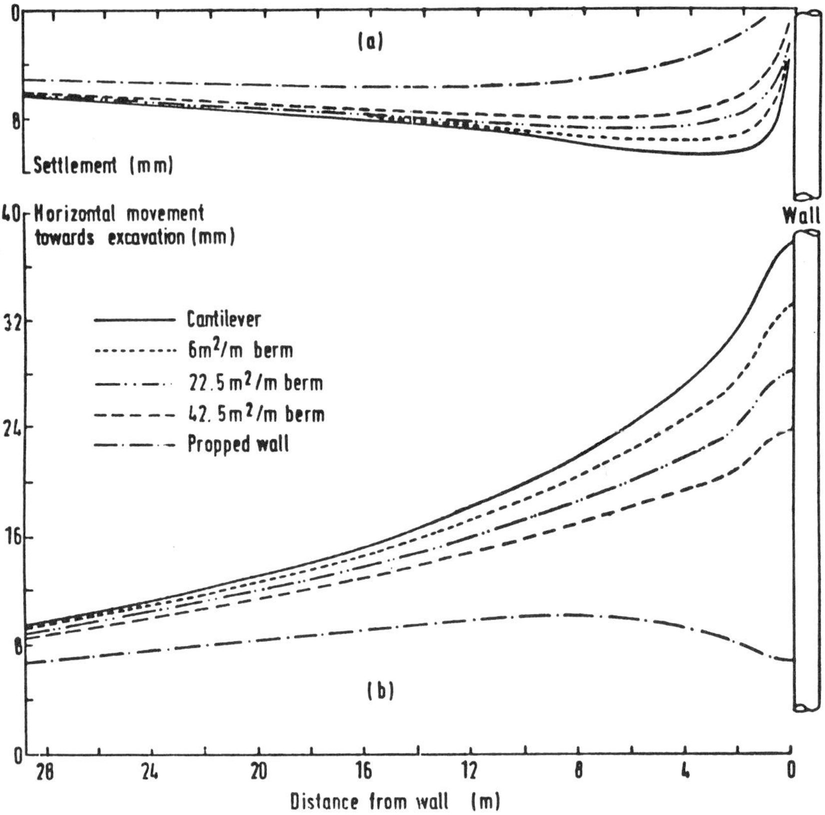

Fig. 6: Effect of berm volume on the movement of the retained ground surface: (a) Settlement; (b) Horizontal movement

present investigation.

For berms with a height between 2.5m and 5m it is the volume of the berm not its specific geometry that dictates soil movements adjacent to the excavation and the final displaced shape of, and bending moments in, the wall it was used to temporarily support. As the berm height reduces below 2.5m, berms of equal volume result in different wall deflections and bending moments; deflections increase and the berm becomes less efficient.

The horizontal resistance provided by a berm has a substantial effect on wall behaviour. It is therefore not accurate to use an equivalent surcharge load to model the effect of a berm. If such an approach is used the effectiveness of the berm is likely to be underestimated.

Soil berms are effective in reducing the final maximum and minimum bending moments in the wall. In this respect there would appear to be an optimum berm size.

The greater the volume of the berm used to provide temporary support then the smaller the wall deflections and the movements of the retained ground surface behind the wall. However for the largest berm used in the present study, which represented 30% of the total volume of soil to be excavated, a deflection efficiency of only 61.5% was achieved. The reduction of ground surface movement behind the wall, using a berm as opposed to a cantilever wall (with no berm), was only significant in a region up to 15m from the wall.

References

JARDINE R.J. *et al.* Studies of the influence of nonlinear stress - strain characteristics in soil-structure interaction. *Geotechnique*, 1986, **36**, No 3, pp 377-396.

POTTS D. M. and DAY R. A. The use of sheet pile retaining walls for deep excavations in stiff clay. *Proc. Instn. Civ. Engrs, Part 1, 1990, 88, pp 899-927.*

POTTS D. M and FOURIE A. B. The behaviour of a propped retaining wall - the results of a numerical experiment, *Geotechnique*, 1984, **34**, N.o 3, pp 383-404.

Regroutable ground anchors

D. E. SHERWOOD and R. R. HARRIS , Bachy Group, UK

An empirical method is presented for the design of regroutable anchorages where repetitive and selective injection under pressure is used. Corrosion protection, design, and factors of safety required by the current British Standard are discussed in relation to TMD anchorages of the above type.

Introduction

Post tensioned anchoring systems in ground engineering have reached an advanced stage of development. Experience in their usage and performance has increased very considerably in the last ten to twenty years. National standards or codes have also emerged in principal markets.

Rigid application of standards, however, often retards further development of design and technique. The current British Standard is now 10 years old. It seems to be time to review some of its requirements and omissions in the light of intervening experience.

The thrust of research and experimentation in ground anchorages has been to maximise the load that an anchor may safely transfer to the surrounding soil. It has been long recognised however, that due to progressive failure along the fixed length, the maximum efficient fixed length in most soil types for a conventional tensile anchorage is about 8m. This has lead to a growing interest in regroutable anchorages where the load carrying capacity is enhanced and the anchorage is made less subject to ground conditions by careful repetitive grouting procedures during installation.

Regroutable ground anchorages

Regroutable ground anchorages attain much higher loads than can be predicted by conventional soil mechanics. This lead Ostermayer (ref. 1) and other French researchers (ref. 2) to devise empirical design curves for anchorages in cohesive and non-cohesive soils from studies of the performance of many anchorages subjected to load tests. This accumulation of experience has been continuously augmented and is presented in the third edition of the French anchorage standard (ref. 5) which was first published in 1972. That standard now shows a clear differentiation in anchorage performance arising from the manner in which the anchorage has been installed and grouted. The IRS (repetitive and selective high pressure injection) anchorage is distinctly

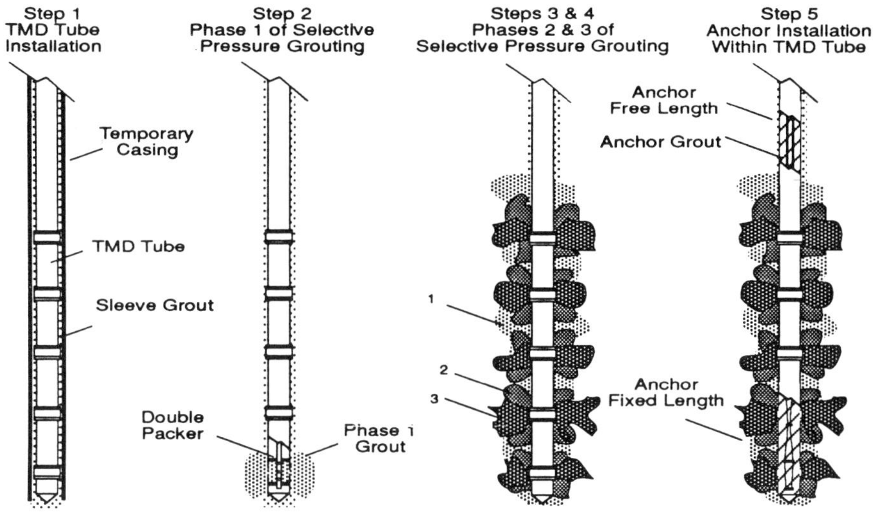

Fig. 1. Steps in TMD anchorage installation

superior to the IGU (global injection at uniform pressure) anchorage.

The British Standard Code of Practice for Ground Anchorages: BS8081 (3), recognises that pressure grouted anchorages are superior to other anchorages but it does not differentiate between IRS and IGU anchorages. In the light of both international and UK experience, this is a serious omission, which leads to both poor anchor design and excessive costs in the UK.

The stages of construction of a TMD anchorage are illustrated in Fig. 1. This is a repetitive and selective high pressure injection (IRS), double corrosion protected soil anchor which has been in use as a proprietary system of SIF Entreprise Bachy, for more than 20 years.

The unique features of the system are capable of overcoming problems inherent in the development of high capacity soil anchorages

- The TMD tube provides a complete protection to the strands over both the fixed and free lengths of the anchorage. It also allows the anchor components to be installed in a fully controlled environment with no danger of borehole collapse.
- The TMD anchorage exploits early observations that the ultimate an-

chor load increases proportionately to the volume of grout forming the body of the anchorage in the fixed length. This is achieved by utilising the tube-a-manchette system to place small volumes of grout in a number of passes, rather than large amounts in a single injection, and leads to controlled recompression of the surrounding soil and enlargement of the grout body itself.

- The fixed length to load ratio of the TMD is lower than for any other soil anchor type. It is therefore possible to minimise anchor lengths.
- The fixed length of the ground anchorage can be greater than the fixed tendon length within the TMD tube. The tendon load may therefore be transferred to the soil by compression over much of the fixed length of the anchorage. This optimises the efficiency of the load transfer and reduces progressive failure along the fixed length noted earlier. The presence of the steel TMD tube also stiffens the anchorage and this also further reduces progressive failure mechanisms.
- The high pressure grouting process 'repairs' loosening or weakening of the soils that the drilling of the anchor hole may have caused. The TMD performance is therefore not dependent on drilling method, a caution expressed in early research on ground anchors (ref. 1). It also improves less compact soil by either some grout permeation into the surrounding soil, or by incremental expansion of the borehole volume under high pressure.

Design of TMD anchorages

It has been shown (refs 1 and 2) that the most significant factor in assessing the satisfactory long term behaviour of a soil anchor is its creep characteristics; that is the gradual relaxation of applied load with time. The terminology used here is defined by Figs 2 and 3, the latter being constructed from the former. This is a graphical presentation of the performance of a typical TMD anchorage test.

The relationship between creep rate and applied load is linear up to the Critical Creep Load (T_c) after which the creep rate increases rapidly until an Ultimate Creep Load (T_1) is reached as shown in Fig. 3.

The Allowable Working Load (T_a) which may be applied to an anchorage is determined as the lesser of the following values

0.67 T_1 or 0.9 T_c for temporary anchorages
0.67 T_1 or 0.8 T_c for permanent anchorages

Tests on more than 260 TMD anchorages have lead to further important observations regarding the creep behaviour of IRS regroutable anchorages

- the value T_c, and the slope of the preceding straight line, are inde-

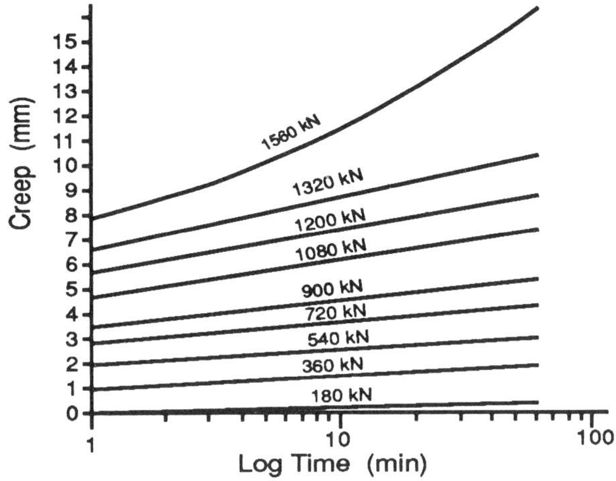

Fig. 2. Creep against log time

pendent of load stage durations i.e. the speed of loading has no effect on T_C.

- cyclic loading or different incremental loading regimes have very little influence on the value of T_C.
- long term performance can be predicted from short term tests.
- the eventual anchor capacity is not affected if T_C is temporarily exceeded by 10 or 15%. Once T_C is reached, however, albeit for a very short period, anchorages loose about 30% of their potential capacity.
- The creep rate (that is the slope of the initial straight line portion of Fig. 3) shows remarkable uniformity irrespective of soil type. The average rate varies from 0.004mm/kN/log cycle of time for gravel anchorages to 0.005mm/kN/log cycle of time for clays.

Figs 4 and 5 present the empirical design curves derived from 15 years of

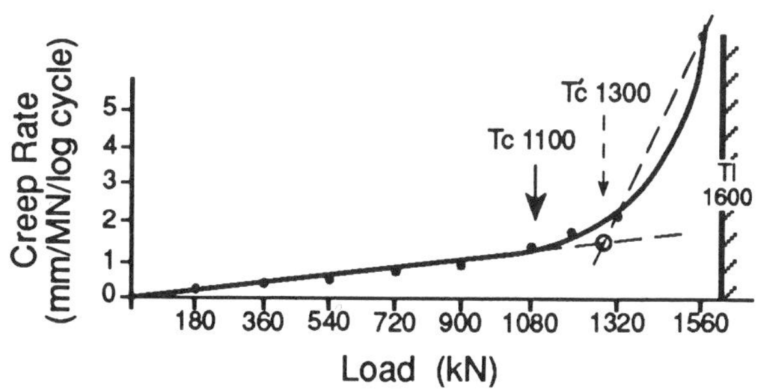

Fig. 3. Creep rate against load

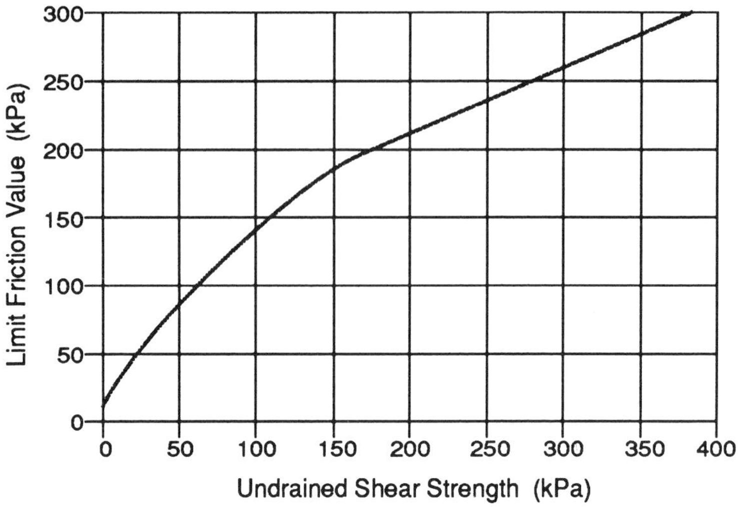

Fig. 4. TMD anchorage in clays and silts

testing of TMD anchorages. These curves, together with data from other European sources, are the basis for the French anchorage standard's (ref. 5) recommendation for IRS type anchorages. This standard is probably the most comprehensive standard currently available for the design and construction of ground anchorages.

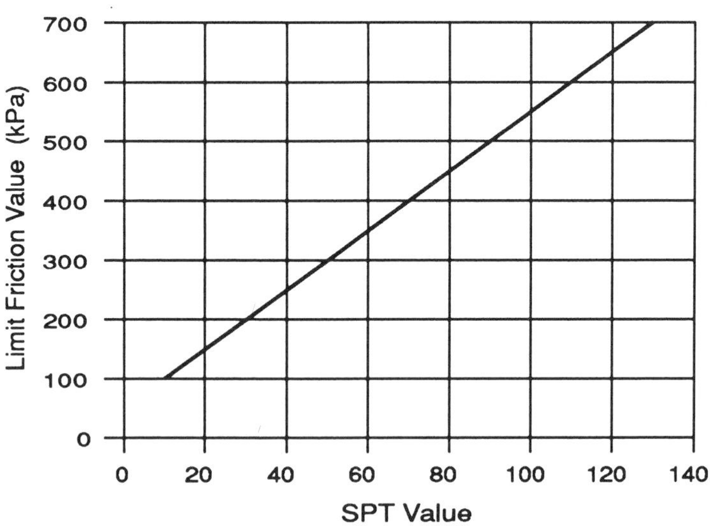

Fig. 5. TMD anchorages in sands and gravels

To apply the Limit Friction (q_s), an appreciation of the soil/grouted body contact area is required. Experience has led to a theoretical anchorage diameter (D_s) being defined as $D_s=nD_d$ where D_d is the known drill hole diameter. For TMD anchorages n may be taken as 1.8 for chalk, marl, clays and gravels and as 1.4 for sands and silts. These values of n presume the injection of pure cementitious grout with a water/cement ratio of not more than 0.5 and an injected volume of at least 50% greater than the related theoretical volume of the drill hole.

In practice, however, much larger grout quantities are normally required to achieve satisfactory grout pressures in the final phase of injection at each level in the anchorage zone.

Preliminary TMD anchorage designs determine allowable working load (T_a) according to the formulae

$$T_a = \pi.\ n.D_d.L.q_s/\text{Factor of Safety}$$

where L is the regrouted length of the anchorage. It is emphasized that this approach is based on limit friction values derived from experience and these are not necessarily ultimate friction values. Limit values have been evolved so that T_a falls in approximately the appropriate relationship to T_c, as set out earlier above.

Corrosion protection

The approach to corrosion protection of permanent anchorages taken in BS8081 is to require that, throughout the anchorage, there should be two continuous, complete, impervious membranes between the tendons of the anchor and the ground. In earlier thinking, a single impervious membrane was considered sufficient but practice showed that this vital membrane was often damaged during the insertion of the anchor assembly and subsequent withdrawal of casing. The double membrane approach demanded by BS8081 was evolved so that during installation, the outer membrane provided protection to the inner membrane which in effect, became the vital element of protection.

The TMD approach is quite different. The principal barrier to corrosion is the steel TMD pipe itself (see Fig. 1). It is installed and grouted to the ground in advance of the anchor tendon installation and thus, if required, its complete integrity can be physically demonstrated. In the UK an inner encapsulation membrane of the fixed length "exposed" strands is sealed to the individual sleeve strands of the free length as a further complete barrier, but its prime purposes are to

- facilitate the use of a short fixed length on the strands themselves thus leading to a fixed anchor length which is substantially in compression

at the anchorage/soil interface outside the TMD tube.
- protect the strands from contamination or damage prior to and during the installation process.

Once the anchorage has been installed the inner encapsulation membrane, although an intellectual comfort, is redundant in the long term protection of the anchorage. This has, clearly, been demonstrated by 20 years of experience in France, where no such inner encapsulations are used except where it is beneficial to have a fixed length substantially in compression as described above. Recently a TMD anchor, which was installed in 1975, was exhumed for examination. Although its environment was severe from a corrosion point of view, the anchor tendons were found to be in perfect condition after 15 years of service, despite using what BS8081 would classify as a single corrosion protection system. Other anchorages have been exhumed after more than 10 years of service, and these also have been found to be in perfect condition.

The standard TMD design for permanent anchorages in the UK market provides double corrosion protection over the entire tendon length as described above. However, from time to time, particularly long life anchorages are required and further protection can be added to the standard TMD system to meet such special requirements. The following sequence of protective layering was provided to the fixed length of anchorages on a recent UK site

1. Factory installed epoxy grout cover to the fixed length tendons
2. Factory installed PVC sleeve installed together with 1 above
3. Anchor grout
4. Steel TMD pipe in fixed length and steel plain pipe in the free length
5. Epoxy coating to the pipe in 4 above
6. Sleeve and consolidation grout outside the TMD pipe.

Anchorages may also be designed using classical BS8081 type double corrosion protection of the anchor strands within the steel TMD pipe protected as described. This in effect gives an eight layer corrosion protection to the tendons.

Whilst such complex multi barrier systems are available, in the view of the authors they are quite unnecessary overkill. They simply reinforce the assurance that the point of failure, when that eventually occurs, will be at the anchor head and not elsewhere in the system.

The background to the definitions of corrosion protection set out in BS8081 are not generally appreciated and this leads to its requirements being enforced in a codified and restrictive manner. This restricts the use of innovative anchor systems and leads to unnecessary cost in many cases. The recommendations of this standard require reconsideration and review.

Table 1. National factors of safety used in anchorage design

Degree of Risk	Recommended Safety Factors for the Grout/Ground Interfaces for Permanent Anchorages		
	UK (1)	Switzerland	France (2)
Low	3	1.6	2.0
Medium	3	1.8	2.0
High	3	2.0	2.0

Notes:
1. Increase Factor of Safety to 4 if ground creep is foreseen.
2. The Safety Factor of 2 may be reduced after site testing.

Anchor design standards

The sophistication of ground anchorages, the degree to which their behaviour is predictable, and their widespread usage in continental Europe is not matched in countries where BS8081 is recognised as the de facto standard. The reason for this is apparent after an appraisal of Table 1.

An anchorage designed in clay in the UK will have about twice the factor of safety of a European anchorage: in reality this means about half the allowable working load.

The BS8081 factors of safety for anchorages owe much to the British approach to safety factors in piling. This may be an appropriate analogy for normal UK anchorages where loosening during drilling and variation in ground parameters from place to place on a site, are ever present factors in the designer's mind. It has been shown by many years of experience that the TMD type of IRS anchorage is much less sensitive, if at all, to drilling procedures because "disturbance" is repaired during careful repeated post installation grouting phases. Experience also shows that this type of anchorage is less affected by normal variations in ground conditions because the high pressure grouting does locally improve weaker soils. Where abnormal or unexpected variations occur, the data recorded during grouting will invariably indicate that something abnormal has occurred.

It is argued therefore that whilst BS8081 factors of safety may be appropriate for simple run of the mill anchorages, they are not appropriate for IRS type anchorages. The Authors advocate the use of a factor of safety of 2.0 on all sites where preliminary or trial anchor data is not available, and a reduction to 1.6 for low risk sites and 1.8 for medium risk sites after trial anchor data becomes available. In their view, all high risk sites, regardless of the size of the project, should incorporate preliminary or trial anchorage tests.

Conclusions

- Experience has shown that TMD anchorages based on the IRS principle mobilise higher working loads than other anchorage types in all soil conditions.
- Because of the influence of controlled grouting in several phases creep performance of TMD anchorages is relatively independent of soil type.
- Corrosion protection definitions have been evolved in relation to conventional ground anchor installation methods, and their background is not generally appreciated. This leads to misapplication of the standard, in practice, in relation to corrosion protection.
- B58081 does not distinguish between the various types of pressure grouted anchorages and its recommendations on design need to be reviewed in relation to IRS and IGU approaches to grouting anchorages.
- Factors of safety recommended by B58081 may be appropriate to simple ground anchorages. They are much too conservative for anchor systems which produce a more standardized and repeatable anchorage performance as a result of their installation procedure.
- The philosophy of reduction of factor of safety after preliminary trial anchorage tests have been made, should be accepted. Such tests would then be perceived as an opportunity to refine design and reduce cost, rather than an expensive formality which normally only confirm the excessive safety of the preliminary design.

References

1. OBSTERMAYER H. Construction, carrying behaviour and creep characteristics of ground anchors.*Proc. Instn Civ. Engrs*, 1975.
2. LACOUR J, DELMAS F and BUSTAMANTE M. Essais de Tirants D'Ancrage Scelles dans une Argile Plastique. *Annates de L'Institute Technique du Batiment et des Traveaux Publics, June* 1978.
3. British Standard Code of Practice for Ground Anchorages. BS8081: 989.
4. SWISS SOCIETY OF ENGINEERS AND ARCHITECTS. SIA Standard 191, Ground Anchors, 1977.
5. Recommendations for the Design, Calculation, Construction and Monitoring of Ground Anchorages. Edited by P Habib, 1989. Translation from Recommendations Concernant la Conception, le Calcul, l'Execution et le Controle des Tirants d'Ancrage. TA86. 3_e Edition, Paris: Eyrolles, 1986.

Ground anchorages to the retaining walls at Peny-y-Clip

M. J. TURNER, Turner Geotechnical Associates, P. J. SELLEY, Travers Morgan and G. M. COOPER, PSC Freyssinet Ltd, UK

The installation of some 1600 permanent ground anchorages to support the major retaining walls associated with the approaches to the Pen-y-Clip tunnel on the A55 North Wales Coast road has been acknowledged as one of the largest, most expensive and technically most demanding of such projects ever to be undertaken in the UK. The retaining walls are constructed through steep, sidelong ground, and support a wide range of very variable materials including deep scree and slope debris, glacial till, weak mudstones and very strong, but often highly fractured, intrusive micro-diorite. The largest of the anchored retaining walls is over 30 metres tall at its highest point and requires more than 750 ground anchorages. The design and construction of the double corrosion protected anchorages, in general accordance with BS8081:1989 (ref.1), the new UK Code of Practice for Ground Anchorages, is described and reviewed.

Introduction

An outline plan of the scheme is shown on Fig. 1. The Pen-y-Clip tunnel scheme is a Welsh Office project which will provide a 1.9 km two carriageway highway, forming part of the A55 North Wales Coast Road, between Penmaenmawr and Llanfairfechan in Gwynedd, Wales. It is one of the final parts of the improvement of this trunk road and negotiates the rocky headland at Pen-y-Clip, just west of Conwy. It forms one of the last stretches of a Euroroute, the E22, which links Sassnitz in Poland with Dublin in Ireland.

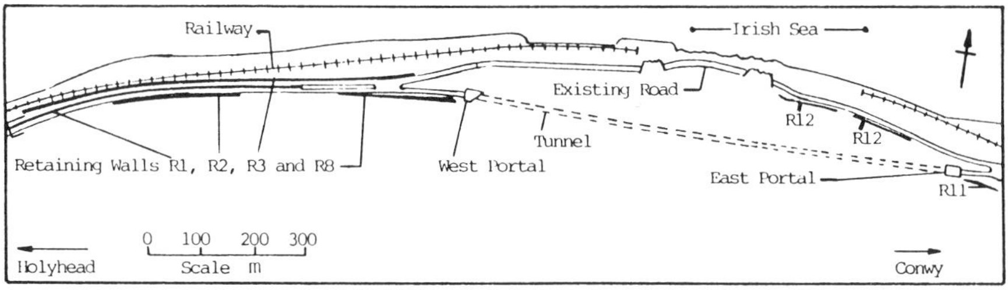

Fig. 1. Site plan of Pen-y-Clip tunnel scheme

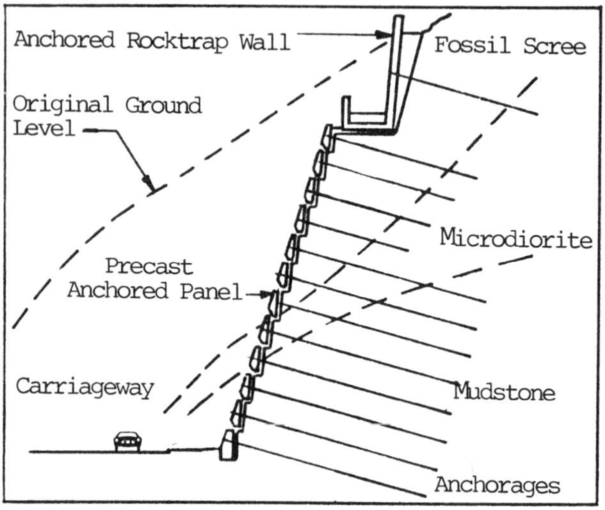

Fig. 2. Typical section through retaining wall R8

Abutting the tunnel portals are two anchored and/or bolted rock cuts, typically inclined at 74°. These abut two major anchored panel walls, R8 and R11, which approach 30m in height. These walls are constructed in bays from the top downwards and are necessary because the upper section of each wall is cut into steep, unconsolidated scree and quarry-related deposits lying at close to their angle of repose. A typical cross-section through one of these walls, R8, is illustrated on Fig. 2.

To the west of the tunnel section, three retaining walls, designated R1, R2 and R3, are necessary: separating and supporting the carriageways where there is a split vertical alignment on the steep sidelong ground between the existing A55 and the railway below. A typical cross-section through this anchored wall complex is shown on Fig. 3. An anchored cantilever wall separates the carriageways, with a retaining wall below supporting the down-slope carriageway and incorporating barette foundations taken down to mudstone bedrock. Up-slope of the anchored wall, the retaining wall above the westbound carriageway is propped underneath the carriageway itself from the top of the anchored cantilever wall. On the western side of the headland, a further anchored panel wall, R12, is required to form a Non-tunnel Traffic Route for westbound traffic.

An extensive monitoring system has been installed as recommended by BS8081, incorporating 110 Glotzl permanent monitored hydraulic load cells transducers. These are connected to central terminal readout units for each structure. Loads being taken by each type of anchorage are then monitored both during construction and after completion of the Works.

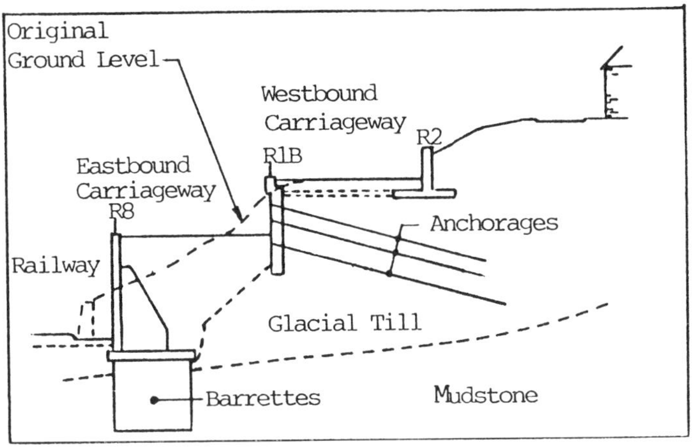

Fig. 3. Typical section through retaining wall R1

Geology and topography

The Pen-y-Clip headland is formed from a laccolithic microdiorite intrusion of Ordovician age. On either side of the headland, bedrock is overlain by scree, fossil scree (in which the interstices of ancient scree slopes have been infilled with an alien matrix of finer material derived from Irish Sea Till), and debris from upslope quarrying operations and boulders.

Further to the west of the headland, glacial till overlies the bedrock, which consists here of narrowly jointed, sheared and altered mudstones, again of Ordovician age, into which the microdiorite was intruded. The microdiorite is also closely jointed and contains sheared zones particularly at its laccolithic margin. The western junction between the mudstone and the overlying microdiorite intrusion daylights within retaining wall R8. The extensive series of site investigations necessary for the project have been described elsewhere (ref. 2).

Design of anchored panel walls

These walls were constructed from the top downwards. The sidelong nature of the existing ground, with slopes between 30° and 40°, made other forms of construction either impracticable or less economic for the conditions at this location. The adoption of permanent ground anchorages to assure the overall slope stability together with precast concrete panels to distribute local anchorage forces thus provided an acceptable and economic solution.

The walls have been designed to function as anchored cuttings in which the precast concrete panels distribute the anchorage loads over the face of each cutting. Panels are not made continuous with their neighbours and are intended to act independently of each other. Separate foundations are not therefore provided for the wall because the facing anchored panels are designed to be integral with the excavated soil.

The working loads for the ground anchorages were determined from established soil and rock mechanics principles. In the case of the largest anchored wall, R8, the analysis takes account of the differing properties of the fossil scree and mudstone strata. The analysis of anchorage forces also took account of the construction method and anchorages were designed to support all temporary heights without relying on the lowest row of anchors.

A factor of safety of 1.5 against global slip failure was achieved at all stages of construction, and this criterion generally governed free anchor lengths. A continuous layer of porous no-fines concrete is provided behind the anchored panels to prevent build-up of water pressure behind the wall.

Ground anchorage design

The ground anchorages were required to be designed in general accordance with the BSI document DD81:1982 (ref.3) and the Contract Specification. DD81 was replaced in mid 1989 by BS8081:1989, the British Standard Code of Practice for ground anchorages. The latter document was therefore published after the tender date of the Pen-y-Clip scheme. The contents of the new Code were, however, known to the Consulting Engineers, and many of its main requirements were incorporated into the Contract Specification.

The Contract Specification sought to clarify the issue of the responsibility for design of the ground anchorages, as envisaged in BS8081. In this case, the Engineer was responsible for defining the overall design parameters for the scheme, including the identification of individual anchorages in terms of anchor force required, orientation and the free anchor length necessary to ensure the overall stability of the anchored structure. The Contractor was then responsible for the design of the fixed anchor length to safely provide the required anchorage force and the design and construction of the fully corrosion-protected tendon which would provide the required force. Thus the Engineer undertook responsibility for the philosophy and design of the global scheme, whilst the Contractor was essentially responsible for the design and performance of each individual anchor.

Prior to finalisation of the overall design for the anchored structures, a test programme was undertaken to investigate the construction and performance of doubly corrosion protected anchorages typical of the type that would be installed in the contract works. Seventeen proving test anchors were installed at four locations throughout the works. Each location was designed to examine a different soil or rock type. The test anchors incorporated shortened fixed anchor lengths and measures to isolate the load transfer lengths, in order to verify the grout/ground bond values developed by the anchorages.

Two anchorages were also exhumed for detailed examination. This work enabled definitive design values to be obtained for ground anchorages within glacial till, fossil scree, and mudstone. Ultimate values within microdiorite were not obtained.

The anchorage loads for walls R1B to R11 varied between 200 and 600 kN. For retaining wall R12, anchorage forces were lower, varying between 50 and 125 kN. The design fixed anchor lengths varied between 3.0m and 12.5m depending upon loads and anchoring stratum.

Within the contract both Type A and Type B anchorages, as defined by DD81/BS8081, were used. Type A anchorages are constructed from straight-shafted tremie-grouted boreholes, typical of those formed in rock or hard self-supporting strata. Type B anchorages employ low pressure grouting techniques utilising a temporary casing or in-situ packers, whereby the fixed anchor grout is caused to penetrate the ground beyond the borehole diameter to enhance the load carrying capacity. Anchorages formed within fossil scree deposits were envisaged as being Type B anchorages. Microdiorite, mudstone and glacial till anchorages were envisaged as Type A.

The ground anchorage tendons were all constructed from multi-unit steel strands encapsulated in cement grout within two corrugated polypropylene ducts over the tendon bond lengths, and enclosed within two plastic sheathings over the free tendon length. The ground anchorages could therefore be classified as permanent ground anchorages with double protection, as identified by BS8081. The construction of a typical contract anchorage is illustrated on Fig. 4.

Design of individual anchorages.

The design anchorage force, T_W, is usually determined by a skin friction equation of the form:

$$T_W = \frac{\pi . d . L . S_{ult}}{F}$$

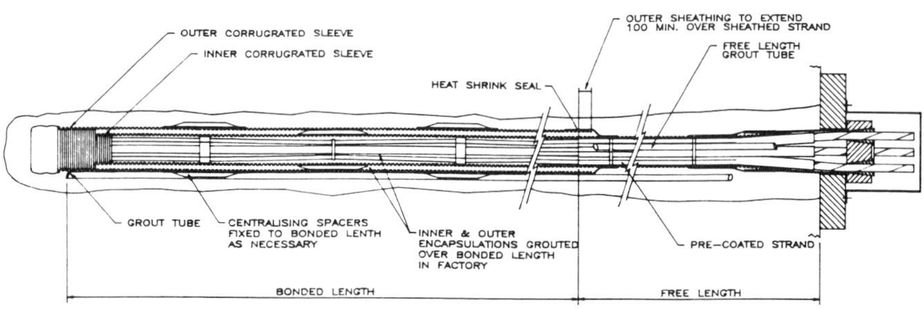

Fig. 4. Details of typical double corrosion-protected ground anchorage at Pen-y-Clip

Table 1. Ground/grout interface design values

Ground Type	Borehole Diameter (mm)	S_{ult} (measured) (kN/m^2)	S_{max} (no failure) (kN/m^2)	S_{ult} (design) (kN/m^2)
Glacial till	160-220	113*-417		140
Mudstone	115-160	416		400
Microdiorite	160-220	-	371-547	540
Fossil scree	160	477		450

* Lowest result assumed to be affected by softened material due to length of time hole was maintained open.

where d is the diameter of the borehole, L is the fixed anchor length, S_{ult} is the ultimate bond or skin friction at the grout/ground interface and F is the required global factor of safety. The values of S_{ult} obtained from the pre-contract proving tests and the values of S_{ult} used in subsequent design are summarised in Table 1, below. The minimum design factor of safety used in conjunction with these ultimate values was 3.0.

The contract specification also stipulated minimum drill hole diameters over the fixed anchor length, based upon the results of the drilling of the proving test anchors. In addition, for certain of the walls, minimum fixed anchor lengths were specified within the contract, again based upon the Engineer's evaluation of the proving tests.

A factor of safety of 3.0 was also used in the design of the tendon/encapsulation grout interface and the interface between the encapsulation and the surrounding anchorage grout.

A minimum design factor of safety of 2.0 was applied to the specified characteristic strength of the tendon steel to calculate the required number of strands forming each tendon (cf Table 2 of BS8081).

The anchor heads were designed to be restressable as defined by both DD81 and BS8081 (i.e a head assembly which permits the tendon, throughout the life of the structure, to be measured by checklifting and small losses, up to 10% of working load to be recovered by shimming or thread-turning). In addition, most anchorages were required to be checked and reset upon completion of the particular structure.

Installation of ground anchorages

Access was extremely limited at the anchored retaining wall sites, and, in particular, walls R8 and R11 are bounded by steep scree covered slopes to the south and an existing road and rail system following the coast line on the northern side. Working access was especially limited in the early stages of these walls, when drilling rigs had to work from 4 to 5m wide benches cut into the previously placed imported fill which provided access to the otherwise unstable scree slopes. At the same time as anchor drilling and installation work was being carried out by the anchoring specialist, the Main Contractor had to continue excavation to subsequent levels and the placing of panels. These working constraints necessitated a high level of coordination of the various activities. In spite of these constraints the works have progressed ahead of programme.

Due to the varying strata envisaged, it was decided that an overburden drilling system (ODS) utilising a down-the-hole hammer technique with simultaneous casing installation would be the most effective. The chosen system formed the drill hole using an eccentric bit to produce a 220mm diameter reamed hole, operating within 203mm OD drill casing. The nominal ID of the casing was 184 mm with a nominal drill through diameter of 165mm. The development of the overburden drilling system at Pen-y-Clip has been described further elsewhere (ref.4).

Because of the congestion caused by the large number of anchorages, the Specification required that anchor drillholes should not deviate from their specified alignment by more than 1 in 30. This is a more onerous tolerance than proposed in BS8081, but the drilling system proved very successful in achieving the specified alignments. Theses were checked using the Reflex-Fotobor dip and direction indicator (ref.5).

Grouting was carried out using high shear colloidal mixers and utilised both neat cement and sand/cement grouts. Extensive grout trials were carried out at the start of the contract, to obtain optimum grout mixes complying with the specification. The on-going quality of grout mixes was constantly checked by site tests consisting of flow cone, wet density, bleed measurements and compression test cubes for each day's grouting.

All contract anchor tendons for the retaining walls were fabricated in factory facilities off site. A feature of such anchorage fabrication is that tendon lengths have to be preplanned and programmed into the fabrication schedule. To cater for varying rock head levels, when scheduling the anchor lengths, the Contract allowed for an additional 3 metres on some of the fabricated tendons.

Problems arose in identifying a supplier of corrugated sheathing to the specified thicknesses for tendon encapsulation. In one case this required special tooling to be set up to manufacture sheathing specifically for this contract. It was found that a sine wave section gave a more consistent wall thickness than the standard square section corrugated sheathing.

Whilst it was not a QA contract, the ground anchorage tendons were fabricated in accordance with PSC Freyssinet's standard Quality Plan to BS5750: 1987 (ref.6). In addition, the Engineer's inspectors regularly monitored fabrication to ensure compliance with the Specification.

Stressing has been carried out using multi-strand jacks fitted with load cells and digital read out units. Jacks, pressure gauges and load cells were regularly calibrated in general accordance with the recommendations of BS8081. In the case of the panel walls, to avoid eccentric loading of the panels both anchors in a panel were stressed simultaneously and panel movements recorded during and after stressing. Upon completion of stressing, each anchor block was wrapped in grease-impregnated tape and the strands resheathed to provide temporary protection until final cutting and capping.

Long term monitoring

A proportion of anchors on each wall have been fitted with permanent load cells and these are read on a regular basis. In addition, on the high panel walls, R8 and R11, overall wall movement and deformation are monitored using targets mounted on the panels. Movements of these targets will be monitored from remote survey stations both during the construction contract and after the completion of the works. Load cell measurements on wall R1B, have already monitored marked load fluctuations in the ground anchorages as the construction works have progressed. It is hoped that the load monitoring methods and results will be the subject of a subsequent paper.

Additionally, a series of ground anchorages have been installed into fossil scree and microdiorite, on behalf of the TRRL. These are double-corrosion protected strand anchorages as installed for the main works at Pen-y-Clip. The research project is intended to monitor the long term performance of anchorages, and will include exhumation at a later date to examine any long-term corrosion effects.

Review and summary

The Pen-y-Clip project was one of the first major ground anchorage projects to be undertaken with the recommendations and requirements of BS8081 in mind. The ground anchorage tendons are double-corrosion protected effectively as defined by BS8081 and comply with all the requirements contained therein, with respect to the corrosion protection of the system.

The project specification set extremely stringent requirements in terms of anchorage design, drill-hole sizes positioning and tolerances. To date all the required tolerances and design requirements have been met. In addition to the stringent performance requirements, the ground conditions have been described as some of the most difficult and abrasive in the country. Boulders within the glacial till and scree deposits, for example, varied from cobble size to several cubic metres in volume. The innovative drilling systems adopted for the works, together with their constant on-site development, and the extensive pre-contract ground anchorage programme were all instrumental in the excellent progress made.

The careful programming of the anchor fabrication from the factory and the delivery of batches of completed tendons to site on purpose-made transporters allowed the high quality, and expensive, double corrosion-protected tendons to be held on-site for the shortest period of time and protected from physical damage and abuse in the harsh working environment.

The anchorages have been installed on steep, potentially unstable, hill slopes, with several construction disciplines and processes vying for possession of the same piece of ground. Despite these difficulties the walls were constructed to an accelerated programme. That the anchored walls have progressed so well reflects to the great credit of all the parties involved with the project.

References.

1. BRITISH STANDARDS INSTITUTION. *BS8081 British Standard Code of Practice for Ground Anchorages.* British Standards Institution, London, 1989.
2. ARBER A. W. and SELLEY P. J. Pen-y-Clip Site Investigations. *World Tunnelling,* 1991.
3. BRITISH STANDARDS INSTITUTION. *DD81 Draft for Development: Recommendations for ground anchorages.* British Standards Institution, London, 1982.
4. ANON. New ODS developed for demanding site conditions. *Geodrilling,* 1991, No. 66 .
5. HARRIS J.S. Ground anchorages: drillhole accuracy determining device - the Fotobor. *Proc. Inst. Civ. Engrs. Part 1,* 1987, **82**, pp 635-638.
6. BRITISH STANDARDS INSTITUTION. *BS5750 Quality systems.* British Standards Institution, London, 1987.

Acknowledgements.

This Welsh Office scheme was designed and is supervised by Travers Morgan and the Main Contractor for the project is John Laing Civil Engineering Ltd. PSC Freyssinet Ltd were awarded the sub-contract for the installation of the ground anchorages to the retaining walls, and Turner Geotechnical Associates were engaged by them for the Contractor-related aspects of the anchorage design. The drilling for the ground anchorages was undertaken by Saxton & Co (Deep Drillers) Ltd as sub-contractors to PSC Freyssinet. The pre-contract test anchorage programme was carried out by Ritchie's Equipment Ltd.

The Authors wish to thank the Welsh Office Highways Directorate for permission to publish this paper.

Stereographic projection methods used in the design of anchor orientations for retention of a joined rock-mass slope

P. M. MAURENBRECHER, University of Technology, Delft

Rock-mass anchor reinforcement design can make use of kinematic admissibility concepts for slope stability in jointed rock masses. By re-orientating the principal stresses within a rock mass using anchor cable reinforcement the danger exists that some joints may be subject to tension forces or that others may fail locally in shear. With the aid of stereographic projections the anchors can be orientated in three dimensions to ensure that the superimposed anchor-reinforcement forces act normally to joint faces so as to take maximum advantage of the shear strength along the joint faces. A design example is given for a retaining structure for a highly jointed rock slope to prevent planar sliding. An anchor design concept is also given for preventing toppling/spalling failures in rock slopes.

Introduction

Stereographic projection techniques are not very familiar to civil engineers in design work. They would recognise such projections in small scale global maps such as of the World. Geologists have been using projection techniques not to draw their geology maps but to describe axes of orientation in the crystals of minerals first developed by Franz Neumann in 1823 (ref.2) and to graphically represent discontinuities such as joints, bedding planes and faulting found in the rock mass.

The concepts of stereographical methods were introduced into rock mechanics in the last two decades (refs 3 and 5). Two similar works on stereographical methods for rock slopes are given in text books compiled for civil engineers (refs 8 and 9).

Stereographic rock mechanics concepts

Projecting not only discontinuity orientations in the ground mass but also the stresses and force directions is considered in all the publications and examined more closely by Priest (ref. 7). References 8 and 9 best summarise

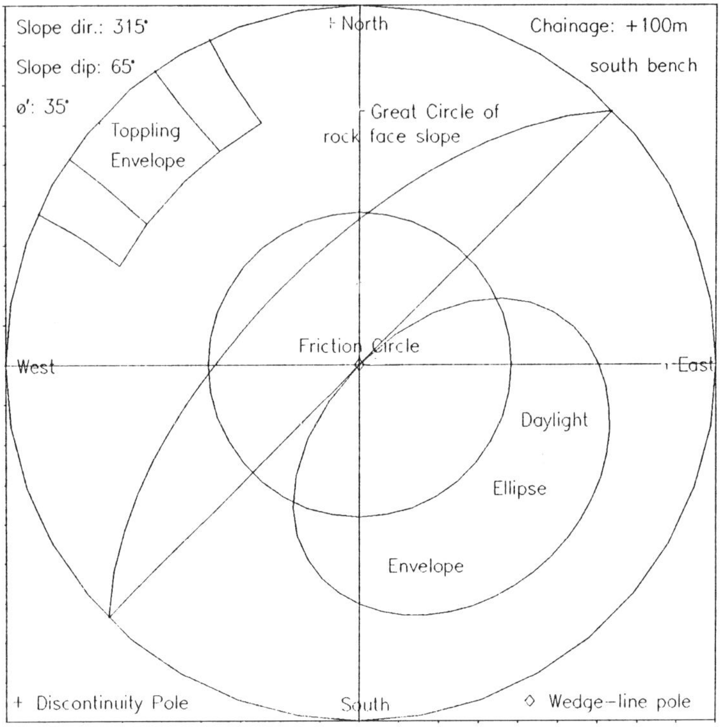

Fig. 1 Stereographic projection of kinematic admissibility envelopes for rock slopes (refs 8 and 9)

the work done on rock slope stability. The authors have produced elegant envelope diagrams enabling both the determination of kinematic admissibility of a rock block to sliding and "toppling" from a rock face. The sliding aspect involves only geometrical orientation of the rock discontinuities with respect to the rock face angle and direction. The rock discontinuities are plotted on the stereographic projection graph as poles, points which both show the direction of dip of a discontinuity and the magnitude of the dip. The rock face is plotted as a series of poles producing an ellipse showing the variation of steepness from the maximum slope angle to a strike or horizontal contour line. Discontinuity poles which fall within this ellipse "daylight" as they are less steep and hence potential sliding could occur. Fig. 1 shows the stereographic projection for a slope of 50°. A further envelope which considers "stress" elements is introduced for further analyzing the sliding mechanism. The envelope consists of a circle which truncates the ellipse into an even smaller envelope. The radius of this circle is the friction angle. Sliding is only considered to occur due to gravity when the discontinuity plane is steeper than the friction angle, stated otherwise: the weight component of a rock mass parallel to the discontinuity exceeds the product of the normal component and the tangent of the friction angle. The circle in Fig. 1 represents a friction angle $\phi = 35°$.

The remaining envelope shown in Fig. 1 also considers stresses in the rock face and is known, rather confusingly, as the "toppling" envelope. The "toppling" stress concept (ref. 4) is not "overturning" as the term suggests, i.e. the weight vector falls outside the base of a rock block such as a slender inclined rock pillar. The term "toppling" describes a model which precedes toppling failure: spalling. The model assumes that the boundary condition at a rock face consists of a major principal stress component parallel to the slope and the minor principal stress is zero (the rock air interface). This condition has been shown in photoelastic modelling (ref. 6) and by analysts working with finite element models. Locally (at the rock face), if the component of the principal stresses along a discontinuity exceeds the product of the normal component and the tangent of the friction angle, sliding will occur (see Fig. 9). The sliding will be local and limited as kinematically admissibility is not satisfied as the discontinuities dip directions are opposite to the rock face slope. Furthermore the principal stress orientation will change within the slope rock mass. If limited sliding does occur it will resemble spalling rather then toppling, though with progressive deterioration of the rock slope due to spalling toppling may ensue. Such slope deterioration has been observed in the field.

Rock anchor orientation

Two types of mechanism can be considered to occur in a rock slope which may require stabilisation by a retention structure, sliding of a large mass or local spalling. Sliding could occur along one plane, ("planar" sliding), or along two or more planes, ("wedge" sliding). Preliminary use is again made by stereographic methods to examine the kinematic admissibility of a potential wedge failure; if the pole of the line of intersection (the normal to the line of intersection in a vertical plane) of two discontinuity planes falls within the daylight ellipse of the slope potential sliding may occur. The friction boundary, however, is representative only on the planar model. Wedge stability analysis can be simply be performed with the aid of stereographic projections to obtain relevant geometrical parameters (ref. 5). The kinematic admissibility envelopes can be used for wedges for preliminary assessment of wedge stability though further analyses for all three mechanisms is required as the projection does not consider the effects of ground water pressure, cementation joint strength, roughness scale effects, continuity and spacing. The type of structures used to stabilize slopes are either a gravity retaining wall or pre-stressed anchoring cables usually in combination with a concrete retaining wall or concrete beams cast flush onto a rock face to spread the anchor loads. In some instances it is conceivable that the foundation abutments of arch structures (bridges or dams) can be designed to cause the abutment loads to prevent a potential block from sliding or toppling. Though such a structure could stabilize a slope, the slope would probably be stabilized separately for safety during construction.

The advantage of using an anchoring system is that retention forces are immediately mobilized through pretensioning whereas a gravity wall would have to allow some active movement of the rockmass before the retention forces are mobilized. This would be undesirable with regard to tunnel portals and bridge foundations. Guidance on the design of anchors is given in BSI DD 81:1982 (ref.1). The section dealing with rock slopes illustrates the various mechanisms of rock slope failure and then through a simple two dimensional mechanical model of a rock block sliding introduces a force P that is required to stabilize the block. The two dimensional model is based on a section through a three-dimensional rock mass.

Most civil engineering works deal with the rock mass near or at the surface. Weathering effects such as chemical or mechanical deterioration can both reduce the shear strength and increase the fracturing. Often such material has to be either self supporting such as a rock cutting or support extra loads from typical structures such as a bridge foundation and abutment, a tunnel portal or embankments. The question remains as to the most ideal orientation for an anchoring system as the orientation brought on to stabilize a block model which is supposed to represent many different possible discontinuity orientations may cause sliding along discontinuities which do not fit the simplified model. To examine this aspect use is made of the kinematic admissibility envelope diagram. The friction cone envelope represents the vertical force due to gravity. Hence by rotating the envelope to an orientation representing an anchor force the effects of the anchor force superimposed on the discontinuities can be examined. In this way the anchor orientations are determined: to ensure as many discontinuity poles as possible are captured by the anchor's friction envelope.

Anchoring for plane failure mode

Field geological-compass readings were made of discontinuities in a peridotite rock slope situated underneath a bridge foundation. Using the compass the dip and orientation of discontinuities are recorded. The individual poles of the discontinuities, indicating their orientation and dip as points, are plotted on a lower hemispherical projection (Fig. 2) and their density distribution, is given Fig. 3.

In Fig. 4 the kinematic admissibility diagram is superimposed on the polar distribution diagram indicating potential sliding failure along a major dis-

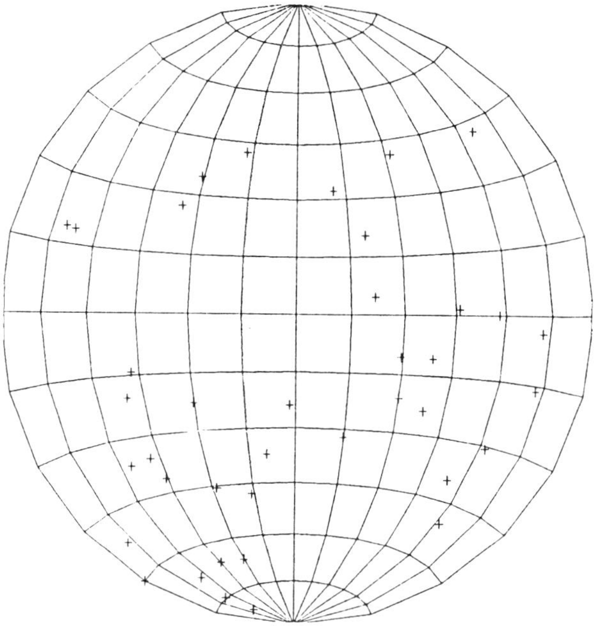

Fig. 2 Stereographic projection of discontinuity poles of peridotite rock

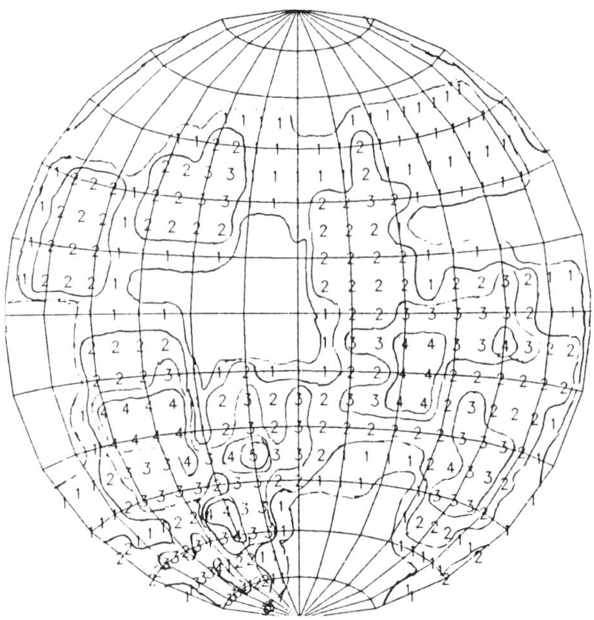

Fig. 3 Density plot of projection of discontinuities (poles per 30° grid at 7.5° centres)

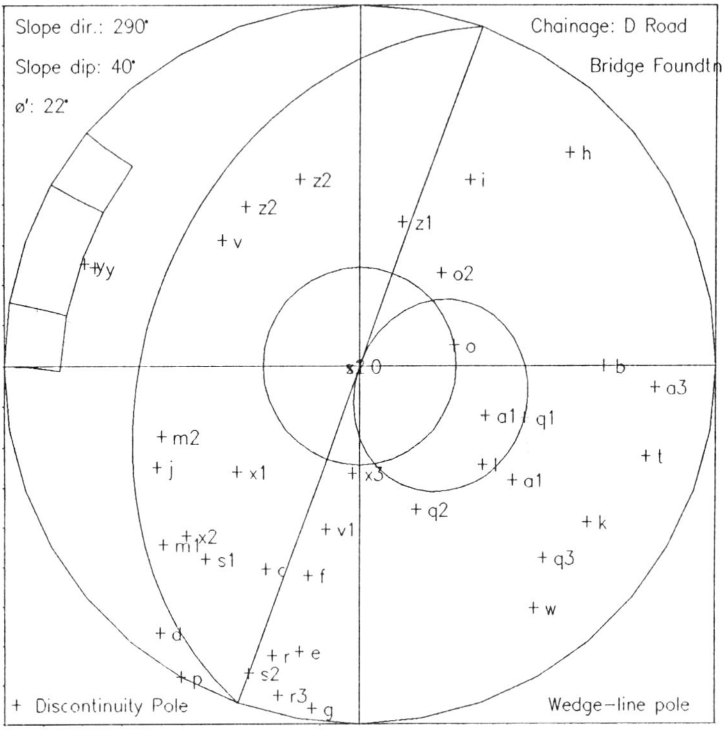

Fig. 4. Kinematic stability envelopes superimposed on pole survey

continuity, "al" and significant discontinuities "o" and "ql". A very low friction angle ϕ of 22° was chosen for determining rock anchor forces, though over-all friction angles of 30 ° probably prevented the existing slope from failing. The low friction was found for highly weathered discontinuities, attributed to weathering of the peridotite rock along the joint planes into talcum.

A simple stability analysis in Fig. 5 suggests the optimum orientation of the anchor would be near horizontal though with a drop in the factor of safety F_2. F_2 is defined as the ratio of restraining forces divided by the disturbing forces, (see ref.5 on discussion on factors of safety Appendix A).

This would result in drilling holes underneath an existing bridge foundation. A steeper orientation would require a larger force (Fig. 5) to satisfy the simple stability model. The steeper anchors, though, have the effect of rotating the friction circle. The central position of the friction circle is based on vertical principal stresses induced by the force of gravity. The applied stress field of the anchors have the effect of rotating the principal (near vertical) stresses by the appropriate amount of rotation resulting from the anchor alignment, the normal stresses acting on the critical sliding planes are in-

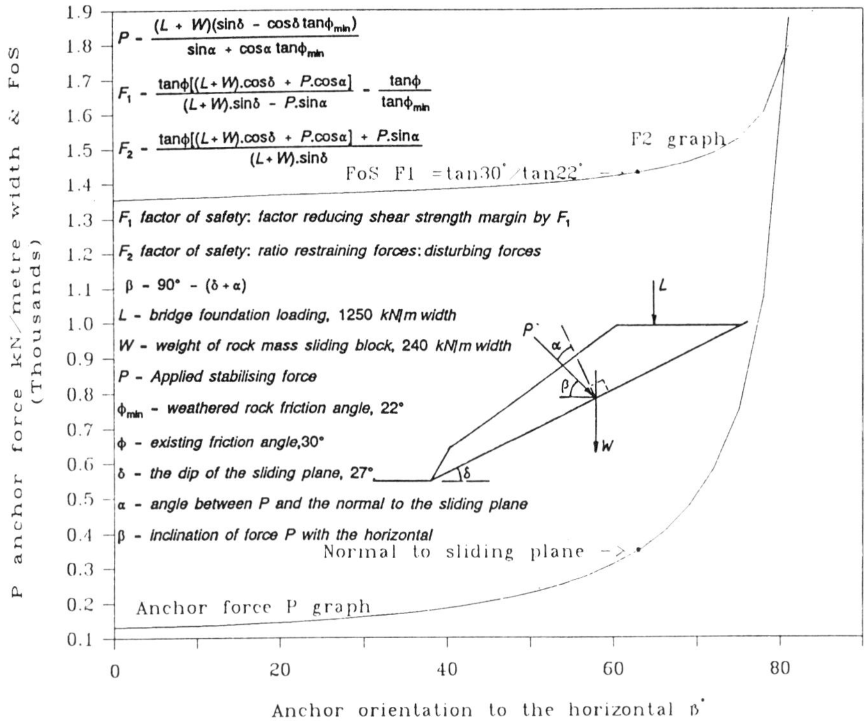

Fig. 5. Variation of retention force P and the resulting factor of safety F2 with anchor orientation

creased (increasing the shearing resistance) without the attendant increased shearing stress.

Appropriate optimum rotation in three dimension can be simply achieved by rotating the friction circle envelope to "capture" the critical discontinuity poles (see Fig. 6.). The horizontal anchors, shown as split ellipses in Fig. 6 would not increase the shear strength of the critical planes, and could conceivably cause some shear deformation resulting in tensioning relaxation. The final design is sketched in Fig. 7; anchors have been designed in four different directions to increase the shear resistance along the discontinuities and they werestressed to resist the loading of the slope and bridge foundation.

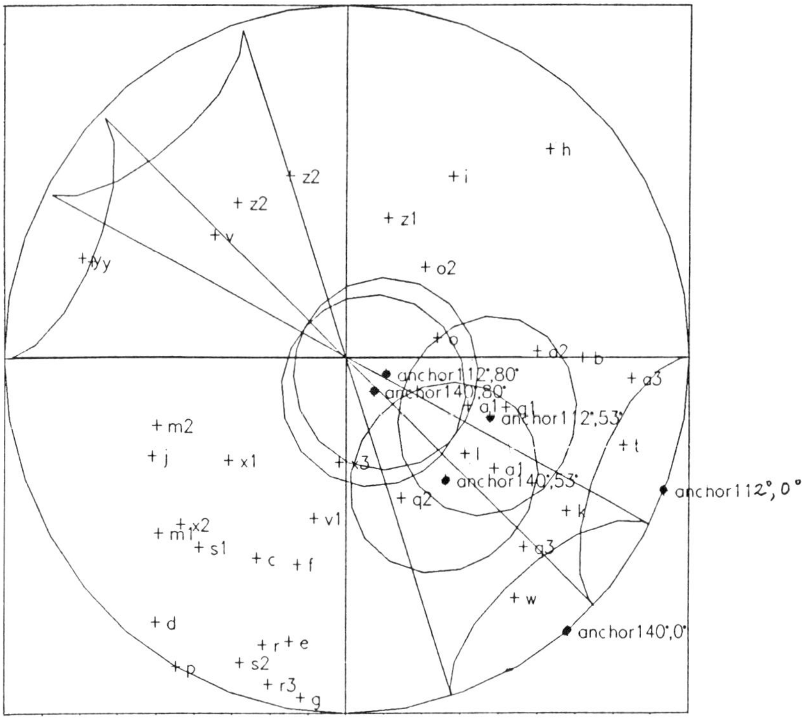

Fig. 6. Anchor friction envelopes superimposed on pole survey

Anchoring for spalling/toppling failure mode

The anchor design approach is based on sliding stability. Often rock slopes can also spall and subsequently topple. The spalling mechanism occurs initially locally at the surface and hence only local reinforcement such as rock bolts would prevent further progressive deterioration. (Note: Often rock masses can have locked-in or residual stresses resulting from previous tectonic movement, unloading due melting of ice or erosional processes. Stress relief can cause spalling. The spalling mechanism described here assumes stresses induced by gravity only.) Spalling cannot be investigated in a similar manner as the sliding mechanism by rotating the friction envelope. Stability occurs when

$$\delta < \phi + (90° - \alpha) \qquad (1)$$

where δ = dip discontinuity, ϕ = the friction angle and α = the slope of the rock face. By introducing a normal force on the rock face through a rock bolt exerting a normal stress causing a ratio μ with the major principal stress (see

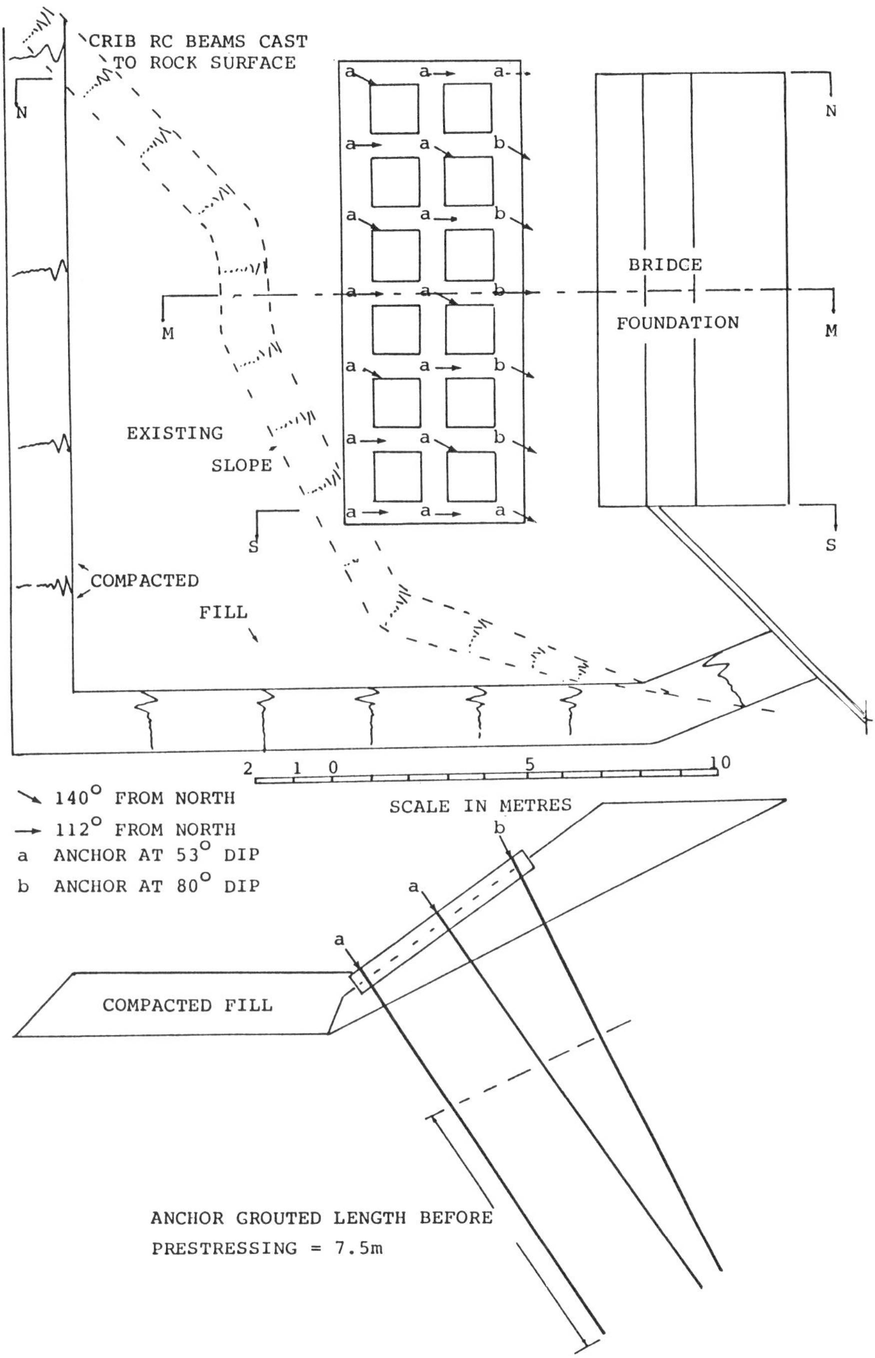

Fig. 7. Sketch of final anchor orientation design

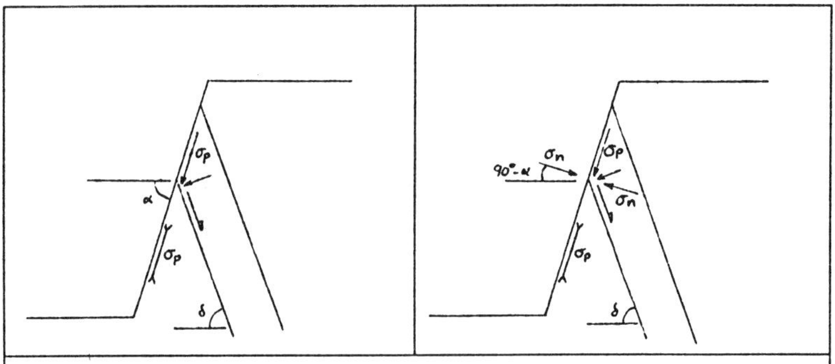

δ = dip of discontinuity
α = slope of the rock face
ϕ = friction angle
Θ = tangent angle of ratio μ
μ = ratio normal stress on rock face with parallel stress on rock face.
σ_n = normal stress
σ_p = parallel stress

from Goodman and Bray, 1977:	Introducing a normal stress on the rock face:
Stress component initiating slip:	$\sigma_p.\sin(\alpha-(90°-\delta))-\sigma_n.\cos(\alpha-(90°-\delta)) <$
$\sigma_p.\sin(\alpha-(90°-\delta))$	$(\sigma_p.\cos(\alpha-(90°-\delta))+\sigma_n.\sin(\alpha-(90°-\delta)))\tan\phi$
Stress component resisting slip:	$\sin(\alpha-(90°-\delta))-\mu.\cos(\alpha-(90°-\delta)) <$
$\sigma_p.\cos(\alpha-(90°-\delta)).\tan\phi$	$(\cos(\alpha-(90°-\delta))+\mu.\sin(\alpha-(90°-\delta)))\tan\phi$
Hence if	let $\mu = \tan\Theta$ ($= \sigma_n/\sigma_p$)
$\sigma_p.\sin(\alpha-(90°-\delta))>\sigma_p.\cos(\alpha-(90°-\delta)).\tan\phi$	$\sin(\alpha-(90°-\delta)).\cos\Theta - \cos(\alpha-(90°-\delta)).\sin\Theta <$
or	$(\cos(\alpha-(90°-\delta)).\cos\Theta + \sin(\alpha-(90°-\delta)).\sin\Theta).\tan\phi$
$\tan(\alpha-(90°-\delta))>.\tan\phi$	above has the form $\frac{\sin A.\cos B - \cos A.\sin B}{\cos A.\cos B + \sin A.\sin B} =$
$\alpha-(90°-\delta)>\phi$	$\tan(A+B)$, hence
$\delta<\alpha-(90°-\phi)$ slip will occur	$\tan(\alpha-(90°-\delta)+\Theta)<\tan\phi$
	$\delta>\alpha + \Theta - (90°-\phi)$ to prevent slip

Fig. 8. Anchoring design concept to prevent spalling

Fig. 8) then tangent angle Θ of this ratio will influence the relation by

$$\delta < \Theta + \phi + (90° - \alpha) \tag{2}$$

The stress-ratio angle Θ in equation 2 would reduce the size of the "toppling" envelope. The magnitude of the normal stabilizing force depends on the magnitude of the slope principal stress. This can be assumed to be equal to the product of the density of the rock material and the height of slope above the anchoring position; ideally the stress field should be determined from in situ strain measuring systems.

Conclusions

Three modes of failure in jointed rock slopes: planar sliding, wedge sliding and toppling can not only be examined for their stability using stereographic techniques commonly used in geological mapping but can further be used to aid design of anchor retention methods for the three modes of failure. The techniques are especially useful in understanding mechanisms in three dimensions.

References

1.BRITISH STANDARD INSTITUTION. *DD 81:Draft for development: Recommendation for ground anchorages*. British Standard Institution, 1982, 123pp

2. ESCHER B.G. *Algemene Mineralogie en Kristallografie*. G.Naeff Gravenhage, 1935, p 50.

3. GOODMAN R.E. *New Introduction to Rock Mechanics*. John Wiley and Sons York, 1980, pp 478.

4. GOODMAN R.E. and BRAY F.W. Toppling of Rock Slopes. *Proc. Rock Engineering for Foundations*, Boulder, Colorado, 1977, Vol. 2, pp 201-234

5. HOEK E. AND BRAY F.W. *Rock Slope Engineering. 3rd Edition*, Butterworths, London, 1981, pp 358.

6. LA ROCHELLE P.*The short term stability of slopes in London Clay*. PhD Thesis, University of London, 1960.

7. PRIEST S.D. *Hemispherical projection methods in rock mechanics*. George Allen and Unwin, London, 1985, pp 124.

8. RICHARDS L.R. , LEG G.M.M. and WHITTLE, R. A. Appraisal of stability conditions in rock slopes. in *Foundations in difficult ground*. Butterworths, London, 1978, pp 449-512.

9. RICHARDS L.R. and ATHERTON D. Stability of rock slopes. in *Ground Engineer's reference book*, Butterworths, London, 1987, pp 3-16.

A temporary secant mini-pile wall propped by a deep jet grout raft

R.L. NEWMAN and C.S. COVIL, Arup Geotechnics, UK

Introduction

The Midland Bank in St. Helier, Jersey, required an extension to their existing building which included a new four storey building with a 4m deep basement. The paper describes the problems associated with basement construction in an area with difficult ground conditions comprising soft compressible soils and a high sub-artesian water pressure. The site is surrounded by traditional granite block type buildings constructed on shallow foundations, as well as roads and buried services.

The temporary works for the construction of the basement had to achieve the following objectives:

(a) provide a retention system which would minimise ground movements;
(b) provide a solution to overcome the high water pressures and ensure a relatively dry and stable base to the excavation.

The Paper describes the solution involving the use of a secant mini-pile wall in conjunction with a deep horizontal jet grout raft. The combined system was designed to resist the sub-artesian water pressure, while the jet grout raft provided an in situ prop to the retaining wall prior to excavation which greatly minimised ground movements. Instrumentation readings demonstrate the success of this temporary works retaining wall solution.

Proposed development

The development is situated in the congested centre of St. Helier as shown on Fig. 1. This created severe problems for access of plant and accommodation of equipment on site. Fig. 1 also shows the dimensions of the site which are only 13.5m by 7.5m.

Two buildings which previously occupied the site had been demolished making way for a new building comprising a single level basement, ground floor and four upper floors. The presence of a 'settlement-sensitive' building on the eastern boundary of the site, shown in Fig. 1, required particular attention to ensure that it did not suffer adversely from ground movements during excavation.

Retaining structures. Thomas Telford, London, 1993

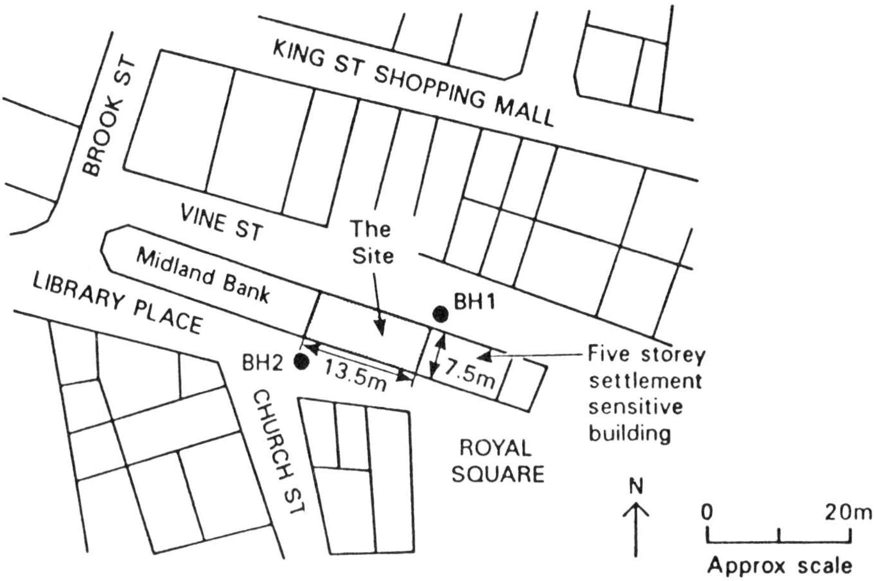

Fig. 1: Site location

Ground conditions

The ground conditions indicated from boreholes 1 and 2 are given below in Table 1.

The made ground consists of a loose brown sand with brick fragments. The windblown sand is a coastal deposit and comprises a loose to medium dense light grey brown fine to coarse sand. The average standard penetration test 'N' value was 10. For design purposes, the material was allocated an angle of a bulk unit weight of 18kN/m^3 and shearing resistance of 30°, a bulk

Table 1: Ground conditions at the site

Stratum	Level to top of stratum (m, local datum)	Thickness (m)
Made Ground	7.7	0.6
Windblown sand	7.1	3.2
Peat	3.9	0.5
Alluvium	3.4	2.7
Head deposit	0.7	1.1
Weathered bedrock	-0.4	1.3
Bedrock	-1.79	

unit weight of $18kN/m^3$ and a drained elastic modulus (E') of $10{,}000kN/m^3$ based on the ratio of $E'/N = 1000kN/m^2$ (ref. 1).

The deposition of the alluvium and peat is related to the Flandrian Interglacial period during which there was a rise in sea level, resulting in the deposition of material in lagoons behind beach bars. The peat is spongy and fibrous with a typical void ratio of 3 and a coefficient of compressibility in excess of $1.5m^2/MN$. The alluvium comprises a reworked loess deposit and is an inorganic clay of low plasticity. It is essentially a fine to medium silt with typically 10% clay content. Vertically orientated fibrous roots of a type characteristic of loess were observed in some of the samples. It was considered that the material would act in a drained manner during excavation, with effective stress parameters represented by a zero effective cohesion intercept and an angle of shearing resistance of 30° based on the clay content (ref. 2). For design purposes a drained elastic modulus of $12{,}000kN/m^2$ was adopted.

The head deposit is derived from solifluction in a periglacial environment and comprises a medium dense to dense clayey, silty fine to coarse angular gravel. Based on standard penetration test data, an angle of shearing resistance of 35°, a bulk unit weight of $20kN/m^3$ and a drained elastic modulus of $45{,}000kN/m^2$ were adopted for design.

The weathered bedrock consists of a strong, slightly weathered fragmented granite rock with fractures often infilled with clay. The underlying unweathered bedrock is a very strong, fresh, closely to moderately jointed fine grained Granophyre.

Piezometric data revealed two groundwater regimes. In the upper windblown sand and alluvium, the water pressure is hydrostatic with a level at 5.6 m (local datum), while in the head deposit and weathered bedrock it is sub-artesian with a phreatic surface at 6.1m (local datum). Permeability tests in the weathered bedrock indicated a permeability of 2×10^{-4} m/s. Data from Littlejohn *et al* (ref. 3) suggest a similar value for the head deposit and average values of permeability of the alluvium from laboratory tests of 2.8×10^{-8} and 5.4×10^{-8} m/s for vertically and horizontally orientated specimens respectively. These laboratory tests did not include the effects of root fabric in the alluvium which were known to increase vertical permeability substantially.

Anticipated problems with basement construction

The anticipated problems associated with the construction of the basement fell into the following categories

(*a*) sub-artesian water pressures;
(*b*) maintenance of existing water levels outside the site;
(*c*) minimisation of ground movements.

The sub-artesian water pressures below the excavation would cause hydraulic uplift pressures which had to be taken into account in both the temporary and permanent works design.

Furthermore as seepage pressures develop in the alluvium liquefaction could occur. This is a common problem with excavations in the St. Helier basin as described by Littlejohn *et al* (ref. 3). These conditions can develop rapidly and as the base of the excavation becomes unstable, the available passive resistance becomes negligible.

Dewatering could not be permitted at the site, due to the presence of peat and the high risk of settlement arising from groundwater lowering outside the site. The temporary retaining wall therefore had to provide an effective cut-off.

Design of the retaining wall also had to ensure that ground movements during excavation did not adversely affect adjoining properties, roads or services. The adjoining five storey building on the east side of the site was known to have poor granite block foundations and would be particularly susceptible to ground movements due to excavation or dewatering.

Temporary retaining wall options

Types of temporary retaining walls

Conventional diaphragm walling or secant pile walling equipment could not be employed because of the small size of the site and the associated problems with access and the presence of flying shores. The installation of sheet piles alongside adjacent 'settlement-sensitive' buildings in such ground conditions was also ruled out.

The retaining wall therefore had to be constructed using highly manoeuvrable, compact equipment with the capacity to install a wall with minimal ground disturbance. At the same time, it had to provide a vertical barrier to groundwater flow and maximise the potential size of the future basement. Mini-piling techniques using a combination of 220mm diameter male reinforced concrete piles installed partly through adjacent non-structural female grout piles using rotary duplex drilling techniques could meet these requirements.

Base stability and passive resistance

A secant mini-pile wall, however, could not cut-off the vertical rise in water pressure due to the sub-artesian condition in the underlying deposits. A temporary retaining wall therefore had to be combined with a further technique to control basal groundwater pressures. Conventional grouting solutions inside the excavation would have required tube a manchette techniques and injections of both cement and chemical grouts to permeate the head deposit and the fractured bedrock as found necessary by Littlejohn *et al*

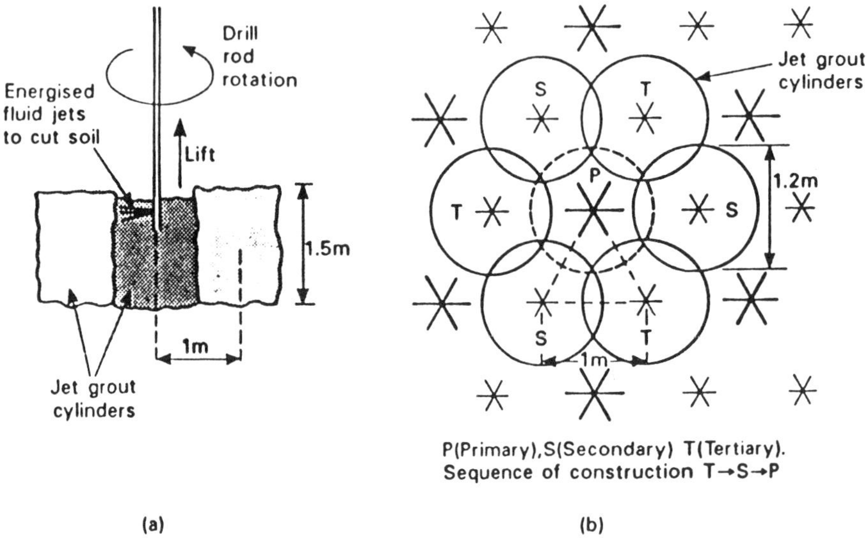

Fig. 2: (a) Section showing construction of interlocking cylinders (b) Plan showing sequence of construction of jet grout raft

(ref. 3). Such measures would have also required extensive testing to demonstrate efficiency and alleviate environmental concerns over the application of chemical grouts.

The solution adopted comprised a secant mini-pile wall in combination with a horizontal structural membrane of relatively low permeability constructed using jet grouting techniques. The process of jet grouting involves the use of high pressure fluid jets to erode the natural soil which is then suspended in the grout mix. For this project the triple tube system of internal rods was employed comprising a water supply line added to the grout and air lines.

The stages of construction of the jet grout raft consisted of the following

(*a*) an initial prebore drillhole to the required depth
(*b*) jetting is initiated with the drill rods rotating at a predetermined rate of 10 rpm and the air-shrouded water jet at a pressure of 40 000kN/m^2 cuts the in-situ soil
(*c*) the drill rods are then lifted at a controlled rate of 10 to 15 cm/min and grout is simultaneously injected to replace the soil. The excess slurry rises up to the annular space in the borehole and is discharged at the surface.

The process produces in-situ cylinders of jet grout material, as shown in

Fig. 2(a). The cylinders are constructed on a triangular grid in a controlled sequence of tertiary, secondary and primary, as shown in Fig. 2 (b) to create a raft of high integrity. The process was designed to create cylinders of 1.2m in diameter and 1.5m in depth with 0.2m overlap between adjacent cylinders.

One possible problem which can arise during jet grouting is heave due to inefficient exhaust of the excess slurry. Survey points on buildings and roads, however, did not detect heave in this case.

The integrated solution was seen to have the following benefits

(*a*) the jet grout raft would provide a low level prop prior to excavation which would substantially reduce ground movements during excavation
(*b*) a mini-pile wall could key into the jet grout raft and so reduce the required length of the piles
(*c*) it would prevent hydraulic uplift
(*d*) the jet grout raft would provide a relatively impermeable horizontal membrane and help ensure that groundwater levels are maintained outside the site during excavation.

Design aspects of the integrated solution

Secant pile wall

If the basal sub-artesian pressures are not controlled, the stability requirements of the secant mini-pile wall necessitate a wall embedment depth into bedrock as shown in Fig. 3(a). Under these conditions, there are two major problems with the retaining wall solution alone

(*a*) installation tolerance of the mini-piles
(*b*) potential ground movements.

To ensure an interlocking mini-pile wall over a depth of 9 m would have required installation tolerances beyond those achievable with mini-piling. To overcome this problem, a double row of mini-piles would have been required with the disadvantage of reducing the available area of the basement as well as a substantial increase in cost.

The horizontal deflection of the mini-pile wall was evaluated using OASYS program FREW and the results are shown in Fig. 3(a) which indicate a maximum horizontal deflection of 28 mm. Analyses of ground movements by Ladd (ref. 4) associated with diaphragm walls in similar ground conditions, suggest that the ratio of maximum vertical displacement (δv_{max}) to maximum horizontal displacement (δh_{max}) is 0.5, giving δv_{max} of 14 mm. Peck (ref. 5) suggests a conservative ratio of δv_{max} to the height of retaining wall as 0.5% for lose sand, giving δv_{max} of 20 mm. Furthermore prediction of deformation of the soil using the simple velocity method described by Milligan (ref. 6) and adopting the profile of wall deflection given by FREW

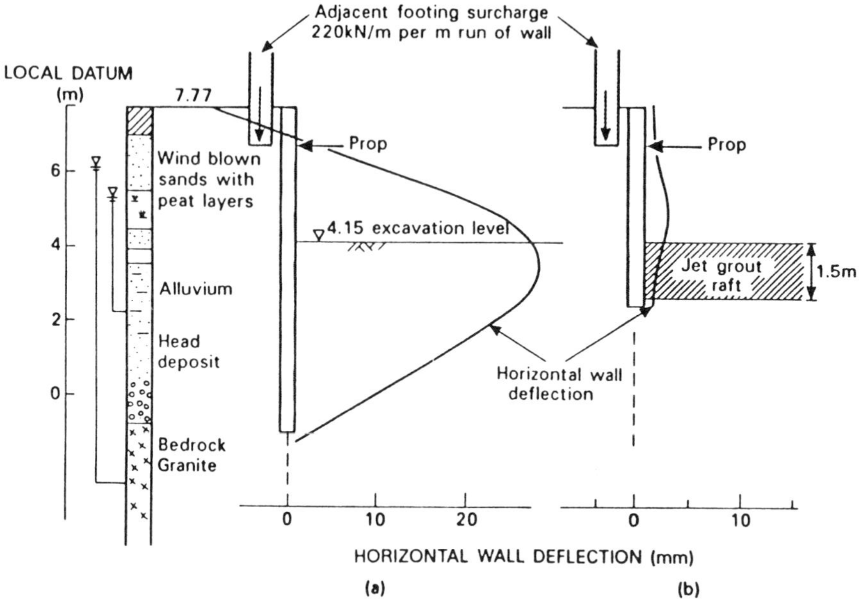

Fig. 3: Required mini-pile wall depth and maximum horizontal deflection, (a) without a jet grout raft (b) with a jet grout raft

gave δv_{max} of 15 mm at the footing level of the adjacent building to the east side of the site. According to Burland and Wroth (ref. 7), the relative deflection of about 1.6 x 10^{-3} arising from the minimum predicted value of δv_{max} of 14 mm would have resulted in severe structural damage of adjacent load bearing brick walls.

Integrated retaining wall solution

To overcome the deficiencies associated with the secant mini-pile wall alone, a 1.5 m deep jet grout raft was installed just below excavation level and was required to exhibit particular design values of strength, stiffness and permeability.

The jet grout raft was designed to prop the base of the mini-pile wall with the mini-piles terminating at the base of 3(b). Consequently, the pile lengths were reduced from 9 m to 5.5 m and, over such a depth, normal positional and alignment installation tolerances ensured a continuous interlock of the secant mini-pile wall negating the need for two rows of piles. The solution also provided a significant reduction in the in the required force and maximum bending moment in the wall as shown in Table 2.

The variation of maximum horizontal deflection as a function of jet grout raft stiffness is shown in Fig. 4. Discussions with practitioners in the field of jet grouting indicated that a minimum mass stiffness associated with jet

Table 2: Maximum prop force and bending moments with and without a jet grout raft

	Without jet grout raft	With jet grout raft
Maximum prop force (kNm)	185	70
Maximum bending moment (kNm)	110	30

grouting an essentially silty material would be 750,000kN/m$^{2.}$ The maximum horizontal deflection of the mini-pile modelled in FREW is about 5mm. For this case, anticipated maximum vertical settlements were likely to small and corresponding relative deflections acceptable.

The jet grout material can behave as a brittle material. It is therefore important to ensure that its operation strength is substantially below its peak strength when used to develop passive pressure. The unconfined compressive strength of the jet grout material, in this case a composition of neat cement grout slurry and natural silt, was expected to be in the range of 2500 to 5000kN/m^2. For the scale of this excavation, however, the mobilised passive pressure proved a small percentage of its ultimate value.

For the jet grout raft to counteract a water pressure of 35kN/m^2 at its base without relying entirely on self-weight, the action of the raft as an inverted arch mechanism with thrusts provided by the active soil forces was investigated. It was found that for active thrusts of 105kN/m^2 and a 1.5m deep raft acting as a two-way spanning slab that the thrust line remained within the middle third of the slab and hence such a mechanism could be theoretically accommodated. As a precaution, however, relief wells were installed directly below the jet grout raft to allow dissipation of water pressure should assumptions not be met.

Seepage analysis using OASYS program SEEP indicated that as long as the jet grout raft exhibited a permeability no greater than 10^{-6} m/s, then water levels outside the site should not be affected.

Instrumentation

Prior to excavation, the following instrumentation was installed

(*a*) vertical settlement gauges were attached to adjoining buildings and on roads

(*b*) an inclinometer was installed in the secant mini-pile wall adjacent to

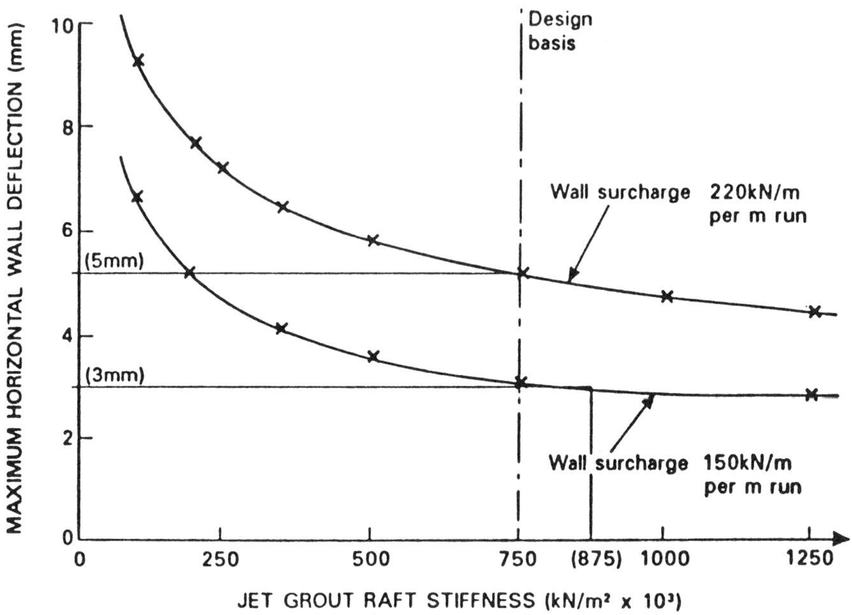

Fig. 4: Maximum horizontal wall deflection as a function of grout raft stiffness

the 'settlement-sensitive' building on the east side of the site
(*c*) Casagrande type piezometers were installed in the alluvium and head deposit outside the site and directly below the jet grout raft inside the site.

Vertical settlement monitoring indicated maximum movements of 2 to 3 mm which are within the accuracy of the surveys.

The inclinometer survey detected a maximum horizontal movement of 3 mm and demonstrated that the deep jet grout raft prop performed according to design requirements. The deflected profile of the secant pile wall was back-analysed using a more accurate assessment of the surcharge loading and according to the relationship in Fig. 4, the stiffness of the jet grout raft was closer to 875,000kN/m^2.

The piezometers outside the site showed no discernible reduction in water level during excavation. The piezometers installed directly below the jet grout raft maintained a head of 3 m which was in agreement with the observation of minimal leakage of groundwater through the raft.

Conclusions

The construction of a 4 m deep basement in the difficult ground conditions of the St. Helier basin was achieved using a secant mini-pile wall with a deep horizontal jet grout raft. Their dual role mitigated the effects posed by the severe ground conditions, and the jet grout raft provided an in situ prop prior to excavation which greatly restricted future ground movements. The solution proved cost effective as it reduced the requisite mini-pile length and number of rows of mini-piles. The instrumentation readings demonstrated the success of the solution.

Acknowledgements

The Authors wish to express their gratitude to the Client Mr. Peter Herrick of the Midland Bank (Premises) Ltd - for his kind permission to publish this Paper.

References

1. STROUD M.A. The Standard Penetration Test - Its Application and Interpretation. *ICE Conf. on Penetration Testing in the UK,* Birmingham, 1988.
2. SKEMPTON A.W. Long Term Stability Of Clay Slopes. 4th Rankine Lecture, *Geotechnique*, 1964, **14**, pp 77 - 101.
3. LITTLEJOHN G.S., NEWMAN R.L. and KETTLE C T. Grouting to control groundwater during basement construction at St. Helier. *Ground Engineering*, January 1989, **22**, No. 1.
4. LADD L. 1989. Unpublished work.
5. PECK R.B. Deep Excavation and Tunnelling in Soft Ground(State of the Art Report). *Proc. 7th Int. Conf. SMFE*, Mexico, 1969, pp. 225-290.
6. MILLIGAN G.W.E. Soil Deformation near Anchored Sheet- piled Walls. *Geotechnique*, 1983, **33**, No. 1, pp 41-55.
7. BURLAND J.H. and WROTH, C.P. Settlement Of building, and associated damage.(State of the Art Review). *Proc. Conf. Settlement of Structures*, Cambridge, 1974, Pentech Press, London, pp. 611-654.

Rock anchors in Southern Brazil - bond strength evaluation

M.C. MACEDO, L. A. BRASSANI and J. MILITITSK,
Department of Civil Engineering, UFRGS, Brazil

The Paper describes a simple technique of evaluation of maximum bond stress at the rock/grout interface which can be used as an indication for the design of rock anchors. The technique involves the use of rock cores recovered from the borings and tested as a composite grout/rock cylinder after being cut at different angles. The results of a series of laboratory tests are compared with field pull tests on rock anchors. The results obtained are encouraging.

Definition of symbols

F_{pu} = tendon yield point;
T_A = initial load;
P_u = rupture load on pull test;
Lef = effective fixed length;
d = diameter of the hole;
UCSn = uniaxial compression strength of grout at n days;
UCS_R = uniaxial compression strength of rock;
DCS - diametrical compression strength of rock (Brazilian Test);
E = average modulus of linear portion of axial stress-strain curve;
CCBS = Composite Cylinder Bond Strength;
S = bond strength determined by proposed method;
α = joint angle on composite cylinders;
σy = uniaxial compressive strength of composite cylinders;
τ_u rock = maximum average bond stress at rock/grout interface;
$\sigma\alpha$ = normal stress on joint inclined at α degrees;
$\tau\alpha$ = shear stress on joint inclined at α degrees;
δ_u = calculated permanent anchor displacement for Pu.

Introduction

Anchored retaining walls have had a large application in Brazil since they were proposed in the late Fifties (ref.6). Nowadays the Dywidag anchor is in widespread use in rock anchoring and a 5 m fixed length is adopted for a

350kN working load. The bond stress under working conditions is assumed to be 0.3MPa at the rock/grout interface.

The tradition established with a large number of successful applications led to no questioning on the bond strength values or minimum fixed length used on routine design. Most of the tests conducted on anchors are acceptance tests only and do not reach failure so that the limit bond strength cannot be verified. The value of 0.3MPa is low when compared with those presented by Littlejohn and Bruce (ref.11). This value is mainly dependent on the rock type and rugosity and grout characteristics. Pull anchor tests to failure may define the maximum bond stress, provided that the tendon does not yield. However, such tests on rock anchors are scarce in Brazil.

There is a great number of anchored retaining walls designed and built on the weathered volcanic rock slopes of South Brazil. The Rio Grande do Sul State Road Department (DAER/RS) partially supported the research in order to improve the knowledge on the available bond strength properties of rock anchors for the local conditions.

Rock Description

Two sites have been chosen for the research. The first is located on the RS235 road between Gramado and Canela towns and the second is on BR116 road near Caxias do Sul town. Both rock masses are part of volcanic events of the Parana Basin which occurred around 130 million years ago (ref.13). Such event had a continental occurrence extending on Northern Argentina, Uruguay, Paraguay and South Brazil.

The rock identification was carried out through microscopical analysis on two thin sections from site 1. Rock texture is glomeroporphyritic with subhedral calco-sodic feldspars present. Silica content is higher than 55% and some micro-fractures are present, with siliceous cementation. In hand specimens the fresh rock has slight grey colour and non-vesicular texture.

Discontinuity structures may be filled or clean with apertures ranging from 1 to 5 mm. Some discontinuities are filled with euhedral calcic mineral. The rock mass was identified as a dacite type.

Deformability parameters were obtained by the uniaxial compression tests according to ISRM (ref.8). The average Young modulus of the rock is 40GPa. The Poisson ratio is close to 0.17 and the modulus ratio (E/UCS_R) is about 330. Tensile strength was obtained from diametrical compressive test (Brazilian Test, ref.9) and varied from 13 to 17MPa.

Thin sections from site 2 presented a porphyritic texture and euhedral pyroxenes when analyzed microscopically. Hand specimens have grey to reddish colours according to the weathering degree. The occurrence of vesicles is common (maximum diameter = 10mm) which sometimes are filled. The rock mass is basaltic and has similar discontinuities to that

Table 1. Geomechanical classification following Bieniawski(ref.5) proposal

Rock Mass Rating - RMR

Conditions	anchor 186	RMR	anchor 2100	RMR
Unconfined Compressive Strength - UCS - (MPa)	142	12	75	7
RQD (%)	95-100	20	90-100	20
Joint Spacing (m)	0.33	20	0.30	20
Joint Condition: Slightly rough surfaces; aperture of 1-5mm (clean)		10		10
Water Condition: moist		7		7
Favorable Joint Orientations		-2		-2
Total RMR (Good Rock)		67		62

described for site 1. Specimens from hole 2100 presented an uniaxial compressive strength of 75MPa and a tensile strength of 14MPa. Table 1 presents the mechanical properties of the rock mass obtained following the ISRM (ref.8) codes and the geomechanic classification proposed by Bieniawski (ref.5) (for foundation cases).

Composite cylinder bond test

A new approach is proposed to evaluate the bond strength at the interface rock/grout. Rock specimens obtained from the site investigation program are used in a simple way to simulate the contact of the grout with the rock mass. The method uses a modified procedure proposed to test the bond strength of concrete/epoxy resins (ref.2).

Sample Preparation

Rock cores are cut at two joint inclinations (30° and 45°). The resulting rock pieces are moistened and put inside cylindrical moulds. Grout is poured on top in order to obtain the composite cylinders which are shown on Fig. 1. Special care should be taken in order to use grout that closely resembles the field material (cement type, water/cement ratio and mixing time). The specimens are taken out of the moulds after one day. Ideally the curing process should simulate field conditions. Considering the usual conditions available in laboratories a moist room is recommended for curing. Further research should be performed in order to evaluate the effect of the curing process on the results. The composite cylinders tested have been cured for 7, 14 and 28 days in a moist room.

Test Description

At the end of the curing period the cylinders were capped as recommended for concrete specimens (ref.1). The test was carried out as an unconfined

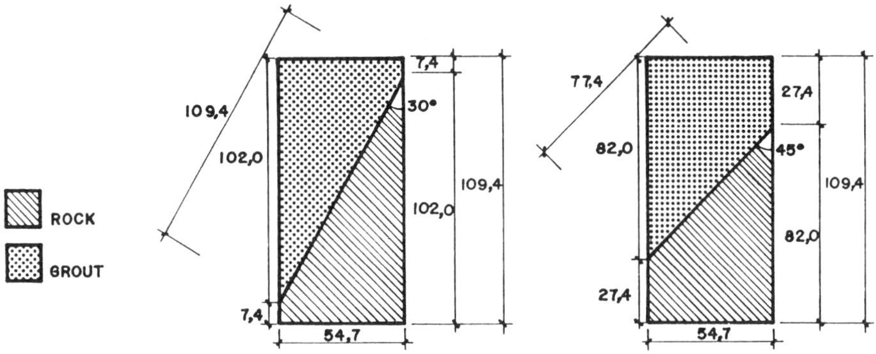

Fig. 1. Test specimens used to obtain rock/grout interface bond strength in laboratory conditions - Composite Cylinder Bond Test

compression test up to failure. The test should have a load application rate of 1500N/s or a strain rate close to 10 mm/min. This strain rate has been used in the tests performed.

A minimum of twelve specimens for each rock/grout and contact angle is recommended. Only the results of test specimens presenting failure at the joint surface should be considered in the analysis.

Test Interpretation

The average uniaxial compressive strength of composite cylinders for each condition is used to obtain the stresses acting on the rock/grout contact surface according to the well known formulae

$$\sigma\alpha = \sigma y \ (\sin \alpha)^2 \tag{1}$$

$$\tau\alpha = (\sigma y/2) \sin 2\alpha \tag{2}$$

Results can also be interpreted graphically in a $\tau \times \sigma$ diagram (Fig. 2). Due to the difficulties in simulating field stress conditions at the contact of rock/grout it is suggested the extrapolation of the results obtained from the 30° and 45° degrees specimens be used to obtain the value S corresponding to a failure condition of $\sigma = 0$ (Fig. 2). This value has been called Composite Cylinder Bond Strength (CCBS).

Test Results

Rock cores from sites 1 (Gramado) and 2 (Caxias do Sul) were tested following the technique described. Fig. 3 presents data obtained from rock specimens taken from anchor 217 and the CCBS value obtained. The experimental points represent the average of twelve results for each joint inclination. Test results on both sites are presented on Table 2.

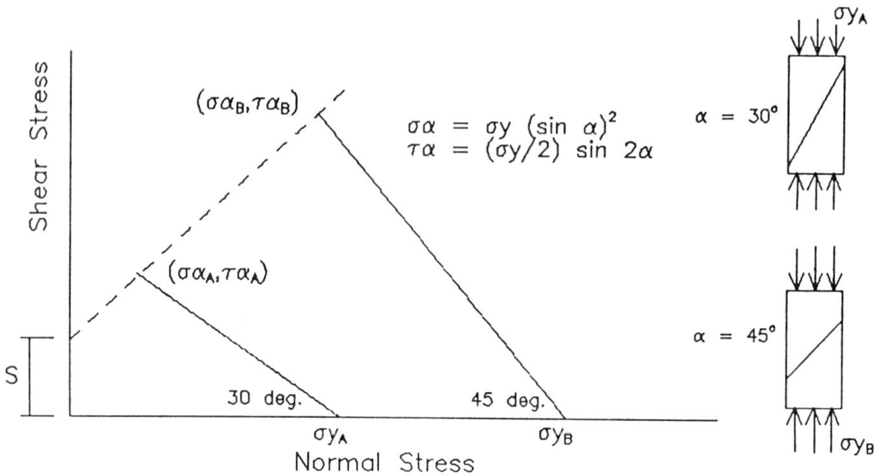

Fig. 2. Suggested approach for interpretation of composite cylinder bond test results

Field Tests

Field pull tests were performed on the rock anchor type usually employed in retaining structures in order to verify the average bond strength and to compare it with the results obtained in the laboratory.

The anchors used locally are of the Dywidag system. The tendon is a 32 mm diameter steel bar with an yield point (F_{pu}) of 683kN and a nominal E tendon = 210GPa. The free length is protected by a coating of cement grout

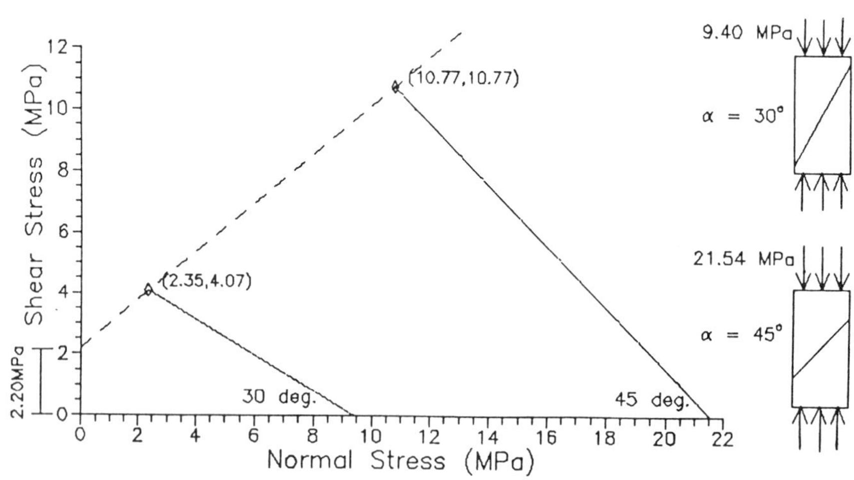

Fig. 3. Composite cylinder bond test results obtained on rock specimens from anchor hole 217. $UCS_7 = 30.84MPa$, w/c ratio = 0.45, $UCS_R = 141.7MPa$, DCS = 16.6MPa

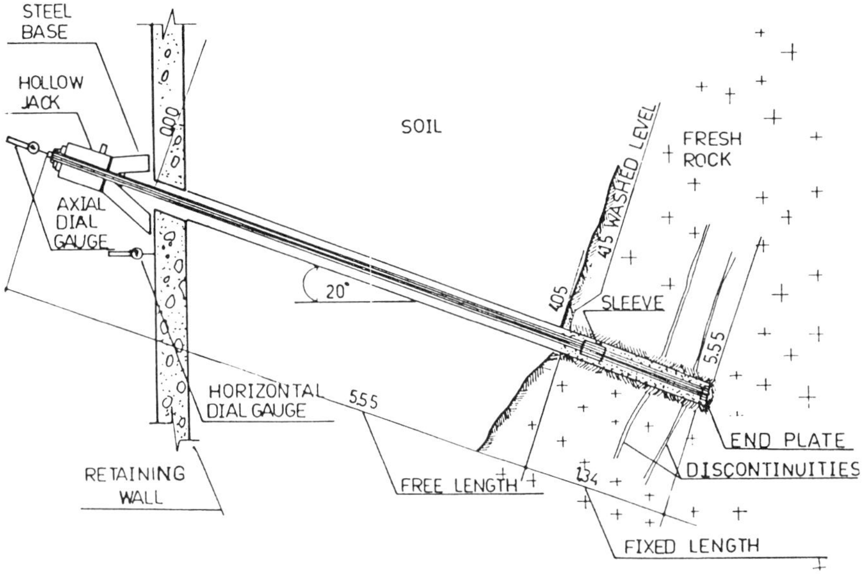

Fig. 4. Details of the tested anchor 217 of site 1 (Gramado, RS235l)

and a polyetylene sheet. The tendon has a variable free length and a fixed length of 500 cm with union steel sleeves as required. The free length is smeared before installation and there is a plate at the tendon extremity. The whole tendon bar and sleeves are pitch painted for protection. The anchor holes had a diameter of 76 mm (NX bits) and inclination of 20 °. No casing was needed for the holes and the average drilling velocity was 0.7 mm/s. After the drilling operation, a Portland pozzolanic cement grout was poured into the hole. The water-cement ratio had a nominal value of 0.6 with excellent control on site 2 and fair control on site 1, according to the Neville's proposal presented by Littlejohn and Bruce (ref.11). The anchors pull tested were similar to this description (Fig. 4) except that they had a shorter fixed length (Lef on Table 2).

The procedure used on operational anchors may leave grout in contact with the theoretical free length. Such a condition was avoided in the tested anchors by careful washing of the free length immediately after grouting.

Four anchors were tested following the methods presented by ISRM (ref.10) and ASTM (ref.3). An hydraulic jack and manual pump were used with a pressure gauge calibrated and accurate to 10kgf/cm^2 ($\approx$10kN) for the pull tests. Dial indicators with 0.0lmm accuracy were used to monitor the axial anchor and horizontal wall displacements. A steel base has been used to

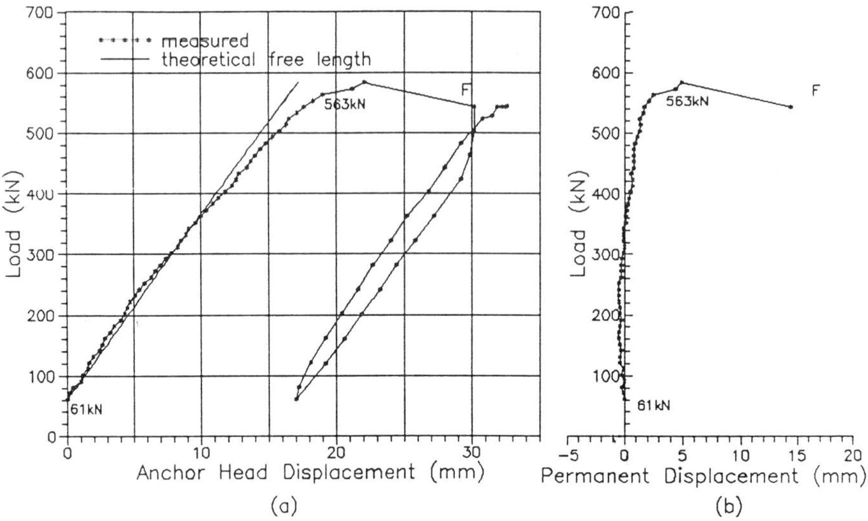

Fig. 5. Pull test results of anchor 217. (a) Load x displacement curve; (b) Permanent displacement curve. P_u *= 563kN (7 days), Lfr = 5.55m,* UCS_7 *= 6.50MPa,* τ_u *rock*

transfer the load from jack to anchor at axial direction.

Anchor 217 was initially loaded to 61kN (Ta) seven days after installation (Fig. 5(a)). Load steps of l0kN were then applied from T_A up to failure (Pu = 563kN). After failure the anchor was unloaded to 61kN. The calculated permanent displacements of the fixed length (without the elastic displacements of the free length are plotted on Fig. 5(b). For loads up to 480kN, the permanent displacements were close to zero in the first loading. The reload was performed to verify the behaviour of the anchor after the rupture has been defined. The anchor showed an ultimate load of 545kN.

The other three pull tests were done on anchors 186, 2100 and 3050 following the same procedure. he main characteristics of these anchors and their rupture loads are summarized on Table 2. A more detailed description can be found in Macedo, (ref.12).

Bond Strength

Instrumented rock anchor tests showed that the stress distribution is not uniform along the anchor fixed length (ref.7). The distribution of stress depends primarily on the relative stiffness of the tendon in relation to that of the rock and grout.

The research carried out by Ballivy and Dupuis (ref.4) on in situ and laboratory pull tests confirmed that stresses at grout/tendon interface are initially exponential but, at rupture loads, the stress distribution becomes uniform. No measurements of the bond stress distribution at the rock/grout interface are known, probably due to the obvious experimental difficulties. For such reasons average bond stress calculated with formula (3) is used in design and have also been used in this paper when comparing results.

$$\tau_u \text{ rock} = \frac{P_u}{\pi \, d \, L_{ef}}$$

Discussion of the results

The composite cylinder bond strength and bond strength calculated from field anchor pull tests are present on Table 2. Laboratory tests on samples from anchor 217 were prepared with the design water/cement ratio. For operational reasons the ratio used on site was higher.The water cement ratio for the other laboratory results and all field anchors was 0.60.

Composite cylinder bond strength (S) obtained in the laboratory conditions are of the same order of magnitude as the ultimate average bond strength (τ_u) calculated from field pull tests using formula (3). Except for anchor 217 from site 1, the CCBS results are lower than the bond strength obtained from field tests. Such result is not surprising and it is probably due to the lack of restraining in the laboratory conditions and the different failure mechanisms. The conditions used to obtain the composite cylinder bond strength value (S) are an over-simplification of the reality. In the authors opinion, however, it gives a strength index for the field conditions of rock/grout interface and it can be used as a rough design indication of the average available field strength. The actual design of such structures should include, in addition, considerations about rock discontinuities and other conditions (e.g. tendon/grout) that can lead to a different mechanism of failure.

Conclusions

A new approach is suggested for the evaluation of the rock bond strength using simple laboratory procedures. Rock specimens are obtained from the site investigation program, cut at two joint inclinations (30° and 45°), used to obtain composite cylinders with grout, tested in unconfined compression up to failure and interpreted in a very simple way to define an index value: the composite cylinder bond strength (S). Based on the comparison of laboratory results so obtained and field bond strength from pull anchor tests, the composite cylinder bond strength (S) proved to be of the same order of magnitude (less than 45% difference, Table 2). The Authors suggest that the CCBS can be used as an indication of bond strength for design of rock anchors.

Table 2. Field and laboratory test results

anchor/ site	laboratory S (CCBS) (MPa)	laboratory UCS grout (MPa)	laboratory w/c ratio	field Lef (cm)	field Pu (kN)	field τu rock (MPa)	field UCS grout (MPa)	field w/c ratio	field δu (mm)	τu/S
217/G[a]	**2.20**[e]	30.80	0.45	134	563	**1.76**	6.50	0.60	1.5	1.27[G]
186/G[b]	**1.27**	16.20	0.60	106	372	**1.47**	9.21	0.60	0.7	
2100/CS[b]	**1.80**	16.24	0.60	100	668	**2.80**	17.20	0.60[d]	0	1.42[CS]
3050/CS[c]	**1.65**	19.58	0.60	50	248	**2.08**	27.10	0.60[d]	1.1	

G - site 1, Gramado. CS - site 2, Caxias do Sul.
a. results at 7 days. b. results at 14 days.
c. results at 28 days. d. control by volume at mixer.
e. not computed for τu/S average.

Acknowledgements

The research was partially supported by FAPERGS (State Research Council), DAER/RS (State Road Department) and Civil Engineering Department/UFRGS (Federal University of Rio Grande do Sul). The contractors SULTEPA and TECNOSOLO provided the field working conditions, workmanship and schedule flexibility necessary for the tests. The first Author has been supported by a grant from CAPES (Brazilian Government Agency).

References

1. AMERICAN SOCIETY FOR TESTING AND MATERIALS. *Practice for capping cylindrical concrete specimens in the laboratory.* ASTM C617-78. Philadelphia, 1978, 4pp.
2. AMERICAN SOCIETY FOR TESTING AND MATERIALS. *Bond strength of epoxy-resin systems used with concrete.* ASTM C882-78. Philadelphia, 1978, 5pp.
3. AMERICAN SOCIETY FOR TESTING MATERIALS. *Rock bolt anchor pull test.* ASTM D4435-84. Philadelphia, 1984, 7pp.
4. BALLIVY G., AND DUPUIS M. Laboratory and field evaluation of the bonding strength of grouted rock anchors. *Canadian Rock Mechanics Symposium*, 1980,**13**, 97-102.
5. BIENIAWSKI, Z.T. Rock mass classification in rock engineering. *Symposium on Exploration for Rock Engineering*, 1976, **2**, 97-106
6. COSTA NUNES A.J., and VELLOSO D.A. Rock anchoring as a technique for construction in slopes, (in Portuguese), *Estrutura*, 1961, **46**, 11-16.
7. HANNA T. H. Foundation in tension - Ground anchors. *Trans. Tech. Publications*, 1982.
8. INTERNATIONAL SOCIETY FOR ROCK MECHANICS. *Suggested methods for determining the uniaxial compressive strength and deformability of rock materials*, Pergamon Press, Oxford, 1981, p. 113-116.

9. INTERNATIONAL SOCIETY FOR ROCK MECHANICS. *Suggested methods for determining tensile strength of rock material* . Pergamon Press, Oxford, 1981, p.119-121.
10. INTERNATIONAL SOCIETY FOR ROCK MECHANICS. *Suggested methods for rock bolt testing*. Pergamon Press, Oxford, 1981, p.161-168.
11. LITTLEJOHN G.S. and BRUCE D.A. Rock anchors - State of the Art, Part 1: Design. *Ground Engineering*, 1975, **13** , No. 3, 25-32 andNo. 4, 41-47.
12. MACEDO M.C. *Anchors in volcanic rocks*, (in Portuguese), MSc Dissertation, (to be published).

Discussion

SESSION 5

Professor C.J.F.P. JONES, University of Newcastle upon Tyne

Dr. Bolton has stated that a check for sliding failure for a mass retaining wall will not be required in the new Code of Practice.

Sliding of a gravity structure leads to active conditions in the backfill; but if the mass structure is a reinforced soil wall, checking for forward sliding on any form of foundation is essential. Reinforced soil structures are internally stabilized, and any forward sliding can lead to a reduction of internal stability, which in the limit can lead to collapse.

D. R. CARDER and A. BROOKES, Transport Research Laboratory

London Regional Office of the Department of Transport commissioned the TRL to undertake measurements of termporary prop loads and wall movements at the A40 Long Lane Improvement Scheme; the results are presented with their permission. The co-operation received from the site staff of Mott-MacDonald Civils and Fairclough Civil Engineering Ltd is also gratefully acknowledged.

Figure 1 shows the temporary raked props which were installed at alternate nominal spacings of 7 m and 5 m to support a bored pile retaining wall during excavation of the berm in front of the wall and during the subsequent construction of a permanent prop slab. The upper end of the prop seating under the capping beam was about 7.5 m above formation level for the permanent prop slab. The props comprised steel tubes 13 m in length which were prestressed to a nominal load of 260 kN. Three props were instrumented with vibrating-wire strain gauges and thermocouples to determine loads and temperatures. Measurements were also made of the lateral movement of the top of the wall using a Geomensor electronic distance measuring system.

Results from all three props were similar and the loads measured on prop B during construction are shown in Fig. 2. Loads rapidly increased during berm excavation and over the following few days. The average loads measured on props A, B and C after completion of the permanent prop slab were 1081 kN, 1188 kN and 1191 kN respectively. Daily fluctuations in recorded prop loads were largely attributable to expansion/contraction of

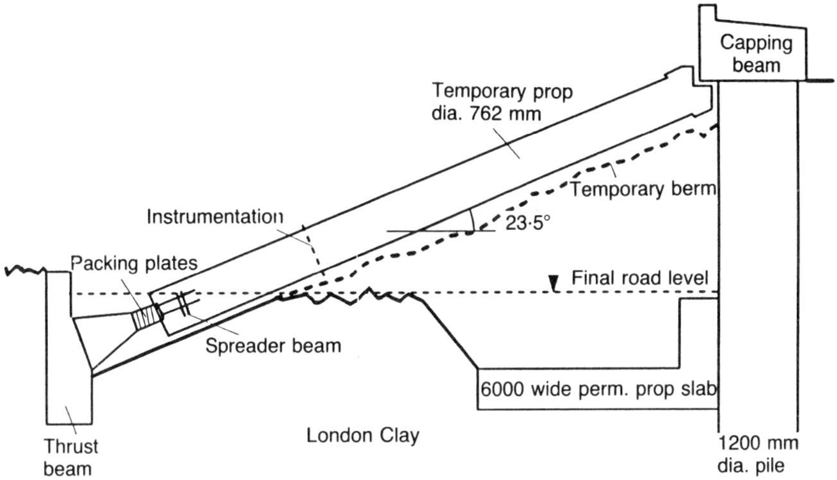

Fig. 1. Cross-section in instrumented area

the steel prop caused by changes in temperature, as illustrated by the results in Fig. 3.

The corresponding load change was of the order of 12 kN/^{o}C. At this site, however, the steel props were only 13 m in length and the variation of load with temperature was not as marked as at another site where 22m long props spanned between the walls of an underpass. A change of temperature of

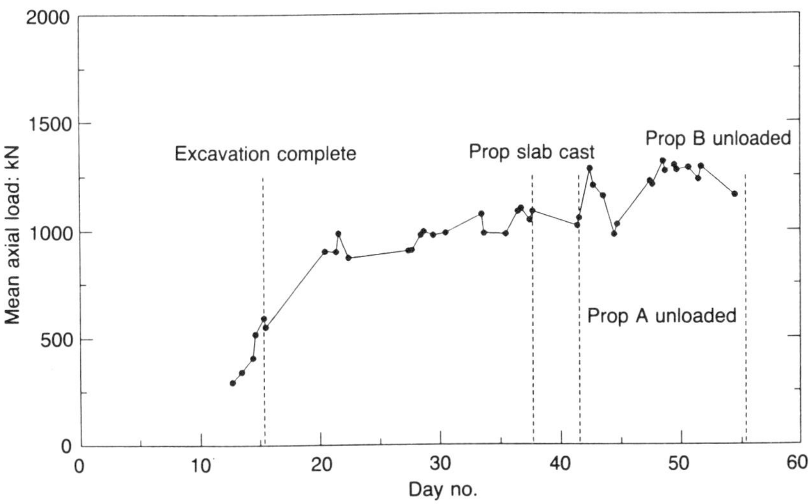

Fig. 2. Variation of load (prop B)

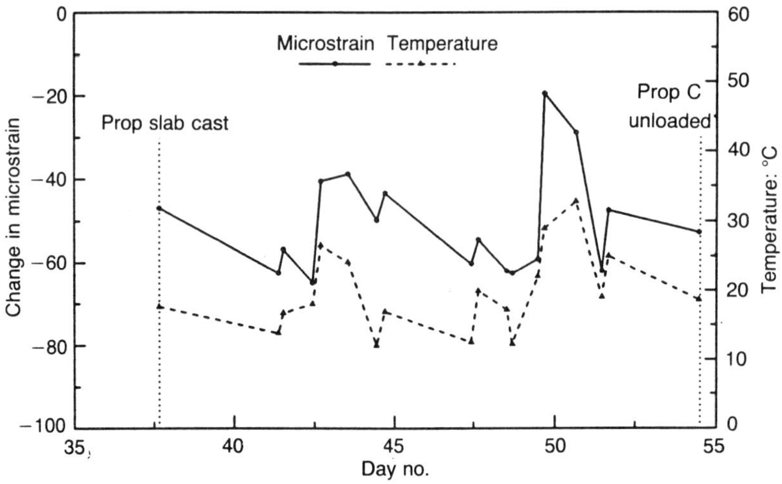

Fig. 3. Variation of prop strain with temperature

about 10°C was recorded over one particular day at this site when no construction was in progress, a prop load increase from about 240 to 430 kN resulted.

The outward movement of the top of the wall whilst the temporary props were in place is shown in Fig. 4. Only small movements of a few mm were measured during excavation of the berm beneath prop B and generally the props appeared reasonably effective in restricting wall movement in the

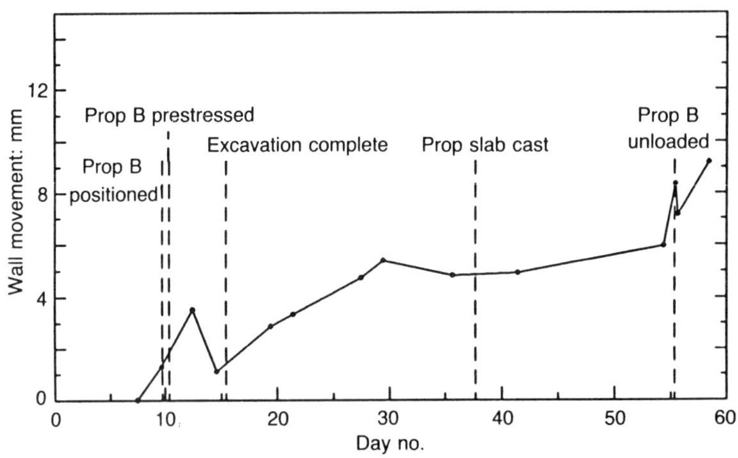

Fig. 4. Top of wall movement at prop B

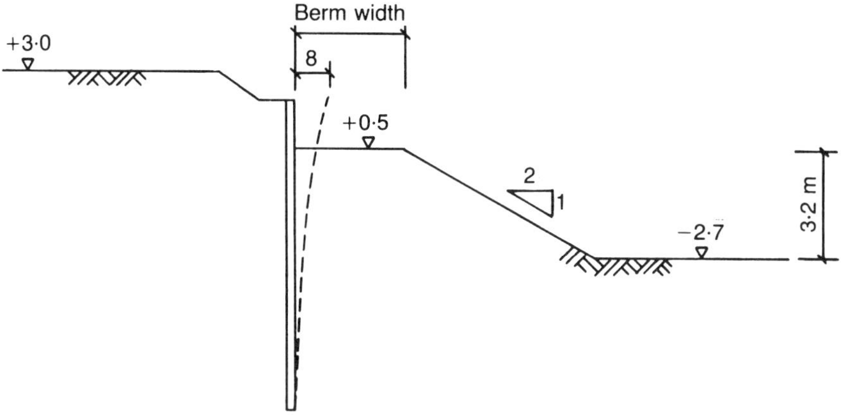

Fig. 5. Construction stage 2

instrumented area. An immediate increase in lateral movement of about 2 mm occurred when temporary prop B was released.

The TRL is continuing to monitor the wall movements in the longer term.

DR E. S. F. LI, Sir Alexander Gibb & Partners

In some sites, such as at the Waterloo International Terminal site, the use of soil berms appears to be the only solution to limit the movement (Figs 5 and 6).

Owing to the enormous span of the excavation at Waterloo, early propping is not possible without first constructing a thrusting block. The anchoring

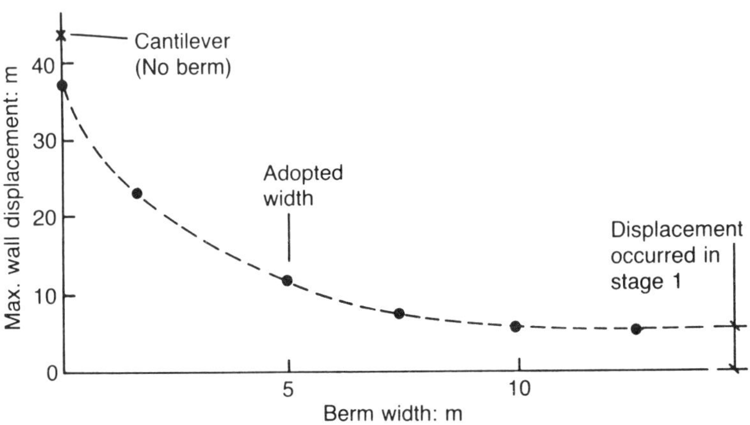

Fig. 6. Displacement against berm width

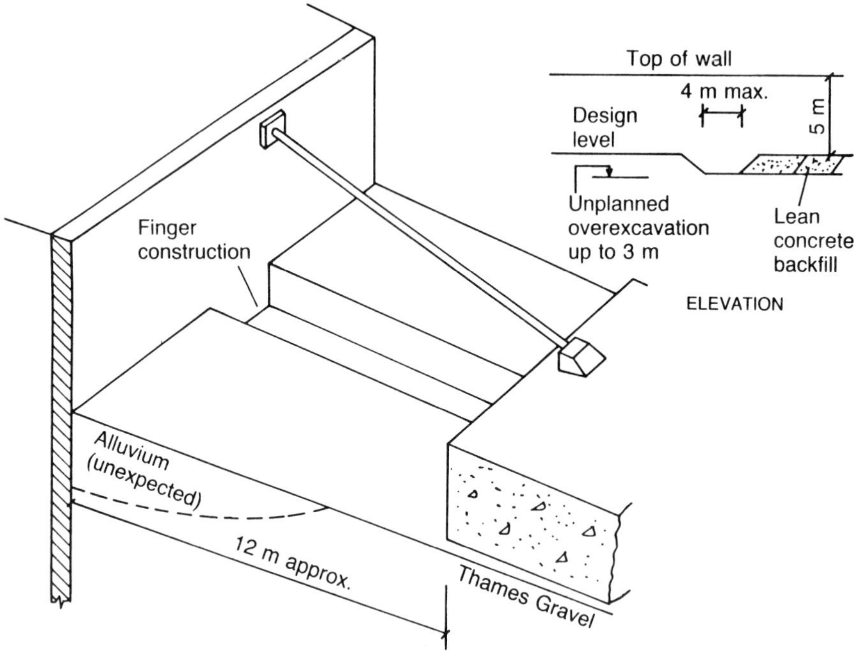

Fig. 7. Finger construction method

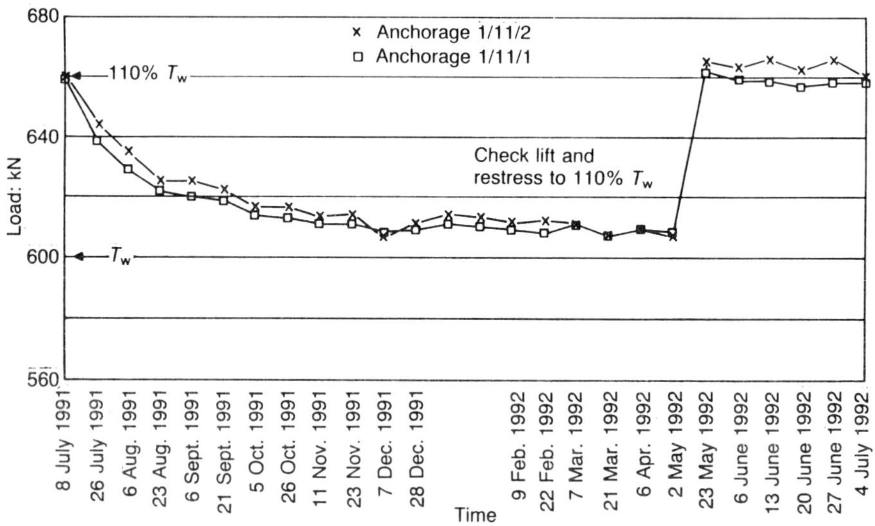

Fig. 8. Retaining wall R8

option is also ruled out on account of the arch foundations behind the wall.

If no soil berm is used - i.e. cantilever - the calculated wall displacement is 47 mm in one case. The results of a parametric study on berm widths show that the displacements vary between 5 mm and 37 mm. In order to select an optimum berm width, it is necessary to compromise between the allowable wall displacement and the restriction to excavation.

There is also another method, namely, finger construction (Fig. 7) which is very effective in limiting movement.

At Waterloo, along the southern boundary, some unexpected soft alluvium was encountered during excavation. To prevent long-term differential settlement, these soft spots had to be removed, which had not been designed for. To do this, the alluvium is removed in the form of 4 m wide trenches, then quickly backfilled by lean mix concrete. The process is started from one end and repeated sequentially to the other end. The finger construction mobilises significant 3D soil strength. The lean concrete backfill can also act as a prop. The movement and prop force were limited to the allowable values in this case.

P. J. SELLEY and M.J. TURNER, Travers Morgan

The anchored panel walls constructed from the top downwards at Pen-y-Clip are frequently referred to as 'Brazilian walls'. In the Paper by Macedo et al. reference is made to a Paper, in Portuguese, by da Costa Nunes. Readers may be interested in a Paper by the same author in English. [1]

The long-term performance of the Brazilian walls and anchored cantilever walls is made possible by 110 glotzl hydraulic load cells with electronic pressure transducers, stainless steel survey pins in many of the panels, and is augmented by additional TRL targets for helicopter-based photogrammetry. The TRL research project in the long-term performance of anchored structures also includes four anchorages, two each into fossil scree and microdiorite which may be exhumed at a future date to determine the nature of corrosion (if any).

A graph showing the behaviour of a typical 600 kN anchorage on a Brazilian wall indicates less than 10% loss in load during excavation of material to the side and below. This was accompanied by a movement of approximately 1 mm which is equivalent to the elastic changes in the tendon. After restressing, the minor fluctuations can be related to temperature.

For the anchored cantilever wall R1, the relationship was measured as $1^{o}C$ = 1 kN which is considerably less than that which has been recorded in props by other delegates. The effect of temperature on load was found to be even less in the Brazilian walls.

Figure 8 shows the topography of the steep mountainside which demonstrates the impracticability of using a conventionally constructed wall. Once trends have developed a more detailed paper will be published. This will be in addition to the TRRL publications on the Pen-y-Clip anchorages.

DR M. M. H. AL-SHAIKH-ALI, DTP Motorway Widening Unit

I cannot agree with Harris *et al.* on their statement in para. 4 of the paper that 'the creep rate shows remarkable uniformity irrespective of soil type'. Given all conditions of loading, grouting and the geometry of the anchors, this statement cannot hold, without qualification. I should like the Authors' to modify this statement in their Paper.

I should like also to ask the Authors the following questions.

(a) How many anchors were tested to failure
(b) Was the F.O.S. based mainly on the grout/soil failures?
(c) How many anchors were tested for creep in varying soil types for a duration of more than three months?

It should be mentioned that anchors may fail as a result of corrosion at the anchor head, with a percentage higher than that recorded for corrosion failure of the tendons at the free or fixed lengths in doubly protected anchors. Would the Authors comment on this.

K. WARD, Sir Robert McAlpine and Sons Ltd

With regard to berms for temporary support, at Holborn Bars, no inward movement of the secant wall occurred while supported by the termporary berm. The berm was 5 m wide across the top, with a volume of about 65 m^3/m run. The top of the wall in cantilever mode actually moved 2 mm away from the excavation, probably as the result of rotational heave.

R. R. W. Harris, Bachy (UK) Limited

Our response to the three elements of Dr Al-Shaikh-Ali's enquiry is as follows.

1. The observation regarding creep rate was one of five statements related to the behaviour of TMD anchors under load. Unfortunately, constrains on the length of published papers precluded further explanation and qualification in listing these behavioural characteristics.

The probable explanation for the observed creep similarity in soils of different strength and composition, is the ability of the re-groutable system to alter the characteristics of soils compaction. It therefore follows that the greater the volume of grout inujected, the more the soil is compacted and the larger the area of grout/soil contact. Thus, for a given limit injection pressure a weak soil will accept a greater volume of grout than a strong soil. The relative deficiencies of the weak soil to act as a load distributive medium will therefore be improved by the pressure injection process.

2. Our reply to the questions raised is as follows.

(*a*) Ten per cent of the anchor tests documented by Bachy were tested to failure; see (b) for further explanation

(*b*) The factor of safety for re-groutable anchors is normally based on the critical creep load (as described in the paper). The critical creep laod is less than the failure laod and it is therefored seldom required to test such anchors to destruction.

(*c*) During definitive trials carried out in 1976, eight anchors were tested for long term creep. Of these, three were observed for more than two years. Obviously, many production anchors completed since 1976 have required intermittent lift-off tests. It is these tests as well as preliminaary test anchors installed on most sites which have contributed to the eventual publication, in 1986, of the French standard for re-groutable ground anchorages.

3. We agree that the anchor head is probably more vulnerable to failure than the fixed or free lengths of the tendons. This vulnerability may be attributed to: variations in temperature and humidity (wetting/drying) not present at depth; the very high concentration of load transfer in the head; the higher probability of construction damage in the exposed tendon after installation and before block off.

Reference

HUNT R.E. and DA COSTA NUNES J. Retaining walls: taking it from the top. Civil Engineering, ASCE, 1975, May.

Introduction to Session 6
Reinforced soils

G.W.E. MILLIGAN, University of Oxford

Introduction

The basic concepts of soil reinforcement are by now well known to most geotechnical engineers, but it may first be useful to review them briefly. Thereafter, the main topics for discussion in this session will be introduced and some points contributing to them selected from the papers submitted to this conference on the subject of reinforced soils.

A vertical or steep slope of soil will not normally be stable, except perhaps in the short term due to cohesion. Such a slope therefore has to be supported in some way by a retaining structure. This can take various forms, as discussed in other sessions in this conference. However in all these cases, the additional forces supporting the soil slope are provided externally; with soil reinforcement they are simply provided internally, by bars, strips or sheets of some suitable material, as shown in Figure 1.

The reinforcement may either be built into soil fill during placement, or inserted into natural ground to stabilise an existing unstable slope or to allow new excavation to take place; the latter technique is normally referred to as soil nailing. The basic principles of both are the same, but the differences in soil types and construction procedures listed in Table 1 may sometimes be significant, as discussed below.

The three main topics chosen for discussion are based on issues raised by

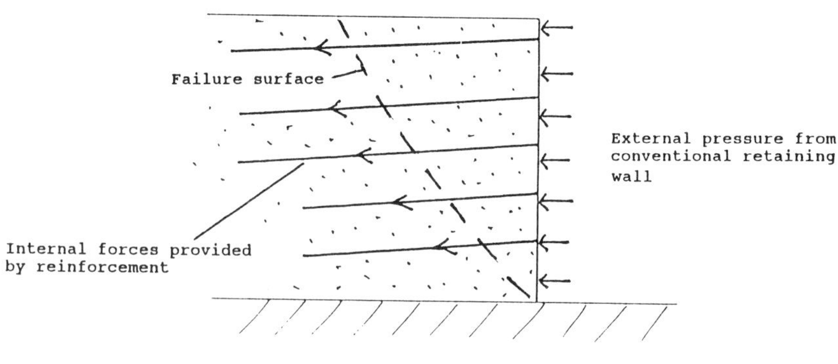

Fig. 1. Basic principle of soil reinforcement

Table 1. Differences between reinforced and nailed soil

SOIL REINFORCEMENT (FILL)	SOIL NAILING (IN SITU)
Soil can be selected Density can be controlled	Soils as found
Drainage easily installed	Drainage less certain
Stresses increase during construction	Stresses decrease
Any reinforcement material Usually palced horizontally Flat strips or sheets - flexible, relatively high bond	Steel (or GRP?) nails only Usually inclined downwards Round bars or angles - stiffer, low bond area
Temporary support possible, allowing control of deformations	Temporary conditions rely on short-term soil strength - potential for local collapse, large movements

the submitted papers or considered by the author to be areas in which uncertainties exist or advances are possible in both design and construction. They are:-

1. Methods of analysis and design: which are most appropriate under what circumstances; and can deformations be predicted accurately?

2. Mechanisms of soil-reinforcement interaction: which structural actions should be included in design; and how is pull-out resistance of reinforcement to be determined?

3. Influence of construction techniques on performance: what techniques are available to control movements or reduce earth pressures; and how can long-term durability of the structure be ensured?

Methods of analysis and design

The basic concept behind most methods of analysis of reinforced soil construction is to consider a volume of soil which would fail if unsupported, and then check that sufficient force is provided by the reinforcement to prevent this failure. The reinforcement must also of course be strong enough to provide this force, and must have sufficient anchorage or bond within unfailing soil not to fail by pulling out. Instability of the complete block of reinforced soil must also be considered. It is also usual to try to ensure local equilibrium between each reinforcing element and the soil around it, so as to reduce the possibility of progressive failure of reinforcements.

In the papers to this conference, McGown et al. (ref.10) discuss potential limit states, taking account of strain compatibility between soil and reinforcement. They show that for relatively inextensible (usually metallic) reinforcement, serviceability limit states involving excessive deformations are unlikely to be critical; failure will be due to reinforcement rupture or pull out. For relatively extensible (polymeric) reinforcement, collapse will result from soil failure or reinforcement pull out, and serviceability requires consideration of limiting strains. These ideas will not be new to those familiar with reinforced soil, but it is useful to have them reiterated in the context of this conference.

Several authors mention limit equilibrium analyses based on various different assumed failure surfaces. Thus Kirkgard and Hammock (ref.8) used a parabolic shape of surface, while Yonan (ref.12) uses more conventional slip circles. De Buhan et al.'s (ref.4) yield theory analysis considers moment equilibrium of a volume of soil; the most critical failure surfaces are found to be logarithmic spirals. A common approach not mentioned here is the use of two-part wedges, which have the advantage of being able to pick up potential failure mechanisms involving sliding along a horizontal plane of reinforcement, particularly when a geotextile is used which may have reduced resistance to sliding of soil along its surface. Such mechanisms cannot be modelled by any analysis involving a curved failure surface.

Segrestin (ref.11) criticises the use of limit equilibrium methods as being unable to reproduce actual failure conditions, particularly once arbitrary safety factors have been included, and presents a design method based on reinforcement forces and soil stresses determined from finite element analyses and observations on model and full-scale walls. However he does admit that such observations must be made for all types of construction since, for example, the line of maximum reinforcement tension depends on the reinforcement extensibility. His Figure 10 is interesting; if the soil is divided into two wedges as shown, the analysis predicts interwedge friction angles which are quite small for typical steep slopes, but equal to the soil friction angle for less steep slopes. This suggests that setting interwedge friction to zero, as is often done to simplify the analysis using two-part wedges, may be rather conservative in the latter case, but good for the former.

Two papers present detailed results of finite element analyses. Harris et al. (ref.6) made comparisons with the standard limit state design by the Department of Transport's method and subsequently with observed performance. Perhaps the major benefits of finite element analyses are that they allow, as here, study of the effects of varying the main parameters within likely limits, and in providing information on local stresses and deformations. The construction modelled in this case was actually of anchored earth, in which the reinforcements are all attached to passive anchors rather than relying on soil-reinforcement friction to resist pull out. Ho and Smith (ref.9) used a 3-D analysis to model soil nailing. Among parameters varied was the initial horizontal stress in the soil. The final nail forces were found to be independent of this, though the equivalent earth pressure was slightly

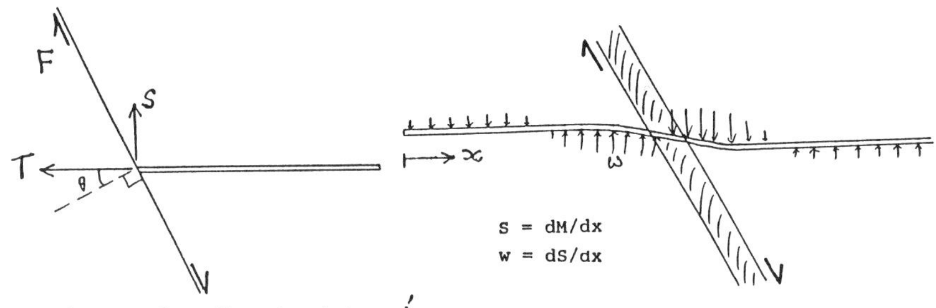

$F = (T\sin\theta + S\cos\theta) + (T\cos\theta - S\sin\theta)\tan\phi'$

Fig.2. Effects of axial and shear forces in nails

greater than fully active, but the deformations increased with initial stress. Note that in both finite element analyses no relative displacement was allowed between soil and reinforcement. This could result in significant underprediction of deformations when the reinforcement is very stiff; it is a better approximation for geotextile reinforcement for which deformations due to reinforcement elongation are likely to dominate.

Mechanisms of soil-reinforcement interaction

In soil reinforcement it has been generally accepted that only the axial force in the reinforcement is of any significance in design. However with soil nailing, perhaps because the typical grouted nail appears to have substantial bending and shear stiffness, there has been considerable debate as to whether these effects should be included. The reinforcement will typically be in a state of contraflexure across the soil failure surface, so that bending moment is negligible. The contributions of axial and shear force to the increase in sliding resistance along the failure surface are then as shown in Figure 2; it might appear that shear force could make a significant contribution. However two effects limit it severely. Firstly, shear force is equal to rate of change of bending moment, and the relatively low maximum bending resistance of typical soil nails could only give rise to large shear forces if the two bending moment maxima of opposite sign on either side of the failure surface were close together. This in turn would require a very rapid gradient of shear force, and hence impossibly high lateral loading from the soil onto the nail.

Furthermore, if such local "kinking" were able to occur and contribute significantly to shear resistance, a nearby failure surface would become more critical with negligibly different resistance to failure. Clearly shear force, unlike axial force, can only occur at a single location corresponding to a narrow failure zone in the soil, which is essentially a post-failure situation. Even in such a case, theoretical analysis, laboratory tests and field measurements have all shown that the contribution of reinforcement shear to overall shear resistance is typically only a few per cent.

In the conference papers, de Buhan et al. include only axial force, stating that to ignore shear and bending is always conservative, and usually only slightly so. The descriptions of actual construction (Kirkgard and Hammock, Barley) make no mention of shear and presumably only axial forces were considered. Ho and Smith's finite element analysis found the shear force in the nails to be small (2-3kN) compared with the axial force (~40kN), and demonstrated local failure in lateral bearing of the soil around the nails.

Davies et al. (ref.5) present results from large shear box tests in which the nails were gauged to measure axial and bending strains. Unfortunately the arrangement of the box means that the soil stresses on the shear plain are not accurately known, and probably change throughout a test as the sheared soil dilates in a manner which is different in different tests. It is therefore difficult to assess accurately the various contributions to the total shearing resistance. Even so it is clear that the shear force is limited by the bending strength of the nails, and that for nails perpendicular to the shear surface (the optimum for shear, but not for axial force), with a localised shear zone enforced by the test arrangement, and a large relative displacement (40mm) between the two halves of the box, the shear in the nails still accounts for less than 20% of the total increase in shearing resistance over the unreinforced soil (Davies et al., Figures 2, 4 and 7). It should also be noted that the absolute value of the axial force is only 7.3kN compared with a yield value of over 100kN, due to the nature of the tests.

The author believes that the best advice for designers is to place reinforcement as nearly as possible to the direction of maximum tensile strain (the most effective) and ignore shear forces in the analysis. Only in exceptional cases, such as when the axial force is very low, due to lack of pull-out length for instance, and a localised failure zone is known to exist, can shear resistance of the nails be significant.

The other problem of soil-reinforcement interaction involves the generation of pull-out resistance. Yonan's designs get round the problem by wrapping the geotextile sheets around the retained soil at the back as well as the front of the reinforced zone, while anchored earth does so by using anchors which can be made as large as necessary to generate sufficient passive resistance. For sheet reinforcement such as geotextiles, pull out resistance may be calculated by multiplying the overburden stress by a suitable interface friction coefficient, though a suitable large-displacement value should be used to allow for progressive development of shear with extensible reinforcement. For flat strips a similar approach may be adopted, though there is some uncertainty about the allowable normal stress, since local dilation can give normal stresses in excess of the overburden stress . Alimi-Ichola et al. (ref.1) suggest a mechanism of pull out from which the bond resistance can be calculated, based on observations in laboratory tests (their Figure 1). However it must be doubted whether such a mechanism actually occurs under field conditions.

Reinforcement sometimes takes the form of ribbed bars or grids, in which some of the pull out resistance is generated by bearing pressures rather than purely by sliding friction. However the statement by McGown et al. that the two effects are simply additive should be treated with caution, as there is some evidence that the generation of bearing resistance can reduce the shear resistance.

With soil nails the uncertainty is even greater, since the normal (radial) stress at the soil to bar or grout interface could be greater or less than the overburden stress, and is likely to be heavily influenced by construction procedures. The calculation is often critical in design, as long bond lengths may be indicated, particularly at shallow depths. In practice bond is often checked by pull out tests on site, and may be found to be considerably higher in dense sandy soils than predicted on the basis of overburden stress. Kirkgard and Hammock report a number of nails failing when tested, which they attribute to ineffective grouting; successful nails on the other hand appear to have generated average bond stresses of over 100kPa. Barley (ref.3) also reports measured pull-out resistances which, for nails at only 2-3m depth in mixed ground conditions, suggest bond of about 60 to 110kPa. However it should be recognised that it is very difficult to perform a pullout test under either laboratory or field conditions which reproduces correctly the actual conditions at failure.

Construction techniques and durability

Typical construction sequences for reinforced soil structures are shown in Figure 3. Reinforced fill is normally constructed incrementally layer by layer, with relatively small articulated units or wrap-around geotextile as the facing (Figure 3a). The facing is essentially flexible and able to deform laterally as required, and can move vertically and hence does not carry any vertical interface forces from the fill. Sometimes full-height facing panels are used which are effectively rigid, usually deform by rotation about the toe, and are

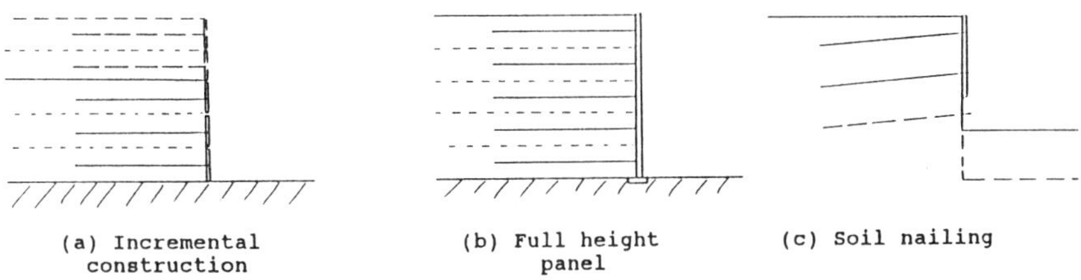

Fig. 3. Construction sequences for reinforced and nailed soil

able to carry vertical forces induced by the settlement of the fill behind the wall (Figure 3b). This reduces slightly the lateral earth pressure against the facing. Soil nailing is installed in a top-down sequence of excavation, nailing and face installation (Figure 3c). The facing may be in the form of precast panels, shotcrete, or flexible mesh retaind by plates, and its main purpose is to prevent local loss of soil between nails.

With soil nailing it is almost impossible to provide temporary support during excavation, and wall movements and local instability are dependent on the short term cohesion of the ground. Successful construction depends on limiting the lift height and installing the nails expeditiously. Filled structures may be supported during construction to control movements; incremental walls are usually supported for only one or two lift heights, but panel walls may be easily propped until filling is complete.

Two papers to the conference offer non-standard techniques of construction. Lareal et al. (ref.9) describe tests on the use of old vehicle tyres incorporated in the fill to reduce the earth pressures, potentially an excellent use of an awkward waste material. One version of this produces lightweight fill, when complete tyres are used and presumably are not completely filled with soil. There must then be some concern about the possibility of uneven settlements occurring in use. Andrawes et al. (ref.2) present test results first to substantiate a method of calculating earth pressure against a propped wall due to compaction, and then to demonstrate the effectiveness in reducing the lateral earth pressures of including a compressible layer between the fill and the wall facing (see their Figure 9). This allows the fill to expand laterally during compaction, inducing the necessary reinforcement forces in the process, so that compaction stresses are relieved and wall movements subsequent to completion of filling are negligible.

Other information on wall movements come from the finite element analyses of Harris et al. and Ho and Smith. Kirkgard and Hammock present site measurements of movements (their Figure 3) which suggest that significant movements did not occur below excavation level, but were concentrated into the currently unsupported zone of the latest stage of excavation. Time-dependent movements are also evident. Barley highlights the potential danger of excessive unsupported height during excavation in reporting a limited collapse in one case.

Barley also includes an invaluable summary of the methods of corrosion protection available for reinforcement; this is summarised in Table 2. It shows a succession of options from simple assessment of loss of steel section, which may be appropriate for reinforcement in well controlled and reasonably homogeneous fill material, through to full double protection systems as considered necessary for ground anchors. His case histories of soil nailing appear to include some in which no protection was used, two in which galvanised bar was used, and one in which a single corrugated plastic sheath,

Table 2. Methods of corrosion protection (after Barley)

METHOD	APPLICATIONS	COST cf. (b)
a) Assess loss of steel section - homogeneous or non-homogeneous soil	Soil reinforcement	
b) Reduction/prevention of corrosion by grout - crack-controlled or randomly cracked	Soil nailing	
c) Sacrificial coating (galvanising) - protection during handling	,,	+60-100% (4-10% of total cost)
d) Epoxy coating - protection during handling	,,	,,
e) Corrugated plastic ducting grouted in situ	,,	,,
f) Stainless steel	,,	x8-9 (+35-50%)
g) Double corrugated ducts, prefabricated - additional haulage and handling costs	Ground anchors	x~5 (+25-30%)

grouted in situ, was used. He appears to favour the last of these for permanent installations, as lacking the potential for local corrosion of simple grout or coated-bar protection, but avoiding the high cost and additional handling problems of preformed doubly-protected nails. An important point to note is that although the corrugated plastic ducting increases the nail costs by up to 100%, this typically only increases the structure cost by about 10%, which will often leave soil nailing competitive on cost with other solutions. The difficulty of providing protection by grout alone is highlighted by Hammock and Kirkgard, who attributed problems with bond to incomplete grouting of the nails. In their case the structure was only temporary, so corrosion protection was not a problem.

For reinforced soil, corrosion worries can be bypassed by the use of non-metallic (polymer) materials, but other concerns such as material damage during placement, long-term creep etc. must then be met. Harris et al. note that corrosion protection is less critical for anchored earth; the anchor ties can be polymeric, or if of steel can be protected more easily since round bar can be used and low-friction plastic coatings are not ruled out. Barley discusses the possibility of using GRP or other non-metallic material for soil nailing.

Concluding remarks

As a result of much high-quality research and development around the world over the last decade and more, the main concepts of soil reinforcement in retaining structures are well understood, and effective and economic construction processes have been developed. In the UK, use of the technique has been limited, while elsewhere (France, Spain, some states in the USA for example) it has become routine. Perhaps the current motorway widening programme will give both soil nailing and reinforced soil the opportunity to show their full value and move from being thought of as exotic to being commonplace.

PAPERS FOR THIS SESSION

1. ALIMI-ICHOLA I., LONG N.T. and LAREAL P. Inclusion pull out tests on small reinforced soil wall model. Results and prediction methods of inclusion strength.
2. ANDRAWES K.Z., YEO K.C. and LOKE K.H. Behaviour of geogrid reinforced soil walls subject to lateral boundary yielding.
3. BARLEY A.D. Soil nailing case histories and developments.
4. de DUHAN P., DORMIEUX L. and SALENCON J. Stability analysis of reinforced soil retaining structures using the yield design theory.
5. DAVIES M.C.R., JACOBS C.D. and BRIDLE R.J. An experimental investigation of soil nailing.
6. HARRIS D.I., POTTS D.M., BRADY K.C., HIGHT D.W. and HIGGINS K.G. Finite element analysis of anchored earth retaining walls.
7. HO D.K.H. and SMITH I.M. Modelling of soil nailing construction by three dimensional finite element analysis.
8. KIRKGARD M.M. and HAMMOCK A.K. Monitoring performance of a soil-nail shoring system in holocene alluvium.
9. LAREAL P., LONG N.T., BENAMEUR F. and COLLAS P. Active earth pressure reducing pneusol: experimental test.
10. McGOWN A., YOGARAJAH I. and YEO K.C. Choice of soil properties for limit state analysis of reinforced soil retaining structures.
11. SEGRESTIN P. Design of sloped reinforced fill structures.
12. YONAN S.J. Geotextile retaining walls for motorway widening.

Modelling of soil nailing construction by 3 -dimensional finite element analysis

D. K. H. HO, Sir Alexander Gibb & Partners Ltd, UK, and
I. M. SMITH, University of Manchester, UK

Introduction

The construction of nailed soil retaining structures has sometimes been regarded as the opposite to reinforced soil wall construction. The reinforcing mechanism in reinforced soil walls is often used to describe the mechanism in a nailed soil wall. Under closer inspection the mechanisms are different. In the reinforced soil wall system the reinforcement and retained fill are subjected to stress increase as the wall is gradually built. In the nailed soil system the retained soil is undergoing stress relief from its in-situ state as excavation commences in front. The reinforcements or nails are introduced into the disturbed ground later. The development of the strengthening mechanism in the two cases may be very different.

There has been much debate on the role of nail bending and shear resistance and the corresponding soil bearing capacity in the strengthening mechanism. So far, theoretical argument and field observation do not give any conclusion on the issue.

In this Paper, the development of deformations and axial and bending stresses in the nails, and the development of strains in the material in the wall and behind are calculated by three dimensional finite element analysis. This gives an insight into the soil-nail interaction as well as into the failure mechanisms of the nailed soil system during excavation. Eight-node 3-D finite elements have been used successfully to model reinforced soil wall construction (ref.1). The process is now reversed to model soil nailing construction. A more advanced 14-node element has been used to give more flexibility than the 8-node brick and to alleviate the storage and computation time problems encountered in the use of 20-node bricks. Two-dimensional plane strain finite element analysis cannot comprehensively model stress transfer between soil and nails.

The results obtained from the 3-D analyses are compared with observations.

Review

Soil nailing is a relatively new technique in stabilizing an excavation. The main differences between the nailed soil and reinforced soil structures are

(1) the construction procedure
(2) the high bending stiffness of the nails, and
(3) the soil types in which the two structures are built (ref.2).

Recently, full scale trial walls and prototypes have been built, but the nail bending and shearing were scarcely monitored.

Numerical analyses have been used to model soil nailing constructions. Very often, composite elements to represent the nails embedded in the soil were used, and the nail's behaviour was recovered indirectly (ref.3). In the discrete element approach, the nails were either modelled by beam or bar elements. However the bar elements will not be able to register any nail bending which may or may not occur. Special slip elements were used to simulate the pullout stiffness of nails. Plane strain conditions were assumed in most cases. Stress-transfer in the soil from above a nail to below was not correctly modelled, and the strengthening interaction between nails in the horizontal direction was ignored. A discrete three-dimensional analysis will give the nail behaviour directly, and the problems mentioned previously will be alleviated. Cardoso and Carreto (ref.4) used a 3-D elastic FE analysis to model a nailed soil structure construction with phased excavation. However theirs was a composite model and any potential collapse mechanisms will not be observed in such an elastic analysis.

Finite element analysis

In the present analysis, an elasto-viscoplastic algorithm was based on the program codes by Smith and Griffiths (ref.5). Modifications were made to simulate a three-dimensional excavation process. Instead of using the traditional 8-node or 20-node brick elements, 14-node brick elements were employed. These new elements require less computational memory storage than the 20-node element. They are capable of reproducing the correct failure mechanisms in a variety of boundary value problems under elasto-plastic conditions (ref.6). Eight sampling points were used to form the element stiffness and for stress sampling. This is enough to almost exactly integrate the stiffness. There are no spurious modes. The advantages of this element and the integration scheme are that the higher number of degrees of freedom means the element will not be as stiff as the 8-node element, and more information can be gained throughout the domain as compared to just a single sampling point at the element's centre. An elastic perfectly-plastic stress-strain behaviour was assumed and the elements obeyed the Mohr-Coulomb yield criterion. The no-tension criterion for soil elements was also

Table l. Properties of nail, shotcrete and soil assumed in the FE analyses.

Properties	Nail	Shotcrete	Soil
E (kPa)	2×10^8	10^5	10^5
υ	0.3	0.3	0.3
c (kPa)	10^{10}	10^{10}	5
ϕ (°)	0	0	30
ψ (°)	0	0	0
γ (kN/m^3)	20	20	20
Length (m)	4	-	-
Width (mm)	50 x 50	-	-
Thickness(mm)	-	100	-

incorporated. Providing the horizontal and vertical separations of the nails were uniform, only a slice of wall was analyzed due to the double symmetry effect. A slice of wall containing nails of half their widths was modelled. The width of the slice was equal to half the horizontal nail spacing. Roller supports which allowed vertical and horizontal movements were placed on both sides of the mesh. Although the symmetrical natural of the problem meant that the behaviour would repeat itself from slice to slice into the plane direction, this would not be the same as a plane strain situation as shown from the results later. This technique was used successfully to model the construction of strip-reinforced earth walls (ref.l). In the present analysis, the nails were modelled by the 14-node brick elements without any special interface elements; a full frictional interaction was therefore assumed. However the nail elements were surrounded by thin soil elements. The nails were 4 m long and they were installed at lm vertical intervals. The benefits of inclined nails have been reported (refs 7-8). However in the present analysis the nails were placed horizontally (providing they were long enough to intercept any potential "slip" surface) without further complicating the 3-D mesh. The properties of the nail are listed in Table 1.

The basic geometry of the wall is shown in Fig.l. The front and back boundaries were situated sufficiently far away from the excavation face. The initial horizontal stresses were set up according to the K_o value. No water was included in the analysis. The top-down nailed wall construction was modelled by excavating lm of soil followed by the placement of nail and shotcrete (model A). The structural elements were modelled by switching the corresponding soil element's properties into those of nail and shotcrete. Therefore soil disturbance due to nail installation was not modelled. The

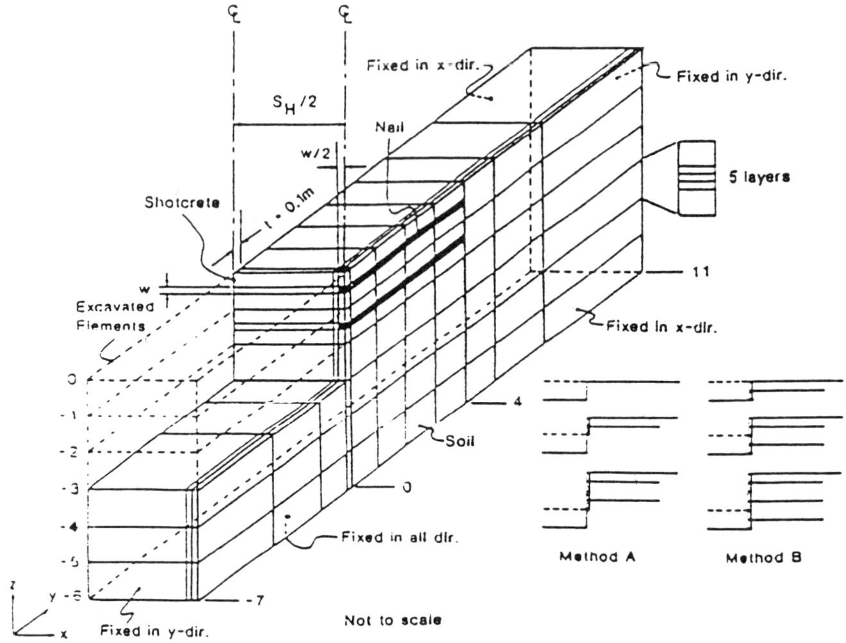

Fig. 1. FE mesh, geometry, boundary conditions and general information

process of soil excavation and nail/shotcrete placement was continued until either the wall failed or the bottom boundary was reached (this gave the final excavation depth of 6m). An alternative sequence of construction (method B), in which the nail and shotcrete were installed ahead of the excavation, was investigated. As the shotcrete facing was not a major stabilizing compo-

Table 2. Parameters studied in the FE analyses

Analysis	Horizontal Spacing, S_H(m)	Coefficient of earth pressure at rest, K_o	Construction method
1	1.0	1.0	A
2	1.0	1.5	A
3	1.0	2.0	A
4	2.0	1.0	A
5	4.0	1.0	A
6	1.0	1.0	B

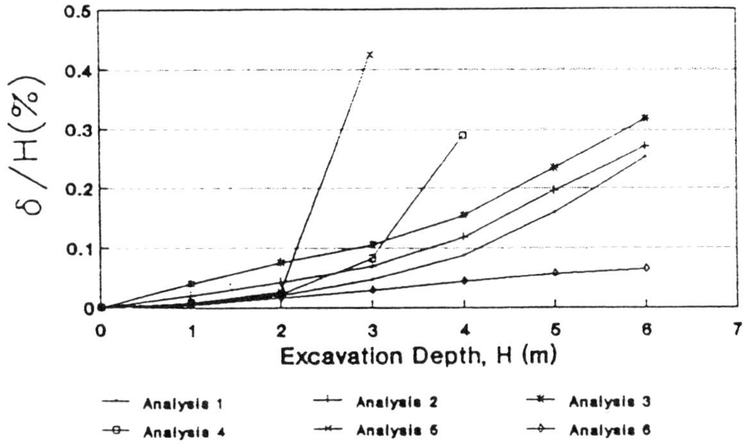

Fig. 2. Graph of wall top deflection as a function of excavation depth

nent in the whole structure, its stiffness was assumed to be the same as that of soil, in order that it would not attract high stresses. In this paper the results of six analyses are reported. The parametric variations were:

(1) the initial stress condition which was governed by K_o,
(2) the horizontal nail spacing, and
(3) the construction method (see Table 2). The soil properties are shown in Table 1.

Results and discussions

Wall Deflections

A normalised plot of the wall top horizontal deflection against excavation depth is shown in Fig.2. The magnitude of wall deflection is proportional to the K_o value. This is consistent with the calculations by Potts and Fourie (ref.9) and Ho (ref.10) of the behaviour of braced excavations. A large horizontal stress relief is associated with a large K_o. The deflected shape of the wall in method A shows the wall was rotating outward about the excavation level at large excavation depths (Fig.3). Similar deflected shapes were also observed in a full scale test (ref.11) and in the field (ref.12). The wall construction by method B led to a translation mode and the movement was much smaller (by about 4 times at the final depth). The small deflection may be partly due to the presence of an extra level of nails at -5.5m depth compared to only five levels of nails in method A. But more importantly, this demonstrates that the early installation of nails played a major role in the control of wall movement which is analogous to the early strutting in a braced excavation. In practice the nailed wall will be constructed under phased excavation and the actual wall deflection will lie between the limits obtained by the two methods.

Axial Force Distributions

The axial force distributions in the nails at the final excavation depth are shown in Fig.3 to Fig.5 (method A), and Fig.6 (method B). The variation of K_o did not affect the nail forces as shown from the maximum tension distribution in Fig.7. The previous stress states in the ground would not be experienced by the nails which were inserted into the soil subsequently when the stresses behind the excavation face would have already reached some "active" condition. In the three cases (S_H=1.0m, method A) investigated, the wall deflections were in accordance with the K_o values suggesting that the total horizontal stress relief was achieved, and the post-excavation stresses were very similar.

The position of the maximum tension line occurred at about half the nail length away from the wall face irrespective of the excavation depth. In the case of method B, the maximum tension took place at a quarter nail length away from the wall face. In method A, the maximum tension increased with depth whereas in method B, the lowest nail experienced less tension. However the tensions were above those associated with the full active earth pressure values (Rankine) except for the lowest nail in method B.

The maximum tensions in method B were lower than those in method A (K_o= 1.0). The extra nail in method B may cause the nails to attract less tensions compared to those in method A. The maximum tensions were doubled by doubling the horizontal spacing (Fig.8).

Bending Moments, Shear Forces and Deformations in Nails

The bending moments and shear forces in the nails at the final excavation depth are shown in Fig.3. to Fig.6. Bending moments are plotted on the "tension" side of the nail. Larger bending moments are observed in the upper nails than those situated near to the wall base. The top nails experienced more deformation as excavation continued. The bending moments were consistent with the way the nails bend, and the position of zero moment was in the vicinity of the point of inflection of the bent nail. The shear force distributions fluctuated from positive to negative without showing a conclusive pattern. However the lines of maximum shear, zero moments, maximum axial tensions and the points of inflection of the deformed nails were close to each other at failure (S_H=2.Om) as shown in Fig.9(b). Their exact positions depend on how finely the mesh is discretized. The bottom nail experienced higher bending moment than the top nails at failure. Although theories regarding the distribution and magnitude of the nail's bending moment and shear force

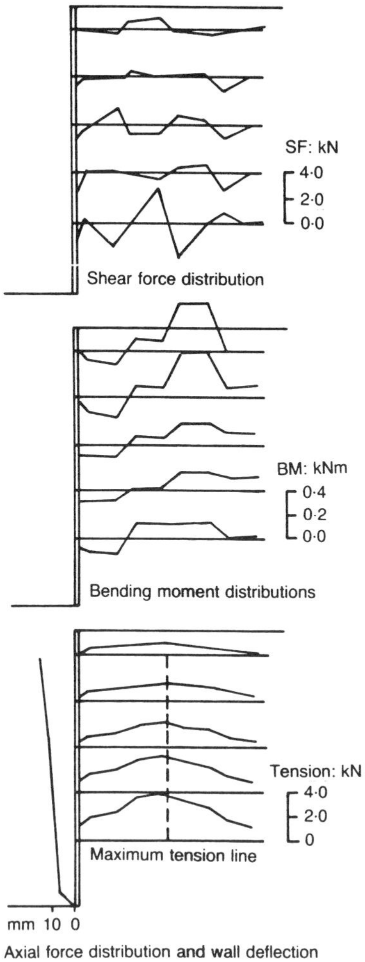

Fig. 3. Wall deflection and nail behaviour at 6m excavation depth (S_H= 1.0m, K_o=1.0, method A)

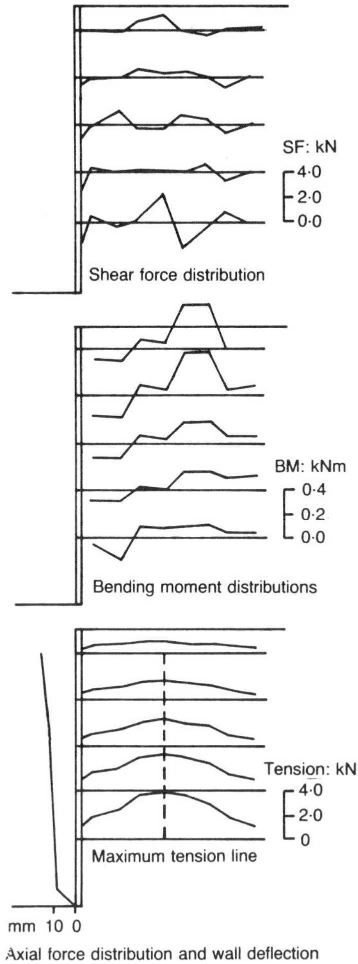

Fig. 4. Wall deflection and nail behaviour at 6m excavation depth (S_H=1.0m, K_o=1.5, method A)

at the limit state were proposed by Bridle (ref.13) and Jewell and Pedley (ref.14), actual laboratory or field measurements are rare. The bending moments at failure of a test wall reported by Gassler show that large moments occurred at the bottom nails and that they were of the same order of magnitude to those (Fig.8) in the present analysis (even though the properties were quite different). It is interesting to note that the maximum bending moment

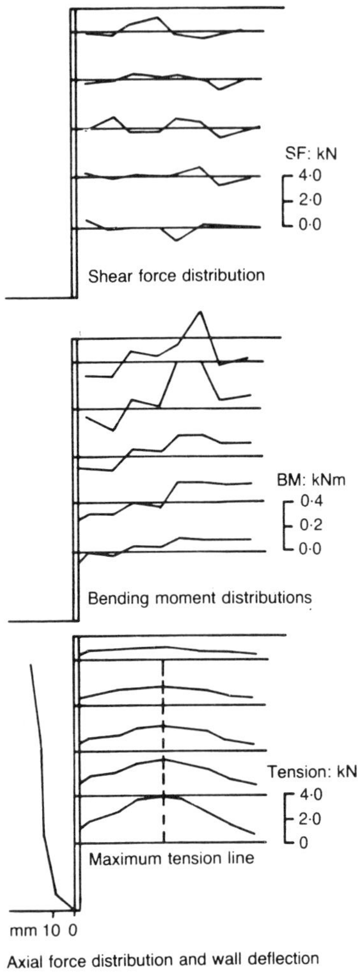

Fig. 5. Wall deflection and nail behaviour at 6m excavation depth (S_H=1.0m, K_o=2.0, method A)

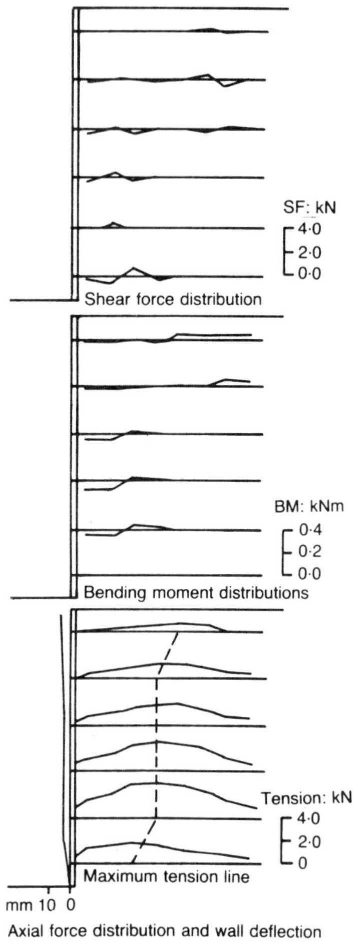

Fig. 6. Wall deflection and nail behaviour at 6m excavation depth (S_H=1.0m, K_o=1.0, method B)

at failure (S_H=2.Om) increased to about six times that of a stable wall (S_H=1.Om) while the maximum tension only doubled.

Failure Mechanism

The excessive wall top deflection with excavation depth as shown in Fig.2 indicates the onset of failure in the walls with wider nail spacings. (S_H=2.0 m and 4.0 m). Plastic deformation was also taking place in the S_H=1 m wall as

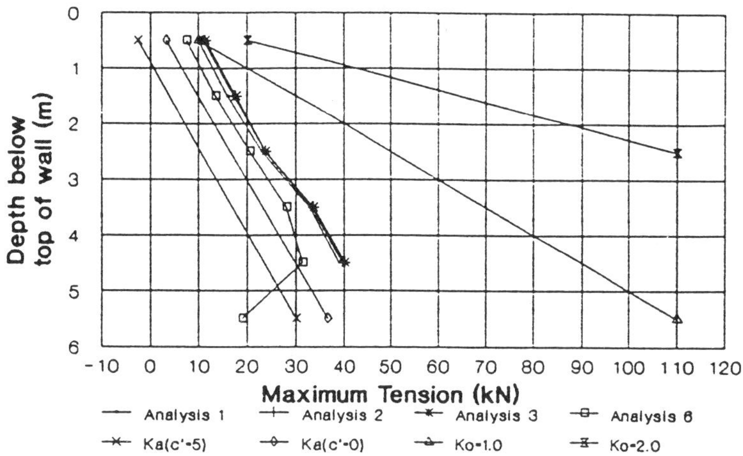

Fig. 7. Maximum tension distribution at 6m excavation depth

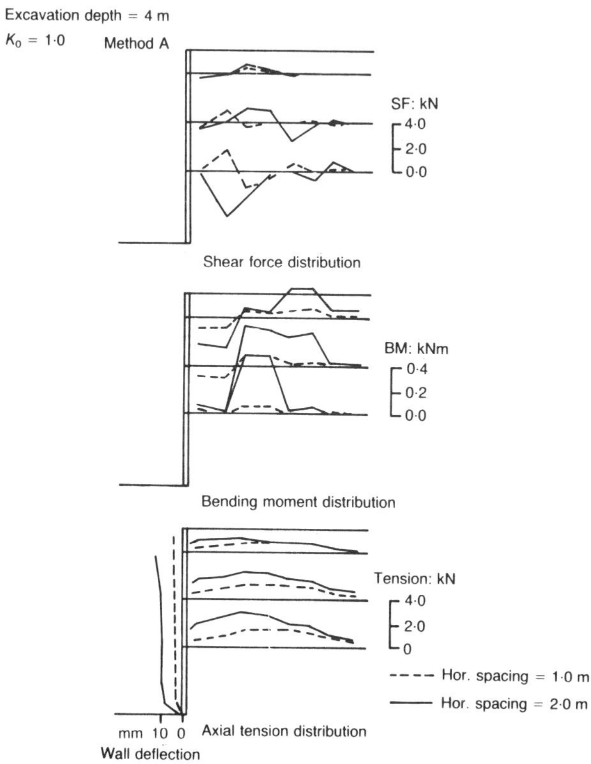

Fig. 8. Comparison of nail tension, bending moment and shear force between a stable wall and a collapsed wall

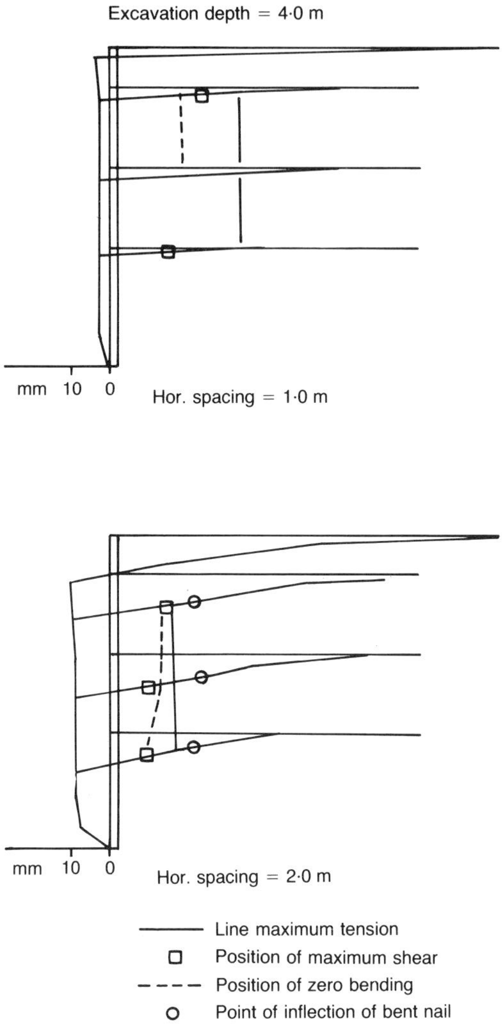

Fig. 9. Comparison of wall movement and nail deformation between a stable wall and a collapsed wall

excavation proceeded (H > 3 m). In the case of S_H=1.0 m, the plastic shear strains in the vertical plane between nails (ie. the least reinforced section) and in the plane containing nails are shown in Fig.10(a) and Fig.10(b) respectively. The high concentration of contours indicates plastic shearing occurred in that region. The effect of nails pulling out of the soil can be seen by the shearing starting from the mid-length of nails to the end of the nails. The top nail was on the verge of pullout as seen by the high intensity of plastic shear. In

Fig.10(a) the shearing was confined to the back of the reinforced mass and at the toe. The wall with S_H=4.0 m failed at H=3m. The corresponding plastic shear plots are shown in Fig.11. The active wall failure mechanism can be seen in Fig.11(a), whereas in the plane containing the nails, the nails were pulling through the soil. In the case of wall constructed by method B (S_H=1.0 m and K_o = 1.0) less shearing was observed compared to method A (see Fig.12). Unlike in a strip-reinforced soil wall where a stable strip-fill block is formed, plastic shearing took place in the soil containing the nails. This may due to the wide spacing of nails and a smaller nail-soil contact area. According to Fig. 9(b), a slip surface passing through the nails was probable.

Conclusions and further research

The 3-D FE analysis yielded direct information on the behaviour of nails

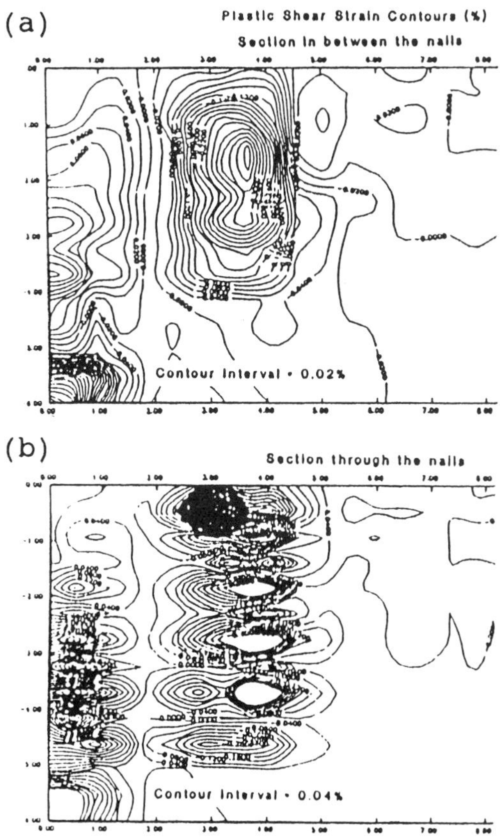

Fig. 10. Plastic shear strain distribution at 6m excavation depth (S_H=1.0m, K_o=1.0, method A)

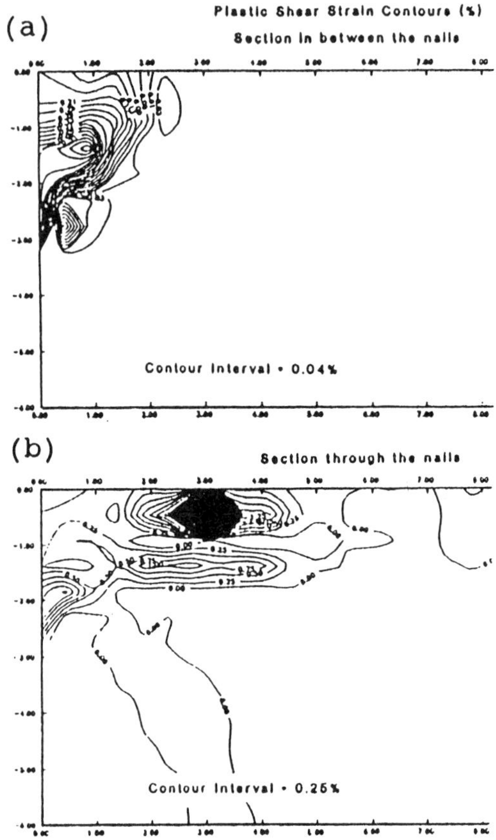

Fig. 11. Plastic shear strain distribution at 3m excavation depth (S_H=4.0m, K_o=1.0, method A)

during a nailed soil structure construction. Besides the nail's axial tension, bending moment and shear force were also obtained. The wall movement and the magnitude of bending moment in the nails at collapse were in reasonable agreement with those observed in the field. The in-situ stress condition did not appear to have any bearing on the nails stresses - only in wall deformation. The advantage of early nail placement in the control of ground movement was demonstrated. Plastic shearing in the soil was observed as nails were pulled through it. At collapse, a probable slip, which was in the proximity of the lines of maximum tension, maximum nail shear and points of inflection in the deformed nails, was computed.

More field information, especially of the behaviour of nails, is still required to substantiate the validity of the present numerical analysis. Future computational work will include the investigation of the role of bearing

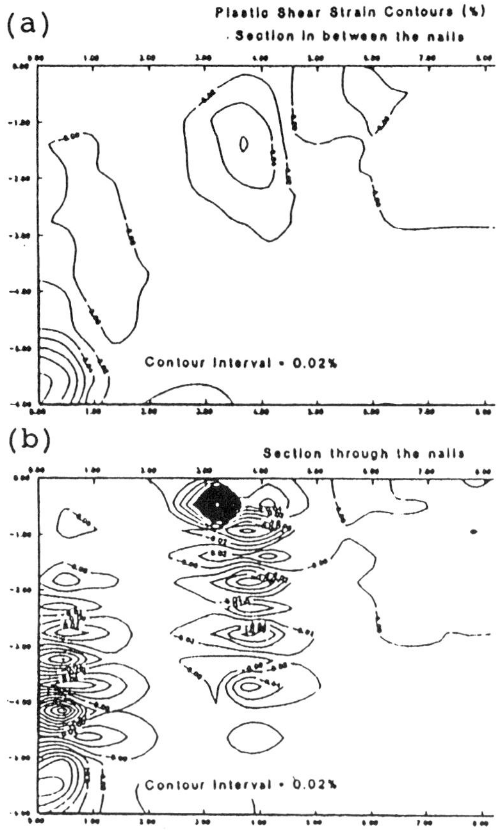

Fig. 12. Plastic shear strain distribution at 6m excavation depth (S_H=1.0m, K_o=1.0, method B)

capacity of the soil as the nails are dragged through it, and the simulation of tension breakage of nails in the computations.

References

1. HO D.K.H. and SMITH I.M. Modelling of Reinforced Soil Wall Construction by the 3-D Finite Element Method. Proceedings at the International Reinforced Soil Conference. University of Strathclyde, Glasgow, UK, BGS, 1990, 335-340.
2. SCHLOSSER F. Mechanically Stabilized Earth Retaining Structures in Europe. *Proc. Conf. Design and Performance of Earth Retaining Structures* . Cornell University Ithaca, New York, Geotechnical Engineering Division, ASCE, 1990, pp 347-378.
3. SHEN M., BANG S. and HERRMAN. Ground Movement Analysis of Earth

Support System. *J. Geotech. Engng Div.*, ASCE, **107**, No.GT12, Dec., 1981, 1609-1624.

4. CARDOSO A.S. and CARRETO A.P. Performance and Analysis of a Nailed Excavation. *Proc. 12th Int. Conf. on SMFE*, Rio De Janeiro, August, 1989, Vol. 2, 1233-1236.
5. SMITH I.M. and GRIFFITHS D.V. *Programming the Finite Element Method*. John Wiley & Sons, Chichester, 1988.
6. KIDGER D. *Visualisation of Finite Element Eigenmodes and Three-Dimensional Plasticity*. PhD Thesis, University of Manchester, 1990.
7. BRUCE D.A. and JEWELL R.A. Soil Nailing : Application and Practice - Part 1. *Ground Engineering*, November, 1986, 11-15.
8. BRUCE D.A. and JEWELL R.A. Soil Nailing : Application and Practice - Part 2. *Ground Engineering*, January, 1987, 21-38.
9. POTTS D.M. and FOURIE A.B. The Behaviour of a Propped Retaining Wall. Results of a Numerical Experiment. *Geotechnique*, 1984, **34**, No.3, 383-404.
10. HO D.K.H. *Analyses of Geotechnical Construction by The Finite Element Method*. PhD Thesis, University of Manchester, 1989.
11. GASSLER G. *Vernagelte Gelandesprunge-Tragverhalten and Standsicherheit*. PhD Thesis, des Institutes fur Bodenmechanik der Universitat Fridericiana, Karlsruhe, 1987.
12. THOMPSON S.R. and MILLER I.R. Design, Construction and Performance of a Soil Nailed Wall in Seattle, Washington. *Proc. Conf. Design and Performance of Earth Retaining Structures* . Cornell University, Ithaca, New York. Geotechnical Engineering Division, ASCE, 1990, 629-643.
13. BRIDLE R.J. Soil Nailing - Analysis and Design. *Ground Engineering*, September, 1989, 52-56.
14. JEWELL R.A. and PEDLEY M.J. Oil nailing design: the role of bending stiffness. *Ground Engineering*, March, 1990, 30-36.

Choice of soil properties for limit state analysis of reinforced soil retaining structures

A. McGOWN, I. YOGARAJAH and K.C.YEO, University of Strathclyde, UK

Introduction

For all civil engineering works, the main aim is to build a cost-effective structure which will behave during and after construction in the manner initially perceived at the design stage. To accomplish this, an appropriate means of predicting the behaviour of the components of the structure during construction and throughout the proposed life has to be employed.

Reinforced soil walls and steep faced slopes have found increasing usage over recent years and a variety of analytical techniques have been employed to design these structures. In general the analytical methods used are based on the Limit Equilibrium approach, e.g. Department of Transport BE 3/78 (1987). Although the finite element analysis has lately been used fairly widely in research, it has not been routinely employed in practice. Presently, Limit State Analysis which takes account of Serviceability and Ultimate Limit States is being introduced.

In this Paper a review is undertaken of the soil properties currently used in the design of reinforced soil retaining structures. Based on the results obtained from several instrumented full scale reinforced soil walls, a comparison is made with the properties obtained by back-analysis. Recommendations for the future choice of soil properties in Limit State Analysis of reinforced soil retaining structures are then given.

Material properties

Soil

There is at the present very little knowledge of reinforced cohesive soil retaining structures. Therefore the paper will only deal with the use of compacted granular soils.

Dense granular soils exhibit two components of shearing resistance: frictional and interlocking. The former is the surface sliding resistance of a granular soil, while the latter is the resistance generated by particles pushing

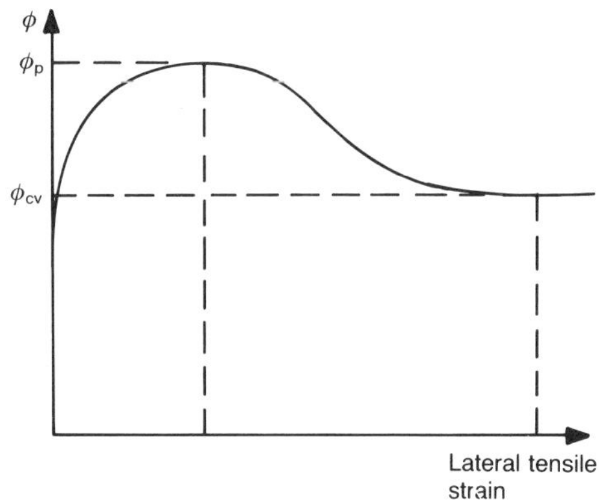

Fig. 1: Typical ϕ *vs strain curve for dense sand*

over each other. Both of these components are mobilised when deformation of the soil takes place resulting in both tensile and compressive strains developing within the soil mass. The rate of mobilisation of the soil shearing resistance is dependent on physical properties such as shape, size and uniformity of size of the soil particles. It is also dependent on the imposed boundary conditions such as plane strain, plane stress or triaxial stress/strain conditions as well as the degree of confinement.

Recognising that most reinforced soil retaining structures act in plane strain, McGown *et al.* (1989) carried out a literature survey on the available data on the behaviour of compacted granular soils in plane strain and found that dense soils reached their peak angles of friction (ϕ'_p) at lateral tensile strains of approximately 3-6% and that lateral tensile strains in the region of 8-12% were required to mobilise the constant volume angle of friction (ϕ'_{cv}).

Bolton (1986), found that for many granular soils the critical angle of soil friction of granular soil was in the range of 32° - 37° and that the peak angle of friction is related to the constant volume angle of friction and the angle of dilation by the equation

$$\phi'_p = \phi'_{cv} + 0.8\Psi$$

where Ψ is the angle of dilation. With this simple correlation, the peak angle of friction can be assessed using the constant volume angle of friction or vice versa.

Based on the above findings, a reasonably close approximation of the shape of the mobilised angle of friction - lateral tensile strain curve for a dense granular soil can be postulated (Fig. 1). Therefore depending on the mobilised angle of friction in the soil, the equivalent strain within the soil at the shear zone can be estimated.

Reinforcing materials

The basic requirements of reinforcements are that they should possess a tensile stiffness over the strain range which is compatible with the tensile strains developed in the soil, and possess sufficient bond with the soil to allow load transfer to occur. McGown *et al* (1978) categorised reinforcements as relatively extensible and relatively inextensible and gave the following definitions.

Relatively inextensible reinforcements

These are reinforcements that have rupture strains which are less than the maximum tensile strains in the soil without inclusions, under the same operational conditions (eg. steel, fibre reinforced plastics,etc.)

Relatively extensible reinforcements

These are reinforcements that have rupture strains which are larger than the maximum tensile strains in the soil without inclusions, under the same operational conditions (eg.synthetic polymeric reinforcements).

Relatively inextensible reinforcements possess a high stiffness and very often strength, while relatively extensible reinforcements have greater deformability but usually have lower strengths and are often affected by time and temperature. This results in differing load strain curves and behaviour in the two categories of reinforcing materials, Fig 2.

Soil - reinforcement interaction

The interaction of soils and reinforcements is governed by two mechan-

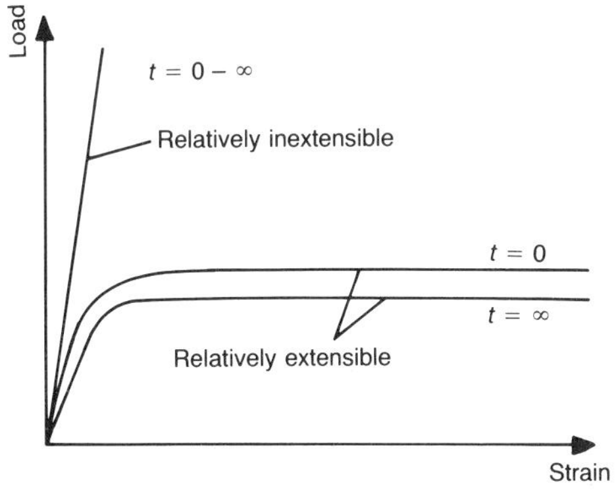

Fig. 2: Typical load strain cure for relatively extensible and relatively inextensible materials

isms

(i) *Direct sliding*, in which the soil moves over the surface of the reinforcements
(ii) *Bearing stresses*, in which transverse bars or protusions on the surface of the reinforcements develop passive resistance in the soil.

The overall bond coefficient for any particular reinforcement is calculated to be the summation of these two mechanisms.

Composite behaviour

Relatively inextensible reinforcements

Relatively inextensible reinforcements will develop much of their tensile strength at low tensile strains, when embedded in the directions of the principal tensile strains in the soil. This allows an overall increase in the load carrying capacity of the system over that of the soil alone at low deformations, or a reduction in deformations compared to the soil alone, under the same stress conditions. These benefits will remain in the system until the rupture of the relatively inextensible reinforcement occurs, after which the behaviour of the system will revert to that of the soil alone, Fig. 3.

Relatively extensible reinforcements

Relatively extensible reinforcementss will also develop their tensile strength when embedded in soil in the direction of the principal tensile strains in the soil but will require larger strains to do so. Nevertheless, the resulting composite system, will develop a smaller strain for a particular load or carry a higher load for the same strain as compared to soil alone. Further, relatively extensible reinforcements have rupture strains much higher than the strain at which the peak angle of friction of the soil develops, indeed they can be higher than the strain at which the constant volume angle of friction is reached. Thus the composite may maintain or even increase its load carrying capacity at strains beyond those at which the peak strength of soil is reached, even though the soil itself is weakening, Fig. 3.

Soil - reinforcement strain compatibility

To maintain strain compatibility between the soil and the reinforcement at all times, the reinforcement must strain the same amount as the soil under the influence of its self weight and externally imposed loads. The induced strains in the reinforcements develop forces which contribute to the equilibrium of the structure. This leads to the definitions of *Required forces* and *Available forces*

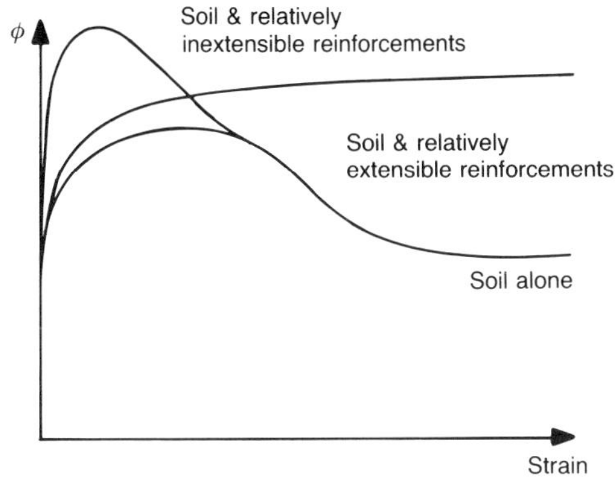

Fig. 3: Typical composite behaviour

Required forces

These are the out of balance forces developed by the soil that need to be maintained by the reinforcement for equilibrium in the structure.

Available forces

These are the forces that are available from the reinforcement to maintain equilibrium in the structure.

For reinforced soil retaining structures, the required force T_{strip} to be sustained by a strip reinforcement at an overburdened depth (h) and unit weight (γ) is calculated as follows:

$$T_{strip} = K_m \gamma h S_v S_h$$

where S_v and S_h are the vertical and horizontal spacing of the reinforcements

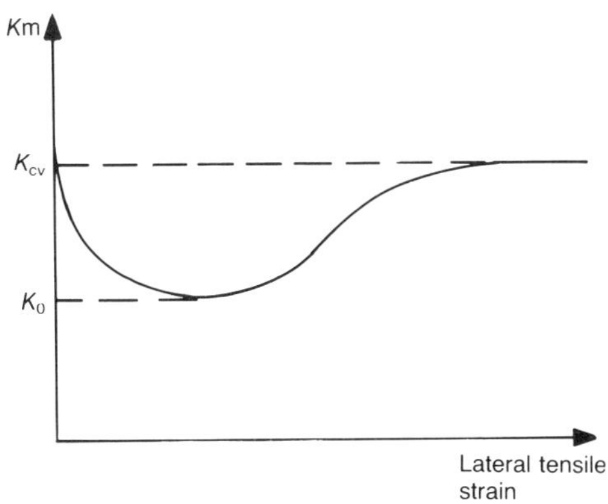

Fig. 4: Variation of required forces with strain soil

and K_m is the mobilised earth pressure coefficient defined as

$$K_m = \left(\frac{1 - \sin\varphi_m}{1 + \sin\varphi_m}\right)$$

where ϕ'_m is the mobilised angle of friction.

The required force (T_{sheet})to be sustained by a sheet of reinforcement at an overburdened depth (h) and unit weight (γ) can be calculated as follows

$$T_{sheet} = K_m \gamma h S_v$$

For a particular layer of reinforcement, the only 'variable' component in the equation is K_m which is dependent on the mobilised soil friction angle (ϕ'_m). Thus for a given h, and γ the required force vs lateral tensile strain curve at an individual reinforcing strip or sheet layer can be drawn as shown in Fig. 4. This is basically a mirror image of the mobilised angle of friction (ϕ'_m) vs lateral tensile strain curve. As shown in Fig. 4, there is a decrease in the required force until the peak angle of internal friction of the soil is reached, followed by an increase in the required force until the strain at which the constant volume angle of friction (ϕ'_{cv}) is reached. Thereafter the required forces remains constant. The configuration of the curve shows that required forces, other than that required at ϕ'_p, can be reached at two different strains levels, one greater and one less than that at which peak angle of friction is reached.

The available force is that provided by the reinforcing material. For relatively inextensible reinforcements, such as steel, this can be estimated from short term constant rate of deformation or stress test data. However, due to the time and temperature dependency of most relatively extensible reinforcements, the available forces have to be estimated by using constant temperature long term sustained load (creep) test data at the operating temperature of the soil. These are presented as isochronous curves and some typical data for a relatively extensible reinforcement may be seen in Fig. 2.

Strain compatibility and load equilibrium in the soil and the reinforcement occurs at the intersection of the required and available force curves, at which point the required force and the available force are in balance. A change in either would compel strains to develop thus allowing a new equilibrium point to be reached.

Limit state and limit equilibrium

Limit state conditions in reinforced soil structures are categorised as serviceability and ultimate limit states and the design conditions can be defined as follows.

Serviceability limit state (SLS)

Under these conditions the out of balance forces generated by the self-weight and the imposed loads would be transmitted from the soil to the reinforcements within limiting acceptable deformations in any of the components and without other forms of distress which would render the structure unsightly, require unforseen maintenance or shorten the expected life of the structure (e.g. deflection in the structure, strains in reinforcements etc.).

Ultimate limit state (ULS)

Under these conditions the out of balance forces generated by the self-weight and imposed loads would be transmitted from the soil to the reinforcing materials, and bring the structure into one or other of a number of previously identified collapse conditions.

Relatively inextensible reinforcements would require to strain more than their rupture strains to develop ϕ'_p in the soil, unless some slippage occurs between the soil and the reinforcements. Even with some slippage it is obvious that only small strains will develop with relatively inextensible reinforcements ensuring that the critical limit states for reinforced structures with Relatively inextensible reinforcements are those of tensile rupture of the reinforcement or pull out of the reinforcement. Tensile rupture will occur when the soil is between the K_o and K_a conditions, however, for pullout the soil may reach large strains and the constant volume angle of friction (ϕ'_{cv}) may be mobilised.

Relatively extensible reinforcements can allow strains greater than those required by the soil to reach ϕ'_p and most likely the ϕ'_{cv}, condition. The critical limit states will be either soil failure or reinforcement pullout, the soil most likely straining sufficiently to reach the constant volume angle of friction (ϕ'_{cv}) in both cases.

In the Limit Equilibrium method, each of the components is taken individually to be in a limiting, failure condition without regard to other components, and the static equilibrium of the system is calculated. Design under these conditions only takes account of forces without considering strain compatibility.

In the design of reinforced soil structures, the British code (BE 3/78, 1987) requires that the peak angle of friction (ϕ'_p) for the soil is used. At this soil condition, the required force for the reinforcements is at a minimum, Fig. 4, and as a consequence reinforcements able to restrain these minimum forces would be employed. The design is based on the concept of Limit Equilibrium and makes no provision for the strain compatibility between the soil and reinforcement components.

All designs, whether based on the Limit State or Limit Equilibrium approach, include either partial or overall factors of safety and sometimes both. As a result, more reinforcement than the amount necessary to sustain the Required forces is used. Thus the strains developed in the composite and the loads carried by individual reinforcements are less than described above.

Instrumented full scale projects

To understand the strain compatibility developed in reinforced soil structures using relatively inextensible or relatively extensible reinforcements, the performance of several instrumented walls at working conditions were analyzed.

Structures with relatively inextensible reinforcements
(Al-Hussaini et.al (1978))

A steel reinforced soil retaining wall 3.66 m high by 4.88 m long and 3.1 m wide, was constructed to study the failure mode by surcharging the wall to failure. The steel strips were 0.64 mm thick, 101.6 mm wide and 3.10 m long and the facing was made of high strength aluminium panels 0.61 m wide, 3.66 m long and 40.64 mm thick. The backfill used was concrete sand with a design angle of friction (ϕ'_{des}) of 36° (it was not mentioned whether this was ϕ'_p or ϕ'_{cv}) and a unit weight of 15.30kN/m^3. The construction was carried out in an incremental manner, by propping both the panel at the level at which the soil was being compacted and the panel immediately below. A surcharge load was then applied until failure occurred. The maximum strains recorded at the end of construction were in the range of 0.02% to 0.04%. At failure, maximum strains in each of the reinforcing layers increased proportionately, and the strains were in the region of 0.1% to 0.12%.

Murray and Farrar (1990)

The 200 m long reinforced soil wall was constructed along the M25 motorway with a height range of 4-8 m. A section of the wall was instrumented to monitor the full scale behaviour of the reinforced wall. The reinforcements used were galvanised steel strips of 40 x 5 mm thick with 3 mm transverse ribs along both surfaces. The backfill used was sand with a unit weight of 23kN/m^3 and a peak angle of internal friction (ϕ'_p) of 56°. The maximum strains recorded in the 5 layers of reinforcement were in the region of 0.10% to 0.23%.

Andrawes et al (1990)

The instrumented 2 m high full-height panel reinforced soil wall was constructed at Transport and Road Research Laboratory. The strain gauged steel reinforcements used were high adherence galvanised steel strips (40 mm x 5 mm) 1.6 m long with an Elastic Modulus of 185E06kN/m^2. The backfill used was uniformly graded Theale sand which had a unit weight of 18.03kN/m3 and an average moisture content of 9.9%. The peak angle of friction (ϕ'_p) at these conditions was 47° measured from triaxial tests. The wall was propped throughout the construction process. After construction, the wall was allowed to tilt 1.5mm about the toe. The reinforcements in the three layers recorded maximum strains between 0.004% to 0.01% respectively.

Structures with relatively extensible reinforcements
Bathurst et.al. (1988)

A 3 m high polymeric reinforced geogrid wall with granular backfill was built at Royal Military College, Kingston, Canada. Four layers of high density polyethylene uni-axially stretched Tensar SR80 reinforcements were used as the reinforcing material. The sand backfill used had a unit weight of 17.6kN/m^3 and a peak friction angle (ϕ'_p) of 43°. The wall was propped throughout the period of construction and the props removed after construction. The wall was subsequently loaded in increments until a 50kPa surcharge was reached and the surcharge load was left for 1000hrs. At the end of this period, maximum strains in the reinforcing layers were in the range of 0.4%-0.9%

Simac et.al. (1990)

The 6 m high polyester reinforced geogrid wall with granular backfill, built at Illinois, USA to evaluate the applicability of existing design methods of the reinforced soil system. The facing of the wall was made up of incremental facing units. The design of the wall was carried out at a very low factor of safety to entice maximum strains and movements in the soil composite. The backfill used was well graded sand with a unit weight of 20.1kN/m^3 and a design friction angle (ϕ'_{des}) of 39°. The wall was constructed in an incremental manner with a vertical batter of 1:20. At the end of construction, the maximum strains in the reinforcements, were in the range of 0.1-0.4%. After the end of construction, a surcharge of 1.5:1 (H:V)was placed over the wall to a height of 2.1m. The strains at this stage reached a maximum of approximately 1%.

Loke K.H. (1991)

A 2 m full-height panel reinforced soil wall was constructed at the Transport and Road Research Laboratory, UK. Uniformly graded Theale sand which had a compacted unit weight of 18.03kN/m^3 and an average moisture content of 9.9% was used for the backfill. The peak angle of friction (ϕ'_p) was 47° measured from triaxial tests. The wall was propped throughout the construction process. After construction, the wall was allowed to tilt 1.5mm about the toe. The reinforcements in the three layers recorded maximum strains between 0.5% to 0.8%.

In all the cases described above, the strains developed at working conditions were very low. In the case of structures reinforced with relatively inextensible reinforcements, the maximum strains up to failure conditions were always less than 0.23% and generally were an order of magnitude smaller. In soil structures, with relatively extensible reinforcements, the maximum strains recorded at working conditions were in the range of 0.4%-1%. None of these structures analysed were taken to failure.

Discussion and conclusion

The Limit Equilibrium approach to design assumes that each of the components have reached a failure condition without regard to the other components, i.e. without regard to strain compatibility. For the soil, the peak angle of friction (ϕ'_p) is generally assumed to be developed and the reinforcements are therefore designed to resist the minimum required forces. At working conditions, the soil strains are smaller than those required to generate ϕ'_p, thus the reinforcements are actually supporting larger required forces than those assumed in the design. These structures maintain their stability due to the presence of excess reinforcements provided as a result of the incorporation of high material factors of safety. Having regard to this it is evident that the Limit Equilibrium method is not an appropriate method for use in the design of reinforced soil retaining structures.

The Limit State approach which takes account of strain compatibility between the soil and reinforcements is appropriate to the design of soil retaining structures. Any particular condition may be modelled by carefully identifying the potential strains in the composite and thereby deducing the loads in the soil and the reinforcements. However, the strain behaviour of structures reinforced by relatively inextensible reinforcements varies from that of structures reinforced by relatively extensible reinforcements, therefore separate general recommendations for the Limit State approach to design are required.

Relatively inextensible reinforcements

At working conditions, the maximum strains will be well below those at which the soil develops ϕ'_p. The mobilised angle of friction for the soil will be less than ϕ'_p and possibly numerically close to the constant volume angle of friction (ϕ'_{cv}). Serviceability Limit States are likely to allow strains approaching those necessary to allow the development of ϕ'_p in the soil. However, these strains will be greater than those required to cause rupture in the reinforcements, if no slippage occurs at the soil, reinforcement interfaces, hence the Serviceability Limit States are not likely to be critical.

For the rupture of the reinforcements, the soil will not have strained sufficiently to develop ϕ'_p. At this failure condition, the mobilised angle of friction is more likely to be numerically close to the constant volume angle of friction (ϕ'_{cv}), Fig. 1.

For the Ultimate Limit State of pull-out of the reinforcements, the strains in the soil will be large and the constant volume angle of friction (ϕ'_{cv}) will be mobilised.

Overall, for structures incorporating relatively inextensible reinforcements, assuming that no slippage occurs between the soil and the reinforcements, the most appropriate numerical value for the mobilised soil angle of friction for use in the analysis of all conditions would appear to be the constant volume angle of friction ϕ'_{cv}.

Relatively extensible reinforcements

At working conditions the maximum strains will be smaller than the strains at which ϕ'_p is developed. Thus at working conditions, a soil angle of friction less than ϕ'_p and numerically close to the constant volume angle of friction would be developed.

To account for Serviceability Limit States, limiting strains will have to be imposed. Depending on these limiting strains, an appropriate mobilised angle of friction can be selected. In general, it is likely that these will be in the numerical range of ϕ'_{cv} to ϕ'_p.

Collapse of these structures will occur at relatively large strain due either to soil failure or reinforcement pull out. In both cases the soil will have strained sufficiently to develop the constant volume angle of friction (ϕ'_{cv}).

Overall the structures incorporating relatively extensible reinforcements, the most appropriate numerical value for the soil angle of friction for use at working condition and Ultimate Limit States, would appear to be the constant volume angle of friction (ϕ'_{cv}). For serviceability limit states the most appropriate values are likely to be in the range of ϕ'_{cv} to ϕ'_p therefore the adoption of the ϕ'_{cv} value would be a conservative choice.

It should be noted that in the Limit State approach, the application of partial factors of safety at various stages of the design may become problematic as they may implicitly violate the concept of strain compatibility which has been explicitly assumed elsewhere in the design. Thus great care in the choice and application of partial factors of safety in the Limit State approach to design is required.

Further it is very often the case that designers do not have control of the final choice of soil material. In such cases the choice of the design angle of friction (ϕ'_{des}) should be made in relation to the above principles linked to detailed specifications of mineralogical, chemical and gradational characteristics of the soil and to compaction requirements.

References

1. Al-HUSSAINI M.M. and PERRY E. B. Field Experiment of a Reinforced Earth Wall. *Proc. Symp. on Earth Reinforcement, ASCE Annual Convention,* Pittsburgh, Pennsylvania, 1978, pp 99-125
2. ANDRAWES K.Z., LOKE K.H. and YEO K.C. Application of Boundary Yielding Concept to Full Scale Reinforced and Unreinforced Soil Walls. *Proc. of the International Reinforced Soil Conference,* Glasgow, 1990.
3. BATHURST R.J., WAWZYCHUK, F.W. and JARRET P.M. Laboratory Investigation of Two Large-Scale Geogrid Reinforced Walls. *The Application of Polymeric Reinforcement in Soil Retaining Wall Structures, NATO Advanced Study Institutes Series,* Kluwer Academic Publishers, 1988 .
4. BOLTON M.D. The strength and dilatancy of sands. *Geotechnique,* 1986, **36**, No.l, pp.65-78.
5. DEPARTMENT OF TRANSPORT. Reinforced Earth Retaining Walls and Bridge Abutments for embankments. Tech. Memorandum BE3/78 1987.
6. Loke K.H.. *Effects of Lateral Boundary Yielding on Large Scale Reinforced and Unreinforced Walls.* Ph.D. Thesis, University of Strathclyde, Glasgow, UK, 1991.
7. McGOWN A., ANDRAWES, K.Z. and Al-HASSANI . Effect of Inclusions on the behaviour of sand. *Geotechnique,* 1978, **28**, No. 3, pp 327- 346
8.McGOWN, A., MURRAY R.T. and JEWELL R. State of the Art Report on Reinforced Soil Walls and Slopes. *Proc. of Soil Conference,* Rio De Janeiro, Brazil, 1989.
9. MURRAY R.T. and FARRAR D.M. Reinforced Earth Wall on the M25 Motorway at Waltham Cross. *Proc. Instn Civ. Engrs, Part 1.* 1990 ,**88**, Apr. pp 261-282
10. SIMAC M.R., CHRISTOPHER B.R. and BONCZKIEWICZ G. Instrumented Field Performance of a 6 m Geogrid Soil Wall. *Proc. 4th Int. Conf. on Geotextiles, Geomembranes and Related Products,* The Hague, Netherlands, 1990, pp 53 -59

Geotextile retaining walls for motorway widening

S. J. YONAN

The improvement of the M3 Motorway would widen the existing 3 lanes carriageway into 4 lanes, with no additional landtake. The option of widening the top of embankments and the toe of the cuttings while maintaining the motorway boundary requires a support from a retaining structure, such as a reinforced soil wall. Due to the difficulty of the traffic management arrangement on the motorway and close proximity of suburban residences, a fast, effective and environmentally acceptable retaining structure would be preferred in this situation. A number of options have been considered as a design solution such as gabion walls, reinforced concrete walls, sheet pile, sheet pile and geotextile reinforced walls. This Paper describes the option of geotextile retaining wall structures.

Introduction

The improvement scheme of the M3 motorway is intended to increase the traffic flow between Junctions 2 at the M3/M20 interchange and Junction 4 at Blackwater Valley interchange by widening the motorway. The sections of motorway under consideration are of dual three lanes standard carriageways.

Retaining walls were chosen as the most effective solution for motorway widening within the existing boundaries, after study conducted by DTP in the 1987 White Paper, Roads for Prosperity. This Paper will present the recommended design options for retaining walls and their advantages and practicality in the M3 Motorway Widening Scheme.

It is important to mention that the ground investigation for the M3 has not yet been undertaken and all the design options are based on the existing information from the original ground investigation undertaken by Soil Mechanics Ltd in 1967 and Soil Engineering Ltd in 1969. Assumptions made for preparing this paper may or may not comply with the actual ground condition.

Site description

The existing M3 was originally intended as the London to Basingstoke motorway and it was then extended to bypass Winchester. The M3 was first

Table 1. Summary of soil parameters

Soil Type	$\phi°$	$C'(kN/m^2)$	$\gamma\ (kN/m^3)$	M/C(mean)(%)
Barton Beds	36	0	19	23
Bracklesham Beds	38	0	19	25
Bagshot Beds	38	0	19	24

Sources: Soil Mechanics Ltd, 1967 and Soil Engineering Ltd, 1969

designed in late 1960s and construction completed in the early 1970s.

The M3 route between Junctions 2 and 4 comprises a total of 10 cuttings and 10 embankments situated on both sides of the carriageway. The proposed scheme would add extra lanes by widening the top of embankments and the toe of cuttings while maintaining the existing slop gradient. Land acquisition is difficult between Junctions 3 and 4, and additional landtake is considered to be necessary between Junction 2 and 3 if more than one lane is required on both sides of the carriageway.

The construction of cuttings and embankments and the resultant slope gradients, heights and lengths are affected by the soil strength, geology, ground water, drainage, land access and the cost of construction. The original recommended design slope gradient for the M3 cuttings was 1 in 2. The final slope gradients adopted range from 1 in 2 to 1 in 3 slope due to the poor ground and high ground water which were found to exist. The maximum height of cutting is 12.2 m and the minimum is 3.3 m. The slope gradient for embankments is 1 in 2. The highest embankment is 9.1 m and the lowest one is 0.9 m.

The geology of the M3 cutting areas, comprises mainly Barton Sand, Bracklesham Beds and Bagshot Beds. The soil strength of these materials vary considerably from weak to firm depending on the presence and location of the groundwater table which has been recorded as artesian in some places. The summary of soil parameters for these materials are shown in Table 1.

Retaining walls

The appearance of the walls and their effects on the environmental impact of the scheme, both in terms of the landscape and the propagation of road traffic noise is an important consideration in option design.

Retaining walls are one of the few options that can be used for motorway widening. Their advantage is that additional landtake is not required. This makes the design of retaining walls critical where ground condition will not allow the steeping of the slope gradient for the cuttings or embankments. The type of wall design is affected by the type and strength of both the retained and foundation soils.

Table 2.

Embankment

Fill material	Reinforcement Strength	Reinforcement Min Spacing	Water Condition	Surcharges
ϕ = 35°	29 kN/m	0.5m	r_u = 0.1	12 kN/m
γ = 20 kN/m³				
c = 0 kN/m²				

Cutting

Fill material	Reinforcement Strength	Reinforcement Min Spacing	Water Condition	Surcharges
ϕ = 35°	29 kN/m	0.5m	r_u = 0.1	12 kN/m
γ = 20 kN/m³			r_u = 0.2	

Traffic management is the most important consideration when dealing with site investigation and construction on motorways. By reducing the construction width and duration for both temporary and permanent works the traffic disturbance and congestion is minimised. Geotextiles offer an effective way of achieving both of the above.

The critical element of design is slope rotational failure passing underneath the foundation of the wall. The type of the retaining structure in this case is not only effected by the soil characteristic of the foundation but also by the groundwater regime to sustain the stability of slopes. The height of the retaining wall is dependent on the width of excavation which is necessary to accommodate the widening and the slope angle. It is important to confirm or review the original soil parameters, and the location and effect of open drains to assess a suitable type of retaining wall and the associated method of the temporary and permanent construction requirements.

For embankments, the wall must accommodate for both the live and dead loads imposed by the widening and also on the adequacy of the wall foundation. The construction of such structures are relatively easy in comparison with the cutting walls because they require reduced excavation for temporary works and it allows the use of good quality fill materials. Provision has to be made for noise fences and lighting columns at the top of the structures.

Geotextile design options

The use of geotextile and polymer yarn fabric for earth retaining structures is relatively new in comparison to the traditional retaining structures. Its use has become widespread in the recent years because of its low cost and simplicity in design. The reinforced earth structure is created form a number of geotextile layers placed within compacted material. Other advantages over traditional walls is the flexibility of the reinforced earth retaining structure to accommodate ground movement associated with the active earth pressure, pore water pressure and wall foundation.

Slope stability analysis based on existing slopes was conducted to examine the validity of the preliminary design and obtain the lowest acceptable factor of safety between 1.1 to 1.2. A typical list of parameters used for design purposes are shown inTable 2.

The design procedure comprises a number of steps. The first was made to determine the factor of safety of the unreinforced structure for the chosen wall profile. A reinforcement layout was designed for the lowest factor of safety. The design was based on the strength of reinforcement, fill material and the reinforcement spacing.

A number of potential circle failures were reconsidered to determine the internal and overall stabilities of the reinforced structures. For each circle, a factor of safety was calculated using Bishops, Janbu, and Swedish methods of analysis. The worst possible slip surface failure is shown in Fig. l.

The idea of creating a retaining structure from weaving a mixture of polymer yarn fabric and soil particles is relatively new. The wall acts as a light weight retaining structure and is similar in section to the mass concrete retaining wall but the concrete is replaced by a mixture of polymer yarn and sand. The system has been used in motorway widening, railway embankments, hard slopes repairs in France but has not been used in this country.

A preliminary design is to be carried out to evaluate the existing ground condition and the anticipated problems for the existing slopes. This will lead to the selection of the appropriate yarn fabric for strength and thickness, type of the sand and method of construction. The wall dimension is then chosen to support the earth pressure and water pressure if applicable.

Typical parameters used for design purposes are shown in Table 3.

The design should be checked for the overall stability of the wall, which includes

stability with respect to bearing capacity of foundation
stability against slip under the base
stability against overturning.

Table 3.

Cuttings and Embankments		
retaining structure	water condition	surcharge
γ = 20 kN/m^3	r_u = 0.1	12 kN/m
c = 10 kN/m^2		
ϕ = 35°		

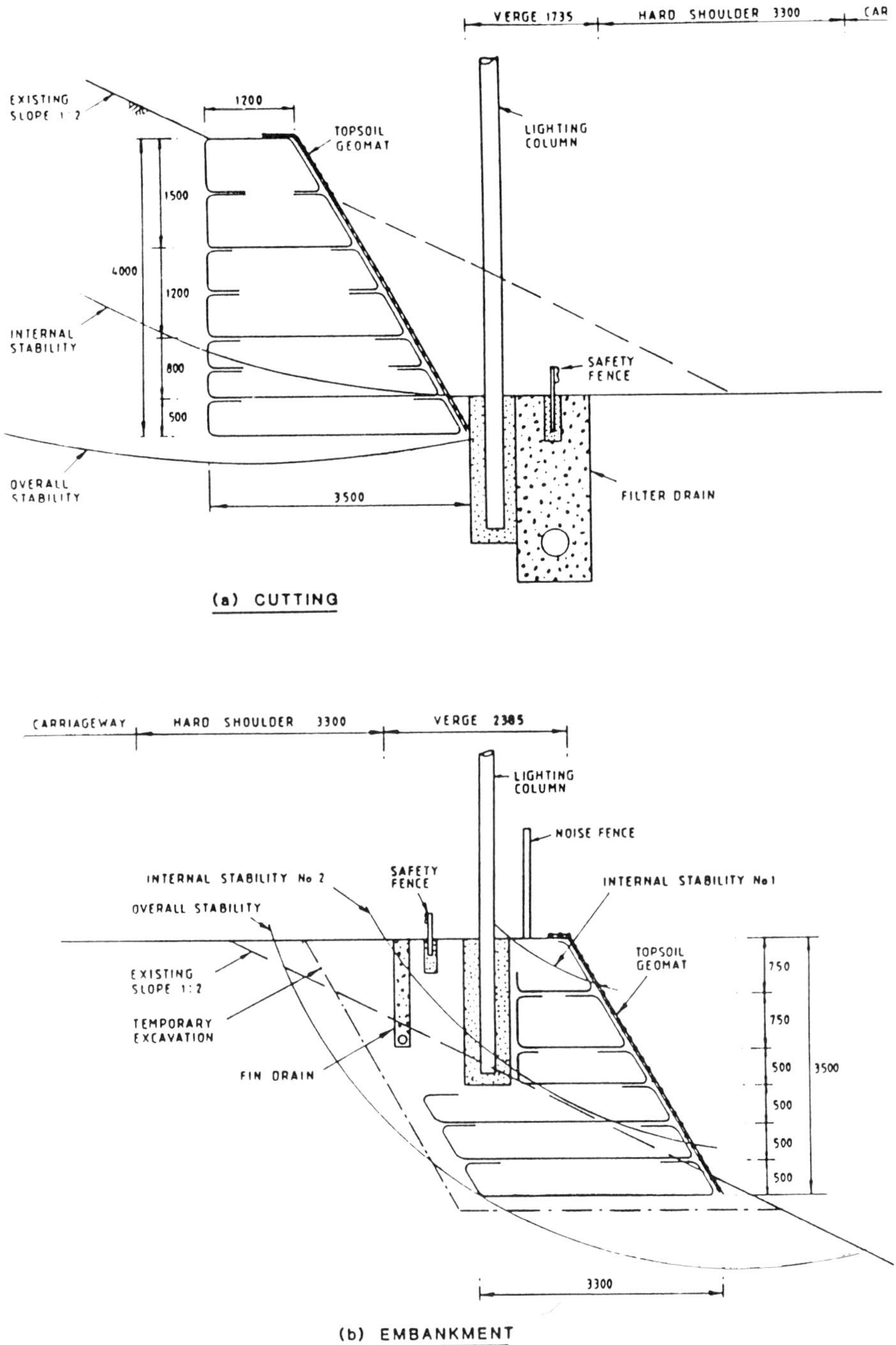

Fig.1. Typical design section for a reinforced earth retaining wall

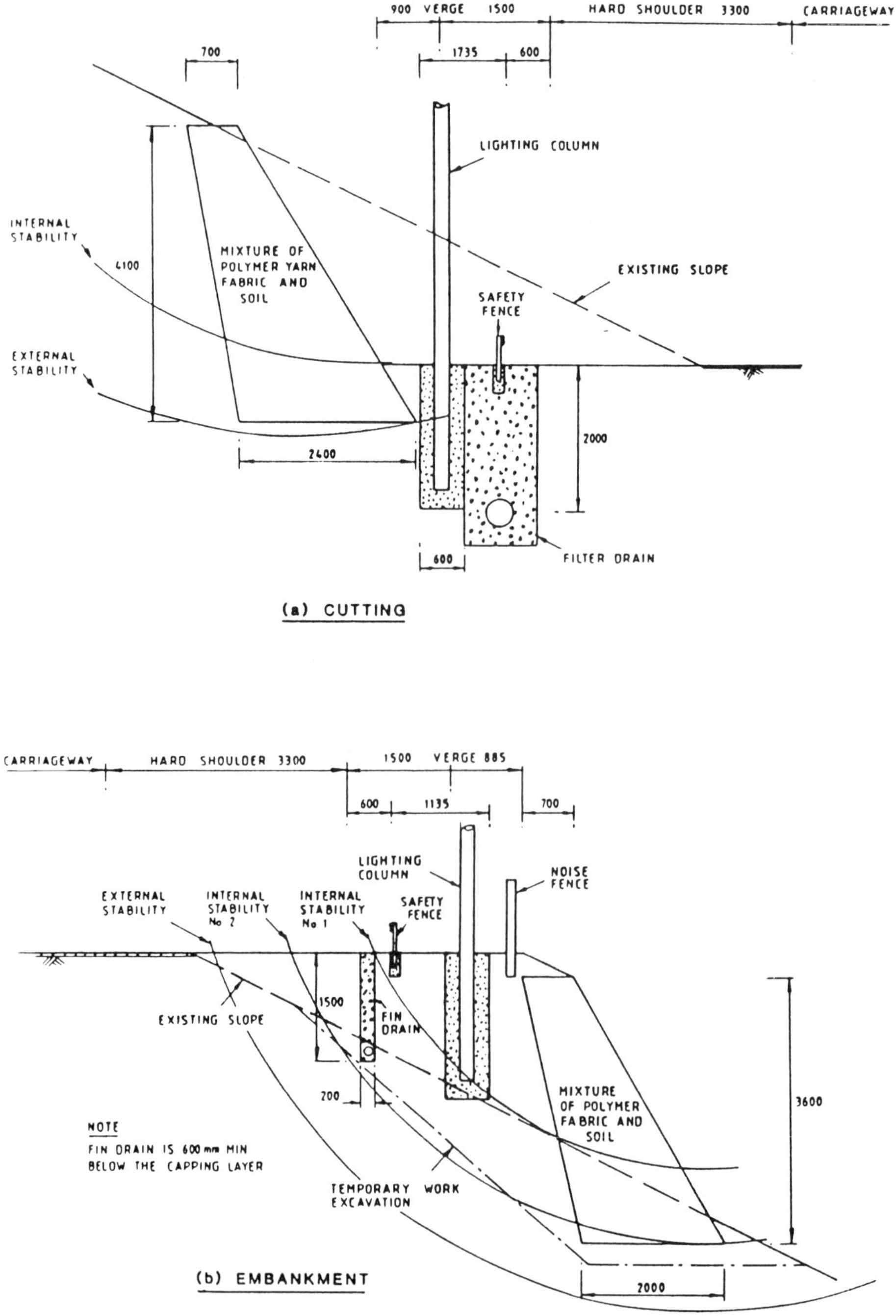

Fig. 2. Typical design section for mixture of polymer yarn fabric and soil retaining wall

The wall stability should be examined for the critical slip generally passing through the base of the front face of a homogeneous wall, and a wall base whose thickness varies linearly with the depth. The study of wall stability should consider the current level of polymer varn fabric knowledge, mechanical properties and an overall strength of polymer yarn fabric and soil mixture. Typical design sections for cuttings and embankments are shown in Fig. 2.

Economics of construction

Construction Cost

The estimated cost for constructing the retaining wall structures will be dependent upon the fill material wall type, the required finish for landscaping purposes and the temporary works. The list of relative cost for retaining walls is established as shown in Table 4 to provide a price comparison between a number of retaining walls options based or a particular set of soil parameters.

Environmental Impact

The impact of the Geotextile retaining structures on the environment would be to restore the natural appearance of the motorway slopes and boundaries, and allow a growth of various types of vegetation on the surface. This would assist in reducing reflected noise in populated areas such as Camberley.

Minimising the traffic disruption will result in the reduction of the social cost that is needed during the construction period.

Conclusion

Due to the nature of the site, the design process should consider the possibilities of installing structures that can accommodate a number of design,

Table 4. Relative cost comparison for various retaining walls

Retaining wall type	Slope 1:2 Wall height 2.0m (£/m run)
Gabion	400
Reinforced Concrete	600
Sheet Pile	1200
Sheet H-pile	1000
Geotextile reinforcement	
Geogrid	500
Polymer yarn fabric	400

environmental and constructional problems. The choice of the reinforced earth or the mixture of polymer and soil will provide the following advantages.

- the structures are flexible and will tolerate some movement where poor founding strata exist
- no special drains are required since they can be designed as self draining
- an environmentally acceptable finish of vegetation growth can be achieved on the surface of the structure
- cheaper than other conventional options such as steel piles or concrete
- minimising the traffic disruption by reducing construction time.

In the case of mixture of polymer and soil, a purpose built machine is used causing minimum earth disturbance. Reinforced earth structures may require temporary support and additional space for the support.

Acknowledgements

The Author would like to express his thanks to the Director of Transport, Motorway Widening Unit for their permission to publish this Paper. It is important to state that the views and recommendations expressed in this paper are not those of the Department of Transport. The Author would also like to express his thanks and gratitude to the M3 Project Management team of Brian Colquhoun and Partners and the staff of Colquhoun Geotechnical Services, especially Sergio Perelberg and David Guy for their valuable advice and comments.

References

1. DEPARTMENT OF TRANSPORT . *Roads for Prosperity*. White Paper (Cm 693), HMSO, London, 1989.
2. BRIAN COLQUHOUN AND PARTNERS. *M3 Internal Reports*. 1990-1991.
3. TEXSOL. *Technical Guide for Retaining Structures*. 1990.
4. SOIL MECHANICS LTD. Site Investigation factual reports on the M3. Soil Mechanics Ltd, 1967.
5. SOIL ENGINEERING LTD. Site Investigation factual reports on the M3. Soil Engineering Ltd, 1969.

Behaviour of geogrid reinforced soil walls subjected to lateral boundary yielding

K. Z. ANDRAWES, K. C. YEO, and K. H. LOKE, University of Strathclyde, UK

Introduction

This Paper describes the influence of two construction methods on the behaviour of reinforced soil walls during and after construction. These construction methods are described and their effects are investigated in terms of lateral and vertical pressure distributions, strains in the reinforcements and lateral boundary yielding. A method of predicting the lateral pressure is proposed.

Experimental set up

Two 2 m high walls were constructed to retain granular backfill reinforced with 3 layers of polymeric geogrids, at 0.7 m vertical spacing. The geogrid was 1.6 m long and 2 m wide. Both walls consisted of three vertical timber facing panels, 0.7 m wide each. A polyurethane compressible layer 75 mm thick was placed between the timber panels of one of the walls and the backfill thus, allowing lateral yielding of the backfill to occur (ref.1).

The walls were propped and the reinforcements left unattached to the facing panels during construction. After completion of the placement of the backfill, the reinforcements were "locked on" to the facing panels. At this stage, it is assumed that the loads on the props are transferred on to the reinforcements.

Instrumentation

Pneumatic pressure cells, 140 mm diameter, were used to measure the lateral earth pressures acting on the walls and the base pressures. Strain gauges and Bison strain coils were used to monitor local and overall strains in the geogrids respectively. Also, the movement of the wall and the compression of the compressible layer were monitored.

Materials

A uniformly graded sand from Theale (Berkshire) was used in the experiments. The sand was compacted in 0.15 m to 0.2 m layers with two passes of a vibrating plate in the direction parallel to the wall. The closest distance between the plate edge and the retaining walls was about 0.1 m. The average moisture content and unit weight were 9.9% and 18.05 kN/m^3 respectively. The drained peak angle of friction of this soil measured in the triaxial tests was 47°. The measured interface angle of friction for soil/timber and soil/compressible layer were 35° and 43.5° respectively.

Prediction of lateral pressure

Compaction induced pressures on unyielding wall

For supported rigid wall retaining reinforced backfill, compaction induces high lateral pressures. In order to assess the effect of compaction, analysis of the compaction pressure using the basic approach proposed by Broms (ref.2) is carried out. The analysis takes into account the number of passes and the direction of the compaction plant. The increase in the horizontal stress due to compaction $\Delta\sigma_h$ is calculated from Boussinesq equation for strip uniform pressure parallel to the wall, as shown in Fig.1, i.e.

$$\Delta\sigma_h = \frac{q}{\pi}[\alpha - \sin\alpha . \cos(\alpha + 2\beta)] \tag{1}$$

where q = uniform strip load at the surface of the fill. The effective lateral pressure due to compaction can be given as:

$$\sigma_{hm} = \sigma_{hi} + \Delta\sigma_h \tag{2}$$

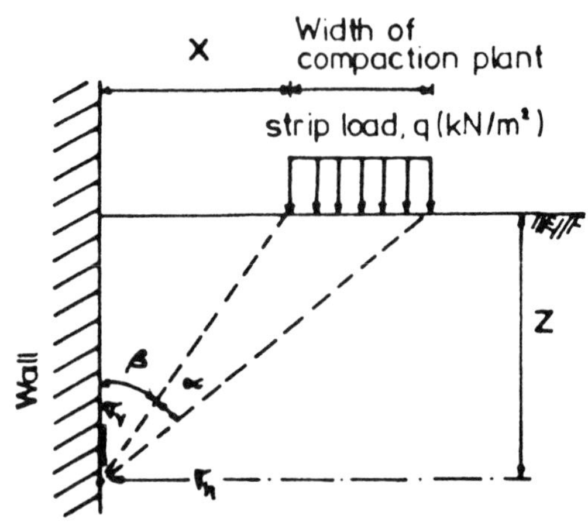

Fig. 1. Stresses due to uniform pressure on the strip area parallel to the wall

Assuming that the wall pressures generated by the self-weight of the fill σ'_{hi} are negligible, thus $\sigma'_{hm} = \Delta\sigma_h$. At critical depth $z_{c'}$ as shown in Fig.2.

$$\sigma_{hm} = +\Delta\sigma_h = K.\gamma.z_c \tag{3}$$

where K is dependent on the number of compaction passes and γ = unit weight of soil. Equating (l) and (3) gives:

$$z_c = \frac{1}{K\gamma}.\frac{q}{\pi}[\alpha - \sin\alpha.\cos(\alpha+2\beta)] \tag{4}$$

For a rigid wall, σ'_{hm} is increased by a factor of 2 to account for the reflected stress on the wall, thus:

$$\sigma_{hm} = \frac{2q}{\pi}[\alpha - \sin\alpha.\cos(\alpha+2\beta)] \tag{5}$$

In addition the value of q is assumed to be increased by 100% over its static value to account for the vibration caused by the compaction plant, thus:

$$\sigma_{hm} = \frac{4q}{\pi}[\alpha - \sin\alpha.\cos(\alpha+2\beta)] \tag{6}$$

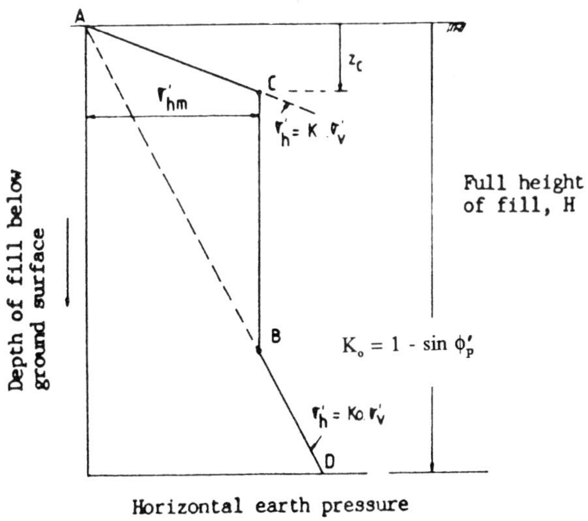

Fig. 2. Horizontal pressure distribution due to compaction of fill

Lateral pressure distribution on yielding wall for reinforced backfills

Two limiting cases can be considered. Firstly, if the lateral boundary is not allowed to yield, the resulting lateral pressure will be equal to that obtained under the "at rest" conditions. This is the upper limit of the lateral pressure acting on the wall. Secondly, if the lateral boundary is allowed to yield sufficiently to mobilise large tensile resistance in the reinforcements and if the reinforcement layers are rigid and rigidly connected to the wall facing, each zone between the reinforcement layers may be considered as a separate wall subject to active horizontal pressures, i.e. at the top of the zone, $\sigma_h = 0$, and at the bottom, $\sigma_h = \cos \delta . Ka. \gamma . z$. This will produce the lower limit of the horizontal pressure distribution on the boundary. If the boundary yielding is not sufficient to reduce the horizontal pressure to a minimum or the rigidity of the reinforcement layers is not sufficient, the resulting pressure will be in between these two limits.

Results

Before removal of props

Figs 3(a) and 4(a) show the lateral pressure distributions on the two walls at the end of construction. For the wall with rigid lateral boundary, lateral pressures significantly larger than the theoretical "at rest" pressure were

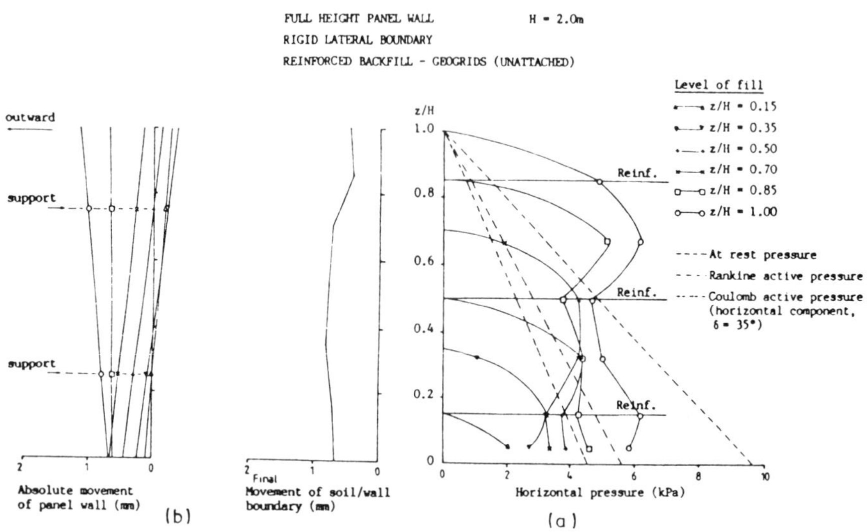

Fig. 3. Horizontal earth pressure distributions, movements of wall facing and movements of soil/wall boundary during construction

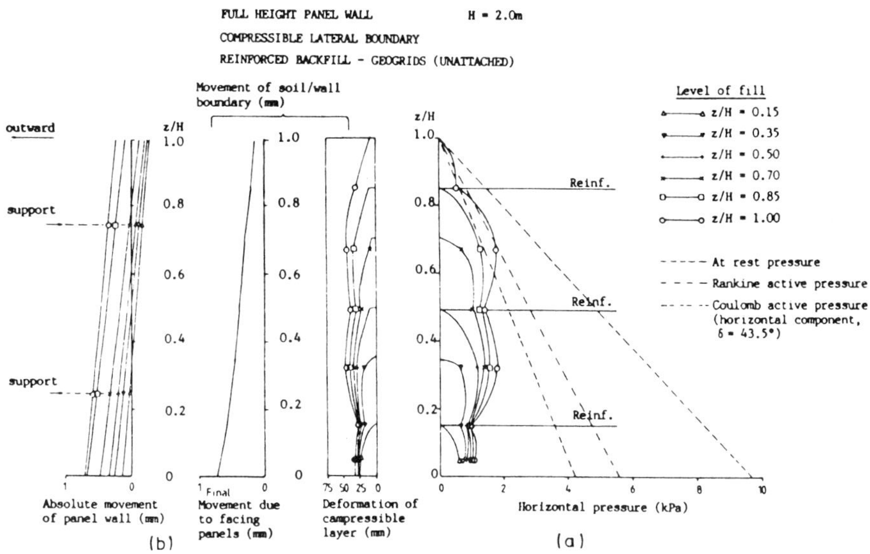

Fig. 4. Horizontal earth pressure distributions, movements of wall facing and movements of soil/wall boundary during construction

recorded at the upper half of the wall. At the lower half, the lateral pressures were between the theoretical Rankine "active" and the "at rest" pressures. However, for the wall with a compressible lateral boundary, the lateral pressure on the wall was less than the Coulomb "active" condition particularly at the lower half of the wall. The lateral pressure distribution conformed closely to that generally associated with rotational wall movements about the top with further reduction in the horizontal pressures near the reinforcements.

For both walls, lateral movements of the facing panels occurred during construction (Figs 3(b) and 4(b)). However, for the wall with a compressible layer, movement of the soil/wall boundary was significantly larger due to the deformation of the compressible layer. At the end of construction, the deformed shape of the compressible layer indicated that wall rotation about the top occurred.

The vertical base pressure distribution, for the rigid wall, was reasonably uniform and equal to the overburden pressure (Fig. 5). For the wall with a compressible lateral boundary an increase in pressure was observed in the zone 0.25 m to 1.0 m away from the wall face. However, for both walls, at close proximity to the wall face, the vertical pressures were lower than the overburden pressures. This is attributed to the friction developed between the wall face and the backfill.

The strain distributions along the geogrids for both walls at the end of the construction are shown in Fig.6. Larger strains were recorded in the geogrids for the wall with compressible lateral boundary, the largest difference being in the second layer of the reinforcements. Overall strain measurements using strain coil in the geogrids are also shown in Fig.6, which agree with the strains measured locally.

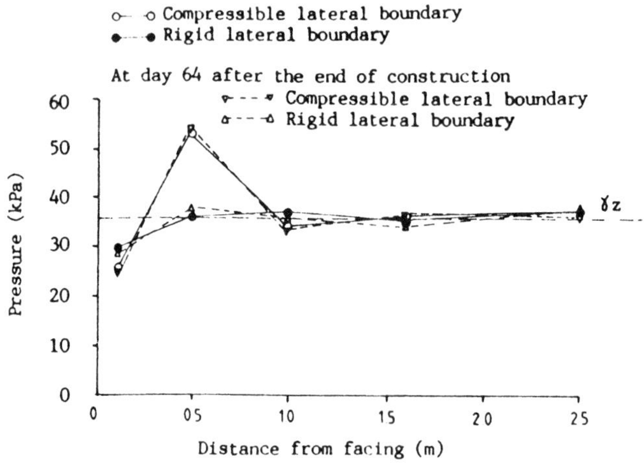

Fig. 5. Vertical pressure distributions at the foundation level

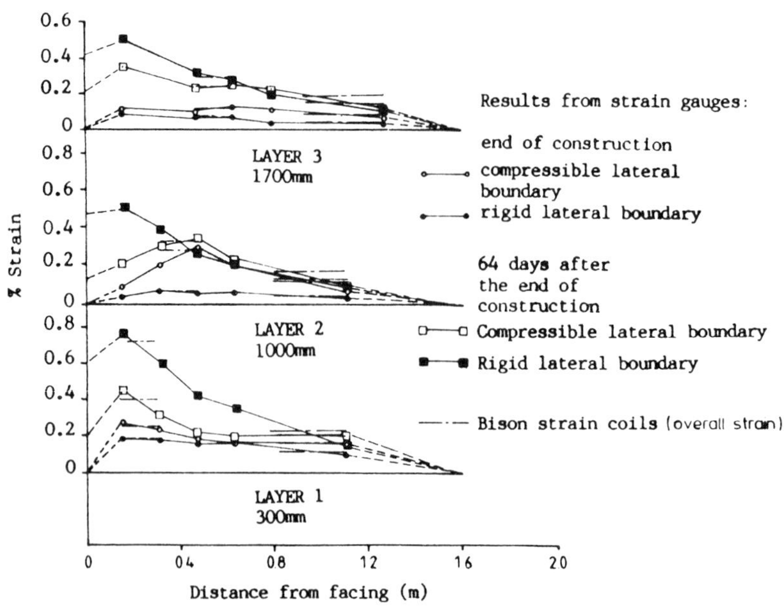

Fig. 6. Strain distributions along the geogrids

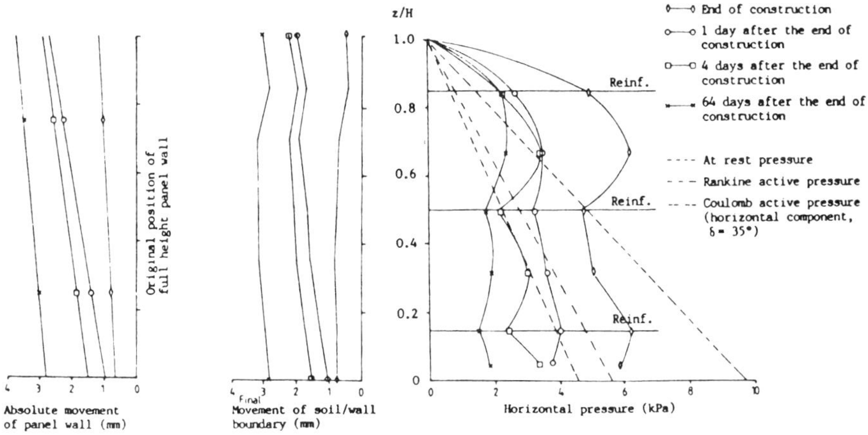

Fig. 7. Horizontal earth pressure distributions, movements of wall facing and movements of soil/wall boundary after the end of construction. (Rigid lateral boundary)

After removal of props

The lateral pressure distributions on both walls at the end of construction and 64 days after the end of construction are shown in Figs 7 and 8. For the rigid wall, significant reductions in the magnitude of the lateral pressure were obtained with time. After 64 days, the area under the pressure distribution was close to that obtained from the Coulomb "active" condition. However for the wall with the compressible layer, little change in the shape

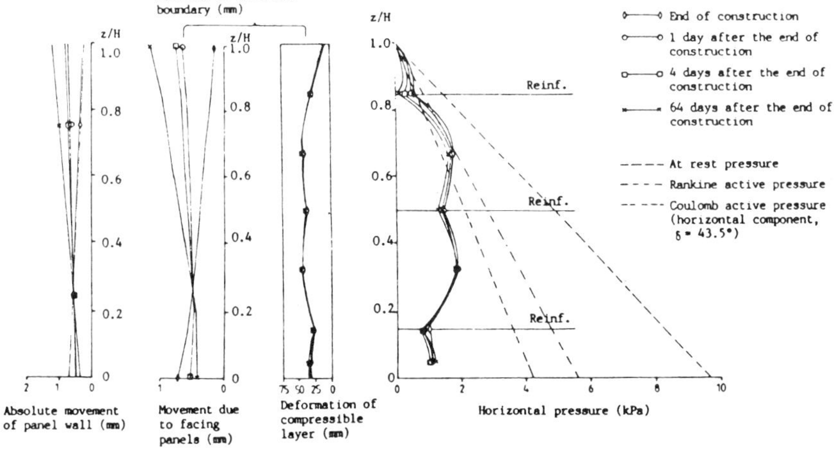

Fig.8.Horizontal earth pressure distributions, movements of wall facing and movements of soil/wall boundary at end of construction (compressible lateral boundary)

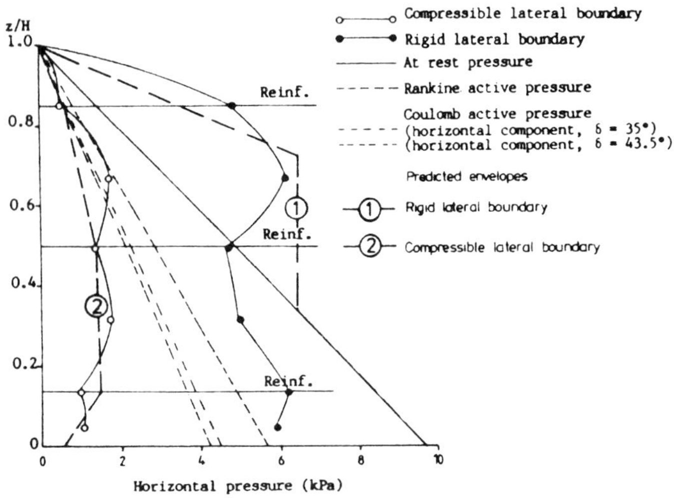

Fig. 9. Comparison of predicted and measured lateral earth pressures

and magnitude of the lateral "active" condition. However, for the wall with a compressible layer, little change in the shape and magnitude of the lateral pressure distributions was observed. In addition, for the rigid wall, a translation movement of the soil/wall boundary of about 4 mm occurred after 64 days (Fig.7). However, for the wall with the compressible layer, little change in the deformation of the compressible layer with time was observed. The maximum movement of the soil/wall boundary due to the facing panels, was about 1.0 mm occurring at the top of the boundary after 64 days (Fig. 8).

The vertical base pressure distributions for both walls after 64 days are shown in Fig. 5. The results indicate that no significant change in the shape and magnitude of the vertical pressure distributions was observed from that obtained at the end of construction.

Fig. 6 shows the post construction strain distributions along the geogrids for both walls. At 64 days, all three layers of geogrids in both walls showed an increase in strains. However, the increase in strains was larger in the geogrids in the rigid wall than that with a compressible layer.

Comparison of predicted and measured results

Fig. 9 shows the comparison of the measured and predicted lateral pressure distributions at the end of construction. For rigid wall boundary, the predicted distribution is calculated (line 1) using q = 6.5 kN/m^2, x = 0.1m, width of compaction plant = 0.46 m and K = 0.2 (number of passes) + Ko. The measured pressure distribution is similar to that predicted shape at the upper half of the wall. At the lower half, smaller pressures were measured than those predicted. This is because of some movements of the facing occurred during construction, resulting in a reduction of the measured pressure along that part of the wall. For the wall with a compressible layer, the predicted distribution is shown in line 2. Since the reinforcements used did not provide a rigid layer along the entire width of the wall, the lateral pressure immediately below the reinforcement layers is not zero but is assumed to be equal to the lateral pressure at the base of the zone above it. This assumption gives good agreement with the measured results.

The calculated loads in the geogrids after removal of the props (Fig.10) were carried out using the measured strain together with the isochronous curves. Comparison of the calculated loads in the geogrids at the facing with the total loads on the wall, measured using pressure cells, showed that the predicted loads were higher than the measured results. These differences may be attributed to the difference in the operational mechanisms of the geogrids in-isolation and in-soil and also due to the effect of compaction.

Conclusion

The above results indicate that the use of a compressible layer allows both the mobilisation of the shear strength of the soil and the straining of the geogrids during construction. As a result, significant reduction in the horizontal pressure was achieved which minimises the post construction strains after the removal of the props.

Thus using a compressible layer placed between the backfill and the wall

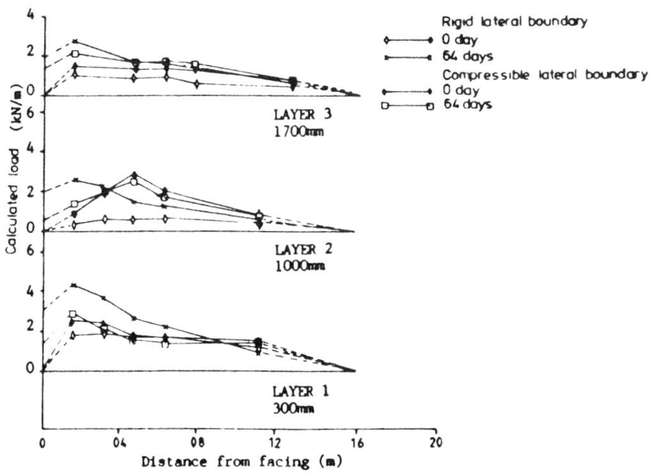

Fig. 10. Calculated tension distributions along the geogrids after the end of construction using isochronous curves at 10°C

facings provides a practical means of reducing the movements of the wall facings, particularly post construction. It optimises the mobilisation of the soil strength and tensile resistance in the reinforcements and minimises 'locked-in' stresses due to compaction.

For reinforced backfills, if sufficient number of reinforcement layers are used, the backfill behaves like a rigid block and when a compressible wall boundary is present, a near trapezoidal vertical pressure distribution at the base is obtained.

The proposed method of analysis for the prediction of lateral pressure takes into account the number of passes and the direction of compaction, and the width of the compaction plant and the distance of the plant from the wall. Comparison with experimental data suggests that the theoretical model can be used to predict the lateral pressures on the wall.

References

1. ANDRAWES K. Z., LOKE K. H. and YEO K. C. Application of boundary yielding concept to full scale reinforced and unreinforced walls. *Performance of Reinforced Soil Structures.* Thomas Telford, London, 1991, pp 79-83.
2. BROMS B.B. Lateral pressure due to compaction of cohesionless soils. *Proc. 4th Int. Conf. Soil Mech. Found. Engng.*, Budapest, 1971, pp 373 -384.

Soil nailing case histories and developments

A. D. BARLEY, Contracts Director, Keller Colcrete

The paper reports on seven soil nailing contracts involving stabilisation of existing walls and both new and existing slopes. Current tendon corrosion protection systems and costs are highlighted and development of new non-corrosive tendons is summarised.

Introduction

The use of soil nails for the stabilisation of slopes has been common in France, Austria, Germany, Canada and the U.S.A. since the mid 1970's. This development followed extensive use of fully bonded rock bolts (passive anchors) for tunnel support during the 1950 and 1960's which lead to the use of rock bolts in weaker rocks such as marls, and weak sandstones. The natural progression continued in the application of the same ground strengthening principles to weakly cemented sands, strong clays and subsequently to mid-strength soils. Soil nailing became an accepted practice for the stabilisation of existing slopes and walls and to allow slope cutting and steepening.

Despite the frequency of such applications of soil nailing techniques in other countries for over a decade, the number of contracts executed in the U.K. during recent years amounted to little more than a handful. Case histories of seven of the earliest soil nailing contracts carried out in the U.K. between 1985 and 1991 are summarised and some useful established design parameters are derived and compared with values previously utilised.

Available corrosion protection systems for steel tendons are assessed from technical, practicable and economic viewpoints. Current developments involving the use of glass fibre and polymeric materials as corrosion free tensile members are also discussed.

Soil nailing to stabilise existing retaining walls

Denholme Clough, Bradford

The use of dry stone walling built more than a century ago to support sloping ground alongside roads through the Pennines is a common feature. At Denholme Clough the road crossed an open hillside with a side slope gradient in the range of 40%. The existing stone walls, between 2.5 and 3

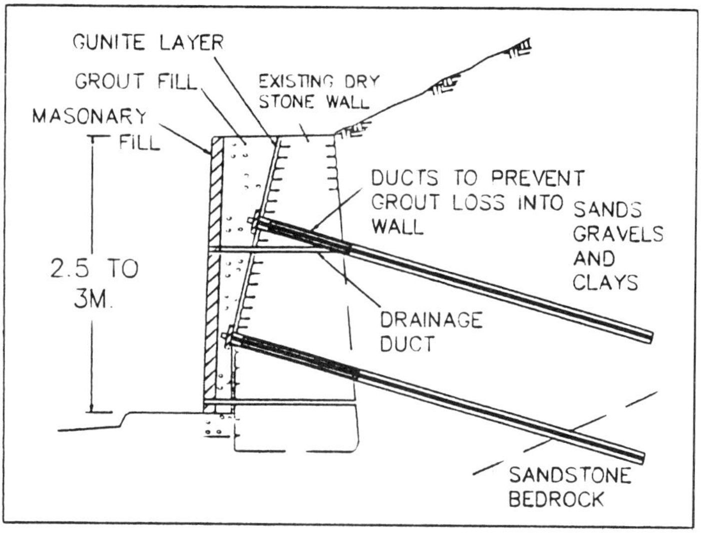

Fig. 1. Soil-nailed wall at Denholm Clough

metres in height, exhibited bulging and short areas of collapse over a total length of 140 metres. Prior to soil nailing, remedial construction works had been carried out in the areas of collapse, some vegetation removed, and a small footing trenching excavated at the base of the wall. Steel mesh (D49) had then been positioned on the face of the wall and a 50 mm layer of gunite sprayed over the entire wall length. The retained soil consisted of a mixture of wet sands, gravels and clays overlying a sloping sandstone bedrock. Some 200 No. 30kN. Load soil nails were installed, generally in two rows at either 1 m or 1.5 m horizontal spacing to depths of 4 m to 5 m at an inclination of

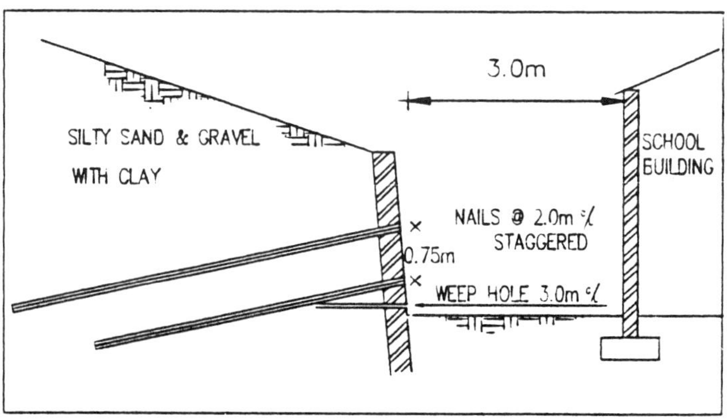

Fig. 2. Soil-nailed wall at Cymsifiog

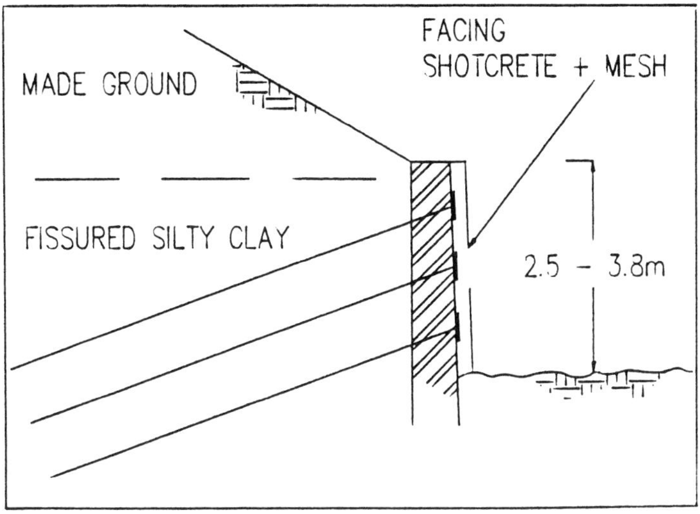

Fig. 3. Soil-nailed wall at Beaufort Road

15°. Drainage holes were installed at base of the wall and at mid-height at 3 m spacing (Fig. 1). During the work, three special trial nails were installed and tested satisfactorily to 80 kN. This proved a ground/grout bond capacity over a 2 m bond length of 110 kN/m^2 without any indication of failure.

After trail nail testing, and nominal loading of all the contract nails to 10kN, a new masonry facing was constructed by the local authority and gaps between the facing and the gunite were filled with sand/cement grout. No indication of movement or detection of cracks has been established during frequent observation throughout a six year period. This would suggest that stabilisation has to date been fully effective.

Cymsifiog Primary School, Glamorgan

A 2.2 m high masonry wall, almost 100 m in length, retains a full range of wet granular materials with some clay, the surface of which slopes upward from the wall at a 35% gradient. This wall is located some 3 metres from a Victorian school building and protects the building and the school playground. The wall is located some 3 metres from a Victorian school building and protects the building and the school playground. The wall exhibited cracking and bulging, and movement of the order of 5 mm had been monitored over a 6 month period. At the time of soil nail installation, temporary strutting was in place to provide additional support to the wall. Ninety soil nails were installed through the wall in two rows at a spacing of 2 m horizontally and 0.75 m vertically (Fig. 2). Working capacities of 30kN. were required of the 5 to 6 m long nails and load was transferred from the nails to the 0.5m thick wall in direct bond via the cement grout. This load transfer over a short length without a bearing plate was aided by the use of a deformed bar (T25) as a soil nail tendon. Corrosion protection to the bar was provided by the use of a 50 mm diameter continuous corrugated plastic duct, identical to that used in anchor encapsulation systems but grouted in situ. Three trail nails were installed and satisfactorily tested to 90 kN., confirming that a factor of safety in the order of 3 was present at the bonded interfaces. Wall movement was monitored for 14 months after soil nail installation and no movement or further cracking was detected.

Beaufort Road, Bristol

This masonry wall was also about 100 years old and retained 2.5 to 3.8 m height of ground plus a rear upward slope at a 50% gradient. However, at this location vegetation was abundant and the upper slope contained shrubs and trees. Retained soil consisted of made ground over fissured silty clay and grey siltstone bedrock. Stabilisation was required prior to building work at the top of the slope. Fifty three nails were installed in 3 rows over the higher section which reduced to 2 rows where the wall height fell below 3 m and where the slope height behind was reduced. Nails were installed on a grid pattern at an inclination of 20° with a 2 m horizontal and 1.5 m vertical spacing, with lengths of 9, 7.5 and 6 m working from top to bottom (Fig. 3). On completion of soil nailing operations, a steel mesh was fitted to the face and a gunite surface applied in a thickness of 50 mm. Detailed observation two years after completion indicates a stable wall with a crack free surface.

Ampthill Road, Bedford

Movement of the 1.5 to 2 m high gravity brick wall supporting the A6 trunk road in the approach to a road over a rail bridge had been causing concern. Although the height of the wall was low, the earth slope behind rose at a 30° to 55° gradient to the road pavement, resulting in a total wall and slope height of approximately 5 m. The retained earth and the embankment generally consisted of firm to stiff silty clay which increased in sand gravel content with depth. The eighty eight soil nails required for temporary wall stabilisation possessed characteristics of both anchors and nails. Although the specification called for an unbonded "free length" equivalent to about one third of the nail length, the nails were unstressed and left in a passive state. Nail lengths ranged from 8.5 to 12.5 m at inclinations of 10° to 20°. Four preliminary trial nails attained 150% design load without failure, and all nails were finally bonded into the 0.5 to 0.6 m thick brick with cement grout without any utilisation of bearing plates. The stabilised wall has been observed at monthly intervals and although accurate monitoring has not been carried out, visual observations over the 2½ year interim period confirm that movement and crack widening has ceased.

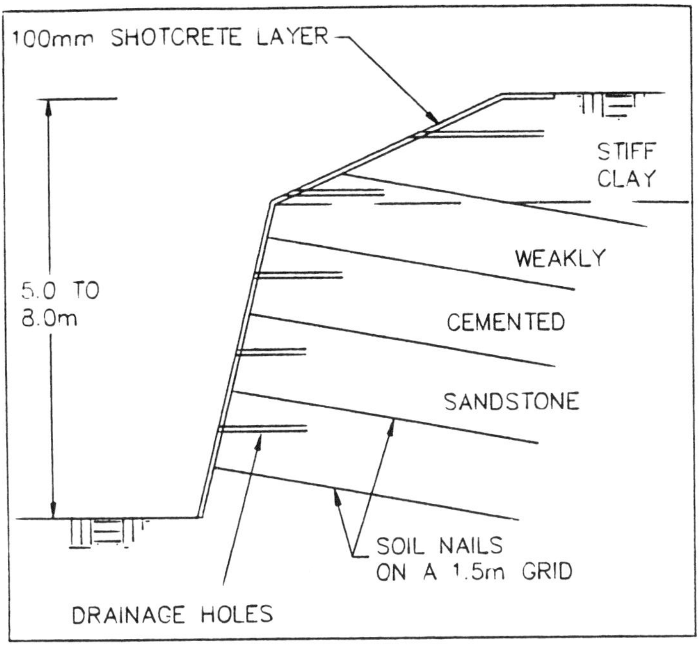

Fig. 4. Temporary soil-nailed slope in Birmingham

Soil nailing to stabilise slopes

Temporary soil nail support in Birmingham

Building construction and site limits demanded the excavation and the accompanying stabilisation of a 150m wide steep face up to 8 m in height. Soils consisted of firm to stiff clay overlying weakly cemented sandstone (Keuper Marl). The upper 2 m in the firm clays could be battered back at a 100% gradient. Some 300 soil nails, in up to 5 rows at 1.5 m spacing both vertically and horizontally, were required to support the shotcreted face. Construction was in a downward progression with the 100 mm thick shotcrete facing, containing 100 x 100 steel mesh, proceeding immediately after face exposure. The nail installation and 50 mm slotted PVC drainage hole construction followed, generally with the unsupported face height not exceeding 2 m. Length ration of the nailed slope generally equated to 1.0, with 8 m nails in a slope of that height and 5 m nails where the slope height reduced to 5 m (Fig. 4). The T25 bars were installed in lengths up to 6 m with the use of couplers being required for the longer nails. The bond capacity over the full length of grouted bar and borehole was supplemented by the positioning of 200 x 200 mm diameter plates on the shotcrete surface.

It was unfortunate that after heavy rain, a collapse at one end of the face

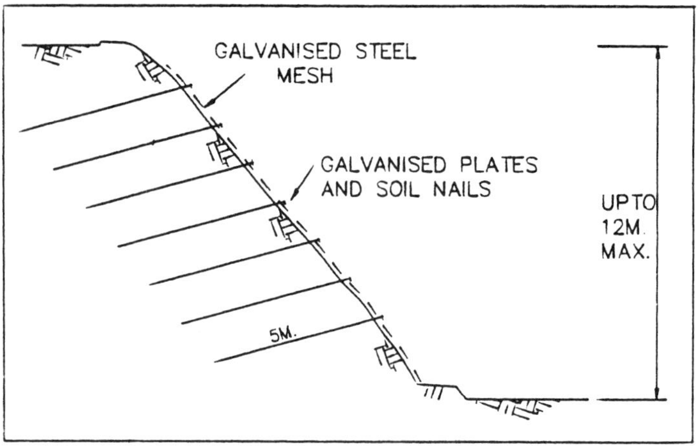

Fig. 5. Soil-nailed slope at Dolywern

took place. It was suggested that this failure could be attributed to insufficient control of the numerous operations being carried out by different working parties.

(a)Excavation proceeding below the 2 m limit of unsupported face,
(b)the omission of a number of soil nails due to over-excavation;
(c)the blockage of a number of drainage ducts with shotcrete.

Subsequent to this collapse, controls were tightened and satisfactory face support to the full height was provided.

Permanent support, Dolywern, Clwyd

A steep embankment alongside the B5400 between Chirg and Glyn Ceirog in North Wales had been providing the County Council with reasons for concern over a number of years due to observations of settlement and cracking of the upper road surface. The vegetated slope between a minor road at the top and the main road below has a gradient in the order of 100% and a maximum height of approximately 10 metres. The soils consist of silty clays with some gravel and cobbles. The slope has now been stabilised by the use of some 180 soil nails in two stages of contract work. Nails were generally required at 1.5 m spacing both vertically and horizontally and at an inclination of 15°. (Fig. 5). Nail locations in up to 7 rows were staggered in adjacent rows. The nail steel consisted of galvanised 25 mm diameter high yield thread bar, 5 m long complete with 250 x 250 x 10 mm galvanised plates and nuts at the head. All major bonding within the soil nail in order to stabilise the soil and the slope was achieved by the sand/cement grout, whilst the plate positioned on the protruding nails provided a mode of retention of the

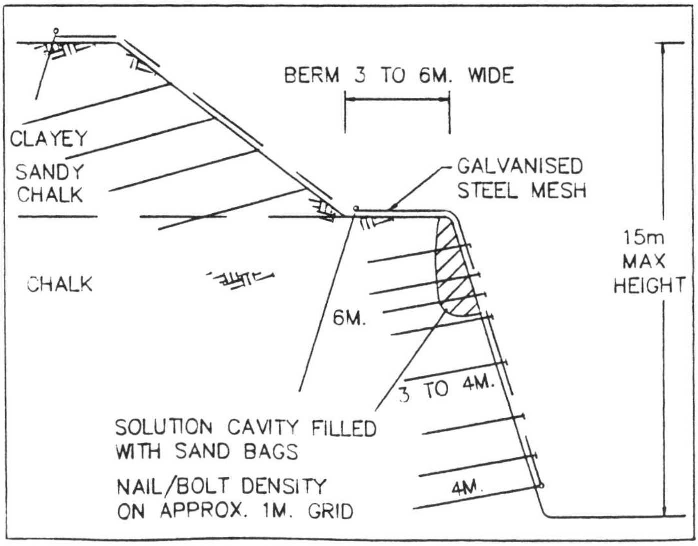

Fig. 6. Rock-bolted and soil-nailed slopes at Lakeside

galvanised steel mesh (A142) which was finally spread over the entire embankment (Fig. 6). This mesh prevented spalling and erosion, and supported the growth of fresh vegetation. On completion of nailing operations, the surface of the minor road was patched up, and from observation of the patching condition there is every indication that further slope movement has to date been prevented.

Lakeside, Thurrock

Recent development of a modern shopping complex within an extensive disused chalk quarry required the trimming and stabilisation of a 500 m long 15 m high face. Initially the exposed chalk face was trimmed to a 70 to 75 degree slope. above the chalk level was a 4 to 6 m wide berm behind which was, for most of its length, a rock armour covered 40 degree clayey, sandy chalk slope. Intermittent within the chalk face were large solution features with a face area as much as 7 m vertical by 5 m wide, where chalk had been washed away and naturally replaced with overlying materials of chalky sands, gravels and flints. Stabilisation was carried out by the installation of some 1500 bolts in up to 8 rows in the required. The 6m long steeply inclined bolts at the back of the berm were threaded with cables to support the suspended galvanised mesh. The intermediate 6 rows ranging from 3 to 4 m long, stabilised the chalk face and held the mesh in face contact by the use of galvanised plates and nuts (Fig. 6). The lower row, complete with threaded cables, retained the lower extremity of the mesh approximately 1 m above the base of the steep face. The granular material content of the exposed solution features was removed and replaced by wet concrete filled bags built into the cavity in a brickwork type formation with a finish flush with the chalk face. 3 to 4 m long bolts in a grid of about 1 m spacing were installed through the filled cavity to tie the bagged facing to the intact chalk behind. At the western end of the face the chalk level and berm level reduced over an 100 m length. This resulted in the presence of a 40^{0} clayey, sandy chalk slope, attaining up to 6 m in height above the berm. The soil nail stabilisation of this slope was only considered necessary where the slope height exceeded 3 m. The 150 nails ranging in length from 3 to 6 m, consisted of galvanised high yield 25 mm diameter steel bars the same as that employed for the bolts. The nails also executed a dual purpose of slope stabilisation and support and constraint of galvanised mesh and tensar mat. Some dozen bolts/nails installed early in the works were satisfactorily tested to 120 kN without failure. On completion, about 7500 m^2 of mesh was installed on the chalk face and 3400 m^2 on the berm and on the overburden slope behind.

Soil nail parameters from slope and wall stabilisation

When summarising the state-of-the-art of soil nailing and reporting on continental practice in 1986, Bruce and Jewell derived four basic parameters each of which provided a useful measure of the characteristics which contribute to the stabilisation or record the slope movement:

(a)length ratio= maximum nail length/excavation height
(b)bond ratio= hole diameter x nail length/vertical face area supported by nail;
(c)strength ratio= (nail diameter) ^{2}/vertical face area supported by nail;
(d)performance ratio= outward movement/excavation height.

Fig. 8 tabulates stabilisation parameters from the three stabilised slopes described, in addition to general parameters produced by Bruce and Jewell. It should be noted that the latter were derived from sites involving steep slopes (80^{0} or steeper) and thus only allow direct comparison with data from the Birmingham slope. In the case of slopes at Dolywern and Thurrock, lower values of length ratio and higher values of bond and strength ratios may be expected, since the shallower slopes would be partly self-supporting and nailing was carried out to increase the factor of safety against slope collapse and to retain a surface mesh to control surface erosion. No data has been published with regard to the application of the three stabilisation parameters of the stabilisation of existing walls. In such an application it is recommended that the effective supported height, and the vertical face area supported by a nail, are taken at a cross section at the mid-length of the longest nail (Fig. 7). On the basis of this consideration, Fig. 9 tabulates values derived from the

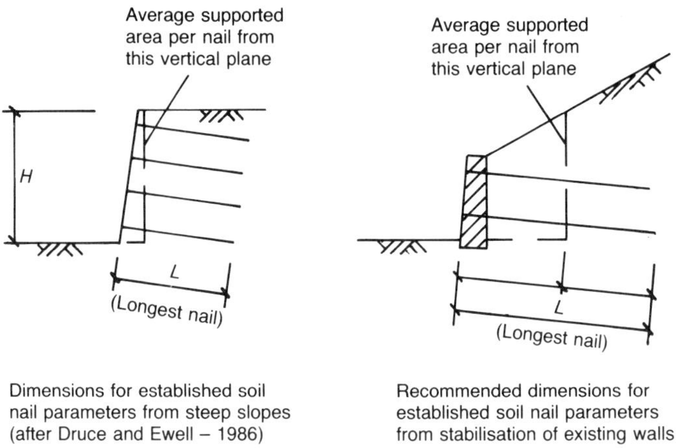

Fig. 7. Mode of dimension analysis for deriving soil nail parameters

four stabilised walls. Owing to the initial conservative approach to wall nailing, length ratios are relatively high. Bond and strength rations are low due to the self-support of the slopes and walls for many years. Again, the recent nailing has been carried out to supplement the stability and not to provide the major component of the resisting force. Although the parameters established from the stabilisation of newly formed steep slopes and those analyzed from early wall stabilisation contracts take the same format, it is important that they are considered separately. The input data for height and nail area is necessarily in each case defined in a different mode as shown on Fig. 7. However, the parameter can be seen to serve a useful purpose for comparing like with like and, furthermore, by contrasting unlike schemes.

Corrosion protection systems for steel tendons

An awareness of the vulnerability of steel tendons due to corrosive action may lead to any of the following considerations in order to establish an optimum balance between cost and risk of loss of steel section in a soil nail:

(a)assessment of loss of steel section in either a homogeneous of a non-homogeneous soil,
(b)the effectiveness of the alkaline environment of cement grout in resisting/reducing corrosion in either a crack controlled or random cracked grout column,
(c)the provision of a sacrificial coating on the steel (i.e. galvanising) the protection of the coating during handling and the estimation of lifespan

STABILISATION OF STEEP OR SHALLOW SLOPES USING DRILLED AND GROUTED NAILS				SUMMARY OF VALUES FOR DRILLED AND GROUTED SOIL NAILS IN STEEP SLOPES (AFTER BRUCE & JEWELL 1986)	
	BIRMINGHAM	DOLYWERN	THURROCK	GRANULAR SOILS	MORAINE AND MARL
LENGTH RATIO	1.0	0.42	1.0	0.5 to 0.8	0.5 to 1.0
BOND RATIO	0.36	0.16	0.15	0.3 to 0.6	0.15 to 0.20
STRENGTH RATIO X 10-3	0:28	0.28	0.20	0.4 to 0.8	0.1 to 0.25
COMMENTS	Steep Slope in Marl	45° slope of Silty Clay	45° upper slope of Clayey Chalk		

Fig. 8. Soil nail stabilisation parameters for slopes

STABILISATION OF EXISTING RETAINING WALLS				
	BRADFORD	CYMSIFIOG	BRISTOL	BEDFORD
LENGTH * RATIO	1.25	1.85	1.50	2.50
BOND * RATIO	0.17	0.18	0.22	0.19
STRENGTH * RATIO X 10-3	0.078	0.19	0.06	0.13
PREDOMINANT SOIL BOND MATERIAL	GRANULAR SOME CLAY SOME BEDROCK	GRANULAR SOME CLAY	SILTY CLAY SOME BEDROCK	SILTY CLAY SOME SAND AND GRAVEL

* Effective height and area per nail determined from ground section at mid-length of the longest nail

Fig. 9. Soil nail stabilisation parameters for existing walls

when in situ,
(d)the provision of an epoxy coating on the steel and the protection during handling and any possible degradation of the protection layer,
(e)the provision of corrugate plastic ducting around the steel tendon, all grouted in situ,
(f)the use of stainless steel and estimation of lifespan when in situ,
(g)the provision of two concentric corrugated ducts around the steel tendon generally prefabricated prior to installation. There are currently two British Standards documents which may be considered appropriate to the design of soil nails: the Code of Practice for Ground Anchorages (BS 8081) is published, whilst the Code of Practice for Strengthened/Reinforced Soils and Other Fills (BS 8006) is drafted for comment.

In soil reinforcement systems where the surrounding soil can be carefully selected and controlled, option (a) may be considered appropriate, whereas in Ground Anchorage practice protection systems at the other end of the range, as highlighted in (g) are recommended for permanent works.

In situ soils into which nails are installed can rarely be considered adequately homogeneous to prevent current differential, or of low enough permeability to eliminate the presence of moisture and a replenishment of a supply of oxygen. Furthermore, due to the general straining of the soil of the nail tendon, and of the grout, the maintenance of crack free or even crack controlled grout cover is improbable. Thus, in the current climate approaches (a) and (b) are generally considered unsuitable for long term permanent stabilisation work. Options (c) (d) and (e) are generally available in a similar price range. Costs increase is in the order of 60 to 100% over and above the cost of the uncoated steel bar. Since the protected steel tendon costs generally range from only 6 to 10% of the total soil nails costs, a variation in preference for any of the three systems would not make a major cost change to the contract value. The Author has no direct experience with the use of epoxy coating system, but with excellent characteristics of adhesion to steel, of impact resistance and of elastic properties compatible with the steel, it clearly deserves due consideration. However, the frequency, the size and the effect of the presence of "holidays" (pin holes in the coating of a few microns in diameter) should be assessed. The use of a single plastic duct is well tried in both nailing and anchoring systems, and the exposure of test samples and the exhumation of a number of ground anchorages, all confirm that the plastic layers are well preserved and remained un-perforated even after being subjected to severe loading conditions. In the Author's experience, galvanising has its limitations. Firstly, the zinc coating is sacrificial and will deteriorate in the long term leaving parts of the steel exposed. Secondly, chipping of the zinc coating during bar loading, unloading, handling and installation appears to occur despite extensive precautionary measures and careful operation controls. Thirdly, within the thread length (if required), the zinc coating will be either of minimal or zero thickness at some point in the male or the female thread.

Although stainless steel (f) is available in a number of base forms, it is not invulnerable to long term corrosion. Furthermore, it has a large cost differential over and above that of normal uncoated steel (800 to 900%) and this would lead to a considerable increase in the cost of the works (35 to 50%). The use of a prefabricated double corrosion protection system, consistent with those generally specified for permanent ground anchorage works, would provide the most efficient corrosion resistance steel tendons currently available. The additional cost of such tendons, of the additional haulage, and of handling considerably heavier units must be offset against the protection benefits available. Additional costs increase a nail tendon cost by some 500% which in turn may lead to an increase in the cost of the works of perhaps 25 to 30%. In terms of the use of prefabricated double plastic ducting as compared to single in situ grouted ducting, the additional cost would be in the order of 20%.

Development of non-corrosive soil nail tendon systems

Continental practice has already generated and proven the use of glass reinforced plastics both in rock bolting and ground anchorages, the later system being first used in prestressed bridge decks. Bonding with resinous grouts has been favoured in rock bolting technology, although published data has also confirmed the adequacy of bond capacity between G.R.P. and cementitious grouts. Research work is currently being undertaken by Keller Colcrete to further investigate bond capacity at this interface when utilising a variety of G.R.P. sections. Considerable investigation work has been completed by Keller Colcrete on the use of folded geotextiles as the main tensile member within of a grouted soil nail borehole. Bond capacities have been well proven, and modes of installation of such flexible members have been established.

Both G.R.P. and geotextiles, when incorporated as tensile members, exhibit a considerable increase in elastic extension over that of steel bars. However, the resulting problems have already been overcome in the rock bolting application, and the investigation work is currently underway to assess the effect of increased tendon elasticity on the stability of soil nailed slopes. The alternative system of prestressing of soil nails to mobilise tensile capacity and thereby reduce soil strain is also being assessed. This operation would require a two phase grouting operation which may prove to be cost prohibitive.

Summary

Soil nailing has been used extensively on a worldwide basis and has now been satisfactorily proven in a number of slope and wall stabilisation works in the United Kingdom. When utilised for temporary support of steep excavated slopes the necessity for tight control of all phases of operation has been highlighted.

In addition to soil nailing works reported in this Paper, slopes have been stabilised using soil nails at a number of other locations in the U.K. New techniques involving the firing of soil nails are also being investigated but these clearly demand a different approach to corrosion protection systems and modes of load transfer. The soil nail parameters originally derived by Bruce and Jewel have provided a useful yardstick for assessment of stabilisation characteristics of both soil nailed slopes and nail stabilised walls. A number of options are available to provide varying modes and degrees of corrosion protection to steel tendons. The Paper has provided guidelines to cost and efficiency of a number of systems available. The use of non-corrodible man-made fibres in the form of glass reinforced plastic or polymeric materials are currently in use in rock bolting, anchoring and soil reinforcement. Research and development is approaching a stage where their introduction into soil nailing should be carried out within the near future.

References

1. Bruce D. A., Stabilisation of Dry Stone Wall at Denholme Clough, Bradford with Soil Nails, n.p.
2. Maddison J. D., Jones, D. B., Spencer I. M., Retaining Wall Stabilisation Using Soil Nails at Cwmsyfiog Primary School, South Wales. Proceedings of the Second International Conference on Foundations and Tunnels - Engineering Technics.
3. Bruce D. A., Jewell R. A. Soil Nailing part 1 & 2 Ground Engineering November 1986, January 1987.
4. Jones D. E. Rock Bolts and Soil Nails for Chalk Face Stabilisation at Lakeside, Thurrock, n.p.
5. B. S. Code of practice for Ground Anchorages BS8081.
6. Draft British Standard Code of Practice for Strengthened/Reinforced Soils and Other Fills - BS8006 BSI publication.
7. Barley A.D. Investigation of the Use of Geotextiles & Glass Reinforced Plastic in Soil Nails. n.p.
8. Barley A. D. Gun Barrel Pull-Out tests and Exposure Inspection of Plastic Corrugated Plastic Ducting n.p.
9. Barley A. D. Exhumation of Test Anchorages at Castle Mall, Norwich, n.p.
10. Weidmann. Fibre Glass Rock Bolts, Rockbolting Technology, Commercial Publication.

11. Strabag, Application of Heavy Duty Composite Prestressing Tendons, Commercial Publication.
12. Barley A. D. Development and Application Discussion Session Proceedings of Conference on Slope Stability Engineering, Isle of Wight, April 1991, Thomas Telford Publication.

Design of sloped reinforced fill structures

P. SEGRESTIN, Terre Armee Internationale, France

Fields of application

More and more sloped retaining walls and slope stabilization structures using reinforced fill techniques are now being built in many countries. Since the first inclined Reinforced Earth (R) structures were built 14 years ago, with hard discrete concrete facing units for high V shaped coal storage slots (McKittrick, 1982), the use of this type of retaining structure quickly extended to other fields of application. They now frequently apply to rustic soft-faced steepened embankments (Smith, 1990), as well as plantable battered walls. These various structures employ a variety of facings according to their purpose and are also built with different types of reinforcing materials, made from steel or from geosynthetics according to the function, the required minimum service life and the tolerable deformations of the structure.

Deficiencies of common design methods

There are a number of methods presently in use for the design of such steep reinforced sloped walls, with regard to their internal stability, i.e. the arrangement of the reinforcements.

Earth pressure

One category of design methods considers the sloping reinforced soil

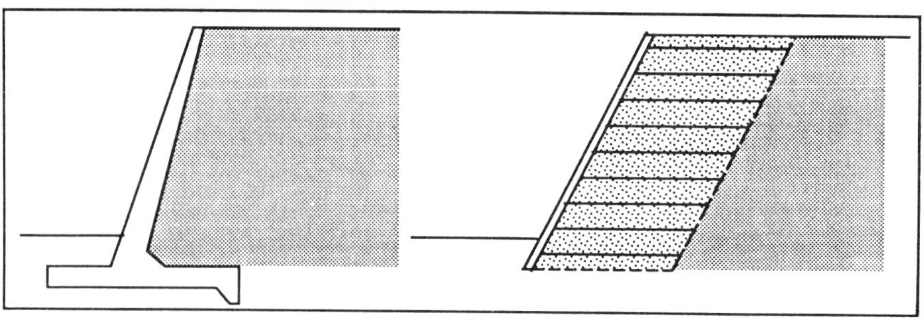

Fig. 1

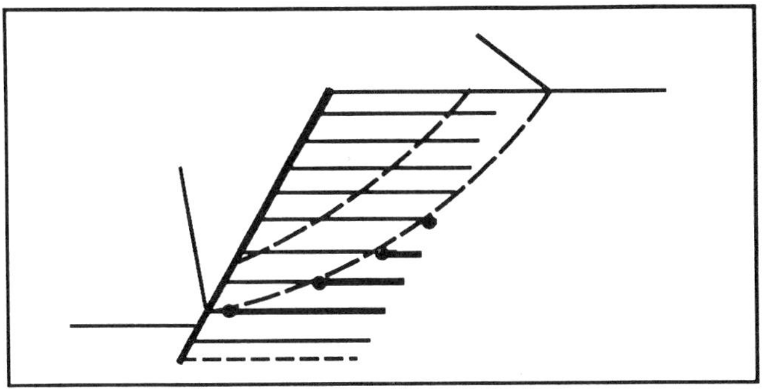

Fig. 2

structure as a coherent gravity wall and derives the internal stresses from the resulting effects of all the external loads applied to the structure, including the earth pressure exerted by the backfill behind. Usually the earth pressure is calculated using the same routine earth pressure coefficients and inclinations as for conventional concrete cantilever walls or masonry gravity walls.

However, when considering examples of significantly inclined walls, it is clear that a concrete wall such as shown on Fig. 1 can be perfectly stable before being backfilled, while the reinforced fill structure (although reinforced) cannot do without the presence of the backfill. Their interaction is not the same. The backfill simply pushes against the masonry wall where it supports a part of the weight of the reinforced soil mass. Therefore the 'earth' pressure has to be estimated in a different way, as will be explained later on.

Potential failure line analysis

Another group of design procedures consists of an extension of methods of analysis - based on bi-linear, circular or non circular potential failure lines- which have been used for a long time for checking the overall stability of unreinforced embankments and natural slopes.

In these modified computation techniques the available resistance of predetermined reinforcing layers, where intersected by the assumed failure line, is combined with the shear resistance of the filler material (Fig.2).

This approach has nevertheless a number of drawbacks

- since the reinforcements have to be predetermined by some other method, the global analysis only confirms that there is no risk of failure along any assumed potential line; it does not guarantee that the arrangement of reinforcements is optimal or most appropriate
- there is no reason for the computed most critical failure line to correspond to the locus where in-service stresses are actually most important
- indeed the critical line depends on the set of safety factors applied to the different components which contribute to the stability (tensile

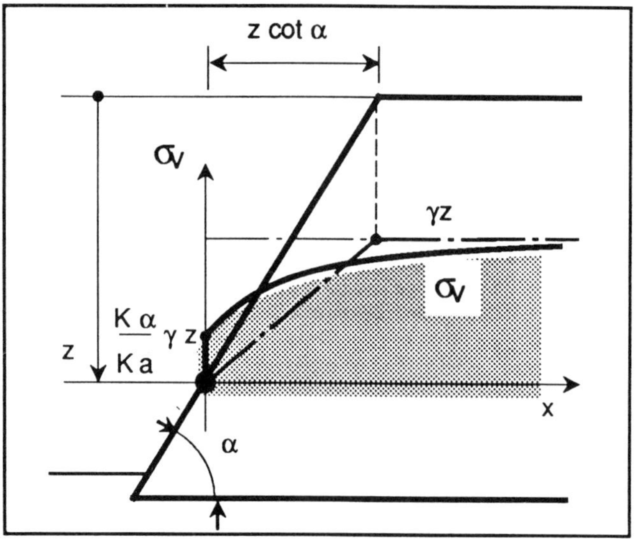

Fig. 3

strength of reinforcing material, friction between soil and reinforcement, shear resistance of backfill). Any other set of values of safety factors leads to another line, which is no more the real one.

In fact these limit equilibrium methods (which correspond to checking rather than design calculations) are essentially theoretical and hypothetical; they clearly do not reflect the actual ultimate state of the structure and certainly not the actual in-service mechanism of the composite reinforced soil material.

Design based on actual behaviour (T.A.I. method)

Experimental basis

On the contrary the design method which is explained below is backed-up, first of all, by the observed behaviour of in-service structures and models and checked by reference to measured stresses. This practical method was developed by Terre Armée Internationale as the outcome of a research program combining three dimensional models, finite element studies and full scale monitoring (Bastick, 1983); the method is currently used for the design of the Reinforced Earth (R) sloping structures built with metal strip reinforcements.

The principle of the method rests essentially on the knowledge of the variations of the vertical stresses inside the reinforced mass on one hand and of the locus of the maximum tension in the reinforcing layers on the other. The former were deduced mainly from finite element studies, the latter from the monitoring of reduced scale models and full scale structures, also confirmed by the FEM analysis (T.A.I.,1989).

Variation of Vertical Stress

The results of the finite element studies showed, as could be expected intuitively, that in a steepened sloping embankment the vertical stress σ_v at depth z increases from σ_0 ($0 < \sigma_0 < \gamma z$) behind the facing to γz well inside (Fig.3). The ratio between σ_0 and γz varies with the slope angle α.

We know that at the macroscopic scale the horizontal stress behind the facing would be $\sigma_h = K_\alpha \gamma z$, where K_α is the coefficient of horizontal active earth pressure on a screen inclined at an angle α, i.e:

$$K_\alpha = \left[\frac{\sin(\alpha - \varphi)}{\sin\alpha + \sqrt{\sin\alpha \cos(\alpha - \varphi) \sin\varphi}}\right]^2$$

At the microscopic scale, in a soil with adequate deformation we also have at the same point :

$$\sigma_h = K_a \sigma_o \qquad \text{where :} \qquad K_a = \tan^2\left(\frac{\pi}{4} - \frac{\varphi}{2}\right)$$

As a result :

$$\sigma_o = (K_\alpha / K_a)\,\gamma z$$

It was found that the following equation could be used for the curve representing the variation of σ_v:

$$\boxed{\sigma_v = \gamma z.\xi(u) = \gamma z.\mathrm{th}(au + b)}$$

where

$$u = \frac{x}{z} \qquad a = 2\tan\alpha \,\mathrm{Log}\frac{2K_a}{K_a + K_\alpha} \qquad b = \frac{1}{2}\,\mathrm{Log}\frac{K_a + K_\alpha}{K_a - K_\alpha}$$

If

$$v = au + b$$

we can also write

$$\xi(u) = \frac{e^v - e^{-v}}{e^v + e^{-v}}$$

As a matter of fact, function $\xi(u)$ (moreover remarkably confirmed by the finite element studies) does satisfy three essential boundary conditions :

1) for $\frac{x}{z} = 0$ $\qquad \xi(o) = K_\alpha / K_a$

2) when $x \to \infty$ $\qquad \xi(\infty) \to 1$

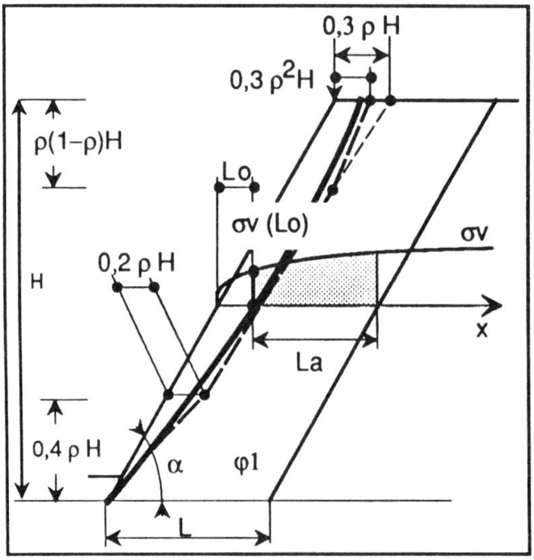

Fig. 4

3) the area between the curve and the asymptote is equal to that of the "missing" triangle between the facing and the vertical :

$$\gamma z \int_{0} [1 - \xi(u)]\, dx = \frac{1}{2}\gamma z^2 \cot \alpha$$

Calculation of tensile loads

From the above, the vertical stress σ_V can be determined particularly at the point of "maximum tensile load" situated at the boundary between the active and resisting zones. For sloped structures with metal reinforcements it was possible to determine the shape of this line experimentally, from the consistent observations on laboratory scale models and real structures as well as finite element models.

The line starts from the toe of this wall, then develops almost in parallel with the facing (at a distance depending on the inclination) before straightening towards the slope of the structure. In practical designs the envelope shape used is as defined in Fig. 4; it allows the width (L_0) of the active zone at each level to be calculated.

Then the maximum tensile load in the reinforcing element is given by :

$$T_m = K_a\, \sigma_h = K_a \sigma_v(L_o) \frac{\Delta H}{n}$$

where ΔH is the vertical spacing of the layers, n the number of reinforcing strips per unit of wall length.

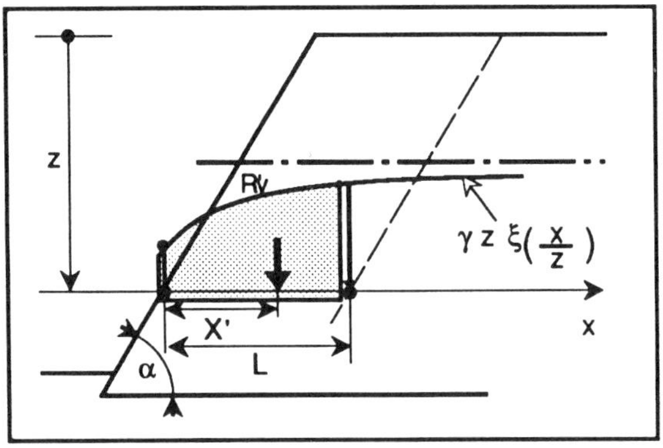

Fig. 5

Adjustment owing to earth pressure

In fact an adjustment of the value of σ_V has to be made before calculating the maximum tension T_m. Up to now reinforcements of unlimited length have been implicitly considered, hence no effect of the earth pressure behind the reinforced mass. This earth pressure (P), inclined at angle δ to the rear face of the structure, first modifies the weight (W) of the structure through its vertical component and then displaces the resultant through the overturning moment (M). With the introduction of the earth pressure the total vertical load at depth z changes from :

$$R'_V = \int_0^L \gamma z \, \xi(x/z) \, dx \quad \text{(Fig.5) to} \quad R_V = W - P \cos(\alpha + \delta) \quad \text{(Fig.6)}$$

It passes through the point with abscissa $X = M/R_V$ instead of X'. The adjustment entails three successive affinities of the cube and we finally obtain (Fig.7):

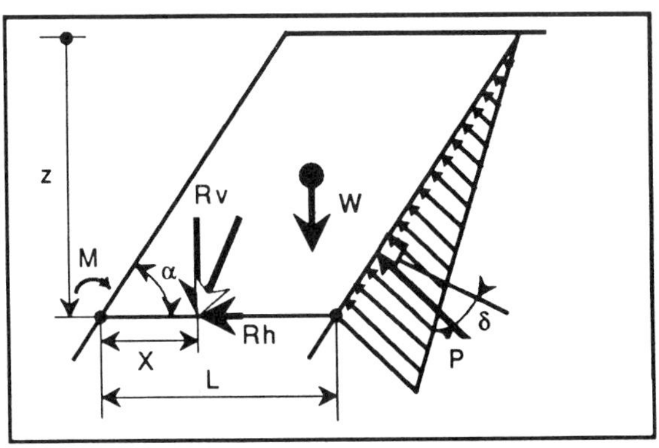

Fig. 6

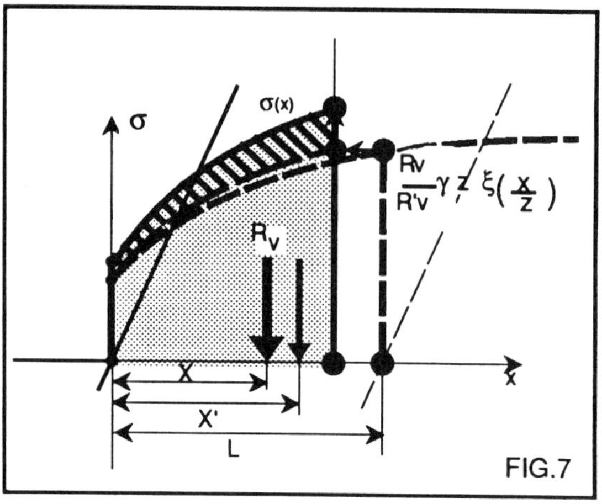

Fig. 7

$$\sigma(x,z) = \frac{R_V}{R'_V}\frac{X'}{X}\ \gamma z\ \xi(x'/z)$$

where

$$R'_V = \int_0^L \gamma z \xi(x/z)dx = \gamma z^2\left[\frac{1}{a}\,\mathrm{Log}\frac{1+e^{2v_2}}{1+e^{2v_o}} - \frac{L}{z}\right]$$

and

$$x' = \frac{X'}{X}x$$

with

$$v_2 = a\frac{L}{z} + b \quad \text{and : } \quad v_o = b$$

Pull-out resistance

Adherence is checked along the resisting length beyond the line of particular tension (Fig.4), taking into account a safety coefficient FS_f with:

$$T_m \le T_f = \frac{1}{FS_f} 2bf^* F_V \quad \text{(b being here the width of the strip)}$$

where

$$F_V = \int_{L_o}^{L} \sigma_V(x)\,dx = \gamma z^2\left[\frac{1}{a}\,\mathrm{Log}\frac{1+e^{2v_2}}{1+e^{2v_a}} - \frac{L_a}{z}\right]$$

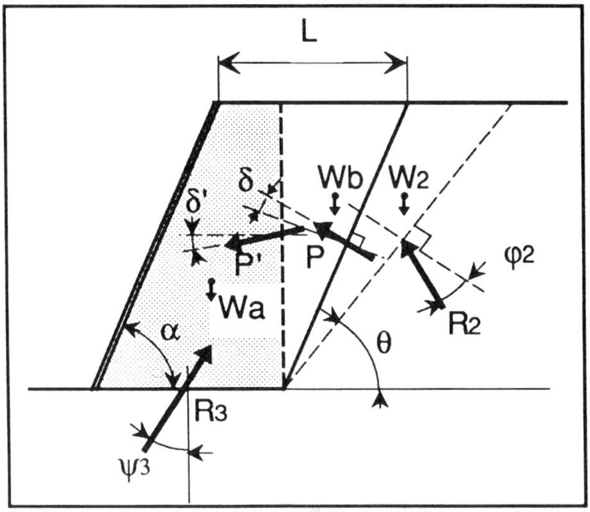

Fig. 8

and

$$v_2 = a\frac{L}{z} + b \quad \text{and :} \qquad v_a = a\frac{L_o}{z} + b$$

When using High Adherence (ribbed) steel -reinforcing strips, the apparent coefficient of friction f* varies with the vertical stress σ_v as a result of dilatancy and the following expression is used (where $\sigma_m = F_v/L_a$ and σ_0=120kPa)

$$f^* = f_0^* \frac{\sigma_0 - \sigma_m}{\sigma_0} + \tan\varphi_1 \frac{\sigma_m}{\sigma_0} \qquad \text{if } \sigma_m < \sigma_0$$

$$f^* = \tan\varphi_1 \qquad \text{if } \sigma_m \geq \sigma_0$$

Interaction between structure and backfill

Analysis

Coming back to the question of "earth pressure", the stability of the structure is analyzed in the same way as for a conventional wall. Behind the structure a Coulomb wedge is formed, bounded by a plane inclined at θ, along which acts the entire internal friction ϕ_2 of the backfill (Fig. 8).

If the structure is steeply sloped, the Coulomb wedge may disappear and the slip plane coincides with the rear boundary of the structure : then θ=α. The structure is appropriately stable so long as the reaction of the foundation is inclined at ψ_3, corresponding at the most to the permissible friction at the interface of the structure and the foundation.

It is useful to subdivide the reinforced fill structure into two "slices" one

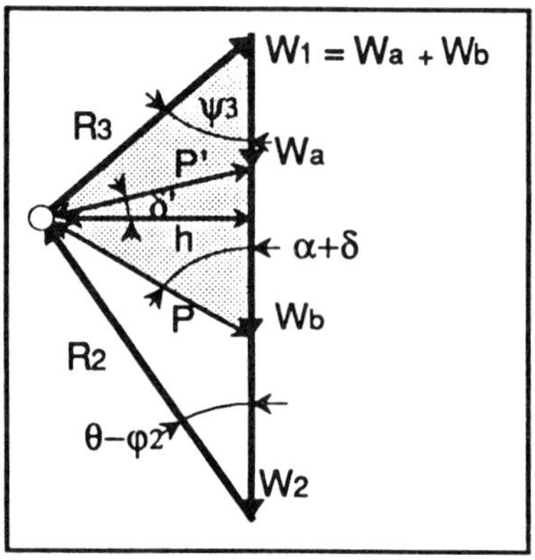

Fig. 9

immediately above the base with weight W_a, the other above the wedge with weight W_b. Their overall stability is represented by the diagram of force vectors shown on Fig. 9. On this diagram the vector representing the interaction P between the backfill and the structure may be identified. The lower triangle illustrates the equilibrium of the wedge, the upper grey tint triangle that of the structure.

Also shown is the "force between slices", P′, inclined at an angle δ′ to the horizontal, which passes through the vertical boundary between the two slices.

This inclination δ′ is clearly correlated with the difference between W_a and R'_v (defined above) at the base of the structure, since $P'\sin\delta' = R'_v - W_a$. This analysis (which cannot be described at length in this paper) leads to charts such as the one shown on Fig. 10 for a flat top embankment $\phi_1 = 36°$.

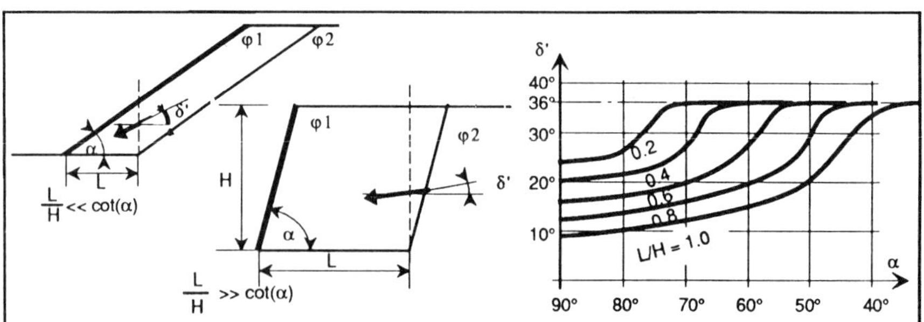

Fig. 10

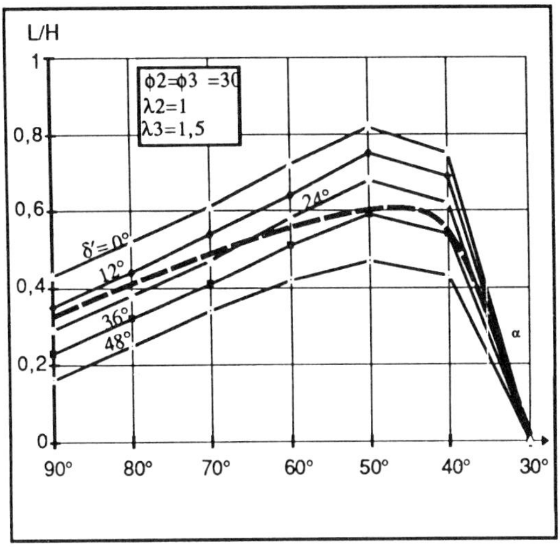

Fig. 11

Computation of the interaction diagram
δ′ being chosen, angle θ is made to decrease from α towards ϕ_2 in order to find the most unfavourable angle θ*, corresponding to ψ_3 maximum. This occurs when the distance shown as "h" on the graph (Fig.9) is itself a maximum called h*)

From the diagram

$$h^* = \frac{[W_b + W_2]\tan(\theta^* - \varphi_2)}{1 + \tan\delta' \tan(\theta^* - \varphi_2)}$$

Then check that

$$\tan\psi^*_3 = \frac{h^*}{h^* \tan\delta' + W_a}$$

complies with the permissible value.

The inclination δ of the total pressure P to the normal to the rear face is drawn from

$$\cot(\alpha + \delta) = \frac{W_b}{h^*} - \tan\delta'$$

Finally

$$P = \frac{h^*}{\sin(\alpha + \delta)}$$

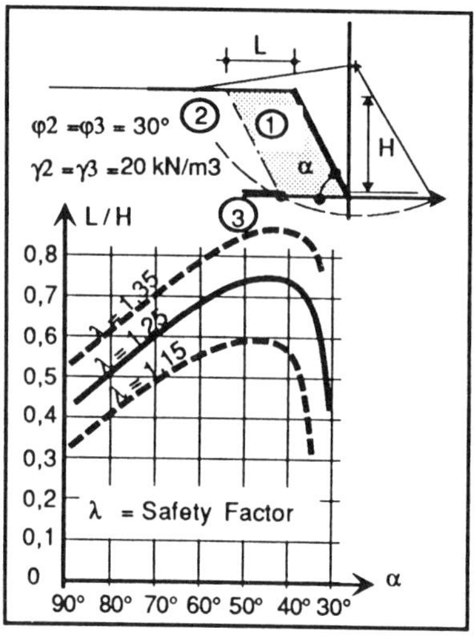

Fig. 12

Minimum L/H ratio
This makes it possible to determine the minimum proportions which ensure the stability of the structure, regarding sliding on the base. As shown on Fig. 11 for the simple case of a purely frictional soil material (and for different possible values of δ′) it is for an inclination of 50 to 45° that the reinforcements should be longest. This is easily and clearly confirmed by a simple slip circle analysis where only circles outside the reinforced fill structure are considered. Again the critical case corresponds to the same range of inclination (Fig.12). It is therefore important to bear in mind that, other things being equal, a sloping structure may be less stable than a vertical one.

Confirmation from experimental and F.E.M. results
Shown in Figs 13 to 16 in graph form are a few examples of the very satisfactory agreement found between the results of the calculations made according to the theory explained here and the results obtained from either FEM models, measurements on lab-scale three dimensional models or monitoring of full scale structures .

Conclusions
The design method used for Reinforced Earth (R) sloping structures are seen to be reliably backed up by both theory and experimentation. Although it was substantiated specifically for metal strips, many of the aspects certainly

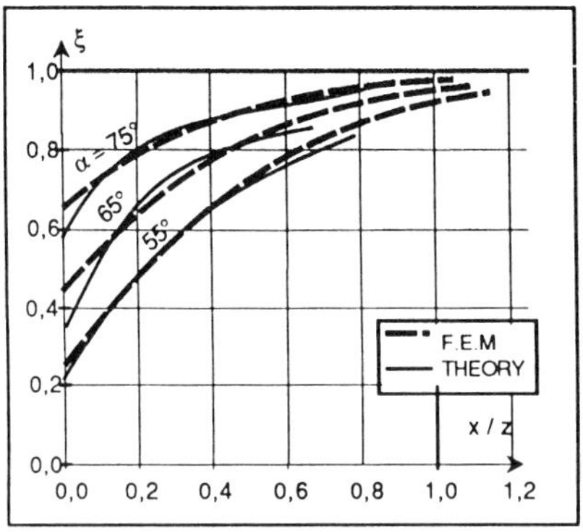

Fig.13. Function ξ(u) = σv/γz. Agreement between theory and FEM analysis

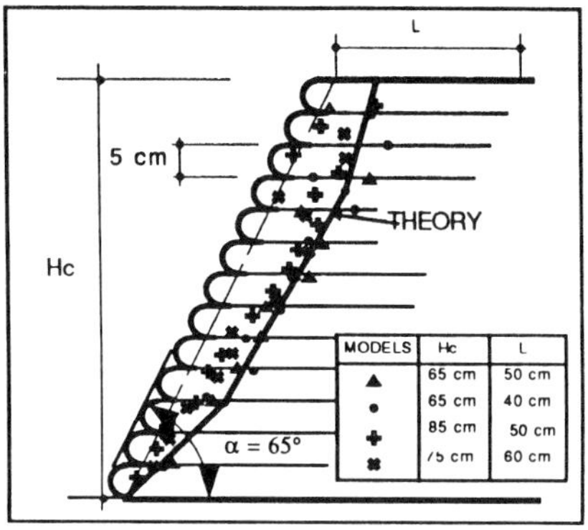

Fig. 14. Line of maximum tension. Agreement between 3D scale models and theory

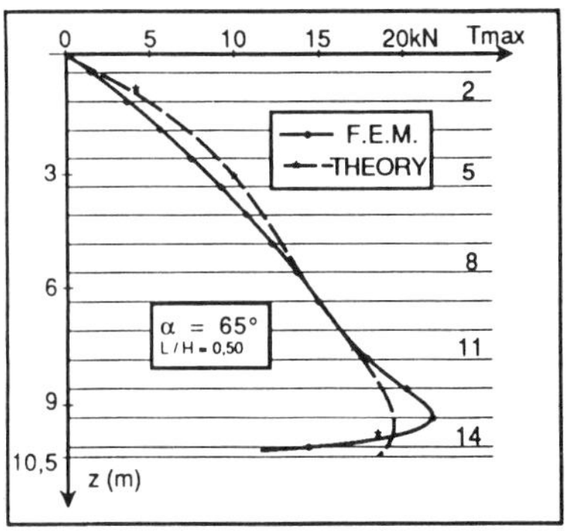

Fig. 15. Maximum tension loads. Agreement between FEM model and theory

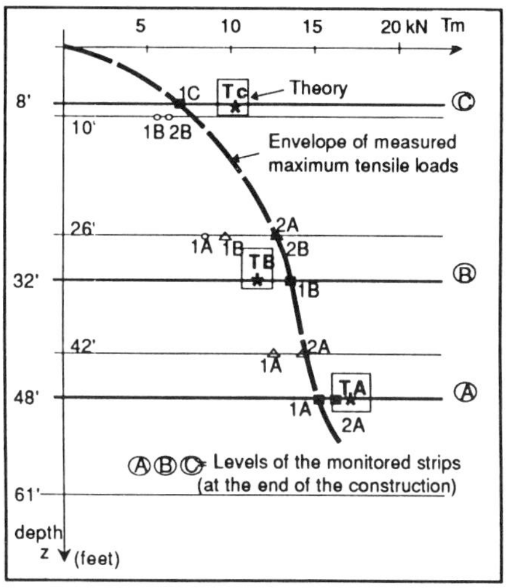

Fig. 16. Maximum tensile loads. Agreement between theory and measurement made on full scale structure(during construction and after completion). Kyanite wall (USA): H = 18.5 m - α = 55°

also apply to other types of reinforcements. However the line of maximum tension, whose shape and location definitely depends on the extensibility of the reinforcements, should firstly be properly established for each type of reinforcing material.

References

McKITTRICK, D.P (1982), Design and application of Reinforced Earth storage slots. Bulk solid handling, Vol.2, n 4, 1982

SMITH, R.J.H. (1990), The development of Terratrel for the construction of steep slopes. Performance of Reinforced Soil Structures, Ed. Mc Gohsn, Yeo, Andrewes, p 135-140 / Thomas Telford

BASTICK, M.J.B. (1983), Murs en Terre Armée à parement incliné; résultats des recherches Internal report, TAI no. R28, 1983

T.A.I.(1989), Industrial applications of Reinforced Earth (R) structures. Technical brochure, Terre Armée Internationale, 1989

An experimental investigation of soil nailing

M. C. R. DAVIES, C. D. JACOBS and R. J. BRIDLE, University of Wales College of Cardiff , UK

An investigation of the behaviour of sand reinforced with soil nails is being conducted using a large shear box. The box has a shear plane 1.5m x 1.5m and is 3m long; permitting prototype scale nails to be tested. The results of experiments investigating the enhancement in shear capacity of soil reinforced using nails and of soil-nail interaction are presented.

Introduction

The relatively recent arrival in the U.K. of soil nailing as a technique for constructing retaining structures has resulted in the requirement for widely acceptable methods of analysis. Various methods exist for placing the nail in the soil, these include drilling and grouting and that of firing the nail into the ground using a compressed air gun. However, once installed the way in which nails perform will be very similar regardless of the insertion technique. In order to assess existing methods of analysis or develop analytical techniques, it is necessary to have a fundamental understanding of the mechanisms involved. To this end an experimental study of the fundamental behaviour of soil nailing is being conducted using a large shear box. A brief description of the apparatus and results of tests conducted to date are described herein.

The shear box is 3 m long and with a shear plane of 1.5 m x 1.5 m - making it arguably the largest shear box ever constructed. A surcharge of up to 150kPa can be applied to increase the effective stress acting on the soil. Because of the large dimensions of the box prototype scale nails - up to 3 m in length - can be incorporated into the soil; removing uncertainties of scale modelling. A bar of 25 mm diameter was used for the experiments described herein. Strain gauges on the nail enable axial and bending strains, and hence stresses and bending moments, to be obtained.

The shear box

The design philosophy behind the shear box was to produce an apparatus in which a nail of prototype dimensions could be tested whilst subjected to

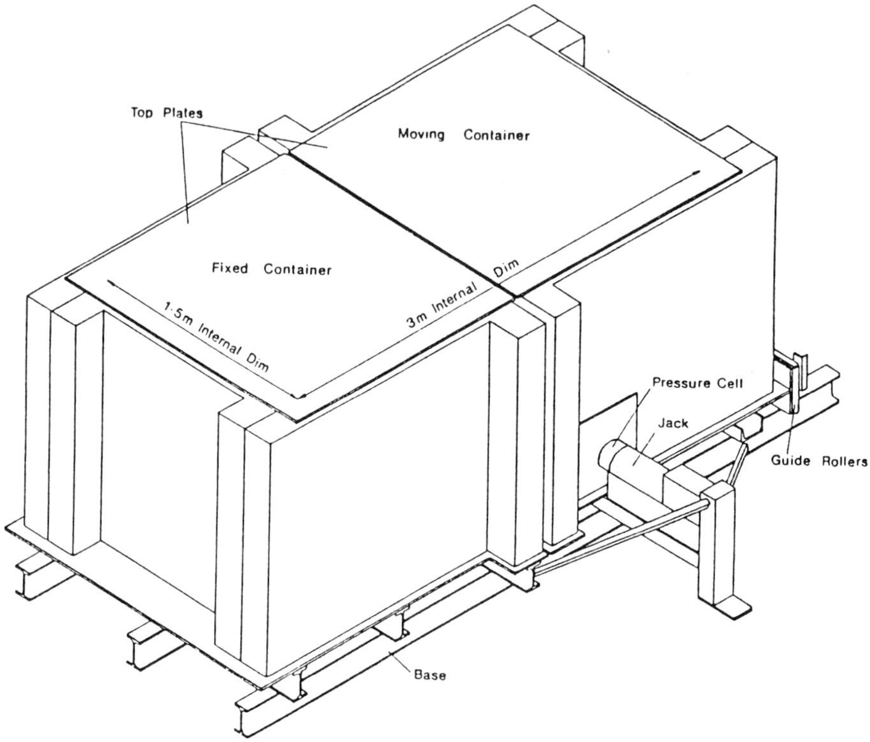

Fig. 1. View of shear box

stresses at typical values experienced in-situ. A detailed description of the shear box used in these experiments has been previously presented by Barr, Davies and Jacobs (ref. 1).

The shear box is illustrated in Figs 1 and 2. Fig. 1 shows the overall arrangement of the box which consists of two containers - one of which is fixed and the other moveable. The containers are identical and each is formed from three large reinforced concrete slabs, 0.2m thick and 1.5m high, as shown in Fig. 1.

From the above description, it is seen that the large shear box is turned through 90° compared to the normal shear boxes used in soil mechanics laboratories. The overall inside dimensions of the box are 3.0 m x 1.5 m x 1.5 m. It is necessary to ensure that the movable part of the box moves in a direction parallel to the interface. This has lead to the development of a system of rigid restraint for the fixed container. The restraint was provided by two series of rollers - located at the base and top of the rear wall of the moving container - which allowed movement to take place only in the required direction i.e. parallel to the interface between the two halves.

To apply pressure to the soil a steel plate is located inside each container

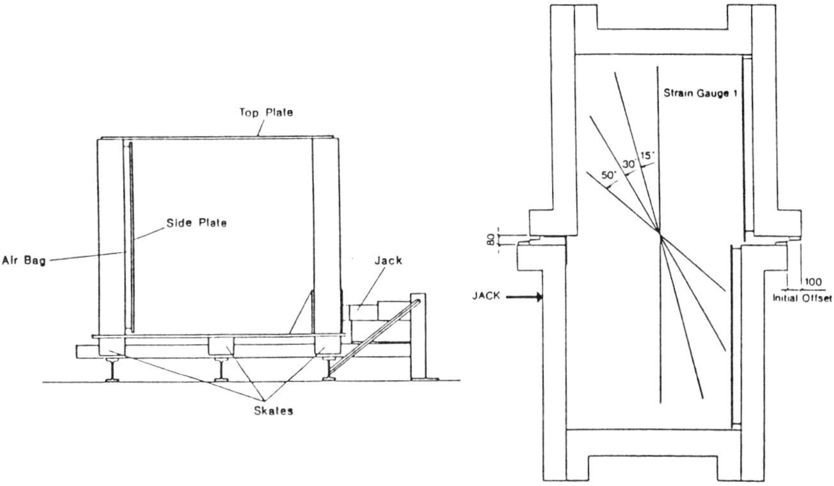

Fig. 2: (a). Section through box; (b). Plan on box showing orientation of nails in tests

adjacent to one side wall with an air bag placed in the gap between the plate and the side wall, Fig. 2. A stiffened top plate is firmly secured to the top of each container to provide confinement for the soil. During filling of the shear box with soil, the air bags are partially inflated to accommodate any dilation that occurs in the soil during shearing. Once the containers are full the surface is struck off level with the sides of the box prior to the placing of the top plate. This is to restrict vertical movement thus providing one dimensional loading boundary conditions during the application of load.

The lateral stress, σ'_x, is applied to the soil by increasing the air pressure in the two air bags. In this configuration the lateral stress is the major principal stress, σ_1', thus the major principal stress direction coincides with the direction of shear. This has the advantage over other large scale shear boxes - which have the same geometry as the standard laboratory direct shear box - in that when nails are located normal to the shear plane the stress distribution along the nail is uniform.

Because of the scale of the box, the above loading configuration results in a drop off in the level of stress from the side of the sample to which the pressure is applied to the opposite face. Miniature stiff diaphragm type load cells, located within the specimens, are used to measure this distribution. The measured stress distribution has been found to agree very well with that predicted using elastic theory. For a lateral stress of 100kPa Fig. 3 shows stresses predicted in this way for the central plane of the box and the face opposite the air bag. Along the greater part of each plane the normal stress is uniform, decreasing at the extremities of the box. Stress distribution results in a reduction in stress to maximum values of 82kPa acting on the central

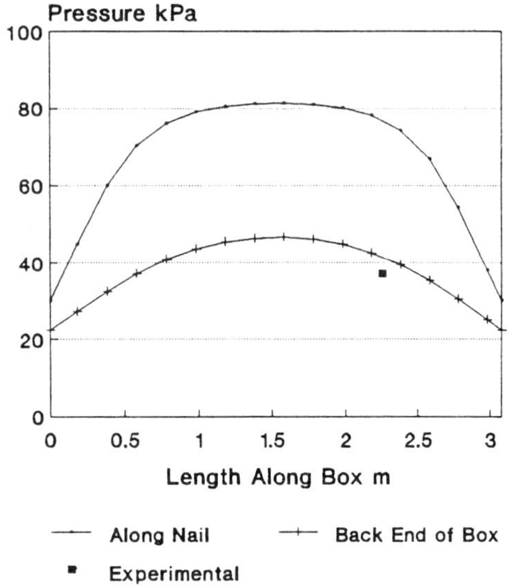

Fig. 3. Predicted and measured stress distribution in box - σ_1 = 100kPa

plane and 47kPa acting on the far side of the specimen. The measured stress on the far face, also plotted in the figure, is seen to agree well with the predicted value.

The shearing force is applied by means of a 500kN capacity jack. Shearing is conducted using stress control, the load being measured by means of a load cell in the jacking system, Fig. 2. The relative displacement between the two halves of the shear box is measured by means of LVDT transducers located at the top and bottom of the side wall of the moving container.

The greatest efforts were made to ensure that frictional forces resulting from end restraint and the rollers were kept to a minimum. In order to assess the effect of friction a series of calibration tests was conducted. This done, the additional shear forces resulting from friction in the system due to the dead weight of the box and its contents during a test were subtracted from the measured values of shear force to obtain the values of shear force across the shear plane in the specimen.

Sample preparation

The sand used in the tests was medium uniform well graded Leighton Buzzard sand (21/75). Sand was placed in the shear box by pluviation from a hopper system. Sand fell through a minimum of 1.0 m into the shear box as the hopper moved at a constant rate across the top of the apparatus. To ensure a uniform density at the boundaries of the specimens knife edges are placed on the tops of the sides of each container preventing sand deflecting from the top of the wall into the box. In this configuration the sample preparation system, initially developed for other applications, produces specimens of medium dense sand with a uniform density of 16.7 ± 0.3 kN/m^3 (ref. 2).

For tests in which nails were to be placed in the sand the pluviation process was halted to position the nail when the sand layer was at the specified depth. Where a single nail was to be tested this was placed at the mid-height of the box i.e. at a depth of 0.75 m. Where 2 nails were tested these were placed with the mid-point of each lying in the central plane of the box at depths of 500 mm and 1000 mm. For three bars the top bar was placed 375 mm below the surface with a gap of 375 mm between each bar. Bars were either placed normal to the shear plane or rotated from the normal to the shear plane by angles of 15, 30 or 50°.

Instrumented nail

An instrumented nail was included in each reinforced soil shear test. Instrumentation consisted of strain gauges located in pairs - one on either side of the nail in the horizontal plane. From the measured surface strains it was possible to deduce axial and bending strains, and hence the axial stresses and bending moments in the bar.

Test results

Tests have been conducted on unreinforced specimens (control tests) and specimens containing one or more bars. To date the programme has consisted of two series of experiments in which lateral stresses of 50kPa and 100kPa have been applied to the specimens.

Shear force/deflection

Fig. 4 displays the plot of shear force versus deflection for a test in which 2 bars were placed in the sand normal to the shear plane, together with the equivalent plot from a control (i.e. unreinforced) test. Volume change during shear was permitted by volume change of the air bags; a typical plot of volume change versus deflection plots is also shown in Fig. 5. Regardless of whether or not the sand was reinforced, and when it was the number of bars present in the specimens and the orientation of the bars to the shear plane, the plots all show very similar general behaviour. This is consistent with the shear force/deflection and volumetric strain/deflection plots obtained from more conventionally sized direct shear box tests.

Shearing of the specimens in these two tests took place after a normal (i.e. lateral) stress of $\sigma_1' = 100$kPa had been applied. The reinforcing effect of the

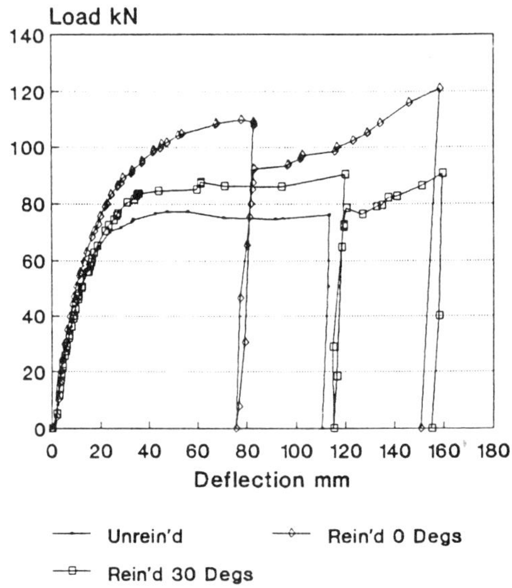

Fig. 4. Shear force v deflection for unreinforced and reinforced specimens

bars is clearly demonstrated by the increased shear stress capacity of the specimen. The observations in tests with differing numbers of bars and with bars located at varying orientations to the shear plane are consistent with test data reported by Pedley (ref. 3), viz: increasing the number of bars increases the shear stress capacity and rotating the bars from the normal to the plane changes the contribution of the reinforcement to the shear stress capacity of the soil. These trends can be seen in Fig. 4 which displays the shear force/deflection behaviour of two tests in which bars were placed at orientations of 0 and 30°, respectively, to the normal to the shear plane. Contrary to the findings of previous studies, e.g. Jewell (ref. 4), Fig. 4 shows that the shear resistance of the specimen reinforced with bars normal to the shear plane was greater than that of a specimen within which the bars were orientated at an angle of 30° to the shear plane. The dimensions of the box necessitated a reduction in the length of the bars by 0.5 m in the latter experiments. This

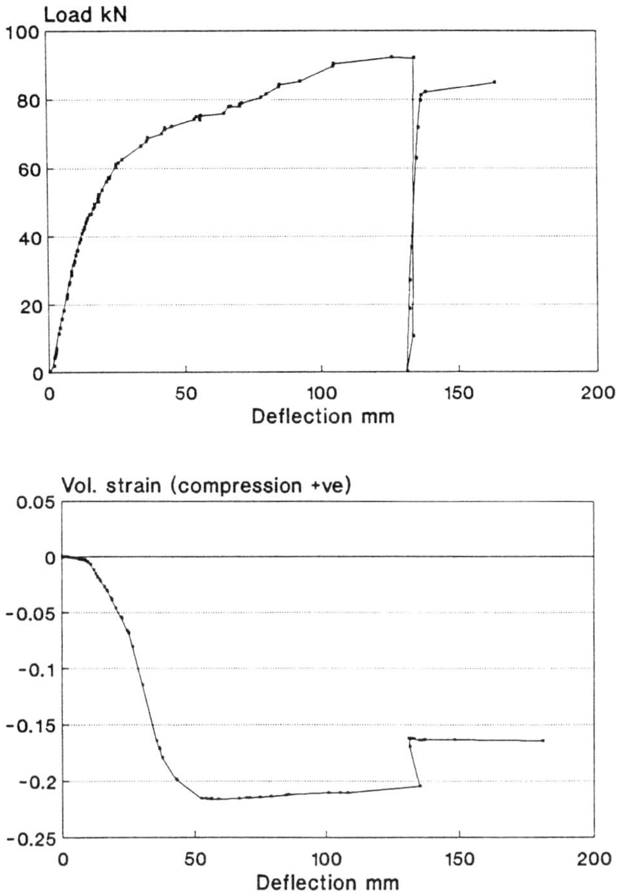

Fig. 5. Shear force v deflection and volume change v displacement (unreinforced specimen) $\sigma_1 = 100kPa$

resulted in a reduction of anchorage length over which frictional forces could develop between the soil and the bar.

By considering the stress distribution within the soil described above, to obtain the normal stress acting on the shear plane, the angle of internal friction of the sand can be calculated for the unreinforced tests. For the test displayed in Fig. 4 . This results in a value of 52°.

Nail instrumentation

The results of the surface strain gauges may be used to obtain axial and

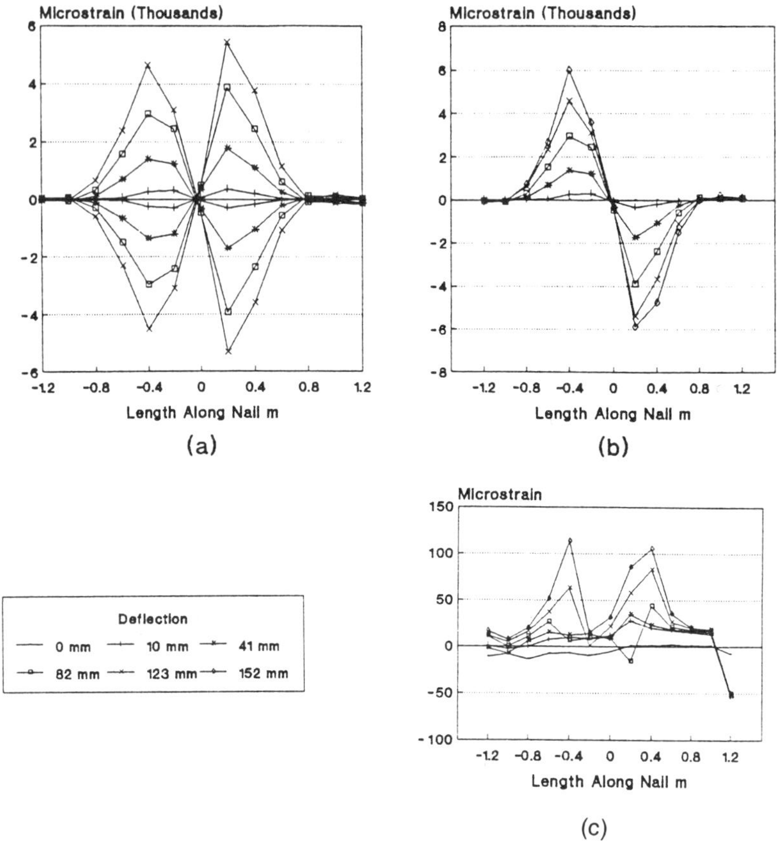

Fig. 6. Measured strains (a) surface, (b) bending and (c) axial (2 bars at 0°, σ_1 = 100kPa)

bending strains in the bar. Fig. 6 displays typical results from the experimental programme, in this case for a bar normal to the shear plane.

Bending strains

The raw strain gauge data, displayed in Fig. 6(a), indicated, as would be expected, that as the bar went into bending one side was in compression and the other in tension - the sides changing at the point of contraflexure at the mid point of the bar. From the initiation of loading the plots of bending strain indicate an antisymmetric pattern of behaviour about the mid-point of the nail (i.e. the location of the shear plane). At the mid-point the bending strain was zero, it then rose to a peak value before dropping to zero once again at the end of the bar, Fig. 6(b).

Following the tests plastic hinges were observed to have formed at the locations of the peak values of bending strain. Tests conducted on elements of the nails and theoretical predictions indicated that a plastic hinge would start to form at bending strains of 2600 μm strain. Typically a displacement along the shear plane of 100 mm to 150 mm was required to produce this value of strain which corresponds to a bending moment of 1.2 kNm.

Strains reducing to zero before the end of nail suggesting that nails were sufficiently long to overcome any problems which could be associated with boundary effects, i.e. modification of bending strain behaviour, in a shear box 3.0 m in length.

Axial strains

During application of the surcharge through the airbags the bar was observed to undergo a small compressive force along its length. Once shearing started this soon disappeared and the bar only experienced tensile axial forces; ranging from zero at the ends to a maximum at the centre of the bar. From Fig. 6(c) it can be seen that after approximately 100 mm displacement of the box anomalies and vast amounts of scatter appear in the axial strain results. This is due likely to the onset of the formation of the plastic hinges resulting in non-linear strains across the bar. Since the axial strains are relatively small compared to the bending strains changes in the compressive side of bending would dramatically affect the calculated value of axial force - which assumes a linear strain distribution across the bar. Indeed, the worst scatter takes place in the results from gauges located in the region of the plastic hinge once it begins to form because of the uneven development of plastic strains on either face of the bar.

However, in general the plots indicate the following behaviour

(i) all bars went into tension from the outset of displacement (N.B. this includes the bars which were initially normal to the shear plane)
(ii) for a particular orientation of bar from the normal to the shear plane greater tensile strains were recorded in tests with the higher overburden pressure
(iii) as the orientation of the axis of the bar from the normal to the shear plane was increased there was a reduction in the peak values of shear stress recorded
(iv) axial forces in the bar varied between 7.3kN when orientated normal to the shear plane and 12.5kN when at an angle of 50° - this is despite a reduction in length of the bar by 1.2 m in the latter case.

Shear force and lateral stress: At any point during shearing the bending moment diagram for the bar may be obtained from the measured bending strains. Bending moments calculated for three stages of the test shown in Fig. 6 are plotted in Fig. 7(a) together with the derived shear force and lateral

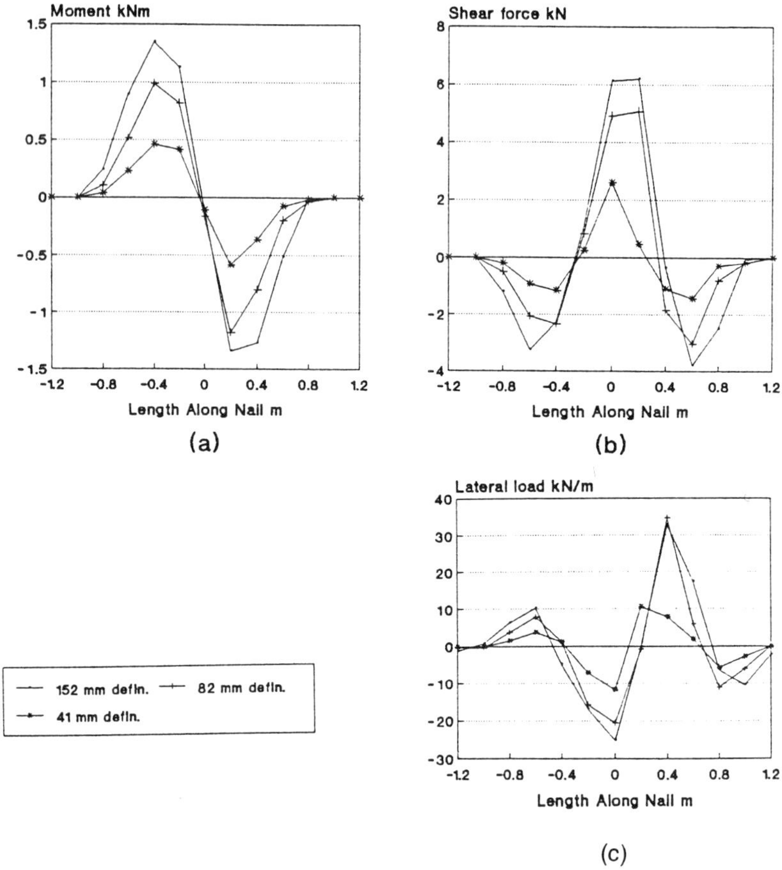

Fig. 7. Diagrams of (a) bending moment, (b) shear force and (c) lateral force (2 bars at 0°, $\sigma_1 = 100kPa$)

stress diagrams in Fig. 7(b) and Fig. 7(c), respectively. Each of the three diagrams remained geometrically similar during a shear displacement of 152 mm. At this displacement the maximum recorded bending moment was 1.35kNm. This compared with the value of 1.47kNm for the fully plastic bending moment - obtained from tests on elements of the nails - indicating that even at large displacement the nail was not fully plastic and still possessed limited extra bending moment capacity should it be possible, or desirable, for this to be exploited. However, if the limiting combination of axial force and bending moment are considered (ref. 3) the value of 1.35kNm is very close to the theoretical lower bound ("safe") fully plastic value of 1.31kNm which would imply that the plastic hinge was, or very nearly, fully developed at this stage.

The maximum derived value of lateral force on the 25mm diameter nail (at a shear displacement of 152mm) was 35kN/m, Fig. 7(c). This results in a bearing pressure of 1400kPa. The maximum shear force in the nail was 6.2kN, Fig. 7(b).

Data of this type - indicating similar behaviour - has been previously reported by Pedley, Jewell and Milligan (ref. 5) from the results of tests conducted by Pedley (ref. 3). This data has been used to support the plastic analysis of soil/nail interaction proposed by Jewell and Pedley (ref. 6). In both the lateral force data presented by these workers and the data obtained in the study currently being conducted at Cardiff there is a certain amount of experimental scatter; which results in part from the indirect method of measurement. This data is of special importance because it is vital for the correct interpretation of the soil/nail interaction - from which the axial (i.e. pull out) and bending capacities of the nail can be determined - to be correctly defined in order to avoid both unsafe or over conservative design.

To complement the above data, a series of experiments is to be conducted in which stresses in the soil at a number of locations adjacent to the nail are to be monitored using miniature earth pressure cells. Derived lateral stress data already obtained will used to determine the optimum locations for this instrumentation. The direct stress measurements will be used to examine the evolution of lateral stresses and assess currently adopted models of soil/nail interaction.

Conclusions

A large direct shear box, having a shear plane of 1.5 m x 1.5 m and capable of testing nails of up to 3m in length, has been constructed. A consequence of the large scale of the box is that there is an uneven distribution in levels of stress inside the box. However, this can be quantified by an elastic stress distribution allowing analysis of the results. The predicted stress distribution has been confirmed by experimental stress measurements.

The shear box has been used to conduct tests on sand reinforced by one or more nails located at varying orientations to the shear plane. There is still work to be done to reach the definitive mechanism of soil/nail interaction and work on measuring pressure and producing a model of its distribution is underway. However, future papers will deal with the soil/nail interaction based on the intermediate results obtained and the tests now being conducted.

Acknowledgements

The Authors wish to acknowledge the financial support to this project provided by Soil Nailing Limited. The second Author is sponsored by the Science and Engineering Research Council and Ryan International Plc.

References

1. BARR B. I. G., DAVIES M.C.R. and JACOBS C.D. A Large Direct Shear Box - Some Initial Results of Tests on Soil Nails. *Ground Engineering*, 1991, **24**, No 2.
2. DAVIES M.C.R. Physical modelling of impact loading on buried structures. in *Structures Under Shock and Impact*, Ed. P.S. Bulson, Elsevier, Amsterdam, 1989.
3. PEDLEY M.J. *The performance of soil reinforcement in bending and shear*. PhD Thesis, University of Oxford, 1990.
4. Jewell R.A. Some effects of reinforcement on the mechanical behaviour of soils. PhD Thesis, University of Cambridge, 1980.
5. Pedley, M.J., Jewell, R.A. and Milligan, G.W.E. A large scale experimental study of soil-reinforcement interaction. Ground Engineering, Vol 23, 1990, No. 6.
6. Jewell, R.A. and Pedley, M.J. Analysis for soil reinforcement: with bending stiffness. Ground Engineering Vol 22, 1990, No. 10.

Finite element analysis of anchored earth retaining walls

D. I. HARRIS, D. W. HIGHT, and K. G. HIGGINS, Geotechnical Consulting Group, D. M. POTTS, Imperial College of Science, Technology and Medicine, and K. C. BRADY, Transport and Road Research Laboratory, UK

Introduction

The construction of anchored earth and reinforced earth walls is similar, as in both systems multiple ties are placed in the fill to transfer tensile loads and ensure the stability of the structure. The modelling of anchored earth is, however, significantly different to that of reinforced earth as the anchors provide resistance in a discrete zone remote from the wall facing and the frictional capacity of the tie between the anchor head and the wall facing is relatively unimportant.

Anchored earth is a potentially more economical method of construction than reinforced earth by virtue of three major differences. The backfill does not need to possess good frictional resistance and so more widely available and cheaper fill materials can be used. Secondly, protective coatings can be applied to anchors, which would otherwise reduce frictional resistance if applied to reinforcing strips, and so the specification controlling the corrosivity of fill may be relaxed. Thirdly, anchors are more efficient at generating pull-out resistance than frictional strips.

The first anchored earth wall built on the UK road network was at Otley in 1983 (Jones *et al.*, 1985). The anchors were fabricated from reinforcement bar, bent into a triangular shape to form the anchor head (Fig. 1). This system was developed at TRRL with a full scale trial wall, following an earlier programme of pull-out tests (Murray and Irwin, 1981) and centrifuge tests (Bolton, 1981).

Many variants of anchored earth are possible. Discarded car tyres, tied together with Paraweb, were used for both the facing and the anchors on a slope repair at Cambridge (Johnson, 1985). At Annan, Scotland, a wall has been constructed, based on an Austrian system, employing semi-cylindrical concrete sections as anchors and "window-box" facing units, tied back by Paraweb loops (Brady and Masterson, 1989). Similar systems have been developed in Japan: A structure consisting of concrete king posts tied back

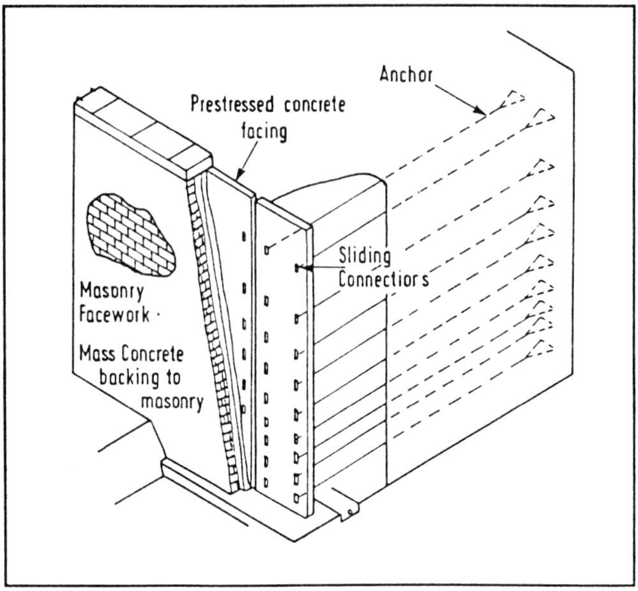

Fig. 1. Anchored earth wall at Otley

to in situ concrete anchor blocks, with fabric facing between the posts, has been described by Fukuoka *et al* (1986).

The current UK design method for anchored earth is an adaption of the method used for reinforced earth and is outlined in BE3/78 (Department of Transport, 1987). TRRL commissioned an extensive parametric study using the finite element program ICFEP to investigate the behaviour of this novel form of construction under working loads. This paper presents results from these analyses and compares these with the provisions within the design method. Further analyses to investigate potential failure mechanisms, the effect of surcharge loadings and compaction are underway; and back-analysis of the structures at Otley and Annan and of the centrifuge models will be undertaken to compare FEM results with observed performance.

Finite element modelling

The concentration of the pull-out resistance at the anchor head precludes the use of an equivalent homogenous material for the numerical modelling of anchored earth. Discrete modelling of anchors is the only valid option for using FEM to analyze anchored earth. The modelling of the two components of an anchor, the tie and the anchor head, are considered separately.

Modelling of the tie

The tie is represented as a continuous plane or sheet if it is included directly into the mesh for a 2D analysis. The properties of the interface between the tie and the soil should produce the correct transfer of load to the tie. The displacements of the soil on either side of a tie are then prescribed to be identical to ensure that the total shear resistance at the elevation of the anchor is unaltered.

The inclusion of the ties in an analysis is expensive in terms of the number of elements required. Conceptually, it would appear to be acceptable to simplify the modelling since the contribution of the tie to the total pull-out resistance of the anchor is theoretically small. The degree of the approximation depends on the nature of the tie, which can vary considerably between different anchored earth systems. In its simplest form the tie could be modelled purely as a connection between the anchor head and the wall unit, making no contribution to the resistance of the anchor.

Modelling of the anchor head

The pull-out resistance of an anchor head has two components: passive resistance in front of the block; frictional resistance on the sides and edges of the block. Each of the components of resistance will be developed at different rates and the importance of each component will vary significantly with the aspect ratio of the different types of anchor.

In a 2D analysis, an anchor would be represented as a continuous member in a discrete finite element formulation. The frictional component of resistance could be reasonably modelled by modifying the interface properties to allow for the difference in area between prototype and model.

The pull-out failure of a deeply embedded anchor head is associated with a bearing capacity failure and the formation of a plastic zone in front of the anchor. The 3D nature of the anchors affects the development of both components of resistance, and therefore the use of a 2D analysis requires the behaviour to be 'fitted' and thus to be estimated a priori. No straightforward method of estimating such behaviour exists and faithful reproduction of this behaviour in 2D FE analyses is, therefore, difficult to achieve.

Simplified modelling

The complexity of the modelling used in an FE analysis should suit the purpose of the analysis and the quality of the data available from which the anchor behaviour must be derived. The analyses described in this paper are drawn from the parametric study referred to earlier. The effects of a wide range of parameters have been considered. The cost of such an exercise using optimal modelling techniques would have been prohibitive and it was therefore necessary to derive simplified methods of modelling which nevertheless capture the essential features of behaviour.

The analyses were intended to model the TRRL anchors (Fig. 1) and to reproduce the working condition. In this context

- failure of anchors was unlikely to control behaviour
- anchor resistance would be mobilised at small relative displacements
- the frictional capacity of the tie was negligible.

The anchors could therefore be modelled, to a reasonable approximation, simply as a tie between the wall and the anchor head. The force in the anchor was applied at a point within the soil and consequently the relative displacement between soil and anchor was neglected. The tie itself has been modelled by one of three methods:

- tied degrees of freedom
- elastic springs
- bar elements.

In the first method, the horizontal displacements of a point on the rear of the wall facing unit and at the anchor head were constrained to be equal, giving an inextensible connection. i.e. the horizontal degree of freedom between these points was 'tied'. In practice the tie would extend as a load was applied, so analyses were also performed with the tie modelled as a linear elastic spring. Yielding of the tie can be included in the FEM model if bar elements are used instead of the spring. In either case there is no interaction between the tie and the soil, except at the location of the anchor head. The failure of a bar element will be ductile i.e. the maximum load is sustained post failure. A brittle failure, if so required, can be modelled by removing the bar element once its capacity is exceeded.

Three modes of failure of an anchor can be identified, Fig. 2

- overstressing of the tie
- flow of the soil around the anchor head
- a passive wedge extending to the fill surface.

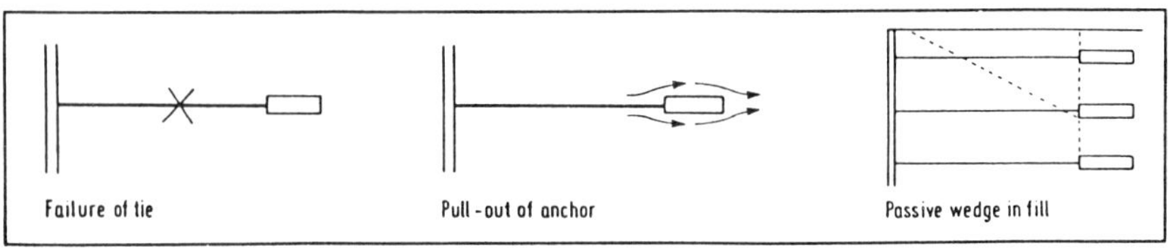

Fig. 2. Modes of anchor failure

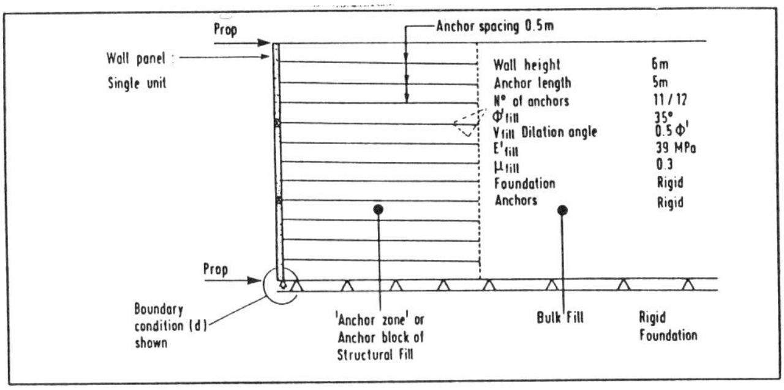

Fig. 3. Standard Case for finite element analyses

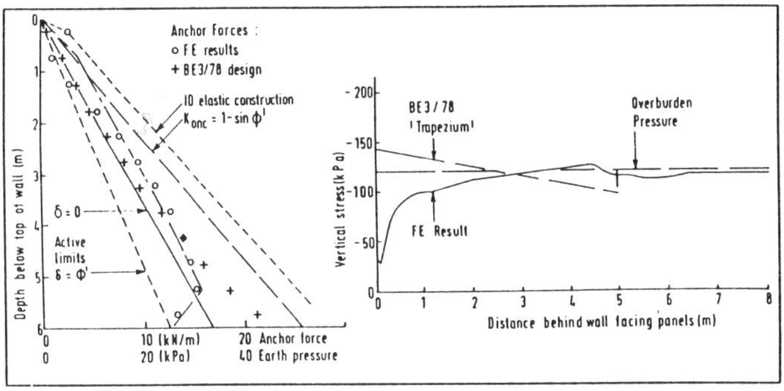

Fig. 4. Standard Case: Anchor forces, earth pressures and vertical stress.

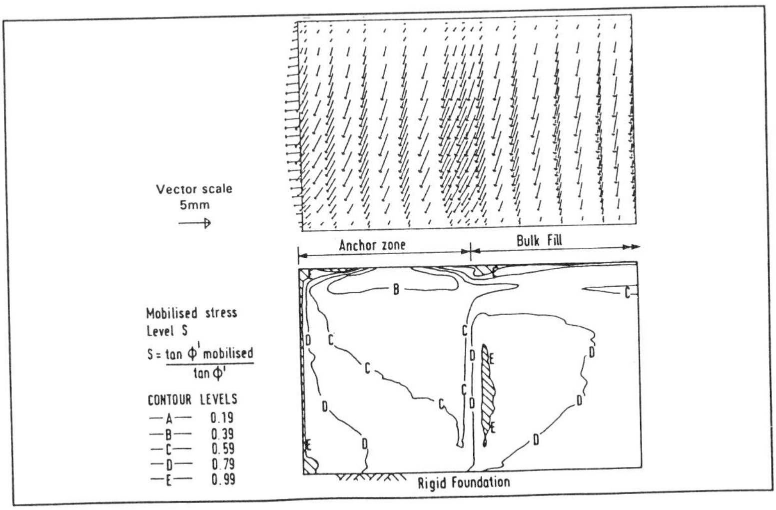

Fig. 5. Standard Case: Vectors of displacement and mobilised stress levels

The yield load of the bar element can be set equal to the calculated tension capacity of the tie or the estimated flow round capacity of the anchor head, whichever is the lower. A classical passive wedge can be formed if the sum of the capacities of a number of individual anchors exceeds the passive limit to the depth of the lowest anchor. This mechanism is more likely to apply at shallow depths of embedment. The possibility of this type of failure is automatically included in the FE analysis.

Standard case

A standard case was established to allow the effect of varying individual parameters to be evaluated. The geometry and properties adopted are given in Fig. 3.

The predicted distribution of anchor forces, earth pressure and vertical pressure on the foundation are compared in Fig. 4 to the design forces calculated by the methods given in BE 3/78 (1987). Fig. 5 shows the vectors of displacement and mobilised stress levels within the fill.

At the top and base of the wall the anchor forces are affected by the temporary props used during construction. The earth pressures on the wall are above the active limit. The similarity of the distributions of anchor force and earth pressures suggest that the concept of local equilibrium is valid. The design anchor forces include the effect of the 'external moment' applied by the bulk fill on the rear of the anchor zone.

The trapezoidal distribution of vertical pressure at the base of the fill assumed in the design method to be generated at failure is not evident in the finite element results. The vertical stress acting at the front of the anchor zone is less than the overburden pressure due to transfer of load to the wall unit in shear. The pressure at the rear of the anchor zone is greater than the nominal overburden due to transfer of load from the bulk fill to the anchor zone. The anchor forces and vertical pressures at the base of the fill suggest that the idealisation of the anchor zone as a rigid block to which external loads can be applied has limited validity when considering serviceability limit states.

The vectors of displacement show the role of progressive construction in determining behaviour. The maximum settlement occurs at about mid-height of the fill and the horizontal wall displacements reflect these movements. The mobilised stress level contours show plastic zones formed adjacent to the wall units, immediately behind the anchor zone and at the surface of the bulk fill. This confirms that active conditions are not generated at working conditions within the structural fill, with yield only induced by the settlement of the fill relative to the wall. The conditions are closer to active in the bulk fill but again the role of progressive construction is clear as yield only occurs at the mid-height of the wall.

Table 1. Description of analysis

	Total Anchor Force (kN/m)	Top Prop Force (kN/m)	Max Wall Movement (% Std. Case)	Anchor Force (% Design Load)	Boundary Condition
RIGID FOUNDATION - 11 Anchors					
STANDARD CASE	68/101	5	100	87	a & c
STANDARD CASE	98	5	100	84	b
STANDARD CASE	106	-	100	91	d
Properties of Fill					
Strength/ Stiffness 25°/26 MPa	70/112	5	338	60	a & c
Strength/ Stiffness 55°/52 MPa	58/ 86	3	67	226	a & c
Bulk Fill Strength 25°	104	-	128	83	d
Bulk Fill Strength 45°	107	-	-	98	d
Structural Fill Strength 25°	118	-	184	60	d
Stiffness 26 MPa	64/101	6	-	87	a & c
Stiffness 52 MPa	71/101	4	-	87	a & c
Poisson's Ratio 0.23	58	4	-	-	a
Dilation angle 0.0	88	6	-	76	b
2m water in fill	109	-	114	79	b
4m water in fill	127	-	214	53	b
Properties of wall and anchors					
Wall Stiffness 3 GPa	98	1	122	84	b
Wall Stiffness 300GPa	97	15	100	84	b
3 unit wall: central hinge	107	-	92	92	e
3 unit wall: rear hinge	105	-	84	91	d
Anchor Length: Decreasing with depth	117	-	132	101	d
Anchor Length: Increasing with depth	115	-	96	99	d
Anchor Spacing non-uniform	110	-	128	95	d
Anchor Stiffness 13.2 MN/m	90	6	136	78	d
RIGID FOUNDATION - 12 ANCHORS					
STANDARD CASE:	109	5	100	94	d
Fill undrained	457	22	932	-	d
Bulk fill undrained	125	21	718	108	d
Anchor Stiffness 13.2 MN/m	91	6	141	78	d
Anchor Stiffness 6.6 MN/m	82	6	177	71	d
3 unit wall	110	-	86	95	d
3 unit wall Anchor E = 13.2 MN/m	90	-	127	78	d
Top prop - no tension capacity	108	2	109	93	d
Joint elements to base	108	6	132	93	d
SAND FOUNDATION - 12 ANCHORS					
STANDARD CASE	119	19	100	103	f
Fill strength 45°	114	19	100	165	f
Fill strength 25°	138	20	143	83	f
Anchor Stiffness 13.2 MN/m	103	18	123	89	f
3 unit wall	125	-	106	108	f
3 unit wall - Anchor E = 13.2 MN/m	108	-	114	93	f
SOFT CLAY FOUNDATION - 8 ANCHORS					
STANDARD CASE: 4m HIGH WALL	72	36	100	153	g
Fill strength 45°	48	30	83	166	g
Fill strength 25	106	41	119	145	g

BOUNDARY CONDITIONS:
a = toe of wall restrained
b = toe of wall allowed to translate (but not rotate) after 1m fill placed
c = as b but toe released at end of construction
d = toe of wall allowed to translate and rotate about rear edge of wall unit
e = toe of wall allowed to translate and rotate about centre of wall unit
f = wall founded directly on foundation soil
g = wall rigidly connected to 450mm wide footing

Parametric study

Details of the various parameters, and the results obtained for the total anchor force, the maximum wall displacement and the force in temporary props are given in Table 1. For comparison, the total anchor load is also given as a percentage of the design load calculated using Rankine active pressures in accordance with BE3/78 (1987). BE3/78 stipulates that a load factor of 2 should be applied to the design load to determine the ultimate capacity required for the anchors.

In this study, the variables which had the greatest effect on the total anchor force predicted by the FE analyses for working load were:

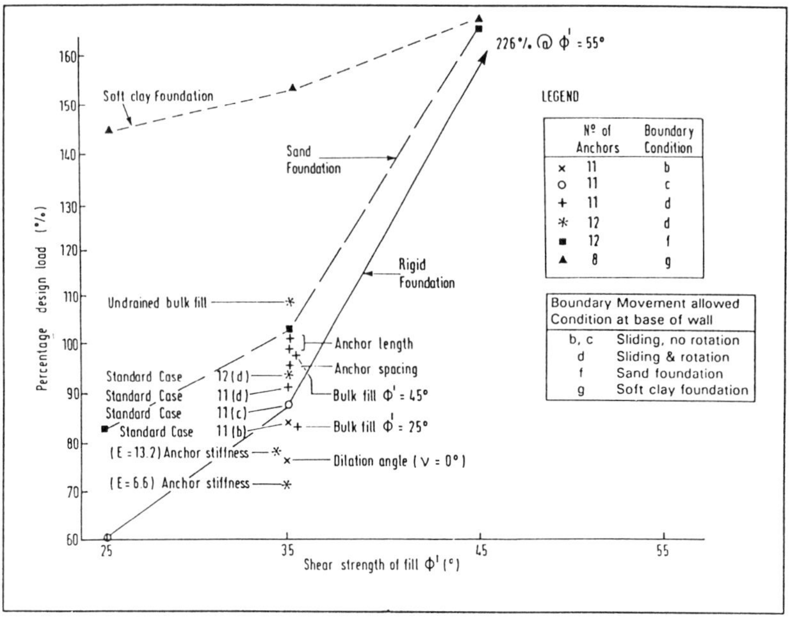

Fig. 6. Total anchor forces compared to Design Load

- the Poisson's ratio of the fill, which determines the stress path direction before yield and hence controls the horizontal stress produced by the construction of subsequent layers of fill
- dilation in the plastic zone at the rear of the facing, which produces an expansive force within the anchor zone
- the anchor configuration; decreasing length with depth produces a concentration of force in the lowest anchor of each length
- anchor stiffness; a reduction in stiffness allows increased strain within the anchor zone allowing the horizontal stresses and anchor forces to decrease
- compressibility of the foundation; the tendency for lateral expansion in a compressible foundation under load increases the force in the anchors in the lower half of the wall
- water pressures within the fill, which increase the pressure acting on the wall as suggested by classical theory.

None of these parameters are explicitly included in the design method for calculating the required anchor capacity. Water pressures are assumed to be zero and drainage is provided to ensure this condition is attained: they can however be included in the design method, and this has been done for the comparison of total anchor force given in Table 1.

The primary variable in the design method is the strength of the fill which is appropriate for consideration of the ultimate limit state. The effect of the strength on the anchor forces at working conditions is however only of the same magnitude as the parameters discussed above. The results of the analyses, summarised in Fig. 6, suggest that at working conditions the anchor force ranges between 60% (ϕ'= 25°) and 165% (ϕ'= 45°) of the design load. For a ϕ' value of 35°, as frequently used for design, the anchor forces were about 85% and 150% for rigid and compressible foundations respectively. The FEM analyses confirm, that from a loading point of view, service conditions may be more onerous than those that apply at collapse.

The greatest influence on the anchor forces under working conditions is the foundation. The anchor forces increase as compressibility of the foundation subsoil increases.

Conclusion

The value of the finite element method in determining the influence of a wide range of parameters on the behaviour of anchored earth structures has been demonstrated. In particular, the relationship between the anchor forces at working conditions given by FEM and those calculated by current design methods are shown to vary over a wide range. The results indicate that anchor forces are strongly dependent on the compressibility of the soil foundation, whereas current design methods assume that they are independent. A factor of safety of 2 would appear to be justified, and the site works at Otley show that the current design method produces safe and serviceable structures. The comparison of the actual and predicted performances of the full-scale and model structures will provide a measure of the ability of FEM to reproduce the essential behaviour of such structures.

Acknowledgements

The work described in this Paper forms part of the research programme of the TRRL and is published with the permission of the Chief Executive.

References

JONES C.J.F.P. *et al.* First application of anchored earth. *XIth ICSMFE*, San Francisco, 1985.

MURRAY R.T. and IRWIN M.J. *A preliminary study of TRRL anchored earth.* TRRL Supplementary Report 674, TRRL, 1981.

BOLTON M. *Centrifuge Model Tests on TRRL anchored earth system*. Report to TRRL, CUED, 1981.

JOHNSON P.E. *Maintenance and repair of highway embankments: studies of seven methods of treatment*. TRRL Research Report 30, TRRL, 1985.

BRADY K.C. and MASTERTON G.G.T. Design and construction of an anchored earth wall at Annan. *Proc. Int. Conf. Reinforced Soil*, British Geotechnical Society, Glasgow, 1990.

FUKUOKA, M., IMAMURA, Y., AND NISHIMURA, J. Fabric faced retaining wall with multiple anchors. *Geotextiles and Geomembranes*, 1986, **4**.

DEPARTMENT OF TRANSPORT. Technical Memorandum BE3/78 (revised 1987). Reinforced and anchored earth retaining walls and bridges. HMSO, 1987.

Monitoring performance of a soil-nail shoring system in Holocene alluvium

M. M. KIRKGARD and A. K. HAMMOCK, Law/Crandall Inc., USA

Introduction

This Paper presents data from measurements and observations of a soil-nail shoring system constructed in Holocene (recent) alluvium in Beverly Hills, California. The soil-nail shoring was designed to retain a 20- to 36-foot-high (6 to 11m) excavation for the subterranean parking levels of a new commercial building. A number of different soils have been supported by this technique since the early 1970s (refs 1-3). However, since soil-nail shoring is a recent innovation in the Southern California area, close control on the performance of the system was desired and was included in the project specifications.

The soil-nailing procedure is a soil improvement concept, where the active soil zone is reinforced with steel "nails," allowing the excavation to be made without external support. The active pressure used in a conventional shoring design is not used. The nails are similar to conventional shoring anchors. However, most of the nails are not preloaded and are attached to a shotcrete facing on the surface of the excavation. The subject project included the construction of approximately 775 soil-nails installed in up to six rows across the four faces of the excavation. Measurements were made periodically during the construction process to monitor the performance.

Inclinometers were used to measure lateral displacements behind the soil-nail shoring. Survey points were established around the project and used to measure horizontal displacement and vertical settlement. Load tests were used to evaluate the capacity of the soil-nails. The various measurements are analyzed, discussed, and presented in plots and diagrams appropriate for indicating the behaviour of the shoring system.

Soil conditions

The natural soils beneath the project are Holocene (recent) alluvium consisting of clay, sand, silty sand, and clayey sand. These soils are generally

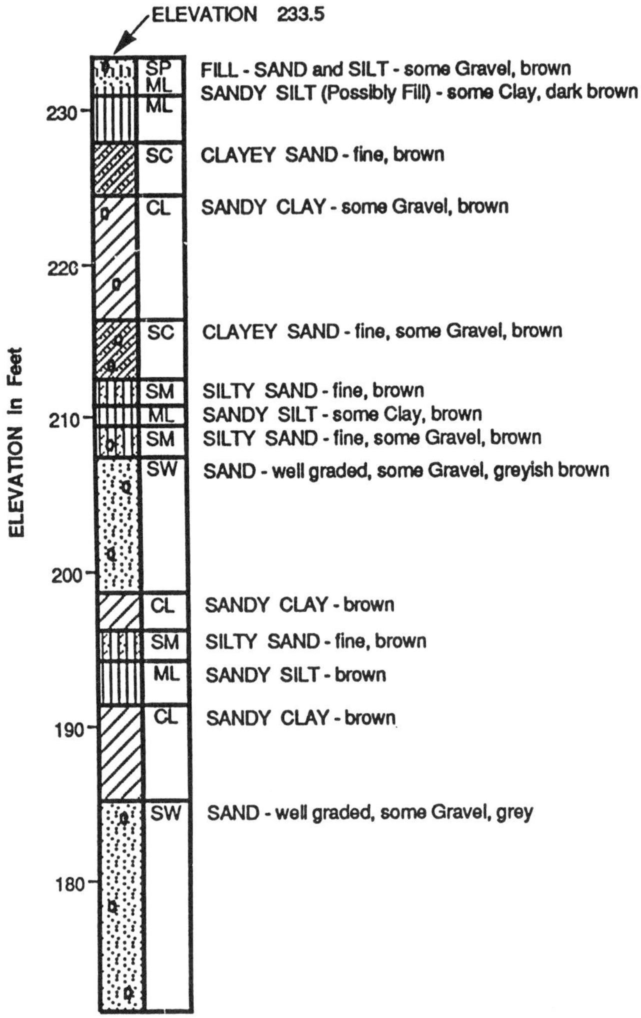

Fig. 1. Typical boring log

firm, with some softer layers. The log from one of the test borings performed on the site is presented in Fig. 1.

The soils above the proposed excavated level had dry densities ranging between about 108 and 124 pounds per cubic foot (17.0 and 19.5 kN/m^3). Measured field moisture contents varied between about 7% and 20%. The strength of the soils was characterized by performing direct shear tests on selected undisturbed samples. The tests were performed at field and increased (soaked) moisture contents and at various surcharge pressures. The yield-point values determined from the direct shear tests are presented in Fig. 2.

Design methodology

The soil-nail shoring was based on a design method developed at the University of California, Davis (ref. 2). This method checks stability using a limit equilibrium analysis. A parabolic failure surface is assumed and two cases are considered. The first case considers a failure surface that extends beyond the reinforced zone, and the second case considers a failure surface that lies entirely within the reinforced zone. Free-body diagrams of the reinforced block and the unreinforced block are drawn and the forces acting

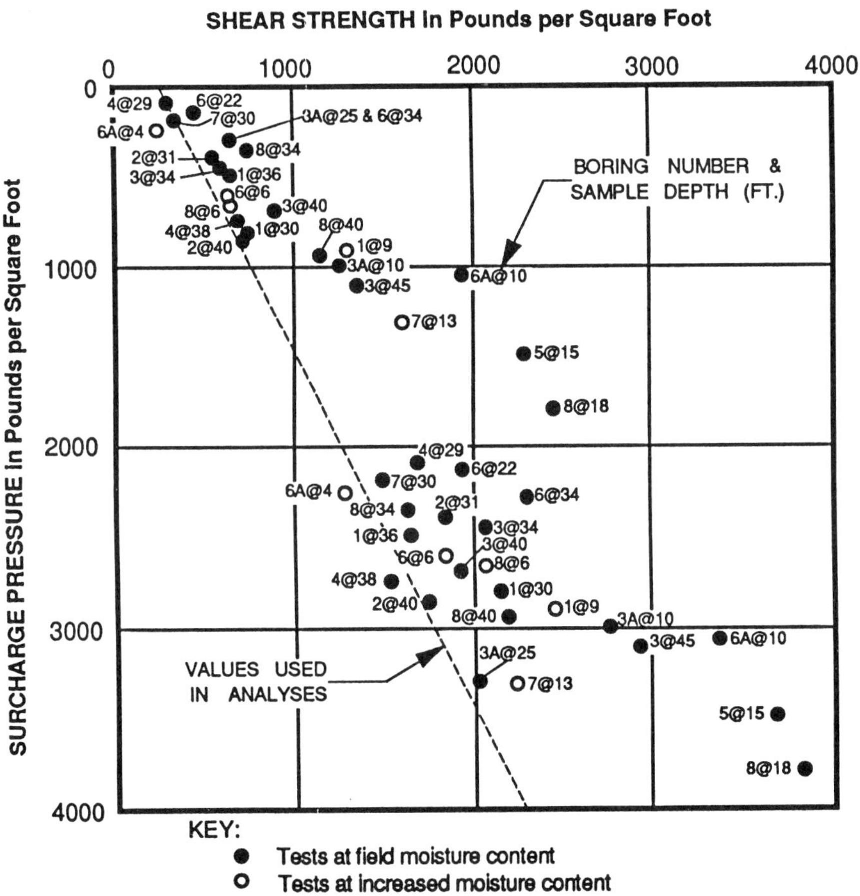

Fig. 2. Direct shear test data

Table 1. Design parameters

Wall Height	= 19 to 36 feet (6 to 11 m)
Friction Angle	= 27°
Cohesion	= 300 psf (14.4 kPa)
Unit Weight	= 125 pcf (19.7 kN/m^3)
Nail Inclination	= 15° to 20°
Shaft Diameter	= 10 inches (254 mm)
Nail Length	= 15 to 30 feet (4.5 to 9 m)
Horizontal Spacing	= 6 feet (1.8 m)
Vertical Spacing	= 4 feet (1.2 m)

on these bodies are calculated. A factor of safety on the shear strength parameters can be calculated from these forces. This process is repeated for various assumed failure surfaces until a minimum factor of safety is obtained. The design method determines the nail capacity by calculating the overburden normal force on the nail and multiplying by the tangent of the friction angle of the soil.

The excavation for the building foundation was roughly 380 feet x 150 feet (116 x 45 m). Excavation ranging between 19 and 36 feet (6 and 11 m) deep was required; the majority of the excavation exceeded 26 feet (8 m) in depth. The site was surrounded on two sides by paved streets. Existing commercial buildings and a paved parking lot were present on the other two sides.

The parameters used in the soil-nail shoring analysis are shown in Table 1.

For design, the soil was assumed to be homogeneous throughout the entire depth. The water table was assumed to be below the bottom of the excavation. A minimum factor of safety for all modes of failure of 1.46 was computed, based on the assumed soil strength. Design nail loads varied between 38 and 63 kips (169 and 280 kN).

The soil nail wall system was constructed in the following manner. First, a 5-foot (1.5m) lift of soil was excavated to expose an unsupported face at an inclination of 0.5:1 (horizontal to vertical). Holes for the nails were augured to the dimension specified above. Soil-nails consisting of Dywidag Grade 60 #8, #10, and #11 threaded bars were placed into the drilled holes and grouted in place. A portion of each nail was left extending out of the hole to allow for the shotcrete facing and the anchor plate/lockoff nut assembly of the nail.

Each nail was suspended in the centre of the hole with a PVC centralizer.

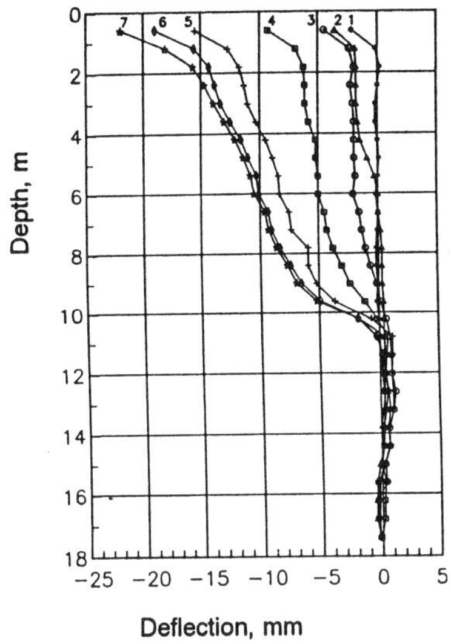

Reading	Date	Depth of Excavation
1	Aug. 9	3.7 m
2	Sept. 6	5.8
3	Sept. 13	8.8
4	Sept. 20	10.4
5	Oct. 4	10.4
6	Nov. 1	11.0
7	Jan. 31	11.0

Fig. 3. Results from Inclinometer #2

A 6 x 6 W1.4 x 1.4 steel wire mesh was attached to the exposed excavation face. Shotcrete was then sprayed onto the face. Reaction plates and nuts were attached to the nails after initial set of the grout and shotcrete. Nails not proof-tested were tensioned to about 5% of the design load and locked-off. This lack of pre-tensioning is the principal difference between soil-nails and conventional drilled tie-backs that are pre-tensioned. After completion of a lift, the process was repeated until the excavation was completed.

Instrumentation

Due to the location of the existing improvements, and the depths of the proposed cut, instrumentation was installed on all four sides of the excavation. Six inclinometer casings were placed around the project, immediately adjacent to the top of the cut. The inclinometers were drilled to depths of between 45 to 60 feet (14 and 18 m) from the ground surface. The slope indicator casing was placed in 8- and 10-inch-diameter drilled holes; the holes were backfilled with #3 Monterey Sand. Measurements were made with readings every 2 feet (0.6 m). Survey points were established around the perimeter of the top of the excavation at 20 foot (6 m) spacing to allow monitoring of both horizontal and vertical movements. Benchmarks for the survey measurements were established at a significant distance from the excavation.

Testing procedures for the nails were not specified by the local building

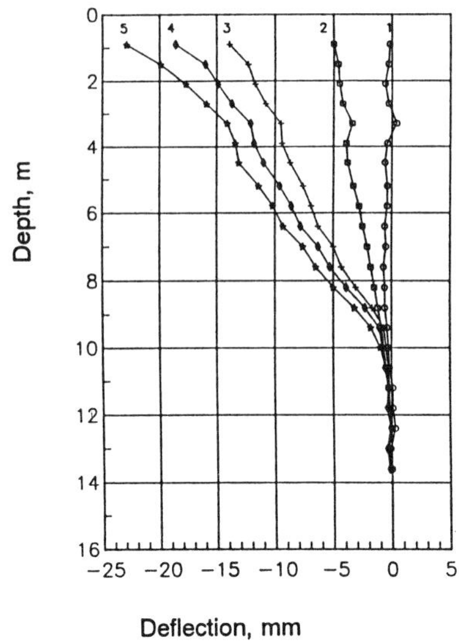

Reading	Date	Depth of Excavation
1	Sept. 13	4.0 m
2	Sept. 20	7.3
3	Oct. 4	8.8
4	Nov. 1	9.8
5	Jan. 31	9.8

Fig. 4. Results from Inclinometer #3

code at the time the excavation was constructed. The geotechnical consultant recommended that at least 10% of the nails in each row be proof-tested to at least 150% of the design load and that one nail in each row be tested to 200% of the design load. The purpose of the tests was to verify the anticipated friction values. As an acceptance criterion, the total deflection in the 150% tests was not to exceed 12 inches (305 mm), and the rate of creep under the 150% tests was not to exceed 0.1 inch (2.5 mm) over a 15-minute period. A lesser number of nails was finally tested during the project, as directed by the shoring designer and contractor.

Measurements

Inclinometers

Representative horizontal deflection profiles generated by Inclinometers #2 and #3 are shown in Figs 3 and 4, respectively. It may be seen from these measurements that the horizontal deflection of the soil-nail shoring decreases with increasing depth from the ground surface, and the horizontal deflection increases as the excavation depth increases. The measurements also indicate that horizontal deflections of the shoring are relatively small at depths greater than the current excavated level. It is also seen that there was some short-term movement following completion of the excavation to the design level; these short-term movements stabilized after a two- to three-month period. The

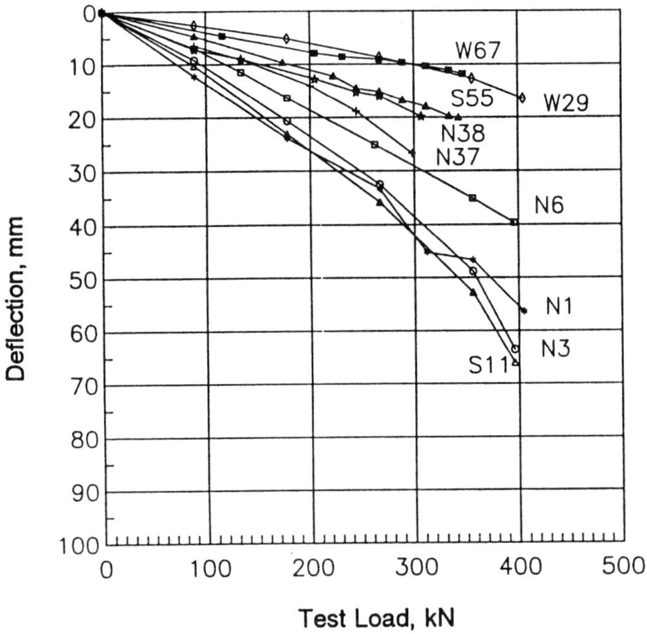

Fig. 5. Typical test results for acceptable soil-nails

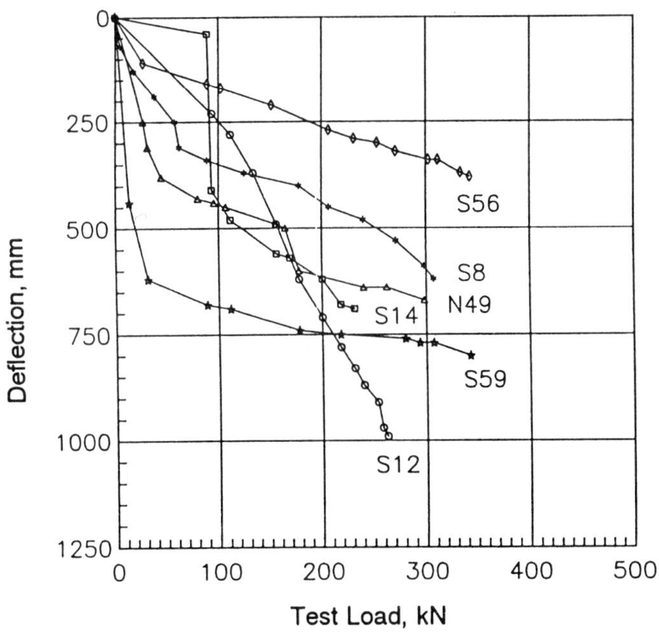

Fig. 6. Typical test results for unsatisfactory soil-nails

maximum measured deflection of the soil-nail system was about 1 inch (25mm), corresponding to about 0.2% of the total excavation height.

Survey points

The monitoring of the survey points showed no significant changes in the elevation of the ground surface surrounding the project. Most of the points changed less than 0.01 feet (3 mm), corresponding to the accuracy of the survey. The horizontal control points were consistent with the results of the slope inclinometer data. Horizontal movements towards the excavation were on the order of 1 inch (25 mm) or less.

Load tests

Of the 775 soil nails, about 2.5% of the nails were finally proof-tested to 150% of the calculated design load. About 70% of the nails held the test load in a satisfactory manner; that is, the observed deflection was less than 12 inches (305 mm). Typical test results for anchors passing the loading criteria are shown in Fig. 5. In contrast, about 30% of the test nails did not pass the acceptance criteria, and had excessive deformations during loading of between about 14 and 39 inches (356 and 991mm) as shown in Fig. 6. These excessive deformations were probably related to the poor bonding behaviour between the reinforcing members and the surrounding grout and soil. The poor bonding likely resulted from an inconsistent grout placement procedure (tremie removed from hole too fast), and the low-slump grout, which was also used to spray the excavation.

For a conventional tied-back shoring system where each anchor is proof-tested, the test results shown in Fig. 6 would require a partial or complete replacement of the failed anchors. However, although concern was raised, for this soil-nail system only the individual nails not meeting the testing criteria were replaced; the other nails that were not tested were left untreated. Based on the test results and inspection of installation procedures, presumably a large number of nails at the project could not hold the design load. Therefore, a questionable or indeterminate safety factor was operative during the life of the shoring. However, the permanent basement wall was completed and backfilled without incident.

Summary and conclusion

This Paper presents the field performance records of a soil-nail shoring system in recent alluvial soils. The system uses a passive earth reinforcement technique to strengthen the soil, and is different from the conventional active systems that retain soils behind a vertical cut, such as a tied-back shoring system. From the displacement measurements taken during construction, the shoring system performed in a satisfactory manner. Lateral displacements were maintained at less than about 1 inch (25 mm) or about 0.2% of the height of the shoring. There were no significant settlements of the ground adjacent to the excavation. The lateral displacement of the inclinometers indicate that the horizontal deflection of the soil-nail wall decreases with the increasing depth from the ground surface, and the horizontal deflection increases as the excavation depth increases.

It is recognized that the design and performance of a soil-nail system are closely connected to the construction method. All of the current design procedures are based on limit equilibrium analyses using soil strength values that need to be verified in the field by testing a certain percentage of the nails. The results of load tests on selected nails for this project indicated an indeterminate safety factor was operative during the life of structure. The adequate measured performance was likely related to the conservative soil strengths used in design of the system. This observation magnifies the need for close monitoring of soil-nail systems by surveying and use of inclinometers. More sensitive monitoring devices that allow continuous tracking of shoring displacements would be required on future projects, particularly to demonstrate the applicability of such systems in this locale under more difficult soil conditions.

Since not all of the soil nails are proof-tested as is the case in a conventional tied-back retaining system, ideally, the contractor should have responsibility for and control of the design, construction, and performance. A performance-type specification would be recommended, which would give the contractor control of both design and construction with a quality and performance level well defined for the final shoring configuration.

References

1. GUILLOUX, A. and NOTTE G. Experiences on a Retaining Structure by Nailing. *Proc. 8th European Conference on Soil Mechanics and Foundation Engineering*, Helsinki, 1983.
2. SHEN C.K., BANG S. and HERRMAN L.R. Ground Movement Analysis of an Earth Support System. *ASCE J. Geo. Eng. Div.*, **197**, No. GT12, 1981.
3. STOCKER M.F. and RIEDENGER G. The Bearing Behaviour of Nailed Retaining Structures. *Design and Performance of Earth Retaining Structures*, Geotechnical Special Publication No. 25, ASCE, 1990.

Stability analysis of reinforced soil retaining structures using the yield design theory

P. DE BUHAN, L. DORMIEUX and J. SALENÇON, Laboratoire de Mécanique des Solides, France

Introduction

In the design of earth retaining structures, geotechnical engineers are primarily concerned with the problem of their stability. Sticking to the spirit of Coulomb's approach (ref.1), the yield design theory (refs.2,3) provides a unified comprehensive framework for dealing with such stability analyses. The basic principle of the method consists in checking the compatibility of the equilibrium requirement of a structure under prescribed loading conditions with the material strength characteristics. Unlike most currently used classical methods (such as the well-known "method of slices" for slope stability analyses), the implementation of the yield design approach does not need the introduction of complementary assumptions and results therefore in clearly interpretable estimates for the stability of soil structures.

Starting from the presentation of the method on the case of a homogeneous soil structure, it is shown that the method is fully applicable to structures made up of several soil layers exhibiting different weight and strength characteristics. Furthermore, any kind of reinforcement of the structure (nails, tie-backs, geotextiles etc.) can be easily taken into account in the analysis through the so-called "mixed modelling" of the composite reinforced soil. All these recent developments of the initial theory have been put in the concrete form of a computational program, whose efficiency is illustrated on the design of a practical example.

Principle of the yield design reasoning

Consider for illustrative purpose the elementary example of a vertical cut of height h, subjected to its own weight γ. The constitutive soil is assumed to be homogeneous and to obey a Mohr-Coulomb strength condition with cohesion C and friction angle φ (Fig. 1).

According to the yield design reasoning, this structure will be termed "safe" or "stable" if a system of internal forces (i.e. a stress field) can be exhibited which complies with the equilibrium equations under the load

 Retaining structures. Thomas Telford, London, 1993

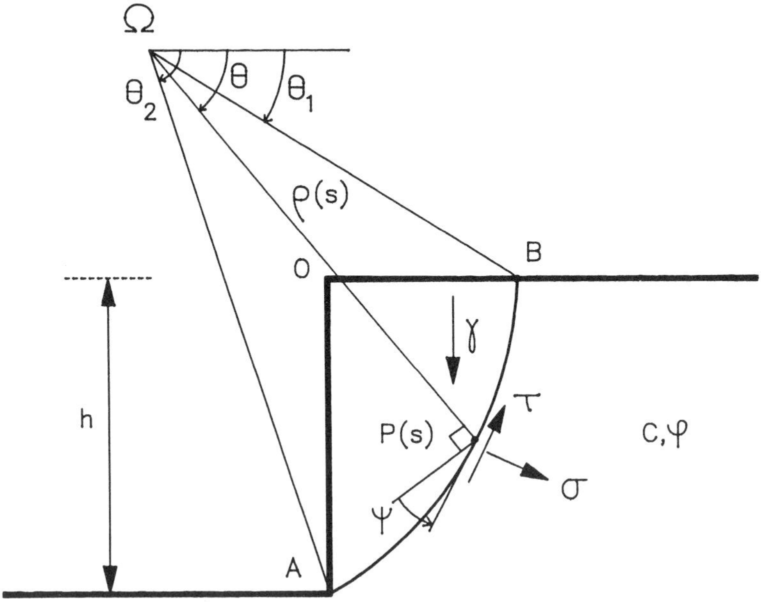

Fig.1 Stability analysis of a vertical cut

parameter γ and with the strength condition of the soil, throughout the whole bulk of the soil. It turns out that the stability of such a structure is governed by the dimensionless parameter γh/C with the following equivalence.

$$\text{stability of the structure} \leftrightarrow \gamma h/C \leq (\gamma h/C)^* \quad (1)$$

where $(\gamma h/C)^*$ denotes the critical value of the parameter γh/C beyond which the failure of the structure will occur.

It is convenient to define the factor of confidence of the structure as

$$\Gamma^* = (\gamma h/C)^* / (\gamma h/C)$$

so that the above statement may be put in the form

$$\text{stability of the structure} \leftrightarrow \Gamma^* \geq 1 \quad (2)$$

The problem being dealt with from a two dimensional standpoint ("plane strain" problem), let OAB be any volume of soil separated from the rest of the structure by an arbitrary line AB. It clearly appears that for the structure to be safe, in the sense stated above, it is necessary that the moment of the weight of OAB with respect to any point Ω (clockwise "driving" moment M^{Ω}_{w}) be balanced by the anticlockwise "resisting" moment developed about the same point by the distribution of normal and shear stress (σ, τ) acting upon OAB along AB, provided the strength condition of the soil

$$|\tau| \le C - \sigma \tan\varphi \tag{3}$$

be satisfied everywhere (tensile normal stresses are reckoned positive).

In other terms, introducing the concept of maximum resisting moment M^{Ω}_{r} defined as

$$M^{\Omega}_{r} = \int_{A}^{B} \mathcal{M}^{\Omega}_{r}(s)\,ds \tag{4}$$

where at every point P(s) of AB

$$\mathcal{M}^{\Omega}_{r}(s) = \rho(s) \sup_{(\sigma,\tau)} \{\sigma \sin\psi + \tau \cos\psi;\ (\sigma,\tau) \text{ satisfying } (3)\} \tag{5}$$

with ψ denoting the angle between the direction normal to the radius ΩP of length ρ, and the tangent to AB at point P (Fig. 1), it comes out

$$\text{stability of the structure} \rightarrow M^{\Omega}_{w} \le M^{\Omega}_{r}$$

and comparing with (2)

$$\Gamma(\Omega, AB) = M^{\Omega}_{r} / M^{\Omega}_{w} \ge \Gamma^{*} \tag{6}$$

The important conclusion which can be drawn from the previous analysis, is that, given any point Ω and line AB, the ratio between the maximum resisting moment and the driving moment due to the weight of the soil, constitutes an upper bound estimate for the factor of confidence of the structure.

Trying to get the best estimate of this factor would therefore require to minimize $\Gamma(\Omega,AB)$ over all possible points Ω and lines AB. Fortunately, this minimization procedure is greatly simplified by the following result (ref.3).

For a given point Ω, it can be shown that the most critical volume OAB, associated with the minimum estimate $\Gamma(\Omega, AB)$, is such that $\psi=\varphi$. This implies that AB is the arc of logspiral of angle φ and focus Ω passing through the toe A of the cut. Its equation in polar coordinates (ρ,θ) attached to Ω (Fig. 1) writes

$$\rho(\theta) = \rho_1 \exp[(\theta_1 - \theta)\tan\varphi]$$

Consequently, denoting by $\Gamma(\theta_1,\theta_2)$ the corresponding value of the ratio $\Gamma(\Omega,AB)$, where θ_1 and θ_2 are the angular parameters defining the position of Ω, the minimization of $\Gamma(\Omega, AB)$ reduces to searching for the minimum value of $\Gamma(\theta_1,\theta_2)$ with respect to θ_1 and θ_2. This is achieved through a numerical procedure, which in the present case yields .

$$\Gamma = \min_{\theta_1,\theta_2} \Gamma(\theta_1,\theta_2) \cong 3.83\ C/(\gamma h)\ \tan\left(\frac{\pi}{4} + \frac{\varphi}{2}\right) \tag{7}$$

This is to be related to a classical result (refs. 2,4). The above analysis may be directly extended to the case when the structure is submitted to any kind

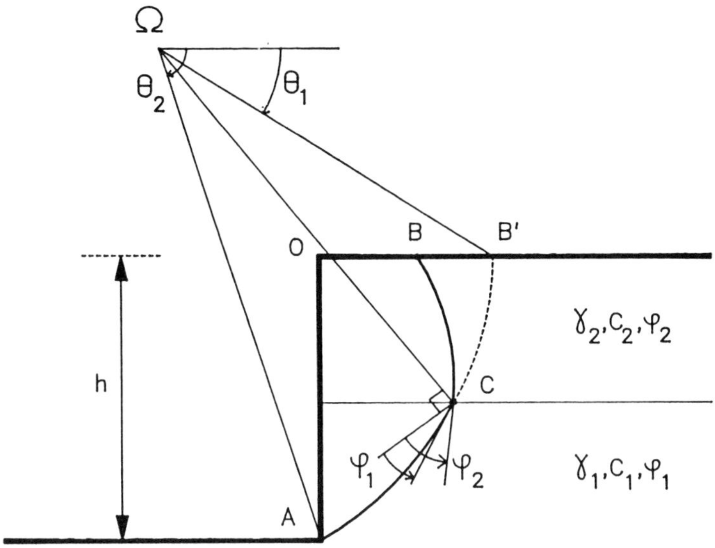

Fig. 2 . Stability analysis of a two-layered vertical cut

of external load applied on its boundary surface. For a given point Ω and line AB , the definition of Γ (Ω, AB) becomes

$$\Gamma(\Omega, AB) = M_r^{\Omega} \;/\; (M_w^{\Omega} + M_L^{\Omega}) \qquad (8)$$

where $M^{\Omega}{}_L$ represents the moment about Ω of the external loads acting upon OAB. It should be noticed that, depending on the sign of $M^{\Omega}{}_L$ relative to that of $M^{\Omega}{}_W$, such external forces may have favourable or unfavourable effects on the stability of the volume under consideration, and hence of the structure as a whole.

Likewise, the method remains fully relevant for dealing with structures where several, layers of soils having different characteristics must be taken into account. Assuming for instance the vertical cut to be made of two such layers with a horizontal interface, one may analyze its stability by checking the moment equilibrium of volumes such as sketched in Fig. 2.

AB′ is an arc of logspiral with angle φ_1 (friction angle of the lower layer of soil) and focus Ω defined through angles θ_1 and θ_2, the curve bounding the volume of soil whose equilibrium is examined is constructed by connecting the arc of logspiral AC in the lower part of the structure, with CB which is drawn as the arc of logspiral of angle φ_2 having the same focus Ω as ACB′. It induces therefore a discontinuity of the tangent to the line ACB at point C where it intersects the interface between the two layers. The estimate of the factor of confidence derived from the global equilibrium check of volume

OACB is defined in the same way as in Eq. (6), taking into account different expressions of $\mathcal{M}^{\Omega}{}_{r}(s)$ along AC and CB. The best upperbound estimate of Γ^* will once again be derived from a minimization with respect to the angular parameters θ_1 and θ_2.

Application to the design of reinforced retaining structure

Various reinforcement techniques are increasingly used today in order to enhance the stability of otherwise unstable retaining structures. Adopting a "mixed modelling" of the so obtained reinforced soil (refs.5,6), that is treating the reinforcing inclusions as strips or beams embedded in the soil modelled as a classical 3-D continuum, makes it possible to apply the same reasoning as that used for natural soil structures.

Fig. 3 outlines such a retaining structure in which several arrays of reinforcements have been placed at each stage of the excavation ("soil nailing" technique).

Resuming the analysis carried out on the corresponding homogeneous structure, the moment equilibrium of volume OAB with respect to a point Ω is considered. While the driving moment $M^{\Omega}{}_{w}$ generated by the weight of the soil remains unchanged (provided the weight of the inclusions might be neglected), the maximum resisting moment $M^{\Omega}{}_{r}$ appears as the sum of two terms

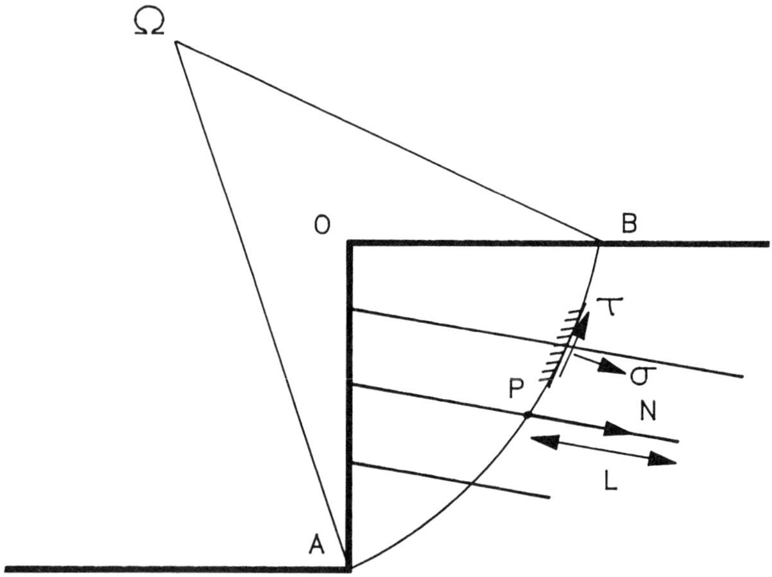

Fig.3. Stability analysis of a reinforced retaining structure

- the contribution due to the distribution of stresses (σ,τ) in the soil along AB, whose expression is given by Eqs. (4) and (5)
- the new term corresponding to the resisting forces developed by the inclusions at their intersection with AB, in account of their strength capacities which are specified as follows.

Failure of the reinforcements may occur according to two modes

- through the breaking of the inclusions themselves
- or by lack of adherence between the soil and the reinforcements due to their limited anchoring length L beyond the failure line AB (Fig. 3).

Either failure mode can be taken into account through a specific condition, namely

- $0 \leq N \leq N_{max}$ which means that the reinforcements can withstand tensile forces up to a maximum value N_{max} corresponding to the yielding of the constitutive material;
- $N \leq f_l L$, where f_l is a lateral friction coefficient characterizing the bonding between the soil and the reinforcement and L the anchoring length.

This deserves two comments

- the adopted strength criterion involves only the axial force in the inclusions, and disregards their resistance to bending and shear forces. Actually, it can be easily shown that such a simplifying assumption is always conversative) as regards the evaluation of the factor of confidence
- the resisting tensile force reckoned for each inclusion is equal to the minimum value between the ultimate load carrying capacity N_{max} and the "pull out" resistance $f_l L$.

$$0 \leq N \leq N_o = \inf \{N_{max}, f_\ell L\} \; . \tag{9}$$

The definition of the ratio $\Gamma(\Omega, AB)$ between the maximum resisting moment and the driving moment relative to a volume OAB and a point Ω is

$$\Gamma(\Omega,AB) = \frac{M_r^\Omega(\text{soil}) + M_r^\Omega(\text{incl})}{M_W^\Omega + M_L^\Omega} \tag{10}$$

where M^Ω_r (incl) is the maximum resisting moment generated by the inclusions, taking their strength condition (9) into account. The two following situations will be encountered when computing the latter term, depending on the sign of the angle δ between the inclusion, oriented outwards with respect to volume OAB, and vector ΩP (Fig. 4).

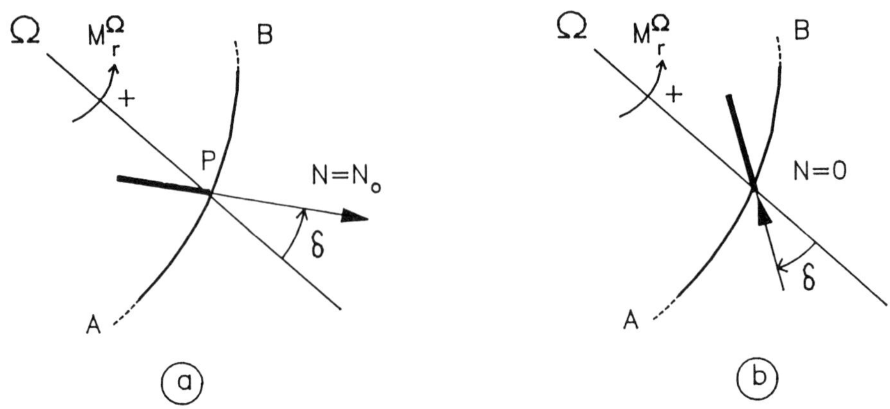

Fig. 4 : Maximum resisting moment generated by a reinforcement

If δ>0 (Fig. 4a) the maximum value of the resisting moment about Ω is obtained for $N=N_0$, that is

$$M_r^\Omega(\mathrm{incl}) = \Omega P\ N_o \sin\delta \qquad (11)$$

Conversely, if δ<0 (Fig. 4b), the resisting moment is positive when the inclusion undergoes compressive forces, and therefore the maximum resisting moment reduces to zero, since the compressive strength of the reinforcements is neglected (it is commonly acknowledged that failure by buckling is likely to occur in such a situation).

It follows that in both cases M^Ω_r (incl) remains non-negative and therefore results in an increase for the evaluation of the factor of confidence given by Eq. (10), when compared with that corresponding to the non-reinforced case.

Practical implementation

Based upon the above-described theory, a computational program named STARS (for Stability Analysis of Reinforced Soils) (refs.6,7) has been developed for analyzing the stability of reinforced soil retaining structures. In its most recent version, this program can handle a large variety of reinforced structures, involving several soil layers, all kinds of surface loadings (concentrated or distributed) and taking possible seismic conditions into account through a regular pseudo-static analysis. The high speed of numerical computations (even when operating with a personal computer) combined with a great conviviality allows any user of such a program to optimize the reinforcement pattern of a structure within a very short period of time, as it will be illustrated on the following realistic example.

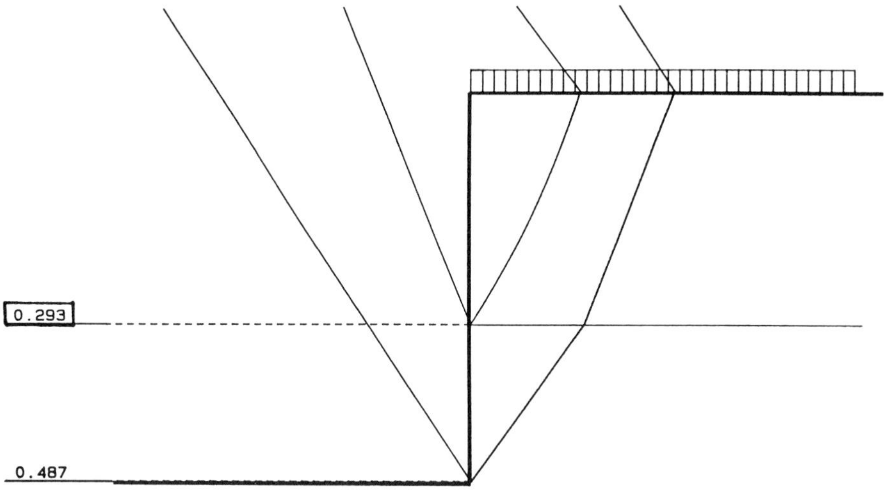

Fig. 5: Stability analysis of the unreinforced structure

The structure under consideration is a 10m high vertical cut subjected to a uniformly distributed load of 10kN/m^2 applied on its upper level (Fig. 5). It is constructed in its upper part (i.e. down to 6 m deep from the upper corner) by a soil with the characteristics γ=18kN/m^3, C = 5kPa and φ=35^0, while those of the lower layer are γ=20kN/m^3, C=30kPa and γ=20°. As shown by a first

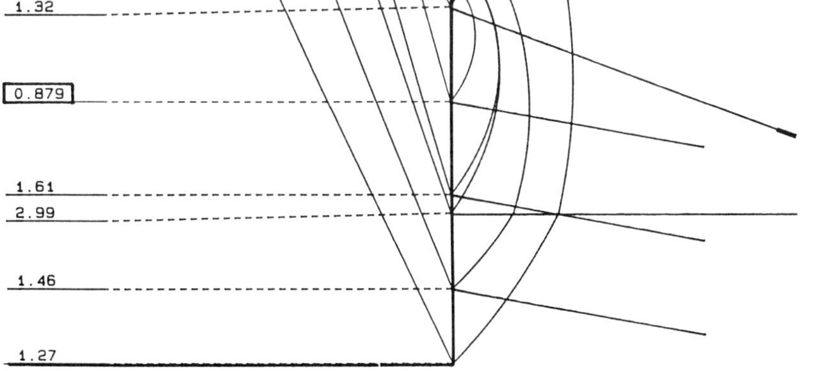

Fig 6: First scheme of reinforcement

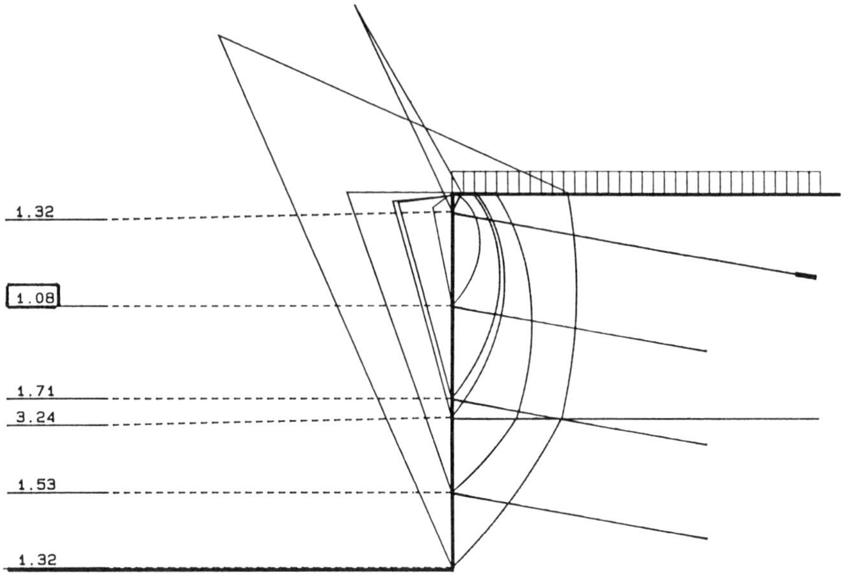

Fig. 7: Modified scheme of reinforcement

calculation, this structure is obviously unsafe, since the best upperbound estimates of its factor of confidence derived from the "global equilibrium check" of volumes bounded by all possible lines made of arcs of spiral passing through either the toe or the emerging point on the interface between layers, remain far below unity (0.49 and 0.29 respectively).

In order to ensure the stability of this excavation, the following reinforcement scheme is proposed (Fig. 6).

Three rows of inclusions (nails) inclined at 10° to the horizontal are placed at respective depths of 3 m, 5.5 m and 8 m, with a regular horizontal spacing of 1 m. The strength parameters of each inclusion as defined earlier, are

N_{max} =100kN; f_l=40kN/m

One layer of tie-backs inclined at 20° and emerging on the facing at 0.5 m below the upper level is fitted, their tension being set equal to 150kN with a spacing of 1.5 m.

The results of the computations are shown in Fig. 6 which , for each family of lines passing either through the toe or emerging above each layer of reinforcement, displays the most critical one with the associated estimate of the factor of confidence. It can be seen that, contrary to what might have been expected , the lowest upperbound estimate (0 . 88) is not given by a line passing through the toe of the excavation, but by a line coming out on the facing at a depth of 3 m. This result emphasizes the fact that, restricting the exploration to those failure lines passing exclusively through the toe could lead to considerably overestimating the stability of the structure (1.27 instead of 0.88 in the present case).

A quite simple modification of the reinforcement scheme may consist in decreasing the inclination of the tie-backs from 20° to 10°. In that case, the calculations presented in Fig. 7 give evidence of a noticeable increase of the estimated factor of confidence from 0.88 to 1.08.

Conclusion

Because it relies upon a clear assessment of the problem along with consistent mechanical arguments for solving it, the yield design approach results in a reliable and versatile method for analyzing the stability of earth retaining structures and especially reinforced ones. Furthermore, it makes it possible to derive quite efficient numerical codes for engineers, as the one presented in this paper. By way of illustration, it should be mentioned that the computations whose outputs are reported in Figs 6 or 7 involve the exploration of nearly 4000 failure lines, which is carried out within a few minutes on any personal computer. One can easily imagine the radical changes that might arise in the design process of structures, when using such numerical tools.

References

1. COULOMB C.A. *Essai sur une application des regles de Maximis et Minimis a quelques problemes de statique relatifs a l'architecture*. Memoire a l'Academie Royale des Sciences, Paris, 1773.
2. SALENSON J. *Calcul a la Rupture et Analyse Limite*. Presses de l'Ecole Nationale des Ponts et Chaussees, Paris, 1983.
3. SALENSON J. An introduction to the yield design theory and its application to soil mechanics. *Eur. J. Mech A/Solids*, 1990, **9**, No.5, pp 477-500.
4. CHEN W.F. *Limit analysis and soil plasticity*. Elsevier, London, 1975.
5. ANTHOINE A. Mixed modelling of reinforced soils within the framework of the yield design theory. *Comp. Geotech.*, 1989, **7**, pp 67-82.
6. ANTHOINE A. and SALENSON J. Une optimisation d'ouvrages en sols renforces. *Proc. XII Int. Conf. Soil. Mech., Rio de Janeiro, Brasil*, 1989, pp 1219-1220.
7. ANTHOINE A. Une methode pour le dimensionnement a la rupture des ouvrages en sols renforces. *Rev Fr. Geotech.*,1991, **50**, pp 5-21.

Active earth pressure reducing pneusol: experimental tests

P. LAREAL, F. BENAMEUR and P. COLLAS, Institut National des Sciences Appliquées, France, and N.T. LONG, Laboratoire Central des Ponts et Chaussées, France

"Pneusol", a combination of old tyres and soils, is a material developed by the LCPC that has a number of areas of application: retaining structures, stiffening of slopes, lightweight embankments arching, etc. Used behind a retaining wall, this material can reduce the active earth pressure on it. Tests on small-scale three-dimensional models have shown how this "pressure-reducing" effect influences the critical height of failure of the structure. A double layered earth fill structure is also studied : a layer of alone sand completed for a layer of pneusol.

Introduction

Pneusol consists of a combination of tyres (truck or car), used whole, partially cut-up (one sidewall removed), or completely cut up (into two sidewalls and tread), with natural or artificial, powdery or slightly coherent soils or various wastes. (Long, 1985, 1990).

Reinforcement with tyres does not impose any special grading requirement on the soil, since the interaction of the tyre and soil is not based primarily on friction, as it is in conventional reinforced structures. There are several types of Pneusol ; broadly (Bailly *et al.*, 1988; Long and Valeux, 1989).

"Lightweight" Pneusol

Pneusol made with whole truck tyres is a "lightweight" material having a density between 8 and 10 kN/m^3. The filling material serves only to level the tyres and make the assembly homogeneous (the size and condition of the tyres varies).

"Heavy" Pneusol

Pneusol made with partially cut tyres (one sidewall removed) or with treads standing on edge constitutes an assembly of gabions, small or large according to the size of the tyres, most often determined by what is available locally.

Other pneusols

Pneusol made with sidewalls arranged in layers mobilizes the soil by the soil- soil contact of the top and bottom layers across the rim area (conventional soil reinforcement function). There are arrangements to tie the various tyre parts together and so constitute a continuous layer. It should be noted that their tensile strength is fairly high: 17 to 52 kN for a car tyre sidewall; 55 to 75 kN far a car tyre tread.

Today, more than 450 000 tons of tyres are discarded every year. Two hundred thousand tons are recovered in one form or another (retreading, incineration, farm use, rubber powder, etc.). The remainder go to official or unofficial landfills. (The quantities are the same in Germany and in England). This waste is fairly uniformly distributed throughout the country and is found in concentrated form. Pneusol is typically a material of the countries of the Northern hemisphere, which are rich countries, but in fact choices in all countries are shaped by economic criteria, and it is far from certain the Pneusol would be without economic and ecological benefits in such Southern countries as Thailand, Singapore, Formosa, etc.

To date, more than one hundred structures are known to have been built with this material (Lareal and Long, 1987 ; Long, 1990)

- retaining walls
- active earth pressure reducing structures
- stiffening of slope
- energy absorbers
- lightweight embankments
- protection of banks and slopes
- stress distribution
- military Pneusol
- special structures

This Paper deals primarily with Pneusol as a means of reducing active earth pressure, and in particular with tests conducted on small-scale three-dimensional models at the INSA at Lyon. We are aware of the drawbacks of such a model, in particular the difficulty of maintaining similarity, but our main aim is to form a qualitative idea of the contribution Pneusol can make to the stability of retaining walls. The test we report simulates the presence of Pneusol behind a gravity and cantilever retaining wall.

Experimental model

The scale-model study is conducted in a rectangular container 80 cm high, 80 cm wide, and 120 cm long (Fig. 1).

The wall, across the width of the container, consists of three identical

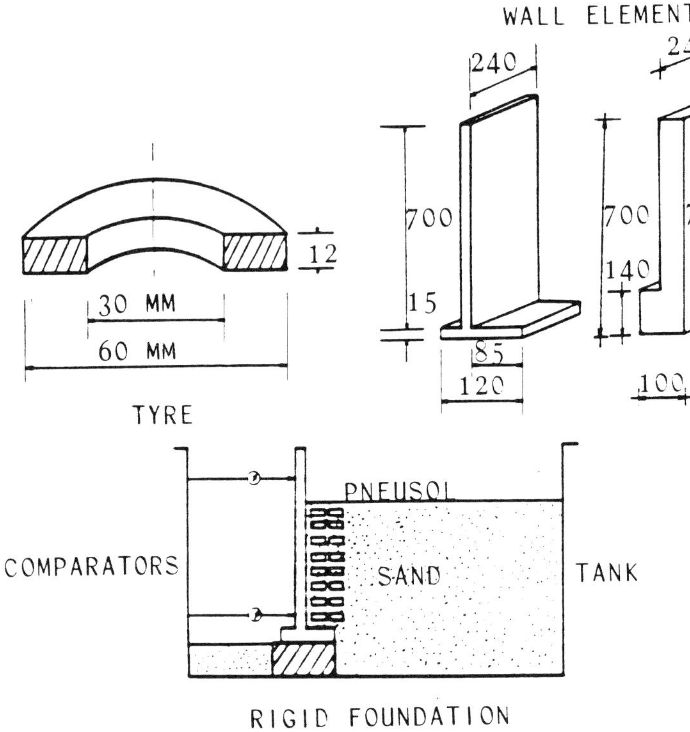

Fig. 1 : The experimental model (dimensions in mm)

elements, made independent to eliminate edge effects. Wall displacement measurements are made only on the centre elements. A strip of polyurethane forms a flexible joint to isolate each element from, and make it independent of the others.

The "tyres" used are made of polyurethane foam. They are rings 12 mm

Table 1. Density of Pneusol

e th (mm)	18	28	38	48	58	68
γ (kN/mm^3)	12,30	12,60	12,90	13,20	13,60	13,90
e th(mm)	78	sand alone				
γ (kN/mm^3)	14,20	14,30				

thick having inside and outside diameters of 30 and 60 mm respectively (about 1/20th scale with respect to a thick tyre).

The soil used in the Pneusol and in the embankment was a crushed sand of uniform grading (D_{min} = 0.3 mm; D_{max} = 1 mm). It was placed at a constant density of about 14.30 kN/m^3.

The density of the Pneusol for each layer spacing th is given in Table 1.

The wall rested on a rigid foundation and its displacement was monitored by two comparators having a precise of 0.01 mm.

Experimental programme

The nature and geometry of the wall, the sand, and individual tyres were constant; the parameters whose influence was investigated concerned the composition of the Pneusol: the spacing "e" between layers of tyres (i.e. the thickness of sand between two layers) and the number "n" of tyres in each layer.

For each case investigated, the displacement of the wall was measured as the placement of the Pneusol and fill progressed.

We also studied a double layered earth fill structure. A double layered ground fill consists of a layer of sand at the bottom up to a height known as h_s completed by a layer of Pneusol up to the failure height known as hr.

The composition of the Pneusol was constant all through the experiments and had the following characteristics (considered as optimal in terms of combined quality and cost)

- spacing between layers, e = 28 mm
- no ties between tyres
- number of tyres per layer, n = 3
- orthogonal tyres disposition.

The parameter whose influence was first investigated was the thickness of the sand layer, h_s, with a uniform distribution of tyres in the upper layer. Then, two trapezoidal distributions were studied.

Tests/result: simple Pneusol fill

The direct measurement results can be used to plot curves h = f (D1) of the height of fill behind the wall versus the displacement of the wall.

Influence of number of tyres

Fig. 2 shows the influence on the number "n" of tyres per layer for a constant layer spacing of 18mm. The simply cubic array of each layer consists of four tyres in contact with each wall element and contains (n x 4) tyres. In

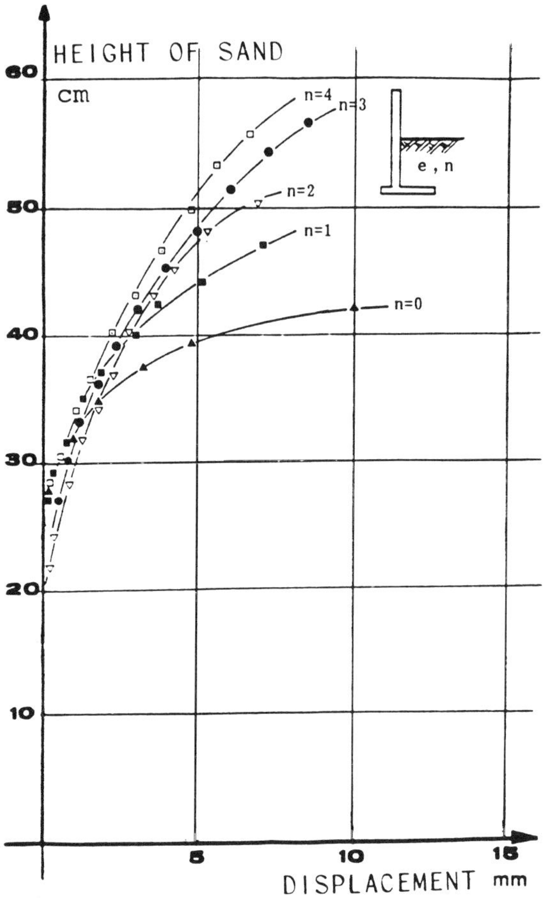

Fig. 2 : Pneusol, e = 18 mm, n = 1-2-3-4 Relation h = f (Δ1)

the tests, n is varied from 0 to 4. With the cantilever wall, comparing the curves h = f(Δ1) to one another, it can be seen that the slope of the cube becomes very small in the range of fill heights tested. Similarly, it can be seen that below a certain height of fill, of the order of 250 mm, the number of tyres has no effect, which is logical because there is very little displacement of the wall without Pneusol. The failure of a wall without Pneusol or with n = 1 is sudden, whereas with n > 1 progressive deformation is observed.

It should be noted that although replacing the fill with Pneusol substantially reduced the weight of material on the heel of the wall and thus its contribution to the stability of the wall, this stability was in fact improved! This is undoubtedly explained in large part by the change from state K_0 (active earth pressure at rest) to state K_a (active earth pressure). The "cohe-

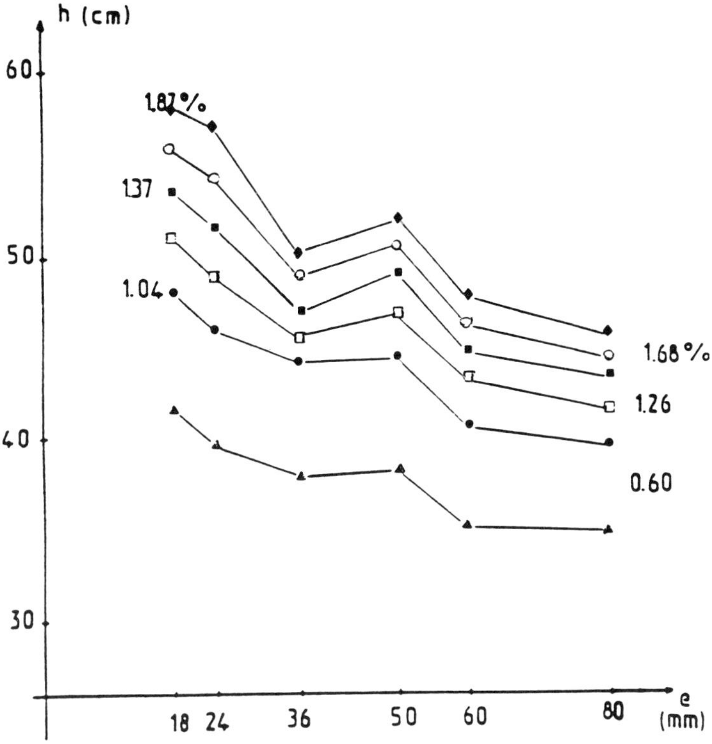

Fig. 3 : Pneusol (n = 3) Relation h = f (e) versus rotation of wall (x %o)

sion" of Pneusol also contributes to stability. A further 10% improvement was observed in the tests on the gravity wall.

Influence of spacing of layers of tyre

Cantilever wall

The spacing between layers, i.e. the thickness th of sand between layers, is varied with the number of tyres per layer held constant (n = 3). Fig. 3 shows the results obtained at different wall displacements. It can be seen that, the smaller the thickness, the greater the height reached by the fill for a given displacement. As th decreases, the Pneusol tends to act like a gravity structure superimposed on the cantilever wall. Again, the larger the admissible displacement of the wall, the greater the effect of reducing the thickness. However, the measurement points show a discontinuity at th = 50 mm that is unexplained; the reproducibility of the test was checked.

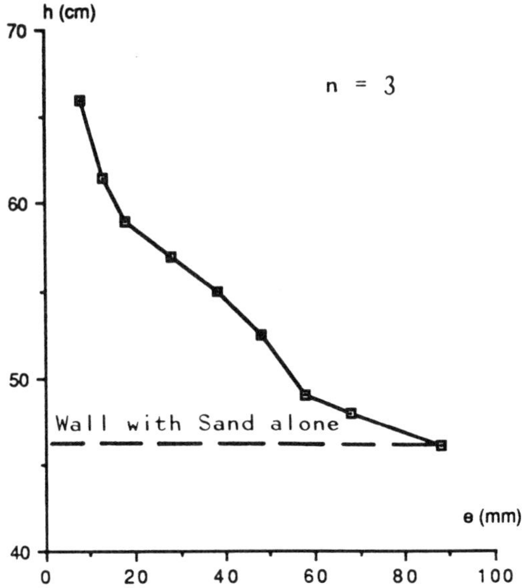

Fig. 4 : Relation between the height of failure and the spacing

Gravity wall

Overturning of the wall occurs at a height of 43 cm of sand alone. Curve no. 4 shows that as layer spacing th increases, the height of failure of the wall with Pneusol decreases, becoming identical to that of a wall without Pneusol at th = 88mm. At this spacing, the tyres have no effect.

Test results: double layered sand-pneusol fill

Uniform distribution

Speed curves [Fig. 5]

All the curves giving the speed of displacement of the wall (dl/ dh) versus total fill height (h) present the same particularities, whatever the value of hs is

- a linear part almost horizontal, similar to that of sand alone
- a breaking point then a vertical asymptote corresponding to the failure.

However, a comparison between those curves (including sand alone and Pneusol alone) shows that the greater the height of sand hs, the earlier the breaking point on the curve and the more sudden the failure. Moreover, the double-layered ground fill behaviour is closer to that of a ground fill made of sand alone than that of Pneusol alone.

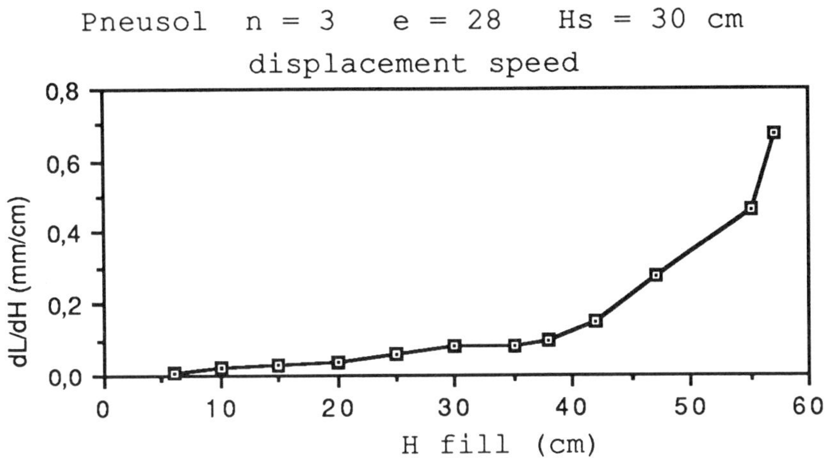

Fig 5 : Speed curves

The hyperbolic part of the speed curves can be changed into a linear one with a mathematical relation. It shows that for great values of displacements, the speed becomes proportional to the inverse of fill height to be added in order to reach failure [Fig. 6].

The breaking point of the curves is approximately situated at 0.2 mm per cm of sand fill added.

Relation between hr and hs [Fig. 7]

For hs inferior to 23 cm, the height of failure is equal to that of the Pneusol

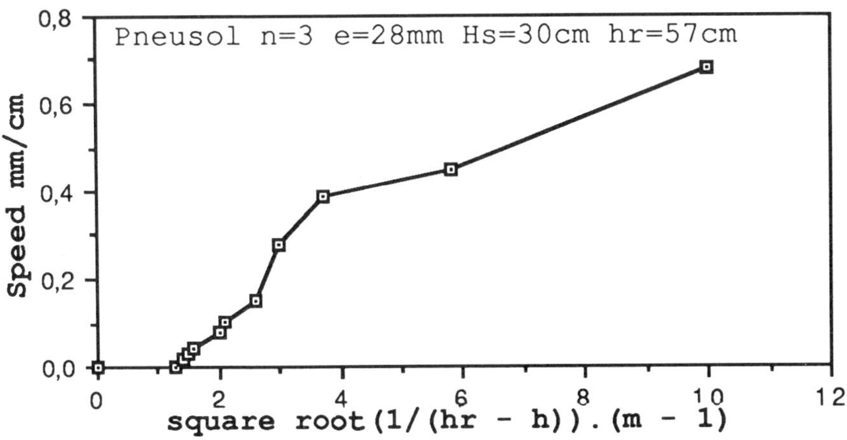

Fig. 6 : Linearisation of the speed curve

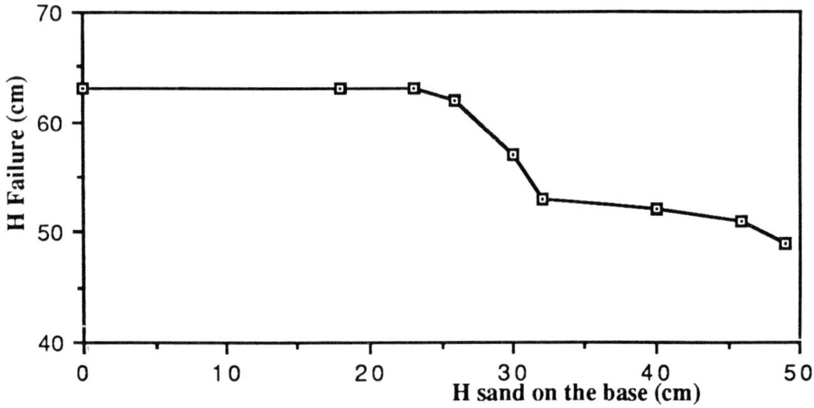

Fig. 7 : Failure of the double layered structure

alone. There is therefore no distinction between the double-layered ground fill and a homogeneous one because the earth pressure is less at the bottom of the structure.

Then, the bigger the height of sand hs, the lower the failure height whose value then reaches the failure height of sand. The reduction of earth pressure induced by cohesion and lightening has a great effect on the increase of hr. The gain in fill height is not negligible even if Pneusol represents a very small proportion of the fill height.

It has also been noticed that the variation of hr with hs (from hs = 23 cm) is a hyperbolic function.

Fig. 8 gives the linear variation of the height of Pneusol (hr - hs) versus the height of sand (hs).

The structure is considered as stable and safe within the triangle limited by the axis and the straight line and unstable beyond it.

These curves, plotted for different values of n and e (ie for different types of Pneusol) would give a useful chart. Knowing n and e and the fill height left, it would indicate the height of Pneusol to be added to reach failure.

Trapezoidal distribution

In both cases the failure height is similar to that of a uniform distribute. This result allows to make a hypothesis on the existence of a Coulomb wedge in the upper layer. Moreover, such a distribution spares at least up to 40 % of tyres.

Before the breaking-point on the speed curves, the trapezoidal and uniform distributions are equivalent but once the break is reached, a trapezoidal distribution is more dangerous because of its greater volumic weight. So it would be advisable not to go beyond the breaking point.

Conclusions

These campaigns of tests on small-scale models have qualitatively demonstrated the "active earth pressure reducing" and "stabilizing" effects of Pneusol. This effect becomes greater as the number of tyres per layer increases and as the thickness of sand between layers is made smaller. There is a maximum spacing beyond which there is no effect. It is consistently greater in a gravity wall when this type of Pneusol, intended to represent lightweight Pneusol made with truck tyres, is used.

The interest of a double-layered ground fill is the ability to give excellent results in terms of fill height at a lower cost than a whole layer of Pneusol (less tyres, easier construction).

Practical advice can be given for the construction of retaining walls using Pneusol whose linear speed curve is known : to a fixed admissible displacement of the wall (i.e.= speed) corresponds to a certain fill height. A reasonable choice would be not to go beyond the breaking-point on the curve at 0.2 mm per cm of added fill height (for example).

Pneusol can also be used in the refurbishment of troublesome retaining walls.

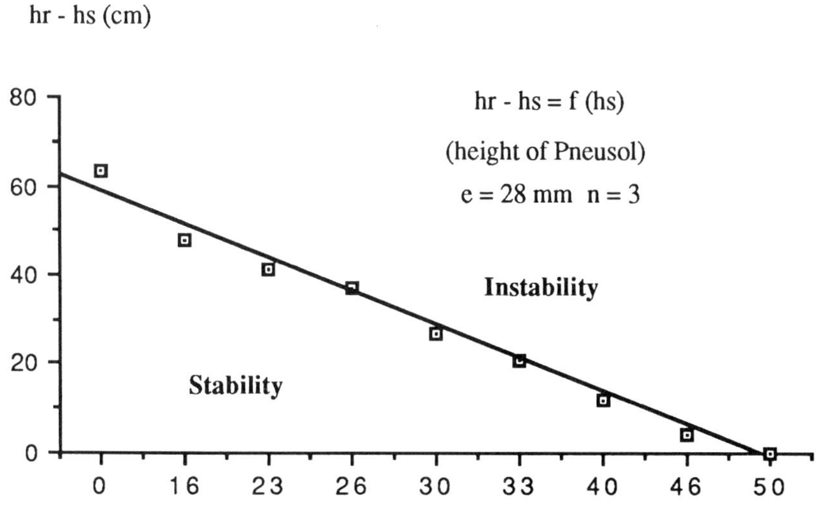

Fig. 8 : Height of Pneusol versus height of sands

References

BAILLY I.C. *et al.* Le Pneusol Recherches et Realisations. *Congres international de la Route de l'An 2000 et de la Securite routiere, Berlin,* 1988.

LAREAL P. and LONG N.T. Le Pneusol reducteur de poussee . *Proc. IV Colloque Franco-Polonais de Mecanique des Sols Appliquee, Grenoble, September,* 1987.

LONG N.T. *Le Pneusol.* Rapport des Laboratoires Serie GT-7, LCPC, Paris, 1985.

LONG N.T. and VALEUX J.C. *Le Pneusol leger.* Rapport des Laboratoires Serie GT37, LCPC, Paris, 1989.

LONG N.T. *Le Pneusol.* Serie Geotechnique GT-44, LCPC, Paris, 1990.

Inclusion pull out tests on small scale reinforced soil wall model: results and prediction method of inclusion strength

I. ALIMI-ICHOLA, Geotechnical Laboratory, Insa-Lyon, N. T. LONG, LCPC, Paris, and P. LAREAL, Geotechnical Laboratory, Insa-Lyon, France

1. Introduction

For internal stability of reinforced soil structure, inclusion characteristics must be chosen correctly. The friction between the soil and the inclusion is an important factor for stability. Inclusion length which gives sufficient resistance strength, must be known. This strength depends on the inclusion type, the soil mass height and the state of inclusion surface.

Pull out tests on inclusions were used to examine the influence of these parameters on inclusion resistance strength. We used fine sand from Fountainbleau.

We present some results of these tests in small scale models and suggest a design method for all results of pull out tests.

2. Model description

The model used in LCPC by Alimi (ref. 1) and Long is a container 1.20 m long, 0.64 m wide and 0.60 m high. The reinforcement consists of thin strips of steel (0.2 mm thick). With some of these reinforcements, gauges were used in order to measure the stress distribution along the inclusions. We built the mass of sand in several layers. These layers were put on in low density or in high density by damping. Different heights of sand mass above the reinforcements were studied.

In this model, investigated parameters were:

the length L and width b of the inclusion
the density γ and the height of sand mass
the stress distribution along the inclusion
the group effect

Another model has been used in INSA-LYON to study large strips of steel (ref. 2).

3. Pull out force

During the extraction test, the inclusion is pulled out at a constant rate (approximately 1 mm/mn). The strength at the head and corresponding displacement are recorded. The maximum pull out force is obtained from the force-deplacement curve. This maximum is the extraction strength, which must be equal to the friction force and is given by

$$T = f^* p L \gamma h$$

where

$p = 2b$ for flat inclusions

f^* apparent soil-inclusion friction coefficient

b width of the flat inclusion

L length of the inclusion

γ the density of soil mass

h height of soil mass above the inclusion

Several tests were done to examine relationships between extraction strength of an inclusion and different parameters (ref. 2).

It was noticed that the extraction strength increases with the inclusion length. The relationship which governs the increasing strength with respect to the length depends on the compacted state of the sand fill above the inclusion. When the density of the sand is low ($D_R = 38\%$), we have a linear relationship between extraction strength and the length. For high density ($D_R = 94\%$) we have a parabolic relationship.

In compacted sand, for a value of b higher than 15mm, we find that the ratio T/b decreases when width of the inclusion increases.

With reinforcements fitted with strain gauges, the tensile force distribution and the friction coefficient along the inclusion are obtained.

The tensile force distribution along the inclusion shows that for a particular value of force applied to the inclusion, the whole length of the inclusion is affected by the force when the sand is loose and the distribution curve is linear. When the sand is compacted, this curve is not linear and when the tensile force is low, only a part of the inclusion length is affected. In this case, the friction force is progressively mobilized along the inclusion when the tensile force increases.

Most of the results are described in another paper (Alimi et al., ref. 2).

4. Pull out force predicting method

According to test results, we noticed that the stress along the inclusion can

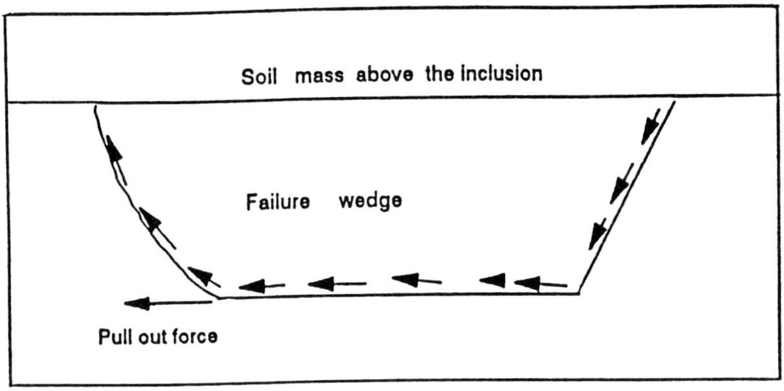

Fig. 1. Failure wedge diagram

be described by a shear motion under and above the inclusion. We have a bidimensional behaviour like a continuous sheet. So the pull out force can be defined for a unit width.

Tests carried out on a group of inclusions, show that the pull out force of an inclusion on a group is less than the pull out force of an isolated inclusion. The pull out force value of the isolated inclusion is found when the space between the layer of inclusions is over three or six times the inclusion width. We concluded that the granular motion above the inclusion did not affect the soil mass over the height of three or six times the inclusion width.

Failure mechanism description

During the pull out test, a granular motion is observed is behind the inclusion, sand grains fall down and their flow defines a concave line which leaves the surface of soil mass and arrives at the inclusion end; in front of the inclusion, sand grains move up and their flow defines a convex line which leaves the inclusion tip and grows up to the soil mass surface. These two lines define the failure block in the soil mass (fig. 1). This block tends to follow the inclusion during the pull out test.

When the height of the soil mass above the inclusion is important, the failure lines in front and behind the inclusion did not reach the soil mass surface. The soil motion is limited to a part of the soil mass height.

Pull out force calculus from test results

To establish a design method, we can use the relationships revealed by test results or we can use a soil mass failure diagram. This soil block makes it possible to define a kinetically deplacement field to evaluate the upper value

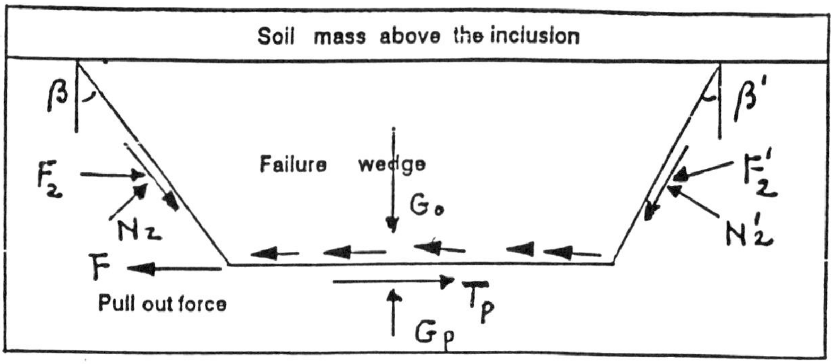

Fig. 2. Simplified failure wedge diagram

of the strength which initializes the motion. In fact these two methods concern the evaluation of the ratio $F/\gamma h^2 b$ (inclusion strength/soil mass weight/soil mass height).

To evaluate this ratio, we consider the simplified failure diagram shown in figure 2. We write the pull out strength as:

$$F = \tfrac{1}{2}K\gamma h^2 b + f\gamma h\, b\, L$$

where f is the soil inclusion friction coefficient.

The coefficient K defines active and passive earth pression behind and in front of the failure wedge. It is evaluated by the formula:

$$K = 2\left(\frac{F}{\gamma h^2 b} - f\frac{L}{h}\right)$$

Relationship between K and the soil mass height

(Fig. 3) shows that, in uncompacted sand, the value of K decreases when the soil mass height increases and becomes fairly constant when the height is over 0.09 m. It is noticed that for a low value of soil mass height, K is overestimated because of a three dimensional effect and for a high value of soil mass height K is underrated, since the soil motion did not affect the whole soil mass height.

When the pull out tests are carried out in compacted sand, the inclusion displacement corresponding to the pull out strength depended on the soil mass height. We show in the figure 4, the variation curves for constant inclusion displacement. We noticed that K decreases as in uncompacted sand but its value increases when the inclusion displacement increases. It is seemed that K becomes constant for a high value of soil mass height.

Referring to the group effect, we can assume that K value does not vary

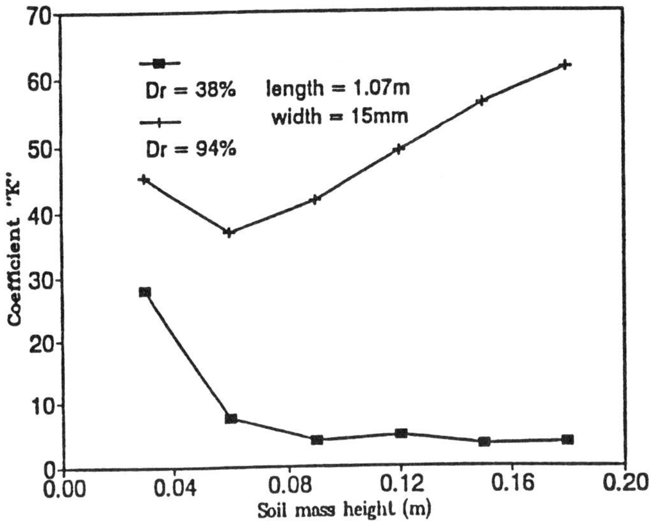

Fig. 3. Influence of the soil mass height on K

when all the soil mass height is affected by the inclusion motion and decreases when the soil mass height exceeds this influenced zone. So the part of soil mass height which did not participate to the inclusion motion was taken as a surcharge in the inclusion pull out strength.

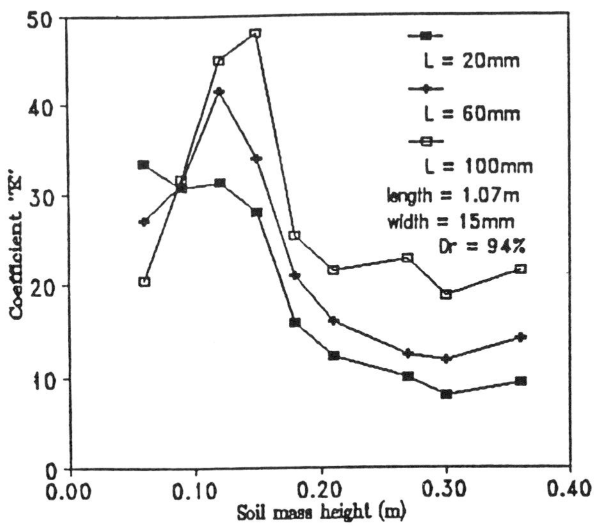

Fig. 4. Influence of the soil mass height on K

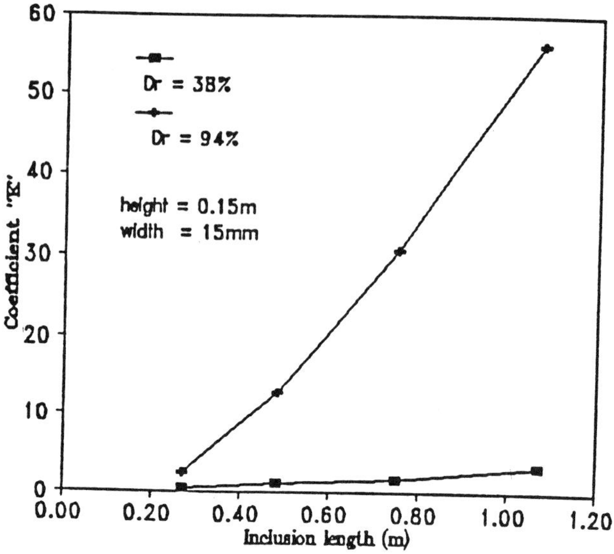

Fig. 5. Influence of the Inclusion length on K

Relationship between K and the inclusion length

As is shown in figure 5, the coefficient K increases with the inclusion length.

The relationship between K and the inclusion length is linear when the density of the soil mass above the inclusion is low. For the compacted sand, the increase of K value depends on the inclusion displacement during the pull out test.

Relationship between K and the inclusion width

Tests carried out in dense sand show that the K value decreases when the inclusion width increases. This behaviour can be explained by the three dimensions effect and by the edge effect. We noticed during our pull out tests that the shear stress is not uniform through the inclusion width.

Evaluation of K from test results

When considering relationships between K and the inclusion length and width, we noticed that the K value depends on the flexibility of the flat inclusion. This flexibility is characterized by the ratio L/b (inclusion length/width). So the variation of the coefficient K is studied as only a function of the ratio L/b, since the influence of the inclusion motion is limited to three times or six times the inclusion width.

We define the relationship between the coefficient K and the ratio L/b as:

Table 1. Variation of α with the soil mass height

h (m)	0.03	0.06	0.09	0.12	0.15	0.18	hc = 3b
h - h (m)		0.015	0.045	0.075	0.105	0.135	Dr = 38%
K	28	7.51	3.94	4.73	3.38	3.55	
α	0.393	0.105	0.055	0.066	0.047	0.05	

$$K = \alpha\left(\frac{L}{b}\right)^3 \quad \text{for compacted sand}$$

$$K = \alpha\left(\frac{L}{b}\right) \quad \text{for loose sand}$$

The coefficient depends on the part of soil mass height which is not influenced by the sand grains motion. As is shown in tables 1 and 2, is fairly constant and the soil mass height affected by the granular motion is greater than four times the inclusion width, when the sand is not compacted; but when the sand is compacted, increases with the soil mass height.

The determination of the relationship between α and the surcharge height will allow K evaluation and the pull out force calculation.

Evaluation of K from kinematically wedge equilibrium

To evaluate K, we consider the failure wedge of the figure 2. The computation of the pull out force can be obtained from the following equations:

$$G_p = G_o + N_2\,Cos\,\beta(tg\beta - tg\varphi) - N_2'\,Cos\,\beta'(tg\beta' + tg\varphi)$$

$$F = T_p + N_2\,Cos\,\beta(1 + tg\beta\,tg\varphi) - N_2'\,Cos\,\beta'(1 - tg\beta' tg\varphi)$$

Table 2. Variation of with the surcharge

h(m)			0.12	0.15	0.18	0.21	0.27	hc = 6b
h -hc(m)			0.03	0.06	0.09	0.12	0.18	Dr = 94%
K	45.287	37.03	41.176	49.209	56.197	61.516	59.76	
$\alpha \times 10^4$	1.248	1.02	1.134	1.356	1.548	1.695	1.646	

G_o is the weight of the failure wedge

$T_p = G_p\, tg\psi$ is the frictional force under the inclusion and $tg\psi$ is soil inclusion friction coefficient.

N_2 and N_2' are related to the height of soil mass affected by the sand grains motion, to β and β′ to the soil internal friction angle. β and β′values depend on the inclusion displacement. That is the displacement necessary to mobilize active and passive earth pressure behind and in front the failure wedge. When we assumed that in front and behind the failure wedge we are in Rankine limit state, the value of β is $\pi/4 + \phi/2$ and the β′ value is $\pi/4 - \phi/2$. In that case N_2 and N_2' have the following expressions:

$$N_2 = \tfrac{1}{2} K_{p\gamma} \gamma r_2^2 \qquad\qquad N_2' = \tfrac{1}{2} K_{a\gamma} \gamma {r_2'}^2$$

where

$$r_2 = \frac{h}{Cos\beta}, \qquad r_2' = \frac{h}{Cos\beta'},$$

$K_{p\gamma}$ and $K_{a\gamma}$ passive and active earth pressure coefficients.

The expressions of F and G_p become:

$$F = T_p + \tfrac{1}{2} K_{p\gamma} \frac{\gamma h^2}{cos\beta}(1 + tg\beta\, tg\varphi) - \tfrac{1}{2} K_{a\gamma} \frac{\gamma h^2}{cos\beta'}(1 - tg\beta' tg\varphi)$$

$$G_p = G_o - \tfrac{1}{2} K_{p\gamma} \frac{\gamma h^2}{cos\beta}(tg\beta - tg\varphi) - \tfrac{1}{2} K_{a\gamma} \frac{\gamma h^2}{cos\beta'}(tg\beta' - tg\varphi)$$

The expression G_p shows that the failure wedge heaves during the pull out test and β must vary from $\pi/4 + \phi/2$ to φ when F_2 becomes parallel to the pull out direction. This expression is positive when the ratio L/h is such as:

$$\frac{L}{h} \geq \frac{1}{2}\Big(\frac{K_{p\gamma}}{Cos\beta}(tg\beta - tg\varphi) + \frac{K_{a\gamma}}{Cos\beta'}(tg\beta' + tg\varphi) - (Sin\beta + Sin\beta')$$

Using $\phi = 30^o$ for loose sand and $\phi = 45^o$ for dense sand, the internal friction angle found during shear box tests, the ration L/h must be greater than 3 when the sand is not compacted, and 10 for dense sand. Supposing that h is the height of the soil mass concerned by the soil grain motion, the expression of the ratio L/h limit value allows the choice of the minimum inclusion length to mobilize earth pressure during the pull out test. This result particularly affects a dense sand in which the friction force is progressively mobilized along the inclusion.

When the soil grain motion ..... not reach the soil mass surface, we introduce the coefficients K_{pq} and K_{aq} to evaluate the coefficient K.

The following expressions give:

$$K_{pq} = tg(\pi/4 + \varphi/2)\, e^{2\epsilon\, tg\varphi} \text{ where } \epsilon = \frac{\pi}{4} + \frac{3}{2}\varphi \qquad\qquad K_{aq} = \frac{Cos\varphi}{1 + Sin\varphi}$$

Table 3. Variation of K with the internal friction angle

					h=0.09m		h=0.15m	
φ °	K p	Kqp	K a	Kaq	Kth	F(N)	Kth	F(N)
20	4	3.703	0.49	0.7				0.86
25	7	6.011	0.406	0.627				16.2
30	13.6	10.63	0.333	0.577		5.78	7.492	43.03
35	30	20.814	0.271	0.521	31.52	45.81	29.34	106.87
40	82	46.442	0.217	0.466	163.2	184.22	100.17	313.58
45	280	122.53	0.172	0.414	777.9	828.14	400.91	1191.59

In that case N2 and N2′ are given by the following relations

$$N_2 = \frac{1}{2}\gamma h^2\left(\frac{x^2 K_{p\gamma}}{cos^2\varphi} + \frac{2x(1-x)}{cos\varphi}K_{pq}\right)$$

$$N_2' = \frac{1}{2}\gamma h^2\left(\frac{x^2 K_{a\gamma}}{Cos^2(\pi/-\varphi/2)} + \frac{2x(1-x)}{Cos(\pi/4-\varphi/2}K_{aq}\right)$$

when the surcharge is given by q = γh(1 - x) and β equal to ϕ.

Tables 3 shows the values of the coefficient K for different values of ϕ, when we consider $K_{p\gamma}$ as Boussinesq coefficient. We noticed that positive value of the coefficient K is obtained when the internal friction angle is greater than 30°. This means that this predicted method is valid when the soil mass above the inclusion is compacted to obtain an high internal friction angle. This table shows that we can obtain the experimental value of pull out force (h = 0.15 m, F = 182.25 N, h = 0.09 m F = 56.6 N; γ = 17.3 KN/m^3, b = 15 mm) for ϕ value between 35° and 40°.

This ϕ value is lower than the value obtained with the shear box test on dense sand. It is possible that we cannot obtain, in our models, a ϕ value as high as that in shear box.

5. Conclusion

When the pull out force is considered as the friction force, it requires a fictitious soil inclusion friction coefficient greater than 1, particularly for dense sand. We propose a computation method which allows the use of the soil inclusion friction coefficient value obtained during the shear box test.

With our test results, we show that it is possible to evaluate the pull out force by assuming the soil mass motion above the inclusion. The computation results are in agreement with the test results and give the pull out force evolution trend with the characteristic parameters of the inclusion and the soil mass. It is noticed that we can find the minimum inclusion length value which initializes the granular motion and give a high soil inclusion friction coefficient. Other test results are needed to confirm these computation methods.

References

1. ALIMI I (1978) Critère de choix de matériaux de terre armée. Etude de la longueur d'adhérence. Thèse L.C.P.C. Paris.
2. ALIMI-ICHOLA I., LONG N. T., BORDEAU Y., LAREAL P. (1991) Soil inclusion friction tests on small and large scale models, 9th Asian Regional Conference on soil mechanics and foundation engineering. December 1991, Bangkok.

Discussion

SESSION 6

Professor A. McGOWAN, University of Strathclyde

As a consequence of previous discussion, three points have arisen:

(a) There are more possible ultimate and serviceability limit states in reinforced walls than in conventional walls.
(b) The level of control/knowledge of soil conditions is not a constant, nor is it always predictable.
(c) Should the present design practice of applying the factor of safety to the reinforcement, soil and overall be maintained, or should it be brought in line with conventional wall design, with the factor of safety applied to the fill or the in-situ soil only.

Professor I.M. SMITH, University of Manchester

3-D finite element analysis can be used not only for nailed soil but also for built-up walls. This shows the difference between extensible and inextensible reinforcements.[1] The former cannot be analysed by the same method as the latter.

DR. M.M.H. AL-SHAIKH-ALI, DTp Motorway Widening Unit

Durability

I feel that there is a need to define durability. At present, life spans of 30,60 and 120 years are mentioned. However, no author at this conference has suggested a definite figure.

Corrosion protection

There are basically two types of reinforcement: first, steel bars or mesh, i.e. galvanized steel bars or PVC coated bars or reinforcement; second, fibre strands or mesh. The conference has not addressed these points.

Defininition of structures

Reinforced soil can be constructed in three types of inclination.

(*a*) 0-45° to the vertical (geotechnical problem)

(*b*) 45-70° to the vertical (grey area)
(*c*) 70-90° to the vertical (structural problem)

I assume that all the cases considered at this conference deal with vertical retaining structures (type (c) above); what about the other two types of reinforced soil ((a) and (b) above)?

Soil types
I agree with Professor McGowan that soil type in all types of reinforcement is paramount. I should like to see a specification for each type of soil to suit a certain type of reinforcement, or vice versa.

Soil nailing
It has been mentioned that soil nailing has not been widely used in the UK. It is my view that performance tests - both short term and long term - should be carried out, especially as all these modern technologies are being considered in the motorway widening industry.

N.T. LONG, Laboratoire Central des Ponts et Chaussees (LCPC)

Pull-out tests on reduced scale models are very difficult to interpret because the respect of similitude laws is difficult. To date, all our guides and recommendations have used data obtained in full scale experiments or from real structures.

Generally speaking, Pneusol, a combination of used tyres and soils, not only helps to consume stocks of old tyres, but also improves the mechanical properties of soils.

So far, more than 200 structures have been built in France, 12 in Algeria, six in the USA, two in Jordan, one in Rumania and one in Rwanda. Some of these structures are lightweight embankments and support roads or highways. The first one (Cannes-Mandelieu) was built in 1985, and was opened to traffic immediately after. Since this date, we have not observed any deflection or cracking on the pavement of the highway. The overall behaviour of all structures is very good.

With regard to the burning of used tyres, when it occurs in the open air, the only countermeasure is to pour sand on to the fire. So according the specialists (Michelin, Goodyear), it is impossible to burn buried tyres.

G. DENBY, GeoEngineers

The soil nail wall completed in October 1991 for the Swedish Hospital Parking Garage set a national record for the deepest excavation using this method.

The project consists of a seven-level underground parking garage. The maximum excavation depth was 74 feet below street level and consisted of a sloping berm at the top and a 68 ft high soil nail wall. The project is located in the Capitol Hill area of Seattle. Several old masonry buildings are located across the streets from the site.

The soil nail wall was installed using the top down method which includes construction of the basement wall as the excavation progresses. The wall thickness is typically 12 in.; however, the first level of the wall on the eastern half of the site and the elevator shaft portion on the south side are 20 in. thick. These portions of the wall required special support which was developed and used for the first time on this project.

Subsurface conditions at the site are glacially overconsolidated and are relatively uniform, as shown in Table 1.

Table 1

Soil type	Blow count N	Depth: ft
Fill	5 - 15	0 - 5
Sandy till	40 - 100+	5 - 30
Advance outwash	40 - 100+	30 - 100

The fill consists of a loose to medium dense silty sand, with zones of medium stiff sandy silt. The sandy till consists of a medium dense to dense silty sand with occasional gravel and the advance outwash of sand and gravel with a trace to some silt and occasional cobbles. The water table is at a depth of approximately 100 ft. These conditions were considered to be ideal for the soil nail shoring system.

A typical section for the deepest portion of the wall is shown in Fig. 1. The steel bar sizes varied from mild steel to 1.25 in. high tensile steel Dywidag bars. Nail lengths varied from 25 ft at the bottom of the wall to 46 ft for the deepest wall section. The maxiumum wall pressure required to counteract the soil nail force on the wall side of the design failure plane was 580 psf.

Shotcrete support

During installation, the soil nail wall is temporarily undermined for each increment of the excavation. Local experience with the soil nail method indicates that shotcrete wall sections up to 12 in. thick may be temporarily supported in favorable conditions by a combination of adhesion between the soil and the shotcrete wall, and the restraint provided by the soil nail. For the 20 in. wall, even though conditions were considered to be favourable, it was decided that additional support of the shotcrete would be required. Three options for shotcrete support were considered: addition of short nails; addi-

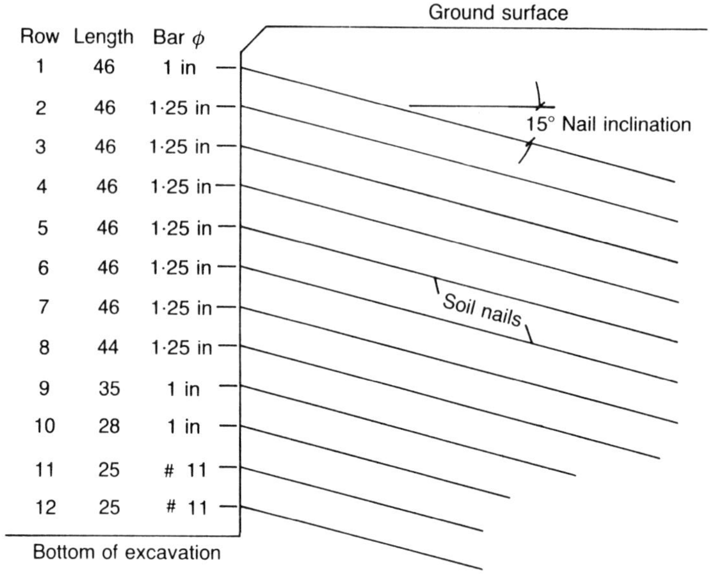

Fig. 1. Soil nail design for deepest wall section

tion of rebar cages to nails; addition of 'strut' nails.

The addition of short nails reduces the vertical load on each nail; however, it relies primarily on the passive resistance of the soil below the nails at the face of the wall. This soil may be disturbed during excavation.

The addition of short rebar cages to the nails at the wall face increases the moment capacity of the soil nail to carry the shotcrete load; however, the success of the option depends more heavily on the passive resistance of the soil below the nails which may be disturbed during excavation.

The strut nail option consists of installing short nails at an angle of about 45 degrees to the vertical and supporting the shotcrete load in compression. The strut nails act in combination with the production nails to form a simple truss which does not depend on the passive resistance of the soil below the nails. The strut nails are located between the production nails and are connected horizontally by the wall. A detail of the strut nail is shown in Fig. 2 along with the resolved forces from the shotcrete load in the production and strut nails. The additional tension component from the shotcrete load is applied at the wall end of the production nail and typically does not cause an increase in the maximum load in the production nail. The load distribution for the shotcrete load component is also shown in Fig. 2.

The strut nail option was selected on the basis of its superior structural mechanism to support the 20 in. shotcrete wall during construction. The strut nails were spaced at 12 ft horizontal centres in the top row of the east half of

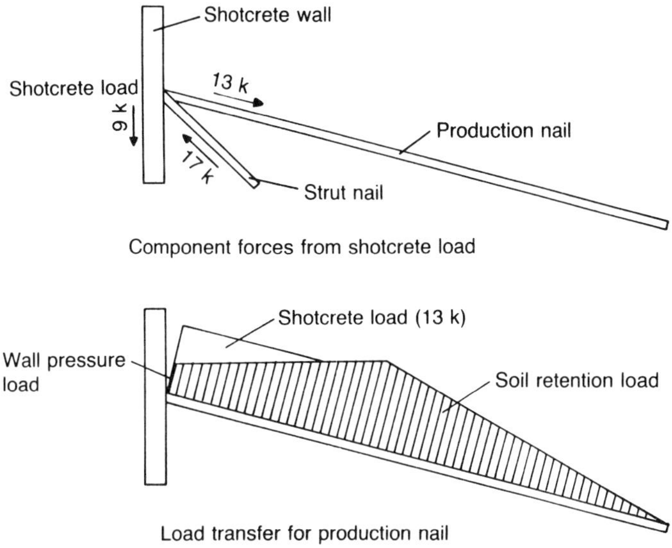

Figure 2: Details of strut nail forces.

the excavation, and at 18 ft horizontal centres in each row for the elevator shaft portion of the wall. The strut nails were 12 in. in diameter and were reinforced with 10 ft long no. 11 mild steel bars.

Face instability

Minor caving of the face started to occur as soon as the outwash soil was encountered in the excavation. The caving appeared to be associated with clean pockets of sand and gravel and seams of organic silt and typically involved less than a cubic yard of material. Below a depth of about 45 ft caving occurred at several places almost on a daily basis and became progressively more frequent with depth. The caving resulted in the formation of voids which extended up behind the shotcrete wall. Further caving up behind the wall appeared to be impeded by arching of the soil between the nails. The typical void dimensions ranged from 5 ft to 25 ft in length, were up to 42 in. deep, and extended up behind the wall a maximum distance of 8 ft. The worst cave occured along the bottom row on Minor Avenue where a 40 ft long by 3.5 ft deep by 4 ft high void developed behind the wall. The caves typically stabilized after the initial collapse; however, some of the caves continued to ravel for a few hours.

Three possible causes of the caving were identified: pockets of cohesionless soil; high in-situ lateral stress; increased vertical stress due to mobilization of the reinforced soil.

All of the observed caving occurred in cleaner layers of the sand and gravel outwash which have cohesion and are more likely to cave than layers with more silt. The high in-situ lateral stress is typically present in a glacially overconsolidated deposit and would also result in spalling when lateral support is removed. It is also possible that the mobilization of the nails in the reinforced soil behind the wall increases the vertical stresses in the soil mass at the base of the wall, which could produce spalling at the face. The relationship of the caving with movements of the walls is discussed in the next section.

Wall performance

Instrumentation of the excavation consisted of an optical vertical survey grid in the adjacent streets, horizontal and vertical optical survey of the top and middle of the walls at about 17 ft centres, and inclinometers 3 - 6 ft behind the mid- points of the four sides of the excavation. The street survey consisted of grid lines at 50 ft centres around the excavation and points about 10ft, 20 ft, 40 ft and 60 ft behind the top of the walls.

The maximum wall movements occurred at the tops of the four walls. The inclinometer profiles for the west and south walls indicated uniform rotation of the wall and maximum movements of 0.5 in. and 0.65 in. respectively. In contrast, the inclinometer profiles for the east and north walls showed distinct breaks in the profile, and maximum movements were greater - 1.65 in. and 1.0 in. respectively.

The inclinometer profiles for five stages of the excavation along the east wall are presented in Fig. 3. The first two profiles indicate rotational outward movement of the wall about the current base of the excavation. At a depth of about 57 ft, the third profile showns a break in the curve. The profile remains near-vertical above the break, indicating translational movement. Below 57 ft the movement reverts to rotational. This movement is considered to be consistent with the mobilization of the nails in the reinforced soil behind the wall.

The maximum horizontal wall movements for the excavation varied from 0.5 to 1.65 in. at the end of construction the maximum wall movement occuring along the east wall, as shown in Fig. 3. This movement is equivalent to 0.0019H where H is the height of the excavation, and is consistent with the average value of 0.002H concluded by Clough and O'Rourke (1990) in their survey of in-situ constructed walls.

The maximum vertical settlements of the east and north walls were 1.25 in. and 0.85 in. respectively. The movements extended behind the excavation a distance of about 40 ft. No offsets were observed in the cracks in the shotcrete that was flashcoated on to the small berm behind the top of the wall, indicating that the wall and retained soil moved together.

A plot of the horizontal and vertical surveyed displacements at the top of the east wall along with the inclinometer data is presented in Fig. 4. All three

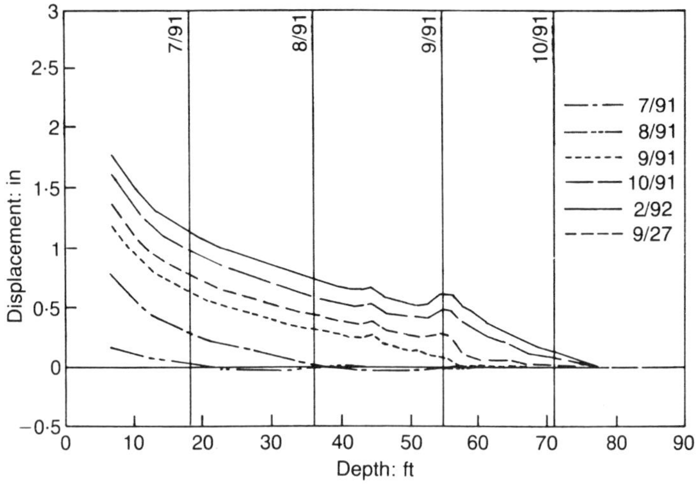

Figure 3. Slope inclomenter profiles for east wall

displacement records indicate a consistent linear pattern (within the scatter of the data) between the horizontal and vertical movement of the top of the wall and the depth of the excavation. The inclinometer data show more displacement than the survey data, most of which is accounted for by the fact that the inclinometer was installed before excavation and the top of wall survey point was installed only after the inital lift of the wall was constructed. Of interest is the fact that the vertical movement is similar to the horizontal movement. This is different from the conclusion reached by Clough and

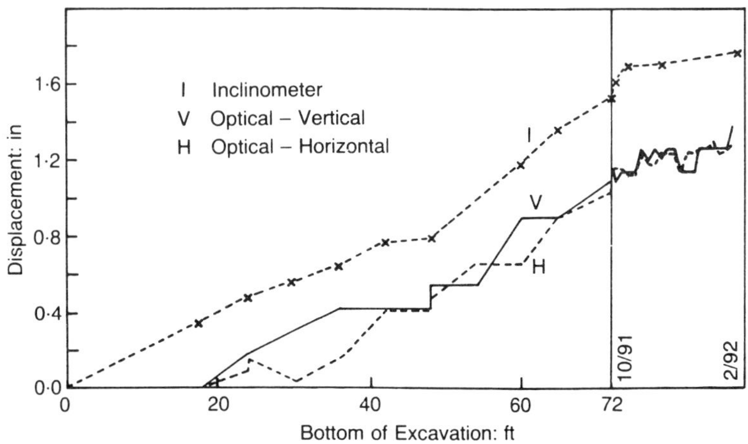

Figure 4. Top of wall displacement

O'Rourke (1990), who indicated that the vertical is typically 25% less than the horizontal movement for in-situ walls.

Referring back to the break in the inclinometer profile discussed in Fig. 3, the additional vertical movement along Minor Avenue may be considered to be related to slippage along the failure plane and mobilization of the nails. There was an inability to correlate the specific spalling events that occurred in the last few rows of the excavation with the vertical movement data: however, it is considered that the spalling is in part caused by an increase in vertical stress due to the downward movement of the reinforced soil necessary to mobilize the nails. This means that the downward movement contributed to the spalling and not vice versa.

All three measurements indicate relatively small displacements on the order 0.1-0.2 in., for four months after the excavation was completed.

Conclusions

The soil nail wall system performed as expected and as predicted by case history data. The maximum displacement of the top of the wall D is in good agreement with the published relationship $D = 0.002H$, where H is the height of the excavation.

Caving of the vertical face became increasingly severe as the excavation depth increased. In the writer' s opinion, the caving was aggravated by the increase in vertical stress owing to the downward movement of the reinforced soil to mobilize the nails.

Caving is a serious problem for a soil nail wall during construction because of the potential for loss of adhesion between the wall and the soil. This may be an inherent weakness of the soil nail system in adverse conditions.

The strut nail concept, introduced for the first time in this excavation, was successful and is recommended where there is concern regarding support of the shotcrete wall during excavation.

References

SMITH, I.M. and SEGRESTIN P. Int. Conf. on Reinforced Earth structures, Kyushbu, Nov. 1992.

CLOUGH G.W. and O'ROURKE T.D. Movements of In-situ walls. Proc. Design and performance of earth retaining structures, Cornell University, Ithaca, New York, 1990, 439-470.

Base stability and ground movement prediction for excavations in soft clay

T.D. O'ROURKE, Cornell University, USA

Introduction

Base stability is of fundamental importance for the evaluation of braced excavation performance in soft clays. Questions about bottom stability have a direct effect on safety and the margin of confidence in design against catastrophic failure. Evaluation of base stability plays a key role in assessing the amount of ground movement generated by plastic deformation of clay at and below the base of excavation. In summarizing measured settlements adjacent to excavations, Peck (1969) recognized that ground movements are affected significantly by the extent of plastic yielding at depth. Accordingly, he distinguished two zones of potential settlement adjacent to braced cuts on the basis of the geometry and dimensionless stability number associated with basal clays. Clough and co-workers (Mana and Clough, 1981; Clough *et al*, 1989) developed a methodology for predicting horizontal wall displacements for deep excavations on the basis of factors of safety against base failure. Henkel (1971) used upper bound plasticity analysis to account for the influence of plastic yielding in basal clays on the apparent earth pressures borne by the overlying support system.

In this Paper, a method is proposed for evaluating the factor of safety against base failure which accounts for the reinforcing effects of wall penetration below the base of excavation. The derivation of the factor of safety is provided, and the assumptions supporting the derivation are tested relative to the measured deformations of excavation walls in the field. Factors of safety are calculated for several deep excavations, which had been near or at failure, and compared with the factors of safety calculated for these cuts according to current stability evaluation procedures.

Simplified methods for estimating horizontal wall movements are reviewed, and a movement prediction method is proposed on the basis of plasticity principles and the factor of safety against base failure. Applications of the simplified predictive techniques are illustrated by means of a case history. Finally, the predictive methodologies for ground deformation are examined in context of a general approach for the estimation of ground movement and its effects on structures adjacent to braced and tied-back excavations.

Factor of safety for base stability

One of the most widely used methods for assessing the factor of safety for base stability of excavations in clay was

proposed by Bjerrum and Eide (1956). Their model for soil behaviour assumed that the unloading caused by excavation was analogous to the performance of a single large footing, such that the bearing capacity expressions put forward by Skempton (1951) could be applied directly to the calculation of a base stability factor. A variation of this approach was used by Eide *et al* (1972) to account for the increase in base stability associated with shear resistance mobilized along concrete diaphragm walls which extend into clay beneath the base of an excavation. These contributions have been very helpful in quantifying the effects of excavation geometry and undrained shear strength on base stability. It is important to recognize, however, that the preponderance of case histories used by Bjerrum and Eide to validate their base stability approach involved shafts with incrementally advanced wall support, free standing cuts, and excavations with walls exhibiting little or no penetration below their bases. The Eide, et al approach takes account of vertical load reduction through shear transfer along the concrete diaphragm walls, but does not consider horizontal deformation nor the lateral resistance afforded by the wall.

Derivation of factor of safety

To account properly for the influence of the excavation wall on base stability, it is necessary to model the basal failure mechanism by means of wall and soil deformations which are compatible and mutually consistent with the assumptions of the base stability, or bearing capacity, factors. Consider the braced excavation in clay and associated wall shown in profile in Figure 1. Soil beneath the lowest support level has been excavated to generate a maximum incremental displacement, δ_m, of the wall and soil. It is assumed that the soil and wall are on the verge of failure. The height of the wall is, therefore, the critical height, H_c, at which failure occurs. If the length and width of the excavation are large relative to the height, then a Pradtl zone of plastic deformation can be assumed for which the following expression for conservation of energy applies:

$$\gamma H D \delta_m = (2 + \pi) D \, S_{ub} \, \delta_m + U \qquad (1)$$

in which γ is soil unit weight, S_{ub} is the representative undrained shear strength of the basal clay, U is the elastic strain energy developed by flexure of the wall, and the other symbols are given in the figure. Unless clay anisotropy is accounted for explicitly, it is recommended that S_{ub} be determined from direct simple shear tests (DSS) or vane shear tests. Eqn. 1 can be

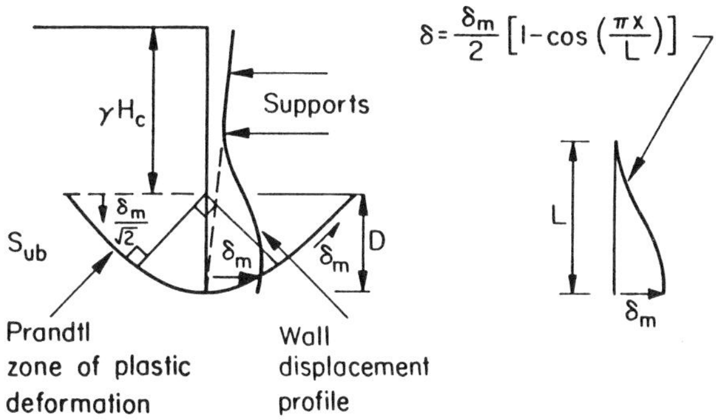

Fig. 1. Cross-section of deformed wall: soil displacements and wall movement profile

rewritten as:

$$\frac{\gamma H_c}{S_{ub}} = (2 + \pi) + \frac{U}{D\ \delta_m\ S_{ub}} \tag{2}$$

For an excavation with width and length of some fraction or moderate multiple of H_c, the base stability factor, N_c, proposed by Bjerrum and Eide (1956) is substituted in Eqn. 2 to yield:

$$\frac{\gamma H_c}{S_{ub}} = N_c + \frac{U}{D\ \delta_m\ S_{ub}} \tag{3}$$

Letting $N_{OR} = (\gamma H_c)/S_{ub}$, one obtains

$$N_{OR} = N_c + \frac{U}{D\ \delta_m\ S_{ub}} \tag{4}$$

in which N_{OR} is the stability factor of Overall Resistance to basal soil failure, including the flexural resistance of the wall. The factor of safety, FS, of an excavation of height H, is:

$$FS = \frac{N_{OR}\ S_{ub}}{\gamma H} \tag{5}$$

If one accounts for surcharge, q, adjacent to the excavation:

$$FS = \frac{N_{OR}\ S_{ub}}{\gamma H + q} \tag{6}$$

The elastic strain energy, U, caused by wall flexure can be evaluated conservatively by considering two dimensional bending (e.g., Timoshenko, 1958) in the form:

$$U = \frac{EI}{2} \int_o^L \left(\frac{d^2\delta}{dx^2}\right) dx \qquad (7)$$

in which E and I are the elastic modulus and moment of inertia per unit length of wall, L is the wall length beneath the lowest level of support, and x is distance relative to the lowest level of support.

Assuming that the wall displacement profile, as illustrated in Figure l, is given by:

$$\delta = \frac{\delta_m}{2} \left[1 - \cos\left(\frac{\pi x}{L}\right)\right] \qquad (8)$$

and substituting Eqn. 8 into Eqn. 7, one obtains:

$$U = \frac{\delta_m^2 \pi^4 EI}{16L^3} \qquad (9)$$

The yield moment per m of wall, My, can be equated to the maximum wall curvature, as:

$$M_y = \left(\frac{d^2\delta}{dx^2}\right)_{max} EI \qquad (10)$$

Combining Eqns. 8 and 10, one obtains the maximum incremental displacement:

$$\delta_m = \frac{2M_y L^2}{\pi^2 EI} \qquad (11)$$

In this way, the maximum displacement of the soil is compatible with the formation of a plastic hinge in the wall. The wall behaves as a mechanism, rotating about the lowest level of support, to result in unrestrained plastic soil deformation and rotation of the wall.

Combining Eqns. 4, 9, and 11, one obtains:

$$N_{OR} = N_c + \frac{M_y \pi^2}{8DL\ S_{ub}} \tag{12}$$

Eqn. 12 is the base stability factor for a free end wall which has been installed to some depth in clay below the excavation base, but not within an underlying firm stratum. Using the derivations in Appendix A, the overall base stability factor for a fixed end wall is:

$$N_{OR} = N_c + \frac{M_y \pi^2}{2DL\ S_{ub}} \tag{13}$$

This stability number applies for a wall installed into an underlying firm stratum with sufficient penetration to result in full moment restraint.

In some cases, a sheet pile or soldier pile wall is driven to refusal on bedrock. Under these conditions, the wall tends to slide along the bedrock interface without full moment restraint. Using the derivations in Appendix A, the overall base stability factor for a sliding end wall is:

$$N_{OR} = N_c + \frac{9\ M_y \pi^2}{32DL\ S_{ub}} \tag{14}$$

In summary, three expressions for the overall base stability factor, N_{OR}, are given by Eqns 12, 13, and 14 for free, fixed, and sliding end conditions, respectively.

Comparison with field observations

The theoretical derivations in the previous section can be checked in two ways. First, the assumption associated with Eqn. 8, that the wall displacement profile is represented by a cosine function, can be checked by comparing actual incremental displacement profiles with the postulated cosine function. Second, the factors of safety associated with excavations that are known to have been near or at failure can be calculated using Eqns 12, 13, or 14. These factors of safety also can be compared with the safety factors derived by Bjerrum and Eide (1956) to determine the relative improvement, if any, associated with accounting for the influence of the excavation wall.

Inclinometer data for four different excavations, listed in Table 1, were analyzed to determine the incremental displacement profile generated by excavation of the last lift of soil beneath the lowest level of supports. The four excavations summarized in the table were undertaken in soft clays in San

Table l. List of excavations in soft clay used for analysis of Wall Deformation

Excavation	H m	Wall Characteristics	Supports
Embarcadero III, San Francisco, CA	13.7	PZ 32 sheetpile driven approx. 13.7 m below base of excavation	3 levels of cross-lot braces
Levi Strauss Bldg., San Francisco, CA	13.7	PZ 32 sheetpile driven approx. 6 m below base of excavation	2 levels of cross-lot; braces; 2 levels of rakers
Marriott Hotel Chicago, IL	8.5	PZ 27 sheetpile driven approx. 4 m below base of excavation	2 levels of rakers
Sears Tower Chicago, IL	13.7	2.5-m-thick concrete diaphragm wall to approx. 7.5 m below base of excavation	3 levels of rakers

Francisco and Chicago. The excavation walls, comprising either steel sheet-piles or a concrete diaphragm wall, were installed into firm strata at depth. Accordingly, all four cases represent fixed end walls.

Figure 2a illustrates the method of incremental displacement calculation. Inclinometer measurements performed after the excavation of the bottom soil lift were subtracted from those performed before excavation to develop an incremental displacement profile as illustrated in the figure. The thicknesses of soil lifts varied from 2.1 to 3.5 m, with maximum incremental displacements of approximately 50 to 100 mm. The displacements were plotted in dimensionless form and compared with the postulated dimensionless displacement profile for fixed end conditions shown in Figure 2a. Comparisons of the measured and postulated cosine curves are provided in Figures 2b and 2c. In general, the patterns of displacement compare favourably with the cosine function.

The actual wall energy associated with a measured displacement profile was determined by numerically integrating the curvatures associated with the profile over the length, L, from approximately the lowest support level to the fixed base of the wall. The measured wall energy for each inclinometer station was compared with the energy predicted for the same incremental displacement, assuming a cosine function. Measured versus predicted wall energies are plotted in Figure 3. The best fit straight line through the data is

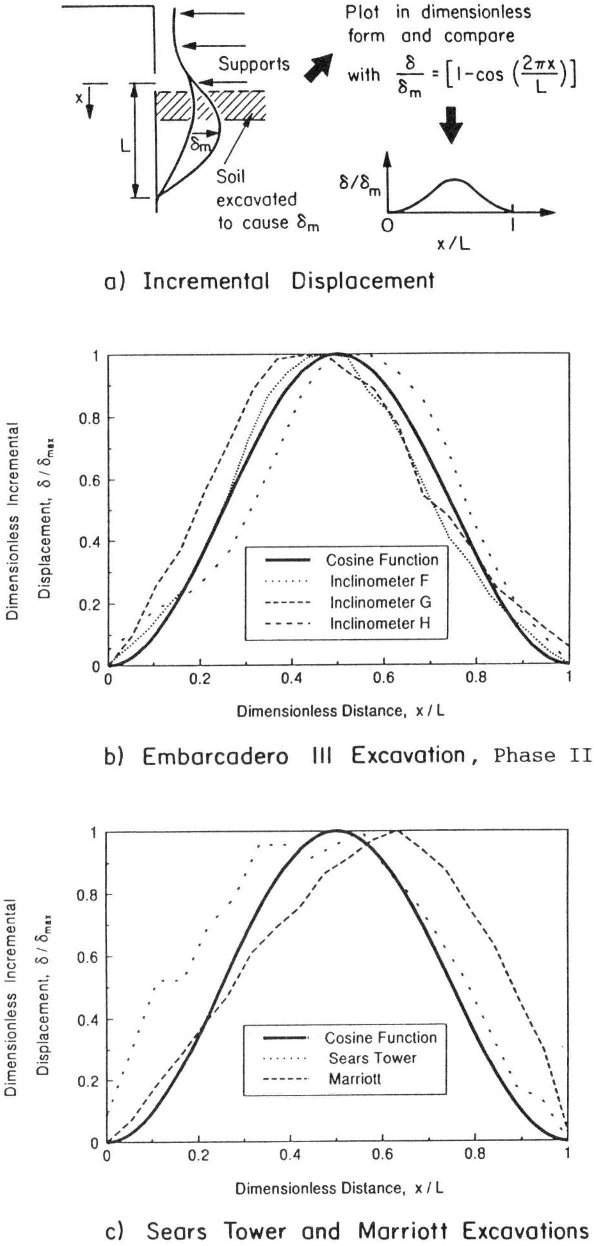

Figure 2. Incremental displacement profiles plotted in dimensionless form relative to cosine function for various excavations

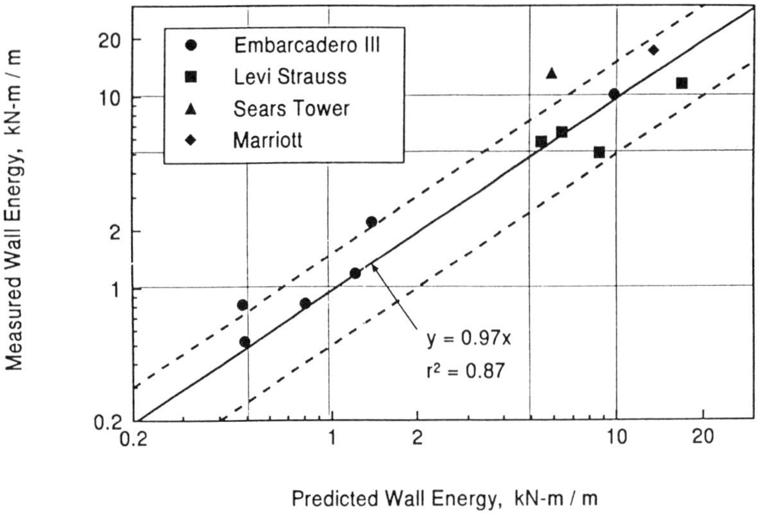

Figure 3. Measured versus predicted elastic strain energy for various excavation walls

shown. Regression analysis indicates a slope of 0.97, when the regression is constrained through the origin, with a coefficient of determination, r^2, of 0.87. A slope near one and a high r^2 are strong evidence that the cosine function is a good representation of the deflected wall shape.

Factors of safety using Eqn. 5 were calculated for seven excavations, each of which either was close to base failure or had failed by excessive deformation and wall movement below the base of excavation. The pertinent data associated with each excavation are summarized in Table 2, with a key diagram for the appropriate dimensions in Figure 4. For comparison, the factor of safety, FS, calculated on the basis of N_c, as proposed by Bjerrum and

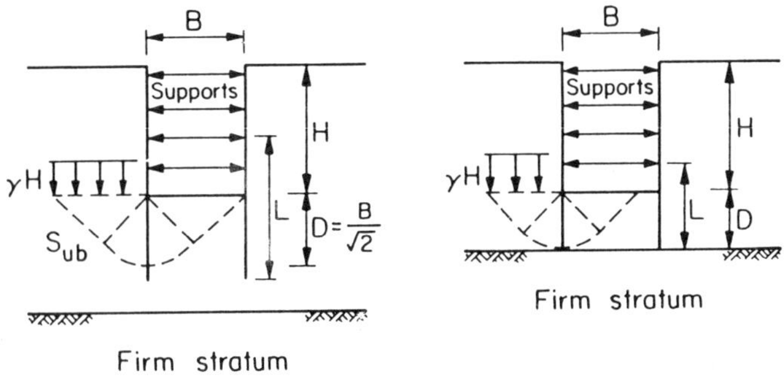

Figure 4. Key diagram for excavation dimensions listed in Table 2

Table 2. Summary of dimensions, soil properties, wall characteristics, and stability factors for soft clay excavations at or near failure

Excavation	H m	B m	D m	L m	γ kN/m^3	S_{ub} kN/m^2	Wall	End Conditions	σ_y Mn/m^2	S cm^3/m	N_c	N_{OR}	FS based on N_c	FS based on N_{OR}
Vaterland 1 (NGI, 1962a)	11.0	11.0	5.3	6.7	18.6	23.5	sheet-pile	between fixed & sliding	270	2050	6.6	8.4-9.8	0.76	0.96-1.13
Vaterland 2 (NGI, 1962b)	11.0	11.0	6.5	7.5	18.6	20.2	sheet-pile	between fixed & sliding	270	2050	6.5	7.9-8.9	0.75	0.91-1.02
Vaterland 3 (NGI, 1962c)	10.2	11.0	7.8	4.5	18.6	34.3	sheet-pile	free	270	2050	6.3	6.9	1.26	1.38
Mexico City (Rodriguez & Flamand, 1969)	9.0	7.0	5.0	6.3	12.0	12.2	sheet-pile	free	270	1688	7.4	8.7	0.95	1.12
Davidson Ave. 1 (Clough & Reed, 1984)	9.1	7.6	5.4	6.4	16.3	19.2	sheet-pile	free	415	2060	6.5	8.1	0.83	1.04
Davidson Ave. 2[a] (Clough & Reed, 1984)	7.3	7.6	5.4	12.7	16.3	19.2	sheet-pile	free	415	2060	6.2	7.1	0.87	1.00
Washington, DC (Swanson & Larson, 1990)	9.1	21.3	6.7	9.8	17.0	23.0	soldier pile & lagging	fixed	249	1672	5.5	6.9	0.81	1.02

a Note: refers to failed section of excavation

Symbols: σ_y - yield stress of steel S - section modulus of wall FS - factory of safety

Eide (1956) is listed next to the factor of safety calculated with N_{0R}. As mentioned previously, the Bjerrum and Eide approach does not account for the effects of wall penetration below the excavation base. In six of seven cases, FS based on N_c is well below one, even though no visible failure of the wall was documented for Vaterland, Mexico City, and the general conditions along Davidson Ave. In contrast, FS based on N_{0R} is near and slightly above one for these same excavations, thereby rendering values more consistent with actual observations.

Of special interest is the Davidson Ave. excavation, which has been described by Clough and Reed (1984). The Davidson Ave. 1 excavation refers to the general conditions observed along this braced cut, for which maximum horizontal wall displacements varied from 100 to nearly 250 mm. Although no failure was observed under the general conditions, the horizontal displacements were exceptionally high for a 9.1 m-deep cut. These conditions are consistent with FS = 1.04 as predicted with N_{0R}, but inconsistent with FS = 0.83, as predicted with N_c.

At one location along Davidson Ave., referred to as the Davidson Ave. 2 excavation, a slope failure inside the cut exposed a section of the wall at depth. The slope failure resulted in inward deformation of the wall, which caused the second level wales to rotate, popping the braces at this level upward. As a result, the wall was propped only by the upper level of struts. Progressive wall deformations ensued, and the excavation was temporarily backfilled as a stabilizing measure. Inclinometer measurements on one side of this section of the cut showed a cumulative horizontal movement of 400 mm.

It seems reasonable to assume that this section of the excavation had experienced failure. Using the depth, H, at the time of failure and the sheet-pile length from the remaining level of struts to the bottom of the sheets, L, the FS calculated on the basis of N_{0R} is 1.0, which is in excellent agreement with observations.

Another case of braced cut failure has been documented by Swanson and Larson (1990) for a 9.1 m-deep, soldier pile and lagging excavation in soft clays in Washington, DC. It was reported that, at 9.1 m, lateral squeezing of soft clay caused 3 m of lateral soil movement, with soldier piles bowing as much as 450 mm into the excavation. A FS = 1.02 is calculated on the basis of N_{0R} for this situation, which again is consistent with the failure conditions observed.

The Vaterland excavations (NGI, 1962a, b, c) have long been of special interest. Although no failure was observed at these excavations, horizontal wall displacements as high as 200 mm were measured. In addition, extremely high strut loads were measured, which exceeded those predicted with conventional earth pressure envelopes (e.g., Terzaghi and Peck, 1968) by a considerable margin. The sheet-piles at Vaterland 1 and 2 (NGI, 1962a and b) were driven to refusal on bedrock, and their end conditions were intermedi-

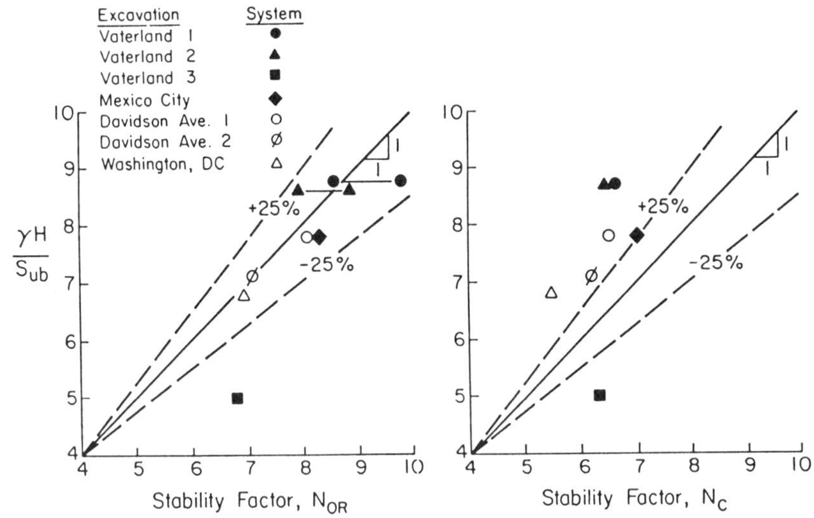

Figure 5. Comparison of base stability numbers for excavations at or near failure

ate between fixed and sliding. Accordingly, the FS based on N_{OR} was calculated for both fixed and sliding end conditions, and is expressed as a range in Table 2. The resulting range is in much better agreement with the observed excavation performance than the extremely low FS values calculated on the basis of N_c.

Figure 5 shows the values of $\gamma H/S_{ub}$ plotted relative to N_{OR} and N_c for the excavations under consideration. For excavations at or near failure, the data points should plot near and slightly below a line with slope equal to one. This is apparent in the N_{OR} plot. In contrast, most data points in the N_c plot are located well above a slope of unity. This grouping of data indicates that FS assessments, which do not account for wall penetration effects, tend to underestimate the actual safety margin.

Simplified procedures for estimating lateral wall movement

In this section, two simplified procedures are described for estimating the maximum horizontal wall movements of braced excavations. The first procedure to be discussed was developed by Clough and co-workers (Mana and Clough, 1981; Clough, *et al*, 1989). The second procedure is proposed as an outgrowth of the base stability considerations in the previous sections of this Paper.

System stiffness approach

With the aid of field measurements and finite element simulations,

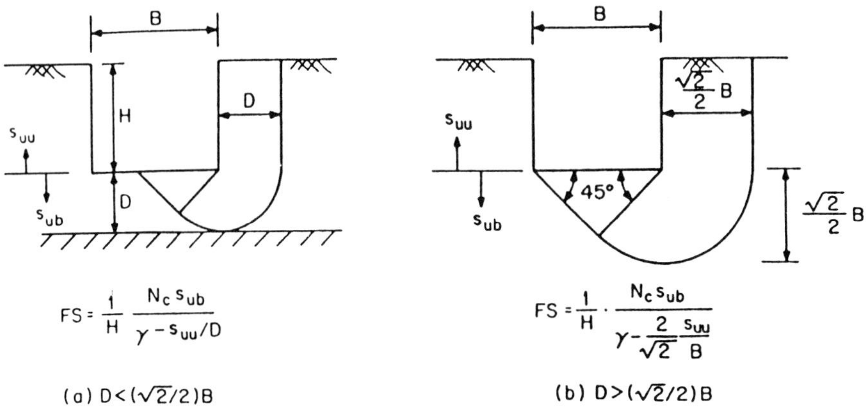

Figure 6. Excavation geometry and soil strength parameters for factors of safety against basal heave (after Terzaghi, 1943)

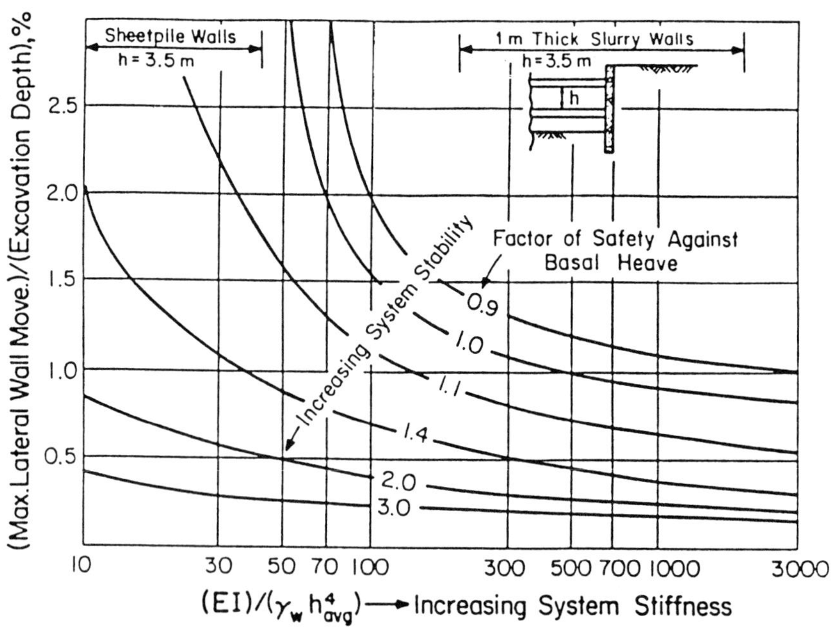

Figure 7. Lateral wall movement as a percentage of excavation depth versus system stiffness (after Clough, et al, 1989)

Clough, *et al* (1989) proposed a semi-empirical procedure for estimating movements at excavations in clay in which the maximum lateral wall movement, δ_{LM}, is evaluated relative to FS and a system stiffness term. The FS is defined according to Terzaghi (1943), as shown in Figure 6. It should be emphasized that the FS is used as an index parameter in this context, and is not intended to provide a direct measure of base stability. The system stiffness is defined as a function of the wall flexural stiffness, average vertical separation of supports, and unit weight of water, which is used as a normalizing parameter. Figure 7 presents δ_{LM}, plotted relative to system stiffness for various FS. The family of curves in the figure is based on average conditions, good workmanship, and the assumption that cantilever deformation of the wall represents a small fraction of the total movement. Clough, *et al* (1989) recommend a method to account for cantilever wall movement in which cantilever displacements are estimated and added directly to those predicted by means of the figure. It is recommended that modification parameters given in previous work (Mana and Clough, 1981) be used, if more accurate adjustments for time effects, preload, and excavation width are required. Other references are suggested (e.g., Clough and Hansen, 1981) to ascertain the effects of clay anisotropy.

Base stability approach

In studies of diaphragm walls (Bolton and Powrie, 1988) and bridge abutments (Bolton, *et al*, 1990), simplified expressions for strains in the basal clays affecting these structures have been proposed and related to the mobilized shear strength of the soil. In this way, strain and associated deformation of the clay can be linked to its mobilized shear strength to make rational decisions about the potential soil displacements associated with a given wall or abutment height. As Bolton, *et al* (1990) point out, such simplifications are consistent with the conventions of structural design, in which idealizations of performance have allowed engineers to design structures which deform tolerably.

This line of reasoning can be adapted to the prediction of wall deformation for excavations in soft clay through the use of the base stability factor, N_{OR}. In applying these concepts, it is assumed that the maximum horizontal wall deformation for a cut in soft clay results predominantly from the accumulation of incremental basal movements, each of which occur beneath the lowest support level at each stage of the excavation. This process of deformation is consistent with measured performance in the field, provided that the upper levels of support are stiff and are installed early in the excavation and support sequence. O'Rourke (1981) has shown that cantilever movements can play an important role in the development of soil and wall displacements, especially when the upper level supports are flexible or are installed after significant

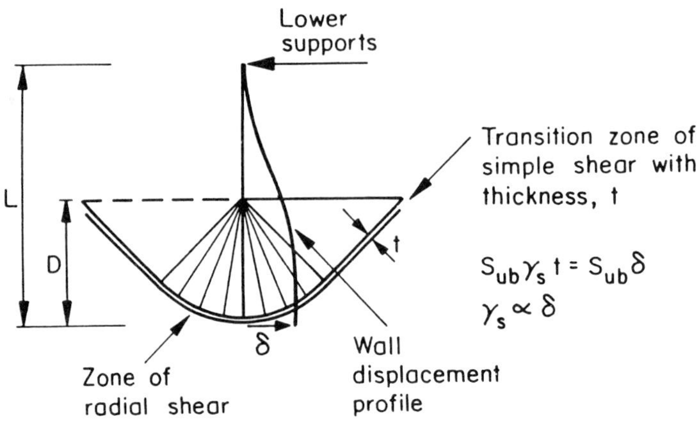

Fig. 8. Shear strain and displacement associated with plastic deformation of basal clay

excavation has occurred. Modifications to the predictive procedure can be made to reflect the influence of cantilever wall movements, following the recommendations of 0'Rourke (1989), Clough, *et al* (1989), and Clough and 0'Rourke (1990).

Figure 8 shows a cross-section of the Prandtl zone of soil deformation for a braced excavation near its critical height, H_c. Shear strains are assumed to be concentrated within a narrow transition layer of thickness, t. The mode of deformation is simple shear, and the rate of energy dissipation within the transition zone is:

$$S_{ub} \gamma_s t = S_{ub} \frac{\delta}{t} t = S_{ub} \delta \qquad (15)$$

The energy dissipation is seen to be independent of t, which indicates that the simple shear strain, γ_s, will correlate directly with the shear displacement, δ.

Following the approach of Bolton and coworkers (Bolton and Powrie, 1988; Bolton, *et al* 1990), the mobilized shear stress, (S_{ub}) mob can be related directly to the soil and wall deformation δ, by means of normalized DSS test results, as illustrated in Figure 9b. The test results have been plotted by-expressing (S_{ub}) mob as a fraction of $(S_{ub})_{max}$ and γ_s as a fraction of a reference strain, γ_{sm}, which is taken as the strain at which (S_{ub}) max first occurs.

From Equation 5, the maximum shear strength of the clay can be related to the excavation geometry and wall characteristics as:

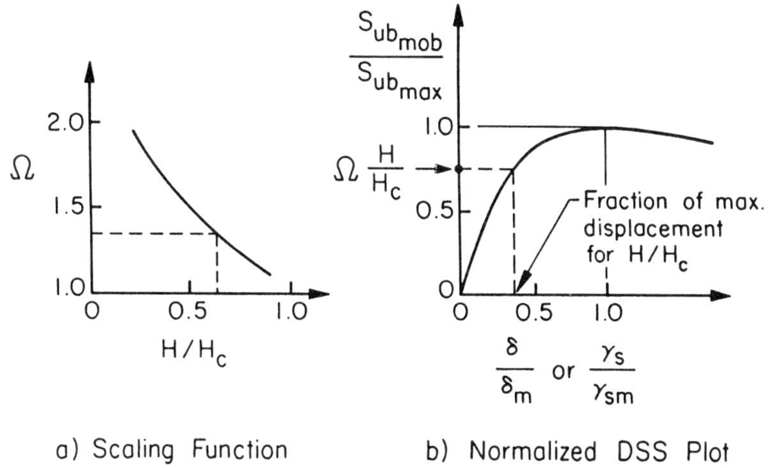

Figure 9. Simplified procedure for estimating wall movement on the basis of scaling factor and normalized DSS plot

$$(S_{ub})_{max} = \frac{\gamma \ H_C}{N_{ORc}} \tag{16}$$

in which N_{ORC} is the stability factor for the critical height, H_C. In a similar fashion, the clay shear strength mobilized at any depth, H, where $H < H_C$, can be expressed as:

$$(S_{ub})_{mob} = \frac{\gamma \ H}{N_{ORH}} \tag{17}$$

in which N_{ORH} is the stability factor pertaining to H. Combining Eqns.16 and 17 results in:

$$\frac{(S_{ub})_{mob}}{(S_{ub})_{max}} = \Omega \ \frac{H}{H_C} \tag{18}$$

in which Ω is a scaling factor which accounts for the variation of shear strength and stability factor from H_C to H. The scaling factor can be calculated for each height of excavation as:

$$\Omega = \frac{(S_{ub})_C \ N_{ORC}}{(S_{ub})_H \ N_{ORH}} \qquad (19)$$

in which $(S_{ub})_C$ and $(S_{ub})_H$ are the representative undrained shear strengths pertaining to the critical height, H_c, and any other height of interest, H, respectively.

The prediction of maximum horizontal wall movement is illustrated by Figure 9. The critical height of the excavation, H_c, is determined first. The scaling factor, Ω, at the wall height of interest, H, is selected according to Figure 9a. The product of Ω and H/H_c then is used to obtain the corresponding fraction of maximum wall displacement, δ/δ_m, as illustrated in Figure 9b. This fraction is multiplied by the best estimate of maximum wall deflection at the critical depth, using Eqn. 11 for free end conditions, or Eqns A.8 and A.10 for fixed and sliding end conditions, respectively (see Appendix A).

When evaluating the maximum wall deflection, the value of L must reflect the entire excavation and support process, not just the final incremental movement before critical height is reached. It seems reasonable to choose L for this calculation as the distance, or wall length, from the highest level of support to the bottom of the wall or depth at which the wall is fixed by embedment in a firm stratum. In this way, the entire length over which deflections accumulate is accounted for, and the selection of L is consistent with the cumulative displacements measured for actual excavations in soft clay.

Application of simplified predictive methods

In this section, the simplified prediction of maximum lateral wall movement caused only by the excavation and bracing process is illustrated by comparison with actual field measurements. The application of the system stiffness approach given here is essentially the same as explained previously by O'Rourke (1989).

The excavation chosen for the comparative example is the Embarcadero III Project in San Francisco. Construction involved twin excavations, each approximately 43 x 85 m in plan. Figure 10 provides a profile view of the Phase 1 excavation, which was 13.7 m deep and supported by three tiers of preloaded cross-lot braces. Initially, the cut was excavated to a depth of approximately 5 m to clear previous foundations as PZ-32 sheet-piles were installed around the perimeter. About 30% of the sheet-piles were driven into the alluvium at a depth of 13.7 m below the bottom of excavation, and the remainder were driven to elevations roughly 1 to 3 m higher. Dewatering was performed during construction to lower the water table in the upper fill and underlying alluvium.

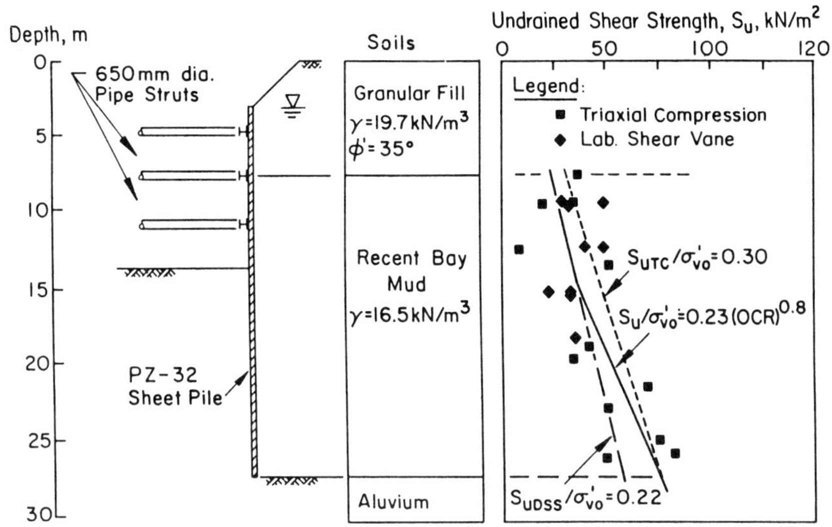

Figure 10. Profile view of embarcadero III phase 1 excavation and soil conditions

Figure 10 also shows the soil profile in which the work was performed. Undrained shear strength data are plotted, based on the results of triaxial compression and laboratory vane shear tests (Harding, Miller, Lawson and Associates, 1971). More recent soils investigations within two to three blocks of the site have disclosed that the Bay Mud has anisotropic strength characteristics (Dames and Moore, 1989). Ratios of undrained shear strength to effective vertical stress for triaxial compression (S_{uTC}/σ_{vo}') and direct simple shear (S_{uDSS}/σ_{vo}') from the more recent work are plotted in the figure. Also shown is a normalized shear strength projection [$S_u/\sigma_{vo}' = 0.23(OCR)^{0.8}$] below a depth of 15 m, which is intended to reflect the average anisotropic strength relationship with overconsolidation ratio, OCR, as determined from one-dimensional consolidation tests on specimens of the Bay Mud (Dames and Moore, 1989). It is postulated that deep well pumping from the alluvium has resulted in some localized preconsolidation of the overlying clays and silts.

Application of system stiffness approach

The complexity of the strength profiles and attendant shear strength determinations leads to questions regarding how FS should be calculated so that the system stiffness plots in Figure 7 can be utilized properly. For this Paper, Peck's method (1943) of calculating an equivalent undrained shear strength of the upper fill was used and combined with the average undrained shear strength in triaxial compression, S_{uTC} for Bay Mud above the base of

Table 3. Summary of shear strength parameters, factors of safety, and predicted lateral wall displacements

Case Number	Upper Shear Strength,[2] S_{uu}, kN/m^2	Basal Shear Strength, S_{ub}, kN/m^2	Factor of Safety, FS	Predicted Max. Wall Movement δ_{LM}, mm
1	35.6	42.0	1.13	200
2	35.6	51.5	1.38	128
3	35.6	57.4	1.54	109

1 - Refer to Figure 6 for definition of parameters
2 - Determined on the basis of Peck's method for calculating equivalent undrained shear strength of granular fill (1943) and triaxial compression data for clay

excavation. This approach provides a rational and consistent basis for determining S_{uu} (see Figure 6).

Below the base of excavation, three separate strategies were pursued to explore the sensitivity of wall movement prediction to variations in the determination of S_{ub} (See Figure 6). In Case 1, the direct simple shear strength, S_{uDSS} was taken as representative of the average anisotropic strength, assuming that the basal clay is normally consolidated (OCR = 1). In Case 2, the direct simple shear strength, S_{uDSS} was calculated on the basis of its relationship with OCR, as illustrated by the middle line projection in Figure 10. In Case 3, the triaxial compression strength relationship was used to estimate S_{ub}. In each case, the weighted average shear strength over the full thickness of the underlying Bay Mud was calculated, and the depth to firm soil beneath the base of excavation, D, was taken as 13.7 m for determining FS.

Table 3 summarizes the shear strength parameters, FS, and predicted wall movements for each case. Maximum lateral wall movements were determined from Figure 7, assuming an average 3.4 m vertical spacing of the support levels.

Figure 11 presents a plot of the final horizontal displacement profiles, measured by five inclinometers at locations shown in the inset diagram. A histogram of the maximum lateral wall movements relative to the predictions for Cases 1 and 2 also is shown. The predictions are in good agreement with the field data, although the Case 1 prediction exceeds the largest horizontal displacement that was measured during this phase of construction.

This case history provides some valuable insight regarding the prediction methodology and behaviour of real walls. First, it is necessary to scrutinize and carefully characterize the shear strength parameters. Anisotropic strength characteristics are likely to be important for soft to medium clays.

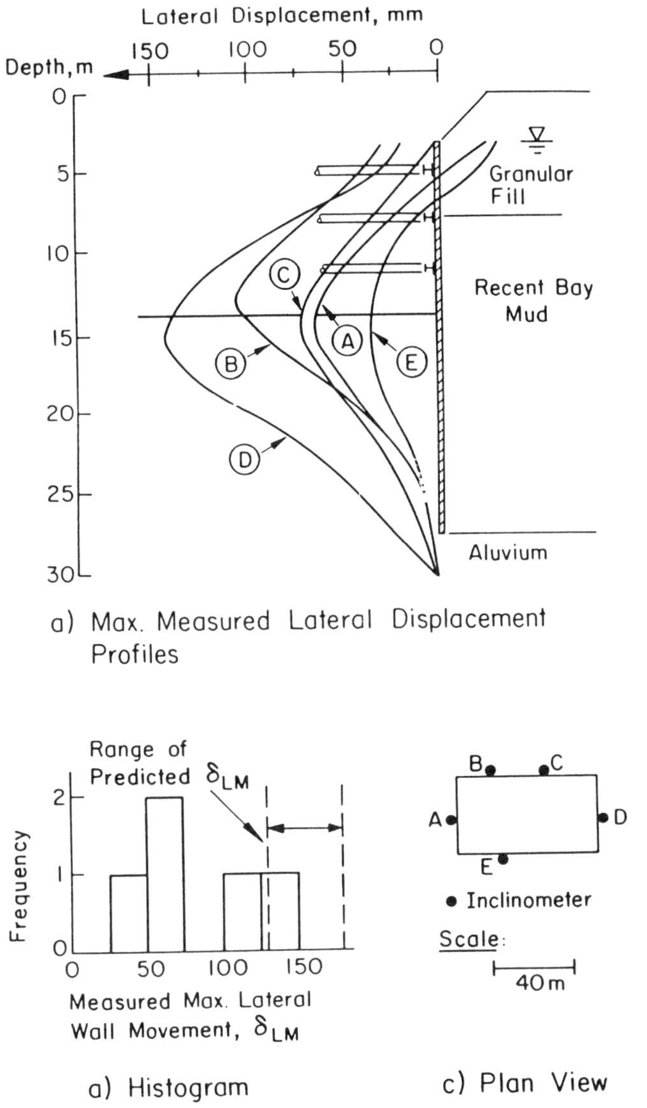

Figure 11. Measured horizontal displacement profiles at the Embarcadero III, phase 1 excavation compared with the predicted displacement

Moreover, the level of preconsolidation will be important. If overconsolidation effects are not accounted for, the wall movement prediction in this example is larger than the measured movement. Second, it is helpful to realize that real wall movements are likely to be variable. In this example, the mean measured displacement was 84 mm, with a standard deviation of 41 mm. The predicted displacement of 128 mm for Case 2 is slightly higher than the mean plus one standard deviation.

The iterative nature of the design process also may be illustrated with this example. If the maximum predicted displacement were deemed inappropriate, then variations in the system stiffness could be explored. For example, if a 900 mm thick slurry wall with braces on 3 m vertical spacing were used, reference to Figure 7 indicates that there could be a two to three-fold reduction in movement. Iterations also may involve alternative interpretations of the shear strength profile. To facilitate the iterative process, a personal computer program has been developed (Clough, *et al*, 1989) to test automatically for the effects of different soil strength parameters, geometries, and support and wall characteristics.

Application of base stability approach

The results of DSS tests for Bay Mud were normalized and plotted in accordance with the method illustrated in Figure 9. The resulting relationship between $(S_{ub})_{mob}/(S_{ub})_{max}$ and δ/δ_{max} is shown in Figure 12a. Scaling factors, Ω, were evaluated in accordance with Eqn. 19, and plotted as a function of H/H_c in Figure 12b. It should be noted that the determination of Ω requires a reassessment of N_{OR} and S_{nb} at each excavation height of interest relative to those at the critical height, H_c. This involves considerable computation, which is facilitated by means of a personal computer employing spreadsheet software. Two cases were evaluated in a manner similar to the application of the system stiffness approach. In Case 1, the direct simple shear strength, S_{uDSS} was taken as representative of the average anisotropic strength, assuming that the basal clay is normally consolidated (OCR = 1). In Case 2, the direct simple shear strength, S_{uDSS} was calculated on the basis of its relationship with OCR, as illustrated by the middle line projection in Figure 10.

Figure 13 shows the factor of safety, FS, determined according to Eqn. 5, plotted with respect to the height of excavation, H. Both Cases 1 and 2 are represented. For each height, the representative S_{ub} was determined at a depth equal to one half the distance from the excavation base to the top of the underlying alluvium. The FS was calculated in each case by assuming that the lowest level of support was 3 m above the bottom of the cut.

For Case 1, FS drops below one at a depth of 16.1 m, which is H_c. For Case 2, the increase in shear resistance relative to Case 1 associated with overconsolidation effects results in FS > 1 for all excavation depths to the alluvium. The FS increases for H > 20.2 m, primarily because the strength contribution of the wall increases as the depth to alluvium and the span between lowest support and fixed wall end decline. For Case 2, the 20.2 m depth, at which the lowest FS occurs, was taken as H_c the purposes of wall movement prediction. The relatively low FS associated with this depth implies that it

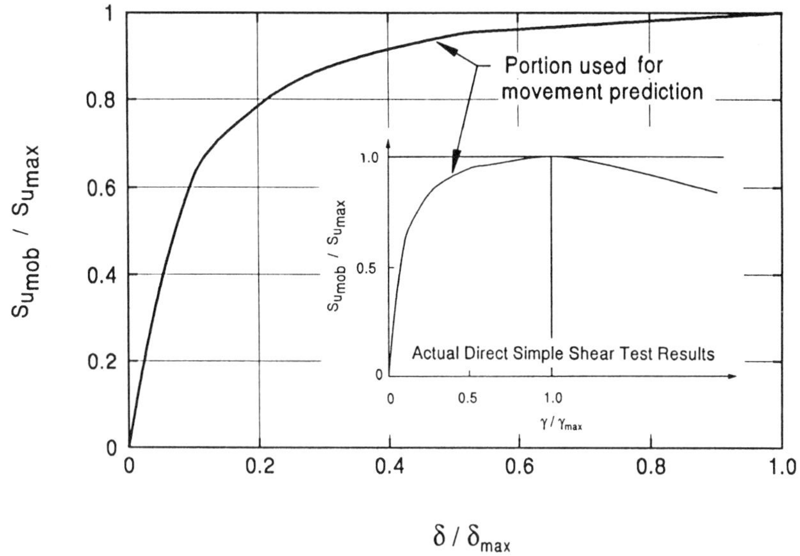

a) Normalized DSS Results

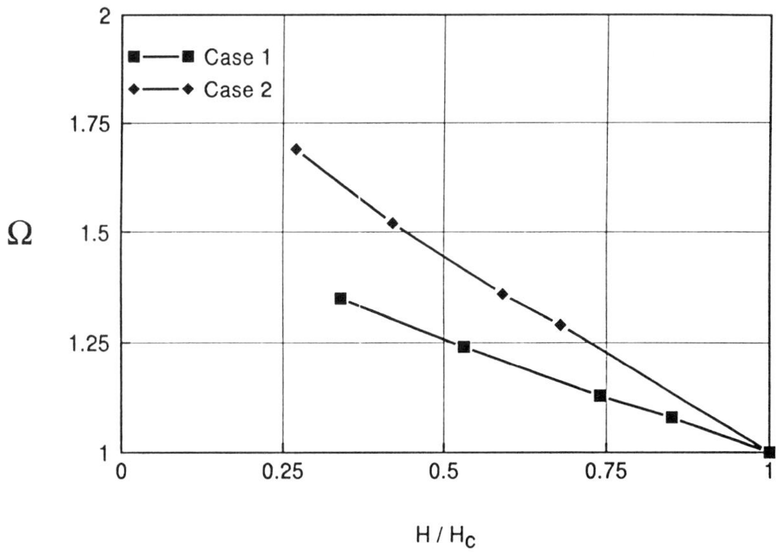

b) Scaling Factors

Figure 12. Normalized DSS test results for recent bay mud and scaling factors

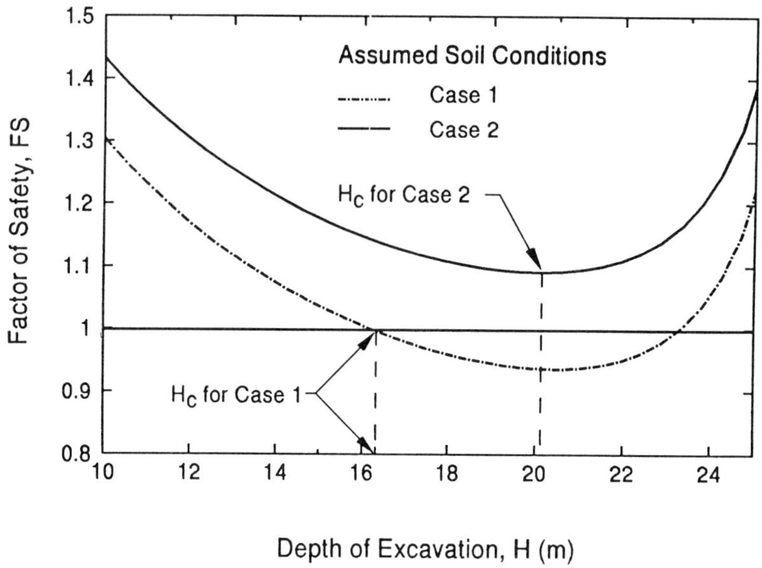

Figure 13. Factor of safety for cases 1 and 2 as a function of excavation depth

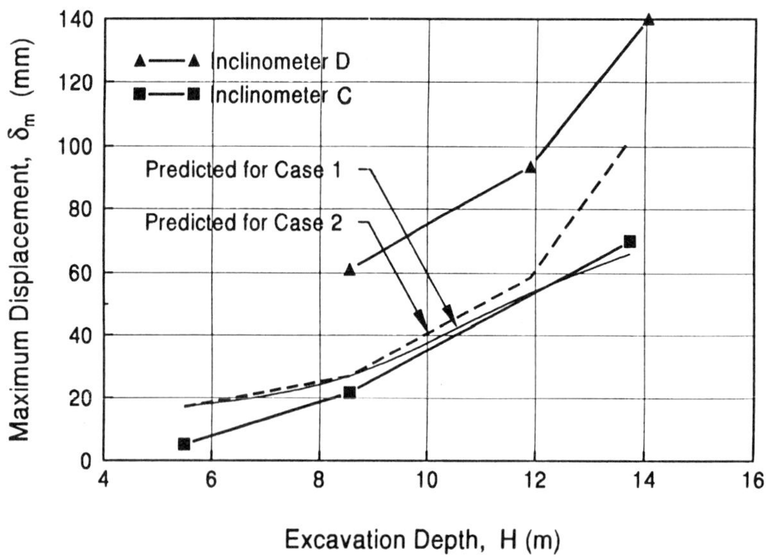

Figure 14. Comparison of predicted and measured horizontal wall displacements for various depths of the Embarcadero III, phase I excavation

will serve as a good reference height for conditions under which $(S_{ub})_{max}$ is developed throughout the underlying clay.

Figure 14 compares the maximum wall displacement predicted for Cases 1 and 2 with the actual maximum wall movements measured at Inclinometers C and D. Because the base stability approach is well suited for estimating displacements at any wall height, it can be applied easily to predict the increase in displacement with excavation depth.

The predicted displacements compare favourably with the measured movements. The predictions for Case 2 compare remarkably well with the average wall displacements for each stage of excavation as determined on the basis of data from all five inclinometers. The Case 2 prediction is only about half the amount of the largest displacement, which was measured by Inclinometer D at the maximum height of excavation. In contrast, the predicted movement for Case 1 is significantly larger than that for Case 2 at the maximum excavation depth of 13.7 m.

Predictions for Cases 1 and 2 are in general agreement until a depth of 12 m, after which the displacement for Case 1 increases more rapidly. The critical depth for Case 1 is close to the final excavation depth of 13.7 m. Accordingly, the final increment of excavation from 12 to 13.7 m for Case 1 generates a mobilized shear strength along the highly curvilinear section of the DSS plot in Figure 12a, resulting in rapidly increasing δ/δ_m. The same increment of excavation for Case 2 conforms more to the initial portion of the DSS plot in Figure 12a, where nonlinearity is not as pronounced, and the resulting δ/δ_m is correspondingly lower.

General approach

The prediction of wall movements, either by simplified procedures or through numerical simulation, needs to be viewed as part of a general process. Although there has been substantial literature written on this subject, the prediction of wall deformation and associated ground movements is only one facet of a more comprehensive and integrated approach which should be adopted by the design engineer. A general design methodology for evaluating and controlling ground movements has been proposed by O'Rourke (1989) which involves three sequential steps.

First, it is necessary to determine what the consequences of ground deformation will be. This means that the structures potentially affected by movement should be identified, characterized, and assessed in terms of permissible levels of deformation. At times, there may be a lack of sensitive structures within the zone of influence. In these cases, considerable economies may be realized by designing excavations for structural stability, rather than the control of movement. In other cases, the presence of historic buildings or very sensitive structures may call for special construction methods and support systems.

Second, it is necessary to identify clearly the different activities involved in the construction at site. Deep excavations often are pursued concurrently with dewatering, removal of existing piles, and the installation of deep foundations. Each of these activities is a potential source of ground movement which must be evaluated and added to the displacements generated by the actual excavation and support process.

Third, ground displacements caused by soil excavation and the installation of wall support are estimated and combined with movements predicted from the other activities. The combined deformations then are compared with the levels of acceptable distortion for adjacent structures. If the predicted deformation exceeds permissible limits, then adjustments can be made in the stiffness of the support system and related construction procedures. It may be shown that ground movements are likely to exceed acceptable thresholds, irrespective of special adjustments, in which case protective measures for the adjacent structures would be in order.

It is imperative to think of the construction project in a broad and systematic way. Sources of ground deformation, not explicitly related to the excavation and bracing process, have been identified and discussed in detail elsewhere (e.g., 0'Rourke, 1981; Clough and 0'Rourke, 1990). They include relocation of utilities, dewatering, removal of existing basement or other subsurface structures, removal of existing piles, deep foundation construction, pre-auguring or driving soldier piles, concrete diaphragm wall construction, auguring or percussive drilling of tiebacks, and the installation of sheet-piles. Soil movements from each of these construction activities may contribute substantially and even dominate the ground deformation eventually related to the work.

The identification of separate activities, such as dewatering and the removal and construction of deep foundations, has practical significance. It encourages the designer to analyze each aspect of the construction, estimate potential movements for each phase, and propose corrective measures for controlling movements which are tailored to a specific activity. These distinctions also encourage the designer to evaluate carefully the entire excavation and support sequence, select appropriate specifications for construction, and promote quality assurance and inspection procedures consistent with the design assumptions.

Concluding remarks

It has been shown that the base stability evaluation for a braced excavation in soft clay can be improved significantly by accounting for the flexural capacity of the wall which extends below the base of the excavation. Derivations, based on plasticity principles and conservation of energy, have been given to show how the flexural effects of the wall can be used to evaluate the factor of safety against base failure through the use of a dimensionless stability factor, N_{OR}. Factors of safety determined by means of the proposed method are in excellent agreement with the performance of real excavations, which either were at or on the verge of base failure. A methodology also is proposed for estimating the maximum horizontal wall movement of an excavation in soft clay, based on an assessment of the base stability factor, N_{OR}. Good agreement is shown between the measured and predicted wall displacements for the case of a deep, braced cut in soft clays and silts.

It is emphasized that the evaluation of base stability and estimation of ground movements caused by the excavation and bracing process is only a part of a more comprehensive and integrated aprlroach that should be adopted by the design engineer. A three-fold process is proposed. First, structures adjacent to the construction site are characterized according to permissible levels of deformation. Second, ground movements are estimated for construction activities performed separately from the excavation and support process. Third, ground movements estimated for the excavation and support process are combined with those from other activities and used to evaluate how adjacent structures are likely to respond in light of the previously established levels of tolerable distortion. The process is iterative and emphasizes constructibility. Concern for the full complement of construction activities is a hallmark of good design, and quite arguably the most important factor which contributes to the control of ground movements.

Acknowledgments

The Author gratefully acknowledges B. Kelly, who helped with the analyses and development of the simplified procedure for wall movement prediction using base stability assessments. Thanks are extended to L. Mayes, who typed the manuscript and A. Avcisoy, who drafted the figures.

References

1. BOLTON, M.D. and POWRIE, W. "Behaviour of Diaphragm Walls in Clay Prior to Collapse", Geotechnique, June 1988. Vol. 38, No. 2, pp. 167-189.
2. BOLTON, M.D., SPRINGMAN, S.M. and SUN, H.W., "The Behaviour of Bridge Abutments on Clay", Design and Performance of Earth Retaining Structures, Geotechnical Special Publication No. 25, P.C. Lambe and L.A. Hansen, Ed., ASCE, New York, NY, 1990, pp. 292-306.
3. BJERRUM, L. and EIDE, O. "Stability of Strutted Excavations in Clay", Geotechnique, 1956, Vol. 6, No. 1, London, U.K., pp. 32-47.
4. CLOUGH, G.W. and HANSEN, L. "Clay Anisotropy and Braced Wall Behaviour", Journal of the Geotechnical Engineering Division, ASCE,

1981,107, No. 7, pp. 893-914.

5. CLOUGH, G.W. and 0'ROURKE, T.D. "Construction Induced Movements of In situ Walls", Design and Performance of Earth Retaining Structures, Geotechnical Special Publication No. 25, P.C. Lambe and L.A. Hansen, Ed., ASCE, New York, NY, 1990, pp. 439-470.
6. CLOUGH, G.W. and REED, R.W. "Measured Behaviour of Braced Wall in Very Soft Clay", Journal of Geotechnical Engineering, ASCE, Jan. 1984, Vol. 110, No. l, pp. l-19.
7. CLOUGH, G.W., SMITH, E.W. and SWEENEY, B. P. "Movement Control of Excavation Support Systems by Iterative Design", Foundation Engineering Current Principles and Practices, Vol. 2, ASCE, New York, NY, 1989, pp. 869-882.
8. DAMES and MOORE, "Factual Report Site Investigation for Completion of Preliminary Design - Phase I, MUNI Metro Turnaround Facility , Vols. I, II, and III, Job No. 185-21503, Dames and Moore, San Francisco, CA, Nov. 1989.
9. EIDE, 0., AAS, G. and JøSANG, T. "Special Application of Castin-Place Walls for Tunnels in Soft Clay in Oslo", Proceedings, 5th European Conference on Soil Mechanics and Foundation Engineering , Madrid , Spain, 1972, Vol. 1, pp. 485-498.
10. HARDING, MILLER, LAWSON, and Associates, "Foundation Investigation", Embarcardero Center Office Building Block 231, San Francisco, California , HMLA, San Francisco, CA, Apr. 1971.
11. HENKEL, D.J., "The Calculation of Earth Pressures in Open Cuts in Soft Clays", The Arup Journal, 1971, Vol. 6, No. 4, pp. 13-15.
12. MANA, A.I. and CLOUGH, G.W. "Prediction of Movements for Braced Cuts in Clay", Journal of the Geotechnical Division ASCE, June 1981, Vol. 107, No. GT6, pp. 759-777.
13. Norwegian Geotechnical Institute, "Measurements at a Strutted Excavation , Oslo Subway , Vaterland 1" , Technical Report, Blindern , Norway , 1962a.
14. Norwegian Geotechnical Institute, "Measurements at a Strutted Excavation, Oslo Subway, Vaterland 2", Technical Report 7, Blindern, Norway, 1962b.
15. Norwegian Geotechnical Institute, "Measurements at a Strutted Excavation, Oslo Subway, Vaterland 3", Technical Report 8, Blindern, Norway, 1962c.
16. 0'ROURKE,T.D., "Ground Movements Caused by Braced Excavations",Journal of the Geotechnical Engineering Division, ASCE, Sept. 1981, Vol. 107, No. GT9, pp. 1159-1178.
17. 0'ROURKE, T.D., "Predicting Displacements of Lateral Support Systems", Proceedings, Seminar on Design, Construction, and Performance of Deep Excavations in Urban Areas, Boston Society of Civil Engineers, Boston,

MA, 1989, pp. 1-36.
18. PECK, R.B., "Earth Pressure Measurements in Open Cuts, Chicago (IL) Subway", Transactions, ASCE, 1943, Vol. 108, pp. 1008-1036.
19. PECK, R.B., "Deep Excavations and Tunnelling in Soft Ground" Proceedings, 7th International Conference on Soil Mechanics and Foundation Engineering, Mexico City, State-of-the-Art Volume, 1969, pp. 225-290.
20. RODRIGUEZ, J.M. and FLAMAND, C.L. "Strut Loads Recorded in a Deep Excavation in Clay", Proceedings, 7th International Conference on Soil Mechanics and Foundation Engineering, Mexico City, Vol. II, 1969, pp. 459-467.
21. SKEMPTON, A.W., "The Bearing Capacity of Clays", Building Research Congress, London, U.K., papers presented in Division l, 1951, pp. 180-189.
22. SWANSON, P.G. and LARSON, T.W. "Shoring Failure in Soft Clay", Design and Performance of Earth Retaining Structures, Geotechnical Special Publication No. 25, P.C. Lambe and L.A. Hansen, Ed., ASCE, New York, NY, 1990, pp. 551-561.
23. TERZAGHI, K., Theoretical Soil Mechanics, John Wiley and Sons, New York, NY, 1943.
24. TERZAGHI, K. and PECK, R.B. Soil Mechanics in Engineering Practice, 2nd Ed., John Wiley & Sons, Inc., New York, NY, 1968.
25. TIMOSHENKO, S., Strength of Materials, Part II, 3rd Ed., Robert E. Krieger Publishing Co., Huntington, NY, 1958.

Appendix A. General derivation of wall energy and deflection expressions

Figure A.l shows a profile view of wall and soil deformation at the critical height, H_c. It is assumed that the wall deformation pattern conforms entirely

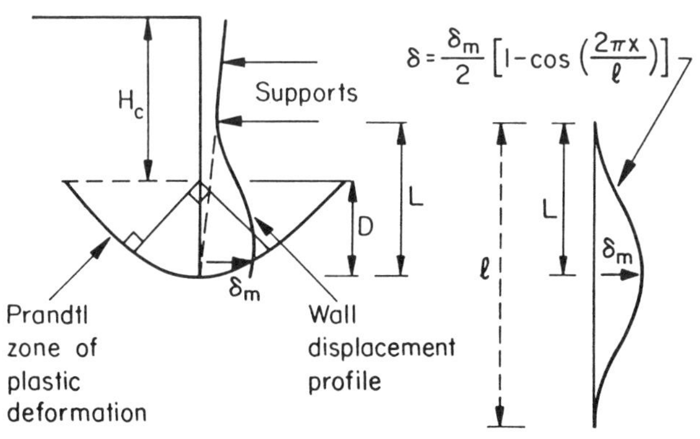

Figure A.1. Wall deformation for free end conditions

or partly to the deformation profile shown in the right side of the figure, as:

$$\delta = \frac{\delta_m}{2}\left[1 - \cos\left(\frac{2\pi x}{\ell}\right)\right] \qquad (A.1)$$

in which δ is the incremental displacement at any distance x from the lowest support level, δm is the maximum incremental displacement, and l is the distance of a full wavelength.

For the free end condition shown, the elastic strain energy, U, caused by wall flexure, is:

$$\Lambda = \frac{EI}{2}\int_{o}^{\ell/2}\left(\frac{d^2\delta}{dx^2}\right) dx \qquad (A.2)$$

Combining Eqns A.1 and A.2 results in:

$$U = \frac{\delta_m \, \pi^2 \, EI^2}{2\ell} \qquad (A.3)$$

The yield moment, My, can be equated to the maximum wall curvature as:

$$M_y = \left(\frac{d^2\delta}{dx^2}\right)_{max} EI \qquad (A.4)$$

Combining Eqns A.1 and A.4, one obtains the maximum incremental displacement:

$$\delta_m = \frac{M_y \, \ell^2}{2\,\pi^2\, EI} \qquad (A.5)$$

Recognizing that l = 2L, one obtains for free end conditions:

$$\delta_m = \frac{2\, M_y \, L^2}{\pi^2 \, EI} \qquad (A.6)$$

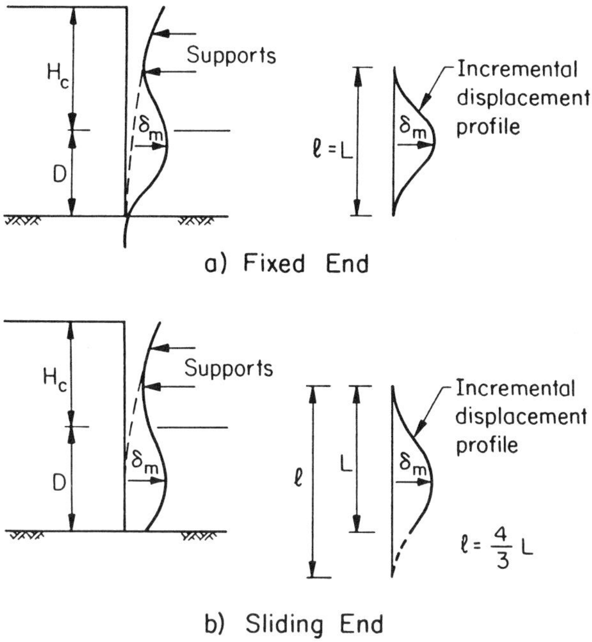

Figure A.2. Wall deformations for fixed and sliding end conditions

Combining Eqns A.3 and A.6, one obtains for free end conditions:

$$U = \frac{\delta_m M_y \pi^2}{8L} \tag{A.7}$$

Similarly, by recognizing that l = L for fixed end conditions and integrating Eqn. A. 2 from 0 to l, one obtains:

$$\delta_m = \frac{M_y L^2}{2\pi^2 EI} \tag{A.8}$$

$$U = \frac{\delta_m M_y \pi^2}{2L} \tag{A.9}$$

The fixed end condition is illustrated in Figure A.2. Also illustrated in the figure is the sliding end condition, which pertains to cases in which sheetpiles or solider piles are driven to bedrock. In response to excavation, the wall deforms by sliding along the bedrock interface. Accordingly, there is shear but little or no moment at this contact, such that the end condition can be modeled as an inflection point at 3/4 of the wavelength, l.

By recognizing that l = 4/3 L for sliding end conditions and integrating Eqn A. 2 from 0 to 3/4 l, one obtains:

$$\delta_m = \frac{8\ M_y\ L^2}{9\ \pi^2\ EI} \qquad (A.10)$$

$$U = \frac{9\ \delta_m\ M_y\ \pi^2}{32\ L} \qquad (A.11)$$

Introduction to Session 7
Gravity structures

Rapporteur: C. J. F. P. JONES, University of Newcastle upon Tyne, UK

Fifteen papers were allocated to the session devoted to Gravity Structures. Of these, 13 can be considered primary papers in respect of the session theme and two are secondary being mainly associated with other sessions in the Conference.

Commonality

There is little commonality between many of the papers in the session, and a number are not primarily associated with gravity structures in the commonly accepted definition. Four of the papers are associated with seismic design or analysis, and the subject of earthquake action can be identified as a sub-theme within the session.

The structural form covered by the papers is extensive, with 5 papers being devoted to tie back walls, 4 to cantilever structures, and only two papers are explicitly related to masonry gravity walls.

All of the papers refer to design or analysis, although the main balance is clearly associated with analysis, only three papers cover the essential subject of costs. The preference of authors for the analytical/design theme is illustrated by the fact that only one paper in the session and the Conference is devoted to maintenance. It is often forgotten that existing retaining walls represent the bulk of all retaining structures and that the evidence of wall failures due to old age and other factors is growing.

The diversity of the papers presented in the session is illustrated in Table 1 which identifies the main theme(s) covered, together with the soil type considered and the analytical procedures preferred by the authors.

Table 1

Author(s)	Structure	Analysis/ Design	Soil	Theme
Bangash Denby Fragaszy USA	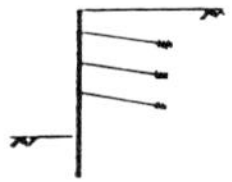	FE parametric	non-linear	seismic design
Bica Clayton Brazil, UK	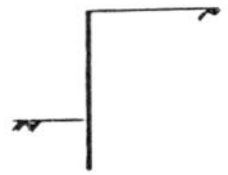	empirical	granular	model tests design
Budhu Neelakahtan Richards USA	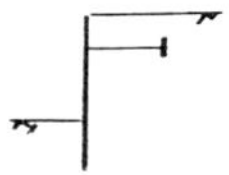	LE	granular	seismic model
Butterfield UK	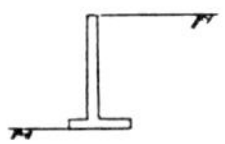	new	sand	design (tilt)
Hagerty Nofal USA	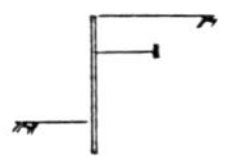	empirical	sand	preliminary design
Hird Djerbib UK	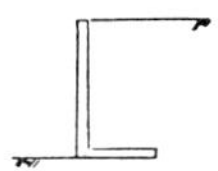	centrifuge	sand	cyclic loading
Hover USA	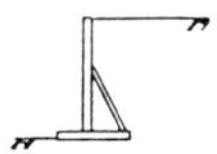	LE		precast structure
Iai Kameoka Japan	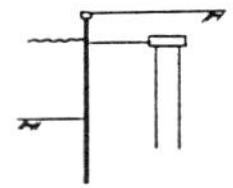	FE empirical centrifuge	sand	seismic design
Jones UK	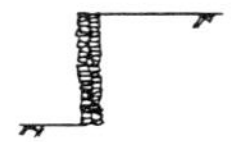	not possible	any	maintenance costs
Lee Butler UK	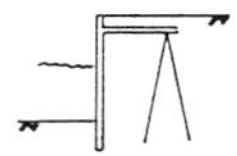	FE empirical	silt sand	design improvement gravel
King McLoughlin UK	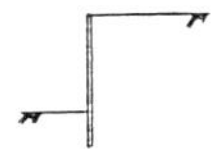	centrifuge	sand	design
Masouri Kastner France	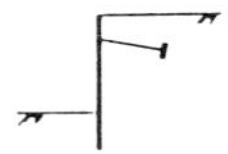	reaction modulus	sand	model test
Moroto Japan	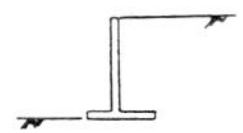	empirical		design costs
Mortensen Denmark		plasticity		
Whitman USA	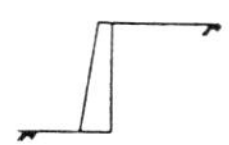	empirical FE centrifuge		seismic effects

Centrifugal model tests of cyclic loading on L-shaped walls retaining sand

C.C. HIRD, University of Sheffield and Y. DJERBIB, Husband and Co., UK

Introduction

Bridge decks are frequently supported on L-shaped walls serving as abutments. Often the deck is seated on a bearing to permit sliding, so that the designer may assume that only a small horizontal force is transmitted from the deck to the wall. In order that the design may proceed with more confidence, the walls are usually backfilled with coarse-grained rather than fine-grained soil. Of course, the designer must attempt to predict earth pressures on the wall which reflect the influence of construction technique, particularly compaction processes, and subsequent traffic loading.

In some cases the deck is used to prop the abutments apart (e.g. Broms and Ingelson, 1971) or the abutment is used to resist horizontal braking forces (e.g. Rathmayer, 1977). No relative sliding is then permissible and the abutment must also be capable of resisting cyclic forces induced by temperature fluctuations in the deck. Such forces may develop unintentionally in an articulated structure due to seizure of the bearing or infilling of the movement joint at the end of the deck.

The analysis of an abutment under the influence of a repeated surcharge (traffic effect) or a cyclic lateral load (temperature effect) would be a challenge for experts in the spheres of constitutive and numerical modelling, and any such analysis would have to be validated before it could be used by a designer. In order to provide data for validation purposes, as well as to improve the understanding of the mechanical behaviour, a series of centrifugal model tests has been conducted. To fulfil their purpose, the tests did not need to be close simulations of prototype events; they can be viewed as events in their own right. On the other hand, it was important to model the essential features of the problem and the styles of loading involved.

Description of tests

The tests were conducted in the geotechnical centrifuge at the University of Manchester Institute of Science and Technology. As is well known, the primary purpose of centrifugal modelling (as opposed to modelling at 1 g) is to create vertical stresses in the model which correspond to those at full scale and, thereby, to permit the soil in the model to display its correct (stress level dependent) deformation behaviour. When modelling soil-structure interaction in the centrifuge, the bending stiffnesses of the model structures must be scaled by a factor of $1/N^4$, where N is the linear scale factor.

Two model walls were constructed for the present tests. With overall dimensions of length 650 mm x height 150 mm x base width 90 mm, the models represented walls 9 m high at full scale (N = 60). They were designed to represent upper and lower bounds to a practical range of stiffnesses, but will be described simply as either "flexible" or "stiff". At full scale the flexible wall stem was roughly equivalent to a reinforced concrete wall 0.4 m thick and the stiff wall stem represented a concrete structure about 3 m thick. The flexible wall was fabricated from 3.2 mm thick mild steel plate and instrumented, on both the stem and the base, with two identical sets of strain gauges positioned symmetrically 50 mm from the centreline in order to measure bending moments. The stiff wall was fabricated from 22 mm thick

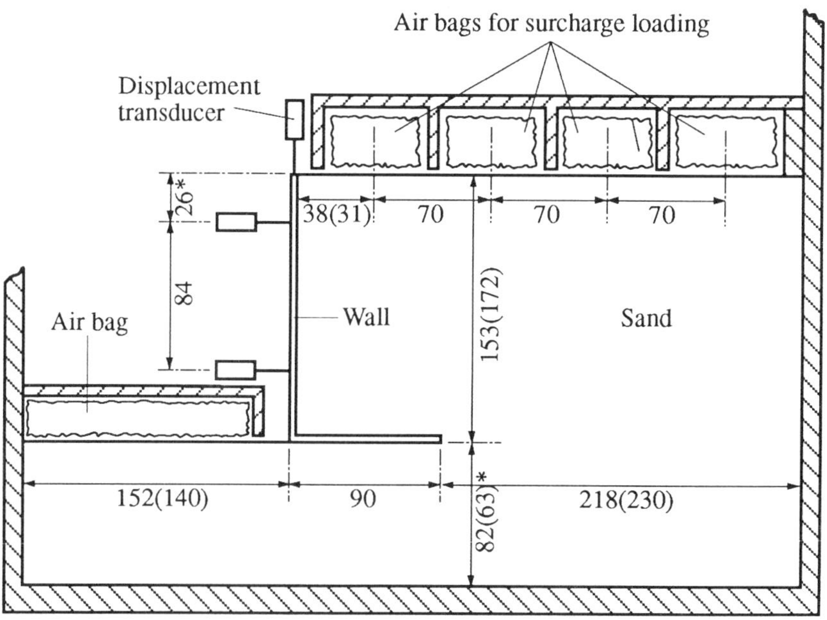

Notes
All dimensions in mm
Figures in brackets refer to stiff wall
Starred dimensions increased by 19mm in lateral loading tests

Fig. 1. Model configuration (cross-section)

mild steel plate and instrumented near the centreline with Cambridge contact stress transducers (Stroud, 1971) to measure both normal and shear stresses on the wall stem (4 transducers) and on the base (4 transducers). In this case the instrumentation was not duplicated.

The configuration of the centrifugal model is shown in Fig. 1, where some construction details have been simplified or omitted for clarity. The model wall was installed on a preformed and carefully levelled bed of dry Leighton Buzzard Sand. This soil was chosen because of the extensive data base on its stress-strain behaviour assembled by researchers at Cambridge University (e.g. Stroud, 1971). It was uniformly graded with all particles passing the l.18 mm BS test sieve and being retained on the 0.6 mm sieve. When placed in the model, by pluviation in air, the sand had a density of 1700 kg/m^3, a void ratio of 0.56 and a relative density of 77% . After the wall had been installed, it was temporarily propped and backfilled with the Leighton Buzzard sand. Other components of the model were then assembled, including transducers to measure the displacements of the wall as indicated in Fig. 1. As the rotor of the centrifuge was not equipped with a swinging arm, the free surfaces of the sand in front of and behind the wall had to be temporarily supported in a vertical plane with air filled rubber bags. After some centrifugal acceleration (5 g) had been applied, the modest air pressures (20 kN/m^2) were released. The model was then accelerated to 60g.

In some tests the rubber bags behind the wall were then used to apply surcharge loading to the backfill. By pressurizing each of the four bags (see Fig. 1) in turn a 65 mm wide strip surcharge moving first towards and then away from the wall could be simulated. In the other tests presently being reported, a double acting pneumatic ram was employed to apply a horizontal pushing/pulling force to the top of the wall. The force was transmitted to the wall via a stiffbeam in order to distribute it along the wall. The beam was equipped with strain-gauged webs, so that it could also function as a load cell, and rested on teflon bearings to restrict friction losses. During the pushing stroke of the ram, the beam made intimate contact with the wall along its entire length and longitudinal bending effects were minimised. During the pulling stroke, however, contact was made through a hook arrangement over the outer parts of the wall to avoid interference with the instrumentation in the central region. Some longitudinal bending in the flexible wall had therefore to be tolerated in this case.

Full details of the experimental apparatus and procedures, including calibration procedures for the instrumentation, are given by Djerbib (1986).

Test results for surcharge loading

Each model was subjected to 20 cycles of surcharge loading during which the bending moments in the flexible wall or the stresses on the stiff wall were

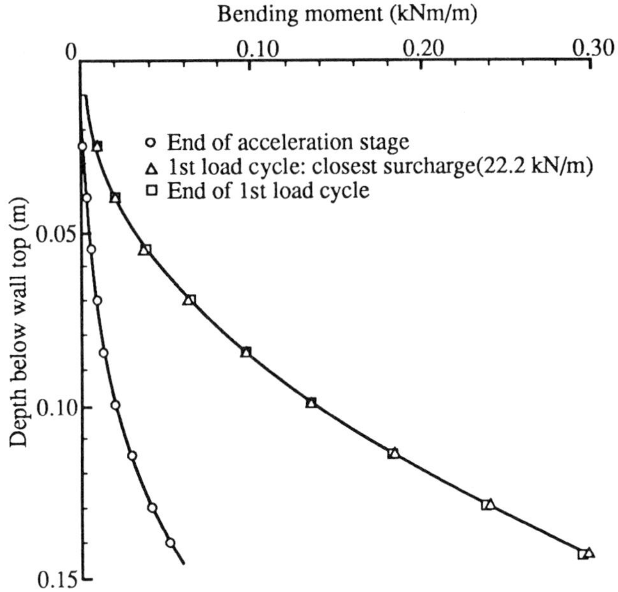

Fig. 2. Bending moment distributions: flexible wall, surcharge loading

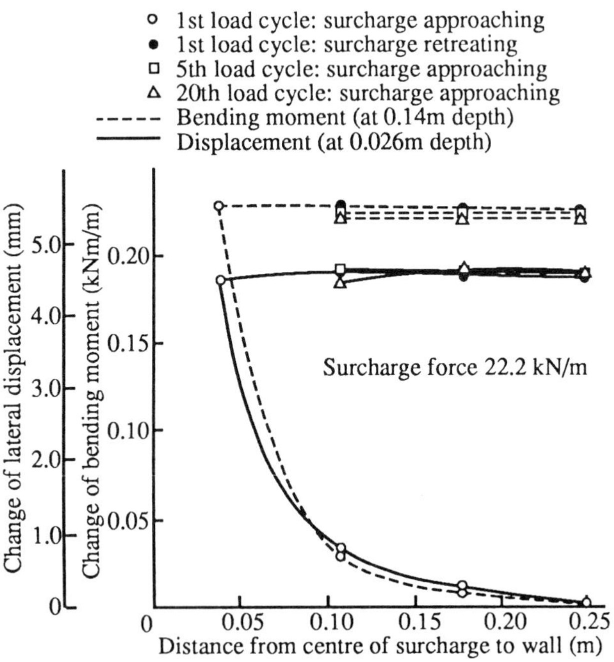

Fig. 3. Changes in condition of flexible wall during surcharge loading

recorded, together with the wall displacements. Since very good agreement was obtained between the bending moments obtained from the two sets of strain gauges on the flexible wall, data from only one set will be presented. For each wall attention will be restricted to results obtained from the wall stem and, furthermore, for the stiff wall only results for normal stress will be given explicitly. The results for the flexible wall will be presented first.

The bending moments in the stem of the flexible wall at selected early stages of the test are shown in Fig. 2. The initial bending moments, before surcharge loading, were shown by Djerbib (1986) to be consistent with those predicted by classical active earth pressure theories in conjunction with a peak angle of shearing resistance of the sand ($\phi = 46.56°$, based on the research by Stroud (1971)). Representative data for the changes of both bending moment and lateral displacement due to surcharge loading are given in Fig. 3, where the variations with surcharge position and number of loading cycles are indicated. The results show clearly that significant deformations occurred during the first loading approach, but that very little recovery took place as the load moved away. The increases in bending moment reflect increases in the lateral pressures in the soil, but the measurements did not allow the distribution of stress to be reliably computed (Djerbib, 1986). Unfortunately, after the first cycle of surcharge loading had been applied, the rubber bag nearest the wall ruptured and the cycling of the load was restricted to the other three positions. Probably for this reason, the subsequent partial cycling of the load produced no measurable changes in the condition of the model. In Fig. 3 it can be seen that, during the first cycle, the changes brought about by applying the surcharge were relatively small until it was applied at the position nearest the wall. In another test, not presently being reported, the same wall was tested with a rigid prop at the top and subjected to 15 complete cycles of surcharge loading. Changes of bending moment and displacement were still greatest during the first load cycle, but smaller changes continued to accumulate throughout the test. After 15 load cycles the maximum bending moment was about 20% larger than it was at the end of the first cycle.

Fig. 4 shows the normal stresses on the wall stem at selected stages of the test on the stiff wall. Again, the results before surcharge loading were compared by Djerbib (1986) with those predicted by classical theories for active conditions, but the agreement was much less satisfactory than for the flexible wall. The measured stresses exceeded the theoretical values at lower levels and, at the bottom of the stem, they also exceeded at rest pressures calculated from the empirical relationship: $K_o = 1 - \sin\phi$ where K_o is the coefficient of lateral earth pressure at rest. The changes of stress and lateral displacement with the progress of surcharge loading are illustrated in Fig. 5. Note that from the beginning (and deliberately) in this test only three of the possible four surcharge positions were used. As for the flexible wall, relatively large changes of horizontal stress and displacement occurred as the

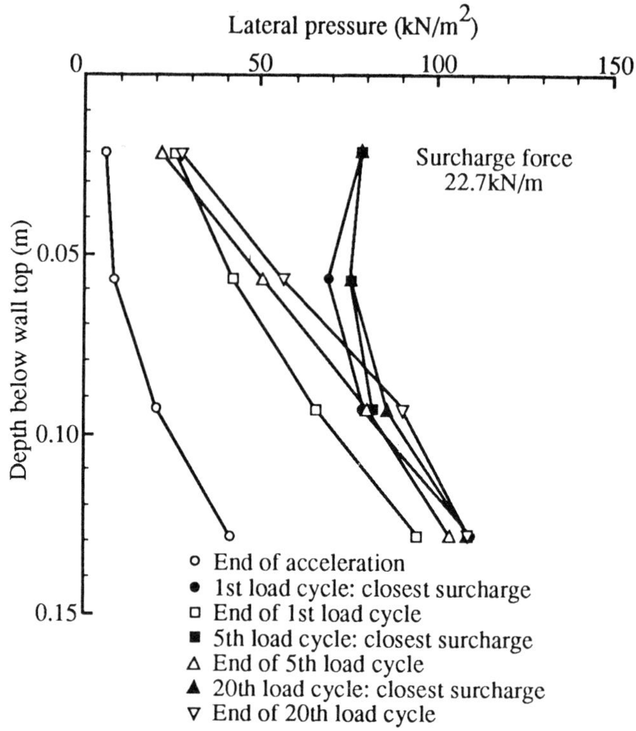

Fig. 4. Lateral pressure distributions: stiff wall, surcharge loading

load first approached the wall, although the displacements were much smaller in absolute terms. By comparison with the previous test, however, the reduction of horizontal stress was more marked as the load moved away. With subsequent load cycles both lateral pressures and displacements continued to increase, as illustrated in Fig. 6, and the model never reached a steady condition.

Test results for lateral loading

Each model was subjected to 50 cycles of pushing and pulling by the horizontal force applied at the top of the wall. In the tests described it was intended that the top of the wall would be displaced between preset limits. However, the capacity of the loading system proved inadequate to achieve this and, on the pushing stroke, the displacement varied during the test. Furthermore, the applied forces did not remain constant; there was a progressive increase in the maximum applied pushing force and a progressive decrease in the maximum applied pulling force (note also that, because of operational difficulties, the pushing force on the first load cycle was very much smaller than on subsequent cycles). The data collected were identical in kind to those obtained during the tests already described, with the addition of the applied force measured by the load cell, and will be presented in a similar manner. Only data collected at the ends of the pushing or pulling strokes will be reported.

For the flexible wall a selection of bending moment distributions in the wall stem is presented in Fig. 8. At the end of the acceleration stage the bending moments were different from those described earlier (see Fig. 2), because of the presence of a modest propping force. During the pushing and pulling stages, the changes of applied force are reflected in the bending moment distributions. A critical examination of the results (Djerbib, 1986) has shown that, during pulling, the longitudinal bending effects were significant. The changes of bending moment and lateral displacement with the number of load cycles are illustrated in Fig. 9. After the first cycle the limits of lateral displacement became progressively closer. On pulling, a fixed displacement limit was achieved at the level of the applied force but, because the transducers were located below this level, the outward measured displacements

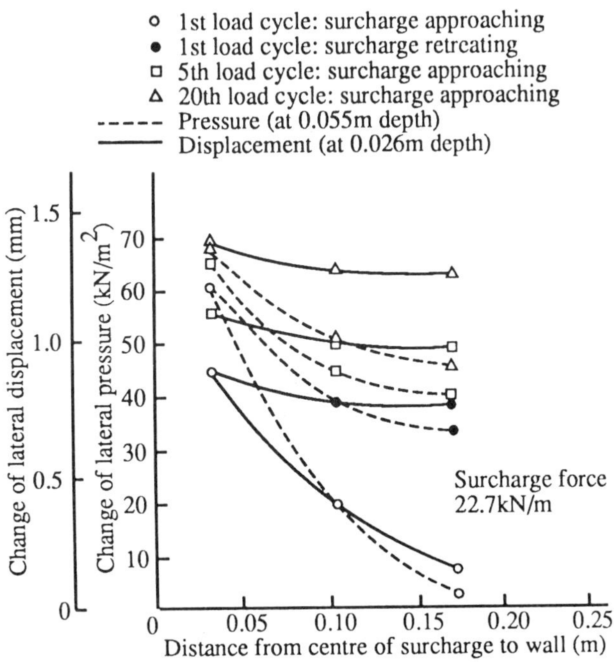

Fig. 5. Changes in condition of stiff wall during surcharge loading

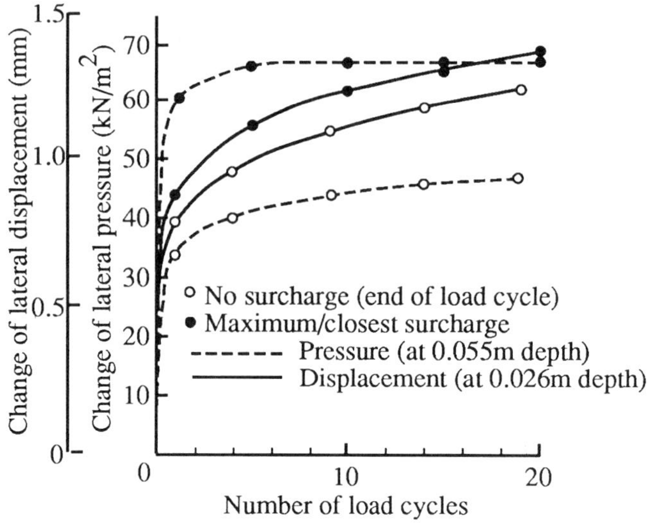

Fig. 6. Changes in condition of stiff wall with no. of surcharge load cycles

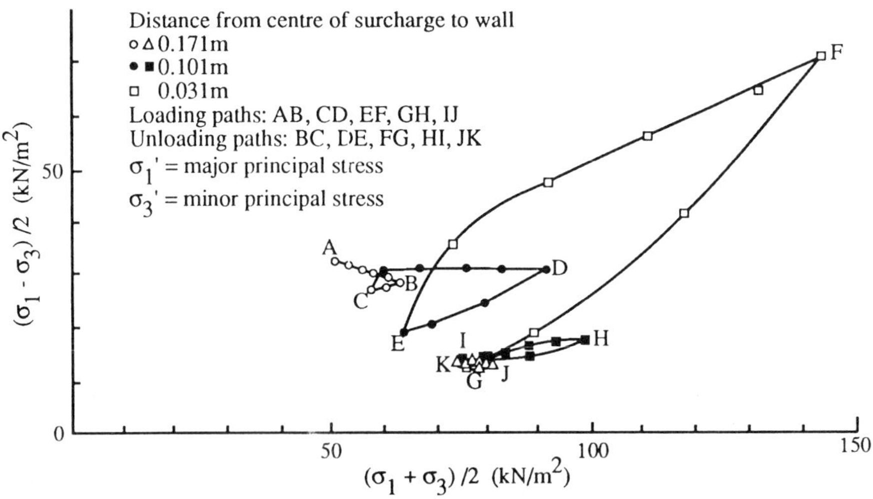

Fig. 7. Stress path for soil next to stiff wall at 0.09m depth during first surcharge load cycle

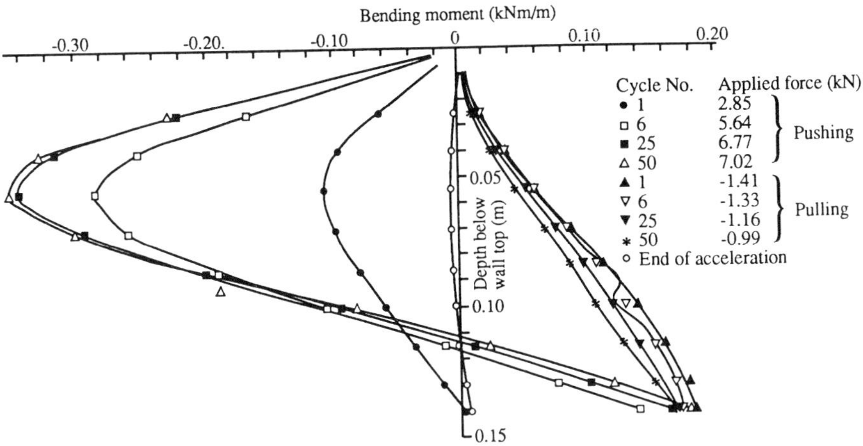

Fig. 8. Bending moment distributions: flexible wall, lateral loading

increased slightly. On pushing, the available force was progressively less able to return the wall towards its initial position.

For the stiff wall a selection of normal stress distributions on the wall stem is presented in Fig. 10. At the end of acceleration the stresses were similar to those already reported (see Fig. 4). At this stage they were only slightly larger than those consistently observed at the end of the pulling stroke. The changes of normal stress and lateral displacement with the number of load cycles are illustrated in Fig. 11, where the behaviour patterns correspond well with those of the flexible wall.

Discussion

During the tests with surcharge loading, horizontal stresses due to the surcharge were largely "locked into" the soil when the surcharge was removed. This important and well known effect is accounted for in several relatively simple theories for the effect of backfill compaction processes (e.g. Ingold, 1979). However, none of these theories gave adequate quantitative predictions for the present tests (Djerbib, 1986). The tests appear to show that the locking in of horizontal stress is more complete as the flexibility of the structure increases, in accordance with the findings of numerical studies by Aggour and Brown (1974).

If further complete cycles of surcharge loading were applied, progressive changes in the models occurred. Some indication of the complexity of model behaviour during the cyclic loading can be gained by plotting the stress paths followed by soil elements in the vicinity of the stress transducers mounted in the stiff wall. A typical result is shown in Fig. 7. In order to plot this path, the

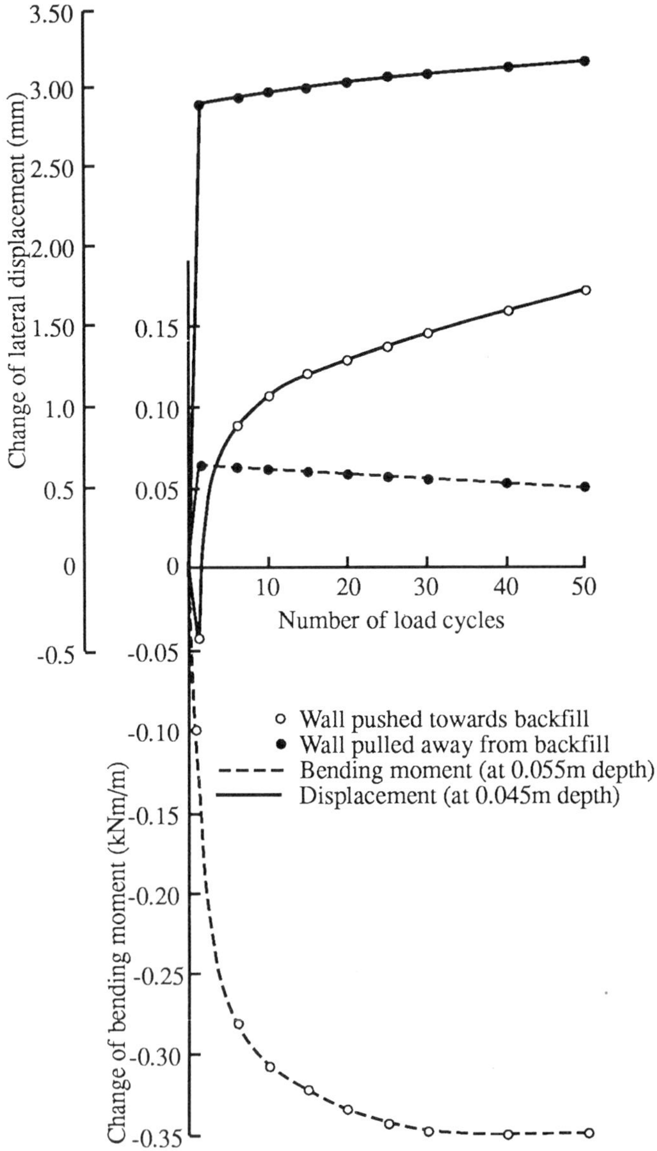

Fig. 9. Changes in condition of flexible wall with no. of lateral load cycles

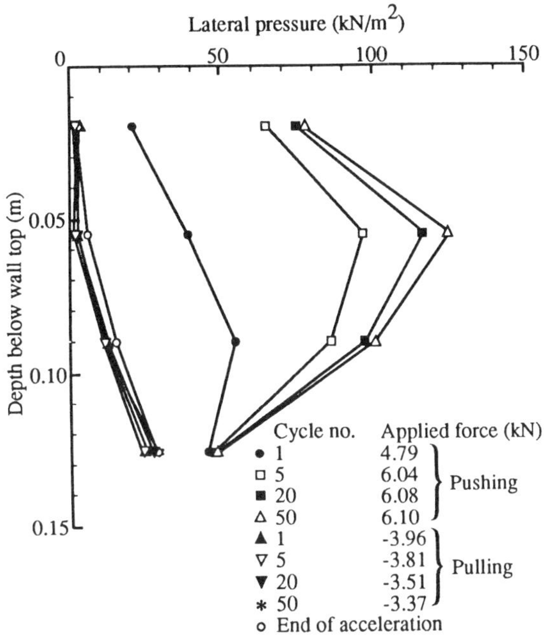

Fig. 10. Lateral pressure distributions: stiff wall, lateral loading

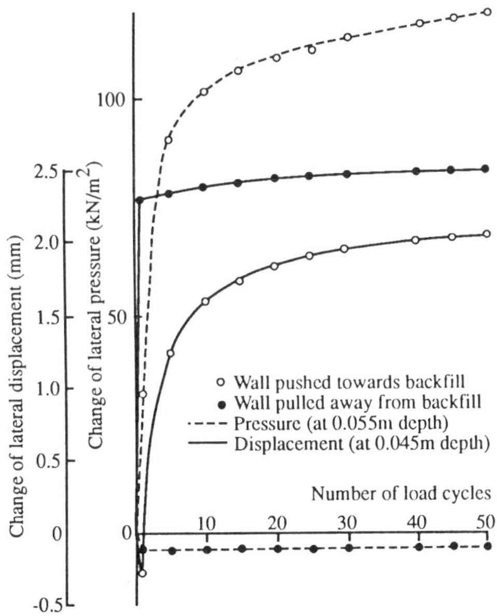

Fig. 11. Changes in condition of stiff wall with no. of lateral load cycles

magnitude of the vertical stress had to be estimated at each stage and both the shear and the normal stress data from the transducer had to be used. Although the Cambridge contact stress transducer is designed to be extremely stiff, it is still possible that deflections of the transducer influenced the stress conditions in the adjacent soil. Nevertheless, the soil is likely to have followed a complex stress path, resembling the one illustrated in Fig. 7 and displaying frequent direction changes. It should be noted that shear stress reversals were a feature of the behaviour at all elevations on the wall stem. At the end of the load cycle the shear stress on the wall acted upwards, but when the full surcharge was applied at the position closest to the wall it acted in the opposite sense.

The changes of stress path direction (and shear stress reversals) discussed above were associated with alternating shear strains in the soil during each load cycle. Soil elements therefore densified and became progressively stiffer. The wall displacements displayed a ratcheting effect, whereby the outward movements of the stem systematically increased. The residual stresses on (or bending moments in) the wall stem after the completion of a load cycle also increased as it became ever harder for the wall to spring back towards its original position. Peak stresses (or bending moments) were also affected, but to a lesser degree.

Only a limited number of load cycles could be applied in the present tests. If further cycles had been applied, changes in residual stresses and displacements would have continued. Some guidance on the effect of a larger number of load cycles can be obtained from the tests of Sherif and Mackey (1977). These tests were carried out at 1 g and involved up to 150 repetitions of a line surcharge at a fixed position. The wall was propped, at both the top and bottom, and its bending stiffness was comparable with that of the stiff wall in the present tests. After 20 cycles of loading, the percentage increases in the residual stresses were of the same order as those described above. However, after 150 cycles, increases of up to 300% were reported. Ultimately, it might be expected that the incremental changes would become negligible.

During the tests with lateral loading, the soil was again subjected to alternating shear strain. The resultant stiffening could be clearly seen in the model's response when the pushing force was applied, since the displacement that could be achieved by a given force steadily diminished. Once more, there was a ratcheting effect for the outward movements of the wall. For a full scale bridge structure in which a cyclic pushing force was being generated by thermal expansion of the deck there would also be a progressive increase in the compressive stresses in the deck and the lateral pressures on the abutment. Indications of such behaviour, accompanying large seasonal changes of pressure and displacement, were obtained by Broms and Ingleson (1972), but their measurements extended for only two years (two load cycles). More useful measurements over a four to five year period were made by

Smoltczyk et al (1977) on the walls of ship hoist structures, which clearly demonstrate the progressive effect of seasonal temperature fluctuations. Over the measurement period the lateral pressures at corresponding temperatures increased by about 50%. Broms and Ingleson considered it possible that the pressures would ultimately approach passive values, but this has not been verified.

Conclusion

Centrifugal model tests have demonstrated the effects of cyclic surcharge and lateral loading on L-shaped retaining walls and have provided data for validating numerical models. Repetition of each type of loading produced significant changes in the stresses on the structure (associated with permanent displacements) which could not be ignored by a designer.

Acknowledgement

The above work was funded by, and published by permission of, the Department of Transport.

References

AGGOUR M.S. and BROWN C.B. The prediction of earth pressure on retaining walls due to compaction. *Geotechnique*, 1974, **24**, 4, pp 489-502.

BROMS B.B. and INGELSON I. Earth pressure against the abutments of a rigid frame bridge. *Geotechnique*, 1971, **21**, 1, pp 15-28.

BROMS B.B. and INGLESON I. Lateral earth pressure on a bridge abutment. *Proc. 5th Eur. Conf. Soil Mech. Fnd. Engng., Madrid*, 1972, **1**, pp 117-123 .

DJERBIB Y. *Centrifugal testing of model L-shaped walls supporting sand*. PhD Thesis, University of Sheffield,1986.

INGOLD T.S. The effects of compaction on retaining walls. *Geotechnique*, 1979, **29**, 3, pp 265-283.

RATHMAYER H. Vertical buffer plates simplify bridge abutments. *Proc. 9th Int. Conf. Soil Mech. Fnd. Engng., Tokyo*, 1977, **1**, pp 695-700.

SHERIF M.M. and MACKEY R.D. Pressures on retaining wall with repeated loading. *Proc. ASCE, I. Geot. Engng. Div. 103*, 1970, GT11, pp 1341-1345.

SMOLTCZYK U. *et al.* Earth pressure variations due to temperature change. *Proc. 9th Int. Conf. Soil Mech. Fnd. Engng., Tokyo*, 1977, **1**, pp 725-733.

STROUD M.A. *The behaviour of sand at low stress levels in the simple shear apparatus*. PhD Thesis, University of Cambridge,1971.

Estimation of earth pressures in the design of a deepened quay wall formed of an existing relieving platform

C.K.T. LEE and F.G. BUTLER, Butler Puller Partnership, UK

Introduction

A programme of modernisation and deepening of the existing berths at Limehouse Wharf, Rochester was envisaged for the bulk handling of newsprint.

Most of the existing wharf structure consisted of a relieving platform with a section as shown in Fig.1. The front wall consisted of sheet piling 15.5 m in

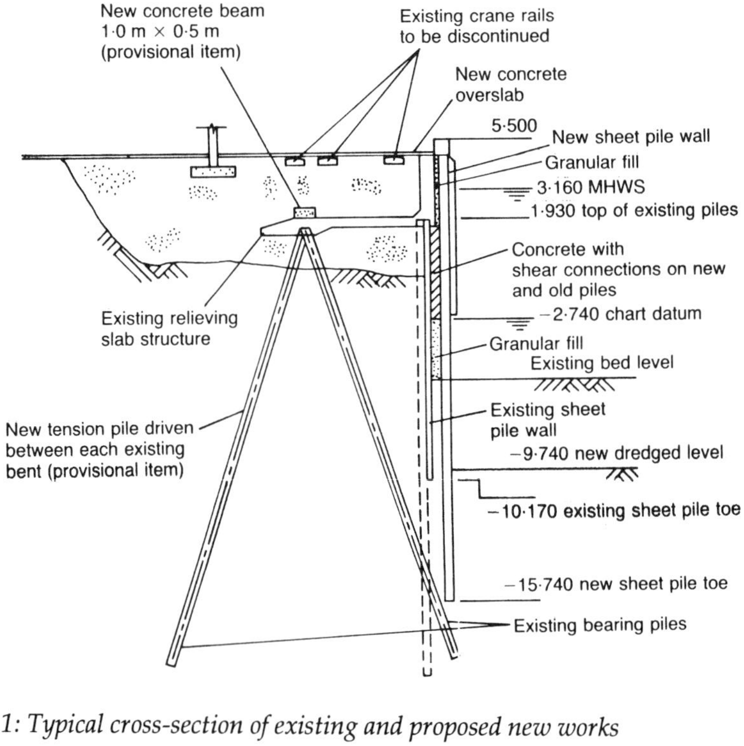

Fig. 1: Typical cross-section of existing and proposed new works

Retaining structures. Thomas Telford, London, 1993

height, attached to an R.C. relieving platform 8 m wide at a depth of 3.5 m below wharf level supported on pile bents. Each bent was believed to consist of three BSP cased piles, two driven to landward and one to seaward, raked at 1 in 2.75. The bents were placed at 4 m centres. From the ground succession and the driving characteristics of BSP cased piles, these piles were estimated to be about 13 m long.

The use of a relieving platform introduces several benefits, particularly in circumstances where the ground conditions limit the efficacy of anchor piles in tension while capacity in compression is good. More particularly since the platform slab carries all the load above to a level at or below dredge, it substantially reduces the height of retained soil. It also provides vertical surcharge to the sheet piles to assist against pull-out forces and to improve passive resistance to the embedded section.

The effectiveness of the relieving platform is a function of both the geometry and the soil-structure interaction. Three empirical hypotheses have been postulated to facilitate analyses of these structures

(*a*) *Fully screened* — The fully screened hypothesis assumes that the concrete slab transfers all the load above to the toe level of the sheet pile wall and raking piles, which are remote and below dredge level. Thus the active pressure diagram against the wall consists of two independent diagrams for pressure above and pressure below slab.
(*b*) *No screening* — The no screened hypothesis assumes that the slab does not carry vertical load and thus the active pressure diagram against the wall is the same as if no slab existed. The wall is in effect a propped cantilever, the prop force provided by the anchor effect of the pile bent supporting the slab.
(*c*) *Partially screened* — The partially screened hypothesis assumes that only part of the self weight above the slab is transferred to toe level of the sheet piles and the platform piles. Thus the earth pressure diagram acting on the sheet piles is larger than in the fully screened hypothesis but less than the no screen case.

Soil conditions

The soil conditions vary somewhat along the length of the wharf but in generalised terms can be described thus

(a) 3.5 m made ground
(b) 3 m silt
(c) 6 m silty sand
(d) 5 m gravel
(e) chalk.

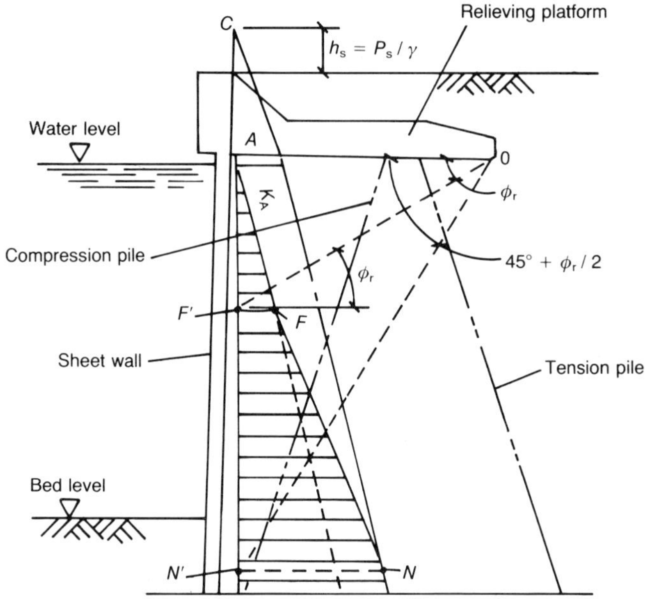

Fig. 2: Theoretical pressure distribution on rear of wall. BS 8349

The Made Ground consisted of loose to medium dense black sandy gravel, ash, with fragments of brick and concrete overlying older soft to firm clay with flints, chalk, glass fragments, shells, timber, ash, etc with Standard Penetration Tests in the range 3 to 19.

Soft peaty clay was sometimes present beneath the made ground.

The Silt was described as soft grey clayey silt with Standard Penetration Tests in the range 8-10.

The Silty Sand was generally loose to medium dense, dark grey or black, often malodorous and slightly organic. Standard Penetration Test values varied from 10-28.

The Gravel was loose to dense sandy and fine to coarse with Standard Penetration Test values generally in the range 25 to 54 and appearing to become more sandy with depth.

The Chalk was Grade V for about 1.5 m increasing to Grade II and then Grade I within 1 to 2.5 m.

Proposed solution

It was proposed to dredge the bed to approximately 15.5 m below cope level, i.e. about the same level as the toes of the existing sheet pile wall, to accommodate larger vessels without grounding and thus allow vessel movements at all states of the tide.

To achieve this it was proposed to drive a new sheet pile wall to greater depth in front of the existing wall, which would also be attached to the existing relieving platform, also shown in Fig.1. To design this required an estimate of the earth pressure on the new wall, the new loadings in the relieving platform piles and a check of the ability of the members in the relieving platform to assume their new duties.

Existing design methods

BS 6349 Part 2: 1988 and the German Code of Practice EAU 1985 adopt the partially screened hypothesis, using empirical methods to determine the extent of screening. Two approaches are described in BS 6349.

The first approach is to define the earth pressure by means of an empirical method as shown in Fig.2. Using this pressure diagram the factor of safety against overturning, bending moment diagram in the sheet piling and the anchor force applied to the rear pile bent can be determined.

The effect of the forces in the pile bent on the sheet pile wall is ignored. "Average" soil parameters are assumed, the method cannot deal directly with stratified soil conditions.

The second approach uses the Culmann Wedge method (Fig.3) to determine the effects of the forces in the raking piles on the sheet pile wall. The disadvantage of this method is its inability to deal with stratified ground. It also does not generate an earth pressure distribution and one is therefore

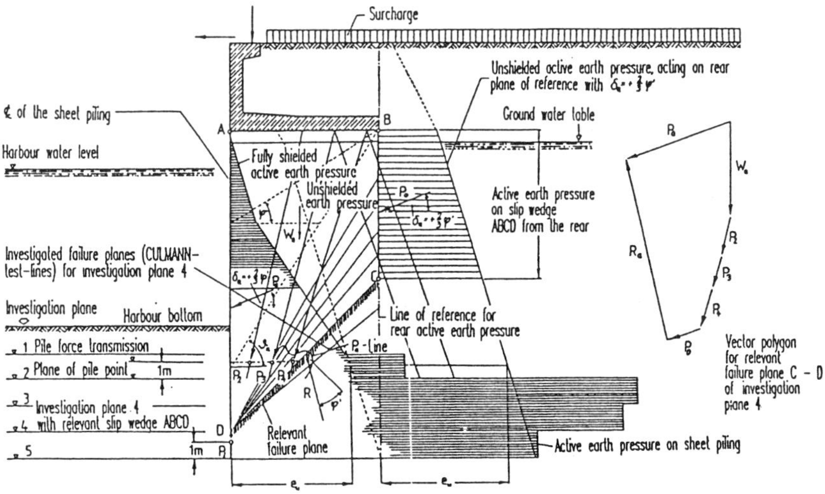

Fig.3. Culmann Trial Wedges

unable to obtain an estimate of bending moment in the sheet piles.

While extensive site investigations were conducted to obtain representative soil parameters for design, it remained central to the proposal to analyze the existing wall structure in order to test the soil parameters we were proposing to use and thereby assess the ability of the existing relieving platform to cope with the forces and moments generated by the deepening of the wharf.

The first of these tasks immediately presented a problem. Even by assuming the most optimistic interpretation of the site investigation it was difficult to justify the existing design using either of the empirical analytical methods. It was essential to resolve this problem before proceeding to the analysis of the proposed new works. The fallacy lay in the empiricism rather than in the soil parameters. It was therefore decided to examine these methods critically in an attempt to improve on them.

Methods adopted

Since each method possesses some merits of its own, it was decided to attempt to combine the principles behind each and to compare the end result with a non-linear plane strain finite element analysis.

Earth pressure distribution method

The procedure for carrying out the analysis based upon earth pressure distribution, but adapted to cater for stratified soil and incorporating the effects of the thrust in the pile bent was as follows

(*a*) The "average" c′ and ϕ' of the soil behind the wall was estimated.
(*b*) Based on the "average" ϕ' value the transition line between the "fully shielded" and "unshielded" earth pressure was found.
(c) The unshielded earth pressure line for the actual stratified soil was then superimposed on the diagram. The transition line representing the stratified soil was assumed to be given by applying the ratio of the transition line intercept obtained in (*b*) to its unshielded intercept as a "shielding factor" to the stratified unshielded line.
(*d*) Using this diagram the depth of penetration of the sheet piles for a required factor of safety can be determined together with the bending moment diagram for the sheet piles and the anchor force pulling on the rear pile bent. This set of results was called Result Set No.l.
(*e*) It was then assumed that the load above the concrete platform is carried by the platform which is simply supported by the sheet piling and anchor pile bent. Hence, by simple statics the vertical load component in the pile bent can be obtained.
(*f*) Combining the horizontal pull from (*d*) with the vertical force from (*e*)

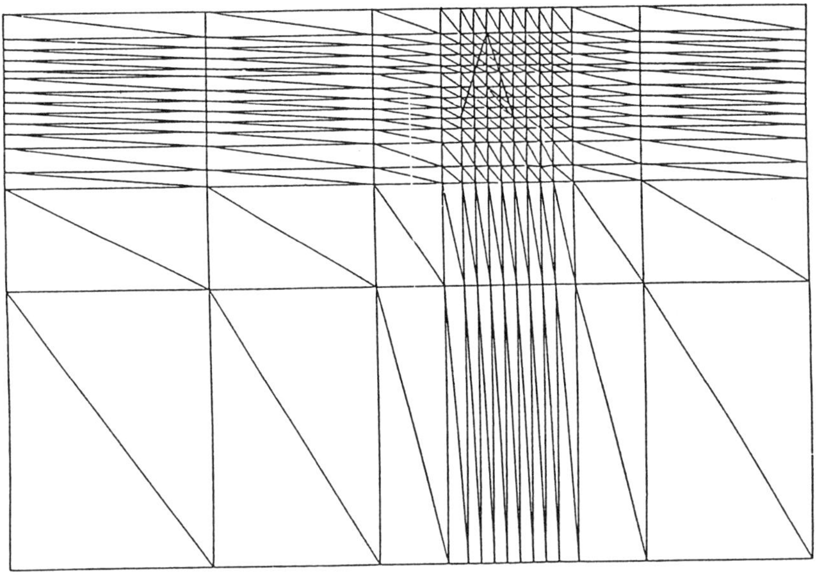

Fig.4. Finite element mesh

the force equilibrium at the top of the pile bent gives the load in each raking pile.

(*g*) Assuming each raking pile transfers all its load from the head to the toe, and that at toe level this force is applied to the soil as an inclined uniform line load, the surcharge on the rear of the sheet piles due to the raking piles can be found using Melans formula.

(*h*) When this additional surcharge is introduced into the re-calculation of the stability of the wall this yielded a new anchor force.

(*i*) Introducing this new anchor force steps (*f*) to (*h*) were repeated.

(*j*) Steps (*f*) to (*i*) were repeated several times until the difference in anchor force became insignificant, i.e. a short manual iteration process. Although this process sounds laborious, convergence is actually quite quickly achieved. The anchor force, factor of safety etc resulting from this process was called Result Set No.2.

A corresponding result was obtained using the wedge analysis based upon average c′ and φ′, with the forces indicated in Fig.3 together with the effects of the surcharge due to the raking piles from analysis of earth pressures. This was termed Result Set No.3.

The finite element approach

The finite element analysis was carried out using CRISP-90 which is able

Table 1.Summary of analysis of old dredge level

	Analysis by Earth Pressure		Analysis by Wedge	Analysis by Finite Element
Result Set No	2	2	3	
FOS	1.2	1.16	1.04	1.86
Active thrust (kN)	383	392	427	495
Prop force (kN)	136	133	N/A	190

to combine two dimensional elements, beam elements, bar elements and interface elements together. In our analysis plane strain conditions are assumed with the concrete slab and sheet piles represented by the beam elements, the raking piles by bar elements and an interface element used beneath the slab and on both faces of the sheet pile. The mesh used is shown in Fig.4. In order to maintain a state of equilibrium at the start of this analysis, the elastic continuum representing the soil initially has a level surface with internal stresses in equilibrium with external constraints. Staged excavation is then carried out in front of the wall, eventually to the old dredge level. The sheet pile is then extended to the new level and further staged excavation carried out down to the new proposed dredge level.

Summary of results

A summary of the results obtained, compared with those from the Finite Element method are shown in Tables 1 (existing dredge level) and 2 (proposed new dredge level).

From these comparisons it appears that the factor of safety obtained by the methods used by BSI and EAU, when extended to allow for stratified soil as proposed, are conservative compared with the Finite Element Analysis,

Table 2. Summary of analysis of proposed dredge level

	Analysis by Earth Pressure		Analysis by Wedge	Analysis by Finite Element
Result Set No.	1	2	3	
FOS	1.48	1.32	1.21	1.98
Active thrust (kN)	771	860	987	1208
Prop force (kN)	291	293	N/A	291

Table 3. Summary of the effect of raking pile

	Old Dredge Level		New Dredge Level	
	Surcharge (kN)	% Increase in Active Thrust	Surcharge (kN)	% Increasein Active Thrust
Wedge analysis	44	11.5%	112	14.5%
Finite Element Analysis	216	56.4%	437	56.7%

although the anchor forces are in reasonable agreement.

When the berth is deep or when the penetration characteristics of the anchor piles is limited (either by the system, or by the density of the soil) the compression piles may be founded above dredge level. The accepted convention is that if the piles are founded more than 1m above dredge, the surcharge effect of the pile load should be included in the analysis of the sheet pile wall. From the results the authors have attempted to assess these effects on the design of the sheet piles. The effect of taking the forces in the raking piles into account was assumed to be the difference between the results from the wedge analysis and Results Set No.l and the difference between the Finite Element Analysis and Result Set No.2.

From Table 1, using the modified earth pressure method the factor of safety in the existing section drops from 1.2 to 1.104 with the inclusion of the effects of the raking pilea by Wedge Analysis.

In the case of the proposed new dredge level the corresponding factor of safety drops from 1.48 to 1.21.

The corresponding factors of safety obtained from the finite element analysis are 1.86 and 1.98.

Conclusions

A method has been postulated, combining the Earth Pressure Distribution and Wedge approaches embodied in BS 6349 to investigate the pressures on a sheet pile relieving platform in stratified soil, in which the pile bents impose load on the rear of the wall. The results of this method have been compared with those obtained from finite element models of the same cases.

The most striking result of this comparison is that whereas using BS 6349 methods the adequacy of the existing wall appears, on paper, to be very marginal (yielding factors of safety ranging between 1.04 and 1.2), the factor of safety against passive failure, from the finite element analysis is 1.86, which would seem unnecessarily high. Needless to say, the existing wall has behaved entirely satisfactorily for some years.

Similarly, when the deepened wall is designed by BS 6349 to provide an acceptable factor of safety against active failure (factors of safety ranging from 1.21 to 1.48) the finite element analysis yields a factor of safety of 1.98.

Prop forces, deduced from the finite element analysis are equal or up to 40% greater in the finite element analysis than from the earth pressure method in which active thrusts are also higher, by 30-57%.

The wide disparity between the results is a matter of some concern, since the conventional methods are yielding a design which falls short of the accepted minimum standards while the finite element approach suggests that the designs are somewhat conservative. In part at least, the difference lies in the inclusion of friction between soil and wall in the latter case.

It seems that the empirical methods currently being used, although apparently performing satisfactorily, are achieving their objective fortuitously rather than definitively.

Since all the methods rely to a greater or lesser extent on empiricism or theory, many more results from model tests and field monitoring are needed before improved methods of analysis can be verified.

Acknowledgements

The work described was carried out as an amplification of analyses undertaken on behalf of Beckett Rankine, Consulting Engineers, responsible for the redevelopment of Limestone Wharf, to whom the Authors thanks are due for permission to publish this paper.

Centrifuge model studies of a cantilever retaining wall in sand

G.J.W. KING, University of Liverpool and J.P. McLOUGHLIN, City of Liverpool Community College, UK

Introduction

Flexible cantilever retaining walls resist the overturning moment, due to the pressure from the retained soil, by developing a fixing moment due to the earth pressures along the embedded portion of the wall. Appreciable re-distribution of these pressures occurs, due to plastic yield in the soil, as the height of retained soil is increased and the depth of embedment reduced by excavation. For a satisfactory design the associated wall displacements and the bending stresses developed in the wall must not be excessive. For a specified depth of excavation in a particular soil, it is necessary to estimate the required depth of embedment of the wall and its section properties.

Cantilever retaining walls are considered to be suitable for low retained heights (<4.5m) and are mainly used for temporary works. According to Head and Wynne (ref. 1), they account for up to 75% of all retaining walls constructed at the present time. Steel sheet-piling is most commonly used with the advantages of strength, variety of available sections, lightness, re-usability, resistance to hard driving and the ability to be extended in length.

Current design methods for cantilever walls are based on classical limit state equilibrium approaches. These approaches, together with guidelines for the measurement and selection of appropriate design parameters for permanent and temporary walls in stiff clay, have been detailed in CIRIA Report 104 by Padfield and Mair (ref. 2). Comparative calculations, using the alternative design procedures suggested for the determination of embedment depths to give certain factors of safety, were carried out by Symons and Kotera (ref. 3). They concluded that pre-selection of a particular procedure was not advisable. These approaches do not give any indications of wall movements and do not consider the influence of wall flexibility on the earth pressures which develop. Conventional laboratory tests on cantilever retaining walls in cohesionless soils have been described by Rowe (ref. 4) and Bransby and Milligan (ref. 5). While these have produced useful quantitative information about stability and movement patterns, quantitative extrapolation to prototype scale is of dubious value. Also bending stresses in the walls

Table 1. Dimensions and properties of model and prototype

			Model	Prototype
Wall	height	mm	120	11000
	width	mm	459	42077
	thickness	mm	2.02	185.2
	second moment of area (I)	mm^4/mm	6.90×10^{-1}	4.93×10^5
	modulus of elasticity (E)	kN/mm^2	193	207
	flexural rigidity	kNm^2/m	1.33×10^{-1}	10.21×10^4
Soil Mass	depth	mm	229	21000
	length	mm	229	21000
	width	mm	460	42170

were not measured. Recent numerical studies by Fourie and Potts (ref. 6) have indicated that limit state approaches over-predict wall bending moments appreciably.

In this Paper centrifuge model studies on a flexible wall, with different surface roughnesses, in both loose and dense sands are described. The results are presented in terms of the behaviour of an 11 m high, Frodingham No. 5 section, sheet pile wall. The magnitudes of displacements are reported and the observed bending moments and critical embedment depths compared to those predicted by limit state analyses.

Centrifuge model tests

The cantilever retaining wall model was a flat stainless steel sheet. Ten 3 mm foil strain gauges were fixed centrally to each side of the wall at 10.9 mm centres. This was used to model an 11 m high Frodingham No. 5 section sheet-pile wall by spinning at 91.67g acceleration in the Liverpool University centrifuge. This facility has been described in detail by King, Dickin and Lyndon (ref. 7). The dimensions and properties of the prototype and model are summarised in Table 1.

Three wall surfaces were used

(*a*) the natural milled surface, subsequently referred to as having intermediate roughness
(*b*) a smooth surface, achieved by coating the wall in silicone grease and placing a single latex rubber sheet over both faces and
(*c*) a rough surface, achieved by applying double sided adhesive tape to the faces of the wall and coating the outer surfaces with sand.

To minimise the influences of the boundary walls of the package, they were coated with a layer of silicone grease and covered with latex rubber sheet. Short greased membranes were also fixed to the sides of the wall to prevent the sand passing between the ends of the wall and the sides of the bucket.

The wall was placed centrally within the package and the sand was rained into the bucket. The sand used was dry Erith sand which is a fine quartz sand with an effective grain size of 0.16 mm and a uniformity coefficient of 1.25. In order to obtain a dense packing, the sand was rained from a box with a perforated base fixed 1.5 m above the bucket. A very repeatable average density of 16.0 kNm^{-3} was achieved. A loose packing was obtained by raining the sand from a box with a perforated base placed directly on top of the bucket. Consistent average densities of 14.2 kNm^{-3} were obtained. Excess sand was scraped off so that the sand surface was flat and level with the top of the wall.

At the start of each test the centrifuge was spun up to the required speed and dynamic datum readings taken for all strain gauges. The machine was then brought to rest and an increment of excavation carried out using an adjustable scraper and brush. The displacement transducer mounting was then fixed in position and a transducer placed in contact with the centre-line of the exposed face of the wall. The distance of the transducer from the top of the wall was noted and a static datum reading taken. The machine was again brought up to speed and strain gauge and transducer readings taken. This procedure was repeated for each increment of excavation until catastrophic wall displacement occurred. The displacement transducer mounting was adjusted and additional transducers introduced at greater depths as excavation proceeded. Static zero datums were assumed to be appropriate for new placed transducers in conformity with the observed behaviour of existing ones. Ideally the excavations should have been carried out in flight but no suitable mechanism for doing this has been developed. However, it is felt that the unload-reload stress history applied did not have a significant influence on the in-flight behaviour at any stage. In all tests, a further increment of 0.5 m resulted in excessive walls movements before the test speed was reached.

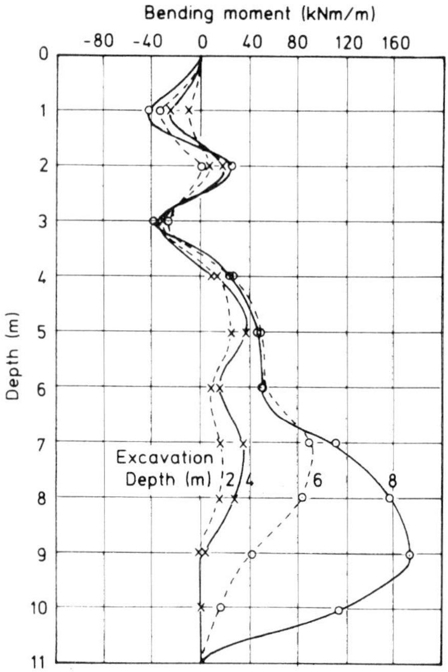

Fig. 1: Typical bending moment distributions

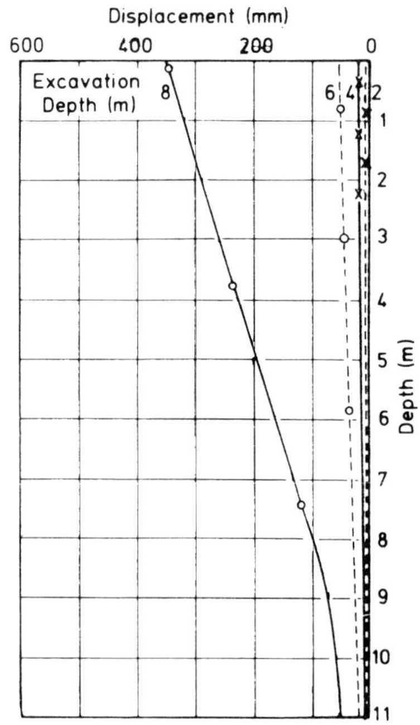

Fig. 2: Typical wall displacements

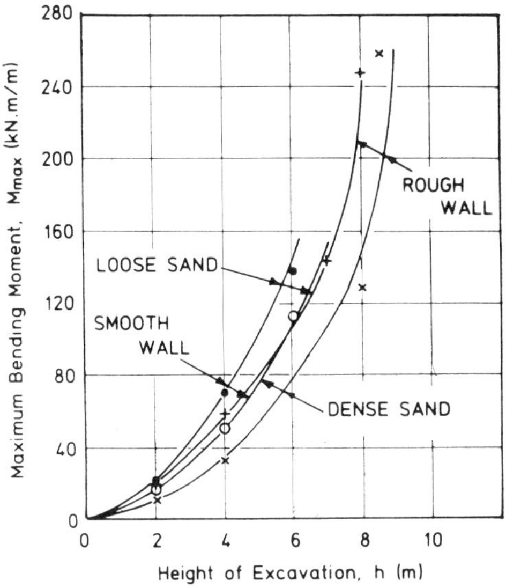

Fig. 3. Variation of maximum bending moment with height of excavation

Test results

The results of all tests have been recorded by McLoughlin (ref. 8). The bending moment distributions and horizontal wall displacements at each stage of excavation were typically as shown in Figs. 1 and 2. These are for an intermediate rough wall in dense sand.

Oscillating bending moments, often with small negative peaks, were recorded in the top part of the wall in all tests. When those moments were negative, the corresponding displacement diagrams showed slight negative curvature. The maximum bending moments occurred at depths of less than

Table 2. Maximum observed values close to failure

Soil	Dense			Loose		
Surface	Smooth	Intermediate	Rough	Smooth	Intermediate	Rough
h(m)	7	8	8.5	6	7	8
M_{max} (kNm/m)	152	168	259	137	166	248
s(m)	1	1	0.5	1.5	1	0.5

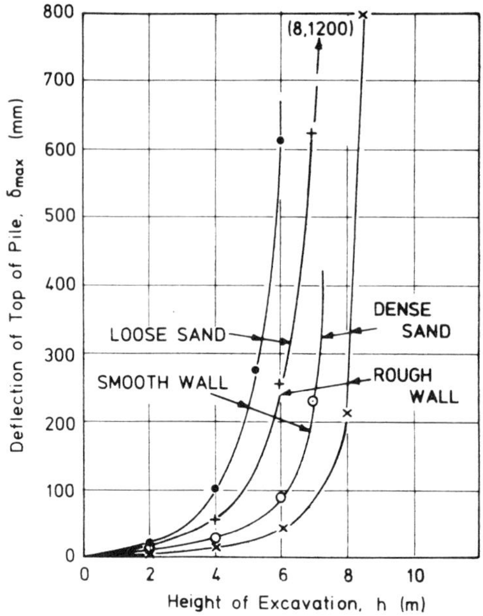

Fig. 4: Variation of maximum deflection with height of excavation.

2m below excavation level for all depths of excavation. These are considerably shallower than the depths obtained, by numerical studies on a different wall, by Fourie and Potts (ref. 6). It is interesting to note that these studies also yielded negative bending moments in the top part of the wall.

The variations in maximum bending moment and maximum deflection with height of excavation, for rough and smooth walls in loose and dense sands is shown in Figs. 3 and 4. These curves also give a good indication of the limiting heights of excavation. Clearly at any stable depth of excavation both maximum bending moments and deflections reduce with increasing surface roughness or soil density. However, because the limiting height of excavation increases with increasing surface roughness and soil density, the greatest bending moments can develop in rough surface piles in dense soil. The maximum deflections observed without catastrophic failure were between 0.5 and 0.75m, for smooth surface piles and between 1.0 and 1.25 m for rough surface piles.

A summary of the average maximum bending moments and their positions recorded at the maximum stable depths of excavation achieved, is given in Table 2 for comparison with values predicted by limit state analyses.

Limit state analyses

The simplified active and passive earth pressure diagrams and equivalent forces ususlly assumed are as shown in Fig. 5. The wall is considered to rotate about point O close to the bottom of the wall. The piling below this point is initially ignored and the pressures on this section replaced by the concentrated force R.

Moment equilibrium about O yields

$$d_o = h \Big/ \left(\sqrt[3]{K_P/K_A} - 1 \right) \tag{1}$$

in which K_A and K_P are appropriate earth pressure coefficients.

The actual penetration depth is then taken as

$$d = 1.2\ d_o \tag{2}$$

It is then checked that $\Delta P_P - \Delta P_A > R$

The depth s below excavation level of zero shear and maximum bending moment is given by

$$s = h \Big/ \left(\sqrt{K_P/K_A} - 1 \right) \tag{3}$$

and the maximum moment by

$$M_{max} = K_A\ (h+s)^3 - K_P s^3)\ \gamma/6 \tag{4}$$

Note that s and M_{max} are not dependent on d_o.

For a fixed overall height of wall, $H = h + 1.2d_o$, Eq. 1 yields the limiting excavation depth

$$h = H \left(\sqrt[3]{K_P/K_A} - 1 \right) \Big/ \left(\sqrt[3]{K_P/K_A} + 0.2 \right) \tag{5}$$

Design with a factor of safety can be achieved by factoring the passive earth pressure coefficient K_P.

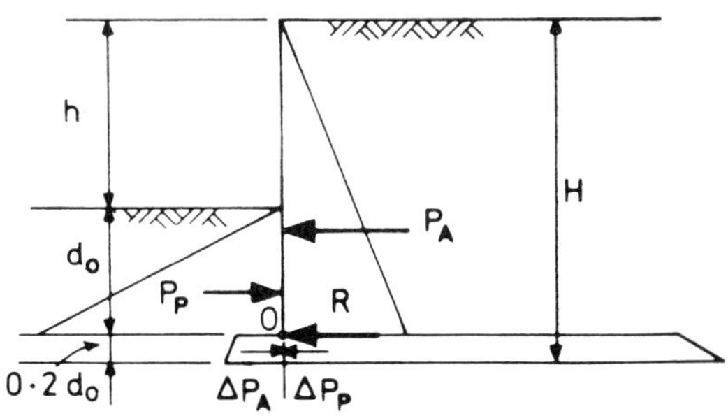

Fig. 5: Simplified pressure diagram

Table 3. Earth pressure coefficients

Soil	Dense (γ - 16 kNm^{-3})			Loose (γ - 14.2 kNm^{-3})		
Surface	Smooth	Intermediate	Rough	Smooth	Intermediate	Rough
ϕ °	49.5	49.5	49.5	40	40	40
δ °	0	17.7	49.5	0	15.8	40
K_A	0.136	0.121	0.102	0.217	0.197	0.168
K_P	7.35	13.77	48.06	4.61	8.61	13.41

Table 4. Limit state values

Soil	Dense			Loose		
Surface	Smooth	Intermediate	Rough	Smooth	Intermediate	Rough
h_{max} (m)	7.7	8.4	9.3	6.6	7.5	8.1
M_{max} (kNm/m)	220	232	244	236	268	265
s(m)	1.2	0.9	0.5	1.8	1.3	1.0

Table 5. Design values (F - 1.5)

Soil	Dense			Loose		
Surface	Smooth	Intermediate	Rough	Smooth	Intermediate	Rough
h(m)	7.2	8.0	9.1	6.0	7.0	7.7
M_{max} (kNm/m)	197	213	231	202	238	241
s(m)	1.4	1.0	0.5	2.2	1.6	1.2

Equations 3, 4 and 5 were used to calculate the maximum excavation depth and the position and magnitude of the maximum bending moment, both at limiting equilibrium and with a factor of safety of 1.5, for each wall surface embedded in dense and loose sand. The earth pressure coefficients used are shown in Table 3.

The friction angles ϕ and δ were determined from plane strain and direct sliding tests respectively and the coefficients from tables by Caquot and Kerisel (ref. 9).

The results are shown in Tables 4 and 5 for comparison with the figures in Table 2.

Since all of the retaining walls failed when subjected to a further O.5m of excavation, the limit state predictions of the critical heights are very good.

By chance the last stable heights of excavation achieved were close to the design values calculated using as factor of safety of 1.5 on the passive earth pressure. At this stage the pile displacements were quite large and their rate of change with excavation depth was increasing rapidly. This does not inspire confidence in this method of applying what would normally be considered a reasonable factor of safety.

The design values of maximum bending moment agree reasonably well with those observed for rough piling but are significantly larger than those observed for intermediate rough and smooth piling.

The calculated depths to maximum bending moment agree with the values observed for the rough pilings in dense sand but, otherwise, they are overestimated.

Conclusions

Centrifuge model studies on a flexible cantilever retaining wall with a range of surface roughness, in both loose and dense sands have been described. The results have been presented in terms of the behaviour of an 11 m high Frodingham No. 5 Section wall.

At stable depths of excavation both maximum bending moments and deflections reduced with increases in surface roughness and in soil density. However, because the limiting height of excavation increased with increasing surface roughness and soil density, the greatest bending moments developed in rough surface piles in dense soil. At all depths of excavation the maximum bending moments occurred at depths of less than 2 m belowthe excavation level.

Limit state predictions of critical heights were very close to those observed. However observed maximum displacements, and rates of change of displacement with heights of excavation, corresponding to design with a factor of safety of 1.5 on passive pressure, were alarmingly high. The maximum displacements were generally in excess of 0.5 m for smooth surface piles and l.0m for rough surface piles. Design bending moments were significantly larger than those observed for all but very rough piling and were predicted to occur at greater depths then those observed for all but the rougher pilings in dense sands.

The simple limit state approach was found to be adequate in many respects. However, modifications to this approach are being considered where it has been observed to be deficient.

References

1. HEAD J.M. and WYNNE C.P. Designing retaining walls embedded in stiff clay. *Ground Engineering*, 1985, **18**, No. 3, pp 30-33.
2. PADFIELD C.J. and MAIR R.J. *The design of propped and cantilever walls embedded in stiff clays*. CIRIA Report 104, London, 1984.
3. SYMONS I.F. and KOTERA H. *A parametric study of the stability of embedded cantilever retaining walls.* TRRL Report RR116, Transport and Road Research Laboratory, Crowthorne, Berks, 1987.
4. ROWE P.W. Cantilever sheet piling in cohesionless soil. *Engineering*, 1951, September, pp 316-319.
5. BRANSBY P.L. and MILLIGAN G.W.E. Soil deformations near cantilever sheet pile walls. *Geotechnique*, 1975, **24**, No. 2, pp175-195.
6. FOURIE A.B. and POTTS D.M. Comparison of finite element and limiting equilibrium analyses for an embedded cantilever retaining wall. *Geotechnique*, 1989, **39**, No. 2, pp 175-188.
7. KING G.J.W., DICKIN E.A. and LYNDON A. The development of a medium sized centrifuge testing facility. *Proc. Int. Symp. on the Application of Centrifuge Modelling to Geotechnical Design*, Manchester, 1984, pp 25-46.
8. McLOUGHLIN J.P. *The behaviour and design of cantilever retaining walls in sand.* M.Sc.(Eng) Thesis, University of Liverpool, 1989.
9. CAQUOT A. and KERISEL J. *Tables for the calculation of passive pressure, active pressure and bearing pressure of foundations*. Gauthier-Villars, Paris, 1948.

The load displacement response of rigid retaining wall foundations

R. BUTTERFIELD, University of Southampton, UK

Introduction

A designer of a rigid retaining wall is frequently faced with the problem of predicting the displacement of the combined soil-wall system due to a set of forces determined from limiting value of active earth pressures and other, fixed, applied loads. Such forces, usually assumed to be independent of both time and any subsequent displacement of the system, will be as shown in Fig.l(a). For an unpiled wall, from the point of view of assessing the contribution of the rigid foundation movements to the total displacement of the system, these forces are equivalent to the three loads (V,H,M) shown in Fig.l(b). i.e. this part of the design problem reduces to that of predicting the load -displacement behaviour of a shallow, rigid foundation under combined vertical, horizontal and moment loading. A point to note ie that the 'sense' of H and M for a retaining wall foundation will always be as shown in Fig.l(b) rather than that of Fig.l(c). This is important in that, as is well established (ref.l), the 1(b) case is less stiff than 1(c), for specific (V,H,M,B) values.

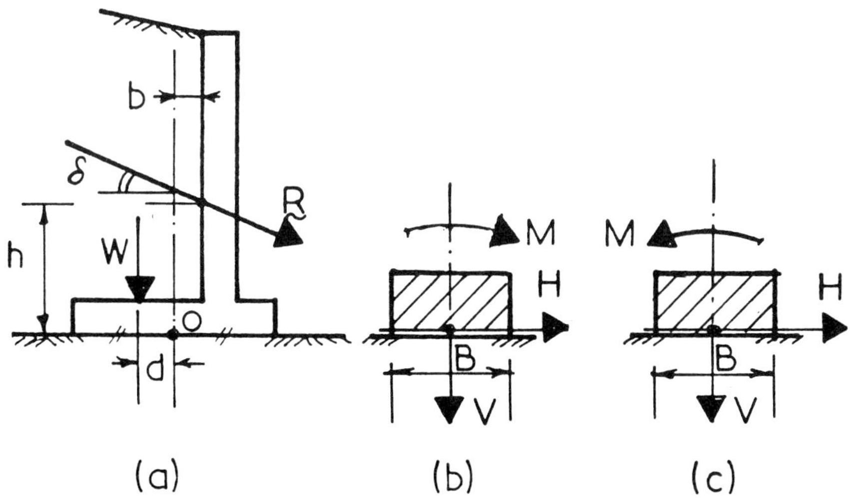

Fig. 1. Equivalence cf force systems on wall and foundation.

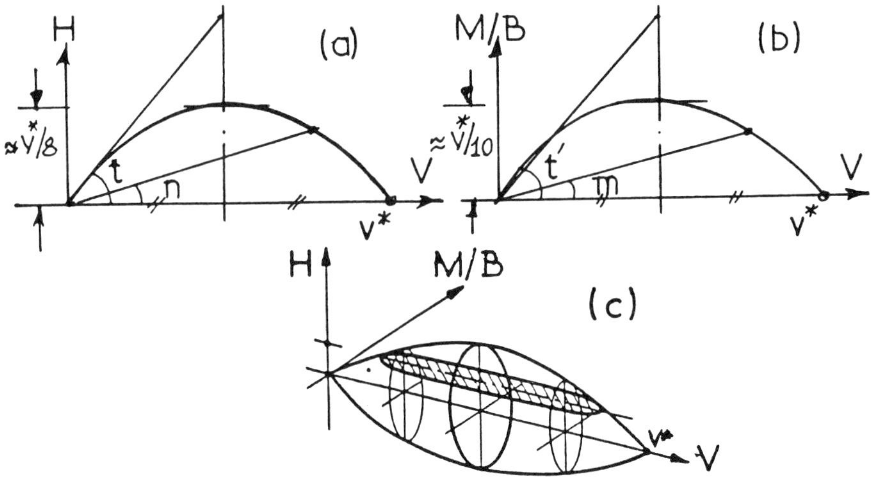

Fig. 2. Plots of parabolic failure/yield surfaces

Outline of analysis

Some years ago Butterfield and Ticof, (ref. 2) proposed that the maximum load capacity of a shallow, rigid, strip footing, of breadth B, supporting loads V and H and bearing on a frictional soil material (i.e. one failing on any plane on which $\tau = \sigma' \tan\phi'$) would, when plotted on V-H axes, lie on a simple parabolic curve with the equation $H = V.t.(V^* - V)/V^*$, as in Fig.2(a). V^* is the, assumed known, maximum central vertical load capacity of the footing and $t = \tan(\delta^o)$ represents the soil-footing friction angle. A similar argument leads to another parabola, relating V with M/B at failure under combined vertical load and moment loading, with the equation, Fig.2(b), $(M/B) = V.t'.(V^* - V)/V^*$. Meyerhof's widely used equivalent breadth concept (ref.3) when interpreted this way, generates a similar curve with $t' = 0.5 = \tan(26.5^o)$. Much evidence has now been accumulated supporting failure envelopes of this form (refs.4-5). A combination of the two figures provides a 3D plot in (V,H,M/B) space ($V > 0$), generating a 'cigar-shaped' surface with an elliptical cross-section in (H,M/B) planes Fig.2(c), (ref. 4). The load capacity of a retaining wall foundation will be confined to the +H and +M/B quadrant of such a surface, as explained above. The form of the complete surface, including ±M/B, has now been established by Gottardi (ref. 6).

These ideas were later extended (ref. 7) to propose that the surface in Fig.2(c) might be treated as a yield surface (with V^*, no longer a "failure" load, but a hardening parameter) and if displacement increments ($\dot{v}$, $\dot{u}$, $B\dot{\theta}$), were to be plotted along the corresponding (V,H,M/B) axes, in the manner of classical plasticity, then a plastic potential for the 'foundation -soil system' might be revealed. This has turned out to be a useful idea, for example

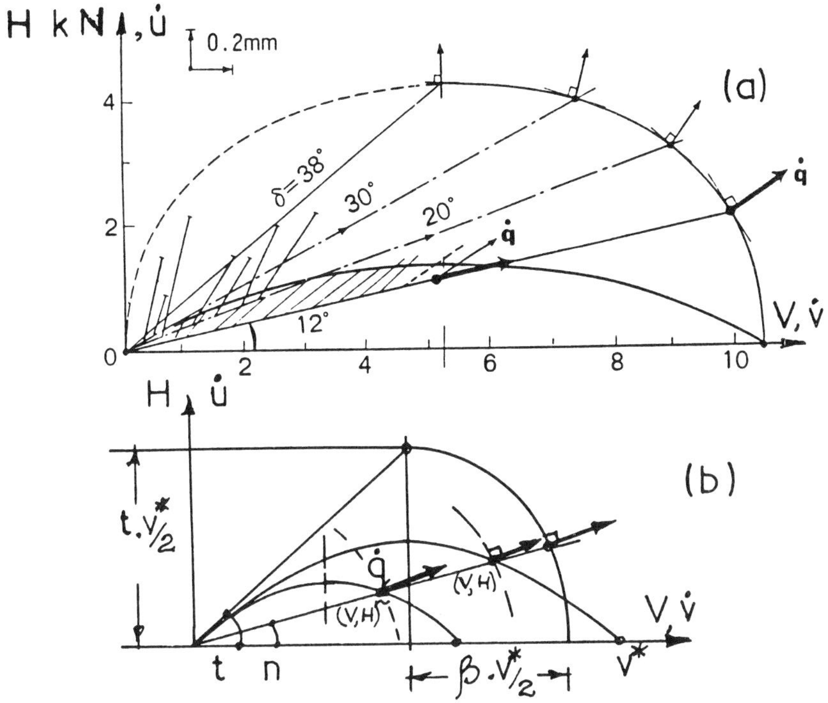

Fig. 3. Displacement increment vector interpretation for monotonic H-V loading: H = N.v

(i) Fig.3(a) which illustrates the near parallelism of the displacement vectors ($\dot{\tilde{q}} = \dot{\tilde{v}} + \dot{\tilde{u}}$) along monotonic, radial load paths in (V,H) space (ref.7). If the displacement increment vector directions are closely normal to an elliptical 'plastic potential' surface, $H^2 = V.t^2(V^* - V)$, as sketched in Fig.3(b), with the shape of the ellipse governed by its axis ratio (t/β), as shown.

(ii) similar results have been obtained (ref. 8) connecting ($\dot{v}$, $B\dot{\theta}$) along monotonic loading paths in (V,M/B) space with a 'plastic potential surface $(M/B)^2 = V.(t')^2(V^* - V)$. More recently, Nova and Montrasio (ref. 9) have developed the idea further,using rather different algebraic curves for the yield-surface and plastic-potential, in a work-hardening plasticity formulation and applied it to the analysis of model footings bearing on a very loose sand

(iii) Gottardi (ref. 6) performed a series of very precise experiments, using rough model-footings (500 x 100mm, on very dense sand), which included exploration of the ±M/B effect, described above, as well as both the load-path and load-increment-direction dependency of the displacement response in 3D. The parabolic yield-surface, elliptic plastic-potential model described above has now been elaborated to incorporate such developments in a work-hardening, non-associated flow plasticity model (ref. 10), which, despite its fearsome description, can load to surprisingly simple and compact results. It requires, as minimum input, only the two 'friction coefficients' (t,t') together with an assisted or measured, vertical load versus vertical displacement curve for a foundation of breadth B and the specific (V,H,M/B) loading path of interest.

The retaining wall foundation problem provides a relatively straightforward application of such an analysis in that

(a) the relative senses of H and M/B are fixed and
(b) the earth-fill loading forcer follow an approximately monotonic loading path in (V,H,M/B) space.

Typical soil loading forces, acting in addition to the weight of the wall and referred to the base centre-line 0 as shown in Fig.l(a), lead to

$$V = R.\sin(\delta) + W \; ; \;\; H = R.\cos(\delta) \; ; \;\; M = H.h - W.d - R.\sin(\delta).b$$

Any such loading path will be along a radial line ($F = \lambda.V$, say) in the (V,H,M/B) diagram, Fig.4, with the displacement increment, corresponding to $F^2 = H^2 + (M/B)^2$, given by $\dot{w}^2 = \dot{u}^2 + (B\dot{\theta})^2$. The remainder of the paper considers the response of a rigid, shallow, wall foundation to loading of this kind interpreted using the 'parabolas and ellipse' model of Fig.3(b).

Analytical results

A detailed development of the plastic analysis will be found in (ref. 10) and only results, with a minimum of theoretical explanation, will be quoted here. Friction coefficients (t,t') are assumed to be known, together with the

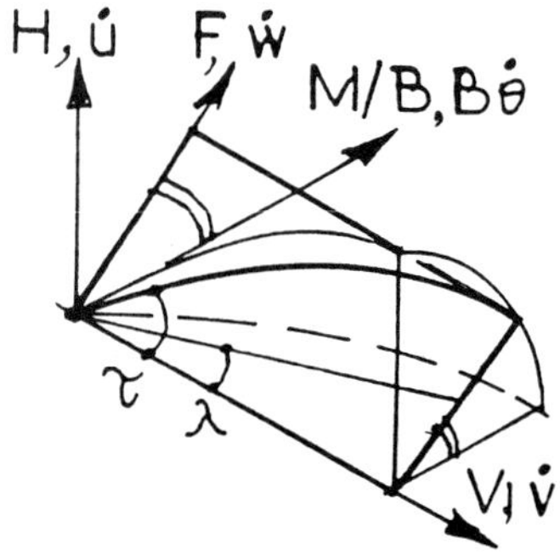

Fig. 4. F plane section of a yield surface

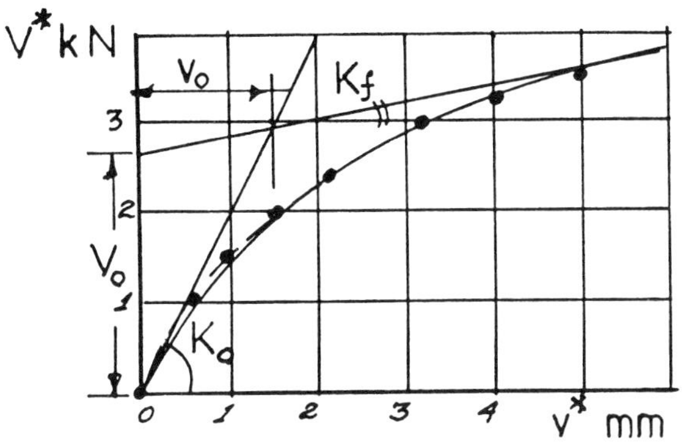

Fig. 5. Footing response for a vertical central load

vertical-load (V*) versus vertical displacement (v*) curve for the foundation, typically as shown in Fig.5, taken from Georgiadis and Butterfield (ref. 11). Such curves can be 'fitted' by an exponential equation of the form

$$V^* = V_o.\{1 - \exp(-v^*/v_o)\} + K_f.v^* \quad ; \text{ with } \quad K = V_o/v_o = (K_o - K_f) \qquad (1)$$

an equation proposed by Butterfield and Georgiadia (ref. 12), a reference which also provides guidance on the extrapolation of plate bearing test results from smaller to appreciably larger plates. In Fig.5 the fitted points were derived from parameter values of V_o = 3200 N; K_o = 2000 N/mm; K_f = 100 N/mm.

Equations (2,3), with these values of (V_o,K_o,K_f), are used in the analysis in which V* is always the leading apex of the expanding parabola along the V axis, shown in Fig.3(b), and plays much the name role in the hardening process as the 'mean effective stress' in critical state soil mechanics. Using the equations given previously for the parabolas and ellipses the following relationships can be established between applied load increments ($\dot{V}$,$\dot{H}$) and the associated displacement increments ($\dot{v}$,$\dot{u}$);

$$\begin{Bmatrix} \dot{v} \\ \dot{u} \end{Bmatrix} = \begin{Bmatrix} t \\ \dfrac{n.\beta^2.(t-n)}{t.(1-\rho)} \end{Bmatrix} (\dot{V} / (t-n).\partial) \qquad (2)$$

in which $\dot{H}$ =n.$\dot{V}$ defines the loading path (Fig.3(b)); $\delta=(\delta V^*/\delta v^*)=K.v^* .\exp(-v^*/v_o) + K_f$ is the gradient of the load-displacement curve relating V* to v* and ρ arises from the requirement that the current yield surface and the current plastic potential must both pass through the current (V,H) load point, Fig.3(b), whence (ref. 10)

$$\rho.t.\{t + \beta.\surd(t^2 - n^2 + \beta^2 n^2)\} = (t^2 + \beta^2 n^2) \qquad (3)$$

From (2) it follows that,

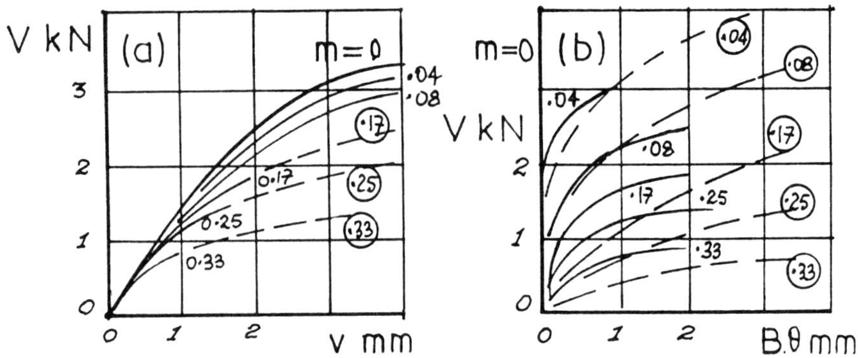

Fig. 6. Eccentric vertical load tests

$$(u/v) = (n/1-\rho).(\beta^2/t^2) \tag{4}$$

Therefore in order to obtain non-linear (u/v) curves, in line with model measurements (Fig.7) (i.e. the displacement vector increments are not quite parallel), β needs to vary throughout the hardening process. In the analysis used here $0.75 \le \beta \le 1.0$, changing at a rate governed by δ, was used ($\beta = 1$ defines an ellipse similar in shape to that originally hypothesised, Fig 3c). By replacing $H = nV$ and t in (2) by either $M/B = mV$ and t′ or by $F = \lambda V$ and τ, as in Fig.4, loading in the (M/B,V) and (F,V) planes can be analyzed similarly.

Experimental results

All the following experimental results are taken from Georgiadis and Butterfield (ref. 11) in which they were interpreted in an alternative, entirely

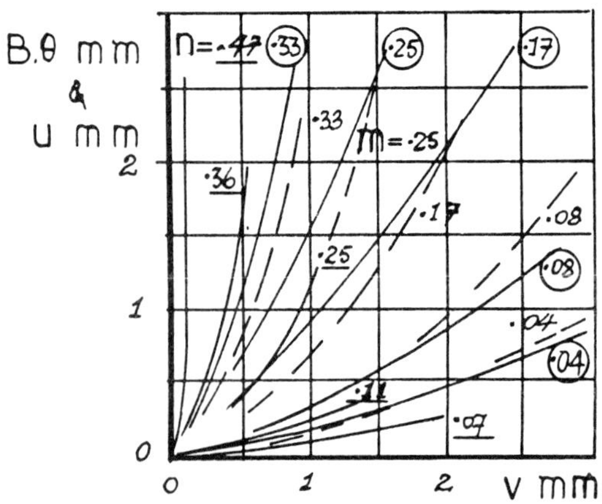

Fig. 7. Comparison between measured and predicted displacement ratios for tests shown in Figs 6 and 8

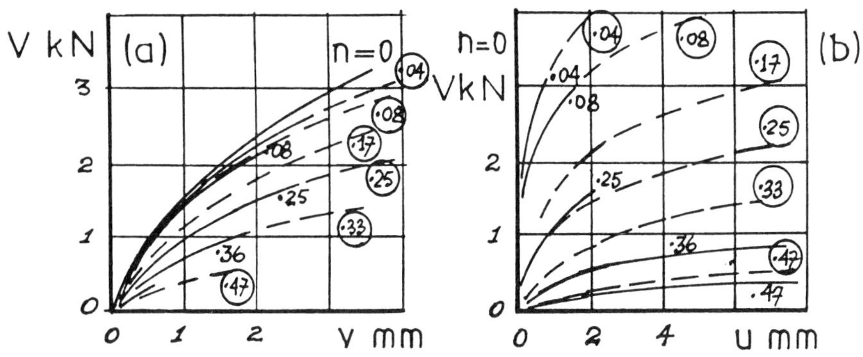

Fig. 8. Inclined central load tests

empirical manner. The measurements were obtained from a series of precise, axial, inclined, eccentric and inclined-eccentric loads tests on a rough, rigid model steel foundation (50 x 400 mm), supported on a dense (e=0.56, I_D≈75%), Leighton Buzzard sand bed ($\phi'= 46^o$ in plane strain). The basic axial central load test is shown in Fig.5 and the bold lines in Figs 6(a,b) are plots of V versus Bθ for a series of eccentric (ε), vertical load tests (M/B = V.ε/B = m.V, with m marked on the curves). If, lacking any better information, t'=0.5, the Meyerhof value, were to be assumed then the dashed curves in Figs. 6(a,b) result from applying equations (2,3). The V versus v predictions are quite good, even with this assumption, and the V versus B.θ curves not unreasonable. Fig.7 shows B.θ plotted against v for these tests (full lines) compared with the values calculated from eqn(4) and again the agreement is reasonable. Generally the ratio (t/t') = 5/4 (ref. 7) and therefore a value of t= 0.6 has been assumed in order to interpret the inclined, central-load tests shown by bold lines in Figs.8(a,b). The experimental results are how reproduced very satisfactorily indeed by the predictions of eqns (2,3), shown dashed.

However, as explained in the introduction, it is the eccentric, inclined loading test which is particularly relevant to retaining wall bases and the results of a series of such tests, Figs 9(a-d), will now be discussed. Figs 9(a,b,c) show, respectively, V versus v, u and B.θ (full lines) from a series of tests in which n ≡ m, (i.e. H/V = V.ε/V.B = ε/B = M/V.B). This means that at any stage of the test H = M/B and the load path lies in a plane passing through the V axis and interacting the H - M/B plane at 45°. Therefore, in Fig.3(b), F =H√2 = (M/B)√2 and if, for convenience, we assume the 'cigar-shaped' yield surfaces to be rotationally symmetric (t = t'), then λ = t√2 and w = u√2 = B.θ√2. If true, the latter prediction, that u = B.θ in such a test, is of considerable

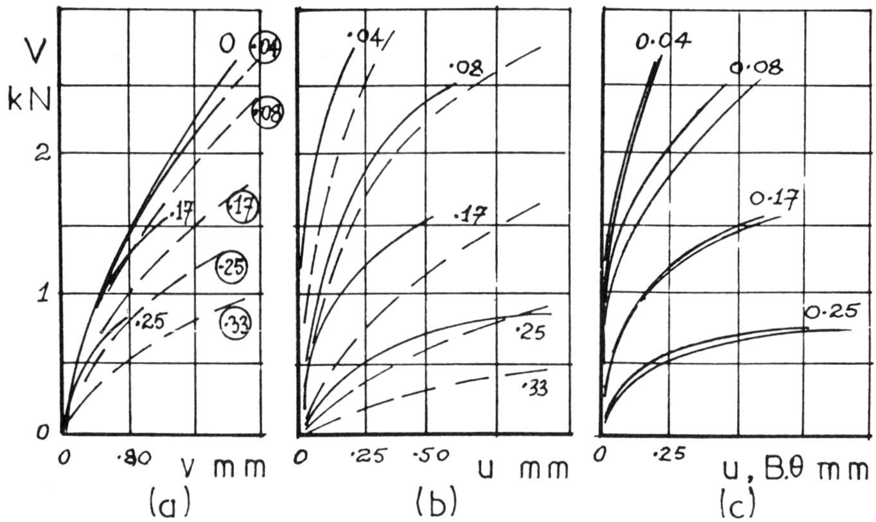

Fig. 9. Combined eccentric inclined loading with n = m

practical importance. Fig.9(c) shows the experimentally measured u and B.θ values superimposed with extremely good agreement between them.

The complete response of the footing is clearly very complicated and we now need to check whether the very simple expressions in eqns (2,3) provide a reasonable prediction of such behaviour (i.e. can the full result of V versus v and V versus u , and thereby V versus B.θ curves, be estimated from t = t′ and the vertical load test of Fig.5).The sense of the +H and -M/B loading in this test was contrary to that occurring in a retaining wall, resulting in a stiffened response as explained previously. This can be approximated in the simple model by arbitrarily increasing the value of t to 0.7, still not an excessive value in relation to $\phi'=46^{0}$. The results of such calculations are shown as dashed lines in Figs. 9(a,b). 'As before, the predicted and measured V versus v values, again quite well whereas the agreement between those relating V to u is less good. Nevertheless all the curves are of the correct general shape and they do provide practically useful guidance on the response of as rigid retaining wall base in a very simple way. The full analysis (ref.10) allows for more complicated 3D load paths, as well as elastic loading/unloading contributions and more sophisticated hardening mechanism, all of which add to both the accuracy of the predictions - and the complexity of the analysis!

Conclusions

The interpretation presented, of a series of model tests on rigid, dense-sand supported, foundations under different combinations of inclined and eccentric loads for which detailed displacement measurements had been made, is based on a new analytical model which possesses the following practically useful features

(i) extremely good predictions can be made of the very complicated, non-linear response of such a base under the general, three-dimensional loading conditions which arise in retaining wall foundations
(ii) the resulting analysis can be explained in a simple form which requires only values of "base-friction" parameters, together with an estimate of the vertical, central-load-displacement response of the foundation, and the applied-loading path to be known
(iii) a striking similarity is demonstrated between the horizontal load (H) versus horizontal displacement (u) response and the moment over breadth (M/B) versus breadth times rotation (B.θ) response of such a foundation when loaded such that n = m (i.e. H =M/B) at a specific value of vertical load
(iv) a graph, Fig.7, is provided which enables the relative magnitude of horizontal displacements and rotating to be predicted for any inclination of the applied load path (i.e. any value of n or m in the expressions H=n.V or M/B=m.V.
(v) despite the simplicity of the solution presented it is demonstrated that reliable results can be obtained from relatively crude estimates of the friction parameters when the foundation response is interpreted in the form shown in the Paper.

References

1. ZAHARESCU E. Sur le stabilite des fondations rigides. *Proc. 5th ICSMFE*, Paris, 1961, **1**, pp 867-871.
2. BUTTERFIELD R. and TICOF I. Design parameters for granular soils. *7th Eur. Conf. SMFE*, UK, 1979, **4**, pp 259-262.
3. MEYERHOF G.G. Some recent research on the bearing capacity of foundations. *Canadian Geotech. J.*, 1963, **1**, pp 16-26.
4. TICOF I. *Surface footage on sand under general planar loads.* PhD Thesis, University of Southampton, UK, 1978.
5. MONTRASIO L. and NOVA R. Assesamenti di una fondazione modello sotto carico inclinato: resultati sperimentali e modellazione matematica. *Rivista Italiana di Geotecnica*, 1988, **21**, No.1, pp 45-49.
6. GOTTARDI G. *The load-displacement behaviour of rigid surface footings on dense sand.* PhD Thesis, Universita de Padova, Italy, 1991.
7. BUTTERFIELD R. Another look at gravity platform foundations. *Proc. IMFE in Offshore Technology*, GSSM, Udine, Italy, 1981.

8. TSUI Y. G. *On the safe design of rigid foundations in granular materials*. MSc Dissertation, University of Southampton, UK, 1978.
9. NOVA R. and MONARASIO L. Settlement of shallow foundations on sand *Geotechnique*, 1991, **41**, No. 2, pp 243-256.
10. BUTTERFIELD R. A work-hardening plasticity model for the analysis of rigid foundations on sand. Submitted for publication to *Soil and Foundations*, Japan, 1992.
11. GEORGIADIS M. and BUTTERFIELD R. Displacements of footings on sands under eccentric and inclined loads. *Canadian Geotech. J.*, 1988, **25**, pp 119-212.
12. BUTTERFIELD R. and GEORGIADIS M. *A general interpretation of plot bearing tests*. Transport Research Board Record No. 810, 1981, pp 60-86.

The preliminary design of free embedded cantilever walls in granular soil

A.V.D. BICA and C.R.I. CLAYTON

Introduction

Free embedded cantilever walls are frequently used to retain relatively small heights of granular soil, for example in river protection works. Cantilever diaphragm and bored-pile walls are increasingly being used for higher walls, as part of routine temporary support during foundation construction.

Free embedded cantilever walls are routinely designed with methods based on limit equilibrium analysis. These methods assume the wall to be a rigid beam free to rotate about a point at some distance above the lower end (Fig. 1(a)). The soil is considered to be at failure everywhere near the wall.

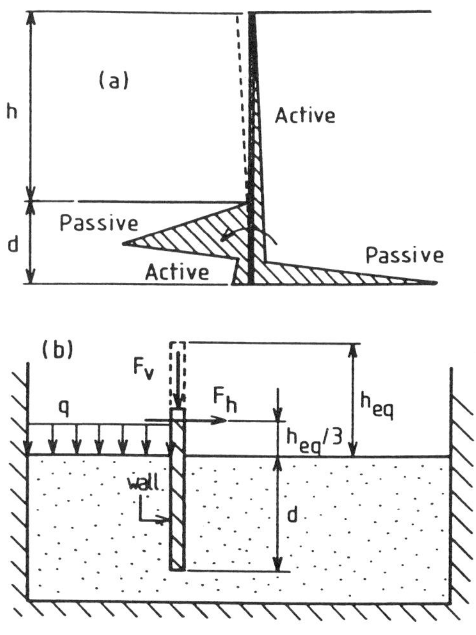

Fig. 1. Free embedded cantilever wall: (a) outline; (b) simplified modelling technique

Active or passive normal stresses are assumed to act on the wall, according to the relative movement with respect to the soil. The design value of depth of embedment of the wall is calculated by requiring equilibrium of both horizontal forces and moments. This calculation usually includes a factor of safety. The design value of maximum bending moment is calculated at the point of zero shear force.

The application of conventional design methods for free embedded cantilever walls in granular soil is not without problems. The details of the design, such as the theory used in the derivation of earth pressure coefficients, the magnitude and orientation of wall friction, the method of obtaining the angle of internal friction of the soil, and the definition and magnitude of the factor of safety, must all be assumed. The spectrum of acceptable assumptions is wide for some of these parameters. Significantly different values of depth of embedment and maximum shear force are obtained from different design assumptions.

A solution for this problem was suggested by Bica and Clayton (1989). It consisted of assessing the performance of each conventional design method by comparison with experimental data. Another solution is developed in this paper. It consists of developing design charts, based directly upon experimental data, without the use of intermediate limit equilibrium analysis.

Previous model wall experiments

Limited experimental data on the behaviour of free embedded cantilever walls in granular soil are available in the literature. Valuable data have been presented by Rowe (1951), Bransby and Milligan (1975), Lyndon and Pearson (1984), and in addition Fourie and Potts (1989) give the results of finite element computations.

Rowe (1951) performed model tests with free embedded cantilever walls in both sand and fine gravel. Earth pressures, displacements of the top of the wall, and depths of embedment at failure were measured in these tests. Bending moments were calculated from measured values of earth pressure. Direct shear tests were used to evaluate the angle of internal friction of the soil. Corresponding values of plane strain angle of internal friction were not given by Rowe (1951), but were estimated by Bica and Clayton(1989). Subsequently, Bransby and Milligan (1975) performed model tests with free embedded cantilever walls in coarse sand. Wall displacements, soil displacements, and depths of embedment at failure were measured, and rupture surfaces were observed in these tests. The value of plane strain angle of internal friction was obtained from simple shear tests. More recently, Lyndon and Pearson (1984) performed centrifuge tests with free embedded cantilever walls in glass ballotini. Earth pressures, wall displacements, and depths of embedment at failure were measured in these tests. Bending moments were evaluated by Bica and Clayton (1989) from reported values of earth pressure. Plane strain compression tests were used to evaluate the angle of internal friction.

Fourie and Potts (1989) performed non-linear finite element analyses of free embedded cantilever walls in clay. Zero pore water pressure was assumed, a condition analogous to model tests using dry granular soil. Excavation at one side of the wall was simulated, and wall displacements, soil displacements, normal and shear stresses on the wall, bending moments, and depths of embedment at failure were calculated.

New model wall experiments

New model wall experiments have recently been completed, at the University of Surrey, to provide further data on the behaviour of free embedded cantilever walls in granular soil. A simplified technique was developed for these experiments. Details are given in Bica (1991).

The simplified technique used in these experiments consisted of partially modelling the free embedded cantilever wall. As shown in Fig. 1(b), only the segment of the wall situated below the excavation level (depth d) was modelled. Both the retained soil and the segment of wall above the excavation level (height h_{eq}) were not modelled in these experiments. Their influence on the behaviour of both the soil and the segment of wall below the excavation level was taken into account in the following way [Fig. 1(b)]. The resultant of active normal stresses situated above the excavation level was simulated by an equivalent horizontal force (F_h) applied to the model wall. Similarly, the resultant of active shear stresses situated above the excavation level was simulated by an equivalent vertical force (F_v). The vertical normal stress applied at the excavation level by the retained soil was simulated by an equivalent uniform surcharge (q). In these experiments, a linear distribution with depth was assumed for active normal stresses above the excavation level. As a result, the equivalent horizontal force was applied to the model wall at a height above the sand surface equal to one-third of the simulated height of retained soil (h_{eq}).

More importance was given in these experiments to the measurement of normal stresses and shear stresses immediately below the excavation level. They were considered to be less well-known, by comparison with both active normal stresses and active shear stresses above the excavation level. Milligan and Bransby (1976) showed, for the case of a model wall with a fixed centre of rotation at a small distance above the lower end, that both active normal stresses and active shear stresses were approximately linear with depth. These stresses were fully mobilized after a small rotation of the wall. As a result, they were not measured in the new model wall experiments, because substantially new data were not expected. Testing procedures could then be simplified, by adopting the technique of partially modelling the free embedded cantilever wall, without significant loss of information.

The model wall was loaded in stages; each new stage corresponded to a larger value of equivalent height of retained soil (h_{eq}). Different values of F_h, F_v and q, were applied at each stage. These values were computed beforehand. In these calculations, Caquot and Kerisel's earth pressure coefficients (see Caquot and Kerisel, 1943) were used, and the value of angle of wall friction was assumed. A value of δ equal to one-half the plane strain angle of internal friction was assumed for the experiments reported in this paper (this assumption is routinely made in practice). For the experiments reported in this paper, the wall was coated with fine sand. Its surface was rough, a condition simulating most real walls. Experiments with other assumptions have also been performed and results are given in Bica (1991).

The model wall was 0.35 m deep. It was located inside a glass-walled tank, 1.25 m long, 0.33 m wide, and 0.47 m deep. The tank was filled with fine Leighton Buzzard sand (mean grain size equal to 0.11 mm). The sand was deposited air-dry from a hopper. Horizontal displacements of the wall, normal and shear stresses on the wall, and equivalent heights of retained soil (h_{eq}) at failure were measured in these experiments. Bending moments were calculated from measured values of normal stress on the wall. Direct shear tests were used to evaluate the plane strain angle of internal friction. Results from these tests were interpreted according to procedures suggested by Bolton (1986), and Jewell and Wroth (1987).

Empirical design charts

As indicated by Bica and Clayton (1989), the comparison between design methods and experimental data is more conveniently performed by expressing both the depth of embedment of the wall and the maximum bending moment in non-dimensional form. By so doing, data from previous and new model wall experiments may be compared. Without this procedure, this would be impossible, because corresponding heights of retained soil were different in each set of published data. The depth of embedment of the wall was normalized with respect to the height of retained soil (d/h). The maximum bending moment per unit length of wall was normalized with respect to the product of the soil density by the third power of the height of retained soil ($M_{max}/\gamma h^3$)

Normalized depth of embedment

Figure 2 shows experimental values of d/h at failure in terms of corresponding values of plane strain angle of internal friction (ϕ). Data from both previous and new model wall experiments are plotted in Fig. 2. As noted by

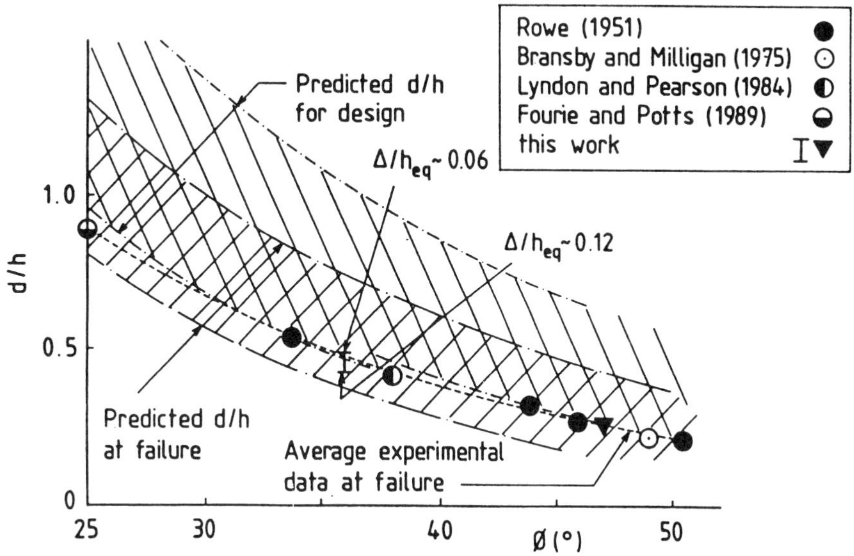

Fig. 2. Experimental values of depth of embedment at failure (Δ: horizontal displacement at the top of the equivalent retained soil).

Bica and Clayton (1989), data from previous experiments, including the one finite element analysis, appear to be consistent within themselves. A virtually smooth average line (also shown in Fig. 2) could be traced connecting these data points. As shown by this line, the larger the value of plane strain angle of internal friction, the smaller is the value of d/h at failure.

For the new model wall experiment performed with dense sand ($\phi = 47°$), a well-defined point of failure was obtained. The experimental value of d/h at failure is represented in Fig. 2 by a data point. Conversely, for the experiment performed with loose sand ($\phi = 36°$), no clear point of failure could be identified. Even for the final stages of the experiment, when very large horizontal displacements were measured, the model wall always reached a stable position, so that the applied horizontal force could be further increased. For this case, a single value of d/h at failure is not indicated in Fig. 2, but a range of plausible values is shown. The upper end of this range gives the value of d/h which corresponds to a horizontal displacement at the top of the wall of about 6% of the equivalent height of retained soil. The lower end gives the value of d/h which corresponds to a horizontal displacement at the top of the wall of about 12% of the equivalent height of retained soil. Similar problems of interpretation were reported by Rowe and Peaker (1965) for their passive earth pressure experiments.

Values of d/h at failure given by the new model wall experiments appear to be consistent with corresponding values from previous experiments (Fig.

2). The experimental relationship between d/h at failure and plane strain angle of internal friction (expressed by the average line in Fig. 2) seems therefore to be robust. This average line may be considered as an adequate reference either for assessing the performance of conventional design methods or for producing a design chart.

Bica and Clayton (1989) adopted the average line in Fig. 2 as the reference for assessing the performance of a number of conventional design methods for free embedded cantilever walls in granular soil. Fig. 2 shows the range of values of d/h at failure predicted by these methods. Also shown is the range of design values of d/h predicted by the same methods. Both ranges are unacceptably wide. Bica and Clayton (1989) suggested conventional design methods which can predict acceptable values of d/h at failure (i.e. similar to the experimental data). Another list was provided, with methods capable of giving satisfactory estimates of design values of d/h (i.e. with a factor of safety on depth of embedment between 1.3 and 1.7). It was noted that few conventional design methods appeared simultaneously on both lists.

A design chart may overcome some of these problems. For example, once the plane strain angle of internal friction of the soil is known, the design value of d/h may be obtained by multiplying the experimental value of d/h at failure (as given by the average line in Fig. 2) by a factor of safety (F). The value of d/h at failure may also be estimated with equation 1. The design value of d/h may then be estimated with equation 2.

$$d/h_{(failure)} \approx (2/3)\ e^{-[(\phi-30)/18]} \qquad (\phi \text{ in degrees}) \qquad (1)$$

$$d/h = F \cdot [d/h_{(failure)}] \qquad (2)$$

As shown by Fig. 3, equation 1 is a good approximation to the average line in Fig. 2. Other curves, determined with equation 2, are also plotted in Fig. 3. Therefore, once the value of plane strain angle of internal friction is known, and a value of factor of safety has been selected, the corresponding design value of d/h may be interpolated from these curves. Fig. 3 can be considered as a design chart.

Normalized maximum bending moment

Figure 4 shows experimental values of $M_{max}/\gamma h^3$ in terms of corresponding values of ϕ and d/h. Data from both previous and new model wall experiments are plotted in Fig. 4. As noted by Bica and Clayton (1989), data from previous experiments show an increase in $M_{max}/\gamma h^3$ as d/h is increased beyond that required purely for equilibrium. In addition, the larger the plane strain angle of internal friction, the smaller is $M_{max}/\gamma h^3$, for a constant value of d/h.

For the new model wall experiments, values of $M_{max}/\gamma h^3$ are smaller for dense sand ($\phi = 47°$) than for loose sand ($\phi = 36°$), for a constant value of d/h.

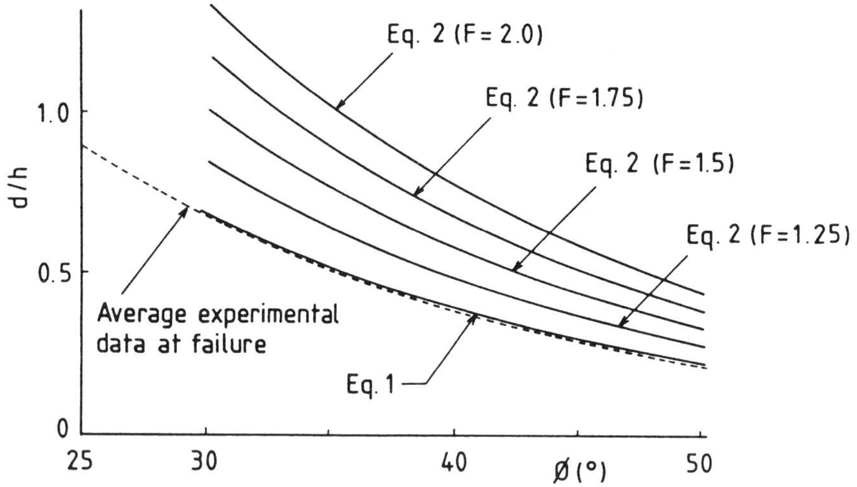

Fig. 3. Design chart for evaluation of depth of embedment

In addition, the value of $M_{max}/\gamma h^3$ increases with increasing values of d/h, for the same value of ϕ, as shown in Fig. 4. For both loose sand (ϕ = 36°)and dense sand (ϕ = 47°), values of $M_{max}/\gamma h^3$ corresponding to d/h equal to 0.70 are consistent with data by Rowe (1951), which corresponded to d/h equal to 0.67. On the other hand, values of $M_{max}/\gamma h^3$ corresponding to d/h equal to 1.06 appear to be somewhat larger than similar data from Rowe(1951), which corresponded to d/h equal to 1.0. There is therefore some uncertainty

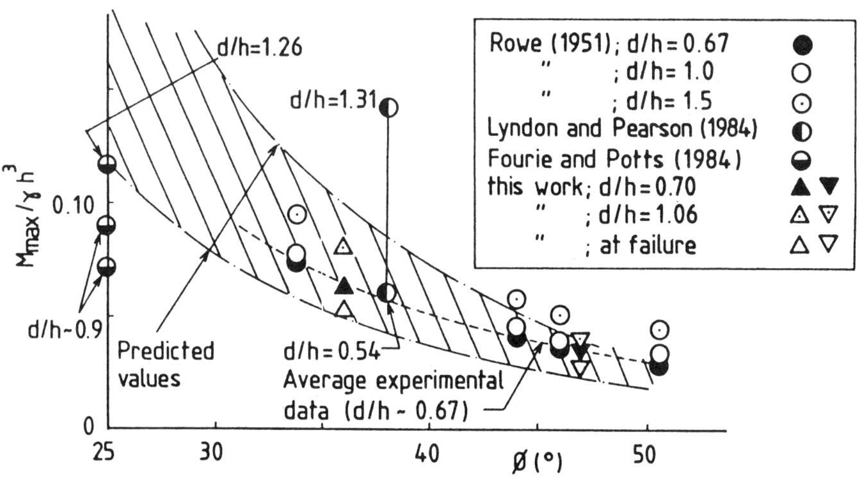

Fig. 4. Experimental values of maximum bending moment

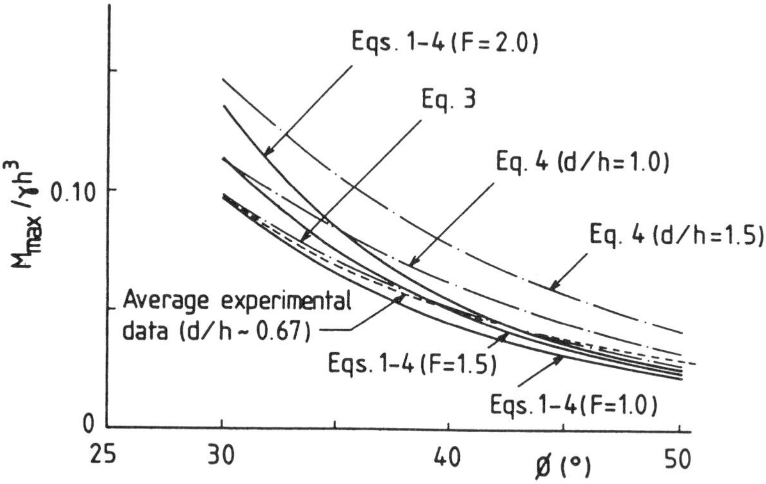

Fig. 5. Design chart for evaluation of maximum bending moment

about the magnitude of the rate of increase of $M_{max}/\gamma h^3$ with increasing values of d/h.

Figure 4 shows the range of values of $M_{max}/\gamma h^3$ predicted by conventional design methods, previously shown in Bica and Clayton (1989). This range is somewhat wide. Some methods appear to underpredict the maximum bending moment, particularly for high values of ϕ. This includes methods which evaluate the maximum bending moment with a factor of safety equal to 1.0, such as that proposed by Padfield and Mair (1984). It is interesting to note that the use of a factor of safety equal to 1.0 is consistent with the assumption (which was contradicted by experimental data in Fig. 4) that $M_{max}/\gamma h^3$ is independent of d/h. Therefore, by considering data from previous model wall experiments, Bica and Clayton (1989) recommended to evaluate the maximum bending moment with a factor of safety larger than 1.0. The new data supported this recommendation.

A design chart may be useful to overcome these problems. An average line can be traced in Fig. 4 through data points corresponding to d/h equal to 0.67 (Rowe's model wall experiments) and 0.70 (new model wall experiments). As shown by Fig. 5, this average line is satisfactorily approximated by equation 3. However, the main difficulty is to establish which rate should be adopted for the increase of $M_{max}/\gamma h^3$ with increasing values of d/h. Examination of Fig. 4 reveals a somewhat large scatter in the measured rates of increase of $M_{max}/\gamma h^3$ with d/h. As a compromise solution, equation 4 is proposed for use in conjunction with equation 3. Equation 4 is an approximation to rates of increase of $M_{max}/\gamma h^3$ observed in the new model wall experiments. Fig. 5 shows curves representing equation 4 for different values of d/h.

$$M_{max}/\gamma h^3{}_{(d/h \sim 0.67)} \approx 0.095 \cdot e^{-[(\phi-30)/16]} \qquad (\phi \text{ in degrees}) \quad (3)$$

$$M_{max}/\gamma h^3 \approx (M_{max}/\gamma h^3)_{(d/h \sim 0.67)} \quad e^{[(d/h)-(2/3)]/2} \quad (4)$$

Other curves are also plotted in Fig. 5. These curves were determined with equations 1 to 4. Once the value of plane strain angle of internal friction is known, and a value of factor of safety has been selected, the corresponding design value of $M_{max}/\gamma h^3$ may be interpolated from these curves. Fig. 5 can therefore be considered as a design chart.

Figures 3 and 5 may be used for preliminary design. Both figures assume a homogeneous, dry or completely submerged (but without seepage) granular soil, and a stiff wall. The value of plane strain angle of internal friction of the soil is required. It may be determined by plane strain shear strength tests or, alternatively, by means of correlations with values from other tests (such as the triaxial compression test). Provided the magnitude of wall displacement is not critical, a factor of safety equal to 1.5 may be adopted, thus producing designs consistent with previous experience. Otherwise, an explicit calculation of wall displacements is required (Fourie and Potts, 1989; Simpson, 1984).

Conclusions

New model wall experiments have been completed, at the University of Surrey, to provide further data on the performance of free embedded cantilever walls in granular soil. These results fit well with previously normalized data set. The trend of increasing bending moment with depth of embedment, for a constant plane strain angle of internal friction, which was postulated on the basis of the previously available experimental results, is confirmed. These data have been re-interpreted in the form of charts, which the Authors suggest may be used for preliminary design purposes.

References

BICA A.V.D. *A study of free embedded cantilever walls in granular soil.* PhD Thesis, University of Surrey, Guildford, 1991.

BICA A.V.D. and CLAYTON C.R.I. Limit equilibrium design methods for free embedded cantilever walls in granular materials. *Proc. Inst. Civ. Engrs, Part 1*, 1989, pp 879-898.

BOLTON M.D. The strength and dilatancy of sands. *Geotechnique*, 1986, **36**, 1, pp 65-78.

BRANSBY P.L. and MILLIGAN G.W.E. Soil deformations near cantilever sheet pile walls. *Geotechnique*, 1975, **25**, No.1, pp75-195.

CAQUOT A. and KERISEL J. *Tables for the Calculation of Passive Pressure, Active Pressure and Bearing Capacity of Foundations*. Gauthier-Villars, Paris, 1948.

CLAYTON C.R.I. and MILITITSKY J. *Earth pressure and earth retaining structures*. Surrey University Press, London, 1986.

FOURIE A.B. and POTTS D.M. Comparison of finite element and limit equilibrium analyses of an embedded cantilever retaining wall. *Géotechnique*, 1989, **39**, 2, pp175-188.

JEWELL R.A. and WROTH C.P. Direct shear tests on reinforced sand. *Géotechnique*, 1987, **37**, 1, pp 53-68.

LYNDON A. and PEARSON R.A. Pressure distribution on a rigid retaining wall in cohesionless material. *Proc. Symp. Application of Centrifuge Modelling to Geotechnical Design*, Manchester, 1984 , pp 271-280.

MILLIGAN G.W.E. and BRANSBY P.L. Combined active and passive rotational failure of a retaining wall in sand. *Géotechnique*, 1976, **27** , 2, pp 473-494.

PADFIELD C.J. and MAIR R.J. *Design of retaining walls embedded in stiff clays*. CIRIA Report 104, Construction Industry Research and Information Association, London, 1984.

ROWE P.W. Cantilever sheet piling in cohesionless soil. *Engineering*, 1951, 7, pp 316-319.

ROWE P.W. and PEAKER K. Passive earth pressure measurements. *Géotechnique*, 1965, **15**, pp 57-78.

SIMPSON B. Computational analysis of flexible retaining walls. *Proc. Symp. Soil Structure Interaction*, Institution of Structural Engineers, London, 1984.

Normalization of analytical results for anchored bulkhead design

D. J. HAGERTY, University of Louisville, USA, and
M. M. NOFAL, ERCE Corporation, Knoxville, USA

Introduction

A number of design methods are in widespread use for anchored bulkheads (ref. 1) and computer programs for analysis are available (ref. 2-4). However, these programs are not intended to make the design most economical or efficient. No guidance is given on the selection of anchor rod placement, for example. No information is generated on the interaction of the input design variables and the output (anchor rod force, required depth of embedment, moment and deflection in the wall). For preliminary design and for cost estimating, extensive computer-based analysis may not be justified. This paper provides guidance for preliminary design and illustrates the interactions among design input parameters and resulting design predictions of behaviour, for bulkheads in sand.

Parameter groupings

Figure 1 shows a schematic diagram of a dredged anchored bulkhead. In the cases evaluated herein, the soil retained by the bulkhead is assumed identical to the foundation soil, a cohesionless sand. Above the water table in the retained soil, a moist unit weight characterizes the sand, but values of shear strength parameters in that zone are no different from those for the saturated soil. In these analyses, the angle of friction between the bulkhead and the retained soil has been assumed in all cases to be equal to one-half the angle of shearing resistance.

Also, since these guidelines are intended only for bulkheads in sand, the water level in the retained soil is assumed to be equal to the water level above the dredge line. To facilitate comparison of various situations, the site geometry for each analysis has been formulated in terms of dimensionless ratios: Z_a/Z_d and Z_w/Z_d.

The input parameters for any situation thus consist of values of Z_a/Z_d, Z_w/Z_d, surcharge pressure (Q), phi, and unit weight values for the soil above and below the watertable. The predicted output parameters include: bending moment in the bulkhead wall per unit of wall length, M; anchor force per unit

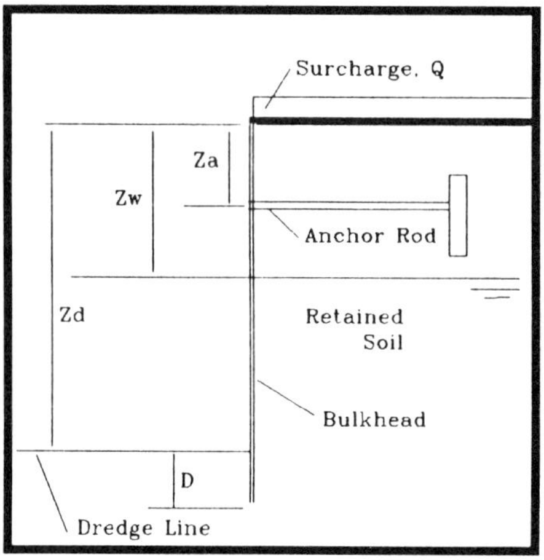

Fig. 1. Bulkhead

of wall length, A_p; required depth of bulkhead embedment below the dredge line, D; and deflection in the bulkhead. It is helpful to express the deflection as scaled deflection multiplied by the modulus and moment of inertia (per unit length) of the bulkhead section.

To simplify computations, the retained soil can be characterized by an average unit weight which would produce the same total lateral force on the bulkhead as would the actual unit weights of the two layers of soil:

xx $$\gamma_a = [\gamma_m Z_w^2 + \gamma_{sub} h^2 + 2\gamma_m Z_w h]/Z_d^2 \quad [1]$$

where h= Z_d - Z_w.

The output parameters likewise can be normalized. Three of the four output parameters can be divided by average unit weight, and dredge depth raised to a power, to yield generalized parameters:

$$\text{generalized moment} = GM = M/[\gamma_a(Z_d)^3] \quad [2]$$

$$\text{generalized anchor pull} = GAp = A_p/[\gamma_a(Z_d)^2] \quad [3]$$

$$\text{generalized deflection} = GDef = Def/[\gamma_a(Z_d)^5] \quad [4]$$

For embedment depth, only division by dredge depth is required:

$$\text{generalized required depth} = GD = D/[Z_d] \quad [5]$$

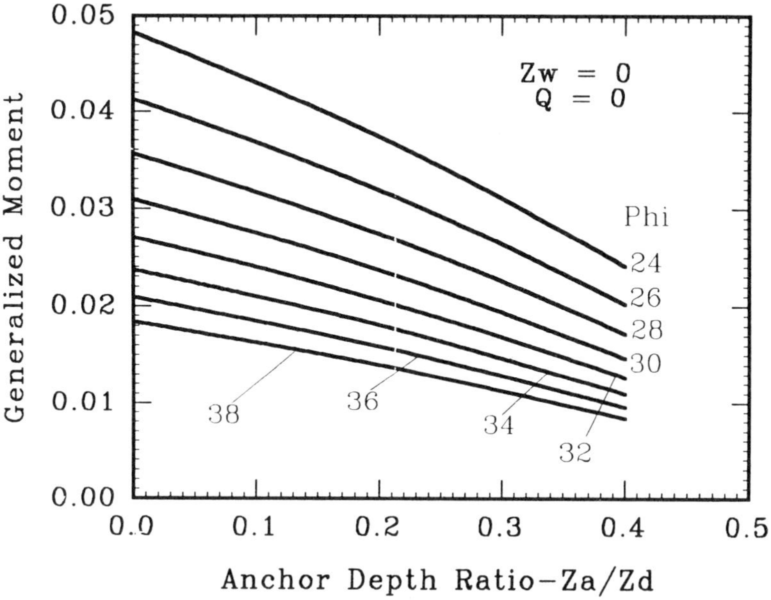

Fig. 2. Generalized Maximum Moment in Bulkhead

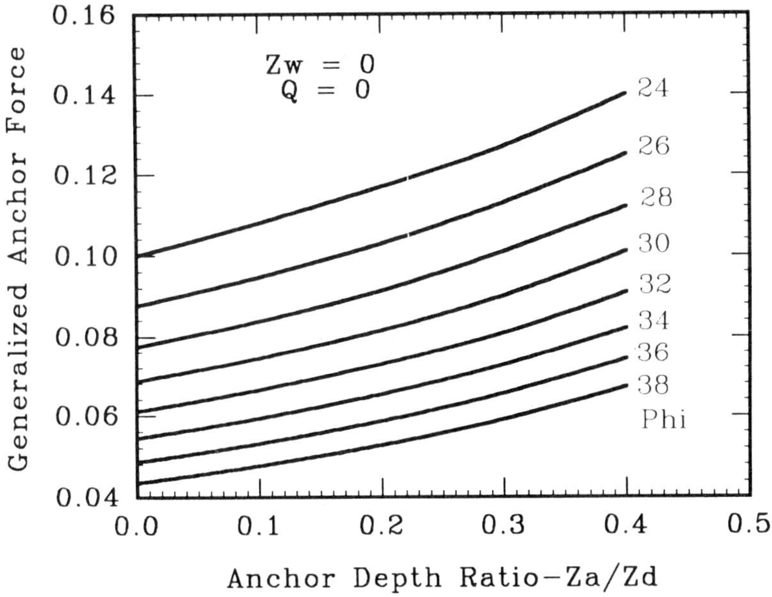

Fig. 3. Generalized anchor force/unit length

Methods of analysis

Methods for analysis of bulkheads can be classified as either a) the free earth support method or b) the fixed earth support method or a variation on that method (ref. 1). In the free-earth support method, the soil below the dredge line is assumed to have reached its limiting shear strength throughout the depth of bulkhead embedment. The bulkhead is considered to be stiff and unyielding relative to the soil; failure is assumed to occur by rotation about the anchor location. The required depth of embedment is found by setting the sum of moments due to pressures on each side of the bulkhead equal to zero. Then the anchor force is found by setting the sum of horizontal forces equal to zero. Since the full shear resistance of the soil in front of the bulkhead is unlikely to be mobilized without unacceptable deflections and settlements; factors of safety are applied to the depth of embedment and to the anchor force. After the required depth and anchor force are found, shear and bending moment in the bulkhead can be found as functions of depth. Deflection can be found by assuming a point of bulkhead fixity or the deflection at a given point, or by using calculated soil pressures in a finite element model of the system. The moment computed from free-earth support analyses may be overly conservative since many bulkheads are not inflexible compared to the soil into which they are installed. Rowe (ref. 5) developed a technique for reducing the maximum bending moment computed according to free-earth support analyses, based on the flexibility of the bulkhead.

The procedure for using Rowe's technique has been given by Bowles (ref. 6). In the fixed-earth support method, it is assumed that the soil beneath the dredge line restrains the bulkhead deformation. The user of this method assumes a point of fixity below the dredge line and the net soil forces below that point are replaced by a force acting away from the retained soil. The resultant bending moment diagram is constructed and the tangential deviation of the point of fixity is computed. The point of fixity is relocated until computed deviation is within a tolerable error from zero. Many designers avoid this cumbersome procedure by using the simpler free-earth support method and using Rowe's method to reduce the computed bending moments, based on the properties of the bulkhead member and the height of the bulkhead (as combined in Rowe's flexibility number). In this study, only the free-earth support method was used. Passive resistance was calculated on the assumption of a logarithmic spiral failure surface. Active pressures were calculated using Coulomb coefficients.

Input-output relationships

A large number of analyses were done to find the influence of input parameters on predicted bulkhead behaviour and requirements. In order to take all of the input parameters into account in obtaining an estimate of bulkhead

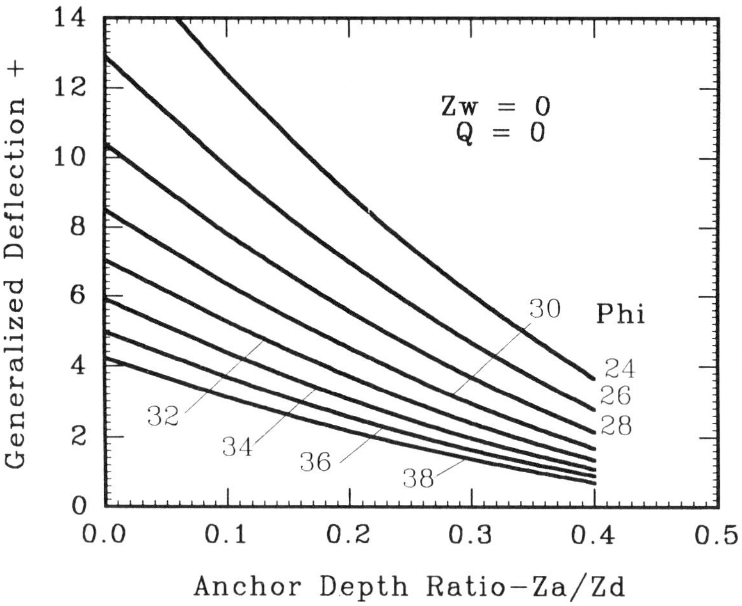

Fig. 4. Generalized Maximum Positive Deflection in Bulkhead

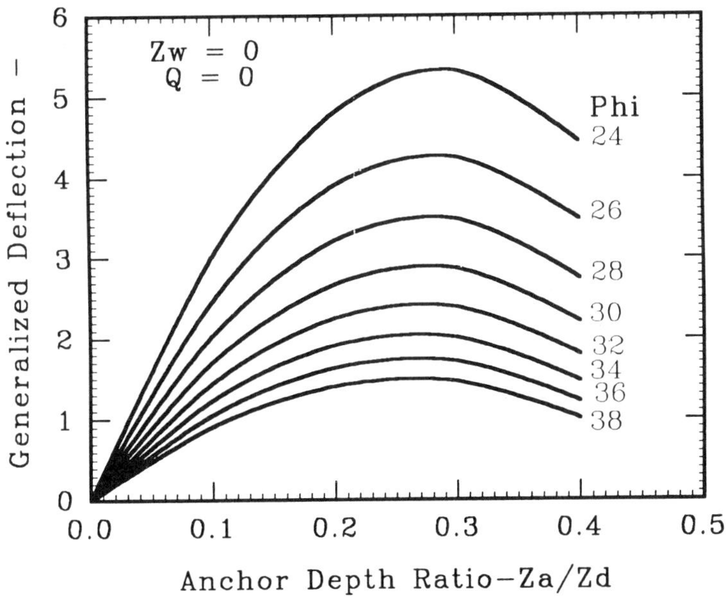

Fig 5. Generalized Maximum Negative Deflection in Bulkhead

behaviour and requirements, the following procedure was used:

1. The effect of depth to water in the retained soil was reflected partially in the value of averaged unit weight, γa, computed in Eq. 1;
2. The generalized parameter groups defined by Eqs. 2, 3, 4, and 5 were used to quantify output; and
3. Figures 2 through 6 were developed to show the variation in generalized output parameters as functions of angle of shearing resistance and location of anchor (given by Z_a/Z_d).

Figure 7 illustrates the influence of surcharge pressure on generalized maximum moment, as a function of angle of shearing resistance, phi, and anchor depth ratio, Z_a/Z_d. The ratio of generalized moment with surcharge to generalized moment without surcharge varied almost linearly with surcharge pressure, as shown. Similar behaviour was found for all output parameters. Space did not permit illustration of these variations in this paper. A detailed account of the effects of surcharge pressure, depth to water and difference in water levels (level within the soil retained by the bulkhead compared to water level in front of the bulkhead) has been presented elsewhere (ref. 7).

If the depth to water within the retained soil is less than the depth to free water in front of the bulkhead, the maximum moment and anchor force can be estimated conservatively by modifying the average unit weight used to develop the generalized output parameters. This modified unit weight is given below.

$$\gamma_a \text{ mod} = \gamma_a + [(\gamma_w(dzw)/Ka)(dzw + 2h)/(Z_d)^2] \qquad [6]$$

where dzw = difference in water levels, Ka = active earth pressure coefficient, and γ_a, h, and Z_d are as defined previously.

If this modified unit weight is used to define output parameters, predicted maximum moment is overestimated by 3 to 4 percent and anchor force is overestimated by 7 to 8 percent, for $dzw/Z_d = 0.1$ (the difference in water levels is hardly likely to exceed this value). Positive and negative maximum scaled deflections are underestimated 1 to 4 percent, but required depth of embedment is underestimated 9 to 10 percent, using this approximation. Obviously, all final design should be done using a method which incorporates the influence of difference in water levels; output parameters predicted using the unit weight defined by Eq. 6 are only crude estimates of the required values.

Units

If unit weight is stated in kilonewtons per cubic metre and depth to dredge line is stated in metres, anchor force is obtained in kilonewtons per meter of

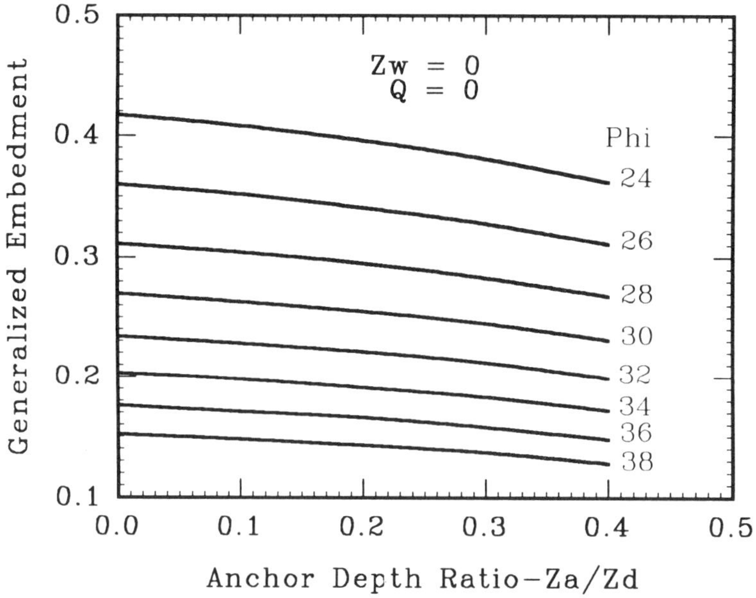

Fig. 6. Generalized Required Embedment of Bulkhead weight is given below:

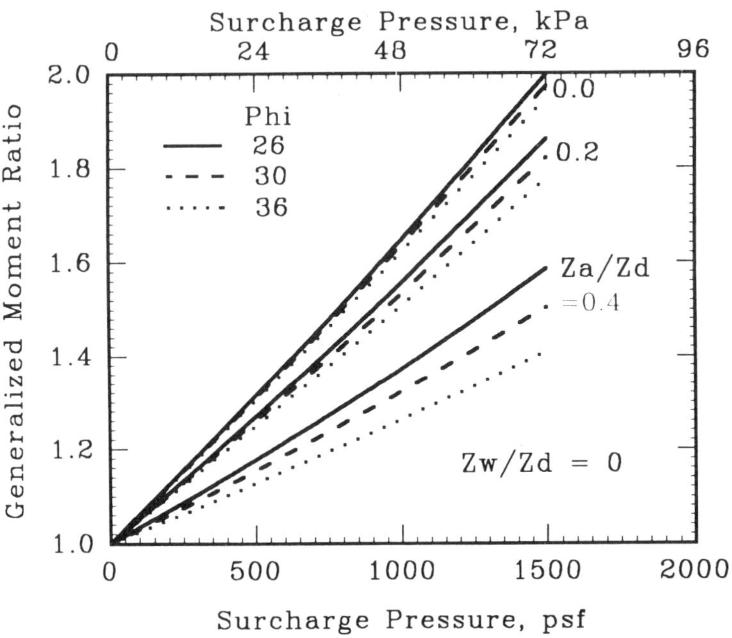

Fig. 7. Effect of Surcharge on Generalized Maximum Moment

bulkhead length when generalized anchor force, GAp, is multiplied by average unit weight, γ_a, times depth to dredge line squared, $(Z_d)^2$. Scaled deflection is estimated in this paper so that division by modulus of elasticity (psi) and moment of inertia (in^4) will yield deflection in inches; bulkhead moment of inertia must be computed as in^4 per foot of bulkhead length. For SI units, divide scaled deflection by modulus of elasticity (MPa) and moment of inertia ($m^4 \times 10^{-6}/m$), and then multiply by 57.88 to produce real deflection in centimetres. Note that moment of inertia must be computed per meter of bulkhead length.

In all of the cases analyzed, it was assumed that the angle of bulkhead-soil friction was equal to one-half the angle of shearing resistance. This assumption may not be valid in a particular case; bulkhead-soil friction varies with bulkhead material and condition. Nevertheless, the relationships shown in Figures 2 through 7 illustrate very clearly the interactions among design input parameters and the values of output parameters generated in analysis. Research on these interactions is being continued.

The effect of surcharge pressure is most important, as indicated in Fig. 7. The effect of depth to water table is much smaller, in a given situation, than the effect of surcharge. As indicated in figure 7, the changes in output parameters caused by a change in surcharge pressure are dependent on the location of the anchor rod and upon the angle of internal friction of the retained soil and the foundation soil. The depth to the water table also has a slight influence on the effect of a change in the surcharge.

The information shown in these figures can be easily used to develop a methosology for optimal design, if material costs, equipment costs and labour requirements are known (ref. 8).

References

1. Lacroix L. and Atmatzidis D. K. Design, construction and performance of anchored bulkheads. Proceedings 11th Ohio River Valley Soil Seminar, Kentucky Geotechnical Groupl ASCE, Louisville, USA, 1980, 3-1 - 3-17.
2. Bowles J. E. Foundation analysis and design. McGraw-Hill, New York, 1988, 4th edn, 592
3. Dawkins W. P. User's guide: Computer program for design and analysis of sheet pile walls by classical methods (CSHTWAL). Report 1 - Computational processes. US Army Engineer Waterways Experiment Station, Vicksburg, MS, USA, 1981.
4. Duncan J. M. FES program for bulkhead analysis. Virginia Polytechnic Institute and State University, Blacksburg, VA, USA, personal communication, 1988.
5. Rowe P. W. Anchored sheet pile walls. Proceedings of the Institution of Civil Engineers, Part 1, 1952, vol. 1, 27 - 70.

6. Bowles J. E. ibid., 629.
7. Hagerty D. J. and Nofal M. M. Design guidlines for dredged bulkheads in sand. Canadian Geotechnical Journal, 1991.
8. Nofal M. M. Computer-aided design and construction cost estimation for anchored bulkheads in sand. M.Engrg Thesis, University of Louisville, USA, 1990.

Predicting earthquake-induced tilt of gravity retaining walls

R.V. WHITMAN, Massachusetts Institute of Technology, USA

Introduction and background

Traditionally, seismic design of gravity retaining walls has been based upon equivalent static forces. These forces include the earth pressure exerted by the backfill, commonly evaluated using the Mononobe-Okabe equation (see Ref.1) plus an inertia force on the wall; see Fig. 1a. The weight of the wall is made large enough to resist the horizontal forces by base friction. Overturning also is checked. A typical design requirement is contact over 80-100% of the base.

Forces are calculated from a design seismic coefficient. Typical values are 0.05, 0.10 or 0.15 (rarely as large as 0.21). Such numbers are small compared

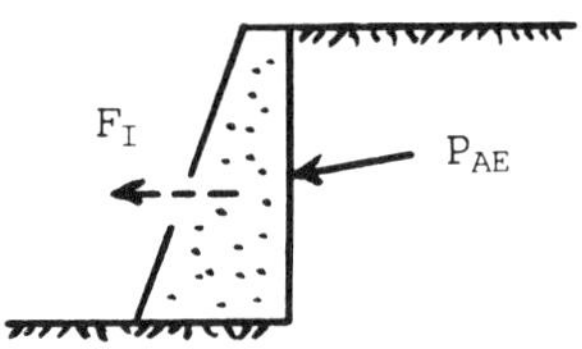

a. Conventional loads, based on seismic coefficients

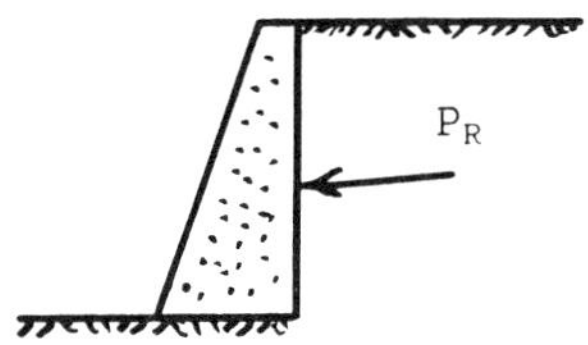

b. Loading considering residual thrust

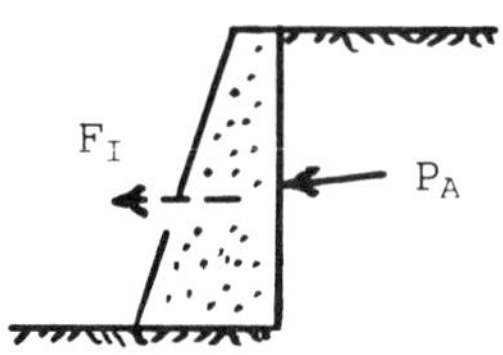

c. Loading considering inertial loading; wall on soil

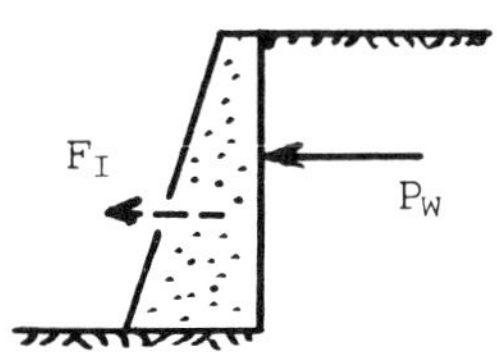

d. Loading considering inertial loading; wall on rock

Fig. 1 Loadings for various cases

to peak horizontal accelerations expected should a major earthquake occur. This situation, which parallels normal practice for seismic design of buildings, implies possible yielding and permanent displacement should the major earthquake occur.

According to modern thinking, the choice of the seismic coefficient should be based upon the permissible permanent outward displacement of a wall. Richards and Elms (ref. 2) proposed an approach for doing just this for walls that slide, based upon an analogy to Newmark's sliding block. Small-scale shaking table tests were performed and validated the approach.

Nadim and Whitman (ref. 3) investigated several complications not considered in the Richards and Elms model, especially the influence of the deformability of the backfill. Inclusion of this deformability can lead to an increase in permanent displacement compared to that computed with the rigid-plastic assumption. Based upon a systematic evaluation of the potential errors and uncertainties in the sliding block analogy, Whitman and Liao (ref. 4) developed an expression for the seismic coefficient necessary to give at least 95% probability that the permissible displacement will not be exceeded if the design earthquake ground shaking were to occur.

The Richard-Elms and Whitman-Liao methods do not explicitly account for permanent outward tilting. Richard-Elms recommend that walls be given sufficient width to ensure that no overturning occurs. A reading of the literature and visits to sites suggests that tilting actually is a more likely form of permanent displacement than sliding. Unfortunately there are very few well-documented case studies where liquefaction was not an issue. The senior author observed tilting of the quay walls in Valparaiso (Chile) harbour as a result of the earthquake in 1985. Grivas and Souflis (ref. 5) describe an example in Greece.

There has been relatively little study of the tilting problem. Prakash (ref. 6) proposed a model for computing dynamic response, including permanent tilt, based upon reasonable but intuitive assumptions for stiffness relationships. Nadim and Whitman (ref. 7) analyzed dynamic response assuming rigid-plastic behaviour for both sliding and tilting.

This Paper presents very preliminary proposals for predicting permanent tilt. The Paper begins with a review of two studies that form the basis for the recommendations.

Centrifuge tests by Andersen

Model gravity walls were constrained to rotate about their heel, and were restrained against rotation by springs (ref. 8). The focus was thus upon the interaction between backfill and tilting wall. At prototype scale, the walls were about 12 m high. The soil was dry sand. The tests involved 80 g centrifugal acceleration and 10 cycles of base motion at 1.5 Hz (prototype)

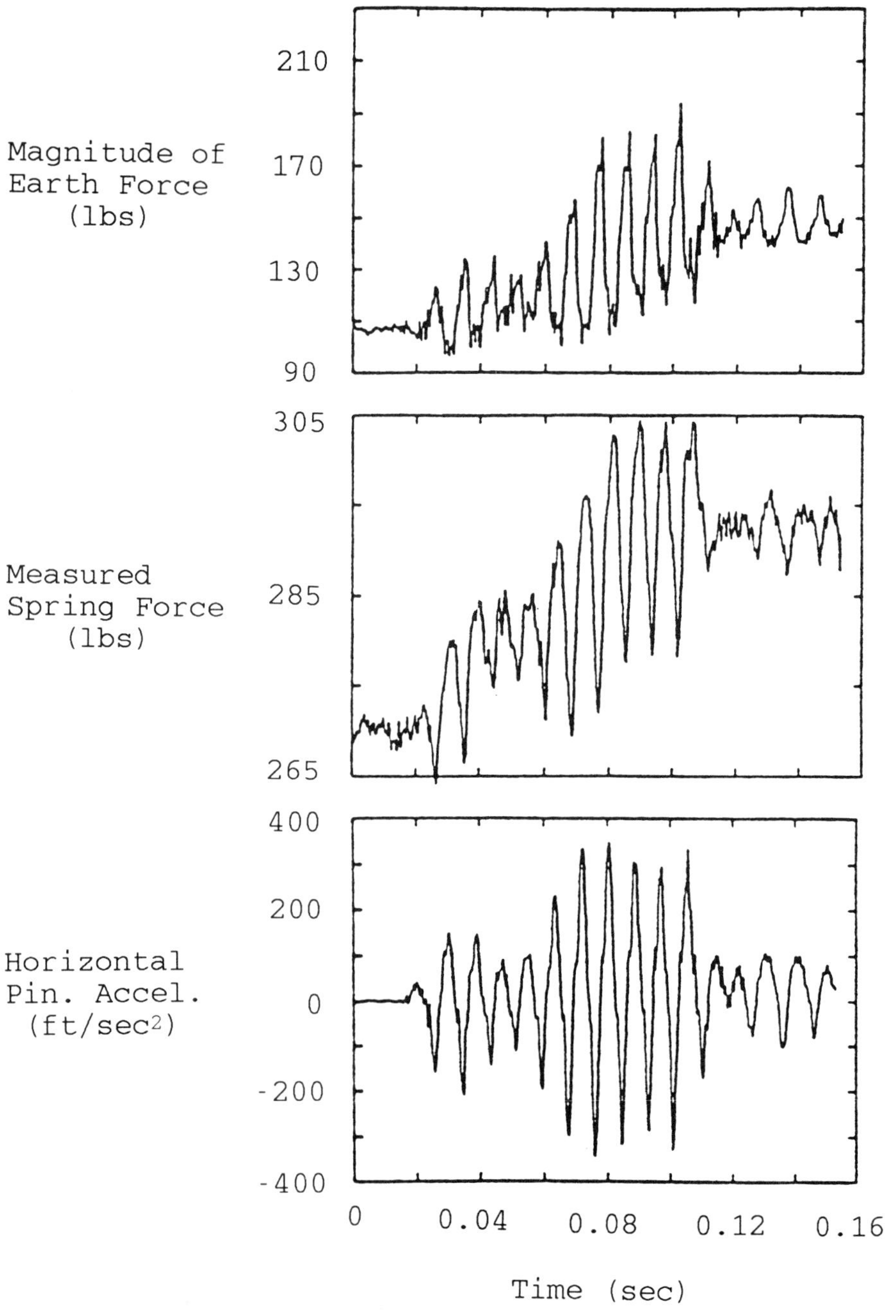

Fig. 2. Typical results from test by Anderson et al.

with peak accelerations of about 20% of gravity. Three stiffnesses of rotation-resisting spring were used, covering the range of stiffnesses typical for walls founded on soil. Typical key results are shown in Fig.2.

The lower plot gives the applied base acceleration, with positive acceleration being in the direction from wall to backfill. The middle plot shows the variation of force in the rotation-resisting spring, and hence rotation. The force exerted by the backfill — see the upper diagram — is at a minimum when the wall is at its greatest rotation away from the backfill, and is not much larger than the initial value. Thus, at the time of maximum rotation, the principal dynamic force is the inertial load on the wall itself and the dynamic component of earththrust contributes little to the amount of rotation. Theoretical studies (ref. 9) using simple mass-spring systems confirmed that the phasing demonstrated in these results is typical for walls founded upon soil.

The second important result is that a residual earththrust develops. The magnitude of this residual thrust is significant. In these tests it was nearly equal to the at rest thrust, and remained at this value with time.

Maxima of earththrust occurred when the wall rotated back against the backfill, and at such times were opposed by the inertial forces on the wall. These peak forces were hence not of concern as regards stability of the wall.

Finite element study by Al- Homoud

Al-Homoud (ref. 10) performed a theoretical study of a gravity wall that could tilt and slide on its foundation, using finite element procedures.

The computer program FLEX (ref. 11) was used for this study. The program provides absorbing boundaries at the two sides of the model, and allows for opening and closing of gaps between the wall and the soil. Soil is modeled as an elasto-plastic material, with an associated flow rule. There were shortcomings to the model. There was too much dilantancy when yielding occurred. The bottom of the model was rigid with no provision for radiation damping. As a result, the stratum tended to amplify the input base motion very strongly. Despite these shortcomings, the general pattern of results are believed to be quite reasonable. The program was used to model Andersen's tests, and giving satisfactory predictions.

The geometry of the wall used for Al-Homoud's study is shown in Fig. 3. The foundation and backfill were dry sand with ϕ of 40 degrees. The wall was designed for static and earthquake loading following usual procedures, using a seismic coefficient of 0.2. Because of the relatively large friction angles, the wall had a narrow base for its height.

The input consisted of either several cycles of sinusoidal motion or an actual earthquake recording, normalized to several different peak accelerations. The behaviour of the wall was influenced both by interaction of wall

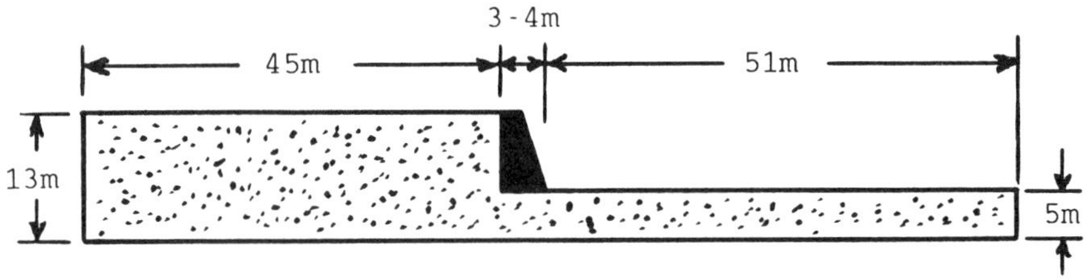

Fig. 3. Case studied by Al- Homoud[10]

and backfill and by the interaction of wall and underlying soil.The general results concerning the magnitude and phasing of earththrust from the backfill were generally consistent with Andersen's experimental data. The dynamic earththrust was small compared to the internal force on the wall at times of maximum outward rotation. As the intensity of shaking was increased, the wall experienced uplifting at its heel. The most significant results concerned the relation between rotation and moment resistance at the base, as this uplifting progressed.

Discussion and proposed procedures

From the standpoint of designing to ensure that permanent tilt of a gravity wall does not exceed an acceptable limit, two significant results emerge from these studies

1. Residual earththrust tends to build-up during a major ground shaking.
2. The peak dynamic overturning moment is determined primarily by the maximum inertia loading on the wall; dynamic earththrust contributes little or nothing.

These results suggest the following approaches.

Focus on residual thrust

Permanent tilt is given the residual thrust and its resultant height (see Fig. 1b) divided by the moment resisting stiffness at the base of the wall. Neither part of this equation can be determined with confidence at this time.

The studies suggest two crude rules for evaluating the residual thrust.

1. In Andersen's tests, the residual thrust was about equal to the thrust (static plus dynamic) calculated from the Mononobe-Okabe equation. This must be a coincidence, but it was true for all stiffnesses.
2. A theoretical study simulating Andersen's tests (ref. 12) found that the earth pressure coefficient for residual thrust builds about 0.6. This is an upper limit, reached during strong, sustained shaking.

In Andersen's tests, the height of the resultant residual thrust acted at about 0.4H (although the test results were not definitive in this regard).

Of course, residual thrust can build-up only if the wall can resist this additional thrust. In Andersen's tests the build-up occurred even for the very soft rotational spring. Obviously, if a wall begins to tip over as a result of the residual thrust, the thrust would decrease to the static active value.

Rotational stiffness at the base of a wall is also difficult to estimate with confidence. A starting point is an equation from theory of elasticity (ref. 13)

$$K_\theta = C\pi GB^2 \qquad (1)$$

where G is shear modulus, B is the width of the foundation in contact with the soil, and C is a function of Poisson's ratio . The product $C\pi$ can reasonably be taken as 0.8. For G, use 80% of the small strain shear modulus. Al Homoud's study indicates this procedure is conservative.

If the moment from the estimated residual thrust cannot be resisted - that is, if there is a net overturning moment about the toe of the wall - a different approach is necessary. This would imply that the wall moves out enough after each pulse of earthquake motion to reduce the stress back to the active level. An intuitive estimate for permanent displacement at the top of the wall is 0.0015HN, where N is the effective number of cycles of motion.

Focus on inertial loading

The peak ground acceleration (and not a reduced seismic coefficient) should be used to evaluate the inertial force. The dynamic thrust from the backfill is assumed to be zero (Fig. 1c). Eq. 1 may be used, assuming that there is very little rebound from an outward rotation.

It is possible that the inertial force will cause a net overturning moment about the toe of the wall, implying total uplift. However, the amount of rotation will be limited by the short duration of the ground motion. The peak outward rotation per cycle may be estimated from

$$\theta = (\text{net overturning moment}) \,/\, I\Omega^2$$

where I is the moment of inertia of the wall about its toe and Ω is the predominant circular frequency in the ground motion.

If a wall is founded upon rock and does not tilt, it is possible that the dynamic earththrust - which should now be evaluated by the method of Wood (ref. 14) - is in phase with the inertial loading on the wall - see Fig. 1(d). Both of these forces should be evaluated from the peak ground acceleration. If this leads to a net overturning moment, the wall must tilt and the earth-thrust will decrease somewhat. The dynamic motion may be estimated using the approach in the previous paragraph.

Combined sliding and tilt

If sliding is predicted to occur, the acceleration transmitted to the wall, and hence the inertial force on the wall, is reduced. However, tilting may still also occur.

An example . Consider a wall 10 m high, supporting a backfill with an unit weight of 16 kN/m^3 and a friction angle of 30 °. The wall rests on a dense sand with a small strain shear modulus of 2.25 MN/m^2. The base friction angle is 40 °. The design earthquake is magnitude 7, causing a peak acceleration of 0.4 g and a peak ground velocity of 1 m/s - a rather severe design environment. A seismic coefficient of 0.15 is used for design. Wall friction will be neglected.

In this example, the required wall width - base width of 6 m and top width of 0.5 m - is governed by moments in the seismic loading case. This wall weighs 780 kN/m. The other forces are

Static active thrust: P_A = 267 kN/m.
Dynamic earththrust: ΔP_{AE} = 90 kN/m, acting 6 m above the base.
Inertia force on wall: F_I = 117 kN/m, acting 3.59 m above base.

The safety factor against sliding is 1.38, which is more than adequate. The resultant force at the base lies 1.62 m from the toe. This is outside the middle third of the base, and there is uplift over about 20% of the base.

Were the design earthquake to occur, the inertia force on the wall and the static plus dynamic earththrust exceed the available base friction. Hence sliding must occur. Using the Richards-Elms approach, the maximum transmittable acceleration is found to be 0.28 g. The computed slip is 92 mm, which is known to be a conservative estimate for the permanent displacement.

With this severe earthquake, it is expected that the residual thrust will reach its maximum value, given by an earth pressure coefficient of 0.6. The corresponding residual earththrust is 480 kN/m, at a height of 4 m above the base. This value is smaller than the static plus seismic force computed by the Mononobe-Okabe equation, and is smaller than the available base shear. With this residual thrust, the resultant force on the base lies 1.52 m inside the toe, and there is uplift over 24% of the base. For this reduced width of base, the rotational spring constant is 94 MIN-m/rad-m. The computed rotation is 0.021 rad, giving a displacement of 210mm at the top of the wall.

With the maximum transmittable acceleration of 0.28 g, the dynamic earththrust is 168 kN/m acting at 6m above the base and the inertia force on the wall is 218 kN/m. Now the resultant on the base lies 1.84 m from the toe, corresponding to uplift over only 8% of the base and a rotational spring constant of 137 MN-m/rad-m. The resulting displacement at the top of the wall is 57 mm. This is the transient displacement during the cycle of ground motion with the largest peak acceleration. Other cycles will add to the total permanent displacement. Hence the estimate of 210 mm residual displacement seems reasonable.

Thus a total permanent displacement on the order of 250-300 mm is expected. This is a rather large movement (compared to a static tilt of 15 mm to reach an active condition), but might be an acceptable outcome of a major earthquake, depending upon the application. Proper consideration of wall friction might reduce the estimated permanent displacement.

Final comments

Except where liquefaction has occurred at waterfronts, gravity walls have generally performed quite well during earthquakes. Hence it can well be argued that the traditional design rules have worked satisfactorily. The reason for the good performance, despite the obvious shortcomings of the traditional procedures, is the considerable conservatism usually inherent in design for static loadings, especially ignoring the beneficial effects of wall friction. When and if less conservative practices are followed for static loadings, it will be essential to be more realistic concerning design against seismic loadings.

Gravity walls are far less popular today than they once were, as mechanically-stabilized walls are now used frequently. While the procedures applicable to gravity walls are not directly transferable to these were walls, many of the concepts discussed above may still be of value.

References

1. SEED H.B. and WHITMAN R.V. Design of earth retaining structures for dynamic loads, in *Lateral Stresses in the Ground and Design of Earth Retaining Structures*, ASCE, 1970, pp 103-147.
2. RICHARDS R.J. and ELMS D. Seismic behaviour of gravity retaining walls, *J. Geotech. Engng Div.*, ASCE, 1979, **105**, April, pp 449-464.
3. NADIM F. and WHITMAN R.V. Seismically induced movement of retaining walls, *J. Geotech. Engng*, ASCE, 1983, **109**, July, pp 915-931.
4. WHITMAN R.V. and LIAO S. Seismic design of gravity retaining walls. *Proc. 8th World Conf. Earthquake Engineering*, EERI, Oakland, CA, USA, 1984, **IIS**, pp 533-540.
5. GRIVAS D.A. and SOUFLIS C. Performance of the Plasteas wingwall during earthquakes. *Proc. 8th World Conf. Earthquake Engineering*, EERI, Oakland, CA, USA, 1984, **III**, pp 509-515.
6. PRAKASH G. Analysis of rigid retaining walls during earthquakes. *Proc. Int. Conf. Recent Advances in Geotechnical Earthquake Engineering and Soil Dynamics*, Univ. Missouri, Rolla, MO, USA, 1981.
7. NADIM F. and WHITMAN R.V. Coupled sliding and tilting of gravity

retaining walls during earthquakes. *Proc. 8th World Conf. Earthquake Engineering*, EERI, Oakland, CA, USA, **III**, pp 477-484.
8. ANDERSEN G.A., WHITMAN R.V. and GERMAINE J. T. Seismic response of rigid tilting walls. *Proc. Centrifuge '91,* Balkema, 1991, pp 417-424.
9. WHITMAN R.V. Seismic design of gravity retaining walls. *Proc. Int. Conf. Earthquake Resistant Design and Construction*, Balkema, 1989, pp 223-244.
10. AL-HOMOUD A. *Evaluating tilt of gravity retaining walls during earthquakes*. ScD thesis, Dept. Civil Engng, MIT, Cambridge, MA, USA, 1990.
11. VAUGHAN D.K. and RICHARDSON E. *FLEX user's guide*. Weidlinger Associates, Palo Alto, CA, USA, 1989.
12. STAMATOPOULOS C. and WHITMAN R.V. Prediction of permanent tilt of gravity retaining walls by the residual strain method. *Proc. 4th US Nat. Conf. Earthquake Engineering*, EERI, Oakland, CA, USA, 1990, **III**, pp 363.
13. DOBRY R. and GAZETAS G. Dynamic response of arbitrarily shaped foundations. *J. Geotech. Engng*, ASCE, 1989, **112**, Feb., pp 109-135.
14. WOOD J.H. *Earthquake-induced soil pressures on structures*. Rpt. EERL 73-OS, Calif. Inst. Technology, Pasadena, CA, USA, pp 1573.

Performance of a precast cantilevered retaining wall

W. H. HOVER, GZA GeoEnvironmental Inc., USA

Post-construction performance of a braced precast retaining wall up to 6.1 m high and 106.7 m long is presented, and is compared to anticipated performance. The proprietary wall installation is the first of its type in the United States and is comprised of precast concrete panels erected on unreinforced concrete footings, which bear in weathered bedrock. Each concrete panel is connected to a reinforced concrete base slab by an inclined precast tension element, and horizontal reinforcing steel. Performance of the wall during backfilling was presented (Hover, Grynkewicz et.al, 1990). Inclinometers indicated 7.6 mm of lateral movement at top of wall, but no movement below grade at the toe. The average axial load in the tension member after backfilling based on strain gage measurements was 133.5 kN, which is 60 to 70 percent of design axial loads based on triangular Rankine and Coulomb active distributions, respectively. After 2 years of service under traffic loading, the lateral movement at the top of the wall increased by 2.7 mm to 10.3 mm. The average axial load in the tension member increased from 133.5 kN to 175.1 kN, which is about 60 to 68 % of the design axial loads based on triangular Rankine and Coulomb active distributions with a uniformly distributed traffic surcharge of 3.88 kN/m^2.

Introduction

Proprietary precast, braced, cantilevered retaining walls have been used in Europe for more than 20 years (Fletcher, Mazza-personal communications, 1988). This paper presents the post-construction performance of a proprietary 6.1 m high wall system, the first of its type in the United States. Behaviour during backfilling, and details of wall design and construction were presented by Hover, Grynkewicz, *et al.*, 1990. This Paper reports behaviour for the first 2 years after backfilling, and focuses on increases in loads attributed to traffic surcharge, and resulting wall deformations.

Wall system description, subsurface conditions, and soil properties

About 106.7 lineal meters of retaining wall was installed to support differential earth cuts of up to 6.1 m between parking areas and roadways at the site. Since September 1989, the wall has experienced traffic loadings from automobiles and trucks. Fig. 1 shows the locus of the wall and site grading. Fig. 2 presents a plan, wall section, and inferred subsurface conditions.

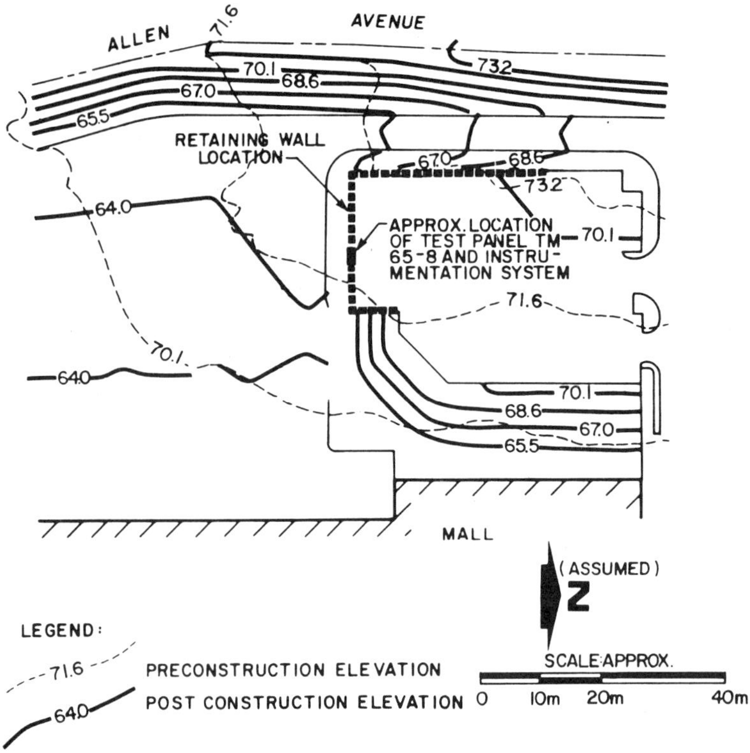

Fig. 1. Locus plan

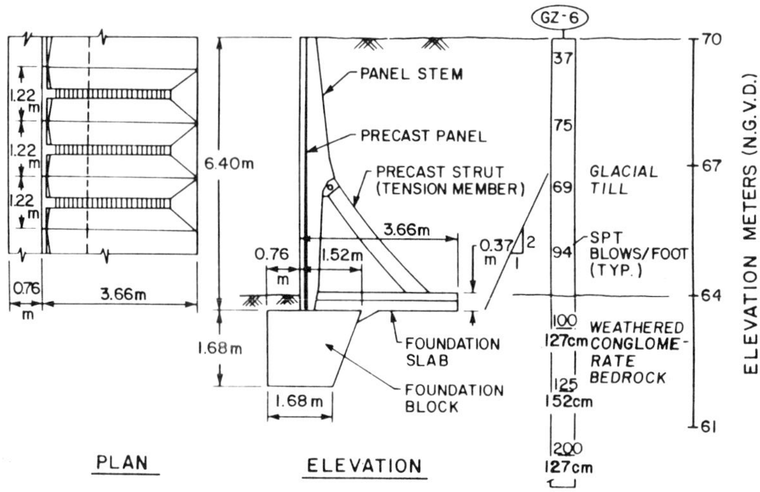

Fig. 2. Plan and typical wall section

The foundation stratum is weathered conglomerate bedrock, and wall backfill consists of glacial till comprised of sand, silt, gravel, and trace clay from on-site excavations. A sand-gravel drainage zone 0.61 m wide with non-woven geosynthetic filter between sand-gravel and glacial till fill extends the full height of the wall panels, and above the base slab.

Laboratory compaction tests performed per ASTM D-1557 Method C (American Society for Testing Materials, 1989) on representative samples of glacial till fill determined a maximum dry density of 21.76 kN/m^3 at an optimum moisture content of 6.5 %. Two consolidated isotropically drained (CID) triaxial tests (U.S. Army Corps of Engineers, 1970) performed during construction on samples of glacial till backfill indicated a drained friction angle ϕ of 37 ° at a dry density γ_d of 20.89 kN/m^3 and a moisture content of 7 %.

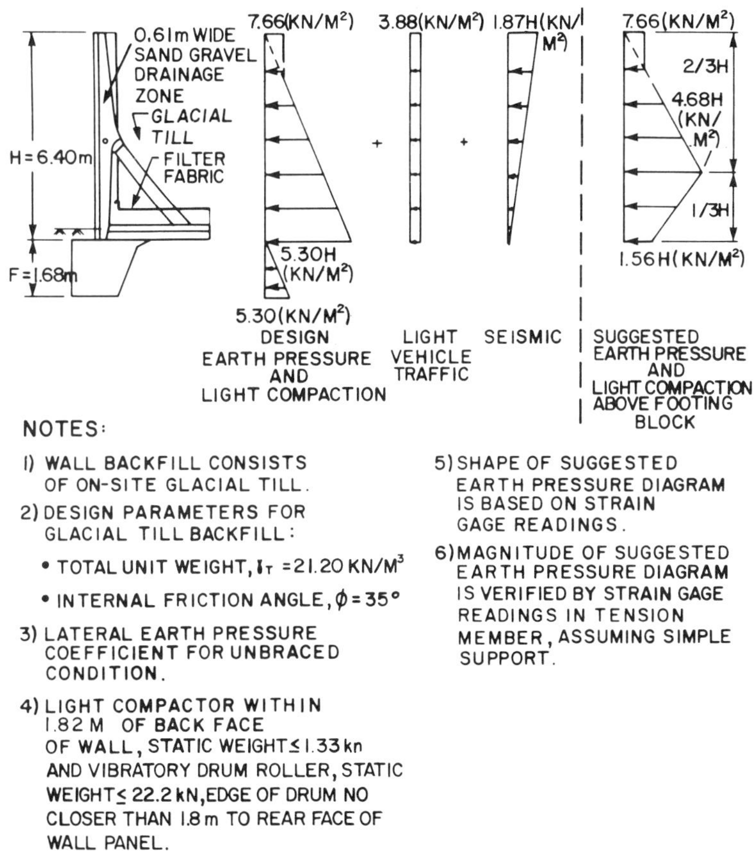

Fig. 3. Design earth pressures

Design earth pressure and theoretical tension member loads

Design earth pressures specified are shown on Fig. 3, and include Rankine active earth pressures (Department of the Navy, 1971) for drained conditions, K_a = 0.25, ϕ=35 °, angle of friction between panel and backfill = 20 °, and backfill total unit weight γ_t= 21.2 kN/m^3. Surcharge loads from light compaction equipment, vehicular traffic, and seismic loads in accordance with local building codes were included.

Theoretical tension member loads for the design Rankine active and Coulomb loading conditions, for the cases of static earth pressure plus compaction only, and for static earth pressure plus compaction plus traffic surcharge, are shown in Fig. 4. For the Coulomb analyses, a friction angle between concrete and sand-gravel of 17° was assumed (Bowles, 1982). Analyses indicate increased theoretical tension member loads of 64.1 and 61.4 kN due to traffic surcharge, for the Rankine and Coulomb cases, respectively.

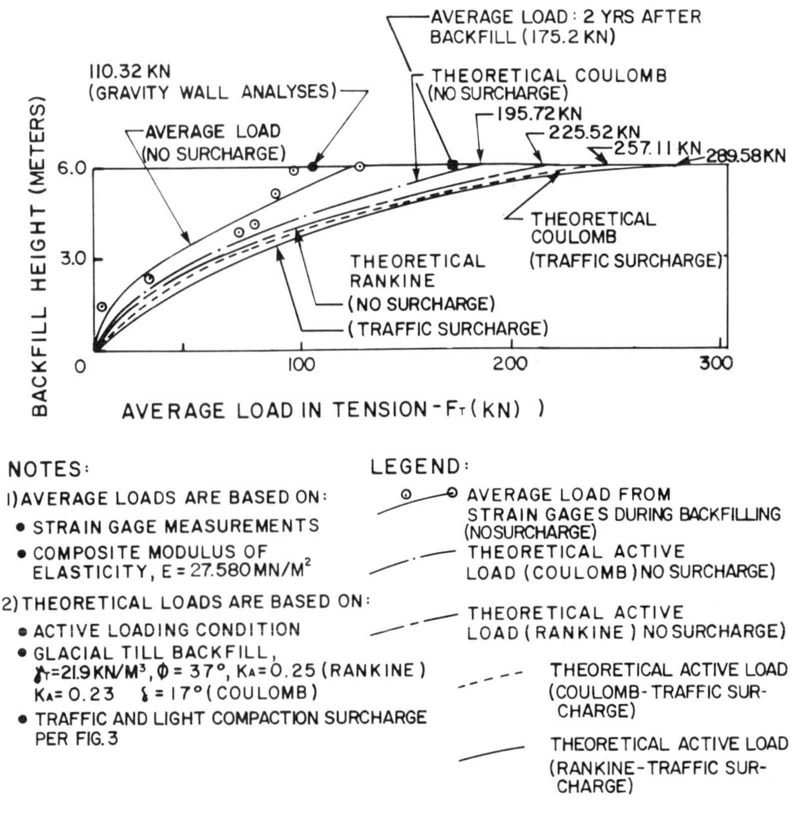

Fig. 4. Tension member load vs fill height

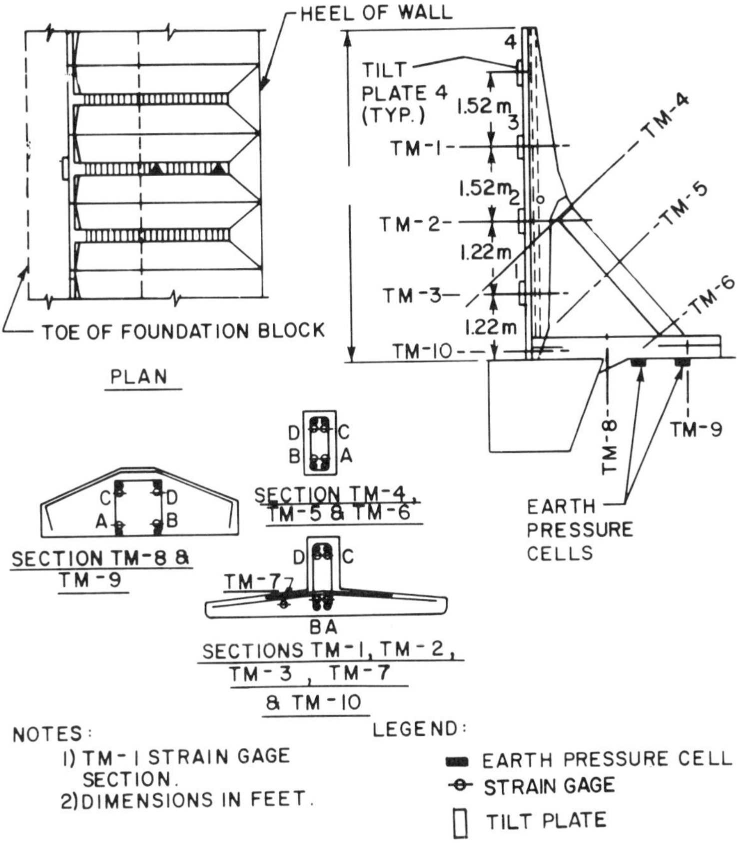

Fig. 5. Instrumentation location plan and section

Instrumentation program

Portions of an instrumentation program undertaken to evaluate a test panel of the wall system during backfilling, were left in place and monitored to provide post-construction data to compare actual performance to anticipated performance. One wall section at the location shown on Fig. 1 was monitored 2 years after construction. Locations of available instruments are shown on Fig. 5, as summarized below. Four inclinometers were deactivated during placement of concrete parapet walls and pavement, and were not accessible.

Four tilt plates were installed on the exterior face of the test panel prior to backfilling, to monitor angular rotation. The tilt plates were epoxied directly to the face of the wall with a temperature stable epoxy to reduce thermal

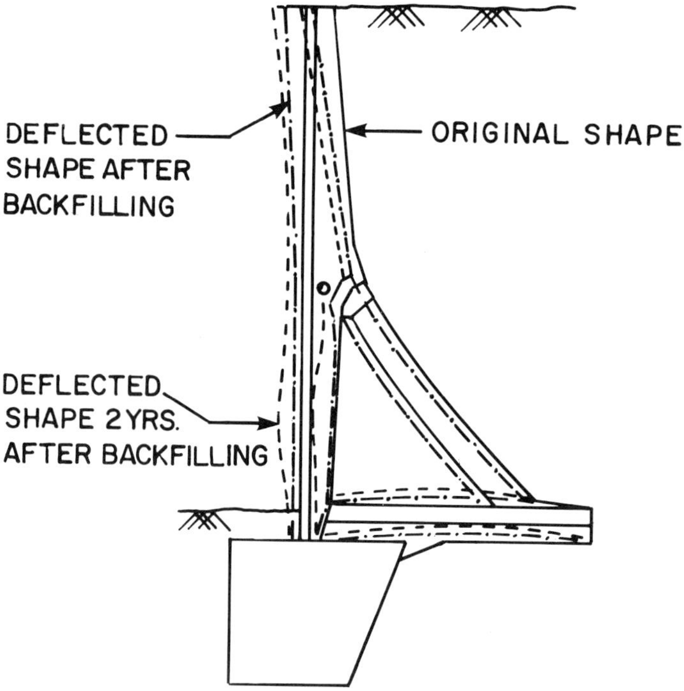

Fig. 6. Deflected shape

effects. The tilt plates were a redundant measurement system for the inclinometers during construction, and were accessible.

Forty concrete embedment vibratory wire strain gages were installed at the locations shown on Fig. 5 and are still functional. The gages are used to monitor strains and to estimate loads in the structural components of the test panel. The strain gages within the wall panel and tension member were installed and calibrated at the precasting facility.

Two earth pressure cells were installed to measure trends in vertical pressures below the base slab of the test panel. One of the earth pressure cells was located near the heel of the wall and one was located near the intersection of the foundation slab and foundation block. Each cell was installed on a levelling pad of 25.4 mm of fine concrete sand. Earth pressure cells were not installed below the footing block, given that it bears in bedrock of very high bearing capacity.

Wall performance 2 years after construction

Wall movement

At completion of backfilling, 7.6 mm of outward lateral movement had taken place at the top of the wall, based on inclinometers, assuming fixity at their bases in bedrock. Most of the deformations were within the upper 3.0 m of the wall, with essentially zero cumulative movement in the lower half of the wall. Refer to Fig. 6 for the deflected shape of the test panel during backfilling, based on deformations of inclinometers assuming base fixity in bedrock and tilt plates.

After completion of backfilling, the net rotation of the test panel ranged from 0.03 ° outward (counterclockwise-towards toe) at the lowest tilt plate, to 0.05 ° at the uppermost tilt plate. Rotation at the tilt plate located at mid-depth of the panel was -0.04 ° (clockwise towards heel), due to the restraint provided by the tension member.

Tilt meter data collected 2 years after backfilling and imposition of traffic surcharges, indicated a rotation of 0.11 °(towards toe) at the upper tilt plate, and 0.06 ° at the lowest plate. The tilt plate located at mid-depth of panel, just below the connection to the tension member, indicated a rotation of 0.02 °. This data suggests that since backfilling was completed, the rotation of the upper tilt plate has increased by 0.06 °; the rotation of the tilt plate located just below the connection of the panel to the tension member has increased by 0.06°; and the rotation of the lowest tilt plate increased from 0.03° to 0.06°. The increase in rotational movement is consistent with increased loads in the tension member, and results in a total lateral movement at top of wall increasing from 7.6 mm (value indicated by inclinometers at completion of backfilling) up to 10.3 mm. Refer to Fig. 6 for a sketch illustrating the deflected shape of the structure 2 years after construction, inferred from tilt plate data and earth pressure cell data.

Tension member loads

The inclined tension member is the key structural component of the wall system. Four internal strain gages were installed at each of three cross sections located near the upper, middle, and lower end of the tension member. Gage locations are designated TM-4, 5, and 6 on Fig. 5. At each section, gages TM-A and B are located near the lower face of the tension member, and gages TM-C and D are located near the upper face.

The average tensile stress in the member estimated from strain gage readings was about 2.97 MN/m^2 at completion of backfilling, which corresponds to an estimated average tensile force of 133.5 kN. Two years after construction, the estimated average tensile stress in the member is 3.90 MN/m^2, which corresponds to an estimated average tensile force of 175.1 kN. This represents an increase of 41.6 kN, which is attributed to effects of traffic surcharge. Data from gage locations where the recorded strain was sufficient to develop the cracked section were excluded.

Fig. 4 compares tension member loads back - calculated from strain gage data to theoretical tension member loads based on Rankine and Coulomb analyses for cases with and without traffic surcharge. Theoretical tension member loads of 225.5 kN and 289.6 kN were estimated for Rankine active cases without and with traffic surcharge, respectively. Theoretical tension member loads of 195.7 and 257.1 kN were estimated using Coulomb analyses for cases without and with traffic surcharge, respectively.

After 2 years of service under traffic loading, the average axial load in the tension member increased from 133.5 kN to 175.1 kN, which is about 60 to 68 percent of the theoretical axial loads based on triangular Rankine and Coulomb analyses for the case of light vehicle traffic surcharge, using a uniform lateral surcharge of 3.88 kN/m^2 applied over the height of the wall above the footing. The data suggest that the 41.6 kN increase in tension member load is about 76 percent of the axial component of the uniform design lateral surcharge load.

Conclusions

Lateral movements of 10.3 mm occurred at top of wall 2 years after backfilling, under traffic surcharge loading, as inferred from inclinometers and tilt plates. The incremental lateral movement is about 2.7 mm since completion of backfilling. Most movement occurred in the upper 3.0 m of the panel, above its connection to the tension member.

Sufficient rotational movement occurred during backfilling to develop the active condition. Further rotational movement occurred in the ensuing 2 years, and additional loads transferred to the tension member are inferred to be from traffic surcharge.

The average calculated axial load in the tension member increased from 133.5 kN at end of backfilling, to 175.1 kN 2 years later. The average calculated load in the tension member is about 60 to 68 % of the theoretical axial loads based on triangular Rankine and Coulomb active distribution with uniform traffic surcharge. The 41.6 kN increase in tension member loading represents 76 % of this axial component of the uniform design lateral surcharge load of 3.88 kN/m^2.

References

1. HOVER W.H., GRYNKEWICZ F.M., HON R. and DARAIE L. Geotechnical Design and Performance of a Precast Cantilevered Retaining Wall, *Proc., Design and Performance of Earth Retaining Structures*, Ithaca, NY, American Society of Civil Engineers, NY, 1990, pp 595-611.

2. Personal communications, Herbert Fletcher and Peter Mazza, The Mazza Consulting Group, 1988.
3. American Society for Testing and Materials. *1989 Annual Book of ASTM Standards, Soil and Rock; Building Stones; Geotextiles*. 1989.
4. U.S. Army Corps of Engineers. *Engineering Manual EM-1110-2-1906, Laboratory Soils Testing, Appendix X*, 1970.
5. Department of the Navy, Naval Facilities Engineering Command. *Design Manual - Soil Mechanics, Foundations and Earth Structures*, NAVFAC DM-7, March, 1971, Sec. 7-10, p.5.
6. BOWLES J.E. *Foundation Analysis and Design*, McGraw-Hill, 1982, p.389.
7. American Institute of Steel Construction. *Manual of Steel Construction*, 1980, pix.
8. LEET K.L. *Reinforced Concrete design*, McGraw-Hill, 1984, p44.

Non-linear dynamic response of tiedback walls

A. A. BANGASH, Bechtel Corporation, G. M. DENBY, Geo-Engineers Inc., and R. J. FRAGASZY, USA

Introduction

The use of tied-back retaining walls as permanent earth retention systems has gained wide acceptance in recent years. A natural consequence of this trend is a renewed interest in the seismic behaviour of such walls. Throughout the seismically active western United States, major highways are protected by these structures. During a major earthquake the failure of such retaining systems would impede critical emergency relief operations. Current design methodologies are based on semi empirical methods originally developed for gravity retaining walls. For example the Mononobe-Okabe method, is often used to account for seismic loading. However, the validity of such methods for tied-back walls has not been determined.

The results shown in this Paper are part of the recent detailed parametric studies conducted by Bangash (ref. 1). In this study, a non-linear dynamic finite element program, FLEX (ref. 2), is used to investigate the seismic response of tied-back walls. Parametric studies are conducted to identify the effects of selected variables. The results of the parametric study are presented in normalized form to provide preliminary design guidelines.

Literature review

Historical accounts of damage sustained by earth retaining structures have been reported over the past five decades. Most of the damaged structures were gravity and cantilever walls. The only reported failures of tied-back or deadman anchor walls occurred in waterfront environments where liquefaction of the soil was a primary cause of failure. Failure of tied-back walls retaining stable material has not been reported. The most common method of incorporating seismic stresses in wall design is to add a pseudo-static component to the static pressures on the wall. This approach was originally developed by Mononobe and Matsuo (ref. 3) and Okabe (ref. 4) and is commonly referred to as the Mononobe-Okabe (M-O) method. Although the M-O method has been validated by field response of gravity walls, its applicability to the seismic design of tied-back walls is yet undetermined.

Retaining structures. Thomas Telford, London, 1993

Since the fundamental design and behaviour of tied-back walls is quite different from gravity walls, the application of the M-O method to the seismic design of tied-back wall is questionable.

Fragaszy *et al.* (ref. 5) performed static and dynamic analyses of a 40 foot excavation supported by a 1 foot thick slurry wall and three rows of anchors. The major conclusions of this study were that the soil, anchors, and the wall move together, therefore negligible anchor loads are introduced in the tendons. However, soil layers above and below the excavation level tend to move relative to each other, thereby causing large bending moments in the wall near the excavation level. They also showed that large vertical accelerations were introduced even though the input consisted of purely horizontal shear waves. Siller (ref. 6) analyzed the static and dynamic behaviour of a retaining wall with a single row of anchors using a time domain finite element program. He assumed one end of the anchor was attached to the wall while the other end was fixed. The input time histories consisted of half sine pulses. Siller concluded that: the wall moves together with the free field; input frequency and acceleration has a significant effect on the response; anchor stiffness can significantly affect the wall displacements; and, large permanent displacements are the most common consequences of poor performance.

Kitajima and Uwabe (ref. 7) performed a large number of shaking table tests and analyzed the response of 110 sheetpile walls. Their conclusions were as follows

(l) the standard design practice needs modification
(2) design methods applied to gravity walls are also applicable to flexible walls
(3) damage was associated with anchor failure
(4) large movements occurred at damaged structures;
(5) most of the damage occurred for input frequencies in the range of 2 – 4 Hz.

It should be pointed out that this report summarized the response of waterfront structures only; therefore, liquefaction of backfill may have been the primary cause of damage. More recently, Steedman and Zeng (ref. 8) performed centrifuge tests on anchored cantilever walls. They demonstrate two modes of failure for such walls

(l) for a backfill with 80% relative density the anchors pull out and the wall rotates outward about its toe
(2) for a backfill with 65% relative density, the anchor again pulls out but the rotation of the wall takes place about the anchorage point.

They also show good agreement between the results of their tests and analytical solutions based on theory of wave propagation.

Denby *et a*l (ref. 9) documented the performance of 16 tied-back walls in

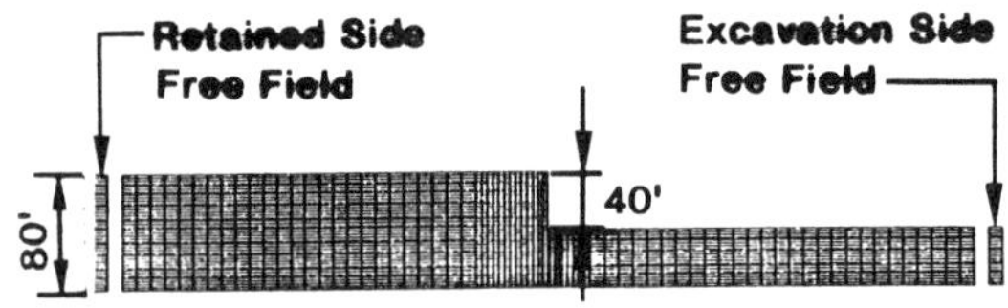

Fig. 1. Finite element mesh used in this study

the Los Angeles area after the 1987 Whittier Narrows Earthquake. The study consisted of interviews with shoring contractors, engineers, workers, and owners in the Los Angeles area. This study concluded that all the tied-back retaining walls performed well during the earthquake; however, the study suggests that the good performance may have been related to overdesign of the walls.

Amplification studies

The program FLEX solves the dynamic equilibrium equations in the time domain. In this study the non-linear soil characteristics were modelled by a modified Drucker-Prager soil model. The typical wall geometry discussed in this Paper is shown in Fig. 1. Transmitting boundaries were used on both the left and right sides to absorb the scattered wave field. The material properties used in this analysis are listed in Table I.The elasto-plastic analyses assume a friction angle ranging from 20 to 30 °. The total design load was based on empirical trapezoidal pressure distribution.

Table I. Material properties

Soil	Density	0.125 kcf
	Bulk Modulus	6540 ksf
	Shear Modulus	3020 ksf
	Damping	10 %
Wall	Density	0.150 kcf
	Stiffness, EI	37,500 kips-ft^2
Anchor	Stiffness, AE/SL	309 kips/ft/lineal ft
	Horiz. Spacing	7 ft

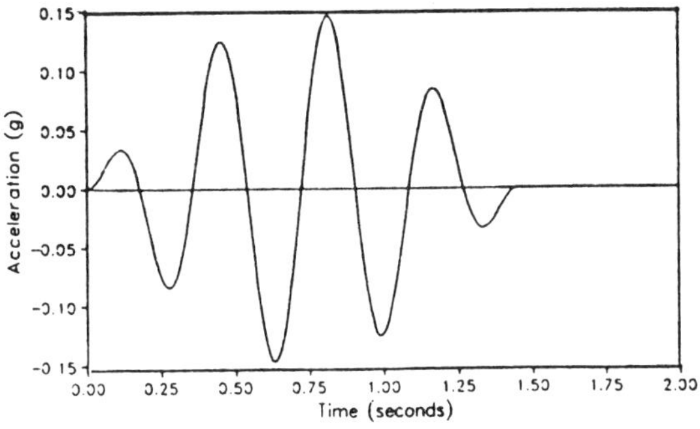

Fig. 2. Typical input motion

The sinusoidal input time history used throughout this study is defined by the following criteria

(1) Four complete cycles of excitation
(2) The amplitude-time relationship is given by

$$A_i(t) = A_1 \sin(\omega_o t) \sin(\omega_i t) \tag{1}$$

where $A_i(t)$ is the input acceleration time history, A_1 is the peak input acceleration assumed as 0.15g in this study, t is time, ω_i is the input circular frequency in rads/seconds, and $\omega_o=\omega_i/8$. The relationship in equation (1) is graphically shown in Fig. 2. The corresponding input frequency in cycles per second, f_i, is equal to $2\pi\omega_i$. All amplification studies were performed at preselected f_i/f_e ratios, where

$$f_e = \frac{C_s}{4H_e} = \frac{882}{4x40} = 5.5Hz \tag{2}$$

where, C_s and H_e represent the shear wave velocity and excavation depth, respectively. Consequently, for the horizontal ground surface configuration the f_i/f_e ratios of 1 and 0.5 correspond to the cases when the input frequency is in resonance with a one dimensional shear beam with heights equal to the depth of excavation and depth to bedrock, respectively.

Fig. 3(a) shows amplification relative to the maximum input velocity for both elastic and elasto-plastic materials. This figure shows two distinct peaks; one at $f_i/f_e = 0.5$ and the other at $f_i/f_e = 1$. The major difference in the two peaks is that it is observed at all locations for the $f_i/f_e = 0.5$ case while only the excavation side free field peaks at $f_i/f_e = 1$. Since the depth of excavation happens to be the same as the height of the excavation side free field, the peak at $f_i/f_e = 1.0$ is simply the result of resonance of the excavation side free field.

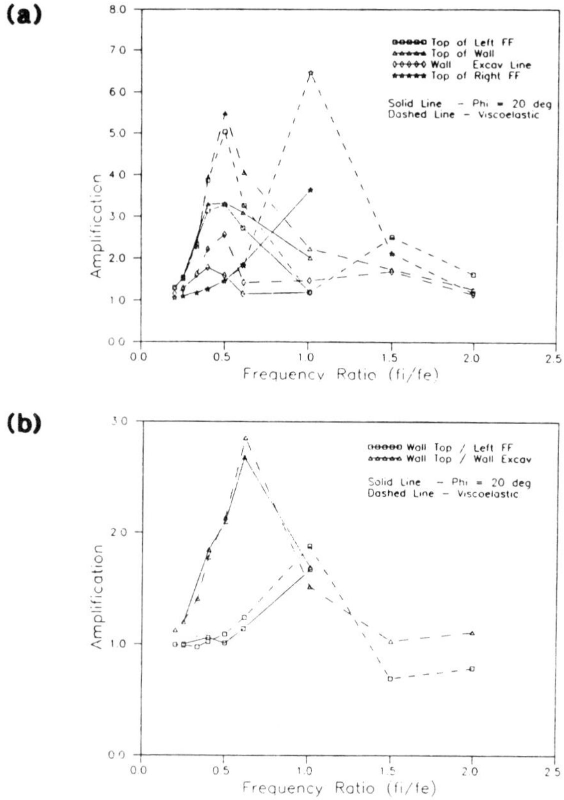

Fig. 3. Amplification curves for horizontal ground surface conditions; (a) relative to base, (b) top of wall relative to retention side free-field and excavation level

The peak at $f_i/f_e = 0.5$ is more significant since it demonstrates the strong influence of the retained side free field response on the overall response of the soil wall system. Clearly, the natural frequency of the soil wall system could be approximated from the characteristics of the retained side free field. This point is further demonstrated in Fig. 3(b) where the maximum velocity amplification is observed at retained side resonance conditions. A smaller peak at $f_i/f_e = 1.0$ shows the effect of resonant conditions for the soil column with height equal to the depth of excavation.

Effect of soil non-linearity on amplification with respect to bedrock is significant because a friction angle of 20 ° limits the maximum shear resistance. On the other hand, amplification at the top of the wall with respect to the retained side free field and the excavation level is practically the same for elastic and elasto-plastic materials, Fig.3(b). Such trends reaffirm earlier

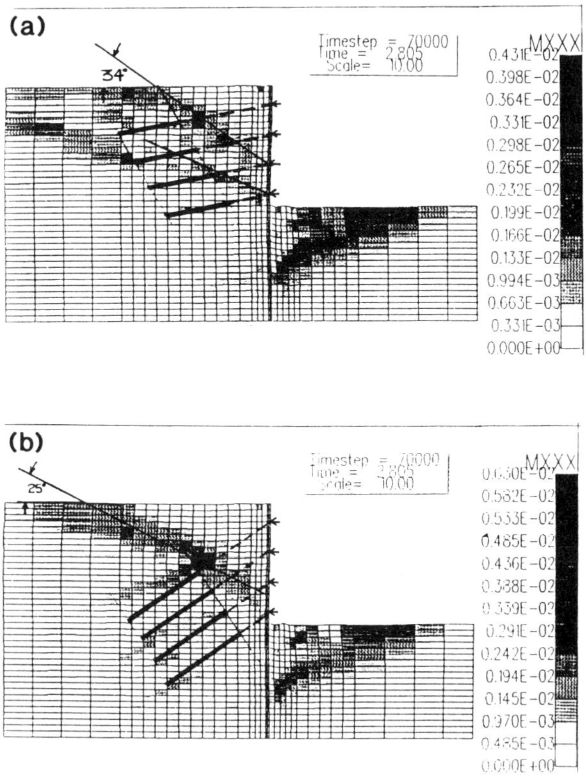

Fig. 4. Tensile failure planes for 30° soil; (a) Anchor inclination = 10 ° (b) Inclination = 35°

observations suggesting that the retained soil and the wall move together with the retained side free field.

Failure mechanism

Figs 4(a) and (b) show the plots of maximum horizontal tensile strains (darkest shade represents the maximum value) for anchor inclinations of 10 and 35 °, respectively. Both plots show one shallow tensile zone behind the wall aligned along the front of the ground anchors and another one just in front of the passive region of the wall. A shallow failure plane develops because of the presence of grouted portion of the anchors, and since significantly larger accelerations are required to force a deeper failure plane behind the anchors. As expected, the failure plane angle decreases as the anchor inclination increases.

The tensile failure plane in front of the wall is primarily the result of interaction between the retained side free-field and the excavation side free-field. The amplification studies described above suggest that wall motions are governed by the motions of the retained side free-field even though the excavation side free-field characteristics could be quite different (due to its shorter height). The soil in the vicinity of the embedded portion of the wall, therefore, experiences large relative motions. When the wall is moving toward, and along with, the retained side free field while the materials below the excavation level are moving toward, and along with, the excavation side free-field, the observed active wedge will develop in front of the wall. During a subsequent reversal of motions, the soil in front of the wall will experience compressive stresses. However, the displacements required to cause active failure are lower than those required for passive failure; therefore, a racheting effect takes place, causing plastic movements toward the wall.

Dynamic earth pressures and anchor loads

The effects of input frequency and soil friction angle on dynamic earth pressures and anchor loads are presented in this section. Fig. 5(a) shows the influence of friction angle on earth pressures. The dynamic earth pressures are roughly twice the static value for the wall section extending from the top of ground to the excavation level. For the embedded portion of the wall, the maximum dynamic earth pressures are three to four times greater than the static pressures, where the higher pressures are associated with the higher friction angle. Similarly, the maximum dynamic pressures in front of the wall are two to two and one half times greater than static earth pressures. Again, the higher pressures are associated with the higher friction angle.

In the passive region in front of the wall, the earth pressure distribution using non-linear soil properties is significantly different from those based on elastic properties. The maximum passive earth pressure is limited to 10 ksf for the elasto-plastic case compared to 20 ksf for the elastic case. This is to be expected, since the elasto-plastic constitutive model imposes a limit on the maximum stress levels.

Fig. 5(b) shows the influence of friction angle on maximum dynamic increment of anchor loads. In this figure, Delta Pi/P is the ratio of dynamic load increment in anchor i, Delta Pi, to the total design load on the wall, P. The parameter D_i represents the distance from the top of the wall to the location of tierod-wall anchorage for anchor i. Fig. 5(b) clearly shows an increase in anchor loads as the friction angle decreases. The maximum anchor load tends to occur in the second and third row of anchors (D_i/H_e = 0.4-0.7).

The relatively small load increase in the lowest level of anchors are the result of large passive resistance and the "racheting" mechanism in front of the wall.

Earth pressures and anchor loads corresponding to the various input frequencies are shown in Fig. 6(a) and (b), respectively. The soil friction angle was selected as 20 degrees. As expected, maximum earth pressures and anchor loads occur near the resonance range of $0.4 < f_i/f_e < 0.6$.

Wall displacements

The effects of friction angle on maximum horizontal wall displacements

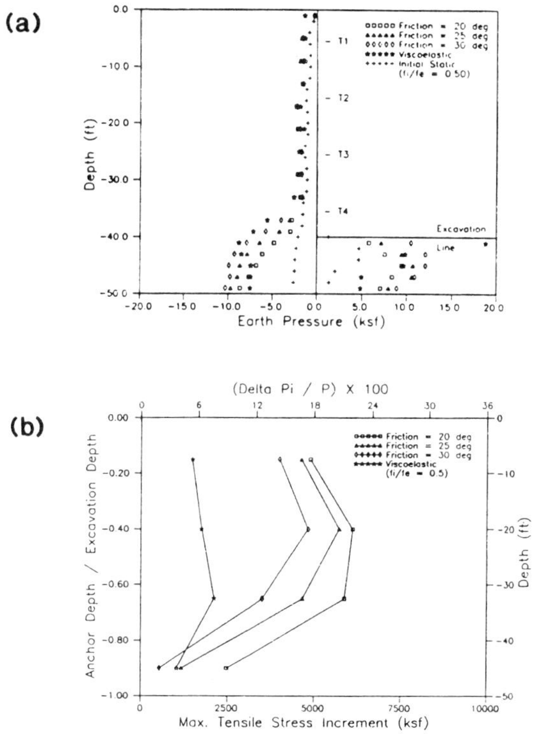

Fig. 5. Influence of friction and horizontal ground on: (a) dynamic earth pressures, and (b) anchor loads

are shown in Fig. 7. The shapes of the displacement curves explain the maximum increase in the top two rows of anchors since maximum relative movement occurs at these locations. Although the maximum displacements occur for the elastic soil, due to maximum amplification, the plastic displacement toward the excavation increases as the friction angle decreases. The plastically accumulated wall displacement toward the backfill are shown to decrease as the friction angle decreases. Another interesting aspect of this figure is the intersection of the elasto-plastic curves with the elastic curve for displacements towards the backfill. The depth at which these lines intersect also represents the approximate depth of a plastic zone, i.e. material above this location has plastically yielded towards the excavation. This observation

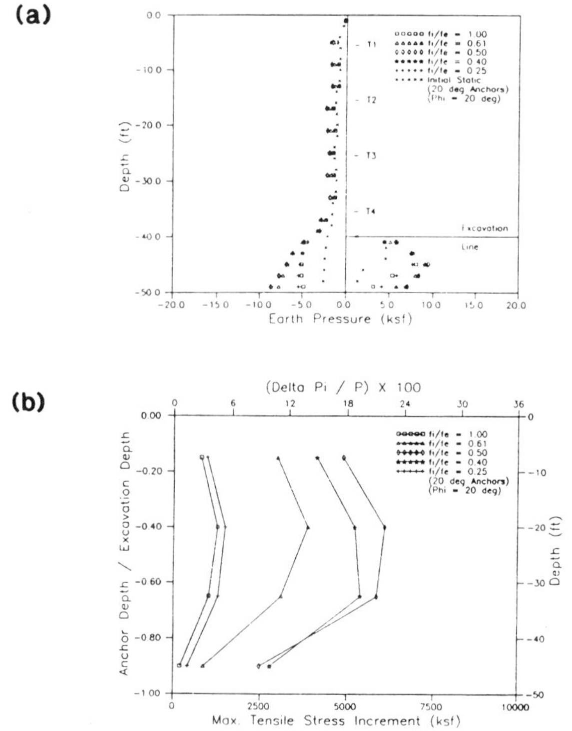

Fig. 6. Effect of input frequency and friction of 20° on: (a) dynamic earth pressures, and (b) anchor loads.

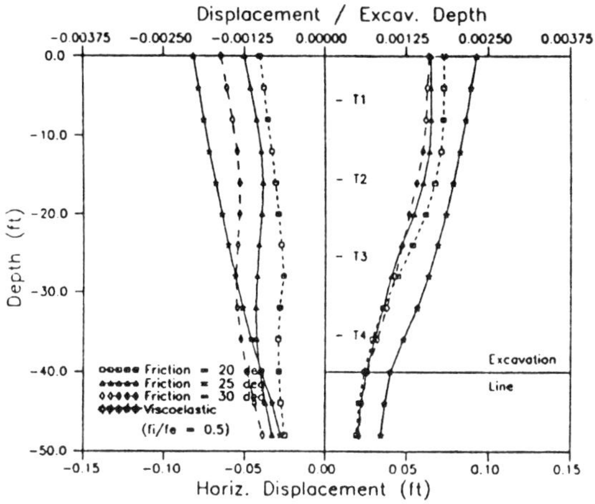

Fig. 7. Effect of friction angle on maximum horizontal wall displacements

is confirmed by the extent and location of tensile failure planes similar to those shown in Fig. 4(a) and (b). The plastic deformation extends all the way down the embedded portion for the weak soil (ϕ=20 °), while for the stronger soil (ϕ=30 °) the plastic zone only extends just below the third row of tiebacks.

The effects of input frequency on maximum horizontal displacement amplifications are shown in Fig. 8. The displacement pattern associated with the higher input frequencies indicate significant flexural bending. Such observations are confirmed by snapshots of wall displacements at selected times presented by Bangash (ref. 1). The maximum horizontal displacements occur along the top 30 feet of the wall. As for the elastic case, the maximum displacements take place when the input frequency is in resonance with the retained side free-field (f_i/f_e = 0.5).

Summary and conclusions

For the specific wall configuration presented in this study, tied-back walls are shown to be remarkably resistant to seismic damage. Their flexibility, low mass, and anchorage to the retained soil allows them to move with the surrounding soil. Such coherent motions result in minimum dynamic anchor loads and dynamic earth pressures during seismic loading. The only location where this motion is impeded in the embedded portion of the wall.

In terms of qualitative behaviour, the most significant aspect of the response is the disparity in displacements toward the excavation versus those toward the retained soil side thereby leading to the so-called interference

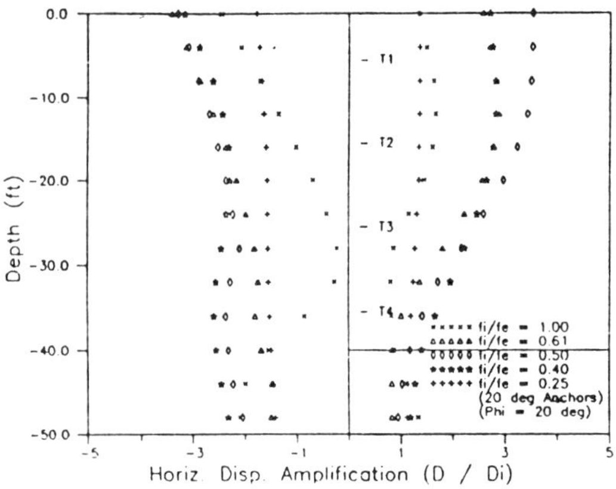

Fig. 8. Effects of input frequency on the amplification of horizontal wall displacements relative to base

zone. Since the amount of movement required for active failure is significantly lower than that required for passive failure, an active wedge tends to form in front of the wall. Another active wedge develops behind the wall. The failure plane associated with this active wedge is located in front of the grouted portion of anchors. A sudden increase in anchor loads occurs when this failure initiates, followed by essentially elastic oscillations.

Another trend observed in the analysis is the transfer of maximum anchor loads to the shallower anchors as the input level increases and/or soil strength decreases. The maximum dynamic loads tends to occur in the top two anchors.

Results of this study show that M-O method may be too conservative when the retained soil ground surface is horizontal and the friction angle exceeds 30 °. It is also shown that the dynamic earth pressure envelope is essentially rectangular for horizontal ground conditions. This represents a fraction of the maximum M-O pressure. However, the wall should be designed for additional dynamic bending moments introduced near the excavation level.

The following quantitative results and conclusions were determined from the parametric study

(1) The overall response of the soil wall system is highly dependent on the response of the retained side free field. For a horizontal ground surface, the response approaches resonance conditions when the input frequency approaches the natural frequency of the retained side free field. The maximum amplification at the top of the wall ranges from 3 1/2 to 6 1/2

times the peak input motion for the weak (ϕ=20 °) soil to a purely elastic medium, respectively.
(2) For resonant conditions, as the friction angle decreases from 35 to 20 °, the anchor loads (expressed as a percent of total design load) increase from 4 to 20%. For the same range of strength parameters, the ratio of maximum displacement at the top of the wall and toward the excavation increases from 0.17% to 0.21%. The maximum permanent displacement of the top of wall is approximately 0.03 feet (0.4 inches) and occurs in the weakest soil.

The analysis in this study shows the development of a shallow active wedge behind the wall as an important consequence of seismic loading. Additional analyses should be conducted with anchors grouted all the way to the wall. Such anchors will essentially act as soil nails and intercept the development of any failure surface. A three dimensional finite element analysis will be required since a plane strain model will be unrealistically stiff.

References

1. BANGASH A.A. *Non-linear dynamic finite element analysis of tied-back retaining walls*. PhD dissertation, Department of Civil Engineering, Washington State University, Washington, 1991.
2. VAUGHAN D.K. and RICHARDSON E. *FLEX Users Guide*. Weidlinger Associates, Los Altos, Ca, 1989.
3. MONONOBE N. and MATSUO H. On the Determination of Earth Pressures During Earthquakes. *Proc. World Engng Conf. 9*, 1929, pp 176-182.
4. OKABE S. General Theory of Earth Pressure. *J. Japanese Soc. Civ. Engrs*, 1926, **12**, No.1.
5. FRAGASZY R.J., ALI A. , DENBY G.M. and KILIAN A.P. *Seismic Response of Tie-back Walls: A Pilot Study*. Transportation Research Record# 1191, National Research Council, 1989, pp 166-175.
6. SILLER T.J. *The Dynamic Behaviour of Tiedback Retaining Walls*. PhD. Thesis, Carnegie Mellon University, Pittsburgh, PA, 1988.
7. KITAJIMA S. and Uwabe T. *Analysis of Seismic Damage in Anchored Sheet-Piling Bulkheads*. Port and Harbour Research InstituteReport 18(1), 1979, pp 68-127.
8. STEEDMAN R.S. and ZENG X. The Seismic Response of Waterfront Retaining Walls. in *Design and Performance of Earth Retaining Structures*, ASCE Geotechnical Special Publication No. 25, 1990, pp 872-886.
9. DENBY G. M., PSCHUNDER H. R. and ARGO D. E. *Response of Tied-Back Walls During the Whittier Narrows Earthquake of October 1*, 1987, Rpt. Grant No. CES- 8804304, NSF, Washington, DC, 1989, pp29.

The maintenance of old masonry retaining walls

C. J. F. P. JONES, University of Newcastle upon Tyne, UK

In 1987 the UK Department of Transport published the results of a census of retaining walls in Great Britain, (ref.1). The results showed that there is a minimum length of 5389 km of walls supporting highways in the United Kingdom. As only half of the highway authorities questioned were able to indicate the length of walls in their area the above Fig. could be underestimated by a significant amount. Of the lengths of walls identified about 50 % were of dry stone construction, 35% were masonry and 15 % of concrete. The fact that approximately 50 % of all the walls in the United Kingdom are formed in dry stone walling indicates that the infra structure is reliant upon structures which were built in the last century before the introduction of modern highway loading criteria and specifications.

The maintenance problems of these structures are growing in importance and in some districts the maintenance costs of dry stone walls now exceeds the cost of bridge maintenance. The 1987 census indicated that between 0.1-0.5 % of the masonry walls need to be repaired annually and that the costs of repair range between £100,000-£800,000 per kilometre.

Most dry stone walls date from the industrial revolution which produced an unprecedented boom in the construction industry. Wall construction was necessitated by the demand for improved transport and the need to build roads and railways on sidelong ground typically represented by the Dales of Yorkshire and the Valleys of Wales and Scotland. In these areas stone was plentiful and today the dry stone wall is a common feature of the landscape. In many cases walls retained land above the highway as well as acting as support structures on the lower slope retaining the carriageway, Fig. 1. The walls were constructed directly onto the topsoil without proper foundations. Only the front face of the structure consists of finished material. The stones used for the face are usually of medium to large size carefully graded and placed by hand to fit. Behind these flat smaller stones were laid in an horizontal plane grading back from the face, Fig. 2. The structures are porous and essentially flexible and would have been built in layers using relatively unskilled labour and little construction plant. The nearest modern approach to this form of construction is a reinforced soil structure.

The use of dry stone retaining walls is not restricted to the United King-

Retaining structures. Thomas Telford, London, 1993

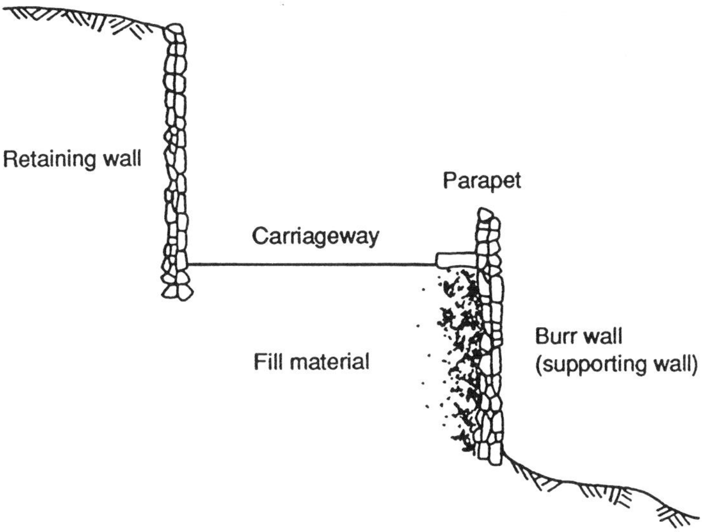

Fig.1. Dry stone walls supporting a highway and retaining load above the highway

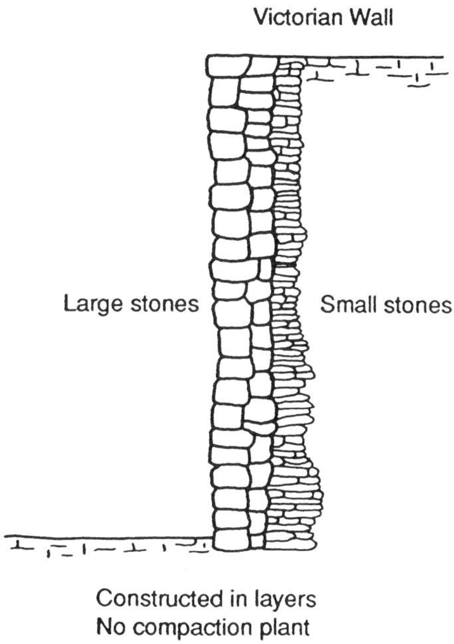

Fig.2. Construction technique

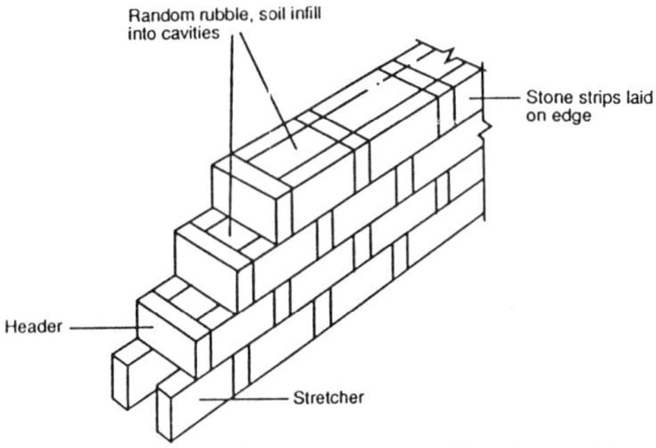

a) Chinese 'box'bonded wall

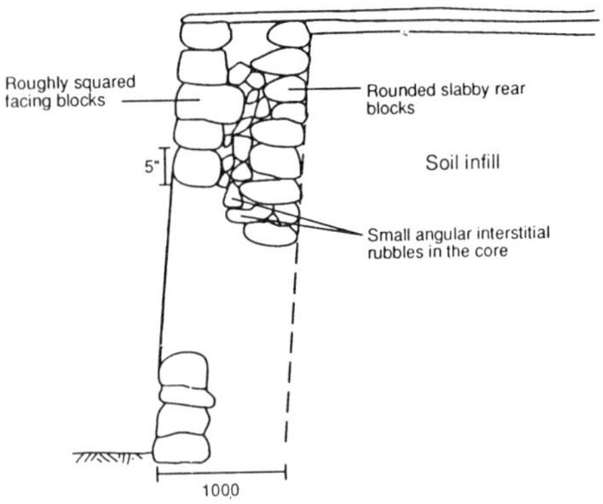

b) Wall at Tung Lung Fort

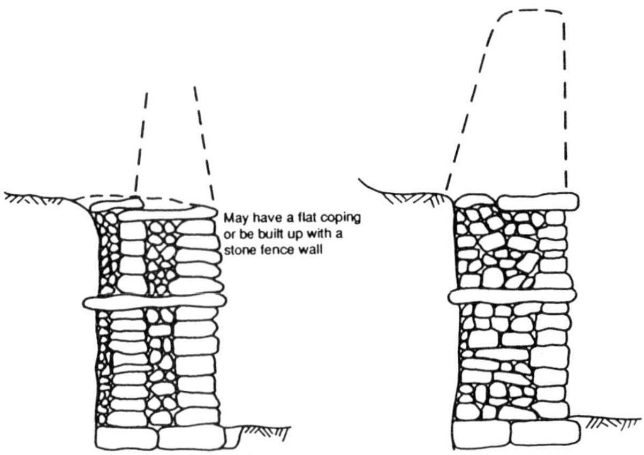

c) Yorkshire style dry stone wall

Scottish style dry stone wall

Fig. 3 Different types of dry stone wall

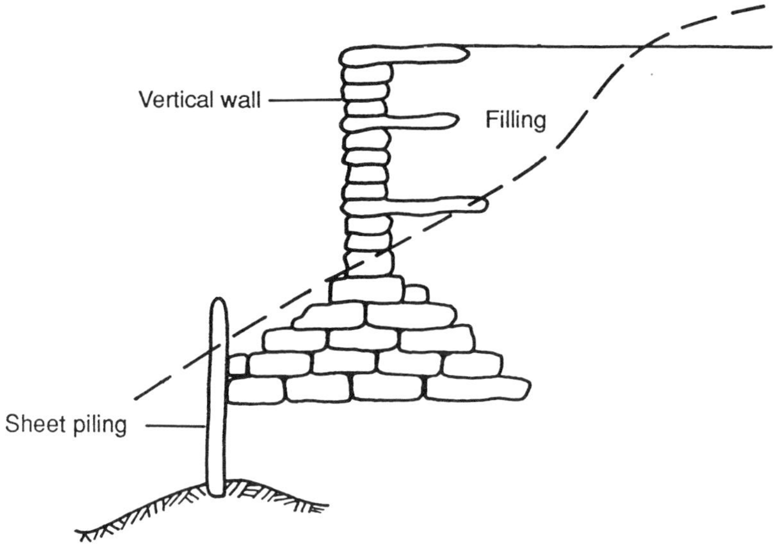

Fig. 4. Use of stone relieving slabs, Columbia

dom. Remarkably similar structures can be seen in China, India, Hong Kong, Japan, Korea, and South America. Observation of the walls in these countries shows certain consistent elements and also regional variations, Fig. 3, (ref.2). In all cases it is apparent that the method of construction is integral to the design ensuring stability of the wall. This is perhaps best illustrated by the

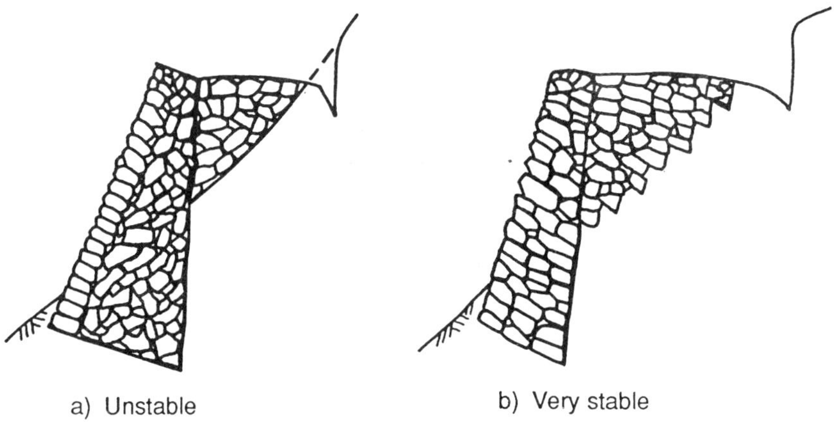

Fig. 5. Importance of orientation of the masonry

Table 1. Factors influencing the deterioration of dry stone retaining walls

Natural/Inherent	Man-made
Form of construction	Traffic loading and vibration
Percolation and removal of fines	
	Accident damage
Material quality	Vandalism
Soil creep	Burst water mains
Storm or flood	Excavation at the toe
Vegitation	New construction (adjacent)
Animals	Removal of support
Old age	Deicing salts
Geological conditions	Air pollution

use of stone slabs as relieving platforms in the dry stone retaining walls in Columbia, Fig. 4 (ref.3.).
Similarly, the orientation of the stones within the wall has been shown to be of critical importance to the stability of the wall, Fig. 5, (ref.4).

The maintenance problem

The growing interest in dry stone retaining walls has been triggered by the increased cost of maintaining these structures. There is a range of factors which influence dry stone wall behaviour and their deterioration, these can be identified in two groups, classified as either natural or man-made, Table 1. The key elements are seen as the nature of the construction, the presence of water, traffic loading and material deterioration. Of these water pressure build up is the most usual cause of collapse.

Material deterioration of the stones forming the wall will lead to failure if not checked. Many walls were built with inferior material prone to weathering. Weathering leads to honeycombing and is accelerated by air pollution and frost action. In badly weathered walls the void ratio of the front face can be as high as 50-60 %, (ref.5). The fact that many structures are so badly weathered is an indication of their age and that the maintenance of many old retaining walls has been ignored for many years.

The most usual failure method is a slip triggered by the collapse of the

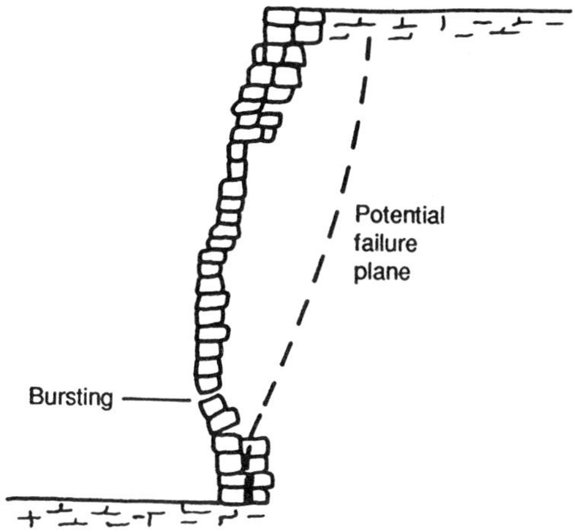

Fig. 6. Failure mechanisms

front face of the wall. This frequently occurs when the front face bulges outwards to a point where it becomes mechanically unstable, Fig. 6. Although a wall may appear to be distressed it may in fact remain in a stable condition for many years. Predictions as to when failure will occur, particularly with a wall which has had a distorted face for some time, is difficult and a matter of experience. As a result the decision as to when to undertake repairs is difficult.

Basing the decision of when to undertake repairs on an analysis of the structure would seem to be a rational approach. However, the design of the old masonry retaining walls is empirical and the application of current design rules frequently suggests that they should not stand up even when in pristine condition, when compared with modern retaining structures.

Modern retaining walls are built with substantial factors of safety incorporated into the design and it is not rational to compare these modern structures with the old dry stone walls. Arguably the use of an empirical design will produce a more efficient design, albeit with a lower factor of safety. Analysis of the stability of the old masonry walls needs to take into account the wide range of factors which influence stability, including the effect of soil structure interaction. The latter can be significant but is not usually considered in modern design, (ref.6). The difficulties of conducting an analysis of the stability of an old wall are compounded by the lack of precise information of the form and nature of the wall and the backfill. No records exist for these structures and even ownership can be a matter of dispute. Walls directly supporting the highway are the responsibility of the highway authority, but walls supporting adjacent property can be private.

In the latter case the cost of maintenance can be beyond the means of the owner.

Maintenance and repair techniques

The maintenance of old masonry retaining walls may be divided into two groups, maintenance prior to collapse and repair following failure. Normally repair is preceded by action to remove any identifiable cause of deterioration or distress. In some cases this may be possible as with the repair of a burst water main behind the structure. Elsewhere the cause of the problem may remain, in which case remedial measures appropriate to the conditions are applicable.

Prior to failure

Where the wall is showing serious signs of material deterioration one of the most common methods of repair is the use of pressure grouting. This method has the effect of transforming the structure and backfill into a gravity structure and the final structure can no longer be classified as a dry stone wall. This technique has two essential elements

(a) the provision of weep pipes to ensure drainage in the future, and
(b) grouting must be undertaken at low pressure so as not to cause collapse of the wall during the grouting process.

With walls showing serious distortion it is sometimes possible to construct buttresses as a support to the face. These usually result in a permanent repair but require additional land in front of the wall, Fig. 7.

Fig. 7. Provision of a buttress to provide stability for a bulging wall

A new form of repair for dry stone walls is the use of soil nailing. The first application of soil nailing in the United Kingdom was for the repair of a dry stone wall in danger of collapse, (ref.7). This technique appears to have considerable potential for the future particularly using such innovative techniques as pneumatic driven soil nails, (ref.8).

After failure

In many places repair is not possible prior to failure, this may be due to a lack of resources or the inability to predict the failure. The most common form of repair following collapse is the construction of a gravity wall faced with masonry. As these are modern structures they can no longer be considered as dry stone walls.

Maintenance strategy

The fact that the repair of many retaining walls in the United Kingdom cannot be undertaken prior to collapse is a growing problem. At present the approach of only responding to collapsed walls may be considered an effective use of limited resources. However, as the rate of collapse of walls increases and the costs of repairs escalates this will no longer be an acceptable solution. In addition the disruption to traffic caused by a collapsed wall is often considerable, as a result the overall costs of the highway users are rising. A different approach to the problem is required.

In Hong Kong the problem of the deteriorating dry stone retaining wall has been tackled in a rational manner. In the past the failure of retaining walls in Hong Kong has had serious consequences including loss of life. As a result, sufficient resources have been made available to study the problem and to devise suitable maintenance procedures. Following a comprehensive survey of the walls a method of assessment has been evolved whereby all walls can be inspected and classified, (ref.9). The need for remedial action is identified, based upon the relative importance of the wall, the consequent of its failure and the degree of distress already evident, (ref.10). The Hong Kong approach is made possible by the relatively small number of retaining walls compared to the available technical and financial resources.

A similar approach to that adopted in Hong Kong is not practical in the United Kingdom without the input of significant additional funding. Inspection of the Hong Kong criteria and consideration of the state of many dry stone walls in the United Kingdom shows a mismatch in that application of the Hong Kong rules in the United Kingdom would require the immediate repair of many hundreds of kilometres of walls.

Because of the scale of the problem and lack of resources a different management strategy is required in the United Kingdom. Many walls become very distorted yet remain stable for many years. With the Hong Kong criteria these could be classified as in need of immediate attention. However where resources are limited it seems rational to retain the structures which remain stable.

It is possible to anticipate that failure of a distorted wall will occur at some time in the future, but with our present state of knowledge it is impossible to predict when this will occur. A better understanding of the failure mechanism of dry stone retaining walls and their behaviour could lead to maintenance engineers being able to predict more accurately when to take appropriate action. However little research is currently being undertaken into this problem and therefore any maintenance strategy in the foreseeable future needs to be based upon current knowledge. Some causes of the failure of old walls can be identified and a rational strategy could be based upon observations associated with previous failures in a particular region. This approach is not labour intensive and could provide potential benefits in the management of resources. The requirement is for a maintenance system which identifies the development of potential critical conditions which can be used as a pre-warning of failure. Such a system is an extension of the sort of information previously supplied by the old highway or bridge inspectors to the local bridge engineer responsible for maintenance. These skills are not always directly available now, particularly where highway maintenance is based upon term contracts and where the employment of the highway and bridge superintendent with an intimate knowledge of their area is no longer possible.

Any new system would need to be based upon a comprehensive survey of the old retaining walls, their previous history and the causes of wall failures. Following this, inspection and assessment procedures could be initiated based upon visual observations aimed at identifying the often subtle changes which can be the forerunner of major collapse. The proposed method follows the Hong Kong approach but modified to accommodate critical condition factors similar to those used in the assessment of masonry arch bridges, (ref.ll).

Coupled with the need for a new assessment procedure is the requirement to develop cost effective repair techniques for application prior to wall failure.

Conclusions

Most of the existing retaining walls supporting highways were constructed in the last century. Many of the structures have deteriorated badly, the incidence of collapse is increasing and costs of repairs are becoming significant. The scale of the problem is such that immediate repair of all of the distressed walls cannot be considered. Predicting when a wall will fail using analytical techniques is very difficult and an empirical approach to the problem appears to offer a logical alternative if maintenance costs are to be kept low.

References

1. DEPARTMENT OF TRANSPORT. *The Assessment of Highway Bridges and Structures, Bridge Census and Sample Survey.* HMSO, London, 1987), p 38.
2. HUGHES K.A. *Tumbling Dry Stone Walls.* MSc Dissertation, University of Leeds, 1987.
3. MALDONADO R and GONZALES M.I. Engineering of the Taironas circa 1200 , Rio de Janeiro. *Proc. 12th ICSMFE,* Balkema, 1989, pp 1273-1275.
4. ARYA A.S. and GUPTO V.S. Retaining Wall for Hill Roads. *Proc. Indian Roads Congress,* 1983, pp 291-324.
5. JONES C.J.F.P. Dry stone walls, in *The Maintenance of Brick and Stone Masonry Structures,* Spon, 1990, pp 349-360.
6. DUNCAN J.M., CLOUGH C.W. and EBELING R.M. Behaviour and Design of Gravity Earth Retaining Structures, in *Design and Performance of Earth Retaining Structures,* ASCE Geotechnical Special Publications No. 25, 1990.
7. BRUCE D.A. and JEWELL R.A. Soil Nailing: application and current practice. *Ground Engineering,* 1986, **19**, No.8, pp 10-16, and1987, **20**, No.1, pp 21-33.
8. MYLES B. and BRIDLE R.J. Fired Soil Nails — the Machine. *Ground Engineering,* July/Aug., pp. 38-39.
9. CHAN Y.C. *Report on the Study of Old Masonry Retaining Walls in Hong Kong.* Geotechnical Control Office, Hong Kong, 1982, p 122.
10. HONG KONG GEOTECHNICAL CONTROL OFFICE. *Engineering Criteria and Guidelines for Old Masonry Retaining Walls,* Circular No. 90, Geotechnical Control Office, Hong Kong.
11. DEPARTMENT OF TRANSPORT. *The Assessment of Highway Bridges and Structures.* Advice Note BA 16/84, HMSO, London, 1984, p 66.

Analysis of earthquake induced damage to quay walls

S. IAI, and T. KAMEOKA, Port and Harbour Research Institute, Ministry of Transport, Japan

The 1983 Nihonkai Chubu Earthquake of magnitude 7.7 in Northern Japan offers precious case history data on the performance of quay walls during earthquakes. A steel sheet pile quay wall suffered damage associated with liquefaction in the backfill sand but a similar adjacent quay wall did not suffer damage because liquefaction did not occur. In order to obtain a full understanding on the performance of the quay walls, a two dimensional effective stress analysis has been conducted. The model used in the present study uses multiple virtual simple shear mechanisms oriented in arbitrary directions. Results of the analysis are consistent with the observed performance of the damaged and undamaged quay walls. The analysis indicates that earthquake induced deformation of quay walls is due to the existence of initial stress and its release in accordance with progressive cyclic mobility of cohesionless soil.

Introduction

The seismic performance retaining structures backfilled with dry sand is well understood based on the Mohr-Coulomb's failure criterion. When the

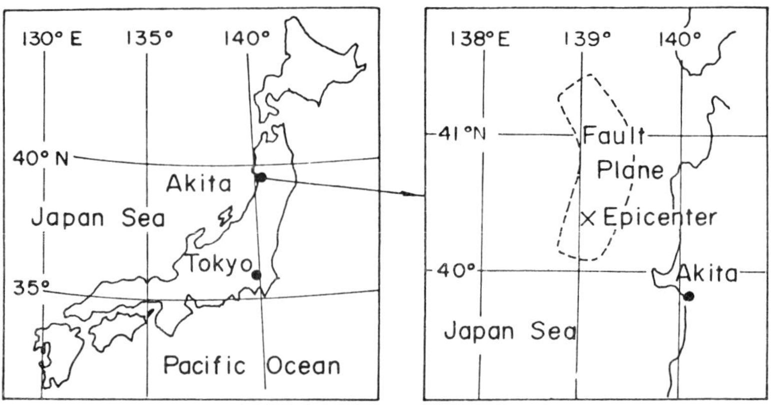

Fig. 1. Location of Akita Port and Epicentre of the Earthquake: (a) Location of Akita; (b) A closer view

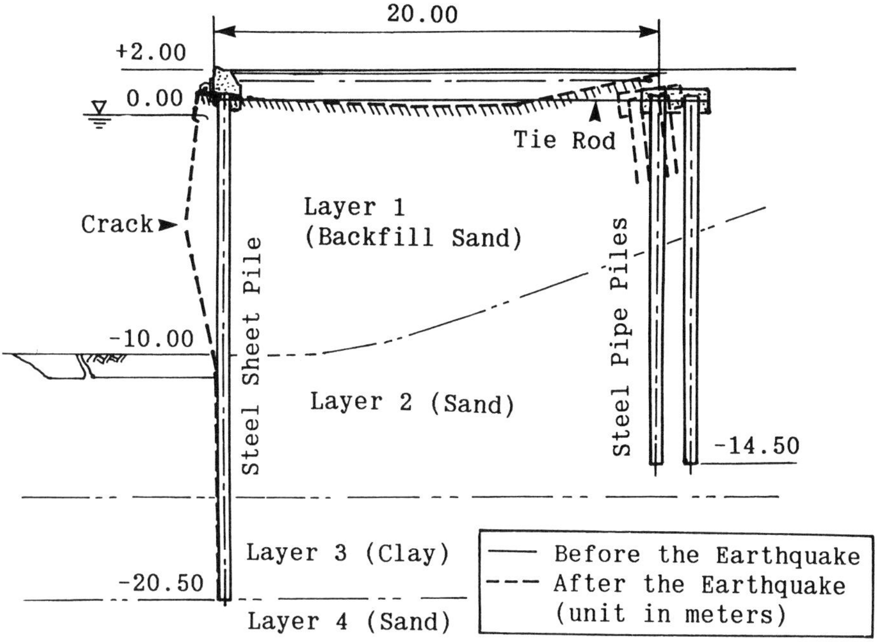

Fig. 2. Cross Section of quay wall at Ohama No. 2 Wharf

Fig. 3. Damage at Ohama No.2 Wharf

sand is saturated with water, as in waterfront structures, the effect of pore water pressure becomes predominant and complicates the performance of retaining structures. Effective stress analysis may be the only means for obtaining a full understanding of the seismic performance of waterfront structures. In particular, a model consisting of multiple shear mechanism offers the prospect of taking into account the essential features in the cyclic behaviour of sand (Iai, 1991).

In May 1983, the Nihonkai Chubu Earthquake of magnitude 7.7 hit the northern part of Japan. The earthquake caused damage to quay walls at Akita Port located about 100 km from the epicentre, the location of which is shown in Fig.l. The damage was associated with liquefaction of the backfill sand. In the vicinity of a damaged quay wall, another quay wall survived without damage. The cross section of this wall was similar to that of the damaged quay wall, but no liquefaction took place in the soil. In addition to the

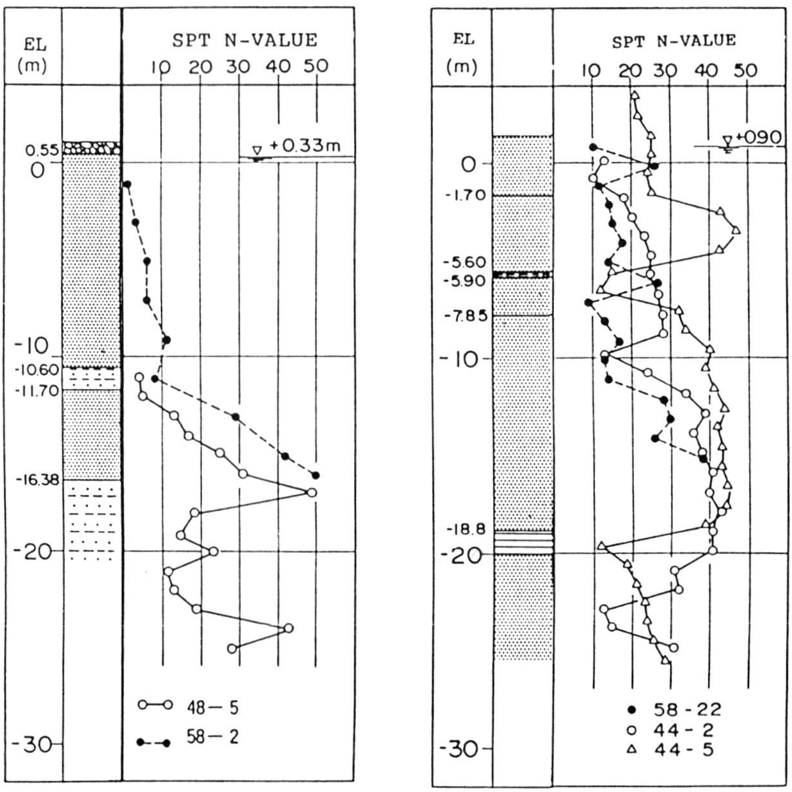

Fig. 4. Ground conditions at Ohama No.2 and No.1 Wharves: (a) Ohama No. 2 Wharf (Damaged); (b) Ohama No. 1 Wharf (No Damage)

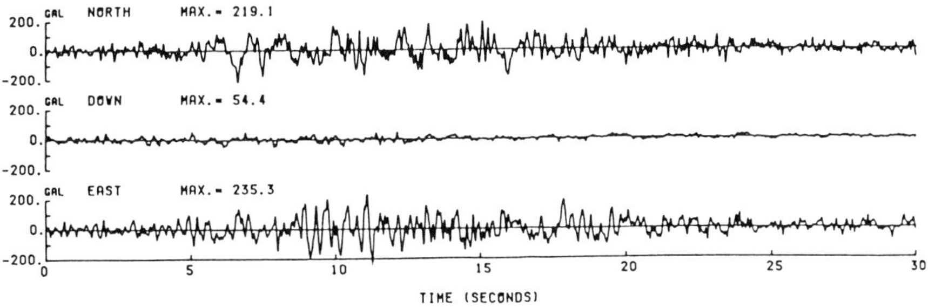

Fig. 5. Recorded Earthquake Motion at Akita Port

geotechnical data, a strong earthquake motion record, which is essential for earthquake analysis, was recovered from Akita Port. This provides a good opportunity to examine the ability of the analytical model mentioned earlier to predict the effects of earthquakes on quay walls.

Performance of quay walls

The quay walls are located at No.2 and No.1 Wharves at Akita Port. The damaged quay wall at Ohama No.2 Wharf had a cross section shown in Fig. 2 and suffered residual displacement toward the sea ranging from 1.1 to 1.8 meters. Settlements of the same order resulted at the apron as shown in Figs. 2 and 3. The anchored steel sheet pile exhibited a yield hinge due to excessively large bending moment, resulting in the opening of a crack at 6 m below the water level. Evidence of liquefaction of the backfill sand was abundant.

The quay wall which survived without damage at Ohama No.1 Wharf had a similar cross section and was located about 400 m from the damaged quay wall. Because Ohama No. 1 Wharf was constructed by excavation without backfilling, the density of the sand at Ohama No.1 Wharf was higher than that at Ohama No.2 Wharf, as indicated by the SPT N-values shown in Fig. 4.

Both of the quay walls were considered to be shaken by the same input earthquake motion recorded at Akita Port, shown in Fig. 5, with a maximum acceleration of 235 Gals in the East - West direction. More details on the earthquake and the damage can be found in a report by Tsuchida *et al*. (1985).

Modelling of soils, structures and sea

As partly mentioned earlier, the constitutive model of the soil used in the present study consists of a multiple shear mechanism with each virtual shear plane oriented in an arbitrary direction (Iai, 1991). The model can take into account fully the effect of principal stress axes rotation, which is essential in analyzing cyclic behaviour of anisotropically consolidated sand. The model has 10 parameters; two of which specify elastic behaviour of soil, two specify plastic shear behaviour, and the rest specify dilatancy.

Table 1. soil model parameters for the analysis

	Layer 1	Layer 2	Layer 3	Layer 4	Layer 5
G_{ma} (kPa)	33800	72200	74970	168200	72200
K_a (kPa)	89930	1921000	199470	447530	19210
$-\sigma_{ma}'$ (kPa)	50	110	140	157	69
Φ_f (deg)	37	41	39	44	43
Φ_p (deg)	30	30	--	30	30
S_1	0.005	0.005	--	0.005	0.005
w_1	0.5	8.5	--	3.8	14.0
p_1	0.4	0.5	--	0.5	0.5
p_2	0.42	0.8	--	0.84	1.0
c_1	1.5	3.3	--	2.4	4.6

In the analysis, the ground at Ohama No.2 Wharf was idealized into four kinds of soil layers as shown in Fig. 2. The parameters of the soil, shown in Table I, were determined from the results of the seismic wave propagation survey conducted at Akita Port and the laboratory test results of the sand under undrained cyclic loading. The ground at Ohama No. 1 Wharf was uniformly idealized using the soil parameters indicated by Layer 5 in Table 1.

As an example, using the parameters for layer 1, a soil element test under undrained cyclic loading condition is simulated as shown in Fig. 6. As seen in this Fig., typical features of the cyclic mobility are appropriately simulated by the present model.

The sheet piles forming the quay walls were type FSP-VI$_L$ manufactured in Japan, having the following properties; the second moment of area 8.6 x10^{-4} m^4, cross sectional area 0.0306 m^2, and density 7.5 t/m^3.

The sheet piles were idealized using linear beam elements. No friction was assumed to be acting between the soil and the sheet piles in order to simulate the effect of water spouting along the sheet piles during liquefaction. The sea water was modelled as an incompressible fluid and was formulated as an added mass matrix (Zienkiewicz, 1977). The side and the bottom boundaries

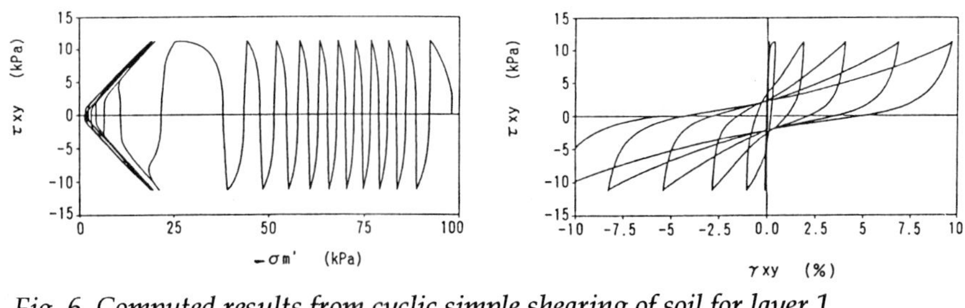

Fig. 6. Computed results from cyclic simple shearing of soil for layer 1

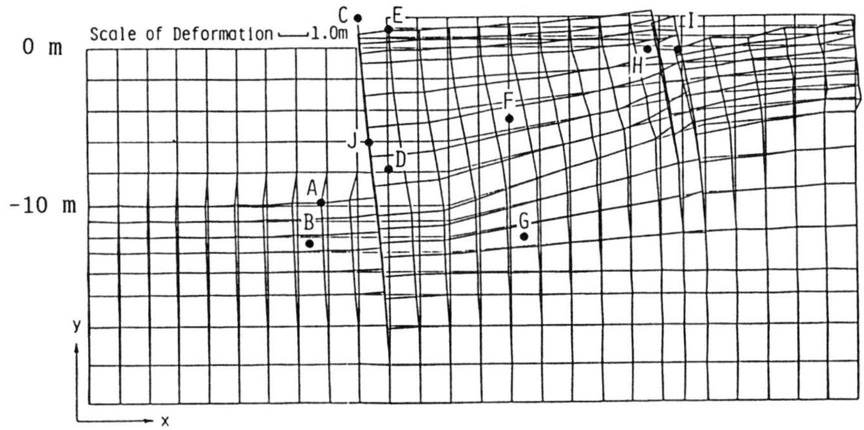

Fig. 7. Computed deformation of quay wall at Ohama No.2 Wharf at 14 seconds

of the cross section analyzed were simulated by viscous dampers to simulate transmitting waves.

The recorded earthquake motion at Akita Port was first deconvoluted to obtain the incident wave and the direction components, perpendicular to the face of the quay walls, were used in the earthquake response analysis. Before the earthquake response analysis, a static analysis was conducted with gravity to simulate the initial stresses acting in-situ. Then the dynamic analysis was conducted under undrained conditions (Zienkiewicz *et al.* 1982). The numerical integration was done by the Wilson- θ method (θ=1.4) using a time step of 0.01 seconds. Rayleigh damping of α=0 and β=0.005 which was reduced proportionally with progress of cyclic mobility was used to ensure stability of the numerical solution process.

Results of the analysis

Damaged quay wall

The effective stress analysis of the quay wall at Ohama No.2 Wharf results in the deformation shown in Fig. 7; deformations towards the sea are mainly seen at the soil wedges in front of and behind the sheet pile wall. The backfill sand behind the sheet pile gradually settles whereas the sea bed in front of the sheet pile gradually swells as the shaking continues. The horizontal displacement at the sheet pile was computed as 1.1 m, being consistent with the observed displacements which range from 1.1 to 1.8 m.

The earth pressures (i.e. normal effective stresses + excess pore water pressures) acting on the sheet pile after 15 seconds of shaking, shown in Fig. 8, tend to increase up to those with the earth pressure coefficient of 1.0. Corresponding to these changes in the earth pressures, the bending moment

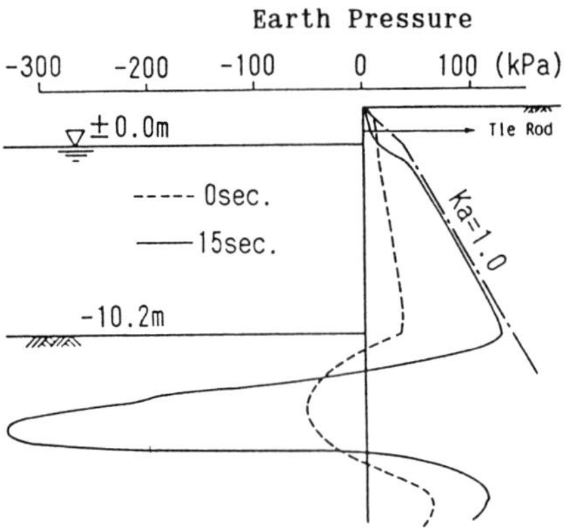

Fig. 8. Earth pressure on quay wall at Ohama No.2 Wharf

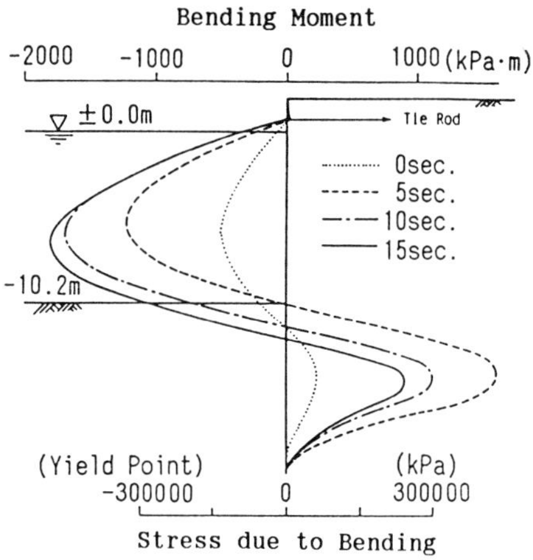

Fig. 9. Bending movements of quay wall at Ohama No.2 Wharf

of the sheet pile changes as shown in Fig. 9. The maximum stress due to bending exceeds the failure strength of the steel sheet pile of 300000 kPa. This is consistent with the observed failure of the steel sheet pile.

Quay wall without damage

The effective stress analysis of the quay wall at Ohama No.1 Wharf resulted in the deformation shown in Fig. 10. The computed maximum horizontal displacement was 15 cm whereas the observed movements were about 5 cm, indicating that the analysis provided a slightly conservative result.

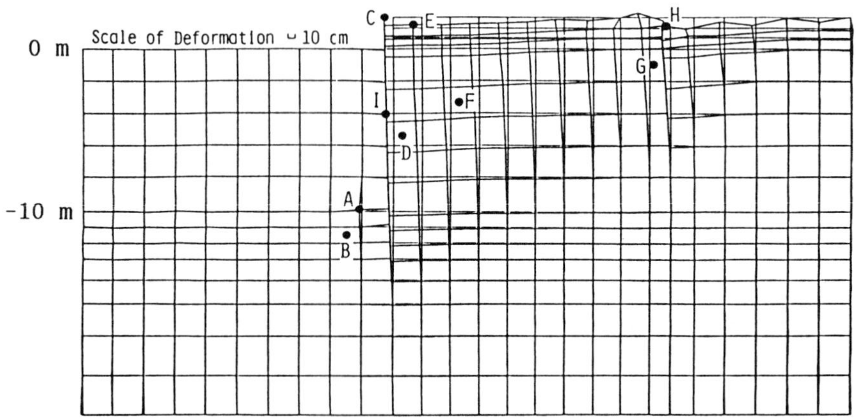

Fig. 10. Computed deformation of quay wall at Ohama No.1 wharf at 14 seconds

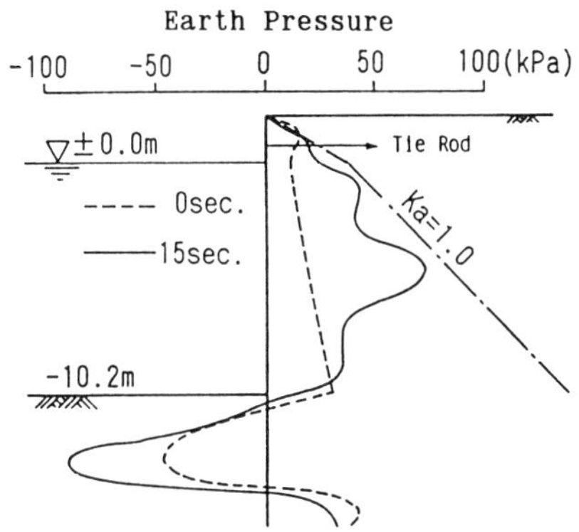

Fig. 11. *Earth pressure on quay wall at Ohama No.1 wharf*

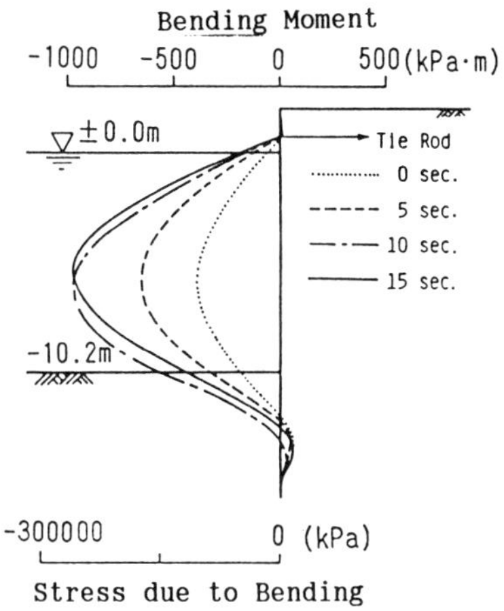

Fig. 12. *Bending moments of quay wall at Ohama No.1 wharf*

The earth pressures, shown in Fig. 11, tend to increase especially at the shallower part of soil but the bending moment, shown in Fig. 12, was less than that of the failure strength of steel, being consistent with the observed performance of the sheet pile quay wall.

Mechanism of deformation

In order to understand the mechanism of deformation of the quay walls, the stress and strain of soil in front of and behind the damaged sheet pile, indicated by B and D in Fig. 7, are shown in Fig. 13. The stress and strain notations are defined as extension being positive.

In front of the sheet pile, the effective stress path, plotted as a relationship between $\tau = -(\sigma_1' - \sigma_3')/2$ and effective mean stress ($-\sigma_m'$) in the upper most row in Fig.13 (a), gradually approaches the failure line with fluctuation around the initial deviatoric stress. In accordance with this, the axial strain difference is gradually induced as shown in the middle row in Fig. 13 (a).

Behind the sheet pile, the effective mean stress ($-\sigma_m'$) rapidly decreases as shown in the upper most row in Fig. 13 (b). In accordance with this, the axial strain difference is gradually induced as shown in the middle row in Fig. 13 (b).

These results indicate that deformation of the quay wall is due to the initial stress and its release in accordance with liquefaction or cyclic mobility of saturated soil. This mechanism of deformation is quite different from the conventional approach based on the sliding block concept.

Conclusions

The effective stress analysis of quay walls with and without damage leads to the following conclusions.

(1) The effective stress model, consisting of a multiple shear mechanism, has a reasonable applicability in the analysis of earthquake induced damage to quay walls. In particular, the model demonstrates the ability to differentiate between serious large deformations (i.e. damage) from negligibly small deformations (i.e. no damage) in similar types of sheet pile quay walls.

(2) The mechanism for inducing deformation in quay walls is due to the existence of initial stress and its release in accordance with cyclic mobility

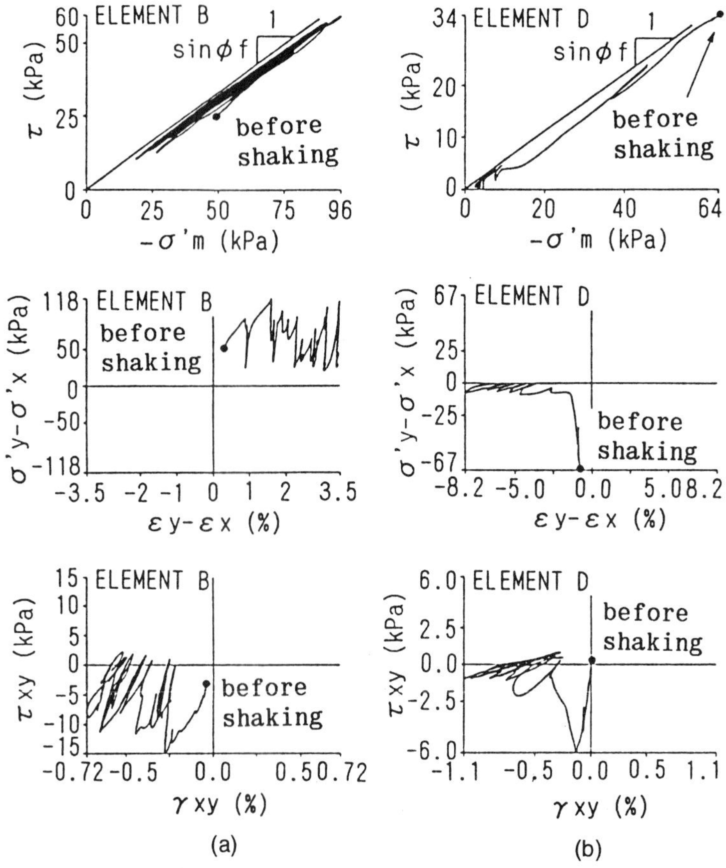

Fig. 13. Stress and strain in soil of quay wall at Ohama No.2 wharf: (a) in front of sheet pile; (b) behind sheet pile

of saturated soil. This is quite different from the mechanism of sliding blocks.

References

1. IAI S. A numerically robust modelling of cyclic mobility. in *Computer Methods and Advances in Geomechanics*, A. A. Balkema, 1991, pp 827-832.
2. TSUCHIDA H. *et al. Damage to port structures by the 1983 Nipponkai-Chubu Earthquake*. Technical Note of the Port and Harbour Research Institute No. 511, 1985, p 447 (in Japanese).
3. ZIENKIEWICZ O.C. *The finite element method*, 3rd edition. McGraw-Hill Book Co., 1970.
4. ZIENKIEWICZ O.C. and BETTESS P. Soils and other saturated media under transient, dynamic conditions, in *Soil Mechanics - Transient and Cyclic Loads*. John Wiley and Sons, 1982, pp 1-16.

Dimension of stable retaining walls and the construction cost

N. MOROTO, Hachinohe Institute of Technology, Japan

Introduction

Retaining walls are popular structures which resist slope instability. When we construct, for example, highway or residential land, retaining walls contribute greatly to the total construction cost. Thus the retaining walls become very important structures, particularly in Japan where flat land is limited.

As everyone knows, many types of retaining wall have been constucted. Typical retaining walls are classified as gravity type, cantilever type and counterfort type. When we design these retaining walls, we must first estimate the width of base slab B and the wall height H together with the construction cost.

In practical civil engineering work, a rough estimation often becomes more important than an exact and complex calculation. This is true for design of structures. An approximate guess is more often requested than performing a detailed analysis, particularly at the first stage of the project. In this contribution, the author focuses on the dimension ratio B/H and the construction cost C in a simplified way.

Dimensions of a stable retaining wall

In the real construction of a retaining wall, it is necessary to secure the bearing capacity of base slab, to set up reliable drainage works and to choose excellent backfill materials. In this contribution, we do not discuss these kinds of practical requirements but try to examine the basic features of the wall dimension. As design loads, we have

- selfweight
- earth pressure
- surcharge
- water pressure
- seismic force

Among these, the primarily important function of the retaining wall is to resist the earth pressure by its own weight. Applying these two external forces to a simple rectangular section as shown in Fig. 1, we examine the

Retaining structures. Thomas Telford, London, Thomas

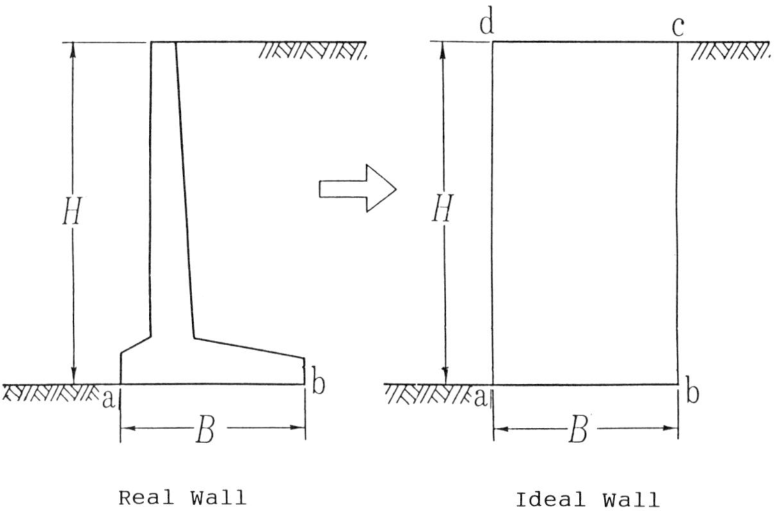

Fig. 1. A simplified retaining wall section

stability of the simplified retaining wall. This ideal retaining wall is considered to have the same unit weight as that of a backfill for convenience. The backfill material is cohesionless. The earth pressure is expressed by Lankin theory.

The stability analysis yields the following two conditions:

$$B/H = \sqrt{K_a} \quad \text{(for the middle third)} \tag{1}$$

$$B/H = \frac{K_a}{2\mu} F_s \quad \text{(for sliding)} \tag{2}$$

where

$K_a = (1 - \sin\phi)/(1 + \sin\phi)$
ϕ: angle of internal friction
μ: frictional coefficient between base slab and foundation soil
F_s: factor of safety toward sliding force

One can see that the stable wall dimension B/H is simply expressed by equation (1) or (2).

Table 1 shows the calculated values of B/H for different angles of internal friction and, $\mu = 0.6$ with $F_s = 1.5$. Old engineers say that they had been taught to use B/H = 0.6.

This empirical evidence fits very well with the case of $\phi = 30$ (in degree) in Table 1. With the values of $\mu = 0.6$ and $F_s = 1.5$, Table 1 shows that the condition of the middle third becomes stronger than that of sliding in the case of ϕ being greater than 15 (in degree). This indicates that the required

Table 1. Stable retaining wall dimension B/H

Angle of internal friction Ø (in degree)	Coefficient of active earth pressure K_a	$\sqrt{K_a}$	$F_s K_a/2\mu$
45	0.17	0.41	0.21
40	0.22	0.47	0.28
35	0.27	0.52	0.34
30	0.33	0.58	0.41
25	0.41	0.64	0.51
20	0.49	0.70	0.61
15	0.59	0.77	0.74
10	0.70	0.84	0.88
5	0.84	0.92	1.05
0	1.00	1.00	1.25
	$F_s = 1.5$	$\mu = 0.6$	

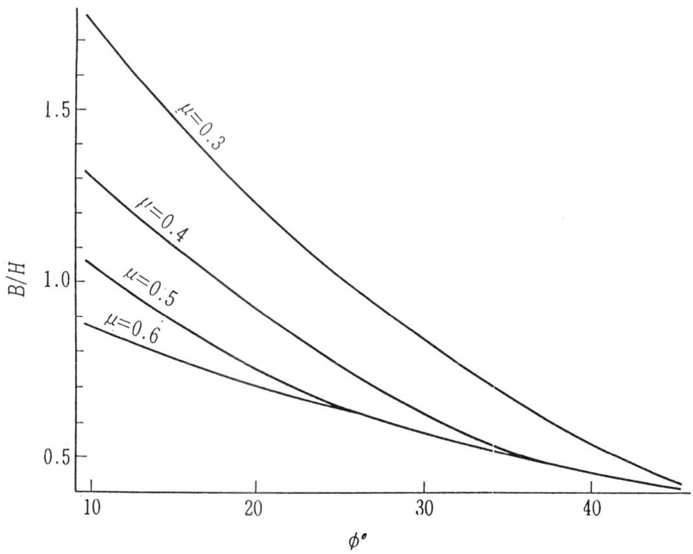

Fig. 2. Stable dimension of ration B/H and ϕ, μ

Table 2. Average values of B/H in a proposed standard design in Japan

Backfill material	Average B/H	Corresponding value of Ø in Table-1 (in degree)
Gravel Gravelly soil Sand	0.5-0.55	35
Sandy soil	0.55-0.6	30
Silt Cohesive soil	0.7	20

dimension ratio B/H can be simply obtained by the square root of K_a for superior backfill and foundation soil. However, as the value of μ becomes smaller, we must judge the value of B/H from both Equations (1) and (2). The relationship between B/H and ϕ of the backfill for the problems of both the middle third and sliding is shown in Fig. 2.

The Ministry of Construction, Japan once presented authorized standard sections of inverted-T type retaining wall (H=3-8 m) with the conditions of $\mu = 0.6$, $F_s = 1.5$, the middle third together with the earth pressure distribution introduced by Terzaghi and Peck. Using the above standard sections, the author has summarized some results in Table 2.

Sufficient coincidence between the soil names and the listed figures in Table 2 can be recognized to be reasonable from our common knowledge. It can be, therefore, considered that this coincidence shows the validity of the author's simplified procedure in this contribution.

Construction cost of retaining walls

The construction cost of the retaining walls consists mainly of concrete work, frame work and re-bar work. It is known that the total cost increases with increasing wall height parabolically.

In order to have a rough estimate for the construction cost, here we consider that the cost C is proportional to the area abcd (B.H) of the rectangular section shown in Fig. 1. Using the relationship given by Equation (1) or (2) into this assumption yields

$$C = b\,H^2 \qquad b\text{: constant} \qquad (3)$$

Table 3 shows a real construction cost estimated for the retaining walls, presented by a highway organization in Japan which covers the whole of the country. The retaining walls are categorized according to their wall height as

Table 3. A real construction cost for retaining wall in Japan

Wall Type	Gravity		Inverted-T			Counterfort			
Wall Height H(m)	2	3	4	5	6	7	8	9	10
1970	10760	30800	55000	78000	122700	163700	197900	250900	301100
1971-1972	18700	32700	57300	80400	127400	170000		260400	312400
1973	21500	37400	59500	84500	132000	174500		278000	337000
1974-1975	28000	49000	77000	110000	172000	227000		316000	438000
1976-1977	34000	58000	99000	139000	184000	239000		375000	450000
1978	37000	64000	109000	153000	202000	263000		413000	495000
1979	35000	60000	121000	166000	221000	312000	404000	490000	583000
1980	44000	76000	146000	201000	267000	367000	476000	577000	686000
1981	47000	81000	158000	218000	290000	377000	489000	529000	705000
1982	47300	81200	158000	217000	289000	375000	487000	590000	703000

(Unit: Japanese yen, ¥)

in Table 3.

This table shows the data for thirteen years from 1970 to 1982. Results of a regression analysis with use of Equation (3) are listed in Table 4 in which an excellent correlation between the real data and the simple equation can be seen.

Table 4. Results of regression analysis

Year	Constant b	Coefficient of correlation r^2	Price index $\frac{b}{b(1970)}$
1970	3109	1.00	1.00
1971-1972	3231	1.00	1.04
1973	3438	1.00	1.11
1974-1875	4297	0.99	1.38
1976-1977	4682	1.00	1.51
1978	5152	1.00	1.66
1979	6074	1.00	1.95
1980	7169	1.00	2.31
1981	7202	0.99	2.32
1982	7378	1.00	2.37

Closure

As has been shown, it can be seen that the dimension ratio B/H plays the most important role for an evaluation of the retaining wall stability.

Teng (ref.l) has shown a standard section of retaining walls such as inverted-T or counterfort type for which B/H= 2/5-2/3 = 0.4-0.67. Teng did not explain the relationship between the dimension ratio and the soil conditions. In this contribution, the author made it possible to relate the stable dimension ratio of the retaining walls to the soil conditions expressed in terms of the coefficient of earth pressure or the frictional coefficient between base slab and foundation soil. In addition, the construction cost of the retaining walls may be simply analysed by the parabolic equation.

Finally, the author's idealized procedure has been clarified to be reasonable. The simpler the better for engineering purposes.

References

1.TENG W.C. *Foundation design*.Prentice-Hall Book Company, 1962.

Appendix

The derivation of Equation (1) and (2) will be shown here.

The horizontal earth pressure P_a is written by

$$P_a = K_a \gamma H^2/2 \qquad \gamma : \text{unit weight of backfill} \qquad (A-1)$$

The resisting horizontal force P_r is expressed by

$$P_r = \mu \cdot \gamma_c H \cdot B \qquad \gamma_c : \text{unit weight of the idealized wall} (= \gamma) \qquad (A-2)$$

The factor of safety toward sliding is given by

$$F_s = P_r/P_a \qquad (A-3)$$

Putting Equations (A-1) and (A-2) into (A-3) gives

$$B/H = F_s \cdot K_a/2\mu \qquad (A-4)$$

Next, the total moment M around the toe point a is expressed by

$$M = \gamma H \cdot B \cdot B/2 - (1/2)\ K_a \gamma H^2 \cdot (H/3) \qquad (A-5)$$

The vertical force V is

$$V = \gamma H \cdot B \qquad (A\text{-}6)$$

The distance x between the toe point a and the point at which the resultant force is acting can be obtained by

$$x = M/V = (B/2)(1-(K_a/3)(H/B)^2) \qquad (A\text{-}7)$$

The eccentricity distance e can be got by

$$e = B/2 - x = K_a(B/6)(H/B)^2 \qquad (A\text{-}8)$$

Inserting e=B/6 into Equation (A-8) yields

$$B/H = \sqrt{K_a} \qquad (A\text{-}9)$$

Anchored flexible retaining walls experiments on models: calculation by the reaction modulus method

F. MASROURI, Laboratoire de Géomécanique - Ecole Nationale Supérieure de Geologie de Nancy, and R. KASTNER, Laboratoire de Géotechnique - Institut National des Sciences Appliquées de Lyon, France

Introduction

The flexible and semi-flexible retaining walls which are frequently used for excavations in urban sites have to ensure the overall stability. These structures must eliminate or limit soil movements toward the excavation, because they can damage adjacent buildings and structures. The anchors, especially those which are prestressed, allow the stabilization of these movements and also stabilize the retaining structures to which they are connected.

A series of experiments on a bi-dimensional small scale model allowed us to study the influence of the anchor specifications: length, angle of inclination and prestressing force on the performance and the behaviour when in service of a retaining wall.

The design by reaction modulus method is compared to the results of experimental tests to clarify the possibilities and limitations of this method. For short anchors, the overall stability of the anchor/wall/ground is also studied and compared with the Kranz method and with french recommendations : TA 86 (ref.1).

Description of apparatus

Our experiments are realized on a model with the bi-dimensional material of Taylor-Schneebeli which is composed of 3, 4, 5mm rustproof steel rollers. ($\gamma_d/\gamma_w = 6.1, \phi = 21^o,\ c = 0$) .

The great density of this material will eliminate in part the simulation distortion due to the volume forces (ref.2). The model pile is made of duralumin and so could be considered smooth. It is in 3 sections (0.08 m width and 0.012 m thick by 0.805 m high) and is placed in a 1 x 2m roller mass (fig. 1). Its flexibility, as defined by Rowe (ref.3), is in the range of the semi-flexible retaining walls such as the retainings to be found in the underground system

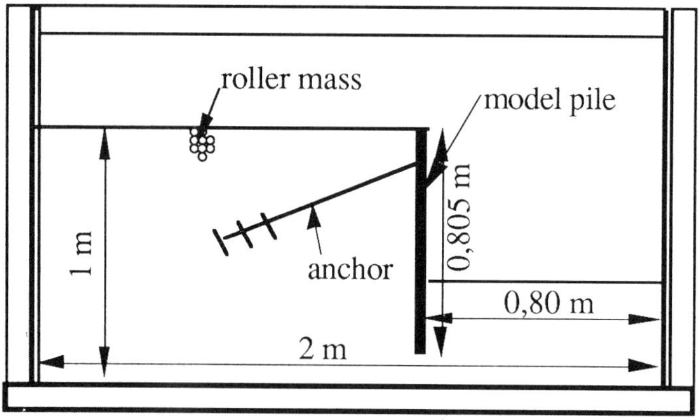

Fig. 1. Diagram of model

in Lyon.

The horizontal movements of the wall at its top and bottom are measured with mechanical dial gauges. The model pile's deflection is measured on 20 levels by 30 strain gauges mounted on both sides of the wall, and permits us to obtain the displacement and the soil differential pressure curves by the least square method, double integration and double differentiation.

The model anchor is composed of a section of steel girder embedded in the rollers (secondary grout) and attached to the model pile by 2 anchor wires equipped with load cells. At every stage of excavation, the deflection of the pile, the displacement at the top and the bottom of the pile and the force in the anchor are measured.

Programme of tests

To study the influence of the different conditions of anchorage on the behaviour and the design of the semi-flexible retainings, we realized 20 experiments. The following parameters could be varied from test to test

- anchor length from 0.55 to 1.17 m
- anchor inclination from 15 to 39°
- anchor prestressing: every test with a passive anchor (maximum anchor load at failure = A_{max}) is repeated with a prestressed one ($A_{initial} = 1/2$ or $2/3\ A_{max}$).

Test procedure

All tests are carried out in a similar manner. First the "soil" on the outside of the piles is dredged away to a depth of 0.15 m. After the fixation of the

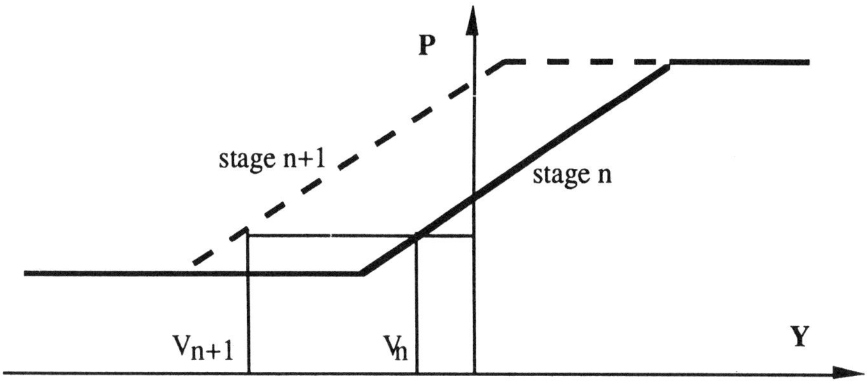

Fig 2. Load-deflection relationship of RIDO programme and its irreversibility

anchor at 0.05m from the top of the pile, the excavation is continued in stages of 0.05m until the collapse caused by a lack of passive pressure at about 0.57m of excavation. The classical design methods dictate an excavation depth of 0.45m for the optimum performance of the retaining wall (in service).

Design of anchored retaining walls

Here we study the performance of the wall in service by the Reaction Modulus Method. The overall stability of the anchor/wall/ground by the Kranz method is also verified.

The Reaction Modulus Design

The software we use is named RID03 (ref.4). It considers the ground action on the every side of the sheet pile with different load-deflection relationships by :

$P = P_0 + K_h (y - v)$
P_0: ground pressure distribution at rest
y: horizontal displacement of the pile at the studied point
v: the hysteresis factor
K_h: modulus of reaction defined by : $K_h = Re + Rp$
Re: constant rate of K_h
Rp: coefficient of increase of the effective vertical stress at the studied point.

P is limited by the active and passive pressures. If P overcomes these pressures the ground reaction will be irreversible, as shown in fig. 2.

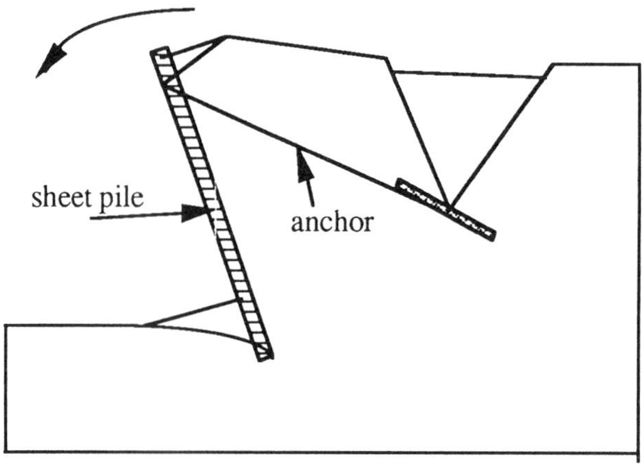

Fig. 3. Overall failure with a short anchor

The Kranz method of overall analysis

A very short anchor can cause an overall failure as shown in fig.3. Kranz (ref.5) proposed to calculate the anchor load in this case with a rectilinear failure plane which passes through the anchor secondary grout) and the assumed pivot point. The french recommendation TA 86 (ref.l) proposes to replace this linear plane by a circular one.

Comparison of calculated and measured values.

Although the failure caused by an insufficiency of embedded length can be correctly predicated by the active-passive calculations, here we are only interested by the optimum performance of the wall in service (excavation depth = 0.45 m).

The Reaction Modulus Design is done with the following values :

$K_{a\gamma} \cos \delta = 0.39$
$K_{p\gamma} \cos \delta = 3.1$
$K_h = R_p \sigma_v$
$R_p = 1500\ m^{-1}$
$\delta/\phi = 1$

The soil reaction modulus (K_h) cannot be measured, so we determined it by retro analysis of two preliminary tests. Independently of test conditions, this value is kept constant (ref.6).

Differential earth pressure

With passive anchors, there is a high pressure concentration at anchor level

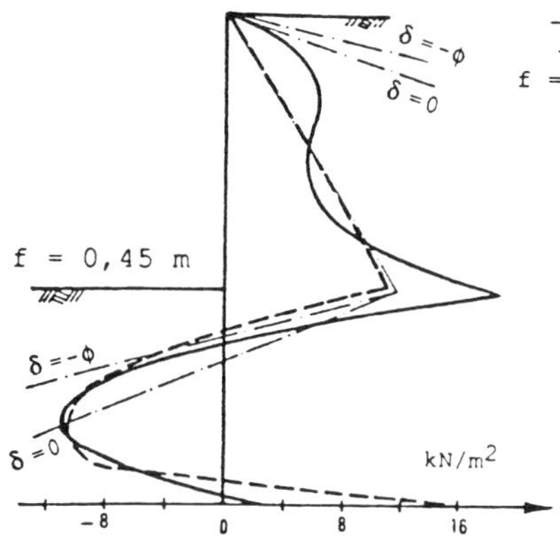

Fig 4. *Differential earth pressure (passive anchor)*

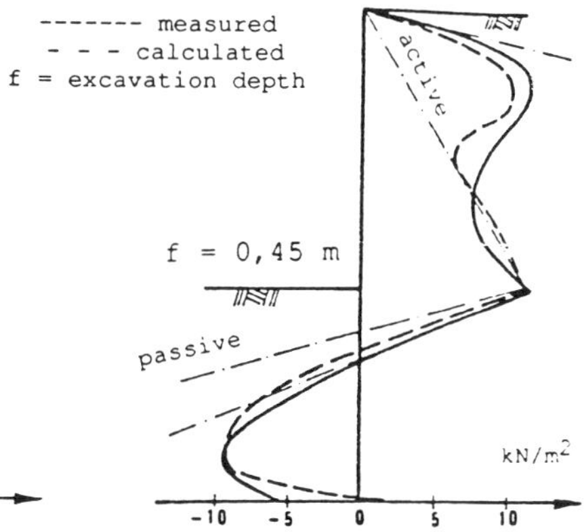

Fig 5. *Differential earth pressure (prestressed anchor)*

(soil arching). It is slightly less for short anchors, or very inclined ones, because of the great displacement of the model pile for such anchors. With the Reaction Modulus Method, this concentration is completely neglected and the distribution of soil pressure is triangular (fig. 4).

When a prestressed anchor is used, a considerable concentration of pressure acts at the anchor level, but it has nothing whatsoever to do with the arching on the active pressure side of the wall. This concentration is due to the passive reaction of the soil, because the anchor's prestressing directly loads it. This phenomenon is correctly calculated by RIDO to an excavation depth of 0.45 m, as shown in fig. 5, and for greater depth values, pile movements toward the excavation increase and the earth pressure decreases and becomes gradually triangular.

Anchor load

The anchor load is influenced by the earth pressure. For passive anchors, the design anchor load is highly underestimated (up to 42%). This is because the Reaction Modulus Method does not consider the effects of soil arching. The magnitude of this difference appears to be less in prestressed anchors, because of the increase of earth pressure on both the measured and calculated curves.

Bending moments

The difference between the measured and calculated curves are smaller than those we obtained for earth pressure and anchor load because of their antagonistic effect on bending moment curves (figs. 6 and 7).

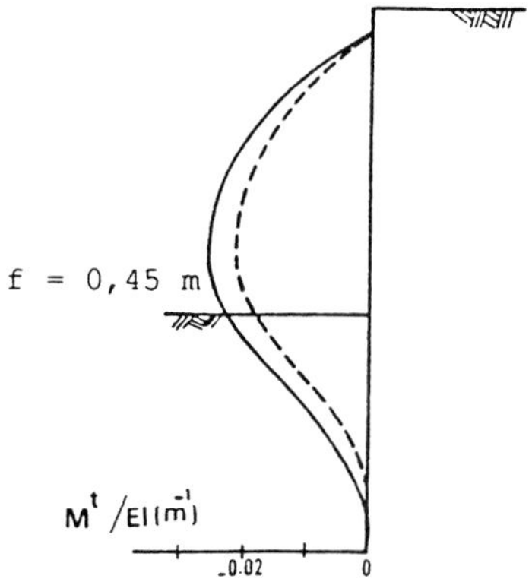

Fig 6. Bending moments (passive anchor)

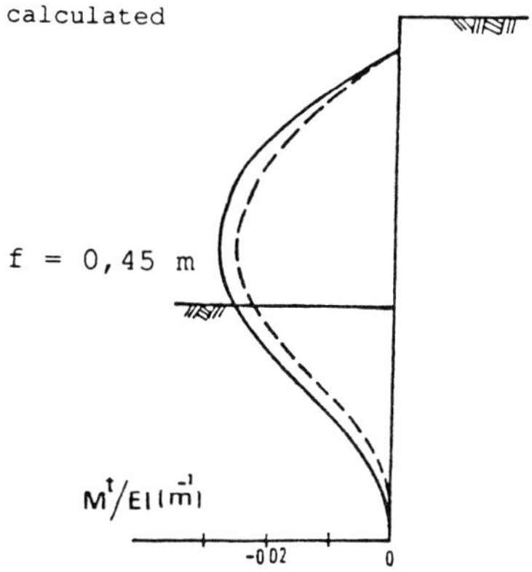

Fig 7. Bending moments (prestressed anchor)

Wall displacements

Both the embedded length and the specifications of the anchor have a profound effect on the wall's movement. For long anchors, the design anchor load is underestimated and the design displacements are slightly less than those measured (fig. 8). When the anchor is short, this difference is larger, because the transposition of the anchor assembly is added to the elastic displacements (fig 9).

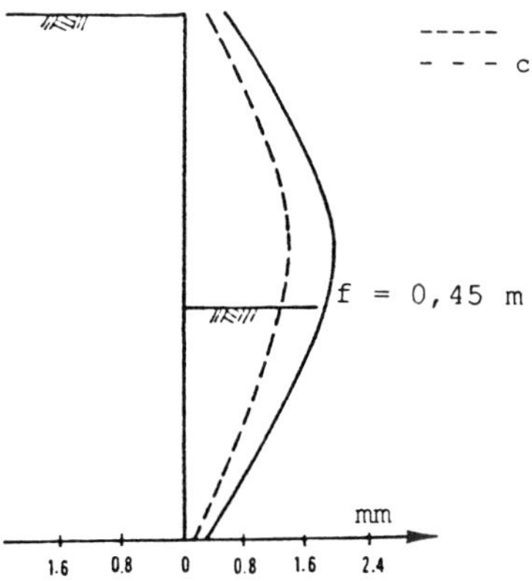

Fig. 8. Deflection curve (long anchor)

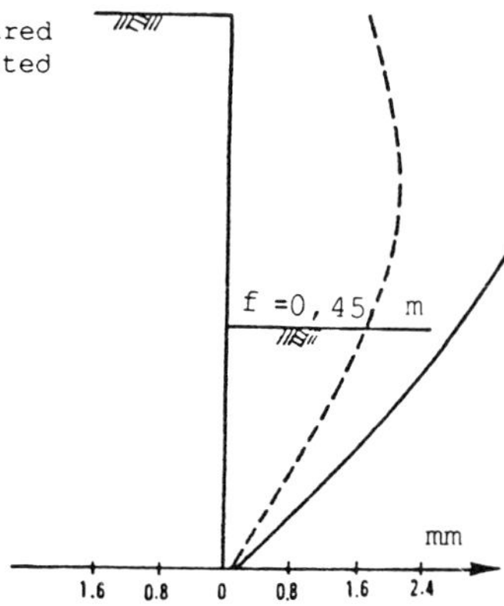

Fig.9. Deflection curve (short anchor)

Overall stability

An experiment conducted with a short anchor (length = 0.65m and angle of inclination = 22°) and continued until overall failure, is studied by stereophotogrammetry method as shown in fig. 10.

The analysis of the displacement field shows that :

- an arching pressure is developed between the secondary grout and the point of fixation of the anchor
- the slip line which limits the sliding wedge is more similar to the linear failure plane proposed by Kranz than the circular failure plane of french recommendations TA 86.

The anchor load at the point of overall failure calculated by Kranz (23 kN) and TA 86 (32 kN) are similar, but the measured load is much greater; 155 kN. To study the reason for this great difference, we analyzed the influence of the different parameters

(*a*) the active pressure distribution on the failure plane (in TA 86)
(*b*) the inclination of the active pressure on the wall (δ)
(*c*) the magnitude and distribution of this pressure
(*d*) the position of the pivot point.

It seems that the third point is the most important. The measured active pressure is only 20% greater than the calculated one, and the substitution of the theoretical triangular value by the real active pressure value decreases this difference. If we substitute all of the above values by the measured and

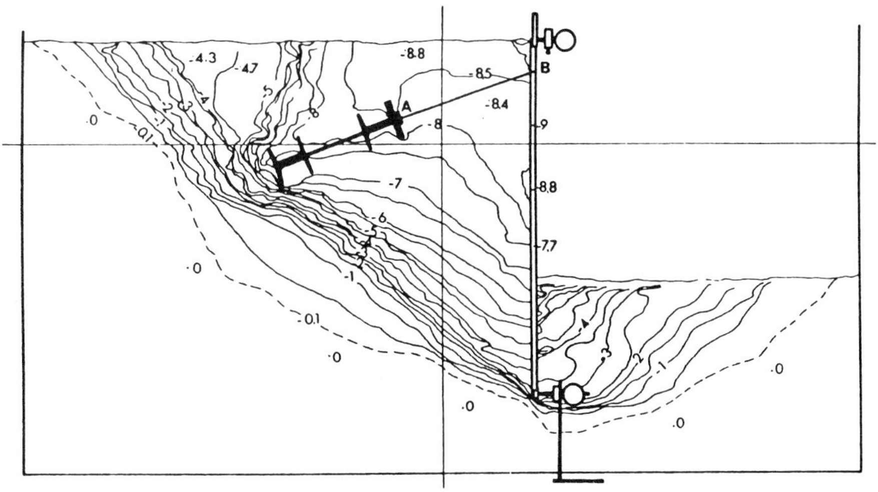

Fig. 10. Stereo photogrammetric image of overall failure (mm)

real values, the anchor loads calculated by the Kranz Method and TA 86 are very similar to the measured values. Therefore the choice of a circular failure plane, that leads to complicated calculations does not seem justified to us.

Conclusions

The bi-dimensional model we used permitted us to study the influence of different anchor specifications on the behaviour of the semi-flexible retaining walls, and also to clarify the limits of the Reaction Modulus Design as well as the overall stability.

However, we recommend that care is taken when the results are translated into full-scale structures, because this model does not take into consideration all of the similitude conditions.

Reaction Modulus Design

When there is no prestess in the anchors, this design method neglects the soil arching and highly underestimates the active earth pressure behind the anchor level. On the other hand, the calculated anchor load is smaller than the measured force. These two effects are antagonistic towards the design bending moment which is very similar to the measured curves. When there is a high prestess value, the Reaction Modulus Design correctly estimates the performance of the structure in service. The differences become important with the increase of excavation depth.

Overall stability:

The exploitation of the displacement field by stereophotogrammetry helps to visualise accurately the phenomenon of overall failure with a short anchor.

The design of the anchor load causing the failure with the Kranz Method represents a high sensitivity to the distribution of the active pressure on the retaining wall, but the choice of a circular failure plane, instead of a linear one, is not justified.

References

1. BUREAUX SECURITAS. *Recommandation concernant la conception, le calcul, l'execution et le controle des tirants d'ancrage. recommandations T.A.86, 3eme edition,* Bureaux Securitas, 1986, Paris.
2. G. SCHNEEBELI. Une analogie mecanique pour l'etude de la stabilite des ouvrages en terre à deux dimensions. *IVeme Congres International de Mecanique des Sols et des Travaux de Fondation*, 1957, vol. 2, Londre, pp 228-232.
3. P.W. ROWE. Anchored sheet pile walls. *Proc. Instn Civ. Engrs*, 1952, **1**, , 27-70.

4. R. FAGES and C. BOUYAT. Calcul de rideaux de parois moulees ou de palplanches. *Travaux*, 1971, 1, 439, pp 21-51.
5. E. KRANZ. *Über die Verankerung von Spundwande.* 2nd edition, 1953, Wm. Ernst und Sohn, Berlin.
6. F. MASROURI. *Comportement des rideaux de soutenement semi-flexibles: etude theorique et experimentale*. These de doctorat, 1986, INSA Lyon, 247pp.

Shaft excavation in soft clay by Caisson construction

T.G. PORTER, B.H. LIEN, H. ABEDI and J.A. RAMOS, Metcalf & Eddy, Inc., USA

Introduction

The City of Detroit Water and Sewerage Department has undertaken a major construction project currently ongoing in Southwestern Detroit. The project involves different types of earth retention systems for excavations in soft to very soft clays. A major structure of the project is a circular reinforced concrete caisson, called PS-2A Caisson, with 140-foot (42.7 meter) outside diameter, 7-foot (2.1-meter) thick walls and approximately 80 feet (24.4 meters) deep. The PS-2A Caisson is one of the largest sinking caissons in the United States. There is also a 44-foot (13.4- meter outside diameter caisson with 7-foot (2.1-meter) thick walls and 50 feet (15.2 meters) deep, called Shaft No.5. Using a self sinking technique, both caissons reached final grade by their own weight and by excavating soils inside the caisson.

The project site is within the Detroit Wastewater Treatment Plant and was previously used as ash lagoons to store ash from the plant incineration operations. Figure 1 shows a plan of the site and the locations of the caissons. Figures 2 and 3 show profiles of the PS-2A and Shaft No. 5 caisson areas. This paper discusses the design considerations and construction experience of these two caissons.

(Note: The elevations indicated throughout the paper refer to the Detroit City Datum, units in feet)

Soil conditions

The soil conditions for the project were determined from three subsurface investigations. In addition, considerable soil data have been collected from previous investigations within the Sewage Treatment Plant. In-site vane shear tests and laboratory unconsolidated-undrained (UU) triaxial compression tests were performed on samples from within the clay stratum. Figure 4 shows the general subsurface profile and undrained shear strengths.

General ground surface on site is approximately elevation 108 (32.9 meters). Surface soils consist of miscellaneous fill overlying fine to medium sand and silty fine sand. The top of the clay stratum is at approximately

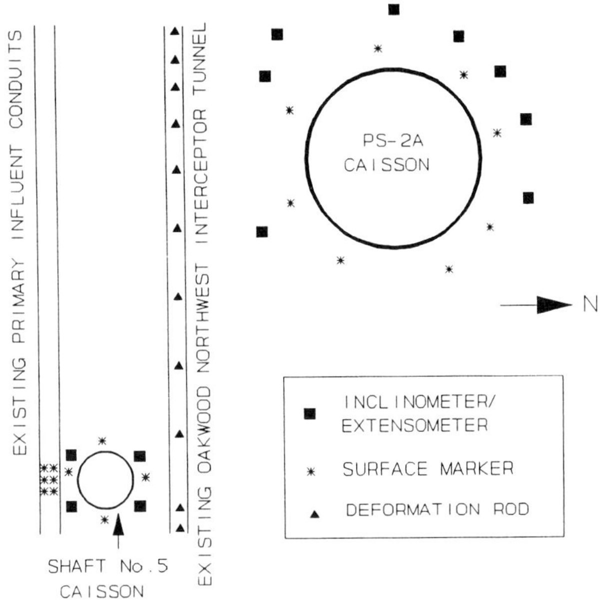

Figure 1.Project site and location of geotechnical instruments

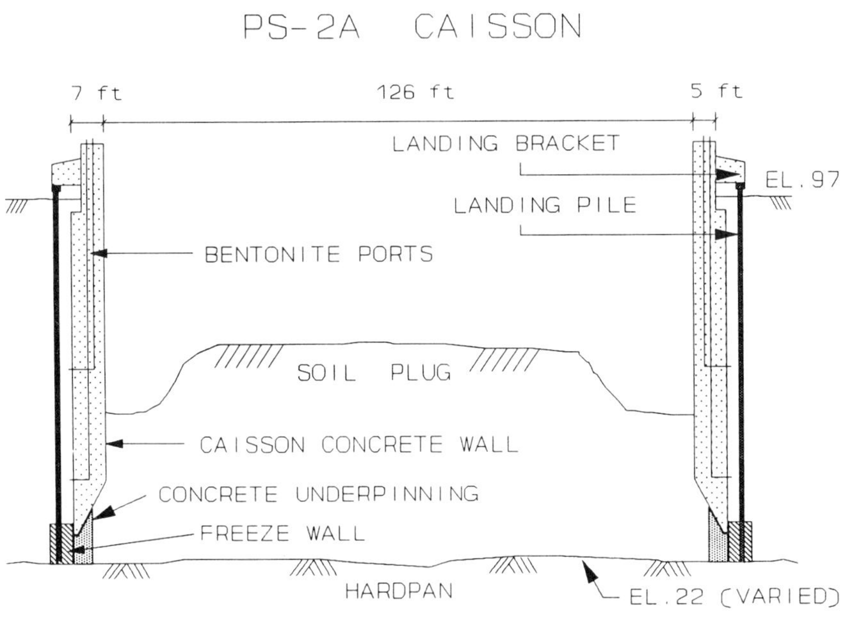

Figure 2. Profile of the PS-2A Caisson construction area

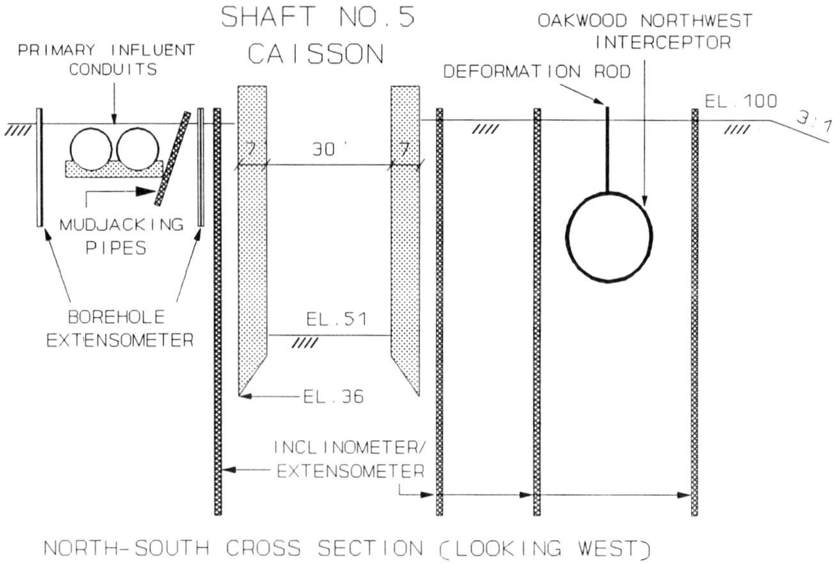

Figure 3. Profile of the shaft No.5 Caisson construction area

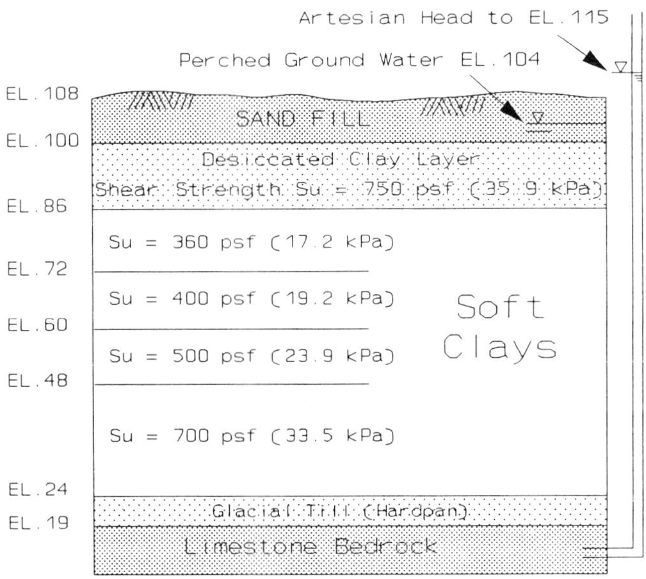

Figure 4.Generalized subsurface profile and test results

elevation 98 (29.9 meters) and extends down to elevation 25 (7.6 meters). The upper 10 feet (3.0 meters) of the clay layer is desiccated and has a medium consistency with an undrained shear strength in the 750 to 800 psf (35.9 to 38.3 kPa) range. At approximately elevation 86 (26.2 meters), the consistency becomes soft to very soft. The undrained shear strength decreases to approximately 350 psf (16.8 kPa) at elevation 72 (21.9 meters) with corresponding increase in moisture content and decrease in dry unit weight.

Below elevation 72 (21.9 meters), the undrained shear strength gradually increases with depth, and reaches a value of about 700 psf (33.5 kPa) below elevation 48 (14.6 meters). The clay is normally consolidated or slightly over consolidated below about elevation 75 (22.9 meters). The clay is classified as low plasticity clay in the Unified Soil Classification System. However, occasional samples are classified as highly plastic clay based on Atterburg limits tests. Stratification or layering was observed in samples taken from the upper clay stratum. The layering typically consists of thin bands of silt or sand.

The clay is underlain by a glacial till, locally referred to as a hardpan, which consists of a very dense mixture of silt, clay, sand, and gravel. A limestone bedrock underlying the hardpan is characterized as a massive competent gray limestone.

Groundwater conditions

Perched groundwater is encountered at elevation 104 (31.7 meters). Also, there is an artesian groundwater condition in the limestone bedrock. The artesian head is at elevation 115 (35.1 meters). Since the hardpan stratum is not continuous across the site, the artesian condition brought up several concerns during the design and construction phases of this project, which will be discussed in subsequent sections of this paper.

General design considerations

Design of the PS-2A and Shaft No. 5 Caissons involved several basic geotechnical considerations. The caissons are to retain the earth around them, and serve as permanent structures. Therefore, in their final position, caissons must be designed to provide adequate soil support for the subsurface conditions encountered at the site. There are several methods by which caissons can be designed and constructed. This section describes the design considerations associated with the sinking caisson method applied at the PS-2A project.

The general principle in the sinking caisson method is to provide a sufficient driving weight to fail the soil below the caisson shoe, and thus sink the caisson through the soft clays. The driving weight includes the steel reinforcement and concrete of the caisson. The required weight is determined by evaluating the soil resistance to caisson sinking which is a function of the soil shear strength. The sinking resistance includes skin friction which acts along the outside and inside walls of the caisson and the soil bearing below the caisson shoe. To control the caisson sinking, the design considerations described in the following sections were investigated and incorporated into the Contract Documents.

Caisson Work Area

The caissons were constructed in areas with varying existing ground surface elevations. Prior to the caissons construction, a pre-excavation was specified to a pre-determined elevation for a distance of 80 feet (24.4 meters) and 45 feet (13.7 meters) around the outside perimeter of the PS-2A Caisson and Shaft No. 5 Caisson, respectively. The extent of these areas was determined by considering an area of influence of 45° extending from the tip of the caisson shoe in its final position, to the ground surface. The work areas were specified to insure a uniform lateral earth pressure acting around the caisson during sinking. Construction surcharges, such as stockpiling of soil and heavy materials were limited within these areas. In addition, construction surcharges from other activities were prohibited from using these areas until the caissons were landed.

Bottom Heave Considerations

To sink the caissons, soil along the inside perimeter of the caissons was excavated. As a result, there was a potential of bottom heave within the excavation area due to the difference in soil elevation outside the caisson as compared to the soil elevation inside the caisson. The potential for bottom heave was reduced due to the penetration of the 7-foot (2.1-meter) thick caisson wall below the excavation. In addition, to further reduce the bottom heave potential, a specified depth of soil plug above the caisson shoe had to be maintained within the caisson during the sinking operation. The required amount of the soil plug was determined by specifying a stability number, which is discussed in the next section.

Stability Number

A stability number can be used to gage the tendency of the clay below the caisson shoe to heave into the caisson. The stability number, N, is calculated as follows:

$$N = (\gamma_1 h + q - \gamma_2 d)/c$$

where

γ_1=Weighted unit weight of the saturated soil layers from the existing ground surface to the tip of the caisson shoe along the outside of the caisson.

γ_2 =Weighted unit weight of the saturated soil layers from the top of the soil plug inside the caisson to the lowest point of the inside caisson wall.

h=Embedment of outer surface of caisson, measured from ground surface to tip of caisson cutting shoe.

d=Depth of embedment of inner surface of caisson, measured from excavated surface inside of caisson (top of soil plug) to tip of caisson cutting shoe.

q=Construction equipment surcharge loads.

c=Shearing resistance of soil at bottom of caisson cutting edge.

In general, stability numbers less than 6 indicate a stable bottom condition within excavations. Stability numbers from six (6) to eight (8) indicate a marginal condition, and stability numbers larger than eight (8) indicate a potentially unstable bottom condition.

For the PS-2A Caisson, a stability number of less than or equal to eight (8) was specified to determine the soil plug height within the caisson. The value of eight (8) was specified due to the relatively large distances between the PS-2A Caisson and the existing structures. This led to a requirement of a maximum 21-foot (6.4-meter) soil plug when the caisson shoe tip was at elevation 62 (18.9 meters). The required soil plug was only 10 feet (3 meters) thick after the caisson had landed. The stability number for the Shaft No. 5 Caisson was limited to six (6), because the caisson is in very close proximity to the existing structures. A smaller stability number reduces the likelihood of adjacent soil movements and damage to the existing structures. Throughout the sinking of the caissons, the actual stability numbers were always less than those specified. The majority of the excavation took place in a trench around the perimeter of the caissons without excavating the cores. The sinking events were usually abrupt with the caisson sinking 2 to 5 feet (0.6 to 1.5 meters) in approximately 5 to 10 minutes.

Landing the Caissons

Due to the uneven topography of the hardpan, the PS-2A Caisson was designed to be landed and temporarily supported upon piles. The hardpan elevations were determined from the geotechnical investigations. The PS-2A Caisson was landed with tip elevation above the hardpan to prevent the caisson from tilting upon encountering the uneven hardpan. Landing piles for Shaft No. 5 Caisson were designed to stop the caisson sinking at a specified elevation.

Underpinning and Ground Freezing

The PS-2A Caisson was landed with the tip of the shoe about four to six feet (1.2 to 1.8 meters) above the hardpan. Underpinning pits below the shoe were constructed to provide support and bearing of the caisson weight on the hardpan. To prevent the clay from squeezing between the tip of the shoe and the top of the hardpan, a ground freezing program was specified to enable the underpinning pits to be constructed (see Figure 2).

The ground freezing specifications required that the frozen earth around and beneath the outside perimeter of the PS-2A Caisson shoe to be watertight, and provide structural support of the soil between the tip of the caisson shoe and the hardpan. At the tip of the caisson shoe, elevation 26 (7.9 meters), the approximate lateral earth pressure was determined to be 7,400 psf (354 kPa). At the hardpan, elevation 21 (6.4 meters), the approximate lateral pressure was determined to be 7,900 psf (378 kPa).

The Contractor designed the freeze wall and utilized a calcium chloride brine system with two mechanical refrigeration units. The brine was cooled and circulated through an eight-inch diameter insulated manifold system leading to vertical freeze pipes around the caisson. There were 102 freeze pipes at 4.5 feet (1.4 meters) centers located approximately 2.5 feet (0.61 meters) from the outside caisson wall. The components of the vertical freeze pipes consisted of an insulated 3.5 inch (88.9 mm.) diameter steel return pipe and a 1.5 inch (38.1 mm.) diameter plastic supply line within the return pipe. The last 10 feet (3 meters) of the freeze pipe was not insulated, which was a closed-end 8 5/8-inch (219mm.) outside diameter steel pipe. The 1.5 inch (38.1mm.) plastic supply line extended to the end of the non-insulated steel pipe. The brine solution was circulated through the system to form the freeze wall required by the specifications. The Contractor anticipated that an average 5.5-foot (1.7-meter) thick freeze wall would be required during underpinning.

From undisturbed samples taken from the PS-2A site, the Contractor froze the sample and determined that the average frozen soil temperature would range between -13°C to -15°C. Laboratory tests conducted by the Contractor on the -13°C samples indicated an unconfined compressive strength (4 months loading) of 30,000 psf (1437 kPa) to 35,000 psf (1676 kPa).

Rock Grouting

As previously indicated, there exists an artesian condition within the bedrock. The artesian head is approximately to elevation 115 (35.1 meters).

To control the artesian pressure within the limestone bedrock, an extensive rock grouting program was specified prior to construction of the PS-2A Caisson. A blanket grouting program was conducted for the bedrock within the interior of the caisson and in the area approximately 15 feet (4.6 meters) outside the caisson wall. Grouting was not conducted in the footprint of the caisson, so as to avoid the possibility of the backfilled grout holes disrupting the sinking of the caisson.

The grouting specifications required both cement grouting and chemical grouting. The cement grouting involved drilling and grouting of over 300 grout holes, consisting of primary, secondary, and tertiary grout holes. The primary grout holes were spaced 20 feet (6.1 meters) on center, and were drilled 20 feet (6.1 meters) into the limestone bedrock. A two-stage grouting approach was specified for each primary hole. Stage one was the grouting of the lower 10 feet (3 meters) of the bedrock, and stage two was the grouting of the upper 10 feet (3 meters) of bedrock.

The secondary grout holes split the distance between the primary holes, and the tertiary holes split the distance between secondary holes. Generally, the bottom 20 feet (6.1 meters) of all grout holes were grouted. Occasionally only the top ten feet (3 meters) of the tertiary holes were grouted depending on the grout takes in adjacent grout holes.

The grout mix consisted of water, cement, and bentonite. The water/cement ratios ranging from 5:1 to 1:1 by volume were utilized depending on the initial grout takes as measured by a flowmeter. Hydrated bentonite was added at a quantity of two percent by volume of grout mix. The grouting pressure was determined to counter the artesian pressure, and had to be controlled to prevent heaving of the overburden and fracturing the bedrock.

It was found that the lower 10 feet (3 meters) of bedrock was competent with very little fracturing. However, the upper 10 feet (3 meters), as anticipated from the geotechnical investigations, was fractured and took the majority of the grouting. A third stage of cement grouting, which was not anticipated, was conducted to seal the zone between the bedrock and hardpan. The cement grouting during the third stage sealed this zone, and therefore, it was decided to delete the chemical grouting program.

Geotechnical Instrumentation

An extensive geotechnical instrumentation program was specified for monitoring soil movements and existing structures around the caissons. Inclinometers, inclinometer/extensometers, and surface markers were installed and monitored during the sinking process. The instruments were located within the theoretical influence zone of both caissons, and measured an insignificant amount of vertical or lateral movement. The only measurable soil movement at Shaft No. 5 was an outward lateral movement associated with the soil moving from beneath the caisson to the instruments as the caisson sank.

Construction of sinking caissons

General

Construction of each caisson was started by erecting a concrete-filled steel shoe over the projected area of the caisson wall. The steel shoe was a truncated triangular section temporarily supported by a series of horizontal and vertical timbers. After the reinforced concrete within the shoe reached 28-day strength, the caisson sinking was launched by pulling out the vertical timbers and excavating soil from inside the caisson. Subsequent sinking was conducted by monolithically placing a 10-foot (3.0-meter) high caisson wall and excavating soils from the inside perimeter of the caisson. There were two levels of bentonite injection pipes installed inside the caisson wall. Bentonite was injected between the outside caisson wall and soil during the sinking operation. This operation intended to reduce the outside caisson skin resistance.

The rate of sinking and plumb of the caisson were controlled by symmetrically excavating soils inside the caisson and by daily surveying. Both caissons were landed within the specified tolerances for plumb [not more than 1 inch (25.4 mm.) in 10 vertical feet (3 meters)], plan and elevation [±6 inches (152 mm.)]. In the PS-2A Caisson area, the rock grouting program was successful. The artesian flow was approximately 80 gallons (303 litres) per minute per grout hole during the drilling phase of the program. After excavation to the hardpan, the flow was reduced to 1 gallon (3.79 litres) per minute in only one local area. The minor leak was controlled by hydrostatic relief wells installed below the caisson base slab.

Shaft No. 5 Caisson

The smaller diameter Shaft No. 5 Caisson was constructed first. It was expected that the experience gained from its construction would be valuable for construction of the larger and more complex PS-2A Caisson. During the sinking operation, minor lateral and vertical soil movements were measured by the geotechnical instrumentation.

Difficulty came after landing the caisson on landing piles. When excavating the inside soil down to one foot (0.3 meter) above the design elevation 55 (16.8 meters), artesian water found a path flowing into the caisson and flooded the caisson to approximately elevation 114 (34.7 meters). Meanwhile, artesian heads in four piezometers installed around the project site dropped down quickly in response to this flow. In order to stabilize the artesian inflow, a 1-foot (0.3-meter) thick, 7-foot (2.1- meter) high wall was added to the top of the caisson. The final solution was to excavate underwater down to below final design elevation and install a pressure relief drainage system and tremie seal the bottom of the shaft. The water was then pumped out of the caisson to construct the base slab in dry condition.

The probable cause of the artesian leak into Shaft No.5 may be due to the combination of fractured hardpan and the driving of the adjacent landing piles into the hardpan.

PS-2A Caisson

Sinking operation of the PS-2A Caisson was similar to that of the Shaft No. 5 Caisson. Only minor lateral and vertical soil movements were observed.

Subsequent constructions, such as landing the caisson, ground freezing, and underpinning the caisson, all went smoothly without any major problems.

Summary and conclusions

The PS-2A and Shaft No. 5 Caissons presented significant design and construction challenges. The combination of construction in very soft clays, the protection of existing structures, and the coordination and implementation of geotechnical tasks lead to the successful completion of the PS-2A and Shaft No. 5 Caissons.

Acknowledgements

The writers would like to recognize the Detroit Water and Sewerage Department for their assistance during the design and construction of this project. The Department provided valuable input and recommendations during all phases of the project. Metcalf & Eddy would like to thank the Department for permission to publish this article. The writers would also like to recognize T. Weber, M. Pochodylo, M. Cameron, and R. Sherman of Metcalf & Eddy, Inc. for their review of the paper, and NTH Consultants, Ltd., Detroit for their assistance during the design and construction.

Discussion

SESSION 7

N.T. LONG, Laboratoire Central des Ponts et Chaussees (LCPC)

I want to give you some information about the improvement of walls. Ten works using Pneusol to reduce active earth pressure are known to have been built in France. The first of these structures, built in, 1986 is a cantilever retaining wall, 54 m long and 5 m high, founded on two rows of piles. Before filling behind the wall, the contractor became aware that the shell of the wall was too thin.

'Heavy' Pneusol was used to relieve the active earth pressure and also to eliminate tensile forces on the piles. It was made of treads arranged in layers and tied together with polyester tapes. Since the fill inside the treads exerts no active earth pressure and that between two layers it is highly confined, the Pneusol mass, provided that the spacing between layers is small, behaves like a gravity wall made up of thousands of tiny gabions stacked one on top of the other.

But the structure most like those used in the tests is that of the Fonderie Wall. One hour before work to widen a heavily travelled road was begun, this 18th centruy wall was classified as a historic monument, so it could not be demolished. It is made up of more or less closely spaced blocks. The Pneusol placed behind the wall consists of layers of truck tyres. Equipped with soil sensors, it gives very good results.

I. PAYNE, Kent County Council

Professor O'Rourke's experience suggested that undrained shear strengths from some tests were better than those from triaxial compression tests. This is not always the case. Published back analysis of failures in soft alluvium at Dartford in Kent suggests good correlations with triaxial compression tests. Vane tests significantly over-estimate undrained shear stength.

We have recorded very low undrained shear strength values inconsistent with exisiting overburden pressures. Clayton's comments regarding sampling disturbance are therefore very valid, and the quality of test data will determine the need for data managing or even the use of more sophisticated testing.

Professor C. J. F. P. Jones, University of Newcastle upon Tyne

Observations on structures in the field and the recent failures of reinforced earth retaining walls, notably in Tennessee in the United States, has shown that some of the current design methodologies used for reinforced soil structures contain weaknesses in analytical procedures and structural detailing.

Analytical methods

Current analytical methods used in the design of reinforced soil structures are based upon two methods, the Tie Back wedge method, developed in the United Kingdom and described in Department of Transport Memorandum BE3/78, and the Coherent Gravity method developed in France, Jones (1984).

The Tie Back method is based upon limit equilibrium methods covering internal and external stability considerations and includes condiseration of wedge stability. The Coherent Gravity method is an empirical method based upon field measurements; a check on wedge stability is not included except in some special loading cases.

In addition, the analytical procedures adopted in United Kingdom schemes sponsored by the Public Sector, require that the external and internal stability calculations for any reinforced soil structure should be linked and undertaken by one designer. Some proprietary systems which use the coherent gravity design method specifically separate responsibility for the external stability and internal stability between different designers. The recent failures of reinforced earth walls in Tennessee in the United States show that this practice of dividing design responsibility can lead to analytical practices which miss the potential failure mechanism based upon wedge instability. In particular, the coherent gravity method of analysis is not valid when the reinforced soil structure supports an embankment or steep slope, usually referred to as the infinitive slope condition, Lee et all (1992a).

Structures crossing ravines

One of the design assumptions usually adopted in the analysis of reinforced soil structures is that the structure can be treated as a 2 dimensional plane strain condition. In the great majority of retaining structure design this assumption is justified and valid.

One of the benefits of reinforced soil is that it can be used to great economic benefit in places when conventional retaining structures would be very difficult to construct. Included in this category of structure is the use of reinforced soil in mountainous regions. Many of these structures built in mountainous regions are constructed across steep valleys or ravines. In this case the assumption of 2-dimensional plane strain conditions may not be

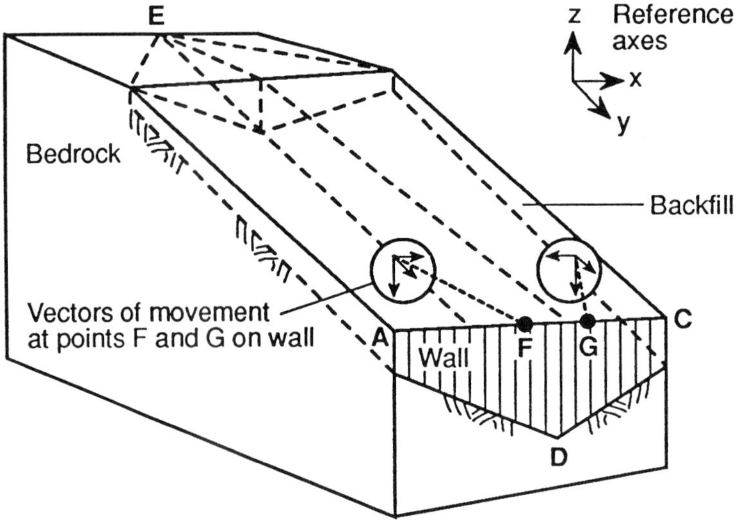

Fig. 1. Isometric view of a reinforced soil wall with in-plane movements conveyed to the wall

valid as the structure can exhibit 3-dimensional characteristics, particularly in respect of the physical movements of the wall facing, figure 1. Although the assumption of a 2-dimensional condition is adequate and usually conservative with respect to the analysis for stability, it is essential to provide sufficient movement capacity in the horizontal plane to cater for the 3-dimensional effects.

Internal Vertical Settlement

Reinforced soil structures are constructed using one of three methods, figure 2. Each of these techniques allows for internal settlement of the reinforced fill which occurs during constructio and use. Field measurements on a range of fill materials and different structural forms has shown that a minimum vertical movement of the fill with respect to the facing can be identified, Jones and Hassan (1992). The minimum movement capacity required to eliminatre the possibility of damage to the facing or reinforcement/facing connections is 1 in 100 (i.e. in a 10m high structure a vertical movement capacity of 100mm is required). To provide an element of tolerance a movement capacity of 1 in 50 is preferred. Some construction systems currently being used do not provide the minimum movement capacity of 1 in 100.

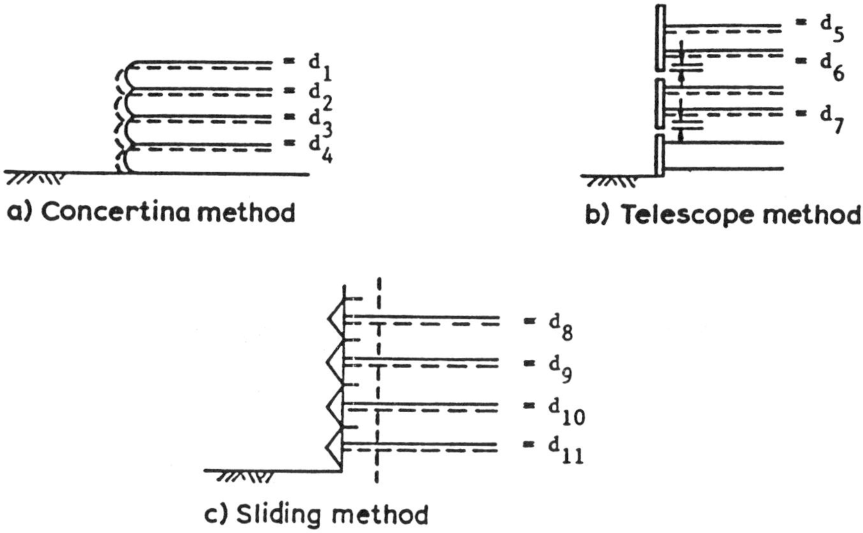

Fig. 2

References

JONES, C.J.F.P. (1984) "Earth reinforcement and soil structures", Butterworths, P. 192

LEE, K., JONES, C.J.F.P., SULLIVAN, R. and TROLLINGER (1992a) "Failure and Deformation of Four Reinforced Earth Walls in Eastern Tennessee, USA", Paper submitted to Geotechnique.

JONES, C.J.F.P. and HASSAN, C.A. (1992) "Compression Movements within Reinforced Soil Structures" Geotropika '92 International Conference in Geotechnical Engineering, Universiti Teknologi Malaysia, April.